Natural Rubber Materials
Volume 1: Blends and IPNs

RSC Polymer Chemistry Series

Series Editors:
Professor Ben Zhong Tang (Editor-in-Chief), *The Hong Kong University of Science and Technology, Hong Kong, China*
Professor Alaa S. Abd-El-Aziz, *University of Prince Edward Island, Canada*
Professor Stephen L. Craig, *Duke University, USA*
Professor Jianhua Dong, *National Natural Science Foundation of China, China*
Professor Toshio Masuda, *Fukui University of Technology, Japan*
Professor Christoph Weder, *University of Fribourg, Switzerland*

Titles in the Series:
1: Renewable Resources for Functional Polymers and Biomaterials
2: Molecular Design and Applications of Photofunctional Polymers and Materials
3: Functional Polymers for Nanomedicine
4: Fundamentals of Controlled/Living Radical Polymerization
5: Healable Polymer Systems
6: Thiol-X Chemistries in Polymer and Materials Science
7: Natural Rubber Materials Volume 1: Blends and IPNs

How to obtain future titles on publication:
A standing order plan is available for this series. A standing order will bring delivery of each new volume immediately on publication.

For further information please contact:
Book Sales Department, Royal Society of Chemistry, Thomas Graham House, Science Park, Milton Road, Cambridge, CB4 0WF, UK
Telephone: +44 (0)1223 420066, Fax: +44 (0)1223 420247
Email: booksales@rsc.org
Visit our website at www.rsc.org/books

Natural Rubber Materials
Volume 1: Blends and IPNs

Edited by

Sabu Thomas, Rajisha K. R., Hanna J. Maria
Mahatma Gandhi University, Kottayam, India
Email: sabuchathukulam@yahoo.co.uk

Chin Han Chan
MARA University of Technology, Selangor, Malaysia
Email: cchan@salam.uitm.edu.my

and

Laly A. Pothen
Bishop Moore College, Mavelikkara, India
Email: lapothan@gmail.com

RSCPublishing

RSC Polymer Chemistry Series No. 7

ISBN: 978-1-84973-610-7
ISSN: 2044-0790

A catalogue record for this book is available from the British Library

© The Royal Society of Chemistry 2014

All rights reserved

Apart from fair dealing for the purposes of research for non-commercial purposes or for private study, criticism or review, as permitted under the Copyright, Designs and Patents Act 1988 and the Copyright and Related Rights Regulations 2003, this publication may not be reproduced, stored or transmitted, in any form or by any means, without the prior permission in writing of The Royal Society of Chemistry, or in the case of reproduction in accordance with the terms of licences issued by the Copyright Licensing Agency in the UK, or in accordance with the terms of the licences issued by the appropriate Reproduction Rights Organization outside the UK. Enquiries concerning reproduction outside the terms stated here should be sent to The Royal Society of Chemistry at the address printed on this page.

The RSC is not responsible for individual opinions expressed in this work.

Published by The Royal Society of Chemistry,
Thomas Graham House, Science Park, Milton Road,
Cambridge CB4 0WF, UK

Registered Charity Number 207890

For further information see our web site at www.rsc.org

Preface

"There is probably no other inert substance, the properties of which excite in the human mind, when first called to examine it, an equal amount of curiosity, surprise, and admiration. Who can examine, and reflect upon this property of gum-elastic, without adoring the wisdom of the Creator" are the words said with reference to natural rubber by none other than Charles Goodyear, who in 1839 became the first American to discover the 'vulcanization' process for natural rubber, the invention that revolutionized the rubber industry and fuelled the Industrial Revolution. Charles Goodyear and Thomas Hancock contributed this indispensable factor of industry, *i.e.* transformation of a material, which is sticky when warm and brittle when cold, to a highly deformable, chemically crosslinked elastic solid. Goodyear wrote: "A man has cause for regret only when he sows and no one reaps." Taking value from these words, the rubber industry and researchers have showed a markedly progressive development over more than 100 years.

Reinforced rubber composites have attracted the attention of many researchers due to their unique properties. In rubbers, fillers are used to achieve products with improved properties for their end use applications. Rubber reinforcement is a very significant issue in rubber science and engineering. Thus, a knowledge of up-to-date developments in rubber reinforcement and the determining factors linked to the final property of the materials are an essential topic. However, many issues in this aspect are still not well understood and explained. Moreover, when rubber reinforcement is obtained by using nanoparticles or when blended with another polymer, it is even more difficult to clearly understand the relationship between the microstructure and properties. The special character of rubber, being a multicomponent system (for example, rubber, vulcanizing agent, accelerants, reinforcements, *etc.*), complicates the analysis of the parameters affecting the final properties. The interest of rubber researchers in macro- and nanofillers, the processing techniques, morphological characterization, and structure–property relationships are growing rapidly, as revealed by the large number of research innovations in the area of natural rubber composites all over the world.

This book on natural rubber presents a summary of the present state-of-the-art in the study of these versatile materials. The two volumes cover all the areas related to natural rubber, from its production to composite preparation, the various characterization techniques and life cycle assessment. Chapters in this book deal with both the science of natural rubber – its chemistry, production, engineering properties, and the wide-ranging applications of natural rubber in the modern world, from the manufacture of car tyres to the construction of earthquake protection systems for large buildings. Although there are a number of research publications in this field, to date, no systematic scientific reference book has been published specifically in the area of natural rubber as the main component in systems. We have developed the two volumes by focusing on the important areas of natural rubber materials, the blends, IPNs of natural rubber and natural rubber based composites and nanocomposites; their preparation and characterization techniques. The books have also profoundly reviewed various classes of fillers like macro, micro and nano (ID, 2D and 3D) used in natural rubber industries. The applications and the life cycle analysis of these rubber based materials are also highlighted.

Volume 1 of this book is comprised of 25 chapters, and discusses the different types of natural rubber based blends and IPNs. The first seven chapters discuss the general aspects of natural rubber blends like their miscibility, manufacturing methods, production and morphology development. The next ten chapters describe exclusively the properties of natural rubber blends with different polymers like thermoplastic, acrylic plastic, block or graft copolymers, *etc*. Chapter 18 deals entirely with clay reinforcement in natural rubber blends. Chapters 19 to 23 explain the major techniques used for characterizing various natural rubber based blends. The final two chapters give a brief explanation of life cycle analysis and the application of natural rubber based blends and IPNs.

This multi-author volume provides a useful summary of current knowledge about natural rubber based materials in general, practically integrating the various aspects of natural rubber research and thereby contributing to the research society a complete book on natural rubber based materials. We anticipate that this book will be of significant interest to scientists working on the basic issues surrounding natural rubber blends and IPNs and their composites and nanocomposites as well as those working in industry on applied problems, such as processing. Because of the multidisciplinary nature of this research, this book will attract a broad audience including chemists, material scientists, physicists, chemical engineers and processing specialists, who are involved and interested in the future frontiers of natural rubber based materials. We trust that this book on natural rubber materials will stimulate new ideas, methods and applications in ongoing advances in this growing area of strong international interest.

We appreciate the efforts and enthusiasm of all the authors for writing their chapters, in spite of their very busy schedules, and we would also like to acknowledge those who were prepared to contribute, but were unable to do so at the present time. We deeply *appreciate* the time and *effort* of the expert reviewers in reviewing the manuscript.

Furthermore, without the tireless efforts of the RSC editorial team, in particular Alice Toby-Brant, Leanne Marle, Lois Bradnam and Sarah Salter, this book would not have been possible. We are also indebted to all others in the editorial wing and the publishing and production teams for their assistance in the preparation and publication of this book. Finally, it is our pleasure to acknowledge our colleagues and good friends for their kind help and cooperation.

Contents

Chapter 1	Natural Rubber Based Blends and IPNs: State of the Art, New Challenges and Opportunities *Gordana Marković, Milena Marinović-Cincović, Vojislav Jovanović, Suzana Samaržija-Jovanović and Jaroslava Budinski-Simendić*		1
	1.1 Introduction and History		1
		1.1.1 Interpenetrating Polymer Networks	9
		1.1.2 History of IPN Development	9
		1.1.3 Properties of Polymer Blends and IPNs	11
		1.1.4 Glass Transition and Viscoelastic Behaviour	12
		1.1.5 Morphology	13
	1.2 Recent Trends and Developments in Natural Rubber Based Blends and IPNs		14
	1.3 Applications and the Potential Market for IPNs		20
	1.4 Environmental Impact and Recycling		22
	1.5 Conclusions		23
	Acknowledgements		24
	References		24
Chapter 2	Natural Rubber: Biosynthesis, Structure, Properties and Application *Jitladda Tangpakdee Sakdapipanich and Porntip Rojruthai*		28
	2.1 Introduction		28
	2.2 Biosynthesis of NR		30
	2.3 Structure and Properties of NR		33
		2.3.1 Initiating Terminal of the Rubber Molecule	33
		2.3.2 Terminating End of the Rubber Molecule	35

RSC Polymer Chemistry Series No. 7
Natural Rubber Materials, Volume 1: Blends and IPNs
Edited by Sabu Thomas, Chin Han Chan, Laly A. Pothen, Rajisha K. R., Hanna J. Maria
© The Royal Society of Chemistry 2014
Published by the Royal Society of Chemistry, www.rsc.org

		2.3.3	Structure of Branch Points, Gel and Storage Hardening	37
		2.3.4	Properties of NR	38
	2.4	Chemical Modification of NR		39
		2.4.1	Hydrogenation	40
		2.4.2	Epoxidation	40
		2.4.3	Chlorination	41
		2.4.4	Grafting Copolymerization	41
		2.4.5	Oxidative Degradation	42
		2.4.6	Cyclization	43
	2.5	Processing of NR and its Applications		44
		2.5.1	Processing of NR	44
		2.5.2	Applications of NR	45
	2.6	Conclusions		46
	References			46

Chapter 3 Non-Rubbers and Abnormal Groups in Natural Rubber   53
Eng Aik Hwee

	3.1	Non-Rubbers in Natural Rubber		53
		3.1.1	Lipids	54
		3.1.2	Proteins, Amino Acids and Other Nitrogenous Compounds	57
		3.1.3	Inositols and Carbohydrates	61
		3.1.4	Ash	62
		3.1.5	Volatile Matter	63
	3.2	Abnormal Groups in Natural Rubber		63
		3.2.1	*Trans*-Isoprene and Dimethylallyl (DMA) Groups	64
		3.2.2	Ester Groups, Fatty Acids and Phospholipids	64
		3.2.3	Epoxide Groups	64
		3.2.4	Aldehyde Groups	65
		3.2.5	Bonded Proteins and Amino Groups	67
	3.3	Future Trends		67
	References			67

Chapter 4 The Production of Natural Rubber from *Hevea brasiliensis* Latex: Colloidal Properties, Preservation, Purification and Processing   73
C. C. Ho

	4.1	Introduction	73
	4.2	Sources of NR	74
	4.3	Evolution of NR as an Industrial Elastomer	75
	4.4	Colloidal Properties of NR Latex and Stability	78

	4.5	Production of Commercial NR Latex from *Hevea*	80
		4.5.1 Preservation of NR Latex	81
		4.5.2 Latex Purification and Concentration Processes	82
		4.5.3 Commercial Concentration Methods for NR Latex	84
		4.5.4 Specialty NR Latices	87
		4.5.5 Chemically Modified Rubber Latices	90
		4.5.6 Recent Advances in NR Latex Technology	93
	4.6	NR Production Methods	94
		4.6.1 Sheet Rubbers: USS, RSS and ADS	95
		4.6.2 Crepe Rubbers: Pale Crepe, White Crepe and Brown Crepe	97
		4.6.3 Block Rubbers: Technically Specified Rubbers and Standard Technical Rubbers	98
		4.6.4 Rubber Products from Field Coagula	98
		4.6.5 Skim Rubbers	100
		4.6.6 Specialty Rubbers and Chemically Modified Rubbers	100
	4.7	Major Industrial Applications of NR and NR Latex	103
		References	104

Chapter 5 Natural Rubber Blends and Based IPNs: Manufacturing Methods — 107

Wanvimon Arayapranee

	5.1	Introduction	107
	5.2	Latex Based Methods	110
		5.2.1 Latex Mixing	111
		5.2.2 Maturation	112
		5.2.3 Latex Curing Processes	112
	5.3	Solution Based Methods	114
		5.3.1 Solution Manufacturing Processes	115
		5.3.2 Solution Mixing	115
	5.4	Solid Natural Rubber Based Methods	115
		5.4.1 Two-Roll Mills	116
		5.4.2 Internal Batch Mixers	118
		5.4.3 Continuous Mixers	122
		5.4.4 Solid Rubber Curing Processes	126
	5.5	Advantages and Disadvantages of Each Technique	128
	5.6	Conclusions	130
		References	130

Chapter 6	**Filler Migration in Natural Rubber Blends During the Mixing Process** *Hai Hong Le, Sybill Ilisch, Gert Heinrich and Hans-Joachim Radusch*	132

    6.1 Introduction 132
    6.2 Theoretical Prediction of Filler Localization in Rubber Blends at an Equilibrium State using the Z-Model 133
    6.3 The Wetting Concept for Experimental Determination of Filler Localization in Rubber Blends 137
    6.4 Equipment and Experimental Methods 141
        6.4.1 Preparation of Blends 141
        6.4.2 Characterization 142
    6.5 Results and Discussion 144
        6.5.1 Silica Localization in Rubber Blends 144
        6.5.2 Carbon Black Localization in Rubber Blends 157
        6.5.3 Carbon Nanotube (CNT) Localization in Rubber Blends 162
        6.5.4 Nanoclay Transfer in Rubber Blends 168
    6.6 Conclusions 171
    Acknowledgements 172
    References 172

Chapter 7	**NR Blends and IPNs: Miscibility and Immiscibility** *Wiwat Pichayakorn, Jirapornchai Suksaeree and Prapaporn Boonme*	177

    7.1 Introduction 177
    7.2 Definitions 178
    7.3 Miscibility and Immiscibility of NR Blends and IPNs 180
        7.3.1 Techniques for Preparing NR Blends 180
        7.3.2 Identification Parameters for Determination of Miscibility 181
        7.3.3 NR Blends and IPNs 182
    7.4 Thermodynamics of NR Blends and IPNs 185
    7.5 Phase Separation and Compatibilization 186
        7.5.1 Achievement of Thermodynamic Miscibility 187
        7.5.2 Addition of Block or Graft Copolymers 187
        7.5.3 Addition of Functional or Reactive Polymers 189
        7.5.4 *In Situ* Graft Polymerization or Reactive Blending 190
    7.6 Techniques for Measuring Miscibility and Immiscibility Properties 191
        7.6.1 Glass Transition Temperature Studies 191

	7.6.2	Scattering Studies	192
	7.6.3	Morphological Studies	192
	7.6.4	Infrared Spectroscopy	192
7.7	Conclusions		193
References			193

Chapter 8 Natural Rubber Based Non-Polar Synthetic Rubber Blends 195
Seiichi Kawahara

8.1	Introduction		195
8.2	Miscible NR Blends		196
	8.2.1	Background	196
	8.2.2	Characterization of NR-Sol and NR-Gel	198
	8.2.3	LCST Phase Behaviour	199
8.3	Immiscible NR Blends		203
	8.3.1	Background	203
	8.3.2	Characterization of NR/SBR Blends	204
	8.3.3	Tear Energy	205
8.4	Conclusions		210
References			210

Chapter 9 Natural Rubber Based Polar Synthetic Rubber Blends 213
Konstantinos G. Gatos

9.1	Introduction		213
9.2	Preparation Methods		214
	9.2.1	Latex	214
	9.2.2	Solution Mixing	215
	9.2.3	Melt Blending	215
9.3	Blend Characteristics		217
	9.3.1	Rheology	217
	9.3.2	Curing	219
	9.3.3	Swelling and Oil Resistance	220
	9.3.4	Morphology	222
	9.3.5	Mechanical and Dynamic-Mechanical Behaviour	225
	9.3.6	Thermal Properties	229
	9.3.7	Dielectric Properties	230
	9.3.8	Infrared Absorbance	232
	9.3.9	Ageing and Other Properties	235
9.4	Applications		236
9.5	Outlook		237
References			237

Chapter 10 Thermoplastic Elastomers from High-Density Polyethylene/Natural Rubber/Thermoplastic Tapioca Starch: Effects of Different Dynamic Vulcanization 242
Mohd Kahar Ab Wahab, Nadras Othman and Hanafi Ismail

10.1	Introduction	242
10.2	Materials and Methodology	246
	10.2.1 Preparation of Thermoplastic Tapioca Starch (TPS)	246
	10.2.2 Dynamic Vulcanization with HVA-2 and Sulfur Curative Agent	246
	10.2.3 Tensile Properties	246
	10.2.4 Gel Content	247
	10.2.5 Fourier Transform Infrared Spectroscopy (FTIR)	247
	10.2.6 Scanning Electron Microscopy (SEM)	248
	10.2.7 Thermogravimetric Analysis (TGA)	248
	10.2.8 Dynamic Mechanical Thermal Analysis (DMTA)	248
	10.2.9 Differential Scanning Calorimetry (DSC)	248
10.3	Results and Discussion	248
	10.3.1 Processing Characteristics	248
	10.3.2 Tensile Properties	250
	10.3.3 Gel Content	253
	10.3.4 Structural Analysis	254
	10.3.5 Blend Morphology	255
	10.3.6 Thermogravimetric Analysis	258
	10.3.7 Differential Scanning Calorimetry	260
10.4	Conclusions	262
	Acknowledgements	262
	References	262

Chapter 11 Natural Rubber/Engineering Thermoplastic Elastomer Blends 265
E. Purushothaman and Mehar Al Minath

11.1	Introduction	265
11.2	Recent Developments in TPEs	267
11.3	Preparation of TPEs	268
	11.3.1 Mixing	268
	11.3.2 Solution Casting	269
11.4	Characterization of TPEs	270
	11.4.1 Rheological Studies	270
	11.4.2 Morphological Studies	271
	11.4.3 Scattering Analyses	273

Contents

	11.4.4	Mechanical Properties	273
	11.4.5	Thermal Analyses	276
	11.4.6	Dielectric Properties	278
11.5	Applications of TPEs		279
11.6	Conclusions		280
References			280

Chapter 12 Radiation Processing of Natural Rubber with Vinyl Plastics — 284
Chantara Thevy Ratnam, Zurina Mohamad and Mohammad Khalid Siddiqui

12.1	Introduction		284
12.2	Radiation Effects on Polymers		285
12.3	Radiation Crosslinking of Polymers		285
12.4	Radiation Sensitizers used as Crosslinking Agents		286
12.5	Radiation Crosslinking of Natural Rubber (NR)		286
	12.5.1	The Properties of Radiation Crosslinked NR	287
12.6	Radiation Crosslinking of Epoxidized Natural Rubber (ENR)		288
12.7	Radiation Crosslinking of NR Based Blends		289
	12.7.1	Radiation Crosslinking of PVC/ENR Blends	290
	12.7.2	Radiation Crosslinking of EVA/ENR Blends	292
	12.7.3	Radiation Crosslinking of PVC/NR Blends	294
12.8	Conclusions		297
Acknowledgements			297
References			297

Chapter 13 Blends and IPNs of Natural Rubber with Acrylic Plastics — 300
Wiwat Pichayakorn, Jirapornchai Suksaeree and Prapaporn Boonme

13.1	Introduction		300
13.2	The History of Natural Rubber–Acrylate Blends and IPNs		301
13.3	Preparation Methods of Natural Rubber–Acrylate Blends and IPNs		302
	13.3.1	Natural Rubber–Acrylate Blends	302
	13.3.2	Natural Rubber–Acrylate IPNs	306
13.4	Natural Rubber–Acrylate Blends and IPNs: Properties and Characterization Techniques		307
	13.4.1	Morphological Properties	307
	13.4.2	Mechanical Properties	313
	13.4.3	Thermal and Thermomechanical Properties	315
	13.4.4	Rheological Properties	318

13.5	Applications of Natural Rubber–Acrylate Blends and IPNs	320
13.6	Conclusions	321
References		322

Chapter 14 Photoreactive Nanomatrix Structures Formed by Graft Copolymerization of 1,9-Nonanediol Dimethacrylate onto Natural Rubber 324
Oraphin Chaikumpollert, Nanthaporn Pukkate and Seiichi Kawahara

14.1	Introduction	324
14.2	Inclusion Complex Formation of NDMA and β-CD	327
14.3	Graft Copolymerization of Inclusion Complex onto DPNR Particles	330
14.4	Conclusions	334
References		334

Chapter 15 Blends and IPNs of Natural Rubber with Thermosetting Polymers 336
Raju Thomas, Ishak Ahmad, Sahrim Hj. Ahmad and Shinu Koshy

15.1	Introduction	336
15.2	Elastomer-Modified Epoxy Resin Systems	337
15.3	Elastomer-Modified Unsaturated Polyester Resin Systems	344
15.4	Conclusions	346
References		346

Chapter 16 Natural Rubber Blends with Biopolymers 349
Silvia Maria Martelli, Carol Sze Ki Lin, Zheng Sun, Nathalie Berezina, Farayde Matta Fakhouri and Lucia Helena Innocentini-Mei

16.1	Introduction		349
16.2	Natural Rubber/Lignin Blends		350
	16.2.1	General Information	350
	16.2.2	Blends and their Applications	351
16.3	Natural Rubber/Protein Blends		353
	16.3.1	General Information	353
	16.3.2	Blends and their Applications	356
16.4	Natural Rubber/Polysaccharide Blends		357
	16.4.1	General Information	357
	16.4.2	Blends and their Applications	358

	16.5	Natural Rubber/Polyester Blends	360
		16.5.1 General Information	360
		16.5.2 Blends and their Applications	361
	16.6	Conclusions and Outlook	366
	References		367

Chapter 17 Clay Reinforcement in Natural Rubber Based Blends: Micro and Nano Length Scales 370
Yamuna Munusamy, Hanafi Ismail and Chantara Thevy Ratnam

	17.1	Introduction	370
	17.2	Recent Developments	371
	17.3	Preparation Methods	372
		17.3.1 Development of Ethylene Vinyl Acetate/ Natural Rubber/Organoclay Ternary Blends	374
	17.4	Characterization of Nanocomposites	375
		17.4.1 Morphology	375
		17.4.2 Mechanical Properties	379
		17.4.3 Thermal Properties	384
		17.4.4 Flammability	386
	17.5	Crosslinking Techniques	386
		17.5.1 Chemical Crosslinking	386
		17.5.2 Irradiation Crosslinking	389
	17.6	Conclusions	391
	References		391

Chapter 18 Rheological Behaviour of Natural Rubber Based Blends 394
Ploenpit Boochathum

	18.1	Introduction	394
	18.2	Rheological Behaviour	396
		18.2.1 Natural Rubber–Thermoplastic Blends	396
		18.2.2 Natural Rubber–Synthetic Rubber Blends	405
		18.2.3 Chemically Modified Natural Rubber Blends	416
	References		439

Chapter 19 Spectroscopy: Natural Rubber Based Blends and IPNs 441
SA-AD Riyajan

	19.1	Introduction	441
	19.2	UV-Vis Spectroscopy	442
		19.2.1 Introduction to UV-Vis Spectroscopy	442
		19.2.2 Sample Preparation and Typical Conditions for UV-Vis Spectroscopy Measurement	442
		19.2.3 Analysis of Polymer Blends	443

	19.3	Fourier Transform Infrared Spectroscopy (FTIR)	445
		19.3.1 Introduction to FTIR	445
		19.3.2 Sample Preparation and Typical Conditions for FTIR	446
		19.3.3 Analysis of Polymer Blends	447
	19.4	Nuclear Magnetic Resonance (NMR) Spectroscopy	460
		19.4.1 Introduction to NMR Spectroscopy	460
		19.4.2 Sample Preparation and Typical Conditions for NMR	460
		19.4.3 Analysis of Polymer Blends	461
	19.5	Raman Spectroscopy	472
		19.5.1 Introduction to Raman Spectroscopy	472
		19.5.2 Sample Preparation and Typical Conditions for Raman Spectroscopy	473
		19.5.3 Analysis of Polymer Blends	473
	19.6	Electron Spin Resonance (ESR) Spectroscopy	475
		19.6.1 Introduction to ESR	475
		19.6.2 Sample Preparation and Typical Conditions for ESR	476
		19.6.3 Analysis of Polymer Blends	476
	19.7	Applications	477
	19.8	Conclusions	478
	Acknowledgements		478
	References		478

Chapter 20 Mechanical and Viscoelastic Properties of Natural Rubber Based Blends and IPNs   481
Wiwat Pichayakorn, Jirapornchai Suksaeree and Prapaporn Boonme

	20.1	Introduction	481
	20.2	Instruments and Techniques for Mechanical and Viscoelastic Evaluations	482
		20.2.1 Mechanical Properties	482
		20.2.2 Viscoelastic Properties	485
	20.3	Mechanical and Viscoelastic Properties of Natural Rubber Blends and IPNs	487
		20.3.1 Natural Rubber/Thermoplastics	487
		20.3.2 Natural Rubber/Thermosets	491
		20.3.3 Natural Rubber/Synthetic Rubbers	494
		20.3.4 Natural Rubber/Biopolymers	496
	20.4	Conclusions	499
	References		499

Chapter 21 Scattering Studies on Natural Rubber Based Blends and IPNs 501
Valerio Causin

21.1	Introduction	501
21.2	Wide Angle X-Ray Diffraction	502
21.3	Small-Angle X-Ray Scattering	507
21.4	Small-Angle Neutron Scattering	517
21.5	Small-Angle Light Scattering	521
References		523

Chapter 22 Transport of Penetrant Molecules Through Natural Rubber Based Blends and IPNs 530
Isaac O. Igwe

22.1	Introduction	530
22.2	Natural Rubber: Properties and Applications	533
22.3	Natural Rubber Based Blends	534
22.4	Natural Rubber Based Interpenetrating Polymer Networks (IPNs)	535
22.5	Transport of Penetrant Molecules through Natural Rubber Based Blends and IPNs	536
22.6	The Effects of Penetrant Absorption on the Properties of Natural Rubber Systems	544
22.7	Conclusions	545
References		546

Chapter 23 Life Cycle Analysis, Ageing and Degradation Behaviour of Natural Rubber Based Blends and IPNs 550
*Cristina Russi Guimarães Furtado and
Márcia Christina Amorim Moreira Leite*

23.1	Introduction		550
23.2	Life Cycle Assessment		551
23.3	Ageing and Degradation of NR Based Blends and Interpenetrating Polymer Networks (IPNs)		553
	23.3.1	Polymer Blends and IPNs	553
	23.3.2	Ageing and Degradation	554
	23.3.3	NR/Thermoplastic Blends and IPNs	555
	23.3.4	NR/Synthetic Rubber Blends and IPNs	558
	23.3.5	NR/Biopolymer Blends and IPNs	561
23.4	Conclusions		564
References			564

Chapter 24 Application of Natural Rubber Based Blends and IPNs in Tyre Engineering and other Fields **569**
Mir Hamid Reza Ghoreishy and Mohammad Alimardani

24.1	Introduction	569
	24.1.1 Properties of Natural Rubber	569
	24.1.2 Elastomer Blends	572
	24.1.3 General Aspects of Compounding of NR and Blends	572
24.2	NR and its Blends for Tyre Components	575
24.3	NR in Seismic Isolation Bearings	577
	24.3.1 Why NR as Seismic Isolation?	578
24.4	Toughened Thermoplastics and IPNs of NR in the Automotive Industry	579
24.5	Membrane Technology	585
	24.5.1 Introduction	585
	24.5.2 Recent Achievements in the Field of NR Blends as Membranes	586
24.6	Miscellaneous Applications of Natural Rubber Based Blends	595
	24.6.1 Retreading of Tyres	595
	24.6.2 NR as an Insulator	595
	24.6.3 Use of NR for Modification of Plastic Properties	596
	References	596

Subject Index **600**

CHAPTER 1

Natural Rubber Based Blends and IPNs: State of the Art, New Challenges and Opportunities

GORDANA MARKOVIĆ,*[a]
MILENA MARINOVIĆ-CINCOVIĆ,[b]
VOJISLAV JOVANOVIĆ,[c]
SUZANA SAMARŽIJA-JOVANOVIĆ[c] AND
JAROSLAVA BUDINSKI-SIMENDIĆ[d]

[a] Tigar, Nikole Pašića 213, 18300 Pirot, Serbia; [b] University of Belgrade, Vinča Institute of Nuclear Sciences, Belgrade, Serbia; [c] Faculty of Natural Science and Mathematics, University of Priština, Kosovska Mitrovica, Serbia; [d] University of Novi Sad, Faculty of Technology, Novi Sad, Serbia
*Email: gordana1markovic@gmail.com

1.1 Introduction and History

The field of polymer science and technology has undergone an enormous expansion over recent decades primarily as a result of chemical diversity. Dilute solution behaviour, elasticity, tacticity, single crystal formation, viscoelastic behaviour, *etc.*, attained the prime interest from past researchers. The concept of physically blending two or more existing polymers to obtain new products is now attracting widespread interest and commercial utilization.

Investigation of polymer blends is one of the most active areas of research and development in the field of polymers at the present time. Polymer blends provide answers to technological challenges posed by the increasing difficulties of synthesizing new monomers and polymers to meet diverse demands, either domestic or industrial. The wide range of properties attainable with these systems were hitherto either impossible to obtain from an individual polymer or would involve costly development of new polymers. In fact, blending of polymers is one of the easiest and most flexible methods of generating new polymeric materials. It has become an important technique for improving the cost–performance ratio of commercial polymers. Polymer blends also provide a platform for scientific investigations in various fields such as newer characterization techniques, processing, molecular engineering to control the blend structure, structure–property correlation and modelling, *etc.*[1–14]

It is important to be able to predict and understand the resultant properties of a blend and its morphology from the properties of the constituent polymers. Predicting the mechanical behaviour of polymer blends and composites with respect to composition, structure (morphology of the blend) and properties of the components covering a wide temperature range are particularly important.[15,16]

Polymer blends and interpenetrating polymer networks (IPNs) form part of the composite materials system. Composite materials may be defined as materials made up of two or more phases.[17] They may be grouped into:

(i) particulate filled: consisting of a continuous matrix phase and a discontinuous filler phase made up of discrete particles;
(ii) fibre filled;
(iii) skeletal IPNs: both as continuous phases.

The factors that influence the properties of composite materials are: properties of the components, shape of the filler phase, morphology of the system, and the nature of the interface between the phases. The mechanical behaviour of composites is greatly affected by the interfacial adhesive bond between the phases. The morphology in polymer blends is indicative of the phase or phases and interrelationships existing in the blend. It reflects the domain size of the dispersed phase, state of aggregation, nature of the interface between the phases, *etc.* phase is nothing but a structurally homogeneous part of a material system. There may be a single continuous phase with one or more disperse phases, or two or more continuous phases which may contain one or more disperse phases. Blends are defined as simply a mixture of two or more polymers or copolymers. They may be miscible (domain size of the order of 0.5 nm) or immiscible (domain size of the order of 100 nm), depending on thermodynamic requirements. Miscible blends are thermodynamically stable, molecular level mixtures. Immiscible blends are separated into microscopic phases with very minimum interfacial adhesion and unstable morphology. Polymer alloy blends (PABs) are a class of polymer blends, heterogeneous in nature, with controlled morphology and properties, achieved by compatibilization and dynamic vulcanization. Compatibilized blends are also macro phase separated (domain phases are between those of miscible and immiscible systems), but the presence of interfacial agents or chemical bonds stabilizes the morphology and increases interfacial adhesion.[18] Domain size is also controlled and reduced, giving a fine dispersion. An IPN,[19] another rapidly growing development in the area of polymer science and technology, is a typical polymer blend in which two or more polymers exist in network form and are synthesized in juxtaposition. When one of the polymers is crosslinked, the product is called a semi-IPN. A general scheme is shown in Figure 1.1.

We can represent the system definition as[32] (Table 1.1).

The advantageous properties of IPNs come from co-continuity and excellent compatibilization (obtained by co-reactions) that result in fine dispersions. Often IPNs do not possess co-continuity of networks; rather one phase forms fine droplets (10–100 nm diameter) dispersed in another phase. IPNs with specific topological network structure provide smaller domains of phase-separated materials. Their miscibility level gives rise to a multiphase polymer system with ordered variety of domain structure ranging from a nanometre to

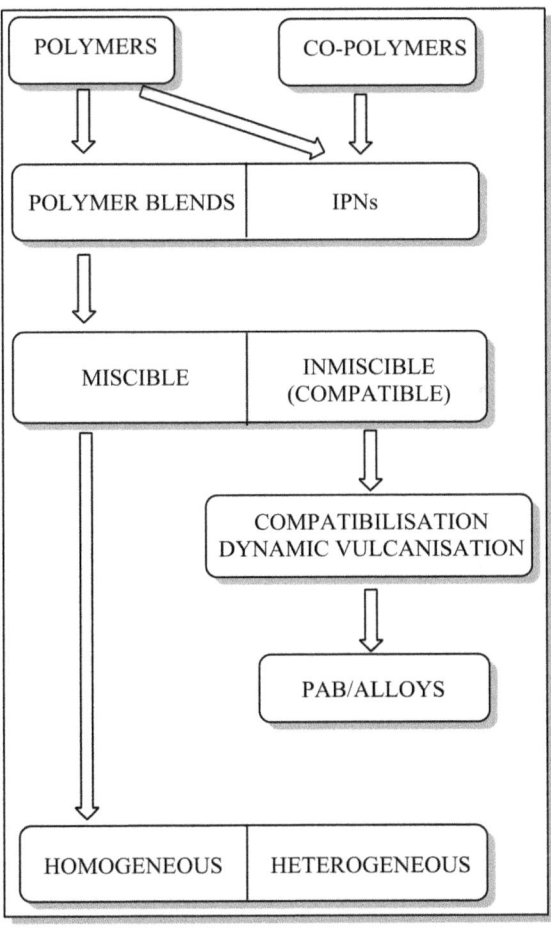

Figure 1.1 Scheme showing various types of polymer and polymer blends.

Table 1.1 Multiphase polymer systems.

Material	Definition	Phases		
		ts	tp	Continuity
Polymer blends	A mixture of polymers and copolymers	a	a	Not required
Polymer alloy blends (PABs)	A compatibilized or modified blend	a	a	Not required
IPNs	A mixture of finely dispersed chemically crosslinked polymers in network form	Yes	No	Required
Semi-IPN	A mixture of at least two polymers in network form where only one polymer is chemically crosslinked: the other is linear	b	b	Required
Thermoplastic polymer alloy blends (PABs)	A mixture of physically crosslinked polymers in network form	No	Yes	Required

[a]The polymer and copolymer may have either thermoset (ts) or thermoplastic (tp) character.
[b]One phase is thermoplastic and the other is thermoset.

micrometre scale.[20] In this respect PABs and IPNs are similar. However there is a difference in the dispersion of the domains. PABs with domain diameter < 100 nm are difficult to prepare, due to the difference in the method of preparation of the multiphase systems. PABs are prepared by melt blending whereas IPNs are obtained by polymerization and crosslinking.

In the preparation of PABs morphological control and stabilization by in situ compatibilization and dynamic vulcanization is gaining importance. The introduction of crosslinks into IPNs restricts the domain size to very small phases and enhances the degree of formulation of micro-heterogeneous structure which results in broad glass transition regions, making them very effective as damping materials over a broad temperature range.[21–25]

Polymer blends may be either homogeneous or phase separated or a combination of both.[26] The factors determining homogeneity or otherwise are mainly: the method of mixing, the kinetics of the mixing process, the processing temperature, and the presence of solvent or other additives, *etc*. However the miscibility of two or more polymers is primarily determined by thermodynamic criteria governed by Gibbs free energy considerations. The relationship between change in Gibbs free energy due to mixing (ΔG_m) and the enthalpy (ΔH_m) and entropy (ΔS_m) of mixing for a reversible system is:

$$\Delta G_m = \Delta H_m - \Delta S_m \tag{1.1}$$

If ΔG_m is positive over the whole composition range at a given temperature, the two polymers in the blend will separate into phases comprising either of the pure components, providing that a state of thermodynamic equilibrium has been reached. For complete miscibility, two conditions are necessary:

$$\frac{\sigma^2 \Delta G_m}{\sigma \phi_2^2} \succ 0 \tag{1.2}$$

ΔG_m must be positive and the second derivative of ΔG_m, also with respect to the volume fraction of component 2 (ϕ_2^2) over the whole composition range.

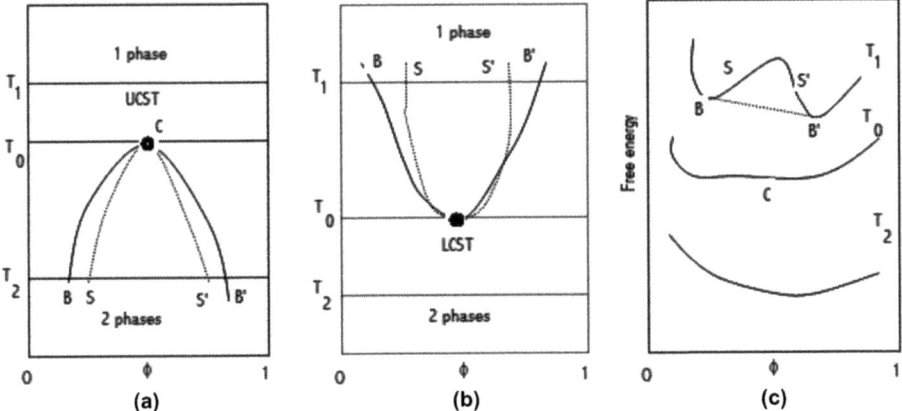

Figure 1.2 Phase diagram for systems showing (a) UCST, (b) LCST and (c) composition dependence of the free energy of the mixture (on an arbitrary scale) for temperatures above and below the critical value.

This ensures stability again phase separation. Figure 1.2(c) shows the free energy–composition dependence at three temperatures. At T_1, Equations (1.1) and (1.2) hold good and miscible single-phase mixtures occur at all compositions at this temperature. At T_2 Equation (1.2) is not satisfied for all composition and mix between the points B and B' separates into two phases resulting in a total free energy falling on the dashed line, which is lower than the homogeneous phase (solid line). The curve at intermediate temperature T_c showed the critical point at C. In Figure 1.2(a) $T_1 > T_2$ and T_c is an upper critical solution temperature (UCST), *i.e.* the temperature above which a homogeneous phase is possible for a particular composition, whereas in Figure 1.2(b) $T_2 < T_1$ and T_c is the lower critical solution temperature (LCST), *i.e.* the temperature below which a homogeneous single phase is possible for a particular composition. In this case miscibility of the component improves with declining temperature. Mixtures that have positive (endothermic) heats and entropies of mixing usually tend to exhibit a UCST whereas mixtures that have negative (exothermic) heats and entropies of mixing usually exhibit a LCST. The area dividing the single phase and two-phase region, *i.e.* the locus of all points B-B' is called the binodal curve. The inflection points S and S' are the free energy curve for T_2 define the spinodal curve, structure and dated line on T–t plane. The critical point at which the binodal and spinodal curve touch may not always lie at the extreme limit of the binodal. In a mix, binodal indicates the boundary between stable and meta-stable compositions whereas spinodal indicates the boundary between meta-stable and unstable compositions. The binodal defines the equilibrium phase behaviour, whereas spinodal is significant with respect to mechanism for kinetic of phase separation processes.

Nucleation is an active process of generating within a meta-stable mother phase the initial fragments, called the nucleus, of a new of more stable phase.[27] Once the nuclei (the critical rate-determining intermediates) are formed, the system decomposes with a decrease in free energy and the nuclei grow. The spinodal decomposition is a kinetic process of generating within an unstable mother phase a spontaneous and continuous growth of another phase. The growth originates not from nuclei but from small-amplitude compositional fluctuations.

The polymer–polymer miscibility is dependent upon a delicate balance of enthalpic (ΔH_m) and entropic (ΔS_m) forces. These are significantly smaller than those observed in the case of mixing of smaller molecules. The combinatorial entropy term is negligibly small because of the very high molecular weight long-chain structure. The heat of mixing is generally endothermic, resulting in positive free energy of mixing (ΔG_m). The polymer–polymer miscibility is primarily due to negative heat of mixing. This can be achieved through specific interactions between the constituent macromolecules comprising the blend.

The morphology of the heterogeneous polymer blends is characterized by two distinct phases and in general depends on thermodynamics and rheology of blend composition, interfacial tension between the component polymers, viscosity ratio and processing conditions and history.[28] In immiscible blends, performance depends upon the interface as well as on the size and shape of the dispersed phase. At equilibrium at low volume fractions of dispersed phase $\phi_d < \phi_m = 0.16$ droplets are expected, with fibres and lamellae usually at higher values. When the dispersed phase concentration increases to the value corresponding to phase inversion (ϕ) a co-continuous morphology results. Ravati[29] has shown that for blends with the same processing history, the melt viscosity ratio and composition determine the morphology. Generally, the least viscous component forms the continuous phase over a large composition range. The size of the dispersed phase will depend upon mixing variables[30] and to an extent on the thermodynamics of mixing, which will also determine the sharpness of the boundary between the dispersed phase and the matrix, for example.

In the case of blending of immiscible polymers (which is generally the case), two structural characteristics have to be met to avoid gross phase separation and ensure consistent good properties. These are proper interfacial tension providing a small phase size (below the micron level) and good interface adhesion providing a strong interface to resist the effects of stress and strain. To be compatible, interfacial tension must be zero or negative.

Morphology of the immiscible polymer blends can be controlled by using compatibilization or dynamic vulcanization, or both.[31,32]

A schematic representation of non-compatibilized and compatibilized blends is given in Figure 1.3.

Mixing, in general, is accompanied by entropy gain and heat absorption (endothermic). Entropy change depends on the number of molecules/unit volume whereas heat of mixing/unit volume is a function of the number of molecules in contact, which remains the same with increasing molecular size. Hence, in the case of polymers, entropy changes/unit volume being exceedingly small, heat of mixing will determine the homogeneity of the mix. Endothermic

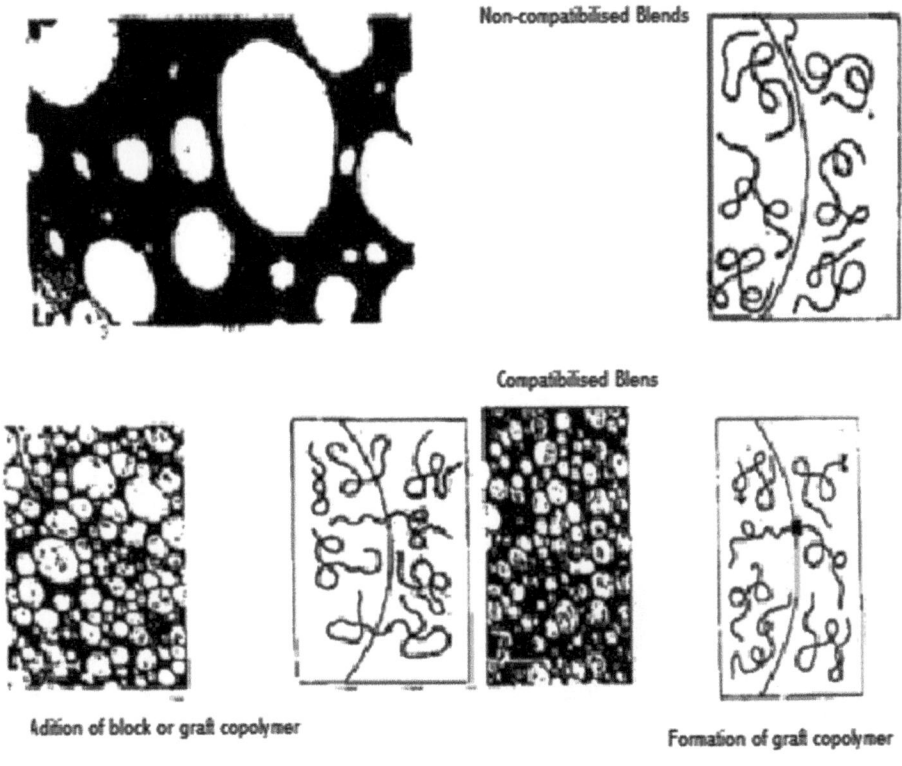

Figure 1.3 Schematic representation of non-compatibilized and compatibilized blends.

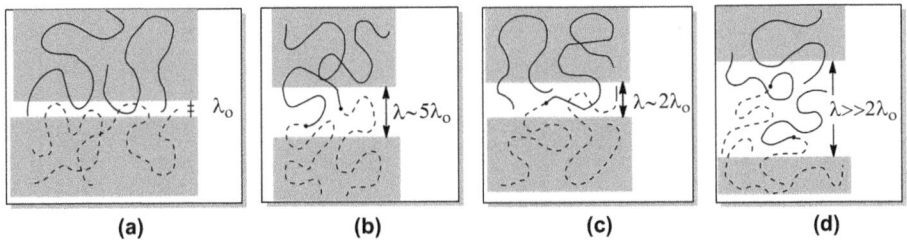

Figure 1.4 Interfaces of blend types: (a) between polymers A and B (non-reactive, *i.e.* physical blends); (b) between micro-domains of neat A-B block copolymers; (c) consisting of polymer A, polymer B and A-B block copolymers; (d) generated by reaction between polymers A and B.

mixing results when the energy of association in the mix is greater than the mean of the energies of association in the pure components. Specific interactions such as H-bonding or stereoisomerism provide favourable association energy in the mix, resulting in homogeneous blends. Interfaces of blend types are presented in Figure 1.4.

Polymer blends are generally made by mechanical as well as chemical methods, as depicted in Table 1.2.

Table 1.2 Methods of polymer blending.

Si no	Type	Method/Principle	Remarks
Mechanical methods			
a	Roll milling	The mixing of polymers can be accomplished by squeezing the stock between rolls	Required that viscosities of the constituents after blending should be such that no gross phase separation or demixing occurs after blending.
b	Melt blending	The polymers are mixed in the molten state	Requirements of high shear rate and temperature (especially in melt blending technique) to maintain low viscosity may lead to degradation of both or either of the polymers. This may result in crosslinking block or graft formation or chain scission. The blended polymers are often physically different from what might be expected for the blend. Hence melt mixing is used for systems where thermal degradation does not normally occur or adequate precaution is taken. Both these techniques avoid the contamination and removal problems of diluents or solvents and assure a system that will not change in moulding operations.
c	Solution blending	The two polymers are dissolved in a common solvent followed by solvent evaporation, freeze drying or polymer co-precipitation	These techniques lower the temperature and shear force requirements for satisfactory mixing thereby eliminating or at least minimizing problems of incomplete mixing and chemical changes caused by heat and shear. The blended stock may contain impurities or contaminations of solvents or emulsifiers, *etc.* Latex blending offers finer scale dispersion than solution blending. In solution blending, attempts to remove the solvent can lead to changes in domain size of the blend and in severe cases can cause complete phase separation.
d	Latex blending	The two polymers, available in the latex form, are mixed followed by drying and/or coagulation, resulting in an intimate mixture of the polymers. Melt mixing is often employed for compounding and pelletizing the latex. Precautions to be taken to avoid degradation of blended stock.	
Chemical methods			
a	Copolymerization	Formed as a result of graft and block copolymerization processes	
b	IPN formation	Synthesized by the polymerization of a monomer in the presence of a crosslinked polymer	This gives rise to a crosslinked polymer completely interwoven with the first polymer. Such precise control of polymer and blend structures offers great promise for novel properties and applications.

There are several characterization techniques for polymer blend systems in respect of mechanical behaviour, structure–property relationships, phase morphology characteristics and their interaction vs. miscibility or compatibility, and so on. These may be categorized as: (i) microscopic techniques; (ii) study of glass transition; (iii) spectroscopic techniques; (iv) scattering methods; and (v) viscosity measurements.

1.1.1 Interpenetrating Polymer Networks

IPNs are chemically homogeneous (on a scientific large scale binary systems) which could be conveniently prepared by first swelling the loosely crosslinked component in a suitably chosen monomer and subsequently crosslinking the latter. This procedure of chemical blending offered a way of avoiding the hazards of imminent phase separation present in the physical blending of linear polymers, due to topological constraints (permanently entangled, chemically different macrocycles) on IPN unmixing. IPN materials have been studied extensively on polymer alloys with synergistic physical properties of technological interest. The polyurethane (PU) with epoxy resin was the first reported simultaneous IPN.[33]

1.1.2 History of IPN Development

It is difficult to pinpoint the origin of IPNs. Goodyear's work on vulcanization, and the development of polymer blends, grafts and blocks, all led to the development of IPNs. The first known IPN was invented by Aylsworth in 1914.[34] This was a mixture of natural rubber, sulfur and partly reacted phenol formaldehyde resins. The term 'interpenetrating polymer network' was coined by Millar in 1960, who carried out a series of scientific studies of PS/PS IPNs, which were used as ion exchange resin matrices.[35] Frisch and coworkers conceived of IPNs as the macromolecular analogue of catanenanes.[36] Sperling and Thomas investigated finely divided polyblends using electron microscopy to understand the phase structure and correlated the mechanical and viscoelastic behaviour with the phase structure.[37] Lastumaki *et al.* considered the use of semi-IPNs as adhesives.[38] The IPN field is not widely explored and it inspires yet more research. The history and development of IPNs and related materials are shown in Table 1.3. Schematic representation of different types of blends is given by Figure 1.5.

Since most IPNs involve the polymerization of one monomer in the immediate presence of the other, they are also called graft copolymers. They form a special class due to the crosslinking of one or both the polymers. The interesting and unique properties of IPNs emerge when the deliberately introduced crosslinks outnumber the accidentally introduced grafts. Some amounts of grafts are still present in IPNs and usually contribute to the IPN behaviour in a favourable manner. The IPNs and graft IPNs are schematically shown in Figure 1.6.

Table 1.3 History of IPNs and related materials.

Event	First investigators	Year
IPN type structure	Aylsworth	1914
Graft copolymers	Ostromislensky	1927
Interpenetrating polymer network	Staudinger and Hutchinson	1951
Block copolymers	Dunn and Melville	1952
Homo-IPNs	Millar	1960
Sequential IPNs	Sperling and Friedman	1969
Latex IPNs	Frisch, Klempner and Frisch	1969
Simultaneous IPNs	Sperling and AMTs	1971
IPN nomenclature	Sperling	1974
Thermoplastic IPNs	Davison and Gergen	1977

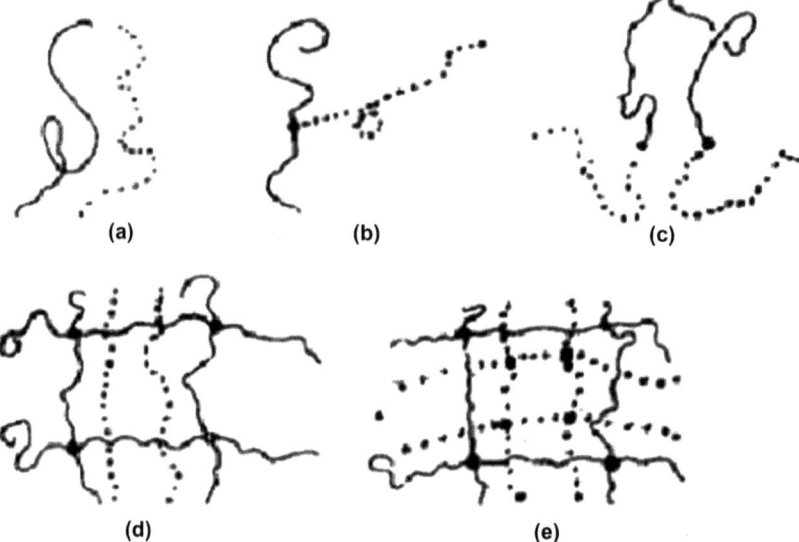

Figure 1.5 Schematic representation of different types of blends: (a) mechanical blends, (b) grafts, (c) blocks, (d) semi-interpenetrating polymer networks, (e) full interpenetrating polymer networks.[63]

The two characteristic features of IPNs that distinguish them from other polymeric mixtures are that (i) IPNs swell but do not dissolve in solvents and (ii) creep and flow are suppressed.[39]

According to the mode of synthesis, IPNs are distinguished into five different types: (i) sequential IPNs, (ii) simultaneous IPNs, (iii) interpenetrating elastomeric networks, (iv) thermoplastic IPNs and (v) gradient IPNs:

> (i) Sequential IPNs, where polymer I is crosslinked initially followed by the swelling of polymer network I with monomer II plus crosslinker and initiator and subsequent polymerization of monomer II, *in situ.*

Natural Rubber Based Blends and IPNs 11

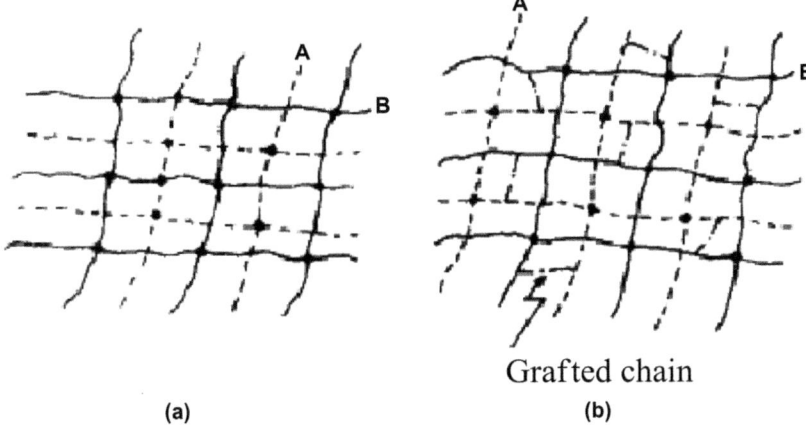

Figure 1.6 Schematic representation of (a) interpenetrating polymer network and (b) graft interpenetrating polymer network.

(ii) Simultaneous IPNs (SINs), where a mutual solution of monomer I, monomer II, crosslinker I, crosslinker II are taken together and then polymerized simultaneously by non-interfering modes. The above two types of synthesis are represented diagrammatically in Figure 1.7.
(iii) There is a third mode of synthesis where two lattices of linear polymers are mixed and coagulated and both the polymers are crosslinked simultaneously. They are called interpenetrating elastomer networks (IENs).
(iv) In the case of thermoplastic IPNs, physical crosslinks are present rather than chemical covalent crosslinks. Frequently, the polymers exhibit some degree of phase continuity. In all such cases, the thermoplastic IPNs behave as thermosets at ambient temperature and as thermoplastics at elevated temperature.
(v) In gradient IPNs the composition is varied within the sample at the macroscopic level. This is carried out by swelling the network in monomer for the required time and polymerizing rapidly, before equilibrium sets in.[40]

Although these types of system are termed IPNs, true interpenetration occurs only at phase boundaries and most IPNs phase separate to some extent. Molecular level interpenetration occurs in the case of total mutual solubility only.

1.1.3 Properties of Polymer Blends and IPNs

The properties of blends and IPNs are dependent on the two-phase nature, phase continuity, domain size and molecular mixing at phase boundaries. A study of morphology, glass transition temperature, modulus, *etc.*, explains the properties of any two-component systems.

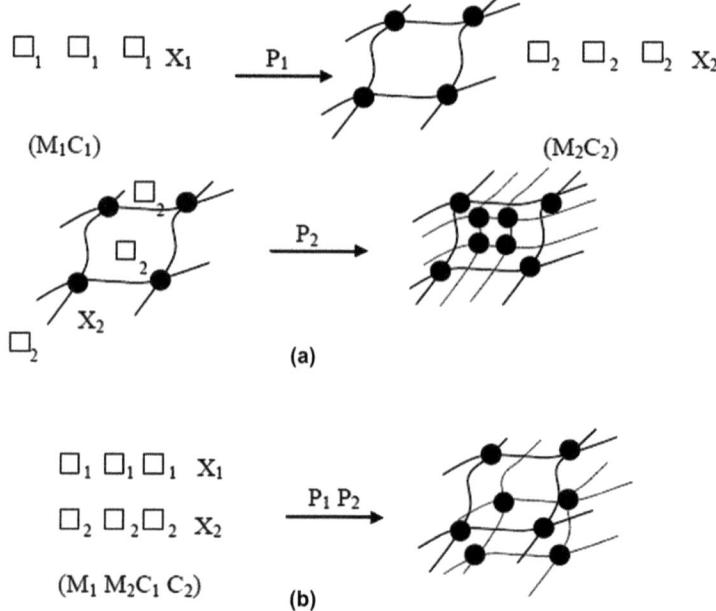

Figure 1.7 Schematic representation of synthesis of IPNs through (a) sequential and (b) simultaneous methods.[62]

The properties of IPNs depend on: (i) properties of the component polymers; (ii) phase morphology; and (iii) interactions between the phases.

Some properties of IPNs are simple averages of the properties of the component polymers. Sometimes the values are higher than expected. Density measurements on PU/PS simultaneous IPNs showed a density higher than expected. This was due to the partial interpenetration of chains of the rubbery and glassy polymer components.[41]

1.1.4 Glass Transition and Viscoelastic Behaviour

Four situations can be observed: (i) two distinct T_g corresponding to each network; (ii) two distinct but inwardly shifted T_g; one broad T_g intermediate to the T_g of each network; (iv) one sharp T_g intermediate to the T_g of each network.[42]

An inward shift or a merging of T_g can be considered as partial evidence of interpenetration. In many cases IPNs do not exhibit a single, broad T_g. Two distinct glass transitions may be observed as a result of phase separation. In the case of acrylic/urethane IPNs, two distinct transitions are observed. The slight shifting and broadening of the peaks indicate moderate degrees of molecular mixing.[43]

Fay et al.[44] evaluated the effects of morphology, crosslink density and miscibility on IPN damping properties for a number of acrylic, methacrylic, styrene and butadiene-based copolymers and IPNs. The differential scanning calorimetry (DSC) curve of 50/50 composition showed a narrow T_g for copolymer IPNs and a broader T_g for multiphase IPNs. The glass transition behaviour of IPNs depends on their morphology.

Nayak[45] studied the glass transition behaviour of polyurethane/polymethyl methacrylate (PU/PMMA) and polyurethane/polystyrene (PU/PS) IPNs. Two T_g were observed in both IPNs showing phase separation. However, an inward shift in T_g is observed, showing some intermixing. The pseudo IPNs and linear blends did not show this inward shift.

Sartor et al.[46] studied the DSC curve of IPN of composition 25% PU/75% PMMA. In contrast to usual observations, the T_g increased over a very broad temperature range, which is due to the distribution of relaxation time. Chen et al.[47] studied pseudo IPNs of linear poly(carbonate urethane) and crosslinked polychloroprene. The single-phase morphology of this pseudo-IPN has been confirmed by DSC measurements.

A large number of IPNs exhibit considerable toughness as measured by stress–strain curves or impact strength. Das and Chakraborty[48] worked on the mechanical properties of epoxy-poly(2-ethyl hexyl acrylate) (PEHA) IPNs. As the elastomeric PEHA content in the blend is increased, properties such as modulus and tensile strength for both semi- and full IPNs decreased, due to the lower mechanical strength of PEHA compared to epoxy resins. Epoxy/PEHA semi-IPNs have higher modulus and tensile strength compared to full IPNs. This may be due to higher interpenetration of linear PEHA molecules with the epoxy network while in full IPNs, PEHA was crosslinked before the formation of the epoxy network.

The styrene-butadiene rubber (SBR)/PS IPN developed by Donatelli et al.[49] showed that at room temperature, yield strength and modulus decrease with increasing SBR content, but increases with the degree of crosslinking of the rubber phase. An optimum degree of toughness is obtained at an intermediate level of crosslinking of SBR. The impact strength was found to be independent of crosslinking in PS.

1.1.5 Morphology

The morphology of IPNs can be determined by electron microscopy. The morphology has a great influence on the physical and mechanical behaviour. The phases differ, however, in amount, size and shape, sharpening of their interface and degree of continuity. The factors affecting morphology are chemical compatibility of the polymers, interfacial tension, crosslink density of the networks, polymerization methods and IPN composition.

The physical, mechanical and morphological behaviour of PBR/PMMA IPNs was correlated effectively by Das et al.[50] The effect of composition and crosslinking of elastomer and plastomer on properties were studied in detail. The morphological studies showed that full IPNs had a more ordered, compact and uniform phase distribution compared to the corresponding semi-IPNs. This can explain the higher tensile strength, modulus and tear strength of full IPNs and higher elongation at break (%) of semi-IPNs.

Frish and Xue[51] used TEM to visualise the network structure in poly(butadiene)/PS semi-IPNs. IPNs of PU and epoxy were studied by Chern *et al.*[52] When the epoxy content is increased, the tensile strength of the IPNs and graft IPNs decreased. The tensile strength of IPNs was lower than that of the graft IPNs. In the graft IPNs, some urethane chains are grafted to the pendant secondary hydroxy group of the epoxy through the reaction with isocyanates, which resulted in higher intermolecular force and high tensile strength.

Pater[53] used an IPN approach to toughen microcracking resistant high temperature polymers, for use in aircraft/aerospace structural components. In this work he combined crosslinked PMR-15 and linear laRC-TP1 to form a new sequential semi-2-IPN called LaRC-RP41. Various techniques were employed to study the phase morphology and phase stability of LaRC-RP41 neat resin and composite.

Most IPNs and related materials show phase separation. The phases however, differ in amount, size and shapes, sharpening of their interface and degree of continuity. All this aspects control the morphology and thereby determine the material properties.

A degree of compatibility between polymers is introduced by IPN formation as two polymers are interlocked in a three-dimensional structure during synthesis. The phase domains are smaller in higher compatibility systems. Most of the early IPN studies were based on immiscible polymers and a small gain in phase mixing was observed. Frisch and coworkers prepared IPNs from PS and poly(2,6-dimethyl-1,4-phenylene oxide), which were miscible.[54] Klempner[55] prepared IPNs that contained opposite charges, and showed better miscibility than IPNs without charge.

IPNs based on SBR/PS studied by Donatelli *et al.*[49] emphasises the control of morphology by varying the crosslinking levels, composition and chemical compatibility. The composition of an IPN affects the morphology and thereby the properties of the product. An increase or decrease in the percentage of the component polymers affects the domain size of the dispersed phase. Chantarasiria *et al.*[56,57] studied the variation of domain size with composition and crosslink density of polymer I for castor oil-urethane/PS IPNs. As the wt% of polymer I increases, the NCO/OH ratio increases and this ratio has an effect on crosslink density, morphology and properties.

1.2 Recent Trends and Developments in Natural Rubber Based Blends and IPNs

Natural rubber (NR) is a typical elastomer, having good resilience, elongation and elasticity, ease of storage, milling and the possibility of using different vulcanization processes.

Das and coworkers studied the morphology and mechanical properties of NR/PMMA IPNs.[58] Mathew studied solid NR and PS, intimately mixed by the sequential method of IPN synthesis.[16] Semi-IPNs based on NR/PS where only the NR phase is crosslinked offer a binary polymer system network having the elastomeric properties of NR imparted to the hard brittle properties of PS. Schematically, the morphology of semi-IPNs can be represented as shown

Natural Rubber Based Blends and IPNs 15

in Figure 1.8, where the chains of the co-continuous components are crosslinked.

Semi-IPNs of NR and PS are prepared in two steps. First, NR is crosslinked with dicumyl peroxide (DCP). The vulcanized sheets of NR are allowed to swell in styrene (inhibitor free) containing 1% DCP. The swollen sheets are heated and then put in a vacuum to get rid of residual styrene monomer. Since the NR phase alone gets crosslinked the resulting IPN is of either semi- or pseudo-IPN type. Figures 1.9, 1.10 and 1.11 depict SEM micrographs of NR/PS rubber blends.

SEM investigation[59] of these IPNs also reveals the morphology to be a predominantly co-continuous structure of the two polymers. It is difficult to justify the findings of the discrete particle model (with PS as matrix) based on this count.

The NR/PS IPN system is a high-performance material with excellent mechanical properties. They possess a fair level of thermal stability and solvent resistance. There are attempts to fabricate useful products using NR/PS IPNs to exploit these materials commercially. The data on impact resistance, volume

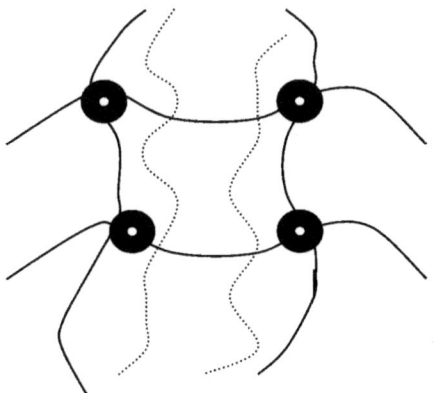

Figure 1.8 Schematic representation of semi-IPN morphology.

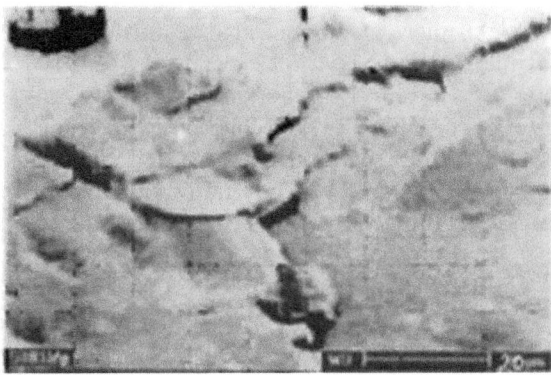

Figure 1.9 Scanning electron micrograph for 30/70 NR/PS semi-IPN.

Figure 1.10 Scanning electron micrograph for 50/50 NR/PS semi-IPN.

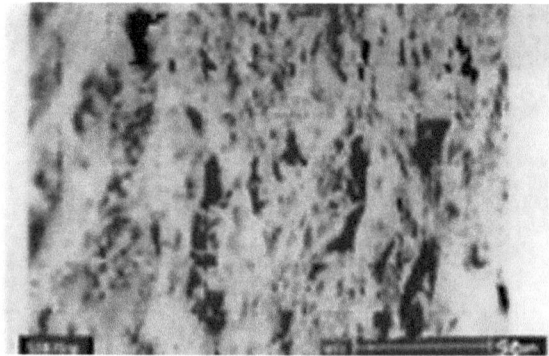

Figure 1.11 Scanning electron micrograph for 70/30 NR/PS semi-IPN.

resistivity, thermal stability and solvent resistance can be made use of while selecting suitable IPN systems for a particular end use.

The T_g of NR and PS were found to occur around −50 °C and 110 °C, respectively, and SEM micrographs revealed mainly dispersed domains of PS particles in a NR matrix.

Physical blends of NR and 1,2-polybutadiene (1,2-PBD), a thermoplastic rubber (Figures 1.12, 1.13 and 1.14)[60] combine the good physical and superior ageing properties of 1,2-PBD with the unique elastomeric/hysteresis characteristics of NR. The NR/1,2-PBD blends find application in films, footwear, adhesives, coatings, *etc.* and also have the potential for use in radiation related fields.

Blends of NR and ethylene-vinyl acetate copolymer (EVA) (physical blends and statically vulcanized)[61] combine the good elastomeric and mechanical properties of NR with the excellent ageing and flex crack resistance of EVA. Blends of NR/EVA copolymer are becoming an important rubber/thermoplastic elastomer blend. Applications of these materials can be found in fields where

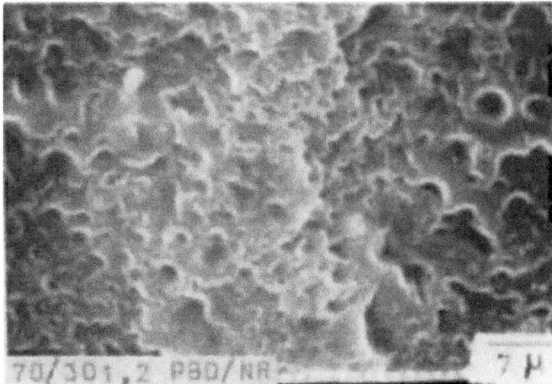

Figure 1.12 Scanning electron micrograph for 70/30 1,2-PBD/NR blend.

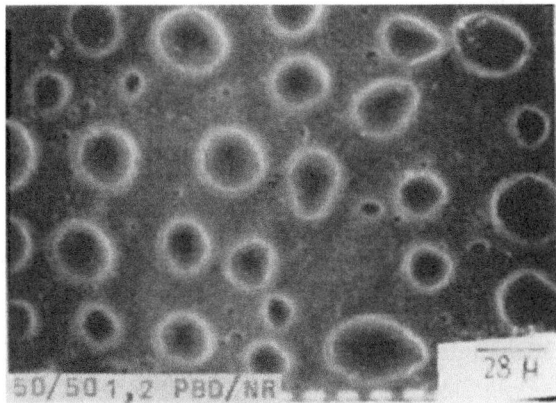

Figure 1.13 Scanning electron micrograph for 50/50 1,2-PBD/NR blend.

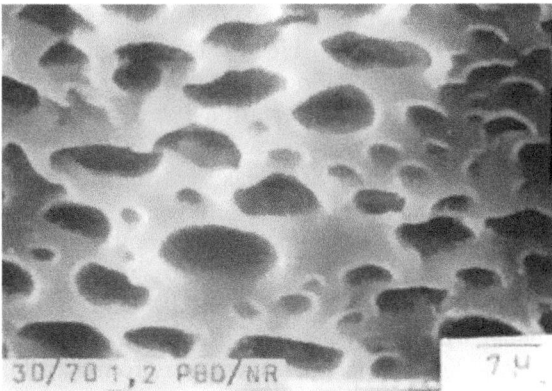

Figure 1.14 Scanning electron micrograph for 30/70 1,2-PBD/NR blend.

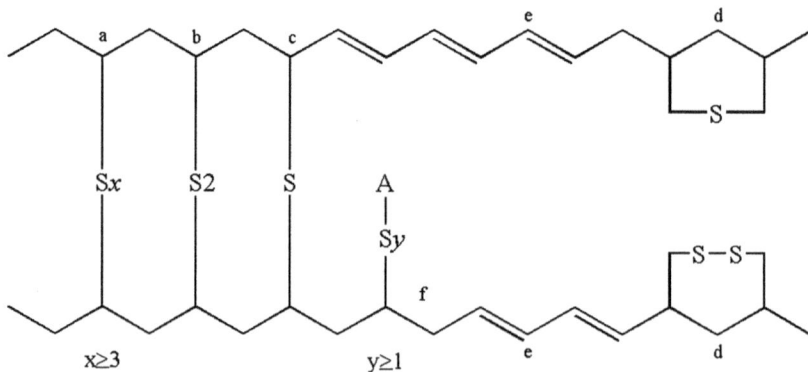

Figure 1.15 Schematic representation of all sulfide link types in a networked rubber: (a) polysulfide links; (b) disulphide links; (c) monosulfide links; (d) cyclic sulfides; (e) conjugated triene; (f) hanging groups originated from the accelerator.

NR is normally used, but where the EVA provides added advantages, such as microcellular soles.

Blends of NR/EVA are conventionally cured after intermix mixing. The blends are cured with a mixed cure system of S + DCP (sulfur + dicumyl peroxide). While sulfur cures only the NR phase of the blend, DCP is expected to cure both the phases. The possible network structure is represented in Figure 1.15. The morphology of the blend is predicted based on uncured blend SEM studies and is schematically represented in Figure 1.16.

When crosslinked using a mixed cure system (DCP and sulfur), good agreement for 30/70 and 70/30 NR/EVA blends in the case of discrete particle model (with EVA as matrix in the former and NR in the composition) is obtained (Figures 1.17, 1.18 and 1.19).

The T_g of NR and EVA were found to occur around −46 °C and −10 °C, respectively. When crosslinked, the T_g of NR was found to increase to −40 °C, while that of EVA remained unchanged. The incompatible nature of the blend was revealed by the presence of individual component tan δ peaks in the blends. Good agreement of 30/70 NR/EVA blend was supported by SEM studies. For 50/50 composition, co-continuous morphology for the blend is observed. For 70/30 NR/EVA rubber blend, the deviations were explained on the basis of larger domains of EVA particles dispersed in the NR matrix. Investigation of T_g and peak temperature from DSC measurements suggested the immiscible two-phase structure of the blends.

NR blended with condensed tannin (polymeric proanthocyanidins) formed semi-IPN composites by haematin-catalysed crosslinking.[62]

Polymerization of two networks, NR and PMMA *via* an IPN technique resulted in a polymer with a wide range of properties. The effect of variation of the DCP as a crosslinker in the NR matrix was found to be prominent. An increase in the DCP concentration result in an increase in the tensile strength, modulus and density.[63]

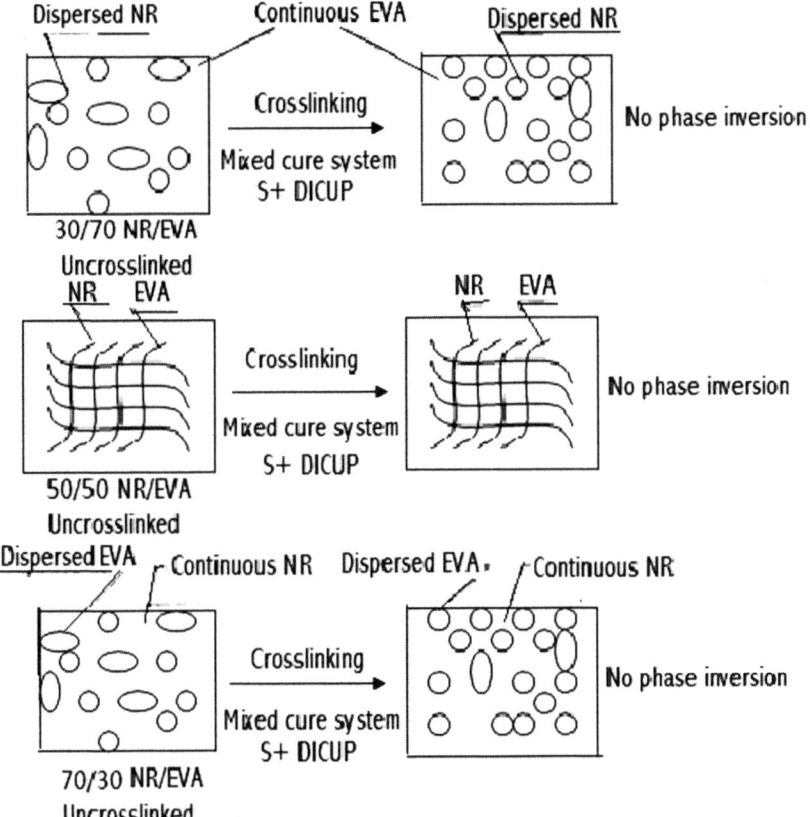

Figure 1.16 Schematic model for the morphology of crosslinked NR/EVA blends.

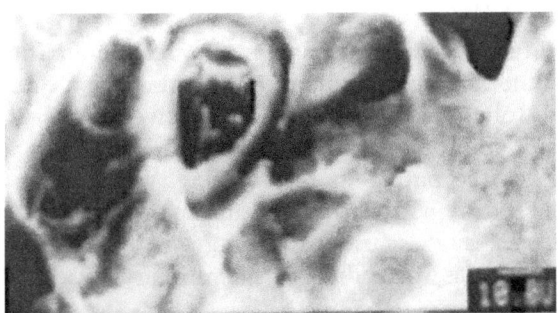

Figure 1.17 Scanning electron micrograph for 40/60 NR/EVA blend.

Full IPNs based on NR and poly(vinyl alcohol) (PVA) were prepared by using glutaraldehyde as the common crosslinking agent. A remarkable improvement was observed in the mechanical strength and thermal stability of the blends by the formation of an IPN.[64]

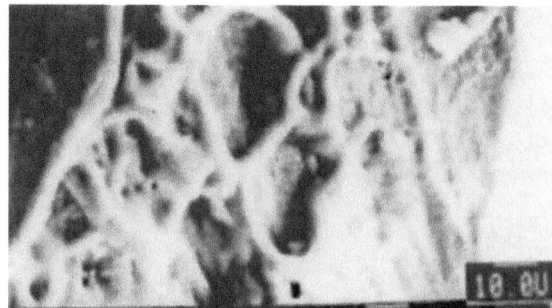

Figure 1.18 Scanning electron micrograph for 50/50 NR/EVA blend.

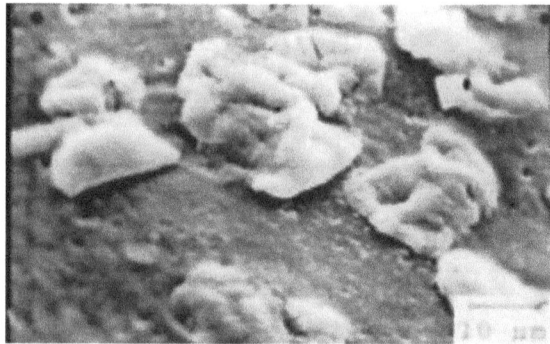

Figure 1.19 Scanning electron micrograph for 70/30 NR/EVA blend.

1.3 Applications and the Potential Market for IPNs

The patent literature reveals that there are many patents utilizing an IPN or closely related materials. Patents mention the production of optically smooth plastic surfaces, tough plastics, pressure-sensitive adhesives, ion exchange resins, noise and vibration damping materials, impact modifiers, contact lenses, *etc*. Besides the patent literature there are many suggested uses in the scientific literature. Predecki[65] mentions arteriovenous shunts and Toushanuent *et al*.[66] suggest uses as casting syrups. Toughened elastomers, impact-resistant plastics, piezodialysis membranes and wire insulation are mentioned.[67]

Hutchinson and coworkers examined a wide range of IPNs, SINs and semi-IPNs that were useful as reinforced elastomers. PU/PMMA semi-IPNs[68] prepared by bulk polymerization are used in shaping polymer articles. Epoxy/PU semi-IPNs[69] are used as adhesives. Ion exchange resins are prepared from chloromethyl PS/sulfonate PS IPNs.[70] PS/PS IPNs are used to prepare optically smooth plastic surfaces.[71] Poly(ethylmethacrylate)/poly(*n*-butyl acrylate) prepared by Sperling *et al*.[72] find the applications as noise damping coatings.

Thermoplastic IPNs have outstanding resistance to shrinkage and distortion on heat ageing, ozone resistance and have better mechanical properties.

It could be noticed that lower melt viscosity, notable elongation at break, high per cent strain to break and high modulus are achieved by chemical blending or IPN formation. Butyl rubber/phenolic IPNs prepared by Tawney et al.[73] exhibited improved high-temperature properties over ordinary butyl rubber. The natural rubber latex was used to synthesize rubber/plastic IPNs.[74–76] Epoxy-based IPNs[77] were found to yield adhesives with high bonding strength to metal. A hydrophilic/hydrophobic SIN suitable for contact lenses was prepared by Falcetta et al.[78] Hydroxy ethyl methacrylate or similar monomers formed polymer I and a polysiloxane formed the network II. Natural leather/rubber IPNs give rise to improved leather, as discovered by Feairheller et al.[79] Plastic/rubber IPNs are useful in preparing optically smooth surfaces,[80] compression moulding compositions[81] and tendture base materials.[82] Plastic/plastic IPNs are used as thermoplastics elastomers and impact resistant plastic.[83] Rubber/rubber IPNs could be used as pressure-sensitive adhesives.[84]

A new application for IPNs involves controlled drug delivery. For many purposes a controlled steady drug delivery is desirable.[85] In general, IPNs and semi-IPNs can be used wherever graft copolymers and polymer blends are used. The crosslinks in IPNs allow a new mode of control over two-phase morphology and hence control their properties. Some commercially used IPN materials are listed in Table 1.4.

Table 1.4 Commercially used IPNs.

Manufacturer	Trade name	Composition	Application
Shell Chemical Company	Kraton IPN	SEBS polyester	Automotive parts
Petrarch Systems Inc.	Rimplast	Silicone rubber-PU	Medical gears
ICI Americas Inc.	ITP	PU-polyester styrene	Sheet moulding compounds
DSM NV	Kelburon	PP-EP rubber-PE	Automotive parts
Shell Research BV		Rubber-PP	Tough plastic
Reichlod Chemical Co.	TRP	EPDM-PP	Auto bumper parts and wire and cable
Rohn and Ilas	–	Anionic-cationic	Ion exchange resins
Monsanto	Santoprene	EPDM-PP	Tyres, hoses, belts and gaskets
Du Pont	Somet	EPDM-PP	Outdoor weathering
BF Goodrich	Telcar	EPDM-PP or PE	Tubing, liners and wire and cable
Exxon	Vistalon	EPDM-PP	Paintable automotive parts
Freeman Chemical	Aopol	Acrylic-urethane-polystyrene	Sheeting moulding compounds
Deutsply International	Trybite Bioform	Acrylic based	Artificial teeth
Hitachi Chemical	–	Vinyl phenolics	Damping compounds

1.4 Environmental Impact and Recycling

Economics coupled with environmental factors of consumer products have been contributing to the growing preoccupation of people involved in industrial nations. Experts have identified three main benefits attributable to recycling:

- Reduction in space for disposal capacity.
- Lowered emissions from landfills and incinerators.
- Reduction in litter and improper disposal.

The three most important benefits resulting from the use of recycled materials are:

- Reduction in energy use and related emissions.
- Reduction in emission of toxic gases due to reusability of polymers, and improvement in extraction and manufacturing processes.
- Long-term value of conservation of raw materials.

Rubber recovery can be a difficult process. There are many reasons, however, why rubber should be reclaimed or recovered:

- Recovered rubber can be half the cost of natural or synthetic rubber.
- Recovered rubber has some properties that are better than those of virgin rubber.
- Producing rubber from reclaim requires less energy in the total production process than does virgin material.
- It is an excellent way to dispose of unwanted rubber products, which is often difficult.
- It conserves non-renewable petroleum products, which are used to produce synthetic rubbers.
- Recycling activities can generate work in developing countries.
- Many useful products are derived from reused rubber and other rubber products.
- If rubber is incinerated to reclaim embodied energy then it can yield substantial quantities of useful power. In Australia, some cement factories use waste rubber as a fuel source.

There is an enormous potential for reclamation and reuse of rubber in developing countries. There is a large wastage of rubber in many countries and the aim of this brief is to give some ideas for what can be done with this valuable resource. Whether rubber is reused, reprocessed or hand crafted into new products, the end result is that there is less waste and less environmental degradation as a result.

In developing countries, there is a culture of reuse and recycling. Waste collectors roam residential areas in large towns and cities in search of reusable articles. Some of the products that result from the reprocessing of waste are particularly impressive and the levels of skill and ingenuity are high. Recycling artisans have integrated themselves into the traditional marketplace and have created a viable livelihood for themselves in this sector. The process of rubber collection and reuse is a task carried out primarily by the informal sector. Rubbers are seen as being too valuable to enter the waste stream and are collected and put to use.

In Karachi, Pakistan, for example, rubber is collected and cut into parts to obtain secondary materials which can be put to good use. The beads of the rubber are removed and the rubber removed by burning to expose the steel. The tread and sidewalls are separated – the tread is cut into thin strips and used to cover the wheels of donkey carts, while the sidewalls are used to produce items such as shoe soles, slippers or washers.

Therefore, nowadays researchers pay more attention to reclaimed interpenetrating rubber. The main reason for this is the stable nature of IPNs in crosslinked rubbers and the strict specifications regarding the quality of products. The disposal problem of crosslinking rubber rejects and possible solutions are available. Another method to reuse large volumes of scrap IPN rejects is to use it as an impact plastic such as PS.

1.5 Conclusions

Polymer blends and IPNs are multiphase polymer systems of technological importance and wide-ranging research interest. Although the multiphase polymer systems (compatibilized blends and IPNs) referred have received much attention in terms of the experimental characteristics of the properties and industrial applications, little work has been done on the use of theoretical models to predict the properties of such materials. The properties of blends and IPNs are dependent on the two-phase nature, phase continuity, domain size and molecular mixing at phase boundaries. A study of morphology, glass transition temperature, modulus, *etc.* explains the properties of any two-component systems. IPNs show great potential as films and coatings in the polymer industry and also in the medical field due to their very good resistance to most common solvents. The NR/PS IPN system is a high-performance material with excellent mechanical properties. They possess a fair level of thermal stability and solvent resistance. Blends of NR and 1,2-PBD (physical blends) combine the good physical and superior ageing properties of 1,2-PBD with the unique elastomeric/hysteresis characteristics of NR. Blends of NR and EVA (physical blends and statically vulcanized) combine the good elastomeric and mechanical properties of NR with the excellent ageing and flex crack resistance of EVA. Polymerization of two networks, NR and PMMA *via* an IPN technique results in a polymer with a wide range of properties. NR and PVA prepared by using glutaraldehyde as the common crosslinking agent.

IPNs have a great future in the rubber and polymer industries as the new polymers in future.

Acknowledgements

Financial support for this study was granted by the Ministry of Science and Technological Development of the Republic of Serbia (Project Numbers 45022 and 45020).

References

1. J. B. Busche, A. E. Tonelli and C. M. Balik, *Polymer*, 2010, **51**, 6013.
2. H. Liang, R. Yu, B. D. Favis and H. P. Schreiler, *J. Polym. Sci.: Part B: Polym. Phys.*, 2000, **38**, 2096.
3. G. Janerfeldt, L. Boogh and J. A. E. Manson, *Polymer*, 2000, **41**, 7627.
4. A. Karim, D. W. Lieu and J. F. Douglas, *Polymer*, 2000, **41**, 8455.
5. B. Meissner, *Polymer*, 2000, **41**, 7827.
6. I. Luzinov, C. Pagnoulle and R. Jerome, *Polymer*, 2000, **41**, 7099.
7. Y. S. Lipatov, *Prog. Polym. Sci.*, 2002, **27**, 1721.
8. A. D. Litmanovich, N. A. Platé and Y. V. Kudryavtsev, *Prog. Polym. Sci.*, 2002, **27**, 915.
9. Y. He, B. Zhu and Y. Inoue, *Prog. Polym. Sci.*, 2004, **29**, 1021.
10. S. Ravati and B. D. Favis, *Polymer*, 2010, **51**, 4547.
11. E. Fekete, E. Földes and B. Pukánszky, *Eur. Polym. J.*, 2005, **41**, 727.
12. R. Mezzenga, L. Boogh and J. A. E. Manson, *J. Polym. Sci. B: Polym. Phys.*, 2000, **38**, 1983.
13. R. D. Siduth and R. Seyfarth, *J. Appl. Polym. Sci.*, 2000, **77**, 1954.
14. Y. Rao and R. J. Farris, *J. Appl. Polym. Sci.*, 2000, **77**, 1938.
15. J. K. Rameshwaram, Y.-S. Yang and H. S. Jeon, *Polymer*, 2005, **46**, 5569.
16. A. P. Mathew, *Interpenetrating Polymer Network Based on Natural Rubber and Polystyrene*, PhD thesis, Mahatma Gandhi University, India, 1999.
17. A. Gungor, *J. Appl. Polym. Sci.*, 2006, **99**, 2438.
18. B. D. Favis, *Polymer Blends Vol. 1: Formulations*, in D. R. Paul and C. B. Bucknall, eds., John Wiley & Sons, Inc., New York, 2000, pp. 501–538.
19. *Interpenetrating Polymer Networks*, Advances in Chemistry Series 239, ed. D. Klempner, L. H. Sperling and L. A. Utracki, American Chemical Society, Washington, DC, 1994, pp. 3–638.
20. R. M. Alcantara, A. P. Pinho Rodrigues and G. Gomes de Barros, *Polymer*, 1999, **40**, 1651.
21. L. H. Sperling, in *Interpenetrating Polymer Networks,* Advances in Chemistry Series 239, ed. D. Klempner, L. H. Sperling and L. A. Utracki, American Chemical Society, Washington, DC, 1994, pp. 3–38.
22. D. Sophia, D. Klempner, V. Sendijarevic, B. Suther and K. C. Frish, in *Interpenetrating Polymer Networks*, Advances in Chemistry Series 239, ed. D. Klempner, L. H. Sperling and L. A. Utracki, American Chemical Society, Washington, DC, 1994, pp. 39–75.
23. R. Hu, V. L. Dimonie, M. S. El-Aasser, R. A. Pearson, A. Hilter, S. G. Mylonakis and L. H. Sperling, *J. Polym. Sci., Part B: Polym. Phys.*, 1997, **35**, 1501.

24. R. B. Fox, J. J. Fay, Usman Sorathia and L. H. Sperling, in *Sound and Vibration Damping with Polymers*, ed. R. D. Corsaro and L. H. Sperling, American Chemical Society, Washington, DC, 1994, pp. 359–365.
25. J. R. Fried, *Polymer Science and Technology*, Prentice Hall, Englewood Cliffs, NJ, 1995, p. 263.
26. L. A. Utracki and M. M. Dumoulin, in *Polymer Materials Encyclopedia*, ed. J. C. Salamone, CRC Press, New York, 1996, Vol. 1, pp. 166–182.
27. D. Scholz, in Polymer Foams: Technology and Developments in Regulation, *Process, and Products*, ed. S.-T. Lee and D. Scholz, CRC Press, New York, pp. 41–67.
28. C. T. Wong, S. S. Lo and L. Huang, *J. Phys. Chem. Lett.*, 2012, **3**, 879.
29. S. Ravati and B. D. Favis, *Polymer*, 2010, **51**, 4547.
30. M. Costoglou, *Ind. Eng. Chem. Res.*, 2012, **51**, 5615.
31. J. George, K. T. Varughese and S. Thomas, *Polymer*, 2000, **41**, 1507.
32. E. Girard Reydot, H. Sauterau and J. P. Pascault, *Polymer*, 1999, **40**, 1677.
33. F. J. Hua and C. P. Hu, *Eur. Polym. J.*, 2000, **36**, 27.
34. L. H. Sperling, *History of Interpenetrating Polymer Networks Starting with Bakelite-Based Compositions*, ACS Symposium Series, American Chemical Society, Washington, DC, 2011, pp. 69–82.
35. W. Lin, *Modeling and Optimization of a Semi-Interpenetrating Polymer Network Process*, PhD thesis, Carnegie Mellon University, Pittsburgh, 2011.
36. H. L. Frisch and D. Klempner, *Adv. Macromol. Chem.*, 1970, **2**, 149.
37. L. H. Sperling, *Interpenetrating Polymer Networks and Related Materials*, Plenum Press, New York, 1981, pp. 265–280.
38. T. M. Lastumaki, L. V. Lassila and P. K. Vallittu, *J. Mat. Sci.: Mater. Med*, 2003, **14**, 803.
39. X. Hou and K. S. Siow, *Polymer*, 2001, **42**, 4181.
40. S. Thawornwisit, K. Charmondusit, G. L. Rempel, N. Hinchiranan and P. Prasassarakich, *Journal of Elastomers and Plastics*, 2010, **42**, 35.
41. Y. Huang and L. Zhang, *Polymer*, 2002, **43**, 2287.
42. K. C. Frisch, D. Klempner, S. Migdal, H. Ghiradella and H. L. Frisch, *Polym. Eng. Sci.*, 1974, **14**, 76.
43. M. Changez, V. Koul and A. K. Dinda, *Biomaterials*, 2005, **26**, 2095.
44. J. J. Fay, C. J. Murphy, D. A. Thomas and L. H. Sperling, *Polym. Eng. Sci.*, 1991, **31**, 1731.
45. P. L. Nayak, *J. Macromol. Sci., Polym. Rev.*, 2000, **40**, 1.
46. G. Sartor, E. Mayer and G. P. Johari, *J. Polym. Sci., Part B: Polym. Phys.*, 1994, **32**, 683.
47. Z. J. Chen, Y. Xue and H. L. Frisch, *J. Polym. Sci., Part A: Polym. Chem.*, 1994, **32**, 2395.
48. B. Das and D. Chakraborty, *Polym. Gels Netw.*, 1995, **3**, 197.
49. A. A. Donatelli, L. H. Sperling and D. A. Thomas, *Macromolecules*, 1976, **9**, 676.
50. B. Das, T. Gagopadhay and S. Sinha, *J. Appl. Polym. Sci.*, 1994, **54**, 367.

51. H. L. Frisch and Y. Xue, *Polym. J.*, 1994, **26**, 828.
52. Y. C. Chem, K. H. Hsieh, C. C. M. Ma and Y. G. Gong, *J. Mater. Sci.*, 1994, **29**, 5435.
53. R. A. Pater, *Polym. Eng. Sci.*, 1991, **31**, 20.
54. H. L. Frisch, D. Klempner, H. K. Yoon and K. C. Frisch, *Macromolecules*, 1980, **13**, 1016.
55. D. Klempner, K. Frisch, H. X. Xiao, E. Cassidy and H. L. Frisch, in *Multicomponent Polymer Materials*, Advances in Chemistry Series 239, ed. D. R. Paul and L. S. Sperling, American Chemical Society, Washington, DC, 1986, pp. 211–230.
56. N. Chantarasiria, N. Farkrachanga, K. Kawnaramitb and A. Eurpermkiatic, *Eur. Polym. J.*, 2000, **36**, 95.
57. M. A. Gómez-Jiménez, J. L. Rivera-Armenta, A. M. Mendoza-Martínez, J. G. Robledo-Muñiz, N. A. Rangel-Vazquez and E. Terres-Rojas, *Lat. Amer. Appl. Res.*, 2009, **39**, 131.
58. B. Das and T. Gangopadhyay, *Eur. Polym. J.*, 1992, **28**, 867.
59. S. Chuayjuljit, S. Moolsin and P. Potiyaraj, *J. Appl. Polym. Sci.*, 2005, **95**, 826.
60. A. P. Mathew, S. Packirisamy and S. Thomas, *J. Appl. Polym. Sci.*, 2000, **78**, 2327.
61. A. T. Koshy, B. Kuriakose, S. Thomas and S. Varghese, *Polymer*, 1993, **34**, 3428.
62. J. Kadokawa, K. Kodzuru, S. Kawazoe and T. Matsuo, *J. Polym. Environ.*, 2011, **19**, 100.
63. S. M. Sanip and N. Katun, *Jurnalteknologi*, 1996, **25**, 1.
64. J. Johns and C. Nakason, *Polym. Plast. Technol. Eng.*, 2012, **51**, 1046.
65. P. Predecki, *J. Biomed. Mater. Res.*, 1974, **8**, 487.
66. R. E. Toushaent, D. A. Thomas and L. H. Sperling, *J. Polym. Sci., Part C: Polym. Lett.*, 1974, **46**, 175.
67. L. H. Sperling, V. A. Forlenza and J. A. Manson, *J. Polym. Sci., Part C: Polym. Lett.*, 1975, **13**, 713.
68. K. C. Jajam, S. A. Bird, M. L. Auad and H. V. Tippur, Development and Characterization of a PU-PMMA *Transparent* Interpenetrating Polymer Networks (*t*-IPNs), in *Dynamic Behavior of Materials, Volume 1, Proceedings of the 2011 Annual Conference on Experimental and Applied Mechanics*, ed. T. Proulx, Springer, New York, 2011, pp. 117–121.
69. Y. Li and S. Mao, *J. Appl. Polym. Sci.*, 1996, **61**, 2059.
70. Y. Li and M. J. Yang, *Sens. Actuators, B*, 2002, **87**, 184.
71. J. J. Staudinger and H. M. Hutchinson, *US Pat.*, 2 539 377, 1951.
72. L. H. Sperling and D. A. Thomas, *US Pat.*, 3 833 404, 1974.
73. P. O. Tawney and J. R. Little, *US Pat.*, 2 701 895, 1955.
74. P. Vijayan, S. Varghese and S. Thomas, in *Handbook of Multiphase Polymer Systems*, ed. P. Vijayan P. S. Varghese and S. Thomas, John Wiley & Sons, Chichester, 2011, pp. 251–310.
75. S. Thomas and W. Yang, *Advances in Polymer Processing: From Macro- to Nano- Scales*, Woodhead Publishing Ltd, Cambridge, 2009, pp. 1–752.

76. S. Thomas, G. E. Zaikov, S. V. Valsaraj and A. P. Meera, *Recent Advances in Polymer Nanocomposites: Synthesis and Characterization*, Brill NV, Leiden, Netherlands, 2009, pp. 1–438.
77. C. N. Cascaval, D. Rosu, L. Rosu and C. Ciobanu, *Polym. Test.*, 2003, **22**, 45.
78. J. J. Falcetta, G. D. Friends and G. C. C. Niu, *German Olien. Pat.*, 2 518 904, 1975.
79. S. H. Fairheller, A. H. Kom, E. H. Harres, E. M. Filachione and M. M. Taylor, *US Pat.*, 3 843 320, 1974.
80. K. C. Frisch, H. L. Frisch and D. Klempner, *German Pat.*, 2 153 987, 1972.
81. Ciba Ltd, *Br. Pat.*, 1 223 338, 1971.
82. B. E. Causton, *J. Dent. Res.*, 1971, **53**, 1074.
83. W. K. Frischer, *US Pat.*, 3 806 558, 1974.
84. B. Volmert, *US Pat.*, 3 055 859, 1962.
85. J. Neelam, K. Sharma, P. Banik, A. Gupta and A. Bhardwaj, *Curr. Drug Ther.*, 2011, **6**, 263.

CHAPTER 2

Natural Rubber: Biosynthesis, Structure, Properties and Application

JITLADDA TANGPAKDEE SAKDAPIPANICH*[a,b] AND PORNTIP ROJRUTHAI[c]

[a] Mahidol University, Department of Chemistry and Center of Excellence for Innovation in Chemistry, Bangkok 10400, Thailand; [b] Mahidol University, Institute of Molecular Bioscience, Salaya Campus, Nakhonpathom 73170, Thailand; [c] King Mongkut's University of Technology North Bangkok, Faculty of Science, Energy and Environment, Rayong Campus, Rayong 21120, Thailand
*Email: jitladda.sak@mahidol.ac.th

2.1 Introduction

Hevea rubber, known as natural rubber (NR), is almost the only commercial rubber from a natural source, even though over 2,500 species of plants have also been found to produce *cis*-1,4-polyisoprene. Guayule has been investigated as an alternative source of NR, which contains resin, rubber-soluble triglycerides and higher terpenes.[1] Guayule rubber is also made of high molecular weight *cis*-1,4-polyisoprene, produced and stored in the parenchyma cells,[2] but having a narrower molecular weight distribution and physical properties slightly different from those of *Hevea* rubber. Among the different plant species producing rubber, *H. brasiliensis* has proved to be the best rubber producer at the present time, owing to the high productivity of the plant and excellent

RSC Polymer Chemistry Series No. 7
Natural Rubber Materials, Volume 1: Blends and IPNs
Edited by Sabu Thomas, Chin Han Chan, Laly A. Pothen, Rajisha K. R., Hanna J. Maria
© The Royal Society of Chemistry 2014
Published by the Royal Society of Chemistry, www.rsc.org

physical properties of the rubber product.[3] In addition, although the cis-polyisoprene synthesized with a Ziegler catalyst has a marked structural resemblance to NR, it shows considerable differences in some properties such as lower tensile strength, lower tear strength and hardness. Due to its outstanding properties, NR cannot be replaced by synthetic rubber in some specific applications such as heavy-duty tyres, gloves, condoms, *etc*. Furthermore, there is currently a rise in the awareness of environmental issues. NR is an interesting material made from a renewable source, while the synthetic rubber is obtained from non-renewable oil-based resources.

NR latex exuded from the *Hevea* trees is a colloidal suspension. Tapping frequencies around the world range from daily to every 4 or 5 days. The amount of latex obtained on each tapping depends on clones, ages, soil, climatic conditions and tapping frequency. The collected latex is usually treated with formic acid to coagulate the suspended rubber particles within the latex. After being pressed between rollers to form the rubber into 3–5 mm thick slabs or thin crepe sheets, the rubber is air- or smoke-dried at 50–60 °C for 2–3 days into a final thickness of 2–3 mm for shipment. These rubbers are known as air-dried sheet (ADS) and ribbed smoked sheet (RSS), respectively. RSS is visually subdivided into six grades based on colour, consistency and observed impurities, as described in the 'Green Book'.[4] RSS No.IX is the top grade and the subsequent grades are RSS No.1 to RSS No.5. RSS is mainly used in automobile tyre manufacturing. Latex is treated with bleaching agents to produce rubber that is pale in colour, called pale crepe. This is used in applications that require light colour and good properties such as footwear, medical supplies and adhesives. The agents used for bleaching rubber are commonly thiols, known as mercaptans, such as xylyl mercaptan ($C_8H_{10}S$), tolyl mercaptan (C_7H_8S) and sodium bisulfate ($NaHSO_3$). The coagulated rubber, after treatment, is pressed to produce 1.2–1.5 mm thick rubber sheets. The sheets are dried at low temperature (35–40 °C) for 3–5 days or air dried for 5–10 days. Other commercial types of NR, known as technically specified rubbers (TSRs) in block form, are systematically graded based on technical specifications specified by the International Standards Organisation (ISO), *i.e.* dirt content, ash content, nitrogen content, volatile matter content, *etc*. The high graded TSR such as TSR L, TSR CV and TSR 5 are made directly from field latex and prepared by a carefully controlled process, while the lower ones such as TSR 10 or TSR 20 are produced from naturally coagulated cup lump. Based on these specifications, major rubber-producing countries introduced their respective standards as Standard Malaysian Rubber (SMR), Standard Thai Rubber (STR), Standard Indonesian Rubber (SIR) and Standard Vietnamese Rubber (SVR).

NR can also be exported as concentrated latex. Fresh field latex consists of 30–40% dry rubber content (DRC), the other 60–70% being mainly water-containing non-rubber substances. The non-rubber components, *e.g.* proteins, carbohydrates, lipids and inorganic salts, vary according to clones, season and age of rubber tree. The latex collected from the plantation is preserved with ammonia (NH_3) and then undergoes continuous centrifugation to produce the concentrated NR latex containing ~60% rubber, which is more uniform in composition than the preserved field latex. For long-term preservation of concentrated latex, the NH_3 content is usually raised to 0.6–0.7%; this is referred to as high ammonia (HA) preserved concentrated latex. Besides preserving the latex and maintaining the colloidal stability, the NH_3 at levels above 0.3% can act as a bactericide. Low ammonia (LA) preserved concentrated latex contains only 0.2–0.3% NH_3 in combination with secondary bactericides, such as tetramethylthiuram disulfide (TMTD) and zinc oxide (ZnO). LA latex is preferred for all applications where a deammoniation process is needed. However, TMTD is known to produce carcinogenic nitrosamines in NR latex products. Other latex grades are commercialized, such as double or multiple centrifuged latex and pre-vulcanized latex. The double or multiple centrifuged latex is highly purified by re-centrifugation for use in hygienic products such as surgical gloves, condoms, catheters, *etc*. Another grade is pre-vulcanized latex, which is a semi-processed latex that does not require a compounding and post-vulcanizing operation. It is particularly used in the manufacture of dipped goods such as balloons, gloves, *etc*. The latex is prepared by mixing centrifuged latex concentrate with vulcanizing agents, which contain sulfur, ZnO and accelerator, under heat to cause partial crosslinking of rubber molecules. The finished products are obtained after drying to produce additional crosslinking.

2.2 Biosynthesis of NR

NR latex is produced and stored in laticifers located in the inner cortex of the *Hevea* rubber tree. The latex is composed of the rubber phase, Frey–Wyssling (FW) particles, bottom fraction and serum. The rubber particles usually range in size from 0.05 μm to about 3 μm.[5] It has been proposed that rubber particles are surrounded by a complex double layer of proteins and lipids.[6] Recent analysis by atomic force microscopy in conjunction with confocal fluorescence microscopy has shown that the surface of rubber particles is surrounded by a mixed layer of mainly proteins and a small amount of phospholipids.[7] FW particles are composite organelles containing lipids and carotenoids bounded by a double layer membrane. The presence of carotenoids in the FW particles, a specialized plastid, suggested that the FW particles might contain enzymes for the isoprenoid and rubber biosynthesis pathways[8] and may be a possible site of rubber biosynthesis. The bottom fraction consists of a large quantity of lutoid particles and non-rubber particles. The lutoids are single membrane-bound vacuoles containing water-soluble substances such as acids, minerals, proteins and sugars. The membranes are very rich in phosphatidic acids, anionic lipids, accounting for 82% of the total phospholipid fraction; thus, the lutoids are negatively charged.[9] The serum is considered to be an important phase because of the presence of rubber biosynthesis pathway enzymes, *i.e.* rubber elongation factor,[10] isopentenyl diphosphate (IPP) isomerases,[11,12] prenyl transferase enzymes,[13] including rubber biochemical intermediates, *i.e.* β-hydroxy-β-methylglutaryl-coenzyme A (HMG-CoA).[14]

The NR molecule is widely accepted to be synthesized by the successive addition of IPP to dimethylallyl diphosphate (DMAPP) to build up isoprene units in the *cis* configuration.[15] Initial biochemical studies presumed that IPP, a precursor of rubber molecules from higher plants, is synthesized via the mevalonate (MVA) pathway,[15] which is operated in the cytoplasm from acetyl-Co A. Later, a non-mevalonate pathway called the methylerythritol phosphate (MEP) pathway has also been considered as responsible for IPP synthesis[16] in the plastids from the condensation of pyruvate and glyceraldehyde-3-phosphate. Mevalonate and 1-deoxy-D-xylulose-5-phosphate (DXP) are intermediates in MVA and MEP pathways, respectively. DMAPP is produced by the isomerization of IPP, catalysed by IPP isomerase, from both pathways. The gene expression of DXP synthase in *Hevea* latex suggested that the MEP pathway for IPP synthesis is involved in rubber biosynthesis.[17] However, an experiment that involved feeding 600 RRIM seedlings with [1-^{13}C] 1-deoxy-D-xylulose triacetate, an intermediate derivative of the MEP pathway, indicated that the MEP pathway is involved only in carotenoid biosynthesis but not in the case of rubber biosynthesis.[18]

Rubber biosynthesis requires three distinct biochemical steps: (i) initiation, the synthesis of allylic diphosphates catalysed by soluble *trans*-prenyltransferase; (ii) elongation, *i.e.* the rubber transferase-catalysed *cis*-1,4-polymerization of isoprene units from IPP; and (iii) termination, *i.e.* the release of the polymer from the rubber transferase.[19] An allylic diphosphate (APP) is needed to initiate the polymerization process.[20–22] The head-to-tail condensation of IPP molecules with DMAPP by soluble *trans*-prenyltransferase leads to the formation of the APP initiators, *i.e.* geranyl diphosphate (GPP, C_{10}), farnesyl diphosphate (FPP, C_{15}) and geranylgeranyl diphosphate (GGPP, C_{20}). The stimulatory effect of the APP initiators on the formation of rubber molecules was reported to increase with the chain length from C_5 to C_{20}.[20,23] The presence of two isoprene units in the *trans* configuration at the initiating terminal of NR analysed by nuclear magnetic resonance (NMR) spectroscopy suggests that FPP is the main initiator for the synthesis of NR molecules, although the dimethylallyl group has not been clearly detected yet.[23,24]

The subsequent chain elongation step was confirmed to proceed via an addition of IPP to polyisoprenyl diphosphate to form isoprene units in *cis* configuration. Rubber polymerization is catalysed by rubber transferase (EC 2.5.1.20), a *cis*-prenyltransferase, and requires divalent cations such as Mg^{2+} or Mn^{2+} as a cofactor.[19,20] The proposed mechanism of rubber biosynthesis is shown in Figure 2.1.

The rubber biosynthesis was found to take place on the rubber surface of rubber particles[20,25] by the incorporation of IPP into a terminal allylic diphosphate group of rubber molecules. It was assumed that the growing hydrocarbon chain of rubber diffuses into the rubber particles, while the hydrophilic diphosphate end group remains in the serum phase where it can react with IPP bound to the active site of the rubber transferase, as shown in Figure 2.2.[15] It has been confirmed that the rubber transferase is tightly bound to the *Hevea* rubber particles even after purification by washing.[19,26]

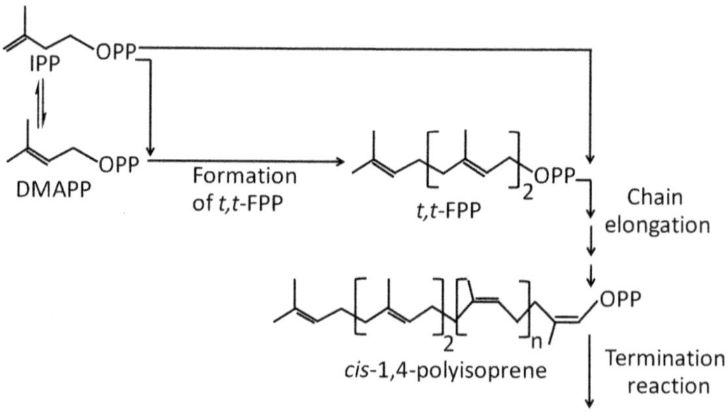

Figure 2.1 Proposed mechanism of NR biosynthesis.

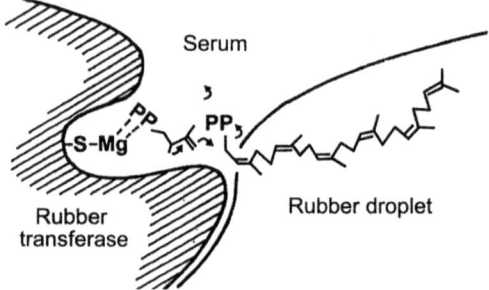

Figure 2.2 Proposed mechanism of enzyme action at the interface between serum and rubber droplet.[15]

Furthermore, the rubber biosynthesis has also been proposed to occur with no requirement of the rubber particles prerequisite site but take place at the washed bottom fraction membrane (WBM), which was prepared from washed bottom fraction particles. This was supported by the higher rubber biosynthesis activity of WBM, compared to the rubber particles. It was interesting that the proteins precipitated at 20% acetone from WBM still had high rubber biosynthesis activity.[27] Nonetheless, the molecular weight of the obtained rubber was not reported.

The mechanism of rubber formation was studied by *in vitro* rubber biosynthesis, by the rubber transferase tightly bound to rubber particles, enzymatically active rubber particles. At least two mechanisms have been postulated for the conversion of IPP to the *in vitro* synthesized rubber.[28] First is the initiation of new rubber molecules on the surface of the rubber particles and the chain extension of pre-existing rubber molecules, the rubber molecules from the rubber particles used for the rubber biosynthesis. In case of *de novo* rubber biosynthesis, the new rubber molecules are initiated by the added APP initiators and/or the APP initiators synthesized by the soluble proteins, *i.e.* IPP isomerase and *trans*-prenyltransferase remained in the system.[29] The second is the formation of completely new rubber particles by a process that does not involve the enzymatically active rubber particles used for the rubber biosynthesis. However, the problem of the mechanism by which new rubber particles originate is still unsolved. In addition, a rubber particle-bound *cis*-prenyltransferase, which catalyses the NR biosynthesis, has not yet been isolated. In *in vitro* rubber synthesis, previous reports almost showed the *de novo* synthesis of new rubber molecules.[20,30] Recently, the chain extension mechanism of pre-existing rubber has been reported in the *in vitro* rubber synthesis with rubber particles purified from *F. elastica*[31] and *H. brasiliensis*.[32]

2.3 Structure and Properties of NR

NR from *H. brasiliensis* is composed of about 94% *cis*-1,4-polyisoprene and 6% non-rubber components, such as proteins, lipids, sugar, *etc*. Even though polyisoprene (IR), prepared with a Ziegler catalyst in 1954, contains higher than 98% *cis*-1,4 configuration, NR exhibits superior mechanical properties such as tack[33] and green strength[33,34] in the unvulcanized state and high tensile strength,[35] high crack growth resistance[36,37] and minimal heat build-up in the vulcanized state. These outstanding properties have been postulated to be due to the non-rubber components, *i.e.* proteins and lipids, and structural characteristics of NR, which are absent in synthetic *cis*-1,4-polyisoprene. However, the physical properties of NR after removing proteins by deproteinization are almost equivalent to those of ordinary NR.[38] This supports the idea that the protein component of NR is not essential to produce its outstanding and characteristic properties. NR purified by removing proteins and lipids by deproteinization and transesterification, respectively, showed stress–strain properties of unvulcanized rubber (termed 'green strength') similar to the synthetic *cis*-1,4-polyisoprene. The fatty acids in NR can be extracted by acetone and completely removed by chemical reactions such as transesterification. The extractable fatty acids are thought to be the fatty acids blended in NR. The fatty acids that can be removed by the chemical reaction are defined as linked fatty acids. The effect of both blended and linked fatty acids in NR was confirmed by the preparation of a model *cis*-1,4-polyisoprene grafted with stearic acid. The similar crystallizability of NR and the rubber model[39] suggests that the structure of chain end groups in NR confers major characteristic properties of NR.

2.3.1 Initiating Terminal of the Rubber Molecule

The low molecular weight (MW) fraction fractionated from NR showed the ^{13}C-NMR signals derived from *trans*-methyl and methylene carbons, suggesting the presence of *trans*-isoprene units in the structure of NR.[40] The absence of *trans*-isoprene units in the *cis–trans* sequence indicated that the *trans*-isoprene units are not derived from isomerization of *cis*-isoprene units.[41] The ^{1}H-NMR spectrum of the low MW fraction of transesterified NR shows the methyl proton signals of *trans*-isoprene units corresponding to those

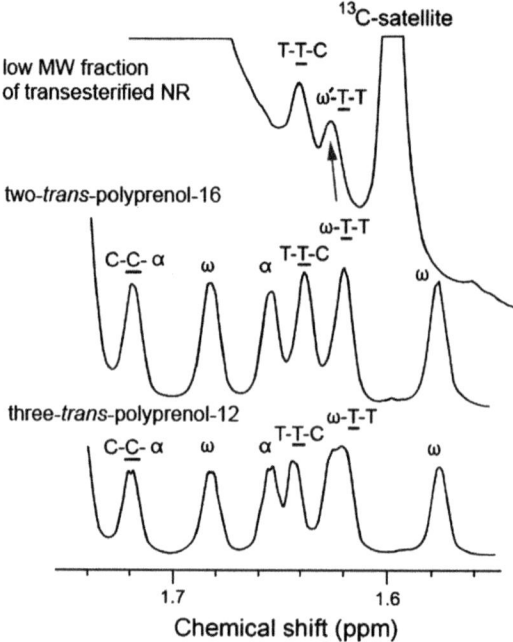

Figure 2.3 ^{1}H-NMR spectra of low molecular weight fractions of transesterified NR, two-*trans*-polyprenol-16 and three-*trans*-polyprenol-12.[23]

of two-*trans*-polyprenol[42] as shown in Figure 2.3. However, ^{1}H- and ^{13}C-NMR studies showed the presence of two *trans*-isoprene units without the dimethylallyl group at the initiating terminal in ordinary NR, NR after deproteinization (DPNR) and also low MW NR from *Hevea* seedlings.[23,43]

Owing to the fact that FPP stimulates the *in vitro* rubber synthesis by washed rubber particles (WRP), the rubber particles purified from soluble enzymes and substrates by washing process, and IPP,[20–22,26] it is difficult to ignore the possibility that FPP is the initiating species in NR. The *in vitro* rubber synthesis by incubation of bottom fraction with IPP in the presence and absence of FPP was carried out.[44] It was reported that only the rubber from the incubation of IPP with FPP clearly showed the dimethylallyl group and *trans*-isoprene units.[45] This suggests that FPP is the initiating species of rubber formation in the *Hevea* tree.

Recently, the ω-terminal group of polyprenol fractions from *Hevea* shoots and fresh bottom fraction contained mainly the middle chain length of C_{60} to C_{80} was analysed using high-resolution ^{13}C- and ^{1}H-NMR techniques.[46] The spectra showed the presence of a dimethylallyl group and two *trans*-isoprene units. This suggests that these polyprenols from *Hevea* rubber consist of a dimethylallyl-*trans-trans* sequence as an initiating end. Moreover, the ω-terminal groups of low MW rubber from various different ages and parts of the *Hevea* tree, as well as that from regularly tapped trees after high purification by washing were also analysed.[46] It was found that the dimethylallyl group at the ω-terminal was not observed in the low MW rubber fractions from *Hevea* shoots, *Hevea* leaves and young *Hevea* trees, while it could be detected in that from WRP. These findings strongly support the evidence that NR from regularly tapped latex contains the dimethylallyl group at the initiating end.

Nevertheless, the absence of the dimethylallyl group in the rubber from various different ages and parts of the *Hevea* tree is owing to the loss or modification of this group during storage in the *Hevea* tree for long periods before tapping. It is noteworthy that the dimethylallyl group at the initiating terminal is observed only in low MW fractions from WRP but not in ordinary NR and DPNR. This indicates that the ordinary coagulation of latex as well as deproteinization with proteolytic enzymes and chemical treatment may cause the modification of the ω-terminal. Accordingly, the ω-terminal of ordinary NR molecules was postulated to be a dimethylallyl group modified by some enzymatic or chemical reactions. Based on the decrease of gel content of NR after enzymatic deproteinizaton,[47] the modified dimethylallyl group at the ω-terminal was assumed to contain a functional group that can link or associate with peptides, leading to the formation of branch points.[24] At present, however, there is no direct evidence to confirm the presence of peptide groups at the initiating terminal of a rubber chain. Accordingly, the fundamental structure of NR characterized by NMR spectroscopy, based on information from model compounds, polyprenols and natural polyisoprenes, was postulated to be a linear rubber chain in NR contains the initiating terminal, *i.e.* ω-terminal, two *trans*-1,4-isoprene units, about 1,000–3,000 *cis*-1,4-isoprene units, and the chain end group, *i.e.* α-terminal.[48]

2.3.2 Terminating End of the Rubber Molecule

On the other hand, long-chain fatty acid ester groups are clearly observed in the ^{13}C- and ^{1}H-NMR spectra of NR, even after purification by deproteinization and acetone extraction.[41] These remaining fatty acids, which are thought to be fatty acids linked with rubber molecules, can be removed by transesterification or saponification with NaOH. The number of long-chain fatty acids linked with the rubber molecule was estimated to be about two, by ^{13}C-NMR and FTIR analyses.[41,43,48] The observed ^{1}H-NMR signals of low MW rubber from 1-month-old *Hevea* seedlings corresponding to –CH$_2$OP and glyceride.[49] The results suggest that most of the chain ends of NR consist of a phosphate group, long-chain fatty acid and glycerol unit, which correspond to phospholipids.[41,49]

The structure of the α-terminal group in NR was analysed in more detail by selective decomposition of the branch points followed by structural characterization.[50,51] The presence of acylglyceride and/or phospholipid at the rubber chain end, assumed by supporting evidence,[41,49] was confirmed by a structural change of a rubber chain after treatment with lipase[50] and phosphatase.[51] These enzymes can selectively hydrolyse acylglycerides, including phospholipids, and monophosphate esters, respectively. DPNR was treated with the enzymes and then extracted with acetone. The ^{1}H-NMR spectra of the rubber before and after treatment are shown in Figure 2.4. The presence of mono- and

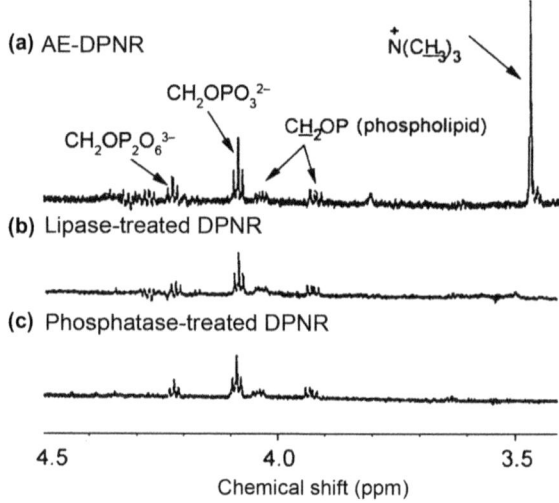

Figure 2.4 ^{1}H-NMR of acetone-extracted DPNR, lipase-treated DPNR and phosphatase-treated DPNR.[50]

diphosphate signals in ^{1}H- and ^{13}C-NMR after lipase and phosphatase treatments clearly indicates that these phosphate groups are directly linked to the rubber molecule.

The selective decomposition of the linkages in phospholipids by the treatment of lipase and phospholipases A_2, B and C affected the change in molecular weight distribution (MWD) and the decrease in the MW of DPNR. These indicate that phospholipids are present in rubber molecules and their acylglycerol group and phosphate are directly concerned with the formation of branch points in NR at the α-terminal.[51] The decrease in MW and the change in MWD of rubber molecules after the addition of 1% v/v ethanol into DPNR solution and FTIR results with DPNR suggested that the formation of branch points was caused by hydrogen bonding between phospholipids through both phosphate and carboxyl groups.[51] It was postulated that the branch points in NR are formed by aggregation of the phospholipids, which are linked to phosphate or diphosphate groups at the α-terminal, to form a micelle structure mainly *via* hydrogen bonding between polar groups in phospholipid molecules,[52] as shown in Figure 2.5. In addition, the Mg^{2+} ions existing in FL-latex were also expected to form branches deriving from phospholipids *via* ionic linkages during storage in the presence of ammonia.[53] However, the removal of Mg^{2+} ions by the addition of diammonium hydrogen phosphate in DPNR latex caused only a slight decrease in the MW of rubber molecules.[51,53]

Accordingly, the structure of the α-terminal of NR has been proposed to consist of a mono- or diphosphate group linked with phospholipids[50,51] *via* hydrogen bonding as a predominant linkage and some parts *via* ionic linkage with divalent metal ions.[51,54] The proposed structure for the α-terminal group of NR is illustrated in Figure 2.6.

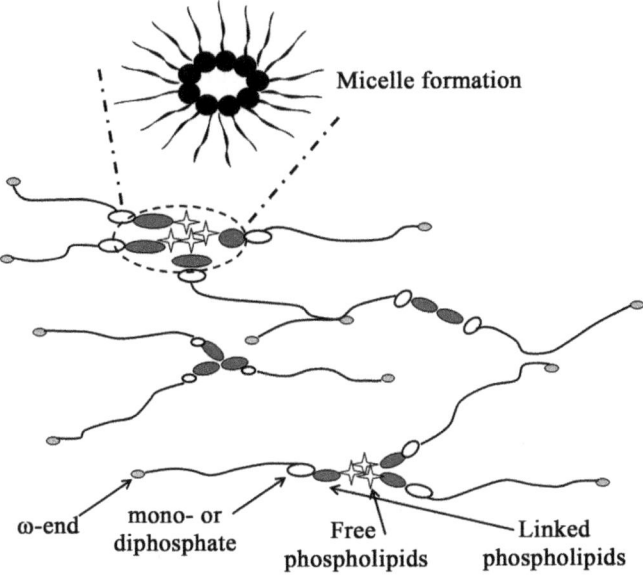

Figure 2.5 Proposed structure of branch points formed at the α-terminal of rubber molecules.[52]

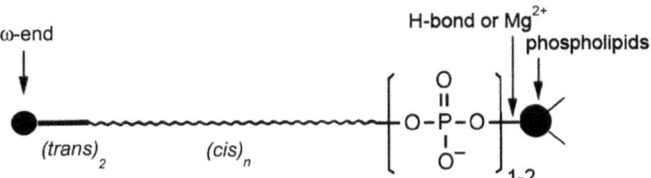

Figure 2.6 Proposed structure for the α-terminal group for NR.[50]

2.3.3 Structure of Branch Points, Gel and Storage Hardening

The formation of branches in NR has been assumed to have a special influence on the green strength of NR. The gel content in NR could be reduced after deproteinization and further reduced drastically after transesterification.[55] Accordingly, the branch points were proposed to consist of two kinds of functional terminal groups, including proteins and phospholipids.[55] The significant effect on green strength and tensile strength of NR, however, was observed only in the case of NR after transesterification.[41,56] This suggests the important role of the branching formed by phospholipids at the α-terminal on characteristic properties of NR. The main phospholipids associated with NR molecules at the α-terminal have been confirmed to be phosphatidylcholine.[50,57,58]

It is recognized that commercial HA latex increases the mechanical stability during storage, but it is always accompanied by an increase in the gel content by as much as 60% after storage for a long period. NR contains both soft gel and hard gel.[47] The content of soft gel fraction in NR decreases by deproteinization or can be partly decomposed in solution by the addition of small amounts of a polar solvent into a good solvent and almost completely solubilized by transesterification. These indicate that branch points originate mainly from functional groups at the ω- and α-terminals.

It is noteworthy that a part of the gel fraction in commercial HA latex cannot be solubilized by transesterification or saponification.[54] This hard gel has been proposed to be formed by radical reactions between rubber chains and tetramethylthiuram disulfide (TMTD), which is normally used as a bactericide preservative, in latex together with zinc oxide (ZnO). The addition of TMTD and ZnO into HA and DPNR lattices resulted in a rapid increase in the gel fraction of hard gel, which is insoluble in good solvents even after the enzymatic or chemical treatments. This suggests that TMTD and ZnO can be another way of increasing the gel content during long storage of latex.[54]

Storage Hardening Phenomenon

The phenomenon of storage hardening in solid NR, which occurs after prolonged storage, has been known to be a factor affecting processing properties[59] such as Mooney viscosity and Wallace plasticity. The change in the properties of NR during storage was thought to be due to the cumulative effects of crosslinking and chain scission, varying with environmental conditions.[60] The branching and crosslinking formations in NR are postulated to derive from chemical reactions of abnormal groups.[59,61] However, it has been reported that these abnormal groups in NR are not major factors for branching and gel formation.[62]

Recently, it was found that the proteins and phospholipids at the chain ends of rubber molecules play the most important role in the storage hardening of rubber under low humidity conditions, but the storage hardening was inhibited under high humidity.[63] This was because the water may disturb the formation of branches by hydrogen bonding. Nonetheless, the centrifuged NR and DPNR showed a significant increase in Wallace plasticity, Mooney viscosity and tensile strength during storage, while NR and DPNR after transesterification showed very low and almost constant Wallace plasticity and Mooney viscosity during storage.[64] This suggests that only phospholipids play an important role in storage hardening. Thus, the interaction of the active functional groups, *i.e.* fatty acid ester groups, in phospholipids at the chain ends of rubber molecules has been proposed to be responsible for storage hardening by the formation of crosslinks during storage.[63,64]

2.3.4 Properties of NR

NR is an important elastomeric material, used in around 70% of all tyres and in a variety of applications in the engineering, medical, sport and household sectors. This is due to its outstanding properties which cannot be matched by the synthetic polyisoprene (IR). Before the vulcanization process, NR is not sufficiently elastic and strong to be used as such. Unvulcanized NR also becomes very soft and sticky with an increase in temperature and hard and brittle at cold temperatures. Vulcanization makes the rubber elastic instead of plastic by forming sulfur links between the rubber chains. The advantages of vulcanized NR are:

- excellent tensile and tear strength;
- low hysteresis and high fatigue resistance;
- high abrasion resistance;
- good rebound and compression set;
- excellent adhesion to metal and fabrics;
- good dielectric strength and electrical insulation.

Vulcanized NR shows a higher tensile strength, tear strength, hardness, abrasion index and cut growth resistance than vulcanized synthetic polyisoprene.[65] These superior properties are more significant at high temperatures. It is commonly accepted that IR cannot be substituted for NR in applications that require high elasticity and high tensile strength. The difference between the properties of NR and IR was suggested to be from the derived MWD, gel content, microstructure and non-rubber components.[65] However, NR also has poor heat aging and resistance to ozone, oxygen, sunlight and oils and solvents.

2.4 Chemical Modification of NR

NR is a renewable natural resource, whereas synthetic polymers are mostly manufactured from petroleum. The use of NR is an environmentally advantageous alternative to synthetic polymers. In order to improve or modify the properties of NR and also extend its use, two major methods have been developed.[66] One is by chemical reaction to modify the molecular structure of NR.[67] Another approach is to blend it with other polymers,[68] with the use of both plastics and rubbers being reported. Only chemical modifications will be discussed in this chapter.

NR has been modified in many ways since as early as 1801, although the first commercial form was not manufactured until about 1915. The term modified rubber can refer to any degree of chemical modification, from a very low mole per cent for the purposes of introducing bound antioxidant functions, crosslinking, bonding, *etc.*, without introducing any changes in the basic physical properties.[69] Another is the reaction of a significant number of the repeated units (say 20–50 mol%), which results in a change in the physical properties of the rubber with higher levels of modification, which tends to alter the nature of the polymer from a rubbery to a more plastic-like or resinous material. Because NR has a fixed *cis*-polyisoprene structure and cannot have its polymerization process tailored like that of synthetic rubbers to provide suitable pendant groups, many of the early types of modified rubber have been prepared with a variety of groups for a whole cross-section of purposes. For instance, the addition of thiols and related compounds to improve low-temperature properties and for crosslinking,[70] for instance reaction of the oxirane ring,[71] or addition of maleic anhydride and maleimides,[72] or the 'ene' addition reaction.[73,74]

Today's environmental concerns demand clean reaction processes that do not use harmful organic solvents.[75] Water is without doubt the most environmentally friendly solvent. NR latex is exuded from the *Hevea* tree as an aqueous emulsion; therefore, it would be desirable to modify the NR latex.[67] Many chemical reactions, such as hydrogenation,[76,77] epoxidation,[78,79] chlorination,[80] graft copolymerization[81,82] and oxidative degradation[83,84] have been performed on the reactive double bonds of the isoprene structure along the molecular chain.

2.4.1 Hydrogenation

Hydrogenation is a potentially useful method for improving the oxidative, thermal and radiation-induced degradation resistance of the unsaturated carbon–carbon double bonds in the NR backbone when exposed to sunlight, ozone and oxygen.[85,86] Hydrogenation of NR can be achieved by both catalytic and non-catalytic methods. The main method used for non-catalytic hydrogenation involves diimide reduction, although undesirable side reactions such as chain scission, cyclization, crosslinking and gel formation may occur.[87] Catalytic hydrogenation can be performed with either homogeneous or heterogeneous catalyst systems. The homogeneous catalyst is a favourable method because of higher selectivity and absence of microscopic diffusion problems. In addition, the role of homogeneous catalysts can be explained and understood at the molecular level.[88] The hydrogenation of NR has been reported in solvent and latex forms.[77,89] The hydrogenation of NR in the presence of Ru[CH=CH(Ph)]Cl(CO)(PCy$_3$)$_2$ was reported by Tangthongkul *et al*.[90] It was found that the impurities within the NR reduced the catalyst efficiency. Hinchiranan *et al*.[77] used the iridium complex, [Ir(cod)py(PCy$_3$)]PF$_6$, to hydrogenate NR in solution form. However, hydrogenation in latex form is of great interest for environmental and economic reasons. The hydrogenation of NR in latex form has been reported to be catalysed effectively by an osmium complex catalyst.[91,92] It was found that OsHCl(CO)(O$_2$)(PCy$_3$)$_2$ is much more efficient than [Ir(cod)py(PCy$_3$)]PF$_6$, Ru[CH=CH(Ph)]Cl(CO)(PCy$_3$)$_2$ and RhCl(PPh$_3$)$_3$ catalysts.[91] The addition of a controlled amount of *p*-toluenesulfonic acid (*p*-TSA) had a beneficial effect on the hydrogenation rate. The hydrogenation improved the thermal stability of NR without affecting its glass transition temperature.

2.4.2 Epoxidation

Epoxidation of NR is a recognized technique to improve the properties of the rubber such as solvent resistance,[93] hydrocarbon oil resistance and gas permeability[94] and to increase damping, while retaining the high tensile strength and tear strength of NR. Epoxidation of NR was first carried out in 1922 by

Pummerer and Bukhard.[95] The reaction proceeds by addition of an oxygen atom to the carbon–carbon double bonds on the NR backbone to form an oxirane (epoxide) ring. The epoxidation has been mainly performed in latex form using a prepared peracetic acid, performic acid, or a mixture of hydrogen peroxide and an organic acid, *i.e.* formic acid or acetic acid. The NR modified by this reaction is called epoxidized NR (ENR). The properties of ENR depend on the degree of epoxidation, which is controlled by the amount of peracid, reaction time and temperature. Two grades of ENR, ENR-25 and ENR-50, are commercially available, with 25 and 50 mol% of epoxide content. The epoxy groups of ENR were reported to be stable under controlled conditions.[78,96] The ring opening and ring expanding of epoxide groups was observed under conditions of excess amounts of formic acid and hydrogen peroxide.[97] Due to the fact that ENR is a reactive elastomer having epoxide groups, it can be used to modify properties of other polymers by blending.

2.4.3 Chlorination

Chlorination is one of the early methods of chemical modification of NR. Chlorinated NR (CNR) shows an improvement over the properties of NR such as chemical resistance, flame resistance and thermal stability. Accordingly, CNR has been used as a raw material for paints, adhesives, inks, coatings, *etc*. The chemical modification of NR by chlorination can be done in solution or latex form. The chlorination of NR in solution form has conventionally been carried out by dissolving in CCl_4. However, this has been prohibited in many countries due to serious problems such as environmental pollution, solvent toxicity and the high equipment investment needed. Preparing CNR in latex is an alternative method. The chlorination of NR can be performed directly in latex form under conditions that stabilize the NR particles with non-ionic surfactant.[80,89,98] Since the NR molecules exist in the rubber particles, the chlorination process in latex form is more difficult than that in solution form, when the reaction occur directly to the NR chains.

Chlorination is also carried out to modify the NR surface of finished products. It produces chemical and physical changes on the modified surface, for example improving oil resistance, and reducing the tackiness and frictional resistance in household NR gloves. Many techniques such as plasma chlorination, acidified hypochlorite solution, chloramines solution and trichloroisocyanuric acid solution have been used to modify the NR surface. The treatment is usually performed using acidified hypochlorite. It was reported that undesirable reactions, *e.g.* cyclization and chain scission, may also occur.[99] However, the cyclized structures of the NR molecules are absent in the case of NR latex films chlorinated in the aqueous phase.[100]

2.4.4 Grafting Copolymerization

The graft copolymerization of NR can be carried out by the polymerization of another monomer on the NR main chain. Graft copolymers are important technological materials used in polymer blends and composites. NR can be grafted with various monomers such as methyl methacrylate (MMA),[101] styrene[102] and acrylonitrile[103] capable of polymerizing into hard plastic materials. MMA and styrene are the most suitable monomers for grafting with NR.[104,105] Graft copolymerization of NR with vinyl monomer provides the toughness, hardness and impact resistance and processability together with the good properties of NR. Grafting copolymerization can be performed either in latex or in solution. Reactions involving free radicals have been most widely applied in grafting processes with a wide range of polymers and monomers.[106] Grafting of monomers onto NR can be carried out by radiation[107–113] and chemical initiator methods. Of these techniques, radiation grafting has the advantage that it is carried out in gaseous and liquid phases of the monomer, and the modified material is free from initiating agent or catalyst.[114] One example of a graft copolymer is poly(methyl methacrylate) or PMMA-graft-NR. It shows improved properties compared to that of unmodified NR, such as hydrophilicity, hardness and modulus, which depend upon the amount of MMA monomer used for grafting. PMMA-graft-NR, containing various MMA contents, has been commercialized in Malaysia since 1950. It can be prepared by polymerization of MMA monomers in NR latex. Cooper et al.[115,116] prepared graft copolymers from NR latex and MMA using visible, ultraviolet and gamma radiation, and found that gamma rays and photo-initiated reactions are of comparable efficiency. Sundardi and Kadariah[117] reported that the increase in the degree of grafting increased the hardness of the grafted polymer of MMA onto NR latex by radiation. George et al.[110] have studied gamma ray-induced graft polymerization of MMA onto NR in field latex. They found that the properties of PMMA-graft-NR prepared by radiation initiation are almost comparable to those prepared by chemical initiation.

Recently, modified NR, such as PMMA-graft-NR (MG) based polymer electrolytes, has drawn the attention of many researchers.[118–124] Modified NR has attractive attributes, such as low glass transition temperature,[119] soft elastomer characterization at room temperature and good elasticity. A suitable elasticity can result in a flat and flexible film, providing an excellent contact between an electrolytic layer and an electrode in a battery system.[125]

2.4.5 Oxidative Degradation

NR is a high MW hydrophobic rubber, which cannot be easily dissolved in several kinds of solvents, leading to limitations in its usage. NR latex containing low MW and reactive terminal groups prepared from the degradation of NR is another interesting form (liquid NR, LNR) to extend the applications of NR, such as compatibilizers,[126,127] adhesives[128] and coatings.[129,130] In addition, the specific functional groups at terminal ends, known as telechelic LNR (TLNR), are potentially reactive with other reagents *via* chain extension reactions to synthesize new polymer structures.[131–134]

LNR can be produced by thermal oxidative degradation using air as an oxidizing agent and phenylhydrazine as a reducing agent. Tillekeratne et al.[135]

used solar energy to depolymerize NR in the presence of nitrobenzene and transition metal complexes. Radvindran et al.[136] used hydrogen peroxide/ methanol and hydrogen peroxide/tetrahydrofuran for some photochemical degradations. However, LNR obtained from the former method is brown in colour and the structure of that obtained from the latter is very complicated. Oxidative degradation for the preparation of LNR and LNR latex was also performed using DPNR latex, which is almost free from proteins,[137] as a starting material, in the presence of $K_2S_2O_8$ and propanal.[138] One reason to use DPNR is because the aging property of DPNR latex is lower than that of the commercial HANR latex, due to removal of naturally occurring antioxidants, *i.e.* proteins, in NR during the process of deproteinization. Recently, a photochemical degradation of NR using H_2O_2, nanometric TiO_2 film and UV irradiation was applied to prepare the hydroxylated LNR.[139] The advantage of this method is that the hydroxylated LNR obtained in either latex or dry rubber form is very clean without the need for further purification.

2.4.6 Cyclization

Of the many NR modification reactions, the cyclization reaction is a simple way of modifying NR to obtain a hard, brittle, thermoplastic derivative of NR. Cyclized products are used in the formulation of adhesives, paints and inks, and also in the compounding of NR to improve its mechanical characteristics.[66–68,140] Sandstrom and co-workers[141] utilized the cyclized polyisoprene polymers incorporated into tyre tread compounds to improve traction, tread wear and tear resistance. Cyclized polyisoprene is further based upon the discovery that blends of cyclized polyisoprene polymers with halobutyl rubber and/or NR can be employed as tyre inner liner formulations.

The cyclization of NR can be carried out in both solution and solid forms,[142] and in a few cases in latex form. Cyclized rubber is obtained from rubber solution by treatment with conventional strong acids such as sulfuric acid or with Lewis acids in solution such as $SnCl_4$, $TiCl_4$, BF_3 or $FeCl_3$, organic acids, such as *p*-toluene sulfonic acid, or a combination of such acids.[141] Dodecyl benzenesulfonic acid is an excellent choice as the catalyst for cyclizing isoprene-containing polymers because it is completely soluble in hexane. Cyclization can be carried out on NR in latex form, but in this case the use of Lewis acids is not feasible. It was found that the cyclization of NR in latex form is effective by treatment with conventional strong acids, in particular, with sulfuric acid.[143–145] In brief, the process consists of stabilizing the latex against acidification, adding sulfuric acid, heating the latex and then separating the cyclized rubber from the latex. Cyclization in the latex state helps to eliminate the need for expensive and hazardous solvents, compared with cyclization in solution. There are other advantages of the cyclization process in latex form.[145]

1. Elimination of the depolymerization step, which is necessary (usually achieved by means of mechanically-induced degradation) if a tractable rubber solution is to be obtained.

2. The possibility of co-coagulating cyclized rubber latex with uncyclized latex to give an intimate mixture of cyclized and uncyclized rubber, which can then be used as a masterbatch for incorporating cyclized rubber into NR compounds.

2.5 Processing of NR and its Applications

2.5.1 Processing of NR

The processing of raw NR into finished products consists of compounding, mixing, shaping and vulcanizing. The compounding involves the incorporation of additives into rubber to provide the desired properties. The additives critically affect the properties with regard to performance, ease of manufacturing, low costs and more attractive appearance. They can be divided into five groups:

(i) Fillers: There are two types of fillers, reinforcing and non-reinforcing. The addition of reinforcing fillers is to improve the mechanical properties, such as modulus, tensile strength or elongation at break. Non-reinforcing fillers, for example clay and calcium carbonate, often called extenders, are used to reduce the production costs.
(ii) Vulcanizing agents: Used to form a crosslinked molecular network to make the elastomeric rubber; sulfur is the most important. The accelerators are normally used in combination with sulfur to increase the rate of vulcanization and the activators, such as zinc oxide and stearic acid, activate the vulcanizing process and help the accelerators to perform more effectively.
(iii) Antidegradants: Antioxidants are used to protect the vulcanized rubber from attack by oxygen, ozone, heat and other factors.
(iv) Plasticizers: Softeners, including oils and paraffin, are used to regulate the hardness and serve as processing aids.
(v) Others: Special ingredients such as pigments in light-coloured compounds, blowing agents for sponge rubber, *etc.*

The additives must be mixed with the rubber to achieve a uniform dispersion within the rubber compound. The mixing is usually carried out in an internal mixer; less frequently in open roll mills. The shaping processes of rubber products can be performed by extrusion, calendering and moulding. An extruder is a machine with a driven screw that forces the rubber through a shaped die under pressure to make a continuous length of rubber product with a preset cross-section. It is preferred for producing long rubber parts of uniform cross-section. The rubber product is vulcanized after it is extruded. A calender is a machine with two or more rolls used to form a rubber sheet of uniform thickness. Compression moulding is the oldest method of making rubber products by compressing an uncured rubber between two heated mould plates. The heat and pressure forces the uncured rubber to mould to the shape of the cavity. The rubber products are vulcanized in mould with the desired shape and are removed from the mould after vulcanization occurs. This process is important in tyre manufacture. The machine used in this process is known as a compression press. An injection moulding machine, also known as an injection press, is an advanced hot-pressing moulding machine for rubber products. The uncured rubber is then forced, by ram or screw, into the shape of the cavity in the closed mould.

A great number of rubber products are made from NR latex, often manufactured by dipping, coating and binding, foaming and extrusion processes. Dipping involves immersion of a former in a latex compound. The latex film formed on the former is dried, cured and stripped from the former. A coating process involves applying latex compound onto fabric or other surfaces. The rubber coated on the fabric is formed after dehydration. NR latex is also used as latex adhesives by adding stabilizers, wetting agents and other components. The latex adhesives can be applied to porous substance such as paper, leather and textiles. The Dunlop process is most often used for making NR latex foam. This involves the beating of air into the compound latex. The latex foam is converted into a semi-solid gel by a gelling agent and turned into a solid rubber after vulcanization.

2.5.2 Applications of NR

Owing to its excellent properties, NR is an essential component of many products used in various applications. Rubber products can be classified into six sectors as follows:

1. Tyre products: Major products of dry NR for over 70% of NR consumption is used in tyres, especially heavy-duty tyres for trucks, buses and airplanes, and tyre products, such as pneumatic tyres and inner tubes for automotive use. The ability of NR to resist heat build-up makes the heavy-duty tyres much safer than those made from synthetic rubber.
2. Moulded products: Prepared by compression or injection of rubber compound into a closed mould. Typical products are sponges, carpet underlay, sandals, shoes, rubber bands, connectors, curing tubes, curing flaps, lining mats, o-rings, rollers, bumpers, heavy-duty pads, seals, gaskets and wheels.
3. Extruded products: Rubber is extruded through dies to form various shapes and profiles. Examples of extruded products are hoses, pipes, electric cables, refrigerator seals, window/door seals, insulators, rollers and erasers.
4. Calendered products: Calendered rubber sheets are widely used in various applications such as conveyer belts, bumpers, gaskets, roll coverings, brake linings, seals and tyre patches.
5. Adhesives: Solution adhesives consist of a mixture of solid rubber and other components dissolved in a solvent. Several adhesive systems are used to produce pressure-sensitive adhesives. The applications of solid

NR are adhesives, electrical insulation tapes, adhesive tapes, packaging tapes, surgical tapes and plasters.
6. NR latex based products: The products can be classified, based on the manufacturing process, into products produced *via* dipping, coating and binding, foaming and extrusion processes.
   - 6.1. Dipped products such as household and medical gloves, nipples, condoms, finger caps, balloons, sporting goods and so on.
   - 6.2. Coating and binding products used for carpet backings, adhesives, adhesive tapes, self-seal envelopes, *etc*.
   - 6.3. Foam products such as foam cushions, foam seating, foam mattresses and pillows, *etc*.
   - 6.4. Extrusion products such as the elastic thread used in undergarments and socks.

2.6 Conclusions

Hevea brasiliensis rubber is a strategically important biomaterial owing to its unique properties. In addition, it is an environmentally friendly polymer from a renewable natural resource. NR is synthesized by the addition of IPP from MVA pathway to FPP initiator and the subsequent addition of IPP into polyisoprenyl diphosphate that catalysed by rubber transferase. Accordingly, the fundamental structure of the NR molecule is a long chain of *cis*-1,4-polyisoprene having two *trans*-1,4-isoprene units at the initiating terminal. This terminal is the modified dimethylallyl group containing functional groups that can associate with proteins. The terminating terminal of NR consists of mono- or diphosphate groups linked with phospholipids *via* hydrogen bonding as a predominant linkage. The branching formation at the initiating and terminating terminals has been postulated to be derived from the aggregation of the proteins and phospholipids, respectively. However, the green strength and tensile strength of NR has been proposed to be the effect of the branching formed by phospholipids. This suggests that the phospholipids existing on NR molecules result in its excellent properties. Accordingly, NR is extensively used in a large variety of applications and cannot be replaced by synthetic polymers in many significant applications. Chemical modifications have been developed to improve the properties and expand the usefulness of NR.

References

1. R. A. Backhaus, *Israel J. Bot.*, 1985, **34**, 283.
2. R. A. Backhaus and S. Walsh, *Bot. Gaz.*, 1983, **144**, 391.
3. L. F. Ramos de Valle and M. Montelongo, *Rubb. Chem. Technol.*, 1987, **51**, 863.
4. *International Standards of Quality and Packing for Natural Rubber Grades (The Green Book)*, 1969. Rubber Manufacturers Association, Washington, DC.

5. C. C. Ho, A. Subramaniam and W. M. Yong, in *Proceedings of the International Rubber Conference, 1975*, Rubber Research Institute of Malaysia, Kuala Lumpur, Vol. 2, 1976, pp. 441–456.
6. E. G. Cockbain and M. W. Philpott, in Chemistry and Physics of Rubberlike Substances, ed. L. Bateman, Maclaren and Sons, London, 1963, p. 73.
7. K. Nawamawat, J. T. Sakdapipanich, C. C. Ho, Y. Ma, J. Song and J. G. Vancso, *Colloids Surf., A*, 2011, **390**, 157.
8. P. B. Dickenson, *J. Rubber Res. Inst. Malaysia*, 1969, **21**, 543.
9. J. Dupont, F. Moreau, C. Lance and J. L. Jacobb, *Phytochemistry*, 1976, **15**, 1215.
10. M. S. Dennis and D. R. Light, *J. Biol. Chem.*, 1989, **264**, 18608.
11. T. Koyama, D. Wititsuwannakul, K. Asawatreratanakul, R. Wititsuwannakul, N. Ohya, Y. Tanaka and K. Ogura, *Phytochemistry*, 1996, **43**, 769.
12. J. Tangpakdee, Y. Tanaka, K. Ogura, T. Koyama, R. Wititsuwannakul, D. Wititsuwannakul and K. Asawatreratanakul, *Phytochemistry*, 1997, **45**, 261.
13. T. Koyama, K. Asawatreratanakul, D. Wititsuwannakul, R. Wititsuwannakul and K. Ogura, in *Biopolymers and Bioproducts: Structure, Function and Applications*, ed. J. Svasti *et al.*, Samakkhisan, Bangkok, 1995, p. 608.
14. R. Wititsuwannakul, D. Wititsuwannakul and P. Suwanmanee, *Phytochemistry*, 1990, **29**, 1401.
15. F. Lynen, *J. Rubber Res. Inst. Malaysia*, 1969, **21**, 389.
16. M. Rohmer, M. Seemann and C. Grosdemange-Billiard, in *Biopolymers: Biology, Chemistry, Biotechnology, Applications, Volume 2, Polyisoprenoids*, ed. T. Koyama and A. Steinbuchel, Wiley-VCH, Weinheim, 2001, pp. 49–72.
17. J. H. Ko, K. S. Chow and K. H. Han, *Plant Mol. Biol.*, 2003, **53**, 479.
18. T. Sando, S. Takeno, N. Watanabe, H. Okumoto, T. Kuzuyama, A. Yamashita, M. Hattori, N. Ogasawara, E. Fukusaki and A. Kobayashi, *Biosci., Biotech. Biochem.*, 2008, **72**, 2903.
19. K. Cornish, *Eur. J. Biochem.*, 1993, **218**, 267.
20. B. L. Archer and B. G. Audley, *Bot. J. Linn. Soc.*, 1987, **94**, 181.
21. S. Madhavan, G. A. Greenblatt, M. A. Foster and C. R. Benedict, *Plant Physiol.*, 1989, **89**, 506.
22. K. Cornish and R. A. Backhaus, *Phytochemistry*, 1990, **29**, 3809.
23. Y. Tanaka, A. H. Eng, N. Ohya, N. Nishiyama, J. Tangpakdee, S. Kawahara and R. Wititsuwannakul, *Phytochemistry*, 1996, **41**, 1501.
24. J. Tangpakdee and Y. Tanaka, *Phytochemistry*, 1998, **48**, 447.
25. F. Lynen, *J. Pure Appl. Chem.*, 1967, **14**, 137.
26. B. G. Audley and B. L. Archer, in *Natural Rubber Science and Technology*, ed. A. D. Robert, Oxford University Press, London, 1988, p. 35.
27. D. Wititsuwannakul, A. Rattanapittayaporn, T. Koyama and R. Wititsuwannakul, *Macromol. Biosci.*, 2004, **4**, 314.

28. B. L. Archer, G. Ayrey, E. G. Cockbain and G. P. McSweeney, *Nature*, 1961, **189**, 663.
29. F. Yusof, K.-S. Chow, M. A. Ward and J. M. Walker, *J. Rubb. Res.*, 2000, **3**, 232.
30. J. Tangpakdee, Y. Tanaka, K. Ogura, T. Koyama, R. Wititsuwannakul and D. Wititsuwannakul, *Phytochemistry*, 1997, **45**, 269.
31. B. M. T. da Costa, J. D. Keasling and K. Cornish, *Biomacromolecules*, 2005, **6**, 279.
32. P. Rojruthai, J. T. Sakdapipanich, S. Takahashi, L. Hyegin, M. Noike, T. Koyama and Y. Tanaka, *J. Biosci. Bioeng.*, 2010, **109**, 107.
33. M. Bruzzone, A. Carbonaro and L. Gargani, *Rubb. Chem. Technol.*, 1978, **51**, 907.
34. L. F. Ramos-De Valle and M. Montelongo, *Rubb. Chem. Technol.*, 1978, **51**, 863.
35. A. G. Thomas and J. M. Whittle, *Rubb. Chem. Technol.*, 1970, **43**, 222.
36. G. J. Lake, A. Samsuri, S. C. Teo and J. Vaja, *Polymer*, 1991, **32**, 2963.
37. G. R. Hamed, H. J. Kim and A. N. Gent, *Rubb. Chem. Technol.*, 1996, **69**, 807.
38. A. A. S. A. Kadir, *Rubb. Chem. Technol.*, 1994, **67**, 537.
39. S. Kawahara, T. Kakubo, J. T. Sakdapipanich, Y. Isono and Y. Tanaka, *Polymer*, 2000, **41**, 7483.
40. Y. Tanaka, H. Sato and A. Kageyu, *Rubb. Chem. Technol.*, 1983, **56**, 299.
41. A. H. Eng, S. Ejiri, S. Kawahara and Y. Tanaka, *J. Appl. Polym. Sci., Appl. Polym. Symp.*, 1994, **53**, 5.
42. A. H. Eng, S. Kawahara and Y. Tanaka, *Rubb. Chem. Technol.*, 1994, **67**, 159.
43. J. Tangpakdee and Y. Tanaka, *J. Nat. Rubb. Res.*, 1997, **12**, 112.
44. J. Tangpakdee, Y. Tanaka, K. Ogura, T. Koyama, R. Wititsuwannakul and D. Wititsuwannakul, *Phytochemistry*, 1997, **45**, 275.
45. Y. Tanaka in *NMR and Macromolecules*, ed. J. C. Randall, American Chemical Society, Washington DC, 1984, p. 233.
46. D. Mekkriengkrai, PhD thesis, Mahidol University, Thailand, 2005.
47. J. Tangpakdee and Y. Tanaka, *Rubb. Chem. Technol.*, 1997, **70**, 707.
48. Y. Tanaka, S. Kawahara and J. Tangpakdee, *Kautch. Gummi Kunstst.*, 1997, **50**, 6.
49. J. Tangpakdee, PhD thesis, Tokyo University of Agriculture and Technology, Japan, 1997.
50. L. Tarachiwin, J. T. Sakdapipanich, K. Ute, T. Kitayama, T. Bamba, E. Fukusaka, A. Kobayashi and Y. Tanaka, *Biomacromolecules*, 2005, **6**, 1851.
51. L. Tarachiwin, J. Sakdapipanich, K. Ute, T. Kitayama and Y. Tanaka, *Biomacromolecules*, 2005, **6**, 1858.
52. L. Tarachiwin, J. Sakdapipanich and Y. Tanaka, *Kautch Gummi Kunstst.*, 2005, **58**, 115.
53. L. Tarachiwin, J. T. Sakdapipanich and Y. Tanaka, *Rubb. Chem. Technol.*, 2003, **76**, 1185.

54. L. Tarachiwin, J. T. Sakdapipanich and Y. Tanaka, *Rubb. Chem. Technol.*, 2003, **76**, 1177.
55. J. T. Sakdapipanich, S. Suksujaritporn and Y. Tanaka, *J. Rubb. Res.*, 1999, **2**, 160.
56. S. Kawahara, T. Kakubo, N. Nishiyama, Y. Tanaka, Y. Isono and J. T. Sakdapipanich, *J. Appl. Polym. Sci.*, 2000, **78**, 1510.
57. P. Rojruthai, L. Tarachiwin, J. T. Sakdapipanich, K. Ute and Y. Tanaka, *Kautch. Gummi Kunstst.*, 2009, **62**, 227.
58. J. Sansatsadeekul, J. T. Sakdapipanich and P. Rojruthai, *J. Biosci. Bioeng.*, 2011, **111**, 628.
59. B. C. Sekhar, *J. Polym. Sci*, 1960, **48**, 133.
60. S. N. Gan, *Trends Polym. Sci.*, 1997, **2**, 69.
61. M. J. Gregory and A. S. Tan, in *Proceedings of the International Rubber Conference, 1975*, Rubber Research Institute of Malaysia, Kuala Lumpur, Vol. 4, 1976, p. 28.
62. J. Yunyongwattanakorn, PhD thesis, Mahidol University, Thailand, 2005.
63. J. Yunyongwattanakorn, Y. Tanaka, S. Kawahara, W. Klinklai and J. T. Sakdapipanich, *Rubb. Chem. Technol.*, 2003, **76**, 1228.
64. S. Ammnuaypornsri, A. Nimpaiboon and J. Sakdapipanich, *Kautsch. Gummi Kunstst.*, 2009, **62**, 88.
65. E. Schoenberg, H. A. Marsh, S. J. Walters and W. M. Saltman, *Rubb. Chem. Technol.*, 1979, **52**, 526.
66. A. D. Roberts, *Natural Rubber Science and Technology*, Oxford University Press Inc., New York, 1988, p. 162.
67. A. Subramaniam, in *Encyclopedia of Polymer Science and Engineering*, ed. F. Mark, N. M. Bikales, C. G. Overberger and G. Menges, Wiley-Interscience, New York, 2nd edn, 1998, vol. 14, p. 762.
68. D. J. Hourston and J. O. Tabe in *Polymeric Material Encyclopedia*, ed. J. C. Salamone, J. Claypool and A. Demby, CRC Press Inc., New York, 1996, vol. 6, p. 4547.
69. P. B. Lindley, Design and use of Natural Rubber Bridge Bearings, NRPRA Technical Bulletin, The Natural Rubber Bureau, 1964.
70. J. I. Cunneen and M. Porter, in *Encyclopedia of Polymer Science and Technology*, ed. H. S. Mark *et al.*, Wiley-Interscience, New York, 1973, vol. 12, p. 321.
71. T. Colclough, *Trans. Inst. Rubber Ind.*, 1962, **38**, 11.
72. C. Pinazzi, J. C. Danjard and R. Pautrat, *Rubb. Chem. Technol.*, 1963, **36**, 282.
73. M. E. Cain, K. F. Gazely, I. R. Gelling and P. M. Lewis, *Rubb. Chem. Technol.*, 1972, **45**, 204.
74. C. S. L. Baker, in *Handbook of Elastomers, New Development and Technology*, ed. A. K. Bhowmick and H. L. Stephens, Marcel Dekker Inc., New York, 1988, p. 37.
75. A. N. Bibi, D. A. Boscott and R. S. Lehrle, *Eur. Polym. J.*, 1988, **12**, 1127.

76. J. Samran, P. Phinyocheep, P. Daniel and S. Kittipoom, *J. Appl. Polym. Sci.*, 2005, **95**, 16.
77. N. Hinchiranan, K. Charmondusit, P. Prasassarakich and G. L. Rempel, *J. Appl. Polym. Sci.*, 2006, **100**, 4219.
78. D. R. Burfield, K. L. Lim and K. S. Law, *J. Appl. Polym. Sci.*, 1984, **29**, 1661.
79. D. Derouet, S. Mulder-Houdayer and J.-C. Brosse, *J. Appl. Polym. Sci.*, 2005, **95**, 39.
80. G. J. Van Amerogen, *Rubb. Chem. Technol.*, 1952, **25**, 609.
81. T. Kochthongrasamee, P. Prasassarakich and S. Kiatkamjornwong, *J. Appl. Polym. Sci.*, 2006, **101**, 2587.
82. J. Saelao and P. Phinyocheep, *J. Appl. Polym. Sci.*, 2005, **95**, 28.
83. H. M. Nor and J. R. Ebdon, *Progr. Polym. Sci.*, 1998, **23**, 143.
84. A. E. Hamdaoui, D. Reyx, I. Campistron and S. F. Tetouani, *Eur. Polym. J.*, 1999, **35**, 2165.
85. N. K. Singha, P. P. De and S. Sivaram, *J. Appl. Polym. Sci.*, 1997, **66**, 1647.
86. M. D. Sarkar, P. G. Mukunda, P. P. De and A. K. Bhowmick, *Rubb. Chem. Technol.*, 1997, **70**, 855.
87. D. N. Schulz, S. R. Tuner and M. A. Golub, *Rubb. Chem. Technol.*, 1982, **55**, 809.
88. S. Bhaduri and D. Mukesh, *Homogeneous Catalysis Mechanisms and Industrial Applications*, Wiley & Sons, New York, 2000, p. 1.
89. A. Mahittikul, P. Prasassarakich and G. L. Rempel, *J. Appl. Polym. Sci.*, 2007, **105**, 1188.
90. R. Tangthongkul, P. Prasassarakich and G. L. Rempel, *J. Appl. Polym. Sci.*, 2005, **97**, 2399.
91. A. Mahittikul, P. Prasassarakich and G. L. Rempel, *J. Appl. Polym. Sci.*, 2006, **100**, 640.
92. N. Hinchiranan, P. Prasassarakich and G. L. Rempel, *J. Appl. Polym. Sci.*, 2006, **100**, 4499.
93. I. R. Gelling, *J. Nat. Rubb. Res.*, 1991, **6**, 184.
94. C. S. L. Baker, I. R. Gelling and R. Newell, *Rubb. Chem. Technol.*, 1985, **58**, 67.
95. R. Pummerer and P. A. Burkard, *Ber. Dtsch. Chem. Ges.*, 1922, **55**, 3458.
96. I. R. Gelling, *Rubb. Chem. Technol.*, 1984, **58**, 86.
97. J. H. Bradbury and C. S. Perera, *J. Appl. Polym. Sci.*, 1985, **30**, 3347.
98. J. P. Zhong and S. D. Li, *J. Appl. Polym. Sci.*, 1999, **73**, 2863.
99. A. F. Halasa, J. M. Massie and R. J. Ceresa, in *Science and Technology of Rubber*, ed. J. E. Mark, B. Erman and F. R. Eirich, Elsevier Academic Press, New York, 2005, p. 497.
100. C. C. Ho and M. C. Khew, *Int. J. Adhes. Adhes.*, 1999, **19**, 387.
101. R. N. Muthurajan, *Plant. Bull.*, 1964, **74**, 131.
102. G. Rajammal, N. M. Claramma and E. V. Thomas, *Proceedings of the International Rubber Conference Session*, 1980, No. 6B, Paper-4.

103. N. M. Claramma, G. Rajammal and E. V. Thomas, *Proceedings of the International Rubber Conference*, Kottayam, India, 1984, **2**(1), p. 13.
104. A. S. Hashim, N. V. Tho and M. O. A. Kadir, *Rubb. Chem. Technol.*, 2002, **75**, 111.
105. W. Arayapranee, P. Prasassarakich and G. L. Rempel, *J. Appl. Polym. Sci.*, 2003, **89**, 63.
106. M. Morton, in *Science and Technology of Rubber*, ed. F. R. Eirich, Academic Press, New York, 1978, p. 25.
107. E. G. Cockbain, T. D. Pendle and P. T. Turner, *Chem. Ind.*, 1958, 759.
108. N. M. Claramma, L. Varghese, K. T. Thomas and N. M. Mathew, *Proceedings of the 18th Rubber Conference on Indian Rubber Manufactures Research Association*, Mumbai, 2000, p. 165.
109. N. M. Claramma, N. M. Mathew and E. V. Thomas, *Radiat. Phys. Chem.*, 1989, **33**, 87.
110. K. M. George, N. M. Claramma and E. V. Thomas, *Radiat. Phys. Chem.*, 1987, **30**, 189.
111. S. Ono, F. Yoshii, K. Makuuchi and I. Ishigaki, *Proceedings of the International Symposium on Radiation Vulcanization of Natural Rubber Latex*, Tokyo and Takasaki. JAERI-M 89-228, 1989, p. 198.
112. M. Utama, Y. S. Soebinto, M. T. Razzak, S. Kusumawati and H. Tunggawiharja, *Proceedings of the Second International Symposium on RVNRL*, Kuala Lumpur, 1996, p. 159.
113. M. Sonsuk and K. Makuuchi, *Proceedings of the Second International Symposium on RVNRL*, Kuala Lumpur, 1996, p. 244.
114. N. C. Dafader, M. E. Haque, F. Akhtar and M. U. Ahmad, *Radiat. Phys. Chem.*, 2006, **75**, 168.
115. W. Cooper, G. Vaughan, S. Miller and M. Fielden, *J. Polym. Sci.*, 1959, **34**, 651.
116. W. Cooper, P. R. Sewell and G. Vaughan, *J. Polym. Sci.*, 1959, **41**, 167.
117. F. Sundardi and S. Kadariah, *J. Appl. Polym. Sci.*, 1984, **29**, 1515.
118. M. S. Su'ait, A. Ahmad, H. Hamzah and M. Y. A. Rahman, *J. Phys. D: Appl. Phys.*, 2009, **42**, 55410.
119. R. Idris, M. D. Glasse, R. J. Latham, R. G. Linford and W. S. Schlindwein, *J. Power Sources*, 2001, **94**, 206.
120. Y. Alias, I. Ling and K. Kumutha, *Ionics*, 2005, **11**, 414.
121. K. Kumutha and Y. Alias, *Spectrochim. Acta A*, 2006, **64**, 442.
122. F. Latif, A. M. Aziz, N. Katun, A. M. M. Ali and M. Z. Y. Yahya, *J. Power Sources*, 2006, **159**, 1401.
123. A. M. M. Ali, R. H. Y. Subban, H. Bahron, T. Winie, F. Latif and M. Z. Y. Yahya, *Ionics*, 2008, **14**, 491.
124. M. S. Su'ait, A. Ahmad, H. Hamzah and M. Y. A. Rahman, *Ionics*, 2009, **15**, 497.
125. M. S. Su'ait, A. Ahmad, H. Hamzah and M. Y. A. Rahman, *Electrochimica Acta*, 2011, **57**, 123.
126. A. Mounir, N. A. Darwish and A. Shehata, *Polym. Adv. Technol.*, 2004, **15**, 209.

127. H. M. Dahlan, M. D. K. Zaman and A. Ibrahim, *J. Appl. Polym. Sci.*, 2000, **78**, 1776.
128. B. Thongnuanchan, K. Nokkaew, A. Kaesaman and C. Nakason, *Polym. Eng. Sci.*, 2007, **47**, 421.
129. D. Derouet, P. Phinyocheep, J.-C. Brosse and G. Boccaccio, *Eur. Polym. J.*, 1990, **26**, 1301.
130. P. Phinyocheep and S. Duangthong, *J. Appl. Polym. Sci.*, 2000, **78**, 1478.
131. N. Kébir, I. Campistron, A. Laguerre, J.-F. Pilard, C. Bunel and T. Jouenne, *Biomaterials*, 2007, **28**, 4200.
132. N. Kébir, I. Campistron, A. Laguerre, J.-F. Pilard, C. Bunel, J.-P. Couvercelle and C. Gondard, *Polymer*, 2005, **46**, 6869.
133. N. Kébir, G. Morandi, I. Campistron, A. Laguerre and J.-F. Pilard, *Polymer*, 2005, **46**, 6844.
134. S. Gopakumar and M. R. G. Nair, *Polymer*, 2005, **46**, 10419.
135. L. M. K. Tillekeratne, P. V. A. G. Perera, M. S. C. DeSilva and G. Scott, *J. Rubber Res. Inst. Sri Lanka*, 1997, **52**, 501.
136. M. R. Radvindran, G. Nayar and D. Joseph Francis, *J. Appl. Polym. Sci.*, 1998, **35**, 1227.
137. A. H. Eng, S. Kawahara and Y. Tanaka, *J. Nat. Rubber Res.*, 1993, **8**, 109.
138. J. Tangpakdee, M. Mizokoshi, A. Endo and Y. Tanaka, *Rubb. Chem. Technol.*, 1998, **71**, 795.
139. J. T. Sakdapipanich, P. Suksawad and K. Insom, *Rubb. Chem. Technol.*, 2005, **78**, 597.
140. J. C. Brosse, I. Campistron, D. Derouet, A. Houdayer, S. Houdayer, D. Reyx and S. Ritoit-Gillier, *J. Appl. Polym. Sci.*, 2000, **78**, 1461.
141. P. H. Sandstrom, J. F. Geiser, J. Chu, D. J. Zanzig, R. G. Bauer, Goodyear Tire and Rubber Company, *US Pat.*, 6303693, 2001.
142. G. J. Van Veersen and B. B. S. T. Boonstra, *Rubber Age*, 1950, **68**, 57.
143. G. J. Van Veersen, *Rubb. Chem. Technol.*, 1951, **24**, 957.
144. G. J. Van Veersen, *Rubb. Chem. Technol.*, 1950, **23**, 461.
145. S. Riyajan and J. T. Sakdapipanich, *Kautsch. Gummi Kunstst.*, 2006, **59**, 104.

CHAPTER 3

Non-Rubbers and Abnormal Groups in Natural Rubber

ENG AIK HWEE

Ansell Shah Alam, Malaysia, Lot 16, Persiaran Perushaan, Section 23, Shah Alam, 40 000, Selangor, Malaysia
Email: aikhwee.eng@ansell.com

3.1 Non-Rubbers in Natural Rubber

Fresh natural rubber latex contains about 30–40% of rubber hydrocarbon that is normally referred to as dry rubber content (DRC). However, the total solid content (TSC) is higher than the DRC due to the presence of non-rubbers in the latex, at around 5%. The DRC and non-rubber content may change due to many factors such as clone, soil and climate conditions, season, type of fertilizers used, and tapping frequency. Most of these non-rubbers are dissolved or suspended in the aqueous serum or adsorbed on the surface of rubber particles. They become trapped, tenaciously held, or co-precipitated during coagulation of the rubber probably due to their poor solubility in the aqueous medium or strong entanglement with the rubber molecule. The major non-rubbers are lipids, proteins and amino acids, minerals, inositols and carbohydrates, as shown in Table 3.1.

When these compounds are chemically bonded to the rubber molecules, they are collectively known as abnormal groups. Based on the biosynthesis mechanism of other naturally occurring cis-polyisoprenes, there are at least two abnormal groups on the main-chain rubber molecule, namely the ω-terminal (*i.e.* initiating end) and α-terminal (terminating end) groups.

RSC Polymer Chemistry Series No. 7
Natural Rubber Materials, Volume 1: Blends and IPNs
Edited by Sabu Thomas, Chin Han Chan, Laly A. Pothen, Rajisha K. R., Hanna J. Maria
© The Royal Society of Chemistry 2014
Published by the Royal Society of Chemistry, www.rsc.org

Table 3.1 Composition of fresh latex.

Component	Percentage (w/w)
Rubber hydrocarbon	36.0
Proteins, amino acids and nitrogenous compounds	1.7
Lipids	1.6
Ash	0.5
Inositols and carbohydrates	1.6
Water	58.6

It is well known that these non-rubbers and abnormal groups are not only of biological significance with regard to the structure and biosynthesis pathway of the rubber, but they also affect the physical and chemical properties of the latex, dry rubber and rubber products. For example, the mechanical stability of the latex increases with storage time as a result of the presence of non-rubbers.[1,2] This is a unique property of natural rubber latex because the stability of synthetic latices would normally deteriorate with storage time. Natural rubber also has a high green strength, which helps to maintain the shape of the rubber compound before curing.[3–9] Natural rubber gloves have been reported to show no leaks in the watertight test after puncturing with a 0.25 mm needle, while 80–90% of vinyl and nitrile gloves leaked when punctured with the same needle. This is due to the ability of natural rubber to reseal when punctured with a small needle.[10] When a larger diameter needle was used, the leak was also markedly less than those of vinyl and nitrile gloves. However, when purified natural rubber was used, the amount of water that leaked through the hole was found to increase significantly, indicating that non-rubbers play a significant role in controlling the aperture of the tear slit.[11] The presence of high volatile fatty acids (VFAs) in natural rubber has been found to significantly reduce the compressive strength of rubber-modified concrete.[12] The non-rubbers have been reported to reduce the tack properties of natural rubber.[13] In addition, non-rubbers that are insoluble in rubber solvents have been found to be responsible for the higher modulus, faster scorch time, heat build-up, and higher hot tear strength of compounded natural rubber. These materials were thought to act as a reinforcing filler and cure activator.[14] The naturally-occurring network, formed by interactions of proteins and phospholipids with the terminal units of isoprene chains, was reported to be mainly responsible for the superior stress–strain behaviour and the strain-induced crystallization of natural rubber in both unvulcanized and vulcanized states.[15]

3.1.1 Lipids

Lipids constitute the largest group of natural rubber non-rubbers by weight. Depending on the clonal origin, the amount of lipids isolated from different latices has been found to vary from 1.3 to 3.5% w/w, as shown in Table 3.2.

There are two types of natural rubber lipids, namely non-polar lipids or neutral lipids, and polar lipids, which consist of phospholipids and glycolipids.

Table 3.2 Lipid content of natural rubber from different clones.[16–19]

Clone	Lipids content, w/w^a			
	Total	Neutral	Glycolipids	Phospholipids
RRIM 600	1.3	0.45	0.30	0.58
PB 28/59	3.4	2.34	0.45	0.57
RRIM 701	3.3	2.32	0.53	0.49
RRIM 730	1.5	0.55	0.53	0.39
PB 255	1.3	0.64	0.28	0.37
RRIM 804	1.9	0.92	0.41	0.54
RRIM 501	1.6	0.88	0.54	0.23

^{a}Based on dry rubber weight.

As shown in Table 3.2, neutral lipids form the major groups of natural rubber lipids. Typically, a whole natural rubber latex contains 54% neutral lipids, 33% glycolipids, and 14% phospholipids.[17] The lipids, such as phospholipids, adsorbed on the surface of rubber particles are normally found in the cream fraction. The rubber particles in the skim latex, on the other hand, have a very low level of adsorbed lipids.[20,21] Smoking, a process used to dry natural rubber, has been found to increase the amount of lipid extract significantly but to decrease the free fatty acid content.[22]

3.1.1.1 Neutral Lipids

The neutral or non-polar lipids can be extracted using hexane/diethyl ether while the polar lipids can be extracted using methanol, followed by methanol–acetic acid.[19] The non-polar lipids consist of more than 14 substances. Among these, triglycerides are the largest component of the neutral lipids. The other components are esters, phenolic compounds (tocotrienols) and sterols. In a study on a latex, free α-tocotrienol and γ-tocotrienol, and esterified γ-tocotrienol and δ-tocotrienol, have been identified.[17] However, in another study on acetone extracts of commercial dry rubbers, four types of tocotrienols have been reported, namely α-, β-, δ- and γ-tocotrienols.[23] The total amount of tocotrienols in the dry natural rubbers varies from 0.02% to 0.15%, depending on the clonal origin of the latex. Tocotrienols, together with other phenolic compounds from the unsaponifiable fraction, are the most important natural antioxidant, preventing the auto-oxidation of the raw rubber.[24] The minor components of the neutral lipids are free fatty acids, fatty alcohols, monoglycerides, diglycerides and carotenoid pigments. The major components of free fatty acids are stearic, oleic and linoleic acids.[16–19] Table 3.3 shows a typical fatty acid distribution from four different clones of natural rubber latex.[16–19] Free fatty acids are believed to be the breakdown products of lipids. These fatty acids could form soaps with cations and at least partially contribute to the colloidal stability of the latex. The free fatty acids and unsaturated methyl fatty esters have been reported to accelerate the oxidation and chain scission of deproteinized natural rubber.[25] Free fatty acids such as stearic and linolenic are activators of sulfur vulcanization. These compounds give better heat build-up

Table 3.3 Distribution of fatty acids in natural rubber from different clones.[16–19]

Components	Composition, % w/w			
	RRIM 701	GT 1	RRIM 600	PR 107
Myristic	0.5	0.5	0.4	0.6
Palmitic	9.4	8.6	9.0	8.1
Palmioleic	2.5	1.6	1.4	1.1
Stearic	16.1	17.7	18.0	17.1
Oleic	15.8	23.6	17.9	16.4
Linoleic	29.2	42.5	49.7	48.3
Linolenic	2.9	2.2	3.2	4.9
Arachidic	0.7	–	–	–
Furanoic	23.0	3.3	0.4	3.5

Table 3.4 Composition of glycolipids in natural rubber from different rubber clones.[16–19]

Component	Compositions, % w/w			
	RRIM 600	PB 235	BPM 24	RRIM 501
DGDG	47	43	51	63
SG	34	30	34	20
ESG	12	19	7	9
MGDG	8	8	8	7

properties to natural rubber.[26] They also function as a crystallization nuclei, i.e. accelerate the cold crystallization of natural rubber by a factor of five-fold at $-25\,°C$ where the rate of crystallization is the highest.[27–30] Bonded fatty acids were found to be more effective than those of unbonded in accelerating the crystallization of the rubber, as demonstrated by the synthetic polyisoprene grafted with fatty acids.[31] In the presence of a large quantity, polyunsaturated fatty acids such as linolenic acid function as a plasticizer to natural rubber.[32] When present in small quantities, a decrease in the plasticity of the rubber has been reported.[33] Nevertheless, when added to natural rubber, the long chain fatty acids do not contribute to the stretch-induced crystallization of the rubber, a property that enhances the physical strength of natural rubber.[34]

3.1.1.2 Glycolipids

As indicated in Table 3.2, the glycolipids in natural rubber vary from 0.3 to 1.0%, depending on the rubber clones. The four main components of glycolipids are digalactosyl diglycerides (DGDG), steryl glucosides (SG), esterified steryl glucosides (ESG), and monogalactosyl diacylglycerols (MGDG). The distribution of these components in the glycolipids is shown in Table 3.4. The sterols are mainly stigmasterol, b-sitosterol and D5-avenasterol. The distribution of covalently bonded long chain fatty acids (acyl components) is very similar to that of free fatty acids, suggesting that these free fatty acids are originated from the hydrolysis breakdown of the bonded fatty acids.[16–19]

Table 3.5 Composition of phospholipids in natural rubber from different rubber clones.[16–19,33]

Component	Compositions, % w/w			
	RRIM 600	PB 235	BPM 24	RRIM 501
PC	66.6	58.2	55.9	58.4
LPC	23.0	31.8	32.0	–
PE	3.2	3.3	4.1	21.0
PI	1.2	1.0	2.2	20.6
LPI	3.4	1.8	1.9	–

3.1.1.3 Phospholipids

Like glycolipids, phospholipids are also a class of polar lipids. Data in Table 3.2 show that phospholipids in natural rubber are normally below 0.6% w/w. A recent study[18,19,33] showed that the major components were phosphatidyl choline (PC) and lysophosphatidyl choline (LPC), as shown in Table 3.5. Other minor components were phosphatidyl ethanolamine (PE), phosphatidyl inositol (PI), lysophosphatidyl inositol (LPI), and metal phosphatidates (MP) or phosphatidic acid (PA). Unlike neutral lipids and glycolipids, the acyl components of phospholipids normally contain very low levels of furanoic acid, except for certain rubber clones. The isolated fatty acids are mainly palmitic, stearic, oleic and linoleic acids.[16–19,33]

The fatty acids containing phospholipids have been postulated to be chemically bonded to the rubber molecule at the α-terminal end and are the branching point of the rubber.[35] Hydrolysis of phospholipids produces phosphate, which could increase the stability of ammoniated latex during storage. Of phosphatidyl choline, phosphatidyl ethanolamine and phosphatidic acid, phosphatidyl ethanolamine has been found to be the most effective antioxidant.[36]

3.1.2 Proteins, Amino Acids and Other Nitrogenous Compounds

3.1.2.1 Proteins

Fresh natural rubber latex contains 1–2% of proteins. The total protein content can be estimated by the semi-micro Kjeldahl method using the formula: nitrogen content × 6.25. When a field natural rubber latex is centrifuged, about one-quarter of these proteins is adsorbed on the surface of rubber particles, another quarter in the bottom fraction and half of them in the serum fraction.[37] The serum phase of natural rubber latex has been shown to contain proteins different from those of rubber particles in terms of their molecular weights, with those in the serum phase fall in the region of 6 to more than 200 kDa, while the major proteins of rubber particles are 14.5 and 29 kDa, which are similar to rubber elongation factor and hevamines, respectively.[38]

α-globulin is the largest component of proteins in the serum of natural rubber latex. The protein has an elemental composition of an unconjugated globulin. It is soluble in neutral salts, alkaline solutions and in acid solutions of pH less than 3. It is insoluble in distilled water and can be denatured and coagulated by heat or by drying. The protein has a low sulfur content of about 0.06% and an isoelectric pH of 4.5, similar to that of rubber particles.[39–41] In contrast to α-globulin, hevein has little surface activity. The isoelectric pH of this protein is 4.7–4.9.[42–44] Hevein is a cystine-rich protein with a molecular weight of 5 kDa. It is a single polypeptide containing 43 amino acids with glutamic acid as the terminal group.[42] The protein has a high sulfur content of 5%. Due to the presence of disulfide groups in cystine, the protein is heat stable. Hevein has been found to have antifungal properties.[43,44] Other natural rubber proteins reported in the literature include microfibrils, which contain as much as 4% carbohydrate, with an isoelectric pH of 4, basic proteins such as hevamines A and B,[39,45] some high isoelectric pH proteins[46,47] and various types of enzymes.[48] The most important enzyme is, of course, the rubber transferase, which is responsible for the biosynthesis of natural rubber. The enzyme was initially thought to be chemically bonded to the rubber particle surface. However, more recently, some enzymes from the bottom fraction have been reported to catalyse the biosynthesis of rubber.[49,50] The molecular weights of proteins in the high ammonia latex concentrate have also been isolated and characterized.[51] Of the six water-soluble proteins, five were found to have molecular weights of 14, 24, 29, 36 and 45 kDa and one greater than 100 kDa. The major protein that associated with the rubber particles was found to have a molecular weight of 14 kDa, the minor protein a molecular weight of 24 kDa.

The rubber proteins exert a strong influence on the latex properties. The carboxyl groups of proteins adsorbed on the surface of rubber particles provide negative charges to the particles, contributing to the colloidal stability of the latex.[52,53] In an early study, proteins were reported to contribute to about 14% of the total negative charges on the latex particles.[53] Recently, based on an atomic force microscopy (AFM) study, the proteins were estimated to cover 84% of the surface area of the rubber particles, while phospholipids covered 16%.[54] Most of the adsorbed proteins are found in the rubber particles, i.e. in the cream fraction. The rubber particles in the skim latex, on the other hand, have a very low level of adsorbed proteins.[20,21] Natural rubber proteins have also been reported to act as an antioxidant, giving a better tensile retention to the rubber. A known antioxidant that is present in natural rubber latex is glutathione.[55,56] It is a tripeptide with a gamma peptide linkage between the amine group of cysteine and the carboxyl group of the glutamate side chain. The cysteine on the glutathione molecule provides an exposed free sulfhydryl group that is a reactive site for radical attack.[56] However, unlike tocotrienols, the effects of this antioxidant on rubber properties have not been widely studied. Since proteins can promote the storage hardening of natural rubber, i.e. spontaneous crosslinking reactions during storage,[57,58] it is not surprising to observe the improvement in tensile retention. This is because storage hardening has been shown to contribute to the plasticity retention index (PRI) value of natural rubber, a measurement of the ageing resistance of natural rubber.[59] Here, PRI is defined as: $PRI = (P_0/P_{140}) \times 100\%$, where P_o is the initial plasticity and P_{140} is the plasticity of the rubber after heating for 30 minutes at 140 °C in an air oven.[60] Proteins could also function as free radical scavengers and retard the chemical modification of natural rubber *via* radical processes. Therefore, in the absence of proteins, a more uniform crosslink distribution along the rubber molecules was obtained in the peroxide crosslinking reaction,[61] more grafting points per chain were found in the grafting reaction,[62,63] and a high speed of degradation to form liquid rubber was observed in the degradation experiment.[64,65] Natural rubber proteins also increase the moisture absorption,[66–68] and dynamic properties such as high modulus,[69–71] creep and stress relaxation.[66] Certain water-soluble proteins from natural rubber latex in various products such as gloves and balloons have been reported to cause Type I hypersensitivity (IgE-mediated allergy) in certain sensitive individuals. In more rare, serious cases, the sensitive individuals may develop angioedema, and even anaphylaxis,[72–75] immediately after exposure to the products. The molecular weights of the allergenic proteins have been reported to be in the range of 11 to 70 kDa.[76] The allergenic proteins that have already been registered by the International Union of Immunological Societies (IUIS) and the World Health Organization (WHO) are shown in Table 3.6.

Of the 13 allergenic proteins, Hev b 1, 3, 5 and 6.02 have been shown to contribute significantly to the total allergenicity of the proteins. The sum of these four allergens was found to be significantly correlated with the validated human IgE-based ELISA-inhibition test.[77] The classification of allergen level for medical gloves is given in Table 3.7. The four allergenic proteins can be quantified using a commercial test kit, known as a FIT-kit, and the test method has been adopted into the ASTM test method, *i.e.* ASTM D7427. The other two test methods are ASTM D5712 for total extractable proteins, based on the modified Lowry, and ASTM D6499 for the antigenic proteins.

Table 3.6 Registered allergenic proteins from natural rubber latex.

Name	Trivial name	Predicted physiological roles
Hev b 1	Rubber elongation factor	Rubber biosynthesis
Hev b 2	Beta-1,3-glucanases	Defence-related protein
Hev b 3	Small rubber –particle protein	Rubber biosynthesis
Hev b 4	Microhelix component	Defence-related protein
Hev b 5	Acidic latex protein	–
Hev b 6.01	Prohevein	
Hev b 6.02	Hevein	Defence-related protein
Hev b 6.03	Prohevein C-terminal fragment	
Hev b 7.01 = Hev b 13 (renamed)	Patatin homologue from B-serum	Defence-related protein
Hev b 7.02	Patatin homologue from C-serum	Inhibitor of rubber biosynthesis
Hev b 8	Latex profilin	Structural protein
Hev b 9	Latex enolase	Glycolytic enzyme
Hev b 10	Mn-superoxide dismutase	Destruction of radicals
Hev b 11	Class I endochitinase	Defence-related protein
Hev b 12	Lipid transfer protein Latex esterase	Defence-related protein
Hev b 13 (= Hev b 7.01)	Early nodule specific protein (ENSP)	Defence-related protein

Table 3.7 Classification of sum of four allergens of medical gloves.[77]

Sum of four allergens content, μg/g	Category
Below 0.03	Very low
0.03–0.15	Low
0.15–0.30	Moderate
0.30–1.15	Moderate to high
Above 1.15	High

Table 3.8 Amino acids of serum from acid coagulated latex.

		Percentage by weight	
No	Amino acid	Reference 55	Reference 80
1	Alanine	10.7	8.3
2	Arginine	4.5	2.0
3	Aspartic acid	14.5	29.8
4	Cysteine	4.7	1.3
5	Cystine	–	1.6
6	Glutamine	–	0.5
7	Glutamic acid	13.7	36.1
8	Glycine	7.9	2.2
9	Histidine	–	0.04
10	Isoleucine	2.6	1.5
11	Lysine	5.2	0.4
12	Leucine	7.5	0.2
13	Proline	–	4.6
14	Phenylalanine	1.9	0.1
15	Serine	6.0	3.1
16	Tyrosine	2.3	0.4
17	Trytophan	–	0.06
18	Threonine	4.8	7.0
19	Valine	7.7	0.8
	Total	100	100

3.1.2.2 Amino Acids

Free amino acids constitute about 0.1% of the latex, of which 80% are in the serum fraction.[78,79] They could be the precursors or hydrolysis products of proteins. A large proportion of amino acids, around 80%, is located in the cytoplasmic serum of the latex. In terms of concentration, both lutoidic and cytoplasmic serums have about the same order, *i.e.* 30 μmol/ml.[79]

The major components are glutamic acid and its amide, alanine, and aspartic acid. Other amino acids include glycine, phenylalanine, tryptophan, leucine, isoleucine, lysine and cystine. Trace amounts of valine, arginine, proline, serine and threonine are also present.[78,79] The results of two recent studies[55,80] on the quantification of amino acids in the serum generated from the acid coagulation of latex are shown in Table 3.8 and confirm that the top three amino acids in the serum of latex by weight are aspartic acid, glutamic acid and alanine.

The antioxidant properties of some of these amino acids were tested on radiation vulcanized natural rubber latex (RVNRL).[81] Among the amino acids, cystine, alanine, asparagine and phenylalanine were found to exhibit high anti-ageing properties on RVNRL. Combinations of these amino acids were observed to improve the antioxidant properties in RVNRL. In another study, the antioxidant activities of the amino acids were studied using 2,2-diphenyl-1-picrylhydrazyl (DPPH) to evaluate the radical scavenging potential. Several amino acids, such as histidine, cysteine, glutamic acid and glutamine, were found to show antioxidant activity.[80] The antioxidant action of amino acids has been proposed to involve two stages: (i) condensation of amino acid under high temperature ageing or vulcanization of the rubber and (ii) chain-breaking donor reactions.[82] The amino acids have also been reported to promote storage hardening of natural rubber.[57] In view of this, if the antioxidant activities are inferred from tensile properties measurement, it could be due to the hardening of the rubber rather than the antioxidant activities of the amino acids.

3.1.2.3 Other Nitrogenous Compounds

Apart from proteins and amino acids, there are other nitrogenous compounds in the latex. These compounds are mainly amines and the corresponding derivatives, such as choline, methylamine, tetramethylenediamine, pentamethylenediamine, ethanolamine and trigonelline.[45] These amine compounds, such as choline, a quaternary saturated amine, have been reported to be an accelerator, and antioxidant for the raw rubber.[83]

3.1.3 Inositols and Carbohydrates

Quebrachitol (2-O-methyl-L-inositol) is the most abundant polyol and it is the largest single non-rubber compound present in the latex, constituting about 1% of fresh natural rubber latex.[84] Crude quebrachitol from dried serum can be purified by a methanol extraction method. The purified compound has a peak melting temperature of 188–189 °C.[85] Its unique chiral structure could simplify the synthesis of many drugs and hence reduce the cost of production. It is, therefore, a base chemical for the synthesis of many pharmaceuticals such as antifungal (E)-β-methoxyacrylate,[86] valienamine[87] and conduritol.[88] The main component of glucid or saccharide in the latex is sucrose. Small quantities of glucose, galactose, fructose, raffinose and two pentoses have also been reported.[89–91] As a result of poor latex preservation, bacteria attack on carbohydrates can lead to the generation of VFAs. The succinic to malic acid ratio could be used to gauge the quality of latex. For a well-preserved latex, values for the succinic to malic acid ratio are normally below 0.6. On the other hand, for a poorly-preserved latex, the values are normally above 0.6.[92] VFA number could affect the physical properties of the rubber films. A low VFA number has been found to give films better physical properties, either unvulcanized or vulcanized, and vice versa.[93]

Table 3.9 Composition of ash from natural rubber latex.[46,94–96]

Component	Percentage by weight
Potassium	0.10–0.30
Magnesium	0.02–0.16
Copper	0.0001–0.0007
Manganese	< 0.0002
Iron	0.001–0.012
Calcium	0.001–0.03
Phosphorus	0.12–0.33
Sodium	0.001–0.10

3.1.4 Ash

A typical mineral composition of field natural rubber latex is given in Table 3.9.

The mineral composition depends on many factors such as clonal origin of the latex, soil conditions, weather, and type of fertilizer used.[95–97] For example, the magnesium level in the latex increases significantly during the winter due to translocation of magnesium from drying leaves.[98] A high magnesium level can cause instability in the latex.[99] It can also lead to an increase in the VFAs and formation of crystals on the surface of the latex. On the other hand, phosphorous compounds, particularly phosphates, are known to make the latex stable. The ratio of phosphate to magnesium is important in determining the stability of the latex.[100–103] When in the tree, most of the magnesium is located in the lutoid particles. However, upon tapping and preservation, the lutoid particles rupture, releasing magnesium ions into the serum. The phosphate can form an insoluble precipitate with the magnesium ions.[104–106] During storage in ammonia, the hydrolysis of phospholipids further releases more phosphates to reduce the magnesium content and leading to an increase in the latex stability.

Copper, manganese and iron are known pro-oxidants of natural rubber, with copper having the highest pro-oxidant activities of the three.[107] However, only free copper can act as a catalyst in the rubber oxidation process.[108] Therefore, copper in fresh latex might complex with proteins and amino acids. This does not have any detrimental effects on the ageing properties of the rubber. However, when the complexes are attacked by microorganisms, free copper is released and this probably explains the general susceptibility of auto-coagulated rubber to thermal oxidative degradation when compared to normal acid-coagulated rubber.[24]

Multivalent metal ions have been postulated to be involved in the ionic crosslinking of natural rubber.[58] Although the addition of these metal ions did not lead to the hardening of the rubber,[109] a microanalysis of the gel fraction of natural rubber by transmission electron microscopy associated with electron energy-loss spectroscopy revealed that the calcium:carbon (Ca/C) and calcium:oxygen (Ca/O) ratios are relatively higher in the gel than in the soluble fraction, indicating strong evidence in favour of ionic crosslinking, probably by

COO⁻Ca^{2+} ⁻OOC bridges.[110] The study also found that the nitrogen:carbon (N/C) ratio is almost the same in both sol and gel fractions, suggesting that protein content seems to have a limited effect on gel formation. This seems to contradict the conclusion from an earlier study using deproteinization, saponification and centrifugation methods to obtain the purified rubber. However, such techniques are rather non-specific.[111,112]

3.1.5 Volatile Matter

Moisture is the main component of the volatile matter. There are small quantities of other volatile acids such as formic, acetic and propionic in the latex.[113] The moisture retention of natural rubber is enhanced by the presence of hydrophilic impurities, mainly inorganic salts and to some extent proteins.[68] Any moisture can promote mould growth and give the dry rubber an undesirable odour.[114] A surgical glove with high moisture absorption properties could cause electrosurgical burns to surgeons during an electrosurgery.

3.2 Abnormal Groups in Natural Rubber

The main component of natural rubber of *Hevea brasiliensis* origin is *cis*-polyisoprene. The rubber also contains a small quantity of chemically bonded non-rubber groups, normally referred to as abnormal groups, on the main-chain molecule. The abnormal groups have been reported to be responsible for the formation of gel and the storage hardening of natural rubber. They have been postulated to give a high green strength property to natural rubber.[3–9] The abnormal groups that have been reported include *trans*-isoprene, ester, aldehyde and epoxide groups, and fatty acid linked phospholipids. Table 3.10 shows a summary of the abnormal groups in natural rubber reported to date.

Table 3.10 Abnormal groups of natural rubber reported in the literature.

Abnormal groups	Fraction	Conc. (mmol/kg)	Method	Reference
Trans-isoprene	Low	10–20	^{13}C-NMR	115
	MW	10	^{1}H-NMR	116, 117
Lactone	Gel	10–15	IR	14
Ester/Fatty Acids/ phospholipids	Whole	10–20	^{1}H-NMR	118–121
Aldehyde	Whole	10–35	H$_2$NOH	122
	Whole	1.5–5.0	2,4-dinitro-phenylhydrazine (DNPH)	123
Amine	Whole	20–35	HBr titration	124
Epoxide	Whole	45–75	HBr titration	
Epoxide	Whole	10–15	Degradation	125
Functionalized epoxide	Whole	30–90	Radiotracer	124, 126

3.2.1 Trans-Isoprene and Dimethylallyl (DMA) Groups

The presence of *trans*-isoprene units was first reported in a ^{13}C-NMR study.[115] The amount of this group was estimated to be 2–4 units per chain of rubber molecule. Subsequently, an investigation into various types of natural rubbers showed that they contained no detectable amount of *trans*-isoprene in *cis–trans* configuration, which is a product of isomerization of *cis*-isoprene units.[116] The *trans*-isoprene units were found to be only in ω-*trans*, or *trans–trans* configurations.[116] Further triad sequential analyses of the isomeric groups by ^{1}H-NMR confirmed that the initiating terminal of the rubber molecule, *i.e.* the ω-group, had a dimethyl allyl derivative followed by two *trans*-isoprene units.[116,117] The ω-group was postulated to be a derivative of the dimethylallyl group, based on the ^{1}H-NMR chemical shifts of the methyl proton in betula-prenol-16[116] and ficaprenol-12[117] as shown in Figure 3.1.

3.2.2 Ester Groups, Fatty Acids and Phospholipids

Based on infrared evidence, ester groups in natural rubber were initially reported to be due to the presence of lactone functional groups.[14] Subsequently, ^{13}C-NMR studies on deproteinized natural rubber showed that the ester groups were due to the presence of fatty acids bonded to the rubber molecule, of which 80% was saturated fatty acids and 20% unsaturated fatty acids.[118] The fatty acids were later identified to be the acyl component of phospholipids linked to the α-terminal end of the rubber molecule.[119–121] The phospholipids were also postulated to be the branching point of the natural rubber molecule.[119–121] The effects of these groups on natural rubber's properties have been discussed above.

3.2.3 Epoxide Groups

The presence of epoxide groups on the main-chain rubber molecule was first inferred from a reduction in the molecular weight of the rubber after treating acid-hydrolysed natural rubber with periodic acid.[124] The amount of epoxide was later estimated by hydrogen bromide (HBr) titration and found to be in the range of 45–75 mmol/kg of rubber, or 0.3–0.5% mol/mol of rubber.[124]

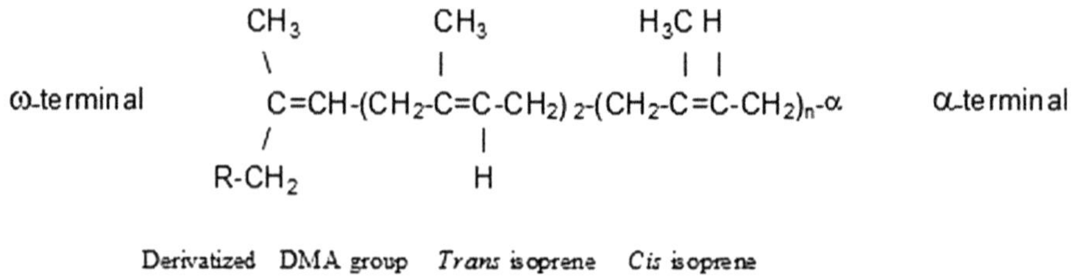

ω-terminal
$$\text{C=CH-(CH}_2\text{-C=C-CH}_2\text{)}_2\text{-(CH}_2\text{-C=C-CH}_2\text{)}_n\text{-}\alpha$$
with CH$_3$, CH$_3$, H$_3$C H substituents; R-CH$_2$ and H below
α-terminal

Derivatized DMA group   Trans isoprene   Cis isoprene

Figure 3.1 *Trans* isoprene of natural rubber.

Table 3.11 *Trans* and *cis* epoxides in degraded natural rubbers.[128]

Degraded rubbers	Temperature (°C)	Trans:cis epoxides
H_2O_2, UV degraded	25	3:40
2% cobalt, thermal degraded	65	10:40
Thermal degraded	120	20:40

The analysis of fractionated natural rubbers indicated that these abnormal groups were concentrated in the high molecular weight fractions.[127]

^{13}C-NMR studies on oxidative degraded natural rubbers showed the presence of epoxides as part of the by-products,[128] of which the *cis* configuration was found to be the major product; the *trans* configuration was the minor product, as shown in Table 3.11. This contradicts the prediction that degradation of natural rubber produces *trans* epoxide groups.[129]

Further ^{13}C-NMR investigations on natural rubbers from fresh latices showed no detectable amount of epoxide groups. However, *trans*-isoprene groups were detected instead in all the samples.[130] The same negative results were also obtained when analysing fractionated natural rubbers of different molecular weights, natural rubbers from different clones, and natural rubbers from different zones of centrifuged latex. Of the six commercial dry rubbers, two were found to contain *cis* epoxide, the amount of which was similar to that of *trans* isoprene units. However, the high molecular weight fraction of these samples contained no epoxide groups when the same analytical technique was used.[129] This contradicts the findings reported earlier.[127] Based on these studies, the *cis* epoxide groups detected in the low molecular weight fraction of rubber from commercial latex by ^{13}C-NMR could be by-products of degradation.[115] This indicates that natural rubber contains no naturally occurring epoxide groups on the main-chain molecule. Due to non-specific methods of detection, the HBr[125] and radiotracer[125] methods are not suitable techniques to quantify epoxide groups in natural rubber. In addition, there has been no direct evidence to indicate that amine and functionalized epoxide groups[124] are present on the main-chain molecule of natural rubber.

3.2.4 Aldehyde Groups

The presence of aldehyde groups in natural rubber was suggested because hydroxylamine was found to inhibit the spontaneous crosslinking reaction in natural rubber[122] and rubber-hydrazone was found when the rubber solution was treated with 2,4-dinitrophenylhydrazine (2,4-DNPH).[123] The estimated level of aldehyde was 1.6–5.4 mmol/kg and the maximum absorption wavelength (λ_{max}) range of 353–357 nm observed for the hydrazone indicates the presence of non-conjugated aldehyde groups in natural rubber.[123] A study on fractionated natural rubbers revealed that all the fractions contain 0.2–0.3 aldehyde groups per linear rubber molecule calculated based on two *trans* isoprene terminal units per chain, or 0.4–1.2 groups per branched chain, as shown in Table 3.12.[20]

Table 3.12 Aldehyde and ester groups in fractionated natural rubber.[20]

Fraction	$M_n \times 10^{-5}$		Ester group/chain[a]		Aldehyde group/chain[b]	
	Linear[c]	Branched[d]	Linear	Branched	Linear	Branched
1	1.8	10.9	1.4	8.5	0.2	1.2
2	1.7	8.7	1.3	6.7	0.2	1.1
3	1.4	6.1	2.1	2.1	0.3	1.1
4	0.4	0.7	1.4	2.4	0.2	0.4

[a]Determined by FTIR.
[b]Determined by UV after converting into rubber-hydrazone.
[c]Molecular weight determined by ^{13}C-NMR.
[d]Molecular weight determined by membrane osmometry.

$$(NR)-CH_2-CHO + OHC-CH_2-(NR) \rightarrow (NR)-(CHO)CH-CH(OH)-CH_2-(NR)$$
$$\downarrow -H_2O$$
$$(NR)-(CHO)C=CH-CH_2-(NR)$$

Scheme 3.1

As discussed earlier, natural rubber ester groups from the bonded fatty acids are located at the branching point of rubber molecules via association with the phospholipids.[35] For branched rubber molecules, the aldehyde content increases in line with the molecular weight of the rubber, as shown in Table 3.12. This indicates that the aldehyde groups are not auto-oxidative degradation products of natural rubber because the degradation would produce a reversed distribution order, *i.e.* the low molecular weight rubbers have a high aldehyde content and *vice versa*. This finding is consistent with the postulation that branching entities of natural rubber are derived from aldo-condensation of the aldehyde groups,[57,122,123] as demonstrated in Scheme 3.1.

However, the reactions in Scheme 3.1 are expected to produce a conjugated aldehyde group, but this has not been observed in the rubber-hydrazone derivative, based on the λ_{max} value.[123] Since the distribution trend of aldehyde groups is similar to that of ester groups, it is more likely that the aldehyde groups are derived from chemical modification of olefinic groups of unsaturated fatty acids bonded to natural rubber,[118] presumably via enzyme-assisted mechanisms during storage of the rubber in the tree, since the presence of oxidizing enzymes such as oxidases has been reported.[48] This is supported by the observation that the removal of bonded fatty acids from natural rubber also reduces the aldehyde content to a level similar to that of synthetic polyisoprene.[20] The aldehyde groups could react with a carbonyl-reactive antioxidant to chemically attach the antioxidant to the rubber molecule, which improved the ageing properties of the rubber.[130] Commercial dry natural rubbers of constant viscosity (CV) grade are produced by adding hydroxylamine to the latex before coagulation, to prevent storage hardening of the rubbers.

3.2.5 Bonded Proteins and Amino Groups

The terminating end, *i.e.* the α-terminal of the rubber molecule, has been shown to be phospholipids, while the initiating end, *i.e.* the ω-terminal, has been postulated to be proteins or amino acids.[35] However, there was no direct evidence to support this idea. Although fractionated natural rubbers have been found to contain 1–2 nitrogen atoms per linear chain,[131] the phospholipids linked to the ω-terminal could also contain nitrogen compounds, such as ethanolamine. This weakens the linked proteins argument.

3.3 Future Trends

With the exception of quebrachitol, most of the non-rubbers are currently being discarded or recovered for low-value applications such as fertilizers[132] and *Bifidobacterium bifidum* growth stimulator.[133] This is mainly because the high value non-rubbers, such as tocotrienols, are present in minute quantities, making it uneconomical to extract them. However, with the development of genetic engineering, such high value non-rubber content may be boosted. In addition, through the use of genetic transformation, transgenic rubber trees could be made to produce proteins and other pharmaceuticals of high commercial value in the latex. Such a system of harvesting high value non-rubbers is non-destructive in nature and highly sustainable as the tree can be tapped every other day for its valuable non-rubbers in the latex for 25–30 years. Other advantages include non-animal origin products, easily propagated via horticultural practices, low cost, and ecologically friendly. Some progress has been made in the genetic transformation of natural rubber trees.[134–136] In the case of chemically bonded abnormal groups, a detailed understanding of their roles in influencing rubber properties would help scientists to design a synthetic rubber with the desired properties. However, due to the complexity of the natural rubber molecular structure and non-rubbers, the detailed structural properties characterization work is expected to take a long time before it can be applied to the design of synthetic rubbers.

References

1. S. F. Chen and C. S. Ng, *Rubber Chem. Technol.*, 1984, **57**, 243.
2. H. Hasma, *J. Nat. Rubber Res.*, 1991, **6**, 105.
3. S. Kawahara, Y. Isono, J. T. Sakdapipanich, Y. Tanaka and A. H. Eng, *Rubber Chem. Technol.*, 2002, **75**, 739.
4. V. L. Folt, *Rubber Chem. Technol.*, 1971, **44**, 12.
5. S. Kawahara, Y. Isono, T. Kakubo, Y. Tanaka and A. H. Eng, *Rubber Chem. Technol.*, 2000, **73**, 39.
6. D. P. Mukherjee, *Rubber Chem. Technol.*, 1974, **47**, 1234.
7. V. A. Grechanovskii, L. S. Ivanona and I. T. Poddubnyi, *Rubber Chem. Technol.*, 1973, **46**, 1234.

8. L. F. Ramos De Valle and M. Montelongo, *Rubber Chem. Technol.*, 1978, **51**, 863.
9. W. S. E. Fernando and M. C. S. Perera, *Kautsch. Gummi Kunstst.*, 1987, **40**, 1149.
10. H. Hasma, A. B. Othman and M. S. Fauzi, *J. Rubber Res.*, 2003, **6**, 231.
11. H. Hasma, A. B. Othman and M. N. Shafie, *J. Rubber Res.*, 2006, **9**, 133.
12. B. Muhammad, M. Ismail, M. A. R. Bhutta and Z. Abdul-Majid, *Constr. Build. Mater.*, 2012, **27**, 241.
13. M. O. David, T. Nipithakul, M. Nardin, J. Schultz and K. Suchiva, *J. Appl. Polym. Sci.*, 2000, **78**, 1486.
14. E. G. Gregg and J. H. Macey, *Rubber Chem. Technol.*, 1973, **46**, 47.
15. S. Amnuaypornsri, J. Sakdapipanich, S. Toki, B. S. Hsiao, N. Ichikawa and Y. Tanaka, *Rubber Chem. Technol.*, 2008, **81**, 753.
16. H. Hasma and A. Subramaniam, *Lipids*, 1978, **13**, 905.
17. H. Hasma and A. Subramaniam, *J. Rubber Res.*, 1986, **1**, 30.
18. S. Liengprayoon, K. Sriroth, E. Dubreucq and L. Vaysse, *Phytochemistry*, 2011, **72**, 1902.
19. S. Liengprayoon, F. Bonfils, J. Sainte-Beuve, K. Sriroth, E. Dubreucq and L. Vaysse, *Eur. J. Lipid Sci. Technol.*, 2008, **110**, 563.
20. A. H. Eng, J. Tangpakdee, S. Kawahara and Y. Tanaka, *J. Nat. Rubber Res*, 1997, **12**, 11.
21. M. M. Rippel, L. T. Lee, C. A. P. Leite and F. Galembeck, *J. Colloid Interface Sci.*, 2003, **268**, 330.
22. S. Rodphukdeekul, S. Liengprayoon, V. Santisopasri, K. Sriroth, F. Bonfils, E. Dubreucq and L. Vaysse, *Kasetsart J. (Nat. Sci.)*, 2008, **42**, 306.
23. M. Morimoto, *Proc. Int Rubber Conf. 1985 Kuala Lumpur*, 1985, **2**, 61.
24. H. Hasma and A. B. Othman, *J. Nat. Rubber Res.*, 1990, **6**, 105.
25. A. R. Arnold and P. Evans, *J. Nat. Rubber Res.*, 1991, **6**, 75.
26. S. Nair, *Proc. Rubbercon'87*, PRI London, 1987, P 2A/a.
27. A. N. Gent, *I.R.I. Trans.*, 1954, **30**, 139.
28. D. R. Burfield, *Polymer*, 1984, **25**, 1823.
29. Y. Tanaka, N. Nishiyama, E. Eijiri and S. Kawahara, *Polym. Prepr. Jpn.*, 1993, **42**, 4324.
30. P. J. Dunphy, K. J. White, J. F. Penock and R. A. Morton, *Nature*, 1965, **207**, 521.
31. S. Kawahara, T. Kakubo, J. T. Sakdapipanich, Y. Isono and Y. Tanaka, *Polymer*, 2000, **41**, 7483.
32. S. Kawahara and Y. Tanaka, *J. Polym. Sci.: Polym. Phys. Ed.*, 1995, **33**, 753.
33. D. Visitnonthachai, *Characterization of Lipids and Their Effects on Mixing Properties of Unvulcanized Raw Natural Rubber*, MSc Thesis, Kasettart University, Thailand, 2005.
34. S. Kohjiya, M. Tosaka, M. Furutani, Y. Ikeda, S. Toki and B. S. Hsiao, *Polymer*, 2007, **48**, 3801.
35. Y. Tanaka and L. Tarachiwin, *Rubber Chem. Technol.*, 2009, **82**, 284.

36. I. S. Bhatia, N. Kaur and P. S. Sukhija, *J. Sci. Food Agric.*, 1978, **29**, 747.
37. S. J. Tata, *J. Rubb. Res. Inst. Malaya*, 1980, **28**, 77.
38. J. Sansatsadeekul, J. Sakdapipanich and P. Rojruthai, *J. Biosci. Bioeng.*, 2011, **111**, 628.
39. B. L. Archer, B. G. Audley, G. P. Sweeney and T. C. Hong, *J. Rubb. Res. Inst. Malaya*, 1969, **21**, 560.
40. S. J. Tata, *Proc. Int. Rubb. Conf. 1975 Kuala Lumpur*, 1976, **2**, 441.
41. B. L. Archer, *Biochem. J.*, 1960, **75**, 236.
42. K. Walujono, R. A. Scholma, J. J. Beintema, A. Mariono and A. M. Hahn, *Proc. Int. Rubb. Conf. 1975 Kuala Lumpur*, 1976, **2**, 518.
43. J. J. Beintema, *J. Rubb. Res.*, 2010, **13**, 265.
44. W. F. Parijs, J. Broekaert, I. J. Goldstein and W. J. Peumans, *Planta*, 1991, **183**, 258.
45. B. L. Acher, *Phytochemistry*, 1976, **15**, 297.
46. B. L. Archer, D. Barnard, E. G. Cockbain, P. B. Dickenson and A. I. McMullen, in *The Chemistry and Physics of Rubber and Rubber-like Substances*, ed. L. Bateman, Maclaren & Son Ltd, London, 1963, p. 41.
47. S. J. Tata and G. F. J. Moir, *J. Rubb. Res. Inst. Malaya*, 1964, **18**, 97.
48. J. d'Auzac and J. L. Jacob, in *Physiology of Rubber Tree Latex: The Lacticiferous Cell and Latex – A Model of Cytoplasm*, ed. J. d'Auzac, J. L. Jacob and H. Chrestin, CRC Press, Florida, 1989, p. 59.
49. J. Tangpakdee, Y. Tanaka, K. Ogura, T. Koyama, R. Wititsuwannakul and D. Wititsuwannakul, *Phytochemistry*, 1997, **45**, 269.
50. J. Tangpakdee, Y. Tanaka, K. Ogura, T. Koyama, R. Wititsuwannakul and N. Chareonthiphakorn, *Phytochemistry*, 1997, **45**, 275.
51. H. Hasma, *J. Nat. Rubber Res.*, 1992, **7**, 102.
52. G. Verhaar, *Rubber Chem. Technol.*, 1959, **32**, 1627.
53. C. C. Ho, *Colloid Polym. Sci.*, 1989, **267**, 643.
54. K. Nawamawat, J. T. Sakdapipanich, C. C. Ho, Y. Ma, J. Song and J. G. Vancso, *Colloids Surf., A*, 2011, **390**, 157.
55. I. Aimi Izyana and M. N. Zairossani, *IIUM Eng. J.*, 2011, **12**, 61.
56. A. R. Ndhlala, M. Moyo and J. V. Staden, *Molecules*, 2010, **15**, 6905.
57. M. J. Gregory and A. S. Tan, *Proc. Int. Rubb. Conf 1975 Kuala Lumpur*, 1976, **4**, 28.
58. D. R. Burfield, in *Development of Plastics and Rubber Product Industries*, ed. J. C. Rajaro, Rubber Research Institute of Malaysia, Kuala Lumpur, 1989, p. 204.
59. M. D. Morris, *J. Nat. Rubber Res.*, 1991, **6**, 96.
60. Rubber Research Institute of Malaysia, *SMR Bull.* No. 7, 1973.
61. P. Tangboriboonrat, D. Polpanich, T. Suteewong, K. Sanguansap, U. Paiphansiri and C. Lerthititrakul, *Colloid Polym. Sci.*, 2003, **282**, 177.
62. Y. Fukushima, S. Kawahara and Y. Tanaka, *J. Rubber Res.*, 1998, **1**, 154.
63. N. V. Tho, M. O. A. Kadir and A. S. Hashim, *Rubber Chem. Technol.*, 2002, **75**, 111.
64. J. Tangpakdee, M. Mizokoshi, A. Endo and Y. Tanaka, *Rubber Chem. Technol.*, 1998, **71**, 795.

65. J. T. Sakdapipanich, T. Kowitteerawut, S. Kawahara and Y. Tanaka, *J. Rubber Res.*, 2001, **4**, 1.
66. K. N. G. Fuller, in *Natural Rubber Science and Technology*, ed. A. D. Robert, Oxford University Press, Oxford, 1988, Chapter 19.
67. A. Stevenson, in *Rubber in Offshore Engineering*, ed. A. Stevenson, Adam Hilger, Bristol, 1984, Chapter 2.
68. D. R. Burfield, A. H. Eng and P. S. T. Loi, *Br. Polym. J.*, 1989, **21**, 77.
69. K. L. Chong and M. Porter, *Int. Polym. Latex. Conf.*, 1978 London, 1980, Prepr. 16.
70. K. L. Chong, and M. Porter, *Int. Conf. Structure – Properties Relations of Rubber*, Kharagpur, 1980, Prepr I.
71. A. V. Chapman and M. Porter, in *Natural Rubber Science and Technology*, ed. A. D. Robert, Oxford University Press, Oxford, 1988, Chapter 12.
72. D. W. Gelfanf, *Am. Radiol.*, 1990, **156**, 1.
73. K. Turjanmaa, T. Reunala and T. Karkkainen, *Br. Med. J.*, 1988, **297**, 1029.
74. J. E. Slater, *New Engl. J. Med.*, 1989, **320**, 1126.
75. J. R. Warpinski, J. F. Folgert, M. Cohen and R. K. Bush, *Clin. Exp. Allergy*, 1991, **12**, 95.
76. H. Alenius, K. Turjanmaa, T. Palosuo, S. Makinen-Kiljunen and T. Reunala, *Int. Arch. Allergy Immunol.*, 1991, **96**, 376.
77. T. Palosuo, H. Reinikka-Railo, H. Kautiainen, H. Alenius, H. Kalkkinen, M. Kulomaa, T. Reunala and K. Turjanmaa, *Allergy*, 2007, **62**, 781.
78. T. S. Ng, *Proc. Nat. Rubb. Res. Conf.*, 1960, **3**, 809.
79. J. Brzozowska, P. Hanower and P. Chezeau, *Experientia*, 1974, **30**, 894.
80. W. Soysuwan, *Antioxidant from Hevea brasiliensis Latex Serum*, MSc thesis, Prince of Songkla University, Thailand, 2009, p. 65.
81. L. V. Abad, L. S. Relleve, C. T. Aranilla, A. K. Aliganga, C. M. San Diego and A. M. dela Rosa, *Polym. Degrad. Stabil.*, 2002, **76**, 275.
82. S. Tuampoemsab, *Control of the Degradation of Natural Rubber: Analysis and Application of Naturally Occurring Anti- and Pro-Oxidants in Natural Rubber*, PhD thesis, Mahidol University, Thailand, 2008.
83. R. F. A. Altman, *Rubber Chem. Technol.*, 1948, **21**, 752.
84. J. Gopalakrishnan, *World Congr. Biotechnol.* India 2011, Abstract.
85. Y. Udagawa, M. Machida and S. Ogawa, *US pat. No. 5041689*, 1991.
86. J. J. Kiddle, *Chem. Rev.*, 1995, **95**, 2189.
87. H. Paulsen and F. R. Heiker, *Angew. Chem. Int.*, 1980, **19**, 904.
88. T. Akiyama, H. Shima and S. Ozaki, *Tetrahedron Lett.*, 1991, **40**, 5593.
89. R. H. Smith, *Biochem.*, 1954, **57**, 144.
90. J. S. Lowe, *I.R.I. Trans.*, 1960, **36**, 202.
91. J. Tupy and W. I. Resing, *Biol. Plant.*, 1968, **10**, 72.
92. V. Galli, N. Olmo and C. Barbas, *J. Chromatogr. A*, 2002, **949**, 367.
93. H. D. Chirinos and S. Carvalho De Jesus, *Periodico Tche Quimica*, 2012, **9**, 38.
94. L. Archer, D. Barnard, E. G. Cockbain, P. B. Dickenson and A. I. McMullen, in *Natural Rubber Science and Technology*, ed. A. D. Robert, Oxford University Press, Oxford, 1988, Chapter 2.

95. E. Yip, *J. Nat. Rubber Res.*, 1990, **5**, 52.
96. E. L. Ong, *J. Appl. Polym. Sci.*, 2000, **78**, 1517.
97. E. L. Ong and A. H. Eng, *The Vanderbilt Rubber Handbook*, R. T. Vanderbilt Co. Inc., USA, 2010, p. 23.
98. J. S. Lowe, *Trans I.R.I.*, 1962, **38**, T208.
99. E. W. Madge, H. M. Collier and J. D. Peel, *Trans I.R.I.*, 1950, **26**, 305.
100. M. W. Philpott and D. R. Westgarth, *J. Rubb. Res. Inst. Malaya*, 1953, **14**, 133.
101. W. L. Resing, *Proc. 3rd Rubb. Technol. Conf. London*, 1954, p. 50.
102. W. L. Resing, *Archs. Rubb. Cult.*, 1955, **32**, 75.
103. L. Karunanayake and G. M. Priyanthi Perera, *J. Appl. Polym. Sci.*, 2006, **99**, 3120.
104. L. Tarachiwin, J. T. Sakdapipanich and Y. Tanaka, *Rubber Chem. Technol.*, 2003, **76**, 1185.
105. L. Tarachiwin, J. T. Sakdapipanich and Y. Tanaka, *Rubber Chem. Technol.*, 2003, **76**, 1177.
106. L. Bateman and B. C. Sekhar, *J Rubb. Res. Inst. Malaya*, 1966, **19**, 133.
107. J. R. Shelton, *Rubber Chem. Technol.*, 1972, **45**, 359.
108. O. Alias and B. L. Chan, *J. Rubb. Res. Inst. Malays.*, 1980, **28**, 109.
109. S. N. Gan and K. F. Ting, *Polymer*, 1993, **34**, 2142.
110. M. M. Rippel, C. A. P. Leite, M. C. M. V. da Silva and F. Galembeck, *Microsc. Microanal.*, 2003, **9**(Suppl 2), 450.
111. J. Tangpakdee and Y. Tanaka, *Rubber Chem. Technol.*, 1997, **70**, 707.
112. J. Tangpakdee and Y. Tanaka, *J. Rubber Res.*, 1998, **1**, 14.
113. R. C. Crafts, J. E. Davey, G. P. McSweeney and I. S. Stephens, *J. Nat. Rubber Res.*, 1990, **5**, 275.
114. M. Nadarajah, M. T. Veerabangsa, G. A. de Silva, S. Senaratne and D. C. R. Perera, *Proc. Rubbercon'87*, PRI London, 1987, p. 3A/1.
115. Y. Tanaka, in *ACS Symposium Series No. 247, 'NMR and Macromolecules'*, ed. J. C. Randall, American Chemical Society, Washington, 1984, p. 233.
116. A. H. Eng, S. Kawahara and Y. Tanaka, *Rubber Chem. Technol.*, 1994, **67**, 159.
117. Y. Tanaka, A. H. Eng, N. Ohya, N. Nishiyama, J. Tangpakdee, S. Kawahara and R. Wititsuwannakul, *Phytochemistry*, 1996, **41**, 1501.
118. A. H. Eng, S. Ejiri, S. Kawahara and Y. Tanaka, *J. Appl. Polym. Sci. Polym. Symp.*, 1994, **53**, 5.
119. A. H. Eng, *Structural Characterization of Natural Rubber*, DEng thesis, Tokyo University of Agriculture and Technology, Japan, 1994.
120. L. Tarachiwin, J. T. Sakdapipanich, K. Ute, T. Kitayama, T. Bamba, E. Fukusaki, A. Kobayashi and Y. Tanaka, *Biomacromolecules*, 2005, **6**, 1851.
121. L. Tarachiwin, J. T. Sakdapipanich, K. Ute, T. Kitayama and Y. Tanaka, *Biomacromolecules*, 2005, **6**, 1858.
122. B. C. Sekhar, *J. Polym. Sci*, 1960, **48**, 133.

123. A. Subramaniam, *J. Rubb. Res. Inst. Malaysia*, 1977, **25**, 61.
124. D. R. Burifield and S. N. Gan, *J. Polym. Sci.: Polym. Chem. Ed.*, 1975, **13**, 2725.
125. D. R. Burfield and S. N. Gan, *Polymer*, 1977, **18**, 607.
126. D. R. Burfield and S. N. Gan, *J. Polym. Sci.: Polym. Chem. Ed.*, 1977, **15**, 2721.
127. D. R. Burfield, L. C. Chew and S. N. Gan, *Polymer*, 1976, **17**, 713.
128. A. H. Eng, J. Tangpakdee, S. Kawahara and Y. Tanaka, *J. Rubber Res.*, 1998, **1**, 67.
129. D. R. Burfield, *Makromol. Chem. Rapid Commun.*, 1988, **189**, 523.
130. W. S. E. Fernando, E. D. I. H. Perera and K. A. R. M. Perera, *Proc. Int. Rubb. Conf. 1975 Kuala Lumpur*, 1976, **2**, 153.
131. D. Mekkriengkrai, J. T. Sakdapipanich and Y. Tanaka, *Rubber Chem. Technol.*, 2006, **79**, 366.
132. C. M. Lau, A. Subramaniam and Y. Tajima, *Proc. Rubb. Growers' Conf.*, 1989, 525.
133. S. Etoh, K. Asamura, A. Obu, K. Sonomoto and A. Ishizaki, *Biosci. Biotechnol. Biochem.*, 2000, **64**, 2083.
134. P. Arokiaraj, H. Y. Yeang, K. F. Cheong, S. Hamzah, H. Jones, S. Coomber and B. V. Charlwood, *Plant Cell Rep.*, 1998, **17**, 621.
135. H. Y. Yeang, P. Arokiaraj, J. Hafsah, M. A. Siti Arija, S. Rajamanickam, H. Chan, S. Jafri, R. Leelavathy, H. Samsidar and C. P. E. Van der Logt, *J. Rubber Res.*, 2002, **5**, 215.
136. P. Arokiaraj, S. Siti Shuhada and E. Sunderasan, *Proc. 22nd Pac. Sci. Congr.*, Malaysia, 2011, Abstract.

CHAPTER 4

The Production of Natural Rubber from Hevea brasiliensis Latex: Colloidal Properties, Preservation, Purification and Processing

C. C. HO

Universiti Tunku Abdul Rahman, Kuala Lumpur, Malaysia
Email: cchoho2001@yahoo.com

CIVILIZATION as we know it today is wholly dependent upon rubber. It is a servant that follows us, literally, from the cradle to the grave.
                                    Ralph Wolf in *Rubber World*, October 1964

4.1 Introduction

Rubber latices are opaque dispersions containing fine latex particles dispersed in a liquid phase. Rubber latexes can either be extracted from plants and trees or they can be synthesized in the laboratory using various monomers. The former is known as natural rubber (NR) latex and the latter is called synthetic latex. So far the only commercial NR latex available as technical feedstock for many industrial applications is obtained from *Hevea* trees (see later). On the other hand, synthetic latexes are manufactured on a large scale from monomers that are the building blocks of polymers making up the dispersed particles of latex. A technique called emulsion polymerization is used to convert the monomers into polymer latex particles in reactors with water as the dispersing medium. This is the most common and important manufacturing route for most of the synthetic latices. NR latex is often converted into solid NR, which is used extensively in many industries, in particular the niche area of aviation tyres, whereas synthetic latices are converted into synthetic rubbers. Emulsion polymerization remains the major industrial process for the production of synthetic rubbers. Of course latexes are sometimes used as such, for example in the production of medical gloves and catheters. Since there are many different kinds of monomers, many different synthetic rubbers can be synthesized. To date nearly all monomers have been derived from crude oil, an exhaustible resource. Hence the price of synthetic rubbers also fluctuates with the price of crude oil. NR is natural, renewable and is considered a green material and green feedstock.

4.2 Sources of NR

Of the industrial elastomers, NR latex from *Hevea* is the largest source of renewable elastomers. *Hevea brasiliensis* (genus *Hevea*, family *Euphorbiaceae*) is a tall softwood tree indigenous to Brazil. However, the major NR-producing country is not Brazil but is the South-East Asia region. In the 1870s the *Hevea* trees germinated from seeds originally from Brazil were introduced to Sri Lanka, Malaysia and Singapore and later found to thrive in surrounding countries, where the weather, terrain and soils were well suited for its growth. In the 20th century the rubber trees in Brazilian plantations were destroyed by serious fungal attack, the South American leaf blight (SALB), caused by the ascomycete *Microcyclus ulei*. This fungus inhibits commercial NR production in South and Central America and signalled the demise of the once flourishing NR industry in Brazil. There is currently intensive and strictly enforced control of international air traffic and freight shipment from South America to other tropical countries to prevent the spread of the fungus to other continents where *Hevea* trees are cultivated. However, disease control is becoming increasingly complex and the probability of uncontrolled spread of this pathogen is high. The spread of SALB into the South-East Asia region is a continuing and serious threat that would have a devastating impact on the economy of these rubber-producing nations and grave repercussions on the world supply of NR with an immediate impact on the aviation and automobile industries. It has been reported that most of the cultivated NR tree genotypes in South-East Asian countries are susceptible to SALB.[1]

The serious threat of disease to the viability of the industry has provided the opportunity for some countries to search for alternative sources of rubber latex from other plants. Guayule (*Parthenium argentatum* Gray) and *Taraxacum koksaghyz* (TKS) are plants in the Compositae family found in the temperate regions and produce latex that contains rubber hydrocarbons. They have been specifically identified as potential commercial sources of NR for the temperate zones that could be exploited to free the EU and USA from importing NR. TKS, commonly known as Russian dandelion, and now Kazak dandelion, is a perennial plant found originally in Kazakhstan and Uzbekistan. The Soviet Union was already actively exploring this as a native source of NR in a strategic programme launched in the then USSR in the 1930s to produce NR.[2] The latex is stored in the roots of the plant and can contain as much as 24% rubber and 35% inulin (a carbohydrate) in addition to proteins and small quantities of fatty acids and resins. The chemical structure of the rubber was confirmed to be >95% *cis*-polyisoprene. This shows that the rubber from Russian dandelion latex is pure, of high molecular weight and very similar to that found in *Hevea* latex. TKS can be viewed as a valuable source of NR latex and solid NR. A program has run since 2006 in the USA and an EU initiative to study the viability of commercial cultivation of alternative NR-producing crops in the temperate regions of the EU was started in 2005.[3]

Guayule (*Parthenium argentatum* Gray), on the other hand, is a desert shrub indigenous to North Central Mexico and the south-western part of the USA. Guayule is the only plant other than *Hevea* that had received serious attention from the US government for the purposes of commercial exploitation. However, commercialization of guayule never really took off successfully as envisioned.[4] Lack of well-established good agronomic practices and viable extraction processes for guayule led to slow uptake by the commercial sector. The rubber in guayule latex is free from rubber proteins that can cause allergic reactions when they come into contact with the skin of certain sensitized users. It was shown that guayule latex contains <1% of the proteins found in highly purified *Hevea* latex. Guayule rubber is almost 100% *cis*-1,4 polyisoprene and identical to *Hevea* rubber,[5] and the two rubbers have identical physical properties. There have been renewed efforts recently to revive the exploitation of the guayule plant as an alternative source of commercial NR.[6] A new strategy was introduced to extract not only guayule NR latex from the shrub but also to recover specialty chemicals and liquid fuel from the guayule biomass as an integrated extraction process.

The road to achieving a fully workable substitute for *Hevea* latex and rubber in the temperate zones is a long and bumpy one. Rubbers from TKS and guayule could at best be useful in the event of supply shortages of NR (complement) as well as providing a suitable alternative for people with allergies to hevein, a major allergen present in *Hevea* latex.[7] In addition, these two plants are not so far threatened by SALB fungus attack in the same way as *Hevea* trees in Brazil and other South American countries.

4.3 Evolution of NR as an Industrial Elastomer

The NR plantation industry has evolved over the years from a largely agro-based industry to that of a multibillion dollar business that provides technically specified industrial feedstock for diverse industries making rubber products.

The introduction of plantation cultivation in the early 20th century had boosted production dramatically and paved the way for high-volume commercial production of dry rubber. This was spurred on by the escalating demand brought on by the revolutionary introduction of pneumatic tyres for automobiles. Better agronomy practices and continuous breeding programmes had succeeded in developing ever higher yielding new clones for the planting industries. Some South-East Asian countries have evolved into very efficient producers of high-quality rubber. Currently, of the total world production of industrial elastomers, about 42% is NR (Table 4.1). The major commercial rubber-producing countries are all concentrated in South-East Asia and part of Asia, led by Thailand, Indonesia, Malaysia, India, Vietnam and China. In fact most of the countries in South-East Asia are current producers of NR (Table 4.2). Amongst them, Vietnam has registered the highest growth in recent years. It is forecast that output from Vietnam will rise further in the future, greatly supported by government incentives. It is estimated that Asia supplies close to 90% of the world's NR, mainly from South-East Asia. It also comes as no surprise that China and India, the two emerging economies, are the two

Table 4.1 Global NR, synthetic rubber and total rubber production.

Year	Elastomer production ('000 tonnes)		Total rubber
	NR	Synthetic	
2000	6762	10 870	17 632
2001	7332	10 483	17 815
2002	7326	10 877	18 203
2003	8006	11 338	19 344
2004	8744	11 977	20 271
2005	8907	12 073	20 980
2006	9827	12 612	22 439
2007	9890	13 347	23 237
2008	10 128	12 711	22 839
2009	9690	12 385	22 075
2010	10 399	14 082	24 481
2011	**10 974**	**15 115**	**26 089**

Source: International Rubber Study Group (IRSG).

Table 4.2 Production of NR in South-east Asian countries.

Countries	2009 ('000 tonnes)	2010 ('000 tonnes)
Thailand	3164	3072
Indonesia	2440	2829
Malaysia	856	939
India	820	851
Vietnam	724	750
China	644	650
Others	1054	1200
Total	**9702**	**10 291**

Source: Monthly Rubber Statistical News, Volume 69, No. 12, May 2011, Statistics & Planning Department Rubber Board, Kottayam, Kerala, India.

largest consumers of NR, followed closely by the USA and Japan. Demand for both natural and synthetic rubbers is expected to continue to grow strongly in the future.

Synthetic rubber is in essence a war product. With the growing demand triggered by the advent of the industrial revolution and the rapid transformation of road transportation, the supply of NR was soon outstripped by accelerating demand. Efforts were started in the USA and Europe to search for a synthetic replacement for NR. The first polyisoprene was synthesized by Fritz Hofmann in 1909 while Bayer was already producing methyl isoprene for making methyl rubber in 1910. These materials behaved like the natural counterpart, but did not quite match it. The first impact of NR on a nation was felt by Germany during the First World War in 1915 when the British blocked all supplies of NR to Germany, hoping to bring to a halt all their war machines. The Germans managed to restart their methyl rubber production plant but eventually lost the War. After the War, production of methyl rubbers stopped but the quest for synthetic rubbers by the USA, Germany and the Soviet Union continued unabated. Progress was slow, however, until Staudinger laid the foundation for macromolecular chemistry and introduced the concept of polymers and the molecular structure of rubber molecules in 1920, when it was recognized these are long-chain molecules with very high molecular weight.[8] Notably, IG Farben was the first to produce styrene butadiene rubber, known as Buna-S then, from butadiene and styrene by an aqueous emulsion polymerization technique. By 1935 Buna-S was already being produced in large quantities in Germany. This was followed by commercial production of Thiokol and Buna-N rubbers (acrylonitrile butadiene rubber) in the 1930s in the USA. Both of these rubbers are oil-resistant. Carothers succeeded, after many attempts, to make chloroprene at about the same time in the USA; by 1940 the Soviet Union had the largest synthetic rubber industry. This was the beginning of the synthetic rubber era.

During the Second World War, the blockage of supplies of NR by the Allies in Europe had prompted the Germans to intensify their search for a good substitute for NR that could be made in the laboratory. At the same time the vital supply of NR to the rest of the world was cut off by the Japanese who were controlling most of South-East Asia, where many of the rubber-producing countries are located. The USA, fearful of how vulnerable to interruption of supply this war material was, started to plan for alternative sources and substitutes, including planting guayule for their latex and rubber. On the other hand, the Soviet Union had already started their search for a NR source from another plant, the Russian dandelion. The War spurred on the rapid development of the synthetic route of manufacture of the NR replacement using chemical methods and accelerated large-scale commercial production of synthetic rubbers. The Rubber Reserve Company was founded in 1940 after President Roosevelt declared rubber a 'strategic and critical material' and the American Synthetic Rubber Research Program began in earnest in 1942, initially producing a Buna-S type rubber following the attack on Pearl Harbour in 1941. The programme succeeded in producing many varieties of synthetic rubbers within a short time under very difficult times. Almost 80% of synthetic rubbers are used in the automobile industry. From an initial 3721 tons of Buna-S produced in 1942, production rose to 756,042 tons by 1945. During the post-WWII period, the demand for elastomers continued to grow rapidly. The USA and Germany intensified their research and developed many synthetic rubber types ranging from SBR to isobutylene isoprene rubbers (IIR) to polysulfides (Thiokol), acrylonitrile butadiene rubber and neoprene rubbers. The discovery of Ziegler Natta catalysts in the 1950s made possible the synthesis of stereoregular polymer rubbers. Synthetic *cis*-1,4-polyisoprene, *cis*-1,4-polybutadiene and ethylene and propylene diene rubber (EPDM) became commercially available from the mid-1950s.

The synthetic rubber industry has never looked back and by the 1960s had overtaken NR in terms of production and usage (Table 4.1). At one stage, it was predicted that SBR could eventually replace NR as an industrial elastomer, in particular in tyre manufacturing. However, that prediction never materialized because NR is still needed to provide the required sturdiness and heat resistance in tyre performance. Radial tyres are typically made up of nearly 50% NR whereas aviation tyres require 100% NR in their formulations. Now NR and synthetic rubber complement each other in many applications, each with its niche properties. In the present drive towards sustainable development, NR is viewed as a renewable raw material and is gradually regaining its dominant role.

4.4 Colloidal Properties of NR Latex and Stability

Rubber latices are colloidal dispersions, which means they contain particles of colloidal size, *i.e.* several nm to several microns. When the particles are polymer particles, they are known as polymer colloids. The size of the NR latex particles is usually no more than 2–3 μm, in fact most of them are smaller than that, while those for synthetic latexes are often smaller, in the hundreds of nm region. The average latex particle size of commercial centrifuged high ammonia latex concentrate (HA) is about 1 μm. The spread of particle size in NR latex is much wider than in synthetic latexes. Bigger polymeric particles of tens of microns can be synthesized but they are no longer colloidal and are known as polymer beads, such as those used as packing materials for chromatographic columns. Most of the concentrated latexes are opaque in appearance due to the ability of particles to scatter light, a typical colloidal behaviour, in addition to their ability to diffuse, exert osmotic pressure and lower vapour pressure. The very dilute latex dispersion will appear milky or turbid.

The scattering behaviour of latex particles is dependent on the particle concentration, the shape and size, and the size distribution of the particles, as in any colloidal dispersion. Thus the size and size distribution of the latex particles can be measured easily using light scattering techniques. In fact both static and dynamic scattering techniques have been employed for this purpose. Another colloidal property of latex dispersions containing highly monodisperse particles is that they are capable of exhibiting a higher order Tyndall spectrum (HOTS) and appear coloured when viewed at different angles.[9,10] In fact regular packing of highly monodisperse synthetic latex particles at high solids on standing in a transparent bottle over time can lead to iridescence, which can be easily seen at the side of the container.

The latex particles consist of polymeric materials of reasonably high molecular weight, in the hundreds of thousands. They exist in particulate form because they are stabilized against aggregation by virtue of the fact that the particle surface is charged and hence they repel each other. All colloidal systems are thermodynamically unstable, and given time they will aggregate to minimize the free energy of the system. Colloid stability is an important property of latex dispersions. In fact the particle size and the size distribution, the composition and structure of the particle surface have a strong influence on the stability, viscosity and other properties of the latex. The bulk properties of rubber, on the other hand, are dependent on the chemical nature of the polymer and its molecular weight and molecular weight distribution.

Chemically the NR produced by the *Hevea brasiliensis* species is almost pure *cis*-1,4-polyisoprene. So far none of the manufacturers of synthetic *cis*-1,4-polyisoprene is able to achieve more than 95% of the *cis* isomer in their commercial products. Jitladda and coworkers[11] have conducted extensive studies on the molecular structures of the NR molecules found in *Hevea* and correlated these to the biosynthesis of this in the trees. This is intrinsically linked to the end groups of the rubber molecules involving the phospholipids and proteins that are linked to the charging mechanism at the latex particle surface and hence has a direct impact on the stability of the latex.

The mechanism of stabilization of the NR latex particles is quite different from that for synthetic latices. They deserve special elaboration here because this is of direct relevance to the production of technical latex concentrate and also rubber. In the case of synthetic latices, the particles are stabilized by adsorbed surfactants, which could be anionic or non-ionic, but are more often than not a mixture of these added for the (emulsion) polymerization process. More surfactant could be added after the polymerization, to enhance the stability. On the other hand, NR latex particles in fresh field latex are stabilized by a mixed layer of proteins and phospholipids.[12] These are the rubber proteins and natural lipids that come with the trees. The latex is susceptible to bacterial and enzymatic attack after leaving the trees, initiating biochemical reactions that generate acidic ions, reducing the pH of the fresh latex to the acidic range and leading to auto-coagulation of the latex. Thus ammonia is usually added to the field latex to preserve it and to raise the pH to increase the stability. In addition, ammonia causes hydrolysis of the phospholipids associated with the rubber particles, converting them into long-chain fatty acid soaps and glycerol. Some of the rubber proteins were also hydrolysed into polypeptides. The long-chain fatty acid soaps and the polypeptides become re-adsorbed onto the latex particle surface. Hence the surface of the latex particles in ammoniated latex concentrate is stabilized mainly by re-adsorbed long-chain fatty acid soaps and polypeptides. The carboxylic groups of these species are known to react with magnesium and calcium ions to form insoluble metal salts that can cause incipient destabilization of latex concentrate. Post-addition of surfactants to the NR latex is sometimes required to boost its stability.

4.5 Production of Commercial NR Latex from *Hevea*

More than 20,000 species of plants from over 40 families produce latex. Of these, over 12,000 species contain rubber.[13] However the vast majority of the latex is unsuitable for commercial exploitation for NR production. Very few of these plants produce latex of quality and quantities that are viable for rubber production on a commercial scale. So far only the rubber tree *Hevea brasiliensis* has been successfully cultivated and commercial production of NR is obtained exclusively from this tree.

The latex comes from the bark of the tree in which the laticifers or the articulated latex vessels are located (Figure 4.1).[14] It is extracted by carefully tapping the bark with a special tapping knife devised by Ridley, who invented rubber tapping and later developed the tapping technique for the industry.[15] Prior to that, latex was extracted from the wild rubber trees in Brazil by chopping them down and slashing their bark to drain off the latex. Ridley's method involves shaving off thin slices of the bark to open the closed ends of the latex vessels and to let the latex flow out along a sloping groove into a collection cup through a spout at the end of the tapping groove (Figure 4.2). The latex will stop flowing after a while due to the coagulation of the latex at the incision. The latex and the self-sealing of the latex vessels is a self-defence

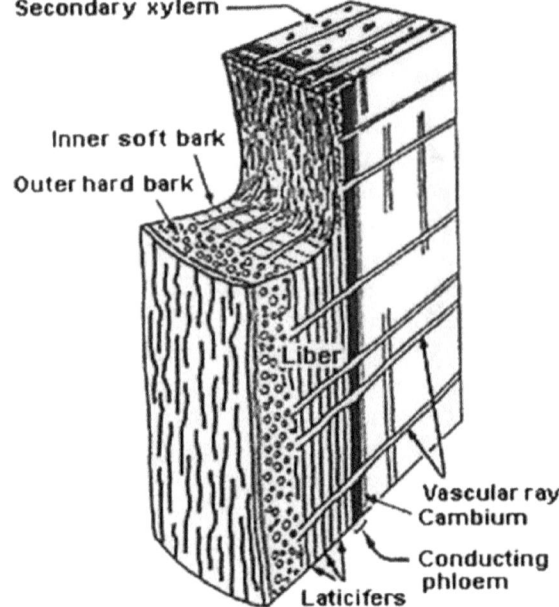

Figure 4.1 The tapping corresponds to a wound in the tree bark. Only a thin slice of the bark is shaved off to cut open the laticifers in the soft bark.

(a) (b)

Figure 4.2 (a) The tapping panel and latex collecting cup at the end of the cut. (b) The actual tapping of the panel by a rubber tapper.

mechanism for the trees. The sloping cut extends to cover half the spiral of the tree. A patch devoid of bark develops on the tree over time. The trees are tapped on alternate days to avoid over-stressing them. After a while the tapped panel becomes covered by newly regenerated bark and this can then be tapped again. With this tapping method the trees are not damaged and can be tapped continuously over their lifetime, in contrast to those wild rubber trees in Brazil that were damaged by indiscriminate multiple cuts on the tree trunk in the early days of the industry. Ridley's tapping method revolutionized the rubber industry by providing an efficient latex extraction technique and helped to establish the rubber plantation industry in South-East Asia. A one-third spiral tapping system has also been developed in conjunction with the application of yield stimulants for older trees that are faced with declining latex yields. Battery-powered tapping knives and auto-puncture tapping knives have also been employed but their adoption by the industry is rather limited. Rubber tapping is done very early in the morning, before sunrise, when the cool weather allows the latex to flow for longer before it stops. Rubber tapping is labour-intensive and hard work. The industry is facing an acute labour shortage, in particular skilled rubber tappers. In the long term, automation of tapping and collection of latex could contribute to great improvements in productivity. However, no immediate solution to resolve this is in sight.

4.5.1 Preservation of NR Latex

The biological nature of the fresh field latex makes it susceptible to enzymatic attack by bacteria and the latex is putrefied within hours of leaving the tree if not preserved. How soon it coagulates spontaneously depends on the ambient temperature and stability of the latex. The latex separates into lumps of rubber and a clear serum, followed by putrefaction and development of the characteristic bad odours. Any preservative for field NR latex to be effective against spontaneous coagulation and putrefaction should be able to destroy the microorganisms present, to aid in the stabilization of the latex and to act against the detrimental effect of heavy metal ions present in the field latex. Ideally the preservative should not be toxic to humans or the environment. It should not cause discoloration of the latex or the products made from it and it should not interfere with the treatment process of the latex. In addition, it should be cheap and easy to handle. Irrespective of the preservatives used, scrupulous cleanliness and hygiene when handling the latex as it exudes from the tree is of utmost importance in ensuring the quality of the latex concentrate later.

Several preservative systems have been developed for NR latex to suppress the proliferation of bacteria. Ammonia is the most common preservative used for NR field latex. It is a known mild bactericide and also functions as an alkali to increase the pH of the latex. The normal practice is to add some ammonia to the field latex upon arrival at the factory to prevent auto-coagulation before further treatment. Since the field latex contains only about 30% solids, it is unsuitable for most technical applications. Thus field latex needs to be further preserved, purified and concentrated to about 60% solids in many commercial technical grade latex concentrates. Long-term preservation requires strong preservatives and/or higher dosages of these. Latex concentrate is preserved usually with about 0.7% of ammonia at a pH of about 10.5. This is known as HA latex concentrate. Alternatively a low ammonia concentration of 0.2% is used in conjunction with tetramethylthiuram disulfide (TDTM) as co-preservative. This latex concentrate, known as LA-TZ, contains 0.025% TMTD/ZnO and 0.05% ammonium laurate. HA and LA-TZ latex concentrates are the two most common latex concentrates marketed.

A whole range of inorganic and organic bases can be used to adjust the pH of NR field latex. However, ammonia remains the most common and the cheapest option and hence is widely used. Additional preservatives are sometimes used to further boost the stability of NR latex concentrates. Very often they function as bactericides, enzyme inhibitors, sequestering agents or stabilizers and are often used in combination. Some early preservative systems that were discontinued mainly because of safety and health hazard concerns are 0.2% sodium pentachlorophenate and 0.2% ammonia; 0.1% sodium pentachlorophenate and 0.1% ethylenediaminetetraacetic acid (EDTA) and 0.1% ammonia; 0.2% zinc diethyldithiocarbamate (ZDTC) and 0.2% ammonia; 0.2% boric acid, 0.05% ammonium laurate and 0.2% ammonia.

The effectiveness of the preservative is evaluated by measuring the volatile fatty acid (VFA) number of the latex, which is defined as the number of grams of potassium hydroxide required to neutralize the VFAs in the latex sample containing 100 g total solids in accordance with, for example, the ASTM D1076-69 test method. Effective preservation ensures the mechanical stability and quality of the latex during storage and processing. More recent developments in latex preservatives with a greater emphasis on safety, such as TMTD-free and ammonia-free systems, will be discussed later.

4.5.2 Latex Purification and Concentration Processes

The latex as obtained from *Hevea* trees by tapping (called field latex) is a dilute aqueous dispersion of latex particles. It contains a host of organic molecules (mainly carbohydrates, phospholipids and rubber proteins) and metal ions (magnesium, calcium), present as minor components collectively known as the non-rubbers. The typical composition[16] of field latex is given in Table 4.3. Most of these are soluble in the aqueous phase; some are adsorbed on the latex particle surface. The phospholipids and some latex rubber proteins are adsorbed on the latex particle surface and the negative charges they carry at the pH of the field latex (about pH 7) is essential in keeping the latex stable. The bulk of the non-rubbers dissolved in the aqueous phase are carbohydrates and soluble proteins. These have to be removed before the latex can be used for industrial applications. Fresh field NR latex can be separated into fractions by high-speed ultracentrifugation. Figure 4.3 shows the separated fractions of

Table 4.3 General composition of NR field latex and HA latex concentrate.

Parameters	NR field latex	HA latex concentrate
Total solids, % w/w	36	
Dry rubber, % w/w	33	59.67
Proteineous substances, % w/w	1–1.5	1.06[a]
Resinous substances, % w/w	1–2.5	–
Soap[b]	–	0.23
Ash, % w/w	<1	0.4
Carbohydrates, % w/w	1	–
Ammonia	–	0.68
Water, % w/w	60	37.49

[a]Includes carbohydrates, amino acids.
[b]Calculated as ammonium stearate.

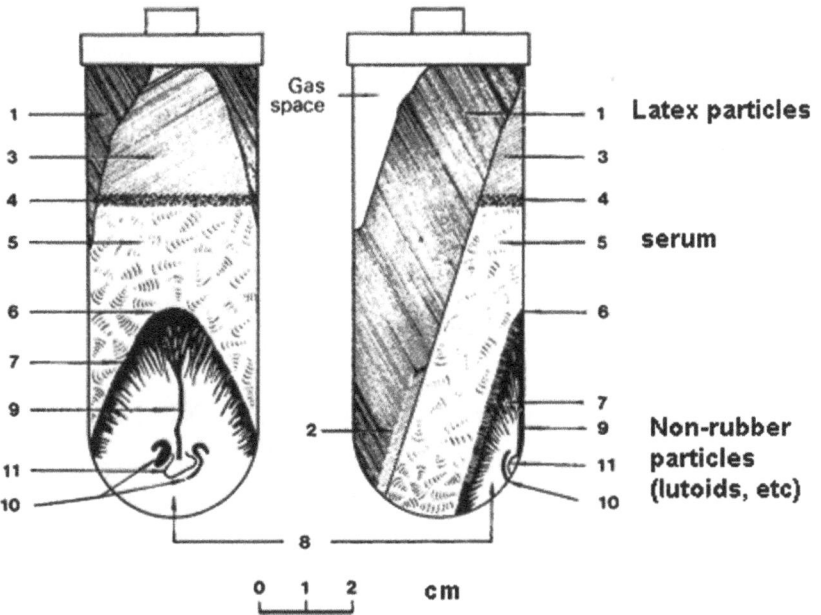

Figure 4.3 Ultracentrifugation of NR field latex into components.[17]

rubber latex particles, non-rubber particles (organelles such as lutoids and Frey–Wyssling particles) in the centrifuge tube.[17]

The non-rubbers and non-rubber particles present in fresh field latex are considered contaminants that are detrimental to the quality of the product made from them. This is particularly true for the presence of residual rubber proteins in NR products, particularly medical gloves. Purification of the field latex is thus absolutely necessary before the latex can be converted into a technically specified feedstock.

To produce technically specified latex concentrate, the field latex is first ammoniated and subjected to centrifugation at moderate centrifugal force to remove the soluble non-rubbers in the serum and also the non-rubber particles such as the Frey–Wyssling particles and lutoids. Centrifugation serves to remove the bulk of the water too. It is a concentration process. It is well known that centrifugal force will compact the denser materials (plant cells, water, *etc.*), resulting in a separation of the lighter rubber particles from the water (Figure 4.3). Basically two product streams are obtained, a latex concentrate stream containing about 60% solids and a skim latex stream containing most of the non-rubbers from the field latex and some very small latex particles that cannot be separated because of their size and stability. The skim latex containing high non-rubber content and about 4–5% rubbers is unsuitable for most latex applications. An option is to recover the residual rubber. However, skim latex is difficult to coagulate due to its high protein and ammonia content. The rubber recovered is of very poor quality due to the high non-rubber content and is difficult to mill (see Section 4.6.5).

It should be noted that not all the rubber proteins and non-rubbers can be removed by one centrifugation under the operating conditions employed at the latex concentrate plants. Residual soluble proteins and those associated with the latex particles are still present in HA latex concentrate (Table 4.3).[18] Hence latex-dipped products such as medical gloves made from this latex would still contain water-extractable rubber proteins. The same applies to the latex obtained from purification by creaming (see Section 4.5.3). NR latex with very low water-extractable rubber proteins can be produced by further treatment to dislodge the proteins associated with latex particles. These are known as low protein or deproteinized NR latex and will be discussed further later. The concern over rubber proteins is that they can cause allergic reactions in certain individuals who are sensitized to them (see Section 4.5.6).

4.5.3 Commercial Concentration Methods for NR Latex

Centrifugation is just one of many techniques used for concentrating latex, to increase the solids content of NR field latex which at about 30% solids is not suitable for many of the industrial applications of latex and is uneconomical in terms of transportation costs. Another advantage of concentrating the field latex is that it offers a more consistent feedstock compared with the dilute field latex.

Centrifuged high ammonia (HA) latex concentrate is the most common technically specified latex concentrate available in the market. A bank of

Figure 4.4 A typical centrifuge used in latex concentrate factories.

commercial centrifuges is used to mass produce NR latex concentrate in a latex concentrate factory with bulking facilities to collect the latex and store the latex concentrate. Figure 4.4 shows the most commonly used commercial centrifuge in the industry, made by Alfa Laval and using disc stack technology. The solids content of the concentrate can be easily adjusted using this machine. Freshly produced ammoniated latex concentrates are kept in storage tanks to mature for 2–3 weeks before being delivered to users. This maturation period is important to complete the biochemical reaction of the latex particles with ammonia and establish the mechanical stability of the latex. Ammonia causes hydrolysis of the adsorbed phospholipids and proteins on the latex particles in field latex and this takes time. The mechanical stability of the latex hinges on the completion of this reaction before usage. In addition, any magnesium ions in the latex will react with ammonia in the presence of phosphate ions to form the sparingly soluble magnesium ammonium phosphate, which will sediment out as sludge together with sand and dirt during the maturation period. Sometimes it is necessary to add diammonium phosphate to the field latex to ensure complete removal of metal ions such as magnesium during centrifugation. High residual concentrations of magnesium in the latex concentrate would be detrimental to its colloidal stability. The long-chain fatty acid soaps thus formed from the hydrolysis of the phospholipids would re-adsorb on the surface of the latex particles. Some of the proteins are also hydrolysed and the polypeptides become re-adsorbed. At the high pH of the latex concentrate, usually above 10, the carboxylic groups of the long-chain fatty acid soaps and the polypeptides are fully ionized, thus rendering the latex particles negatively charged at the surface and hence stabilizing them against aggregation. Both the high pH and fully developed surface structure of the latex particles ensure the latex concentrates remain stable on storage before and during processing.

Instead of using centrifugation to purify and concentrate the latex particles, creaming with the aid of chemicals has been used to produce commercial creamed latex concentrate. Since the density of the NR latex particles is less than that of water, the latex particle will float or cream. However, the process is too slow to be of any practical use, in particular for the smaller particles in the field latex to be separated out. The process can, however, be accelerated using a creaming agent such as ammonium alginate, a hydrocolloid. The field latex is first ammoniated to 1% w/w before addition of about 0.1% w/w of the creaming agent solution, stirred for 1 hour and left to stand for 40 hours. After this period, creaming proceeds almost to completion, the skim is drained off and the cream is ammoniated to 0.8%, bulked and then blended to improve uniformity. More skim can be drained off to adjust the final solids content of the concentrate.

The creaming equipment is simple and the process easy to operate. Not much labour is needed and only minimal energy is required for the process with relatively little loss of skim rubber. However, the process is slow and sensitive to variability in the quality of incoming field latex supplies. Further creaming may occur during storage and delivery. As well as ammonium alginate, other creaming agents can be used, such as sodium alginate, methyl cellulose, locust bean gum, carrageenan and gum tragacanth. However, the current market for creamed concentrate is small.

Another concentration method is to evaporate the water from the field latex by heating it in cylindrical drums. All the non-rubbers are retained in the evaporated latex concentrate because only water is removed. The latex particle size distribution of the concentrated latex is also unaffected by the process, since no latex particles are removed. Two types of evaporated NR concentrate are marketed by Revertex. One has a solids content of 67.5% and is preserved with potassium hydroxide and further stabilized using potassium soap. The other type is preserved with ammonia at 67.5% solids. It is claimed that by careful blending with water, evaporated latex concentrate can be reverted to its original conditions. These latex concentrates have excellent mechanical stability and high tolerance to filler loading, in addition to excellent stability to compounding and storage and good freeze–thaw resistance.

Standards for the various common commercial latex concentrates are available from various standards organizations. For example, ISO 2004:2010 gives specifications for NR latex concentrate types which are preserved wholly or in part with ammonia and which have been produced by centrifuging or creaming, whereas ISO 2027-1990 includes specifications for evaporated NR latex concentrate. ASTM 1076-10 has another set of standards for four categories of latex concentrates, the first three being for *Hevea* NR latex, the fourth for guayule NR latex. For standard specifications for concentrated, ammonia preserved, creamed and centrifuged NR latex from *Hevea*, three types are distinguished:

- Type I: centrifuged natural latex preserved with ammonia only, or with formaldehyde followed by ammonia.
- Type II: creamed natural latex preserved with ammonia only or with formaldehyde followed by ammonia.

The Production of Natural Rubber from Hevea brasiliensis Latex

Table 4.4 ASTM 1076-10 specifications for Types I, II and III concentrated latex.

Parameters	Type I	Type II	Type III
Total solids, minimum %	61.5	64.0	61.5
Dry rubber content (DRC), minimum %^{a}	60.0	62.0	60.0
Non-rubber solids: total solids minus dry rubber content, maximum %	2.0	2.0	2.0
Total alkalinity % (as ammonia % of water in the latex)	1.6 minimum	1.6 minimum	1.0 minimum
Viscosity at 25 °C by Brookfield viscometer	Type I	Type II	Type III
Sludge, max. % of wet weight	0.10	0.10	0.10
Coagulum, max. % of total solids	0.080	0.080	0.080
KOH number, maximum^b	0.80	0.80	0.80
Copper, max. % of total solids	0.0010	0.0010	0.0010
Manganese, max. % of total solids	0.0010	0.0010	0.0010
Mechanical stability, minimum	475–500 s	475–500 s	475–500 s
Colour on visual inspection^c	No pronounced blue or grey	No pronounced blue or grey	No pronounced blue or grey
Odour after neutralization with boric acid	No odour of putrefaction	No odour of putrefaction	No odour of putrefaction

^{a}Dry rubber content by definition and use is the acid coagulable portion of latex, after washing and drying.
^{b}KOH values for latices preserved with boric acid will be higher than normal, by an amount equivalent to the amount of boric acid in the latex.
^{c}Blue or grey colour usually denotes iron contamination caused by improper storage in containers.

- Type III: centrifuged natural latex preserved with low ammonia and with other necessary preservatives.

Table 4.4 gives the ASTM 1076-10 specifications for Types I, II and III concentrated latices.

The preservatives may function as bactericides, enzyme inhibitors, sequestering agents or stabilizers. They are often used in combination. A system widely used today is: zinc oxide/TMTD 0.025%; lauric acid or ammonium laurate 0.05% and ammonia (NH$_3$) 0.2%. This is the low ammonia LA-TZ centrifuged latex corresponding to Type III.

4.5.4 Specialty NR Latices

HA and LA-TZ centrifuged latex concentrates make up the bulk of the commercial supplies of NR latex. They are the general-purpose latex concentrates that meet most of the demands of diverse latex product manufacturers. Some producers also include in their product range a medium level ammonia variation at 0.45–0.50% w/w. The volumes of creamed latex concentrate and the evaporated latex concentrate (*e.g.* Revertex LCS) remain small in comparison, serving the specific requirements of certain manufacturers. In addition, there are several specialty NR latex concentrates available commercially for some niche applications.

There is a market demand for double-centrifuged NR latex concentrate, a purified latex concentrate manufactured by re-centrifuging once centrifuged NR latex that has been suitably diluted. Thin films prepared from this latex exhibit greater clarity, low water absorption and high dielectric properties due to its low water-extractable proteins and non-rubber content. It possesses good mechanical and storage stability. This latex finds application in the manufacture of surgical dipped goods and other niche latex products containing low non-rubbers, in particular low in rubber proteins. These products are targeting users of dipped goods who are sensitized to rubber latex proteins.

Deproteinized NR (DPNR) latex concentrate is a special grade of NR latex concentrate in which most of the proteins that come with the tree have been removed. It can be prepared either from field latex or HA latex concentrate. There are several methods of removing the rubber proteins. The most common is by enzyme digestion followed by centrifugation to remove the contaminants and non-rubbers, and re-stabilization with surfactants. The proteins can also be displaced and replaced by repeated washing with non-ionic surfactant and centrifugation. Chemicals such as urea have also been employed to destroy the rubber proteins from the latex particle.[19] A residual nitrogen content of the rubber down to 0.02% w/w can be achieved under optimum conditions in the laboratory, compared to 0.38% w/w of HA latex concentrate. A recent report explored the combined effect of polar organic solvents and urea in producing protein-free NR latex.[20] Commercial DPNR latex concentrates are usually treated by enzymes and stabilized by added surfactant. Loss of proteins from the latex particle surface will destabilize the latex and surfactants are added to replenish the loss of surface charge.

A common method of obtaining DPNR latex is to break down the rubber proteins enzymatically or chemically. Recently another way of removing rubber proteins from NR latex by binding them with aluminium hydroxide was reported. The precipitated aluminium hydroxide with bound proteins was then separated by centrifugation to yield a low rubber protein NR latex called Vytex. The process is patented and was developed by Vystar, which is actively promoting Vytex for use in various industries.

Low rubber protein latices are developed especially to address the issue of rubber protein allergy for some sensitized users of latex rubber gloves. These latices have very low water-extractable rubber proteins and hence are well suited for manufacturing low-protein gloves. DPNR also finds niche applications in electrical and engineering products (see later).

Pre-vulcanized NR latex concentrate is a specialty latex that can provide small-scale manufacturing operators of latex products the convenience of ready-to-use latex compounds. All the necessary curing agents and accelerators have already been added by the latex supplier and it can be used directly, for example at the dipping plant. The rubber molecules within each latex particle of this latex have already been pre-crosslinked by the supplier. However, the particles remain separate and the latex remains fluid and stable. After dipping, the latex particles in the wet gel film on the mould will start to coalesce and the rubber molecules crosslink between particles to develop the mechanical strength of the film. This occurs slowly at ambient temperature, more rapidly at elevated temperature, to give a fully cured product. Thus the high-temperature curing step of the dipped film is not required apart from drying. This latex has long storage shelf life, unlike ordinary compound latex, and good stability for online dipping processes. It is formulated to be compatible for use in high-speed dipping machines with minimal batch-to-batch variations. Popular for the manufacture of condoms, catheters, baby teats, continuous sheeting, adhesives, carpet backing, cast articles including theatrical masks and toys. Commercial pre-vulcanized NR latex concentrates come in three forms: low, medium and high modulus. Variations of pre-vulcanized NR latex are also available. For example, there is on the market a double-centrifuged pre-vulcanized latex concentrate and a low-nitrosamine pre-vulcanized latex.

Cationic NR latex concentrate is a high solids latex designed to be compatible with a wide range of cationic bitumen emulsions for applications in road construction such as surface dressing, microsurfacing and tack-coats. In ordinary HA latex the particles are negatively charged and hence would be destabilized immediately if mixed with cationic bitumen emulsions. This would not occur if the latex has a similar charge to the emulsion. This latex may cream on standing, and so is best stirred before use. Since the pH of this latex is acidic, it should be stored after dilution in plastic containers or suitably lined drums that are resistant to dilute acid. When the product is to be stored in bulk storage tanks, it should be stirred with a mechanical stirrer or some form of recirculation to maintain latex homogeneity. This is a niche product and there are only a handful of manufacturers.

There are several NR latices which are prepared as a conduit for obtaining the dry rubber from them later by coagulation. This is because it is easier to incorporate solid rubber chemicals into latex in powder form so as to achieve a uniform dispersion of the material. For example, carbon black is often agglomerated and very difficult to disperse uniformly in dry rubber. A lot of energy is usually needed to mix the rubber and carbon black in an internal mixer to obtain a near homogeneous compound.

Carbon black masterbatch latex is a latex manufactured for ease of processing and the convenience of the rubber industry. The process is less energy intensive and can be tailored to the needs of the end users, in particular those small operators who lack facilities and resources in compound formulating. The properly stabilized slurry of carbon black, usually with suitable surfactant, is mixed with NR latex until homogeneous, then acid-coagulated, washed, dried and baled to give a ready-to-use pre-mix that requires a much shorter mixing cycle and hence saves energy. Good integration of the black into the rubber matrix can usually be achieved with high bound rubber content. Silica filler can be incorporated in a similar manner into the NR latex and converted into rubber-filler masterbatch.

Thermoplastic NR (TPNR) or thermoplastic elastomers (TPEs) are materials possessing the processing characteristics of thermoplastics and the properties of vulcanized rubbers at room temperature. Thus TPEs can be softened reversibly with heat like thermoplastics, and yet possess considerable resilience and flexibility, a characteristic of vulcanized rubbers. TPNR can be made from a blend of NR and polypropylene (PP) and in certain compositions will function as a TPE.

By partially crosslinking the rubber phase through dynamic vulcanization, the elastic properties of TPNR can be improved further. TPNR is heat and ozone resistant and finds applications in footwear, sports goods and automobile parts.

Hydrocarbon oils are often used as rubber additives, as processing aids or extenders. As a rule of thumb, when the amount added is <10 phr, it is known as processing oil and when in excess of that it functions in reducing the cost of the product and is then known as an extender. The rubber thus prepared is called oil-extended NR (OENR). The oil can be added to rubber by milling or more conveniently and evenly in the latex stage as an emulsion. When carried out in the latex stage, it is termed OENR latex. The oil imparts to the rubber high skid resistance on wet roads and possesses good wear and groove-cracking resistance. Thus it is widely used in the tyre industry.

4.5.5 Chemically Modified Rubber Latices

Apart from pre-vulcanized latex where the rubber molecules have been chemically crosslinked by sulfur, the chemical nature of the rubber molecules of the other latices described above remain chemically intact during and after the process. There are several other latexes available on the market in which the rubber molecules of the latex have been chemically modified. The chemical reactivity of the rubber molecules arises from the olefinic structure of the *cis*-1,4-isoprene unit within the molecule, which can undergo rapid reactions with, for example, halogens, ozone and hydrogen chloride. Some of these will be described in this section. They are prepared to serve niche applications.

In addition, it is easier to incorporate chemicals in powder form or organic liquids into NR latex to obtain better dispersion of the powder or liquid. It is usually very energy intensive to disperse solids into dry rubber using an internal mixer to obtain a uniformly dispersed powder in the rubber mix. The example of incorporating carbon black into rubber at the latex stage has been described above. By first preparing a slurry of the powder or an emulsion of the organic liquid and adding this to the latex, a rather uniform distribution of the solids or liquid in the latex particles can be achieved and the incorporation process will also be achieved more quickly since mixing is easier in the latex stage. Chemical reactions can then be initiated in the latex particles and then allowed to go to completion. This can be carried out at elevated temperature too. In fact reaction in the latex stage offers better temperature control of the reaction. The dispersing medium (water) is an excellent heat sink for the reaction. The particulate nature of latex offers a high surface area for absorption of the chemical reactants and also a high reaction rate. After the reaction is over, the modified latex can be kept as such for latex applications or coagulated with formic acid to give the modified solid rubber. Many of the chemically modified rubbers are made by this route.

4.5.5.1 Epoxidized NR Latex

Epoxidized NR is prepared by partially epoxidizing the NR molecule to create a new type of elastomer in which the epoxide groups are distributed randomly along the rubber molecule backbone. The reaction is best conducted in the latex stage using performic acid, which is generated *in situ* using hydrogen peroxide and formic acid.[21] Increasing levels of epoxation can be achieved, even up to 50 mol%. The reaction is performed at 50 °C. Non-ionic surfactant is added to help enhance the stability of the modified latex. The modified latex is treated with ammonia to raise the pH to >8 after completion of the reaction. ENR 25 and ENR 50 latices at 25 mol% and 50 mol% epoxidation level, respectively, are both available commercially. Since the modified latex is stabilized with non-ionic surfactants, it is difficult to destabilize during the dipping process. Further, concentration of the modified latex by centrifugation is ineffective, so instead evaporation, steam coagulation or ultrafiltration is needed to raise the solids content from 25–30% to 60% for latex applications.[22,23] ENR 25 and ENR 50 solid rubbers are obtained from these latexes.

Epoxidation increases the polarity and glass transition temperature (T_g) of the modified rubbers. The properties of the vulcanizates change with increasing degree of epoxidation of the rubber molecules. These are reflected in an increase in damping; a reduction in swelling in hydrocarbon oils; a decrease in gas permeability; an increase in silica reinforcement; improved compatibility with polar polymers like polyvinyl chloride; reduced rolling resistance and increased wet grip. ENR is used for products that require hydrocarbon oil resistance and gas impermeability, such as oil seals and tyre inner tubes. It is an important ingredient in making tyre tread compounds in view of its excellent wet grip and rolling resistance properties.

4.5.5.2 Methyl Methacrylate-Grafted (MG) Rubber Latex

Grafting a second polymer to the NR molecule in the latex stage is one of the many routes to chemically modified NR. An olefinic monomer with unsaturated double bonds such as methyl methacrylate (MMA), styrene and acrylonitrile are important monomers used for such grafting.[24] For example, MMA monomer is first converted into an emulsion with some suitable emulsifiers and then mixed with NR latex to copolymerize the monomer in a seeded emulsion polymerization process. It is important to ensure the seed latex particles are saturated with the monomer supplied through diffusion from the emulsified monomer droplets. An oil- or water-soluble initiator can be used to start the reaction. With proper control of the system and reaction conditions, the free radical reaction can be made to propagate within the latex particles as far as possible, so that only grafted NR occurs, without the formation of free homopolymer from the monomer.[25] In this way only chemically modified NR particles are formed, with the exclusion of free homopolymer particles. Both DPNR and HA latex concentrates can be used for grafting. However, DPNR was found to give higher grafting efficiency.[26] It appears that rubber proteins may have some inhibitory effect on the grafting reaction in the latex particles. Depending on the amount of MMA grafted onto the rubber molecules in the latex particles, products with a wide range of compositions can be obtained. The graft copolymerization has been found to be a surface-controlled process. The latex thus obtained is converted into solid MG rubber marketed under the trade name 'Heveaplus MG' rubber. Two commercial types are available, labelled as MG30 and MG49. Various structured NR latex particles can be synthesized with notably core–shell, microdomain and semi-interpenetrating network morphologies. The structured NR latex particles have high modulus and find some applications as toughening agents and impact modifiers for brittle polymers. They are also used in the adhesive and shoe industries.

Alternatively, grafting is effected through an irradiation technique using γ rays on NR latex containing the monomer.[27] In this case, the free radicals needed to initiate the polymerization reaction are generated by high-energy γ rays instead of using a chemical initiator. A higher dosage of irradiation is needed to graft MMA to NR latex particles compared to emulsion polymerization of MMA monomer to achieve the same level of conversion. The viscosity of the NR latex increases with progress of polymerization, due to the increase in solids content of the modified latex.

NR-styrene grafted latex is prepared similarly using styrene monomer. The polymerization reaction can be initiated as in seeded emulsion polymerization or irradiated with γ rays as described above. The solid rubber obtained from this is used as a modifier in microcellular rubber products.

4.5.5.3 Cyclized NR Latex

Early cyclization of NR latex was carried out using a strong acid such as sulfuric acid. The latex was first stabilized against coagulation by the acid with suitable surfactants, followed by heating the latex to 100 °C with sulfuric acid for 2.5 hours.[28–30]

Both non-ionic and cationic surfactants have been used to stabilize the latex. The resulting cyclized NR latex can be coagulated together with untreated NR latex to yield rubber that is a mixture of cyclized rubber intermixed with NR. This modified rubber mix is a hard resinous substance that can be used as a masterbatch for incorporating cyclized rubber into NR compounds. Cyclized NR through the latex route avoids the use of expensive and hazardous solvents in which NR is dissolved before reacting with a Lewis acid such as tin(IV) chloride.

It was reported previously that rubber proteins in NR latex concentrate may interfere with the cyclization reaction; the report suggested the use of DPNR latex or purified NR latex instead.[31] More recent publications have reported studies on cyclization of NR *via* DPNR and purified NR latex using catalysts such as a combination of benzotrichloride and sulfuric acid[32] and a variation of a cationic cyclization system using sulfuric acid[33] or with trimethylsilyl trifluoromethanesulfonate (trimethylsilyl triflate, TMSOTF).[34]

Resinous cyclized NR is a thermoplastic elastomer and finds applications as an adhesive, in paints and in printing inks. Its tensile properties and modulus also make it suitable for use as shoe soles. It is used as a toughening agent and reinforcing resin in, for example, hard mouldings and industrial rollers.

4.5.5.4 Radiation-Vulcanized Latex (RVNRL)

This is a possible alternative to conventional sulfur vulcanization in providing a sulfur-free and accelerator-free crosslinking system to NR. The latex is vulcanized by γ irradiation in the presence of a sensitizer such as N-butyl acrylate in a continuous process. RVNRL is used in the manufacture of dipped products, in particular thick rubber products such as baby pacifiers and teats. It is a form of pre-vulcanized latex with low cytotoxicity, free from the risk of N-nitrosamines generating accelerators. Products made from RVNRL have high transparency and clarity, are soft and with low emission of acid combustion gases. Effluent from the processing of this latex in dipping is free from zinc oxide, which is considered environmentally unfriendly. RVNRL can be used for gloves, condoms and catheters, which involve large contact surfaces with the users. The reasons for poor acceptance of RVNRL by the manufacturers are the inferior tensile strength of the products and high cost of γ irradiation. Much research is still needed to improve the quality of the finished products using RVNRL and mitigation on safety and health issues.

4.5.6 Recent Advances in NR Latex Technology

Most of the recent advances in NR latex technology are in latex preservation systems, with safety and health hazards of the latex chemicals and rubber proteins being the main concerns. The issue of rubber proteins (both water-extractable and rubber-bound) causing allergic reactions in certain sensitized users has dominated the medical glove industry in the 1990s and the first decade of this century. Much research has been carried out on identifying and characterizing the allergen in rubber proteins and on mitigating its exposure to the users of latex products. With extensive research on deproteinization of NR and the now easily available commercial DPNR and double-centrifuged latex concentrates, coupled with vast improvements in latex dipping technology incorporating ample leaching protocol for the dipped gloves followed by chlorination or polymer coatings in recent years, the incidence of such allergic reactions has been vastly reduced. The proactive measures taken by glove manufacturers in following a strict quality control scheme have done much to reduce the risk of exposure of healthcare workers to the allergens.

Another issue was with LA-TZ latex concentrate, where considerable effort has been directed into looking for a benign replacement for TMTD, which is an allergen and can cause dermatitis in susceptible people.[35] TMTD is used both as a preservative for latex and as an accelerator for the vulcanization of NR.

And of course there has been much research into an entirely ammonia-free preservative system for NR. Another pressing issue is the presence of N-nitrosamine in NR latex products and the drive to search for an accelerator system that will lead to nitrosamine-free NR latex products such as baby pacifiers and rubber teats.[36] N-nitrosamines are known carcinogens and it has been found that both nitrosamines and nitrosamine precursors are continuously being formed in vulcanized rubber using TMTD, which turns out to be the source of nitrosamines and secondary amines in rubber products. The other frequently used accelerator, zinc diethyldithiocarbamate (ZDEC), also breaks down to nitrosatable compounds and causes dermatitis in sensitized persons. The search for a friendlier accelerator is a pressing one. A new preservative named BeKa 100 claims to be effective against a whole range of microorganisms in NR latex. The antimicrobial quality of BeKa 100 was reported to protect not only the glove user but also helps to reduce any microbial population on surfaces or persons that are touched.[36]

A very recent US patent[37] disclosed new vulcanization compositions, latex dispersions and elastomeric articles formed using conventional techniques which resulted in reduction or elimination of Type I and Type IV allergenicity. It was claimed the invention is beneficial in enabling both healthcare providers and patients who are exposed to these potential sources of allergens frequently or for prolonged periods of time to avoid problems associated with allergic reactions to medical rubber products. Type I hypersensitivity reactions are mediated by IgE immunoglobulin and the effect is immediate. It has been linked to the residual, extractable rubber proteins present in NR latex products, whereas Type IV delayed hypersensitivity reactions are cell-mediated allergic responses to specific chemicals, such as the rubber vulcanization accelerators.

4.6 NR Production Methods

Most field latex is processed into dry rubber. Latex concentrate production makes up just slightly more that 1 million tonnes or about 10% of the total rubber produced. There are a lot more applications using dry rubber than latex. In fact the bulk of NR produced is used by the tyre industry. There are many types and grades of dry rubbers produced to meet the requirements of diverse applications.

The processing of field latex into rubber changed very little before the mid-20th century. Conventional methods, often based on very old techniques, followed the relatively simple routine of tree tapping, collection of latex, coagulation, sheeting and drying. In the simplest production method for solid rubbers, the field latex collected is first filtered and coagulated with an organic acid such as formic acid to form a coherent gel or soft slab in a shallow tray, squeezed mechanically (for example, through a mill with a pair of rollers) to remove the entrapped liquid (serum) and the contaminants contained therein. The wet gel is further washed to remove the residual acid and the biological residues from the trees while being squeezed through the roller. The washed wet gel, now in thin sheet form, is then air dried, dried under the sun or smoke dried in a smokehouse, depending on the facilities available. These simple steps can be conducted by the smallholders who operate on latex collected from the small number of trees they own. The quality of the dried rubber sheets can vary widely and is usually based on visual appearance and not on technical properties. Therefore the smallholders are unable to sell them for a high premium compared with technically graded solid NR (see below) produced in a rubber factory with accompanying proper technical data.

In fact the NR market is highly fragmented by rubber types and grades within each type, and this is further complicated by sizeable variations in trade flows and price movements. Two main grading systems are in place for NR, namely the conventional visually graded types known as the International Standards of Quality and Packing and NR graded by technical specifications, namely technically specified rubbers (TSR) or the standard rubber scheme. These are in addition to the specialty grades developed separately for rubbers with specialty applications.

The standard rubber scheme has several important features, most important of which is that the rubbers produced under this scheme would be classified on the basis of technical properties and graded according to a set of specifications. The basic specification parameters are dirt, volatile matter, ash, nitrogen, copper and manganese contents. The rubbers are to be presented in standard block form to facilitate palletization, transport and handling in a consumer factory. No skim rubbers are allowed to be included in these rubbers. For example, the Malaysian Standard Rubber (SMR) scheme listed five basic grades, namely SMR5L, SMR5, SMR10, SMR20 and SMR50, where specification for the light colour grade by Lovibond Colour is for L grade only (see Section 4.6.3). Each consignment of the standard rubbers is accompanied by a valid test certificate conforming to the specification scheme. SMR5, the purest grade, is made from latex only whereas SMR50 is the lowest grade.

The rubber industry produces four main types of solid rubbers from field latex, namely block, sheet, crepe and specialty rubbers. The sheet rubbers including ribbed smoked sheet (RSS), unsmoked sheet (USS), air-dried sheet (ADS) and latex grades of crepe, are all graded under the visually graded types. They are made from field latex preserved with sodium sulphite, diluted, coagulated, formed into sheets, dried and often pressed into bales. The production of crepes differs somewhat from that of sheet rubbers. Maintaining hygiene is a very important aspect of rubber processing and is applicable to all the processes. The process for manufacturing latex concentrate products has already been described in Section 4.5.3.

4.6.1 Sheet Rubbers: USS, RSS and ADS

Unsmoked sheet rubbers (USS) are produced with the simplest facilities available to smallholders and so it is difficult to achieve quality anywhere near RSS grade rubbers. Very often they are insufficiently dried and coloured by smoke. The quality of the rubbers could be improved to RSS grades by the intermediaries who further process them in small factories. The smallholders and intermediaries together supply an international market. In some countries they are the main source of exported RSS rubbers.

RSS and ADS rubbers are normally produced mainly in the estates where almost all the latex collected is processed into RSS or ADS. A smokehouse is used to dry RSS rubber whereas ADS rubber is dried in a building similar in design to a smokehouse but without the smoke generation, having instead just a dryer. Nowadays on the large plantations, field latex is converted directly into block rubbers. Almost all the RSS or ADS are produced in the smaller estates where some of the USS rubbers are upgraded to RSS. The industry trend is moving towards production of block rubbers.

ADS rubbers can be considered a more sophisticated version of USS, which is crude and produced by smallholders. It is made in much the same way as RSS rubbers, except that it is dried without the smoke and at lower temperatures. As such it is very much lighter in colour than the RSS grades and may be used as **pale crepe** which has remained the preferred grade when extreme purity and very light colour product is required. It is prohibited to add the fungicide, paranitrophenol to ADS when the rubbers are used in products for food applications. ADS is normally sold as the best grade of RSS 1 or 1X whereas technically specified ADS must meet the TSR 5 specifications.

In the production of RSS, the field latex is first diluted to 15–17% solids before coagulation with acid. The soft and spongy coagulum is easier to wash to remove dirt and to mill into thinner sheet, and is easier to dry but more difficult to handle. The soft coagulum is then passed through a battery of sheeting machines, while being continuously washed, to form the sheet, which is hung on a trolley before being moved to the smokehouse for drying (Figure 4.5). Syneresis takes place during this stage, expelling the serum from inside the sheet and draining off any superfluous water. After drying at temperature between 40 °C and 60 °C, the rubber is inspected, baled and packed.

(a)   (b)

Figure 4.5   a) A trolley loaded with wet sheets ready for smoking in the smokehouse[39] (Processing of Natural Rubber, John Cecil and Peter Mitchell, FAO (AGST) Consultants, EcoPort version by Per Diemer, FAO (AGPC) Consultant and Peter Griffee, AGPC, 2005 ref). (b) RRIM type C smokehouse and trolley.[39]

The RSS are graded as RSS 1X, RRS 1, RSS 2, RRS 3, RSS 4 and RSS 5 in a descending order of quality.

4.6.2 Crepe Rubbers: Pale Crepe, White Crepe and Brown Crepe

Thin (about 1 mm thick) and thick (3–5 mm thick) pale crepes (TPC) and thin white crepe (TWC) are still produced from field latex in Sri Lanka. Sheets of sole crepe, up to 15 mm thick, are made by laminating sheets of white or coloured crepe. The demand for pale crepe is declining and will continue to decline, its place being taken by the technically specified grades of block rubbers.

White crepes are produced either by bleaching or by fractional coagulation, or both. It is best made from latices from selected clones of rubber trees such as PB 86, which have very low carotenoid contents and produce a white crepe without any special treatment. For other clones, the removal of the carotenoids is by fractional coagulation. It consists of coagulating the latex in two stages. In the first stage, a 25% solid field latex is coagulated with acetic acid to form a yellowish agglomerate in which most of the carotenoids and non-rubbers in the latex are concentrated. The agglomerate, which makes up 15% of the rubber, is removed. The remaining latex is then coagulated with more acid and processed in the usual way. The resulting rubber is completely white (Figure 4.6). Fractional coagulation is time-consuming and results in a low yield of quality product (about 85% of the original latex). Bleaching may be used instead of fractional coagulation to lighten the colour of pale crepes. The wet rubber is then milled between heavy rollers to give crinkly lace-like rubber, which is drained and dried.

Field latex for crepe production is coagulated in the same way as latex is coagulated for sheet rubber production and dried in a drying shed at temperatures not exceeding 45 °C.

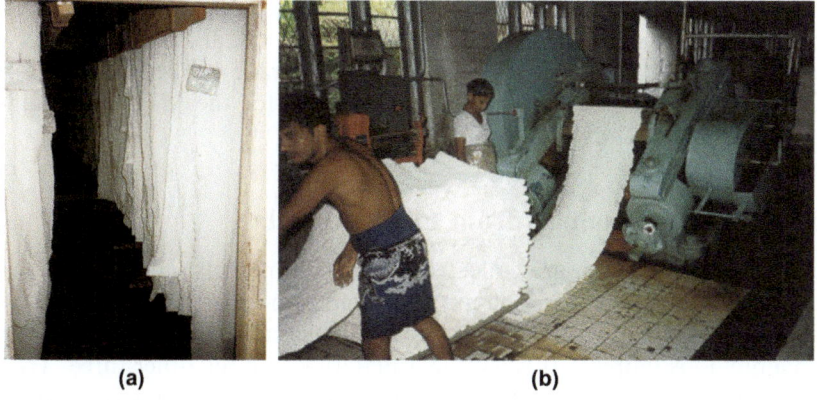

Figure 4.6 a) White crepe festooned in drying shed. (b) Milling white crepe.[39]

4.6.3 Block Rubbers: Technically Specified Rubbers and Standard Technical Rubbers

Block rubber is one of several ways solid rubber is presented for convenience of handling, quality control and marketing purposes. The different types of wet rubber mentioned above are crumbed and compressed into blocks. Hence they can have diverse characteristics and properties. Due to its many advantages, more rubber is traded as block rubber now than all other forms of rubber combined. Block rubber was first introduced in Malaysia in 1965 and by 2000, block rubber of the TSR grade accounted for more than 80% of all the rubber traded internationally. The standard weight of a block of rubber is $33\frac{1}{3}$ kg.

The block rubber scheme converts raw rubber or field latex into granular form in a fast, continuous process in a large factory setting, as opposed to small operators working on sheets. The rubber crumbs are then dried in an oven, weighed, compacted into solid blocks, wrapped and palletized. In the case of field latex, 0.7 phr of castor oil was added to the latex in earlier processes, so that the wet rubber could be broken up into crumbs on passing through the creping machine. With the availability of better comminuting machines such as hammer mills and shredders, castor oil was eventually phased out of the process. The production time from latex to ready-to-ship block rubbers is reduced to less than 12 hours. The block rubber process allows greater mechanization, better and more consistent quality, it simplifies handling, storage and transportation and offers greater convenience for the consumer. Block rubbers are graded to technical specifications on inherent physical and chemical properties. The block rubber process as developed by Malaysia at that time was known as the Hevea crumb process, which evolved into the successful **Standard Malaysian Rubber (SMR)** scheme launched in 1965. However, the use of castor oil was not included in the SMR scheme because of blooming problems. This was the first time a standard rubber scheme graded according to technical specifications was ever introduced to the rubber industry. This revolutionary strategy created a quality system for the industry to uphold and facilitated international trade of this important commodity. Since then Thailand, Indonesia, Vietnam, Sri Lanka and India have also set up similar national standard rubber schemes known as **STR**, **SIR**, **SVR**, **SSR** and **ISNR**, respectively. An example of a packing label from some SMR 5 grade rubber under the SMR scheme is shown in Figure 4.7. The specification for the Malaysian Standard Rubber Scheme is shown in Table 4.5.[38]

4.6.4 Rubber Products from Field Coagula

Four types of coagulum are collected in the field, namely: cup lump, tree lace, earth scrap and smallholders' lump. **Cup lump** is the coagulated material found in the collection cup formed from residual latex after the latex was collected from a previous tapping. It is a cleaner material than the other three. **Tree lace** is the coagulum strip that the tapper peels off from the previous cut before making a new cut. It is often contaminated by manganese from the bark and

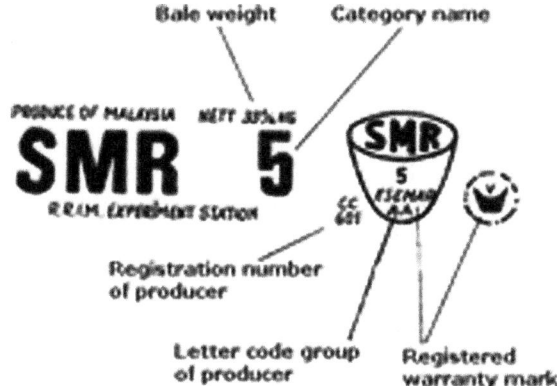

Figure 4.7 Packing label for some SMR 5 grade rubber.[38]

Table 4.5 Specification for the Malaysian Standard Rubber (SMR) scheme.

Criteria	5L	5	10	20	50
Dirt, max. wt %	0.05	0.05	0.10	0.20	0.50
Ash, max. wt %	0.60	0.60	0.75	1.00	1.50
Nitrogen, max. wt %	0.65	0.65	0.65	0.65	0.65
Volatile matter, max. wt %	1.00	1.00	1.00	1.00	1.00
PRI, minimum %	60	50	50	40	30
Wallace plasticity, P_o, minimum	30	30	30	30	30
Colour, Lovibond unit	6.0	–	–	–	–

has a higher copper and manganese content than cup lump. Both copper and manganese are pro-oxidants and decrease the physical properties of the dry rubber. Tree lace is badly darkened because of the manganese. **Smallholders' lump** is rubber lump produced by smallholders who collect rubber from trees a long way away from the nearest factory. They are deliberately coagulated in the field and of very variable quality, often badly adulterated. **Earth scrap** is the rubber coagulum found around the base of the trees. It could be from rain flooding a collection cup or spillage from the tapper's bucket. This is usually contaminated with earth and other non-rubber substances, with very variable quality and rubber content.

It is clear the coagula collected from the field can be highly contaminated and badly discoloured and varies greatly in quality. Obviously not much hygienic consideration has been given to their formation, collection and storage. Earth scrap is of very low quality and should not be used for the production of block or crepe rubbers. Isolation and separation of badly degraded and darkened scrap and lump from clean scrap is essential before pre-cleaning. Pre-cleaning involves washing in water to remove bark debris from the field coagula followed by soaking in a 1% sodium sulphite solution to reduce the colour. They are then milled and crumbed. Soaking in water also softens the rubber and renders it easier to machine.

Scraps are usually made into light brown crepe and dark brown crepe. Estate brown crepes are produced from fresh cup lump and high-grade scraps. Cup lump is normally processed into TSR 10 grades of block rubber. Smallholders' lump, after suitable cleaning, can be made into TSR 20 block rubber whereas tree lace in small quantities can be tolerated in the production of TSR 20 grade block rubbers. To produce the special grade constant viscosity (CV) rubber (see below), the cleaned crumbs are soaked in hydroxylamine solution. In cases where cup lumps are to be blended with field latex to produce block rubbers, the field coagulum has to be first blended, crumbed and spread out on the bottom of the coagulation troughs. Acid is then sprayed onto the crumbs and field latex is run into the troughs to cover the crumbs and fill the trough to produce the wet slab of gel.

4.6.5 Skim Rubbers

A by-product from the latex concentrate factory is skim latex (see Section 4.5.2), which contains only 4–5% solids but very high non-rubbers, mostly proteinaceous materials and dissolved minerals. The quality of the recovered skim rubber is very poor because of high protein, resinous and mineral contents, as much as 15–20% of the dried skim rubber. The skim latex is also more difficult to coagulate and requires prior de-ammoniation. Both acid coagulation and auto-coagulation can be used. More intense water washing of the coagulum is needed to remove the acid. In fact in auto-coagulation much of the proteins and resins are digested during the process and lost in the milling resulting in a lower yield. Auto-coagulation also produces strong smells of putrefaction that can persist in the dry processed rubber. The coagulum can be processed into block rubber or thick crepe. There are no international or national standards for skim rubbers. Skim rubber is low in dirt content and can vulcanize easily. Use is made of these characteristics of skim rubbers in the manufacture of footwear and rubber flooring.

4.6.6 Specialty Rubbers and Chemically Modified Rubbers

4.6.6.1 DPNR

DPNR is a highly purified form of NR with very low nitrogen and ash contents and low water absorption. The preparation of this rubber from DPNR latex is described in Section 4.5.4. DPNR is well suited for electrical and engineering applications and is used in fabricating seismic isolators for buildings in earthquake-prone areas, for example. DPNR exhibits low creep, which is essential for such applications. Removal of proteins from NR reduces its moisture sensitivity, thereby improving its engineering properties.

4.6.6.2 Epoxidized NR (ENR)

ENR is a chemically modified NR with improved properties of oil resistance, low air permeability, better damping and good bonding while retaining its superior high strength advantage over synthetic rubbers. The manufacturing process has been described in Section 4.5.5.1. The degree of improvement is dependent on the level of epoxidation achieved. ENR 25 and ENR 50 are two important commercial grades that find extensive engineering applications such as oil seals, tyre inner tubes, anti-vibration mountings and conveyor belts and floor coverings. ENR can be reinforced with silica filler without any coupling agent in much the same way as carbon black reinforces NR. ENR can bond very well with other materials and so it is used in adhesives and sealants. It is used extensively in tyre tread compounds.

4.6.6.3 Constant Viscosity (CV) and Low Viscosity (LV) NR

NR undergoes hardening or an increase in viscosity during primary processing and subsequent storage under ambient conditions. This phenomenon is known as 'storage hardening' and is enhanced by low relative humidity of the environment. The hardening arises from crosslinking reactions between the 'abnormal' carbonyl groups found randomly distributed on the rubber molecular chain and the amino acids present as non-rubbers. Storage hardening causes processing difficulties and results in high energy consumption. Hardening can be prevented by pretreating the latex with chemicals such as hydroxylamine hydrochloride, hydroxylamine neutral sulphate or semicarbazide before coagulation. These chemicals block the crosslinking reaction of the carbonyl groups with amino acids. Low viscosity (LV) rubber is prepared by adding a small quantity of naphthenic oil to the latex to stabilize the viscosity of the rubber derived from it. Again this helps in better processing of the dry rubber without the need for a pre-mastication step, thus permitting direct mixing.

4.6.6.4 Superior Processing (SP) Rubber

Superior processing (SP) rubber is a special grade of dry rubber originally patented by the Rubber Research Institute of Malaysia (RRIM), in which part of the rubber is vulcanized before it is coagulated.[39] Preparation involves adding a dispersion containing sulfur, zinc oxide and accelerator to a lightly ammoniated field latex slowly with stirring, which is then heated by live steam to 82 °C and held at this temperature for 2 hours to complete the vulcanization. This vulcanized latex is then added to fresh field latex already diluted to 20% solids in a bulking tank, without any added preservative, in a ratio of 4 parts fresh latex to 1 part vulcanized latex. Thorough mixing and blending of the latex is essential. The homogenized latex is then run into the coagulation tank, coagulated, milled, dried and baled as in the manufacture of the equivalent NR.

SP Hevea crumb, SP crepe, SP air-dried sheet and SP smoked sheet are produced by this procedure. Each of these products contains 20% crosslinked rubber and 80% unmodified NR. A masterbatch version is also available that is used as a processing aid for NR processing. PA 80 contains 80% of crosslinked rubber and 20 parts of unmodified rubber, whereas PA 57 is an oil-extended version of PA 80. SP sheet and crepes are normally marketed in 113 kg bales or SP in crumb form is available in 33 kg blocks.

SP rubbers are usually incorporated into compounds to facilitate preparation of complicated extrusions, with low deformation or collapse, and sharp definition of thin sections.

4.6.6.5 Liquid NR (LNR)

LNR is produced by breaking down long-chain high molecular weight NR molecules. Extensive size reduction of the molecular chains by depolymerization can be achieved by heating masticated NR at 220–240 °C. Alternatively depolymerization can be effected through a redox reaction using phenylhydrazine and air or photochemical reaction using irradiation. The process converts the solid rubber into a viscous liquid.

LNR is compatible with NR and is often used as a reactive plasticizer to reduce cycle time and energy usage in rubber compounding in the tyre industry. In view of its compatibility with other elastomers, it is used as a compatibilizer to bond NR to polyolefin blends, NBR to NBR in which the elastomers can be vulcanized together. It can function as an adhesive in bonding NR and NBR to mild steel. It is used as a modifier in making flexible moulds for the printing industry and as a binder for grinding wheels, amongst other applications, in addition to as a polymer base for automotive sealants and sound damping compounds.

4.6.6.6 Chlorinated NR

Chlorinated NR (CNR) can be prepared in the solution stage or by bubbling chlorine gas through the latex. Under acidic reaction conditions, up to 60% chlorination can be achieved.[40] Traditionally the rubber is dissolved in a solvent before chlorination. CNR is resistant to acids and alkalis and to wear and ageing. It is corrosion resistant in sea water and so is used to formulate marine paints. CNR has good solubility in toluene.

4.6.6.7 Tyre Rubber

Tyre rubber is developed specifically for tyre manufacturing. It is prepared by coagulating NR latex containing oil with wet crumb rubber in the following proportion: 30% latex rubber, 30% unsmoked sheet, 30% field coagulum and 10% plasticizer (oil).

The oil could be aromatic or naphthenic. The unsmoked sheet and field coagulum are first converted into wet crumb. Blending of the components could be achieved by coagulating a mixture of latex and oil with the wet crumb. The coagulum is then processed into a uniform blend that requires a minimum of mechanical working. Tyre rubber is marketed in $33\frac{1}{3}$ kg blocks, packed in polyethylene and palletized in 1 tonne units. The attractive features of tyre rubber are a low Mooney viscosity, a low rate of storage hardening, a low rate of crystallization, low heat build-up and easy blending with polybutadiene.

4.7 Major Industrial Applications of NR and NR Latex

The main use of NR is in automobiles. In developed countries nearly 60% of all rubber consumed is for automobile tyres and tubes. The superior heat build-up and tensile strength of NR makes it indispensable in the manufacture of aviation tyres. In heavy-duty tyres such as those for tractors and trucks, the major portion of the rubber used is NR. In addition to tyres, a modern automobile has more than 300 components made out of rubber. Many of these are processed from NR. NR is also widely used in the manufacture of hoses, footwear, battery boxes, foam mattresses, balloons, toys, *etc*. In addition to this, NR now finds extensive use in soil stabilization, in vibration absorption and in road construction. A variety of NR-based engineering products have been developed for use in these fields. Seismic isolators, bridge bearings and dock fenders are niche products relying on the superior properties of NR.

A whole range of application markets have evolved for NR latex in recent years. Most of these are in the dipped goods sector, covering medical gloves, catheters, condoms and medical products and devices. This is an important sector serving the vast healthcare industry, especially in barrier protection of healthcare workers in the wake of the spread of HIV/AIDS and other infectious diseases. The strong growth in demand for NR latex can be traced to its escalating and leading volume consumption in medical gloves (Figure 4.8). Other demand comes from baby teats and soothers, balloons, foam, thread, toys, adhesives and binders for artificial leathers, carpet backing and textiles. Latex

Figure 4.8 Chain dipping lines in a medical glove manufacturing factory.

thread and latex garments have also registered strong growth. However, foam and carpet backing are facing some stiff competition from synthetic latices.

In the current context of sustainable development, NR is receiving renewed and greater attention as a renewable feedstock and could have a leading role in reducing the global carbon footprint in a low-carbon society. Greater emphasis is being placed on expanding and diversifying the usage of NR into new areas or in complementing those of synthetics. ENR is a good example where new properties of NR were created without compromising its superior attributes. In a very recent development, ENR latex is mixed with an emulsion of vegetable oil to produce an oil-extended ENR latex which is then converted into OENR for manufacturing environmentally friendly tyres using a silica filler.[41] The issue is pertinent and important because with the rapid consumption of depleting fossil oil looming on the horizon and the expected escalating cost of extraction, there will be pressure to increase NR usage. Unfortunately world production of NR is currently unable to meet the increasing demand on elastomers. Even though the rubber-producing countries, in particular those in South-East Asia, are increasing their efforts to expand the acreage of rubber plantations, the pace is unable to keep up with escalating demand. R&D underpins the further progress of NR in the upstream plantation and production sector and the downstream applications and usage expansion.

References

1. K. H. Chee, *Ann. Appl. Biol.*, 1976, **84**, 135–145.
2. J. B. van Beilen and Y. Poirier, *Trends Biotechnol.*, 2007, **25**, 522–529.
3. J. B. van Beilen and Y. Poirier, "Guayule and Russian Dandelion as Alternative Sources of Natural Rubber", *Critical Review Biotechnology*, 2007, **27**, 217–231.
4. J. Bonner, in *Guayule Natural Rubber*, ed. J. W. Whitworth and E. E. Whitehead, Office of Arid Lands Studies, University of Arizona, Tucson, 1991, pp. 1–6.
5. E. Campos-Lopez and J. Palacious, *J. Polym. Sci., Part C: Polym. Lett.*, 1976, **14**, 1561.
6. *Bridgestone Targets Guayule Shrub as Rubber Source*, March 9, 2012. Available from: www.environmentalleader.com/2012/03/09/bridgestone-targets-guayule-shrub-as-rubber-source (accessed 18 June 2013).
7. A. Yagami, K. Suzuki, H. Saito and K. Matsunaga, *Allergol Int.*, 2009, **58**, 347–355.
8. H. Staudinger, *Ber. Deut. Chem. Ges.*, 1920, **53**, 1073.
9. R. H. Ottewill and J. N. Shaw, *Colloid Polym. Sci.*, 1967, **215**, 161–166.
10. S. H. Maron and M. E. Elder, *J. Colloid Sci.*, 1963, **18**, 199–207.
11. J. T. Sakdapipanich, *J. Biosci. Bioeng.*, 2007, **103**, 287.
12. K. Nawamawat, J. T. Sakdapipanich, C. C. Ho, Y. Ma, J. Song and J. G. Vancso, *Colloids Surf. A: Physicochem. Eng. Asp.*, 2011, **390**, 157–166.
13. J. E. Bowers, *Natural Rubber-Producing Plants for the United States*, USDA, National Agricultural Library, Beltsville, MD, 1990.

14. K. Kanokwiroon, PhD thesis, "Antimicrobials from latex of *Hevea brasiliensis*", Prince of Songkla University, Thailand, 2007.
15. *Rubber Asia: 20th anniversary issue*, 2007, 28.
16. D. C. Blackley, *High Polymer Latices*, Volume 1, Applied Science Publishers, New York, 1966, p. 159.
17. G. F. J. Moir, *Nature*, 1959, **184**, 1626.
18. K. F. Gazaley, A. D. T. Gordon and T. D. Pendle, in *NR Science and Technology*, ed. A. D. Roberts, Oxford University Press, Oxford, 1988, p. 63.
19. S. Kawahara, W. Klinklai, H. Kuroda and Y. Isono, *Polym. Advan. Technol.*, 2004, **15**, 181–184.
20. O. Chaikumpollert, Y. Yamamoto1, K. Suchiva, P. T. Nghia and S. Kawahara, *Polym. Advan. Technol.*, 2012, **23**, 825–828.
21. S. C. Ng and L. H. Gan, *Eur. Polym. J.*, 1981, **17**, 1073–1077.
22. J. Nambiar, *Concentration of Epoxidised NR Latex by Ultrafiltration*, Proceedings of the International Rubber Technology Conference, Kuala Lumpur, 1993.
23. S. F. Chen, *Types, Composition, Properties, Storage and Handling of Natural Rubber Latex Concentrates., Notes on NR Examination Glove Manufacture,* Rubber Research Institute of Malaysia, Kuala Lumpur, 1988, pp. 1–12.
24. P. Prasassarakich, P. Sintoorahat and N. Wongwisetsirikul, *J. Chem. Eng. Jpn*, 2001, **34**, 249–253.
25. G. F. Bloomfield and P. Mcl., Swift, *J. Appl. Chem.*, 1955, **5**, 609–615.
26. C. Nakason, A. Kaesaman and N. Yimwan, *J. Appl. Polym. Sci.*, 2003, **87**, 68–75.
27. N. C. Dafader, M. E. Haque, F. Akhtar and M. U. Ahmad, *Polym.-Plast. Technol.*, 2006, **45**, 889–892.
28. P. B. Edwards, *India Rubber J.*, 1953, **125**, 334.
29. G. J. Van Veersen, *J. Polym. Sci.*, 1951, **6**, 29.
30. M. Gordon, *Proc. R. Soc. A.*, 1950, **204**, 569.
31. D. J. Hourston and J. O. Tabe, in *Polymeric Material Encyclopedia*, ed. J. C. Salamone, J. Claypool and A. Demby, CRC Press Inc., New York, 1996, p. 4547.
32. S. Riyajan, Y. Tanaka and J. T. Sakdapipanich, *Rubber Chem. Technol.*, 2007, **80**, 365.
33. S. Riyajan and J. T. Sakdapipanich, *Kautsch. Gummi Kunstst.*, 2006, **59**, 104.
34. S. Riyajan, D. J. Liaw, Y. Tanaka and J. T. Sakdapipanich, *J. Appl. Polym. Sci.*, 2007, **105**, 664.
35. K. Vivayganathan, M. Y. Amir Hashim and F. Hanim, "Environmental-friendly natural rubber latex preservation systems" Proceedings, International Rubber Conference 2008, 20–23 October 2008, Kuala Lumpur, Malaysia.
36. J. J. Kabara, J. Lopez and C. F. Robert, *Natuurrubber*, 2006, **41**, 1.
37. S. F. Chen, W. C. Wong and C. Y. Low, *US Pat.*, 2012/0021 021 155A1, 2012.

38. *Natural Rubber: Types, Packaging and Transportation* Version 1.0.0.05, available from www.tis-gdv.de/tis_e/ware/kautschuk/naturkautschuk/naturkautschuk.htm.
39. J. Cecil, P. Mitchell and P. Griffee, *Processing of Natural Rubber*, FAO (AGST) Consultants, EcoPort version by Peter Griffee, FAO, 2003, available from http://ecoport.org/ep?SearchType = earticleView& earticleId = 187&page = -2.
40. J. P. Zhong, S. D. Li, Y. C. Wei, Z. Peng and H. P. Yu, *J. Appl. Polym. Sci.*, 1999, **73**, 2863.
41. T. Sakaki, N. Ichikawa, T. Hattori, C. C. Ho and H. D. Choong, *US Pat.*, 2012/0065324A1, 2012.

CHAPTER 5

Natural Rubber Blends and Based IPNs: Manufacturing Methods

WANVIMON ARAYAPRANEE

Department of Chemical and Material Engineering, College of Engineering, Rangsit University, Phathum Thani, 12000, Thailand
Email: wanvimon@rsu.ac.th

5.1 Introduction

Polymer blending is carried out to achieve commercially viable products having unique properties and/or at low cost. Although blending is a very simple technique, many polymer pairs are immiscible and incompatible, leading to poor mechanical properties. The most important polymeric properties related to rubber/rubber blends are homogeneity of mixing and cure compatibility. Processing rubber into finished goods consists of compounding, mixing, shaping, moulding and vulcanizing. Manufacturing rubber products requires the use of many additives such as accelerators (to initiate the vulcanization process), zinc oxides (to assist in accelerating vulcanization), retarders (to prevent premature vulcanization), antioxidants (to prevent aging), softeners (to facilitate processing of the rubber), fillers (to serve as reinforcing/strengthening agents or reduce the cost), and inorganic or organic sulfur compounds (to serve as vulcanizing agents). It is through compounding that the specific vulcanized rubber gets its characteristics (properties, cost and processability) to satisfy a given application. The additives can be added and mixed into the rubber to form an uncured 'rubber compound'. Compounding requires decisions about which rubber to use, together with the quantities and types of other additives, to achieve after mixing, shaping and curing the properties required for the end application, whereas mixing of the additives with the rubber is to produce a product that has the ingredients dispersed and distributed sufficiently such that it can easily be shaped in the next process.

The manufacture of products directly from latex was developed during the 1920s. The mixing of rubber latex compounds is a simple operation using relatively inexpensive equipment with low energy consumption, but the mechanical stability of the rubber latex mixtures is a serious concern and is reduced by turbulent agitation. All latex products contain several distinct disperse phases and are highly polydisperse with different surface-active species such that a stable colloidal system is maintained until, at the desired time, it is made unstable and converted to a dry product. The drying, which consumes large amounts of energy, restricts latex technology for the manufacture of products generally less than 3 mm in thickness.[1] The dry solid rubber industry has always primarily used batch mixers. The early rubber industry was largely based on mixing with two-roll mills and involved a large quantity of fine particles and hazardous vulcanization accelerators. This made necessary the introduction of the internal batch mixer with isolated mixing chambers. With basic masticating and mixing processes, new methods of mixing that were quicker, consumed less power, and which offered a cleaner working environment than that provided by the 'open' mills become available. Mixing with an internal batch mixer is performed with counter-rotating heavy-duty rotors to efficiently shear the rubber materials, which are confined within a chamber. When dumped from a mixer the mix is a large deformed mass that has to be both cooled and converted into sheet or strips for moulding and vulcanization.

Mixing is the process of thoroughly combining different materials to produce a homogenous product and is the most demanding unit operation in rubber processing because the quality of the final product and its attributes are determined by the quality of the mix. Difficulties in subsequent processes and inadequate end product properties result if either distributive or dispersive mixing is incomplete. Mixing as a general operation may be considered as three basic processes occurring simultaneously, namely incorporation, dispersion and distribution.

Incorporation is a process in which the originally separate ingredients form a coherent mass followed by wetting of the filler with rubber. Mixing is always preceded by a masticating process, introduced in order to soften the rubber so that compounding ingredients can be added; the long polymer chains must be partially broken and stretched by mastication, which results in an increased surface area for filling up the voids within the filler agglomerates. Nakajima[2–4] showed that the rubber undergoes deformation, providing a large surface area for wetting filler agglomerates and then sealing them inside.

Dispersion is defined as a process during which filler agglomerates are broken down to the submicron level, called aggregates, mainly by shear stress. At the end of the incorporation stage the majority of the filler is present as agglomerates. They act as large, rigid particles, whose effective volume is higher than that of the filler alone due to rubber being trapped inside the filler voids and the rubber being immobilized on the surface. The rubber increases the rate of dispersion by increasing the effective size of the filler aggregates.

Distribution corresponds to mixing, which is simple homogenization on the macro scale, during which the various ingredients are randomly distributed throughout the mass of the mix. In the distributive mixing step particles are spread homogenously throughout the polymer matrix without changing their size and physical appearance. Homogeneity is achieved when both the dispersive and the distributive elements of the mixing have achieved a fluctuation in average composition below a certain fixed and acceptable level. The mix reaches its final viscosity as plasticizers effectively internally lubricate the mix. Figure 5.1 illustrates the mixing operation.

A good dispersion is one that evenly distributes filler throughout the polymer with smaller aggregates, whereas a poorer dispersion results in larger aggregates and agglomerates (Figure 5.2).

Vulcanization is the process by which the linear rubber molecules are linked to form a three dimensional network comprising crosslinks formed by one or more sulfur atoms; this is a result of heating the liquid rubber with sulfur. Crosslinking increases the elasticity and the strength of rubber about ten-fold, but the amount of crosslinking must be controlled in order to prevent the formation of brittle and inelastic rubber. The properties of rubber that are improved by vulcanization are tensile strength, elasticity, hardness, tear strength, abrasion resistance, and resistance to chemicals. The traditional

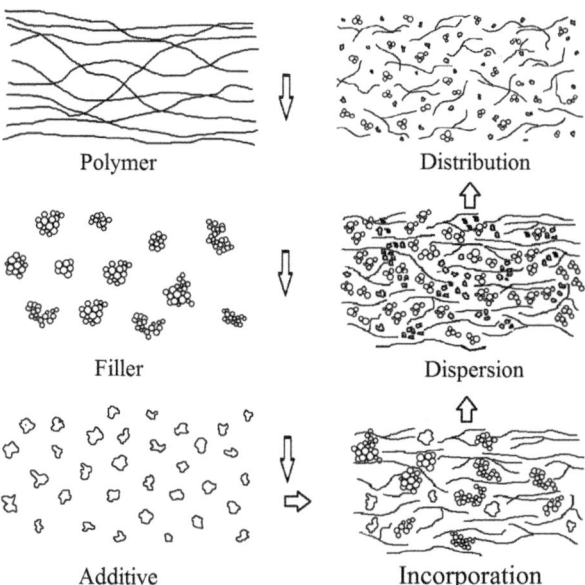

Figure 5.1 The mixing operation.

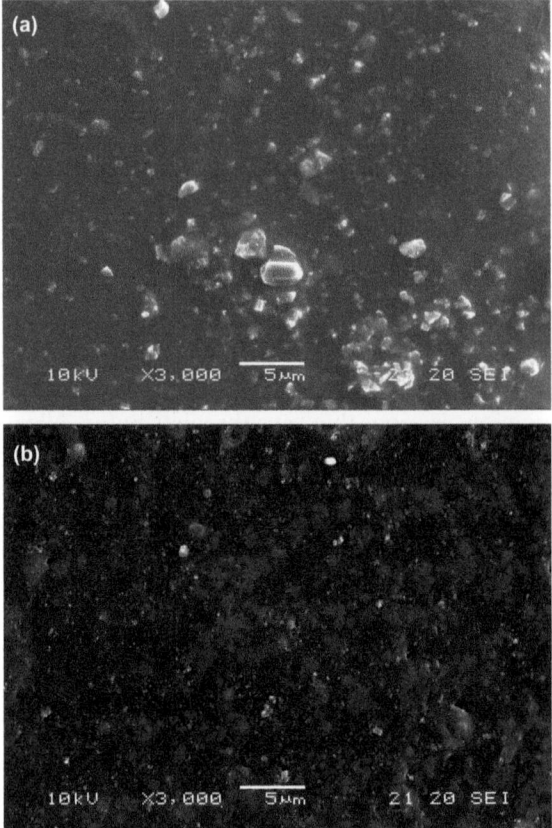

Figure 5.2 Electron micrographs of (a) poor and (b) good dispersions.

simple rubber-plus-sulfur mix is no longer used today, due to long curing times. It has been replaced by systems containing activators and accelerators which greatly increase the rate, efficiency and reproducibility of crosslinking. In the process of vulcanization, the added sulfur allows some C–H bonds to be broken and replaced by C–S bonds. The vulcanized rubber is harder, more durable, and more resistant to chemicals and other damage.

5.2 Latex Based Methods

Today, the term latex applies not only to natural rubber emulsions but also to emulsions of synthetic rubber, since many rubbers have their own latex form and latex can be considered as nanopolymer spheres dispersed in a water medium. In general most latex compounds require additives like stabilizers to ensure adequate processing stability, vulcanizing agents to affect crosslinking of the rubber, and protective agents to ensure adequate service life.

5.2.1 Latex Mixing

Compounding is a process in which special formulations are added to the latex to strengthen it. Chemical additives added to the raw latex must be in the form of emulsions or dispersions. The additives are added in the order: stabilizing agents, vulcanizing agent (sulfur), accelerators (which help control the later vulcanization process), antioxidants (which prevent deterioration of the rubber molecules in the final product by heat, moisture and ozone), pigment, fillers, zinc oxide, and thickener solutions. All solid additives are ground to a fine particle size by milling the substances with distilled or softened water in ball or gravel mills (cylinders half to two-thirds filled with a grinding charge, glass or porcelain balls, or flint gravel) used for dispersing powders into liquids over a few hours up to several days. Mixing is relatively straightforward, *e.g.* by paddle stirring (Figure 5.3(a)). Ball mills (shown in Figure 5.4) are also used, but the process takes longer. The latex itself has to be stabilized with surface-active agents to prevent coagulation, which can be irreversible. The stabilizing agent acts by imparting a charge (negative or positive) to the surface of the minute rubber particles or by holding an envelope of water around the particle, thereby preventing any coagulation. The viscosities of latex, dispersions and solutions (except thickener) are low, so that the mixer is run at a slow rotor speed to achieve a homogeneous mix. Low speed stirring is adequate when the diameter of the agitator blades is sufficient to move the whole of the mix (Figure 5.3(b)). A variable speed motor is recommended for the agitator so that a rotational speed of 50–100 rpm can be used during addition of the ingredients and a speed of 20–40 rpm during the maturation process.[1]

The mixing vessel can be made of stainless steel, epoxy resin lined mild steel, fiber glass-reinforced polyester resin or other materials. Uncoated mild steel should be avoided as it may result in rusting.

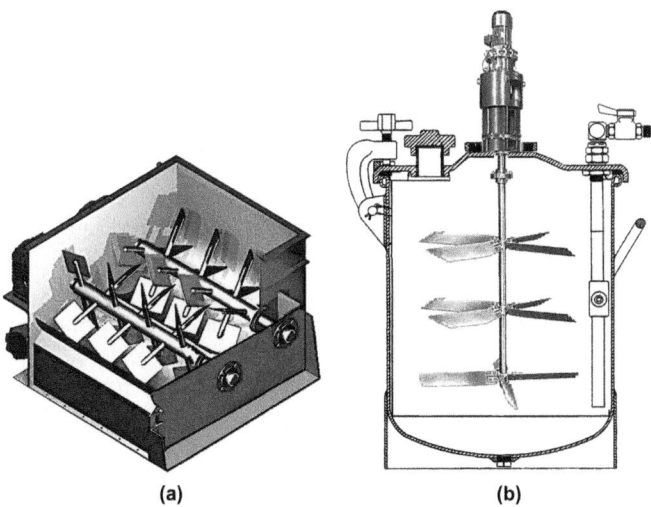

Figure 5.3 Latex mixers: (a) paddle mixer and (b) mixing vessel.

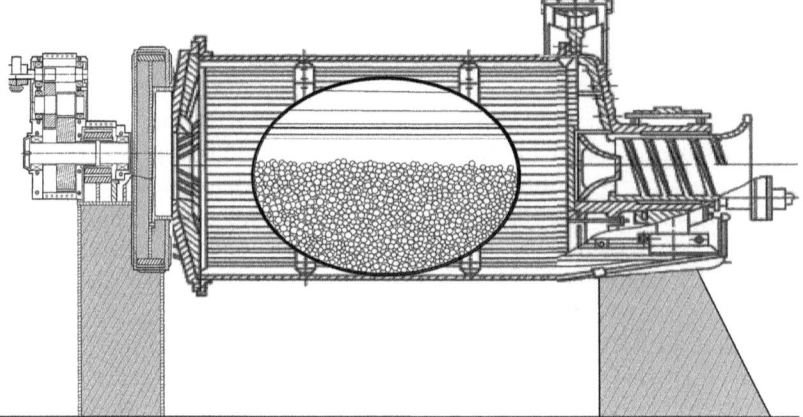

Figure 5.4 A ball mill.

5.2.2 Maturation

Latex compound maturation is the period when the latex compound is stored after mixing, prior to use in the production line. After maturation the latex compound is mixed with a sulfur dispersion to crosslink the rubber molecules to improve their properties. The latex compounds are agitated for a maturation period of 1–7 days, depending on the nature of the production process and on the scale of the mixing operation. During maturation, crosslinking of the rubber molecules takes place inside discrete rubber particles dispersed in the aqueous phase of the latex[1] and air bubbles introduced in the mixing rise to the surface.

5.2.3 Latex Curing Processes

After the pre-vulcanized latex compound has been prepared, the next step is to obtain the shape to be made; set the shape and then vulcanize. The different latex processes classified according to the method of shaping are:

- **Dipping**: A thin skinned article can be prepared from latex by dipping process. After the latex is compounded, it is dispersed in a dipping tank. A "former" the shape of the article to be produced, which is coated with coagulant (calcium nitrate) is immersed slowly in the latex which form a thin layer of latex that coats the former. The formers may be made from glass, aluminium, porcelain, polyester resin or plaster. The coagulant converts the liquid latex film into a wet-gel on the former. After dipping, the mould is dried, washed, vulcanized in circulating hot air, steam or hot water and then peeled from the former.
- **Casting and moulding**: Plaster or metal moulds are used to produce both solid and hollow articles. The latex is poured into a mould containing a hollow cavity of the desired shape, and then allowed to sit in the plaster

mould by partial absorption of water or in a metal mould by using a heat-sensitizing agent. Plaster moulds are generally used for hollow articles produced by forming the designed wall thickness on the inside surface of the mould. In the preparation of the solid articles, the compound with in presence of a gelling agent is solidified and subsequently dried. The final steps include the partial drying of the deposited latex and cooling, removing the deposit from the mould, washing, and then vulcanizing completely in an oven.

- **Foaming**: Air is mechanically beaten into the latex to form a gas bubble filled rubber matrix structure. Latex foam is a flexible cellular material containing an open cell structure or a closed cell structure. Basically, there are two methods in the production of latex, the Dunlop Process (1928) and Talalay Process (1935). In the Dunlop Process, the latex compound is mechanically beaten and/or air is blown into liquid latex to form wet-foam. The density of the foam is obtained by the amount of blowing. After adding a gelling agent, the latex is poured into a mould. The gelled foam is then steamed to vulcanize, removed from the mould, washed and dried. The natural sediments in the latex settle to the bottom of the mould, which makes a latex layer slightly firmer on its bottom side. In the Talalay Process, after foamed latex is poured into a mould sealed under pressure provided by a hydraulic press and a vacuum is applied to remove air and expands to fill the mould cavity with latex completely. The foam is then frozen to stabilize the cell structure by cooling the mould using carbon dioxide as the blowing agent, which penetrates the structure and owing to the formation carbonic acid, which lowers the pH from above 10 to 7, thereby causing gellation. This yields a uniform cell structure. The mould is heated to the vulcanizing temperature to complete the cure. The final steps are the washing and drying processes. The Talalay Process is significantly more complex and costly, but results in a higher quality product.
- **Spreading**: This process has found wide application in the backing of tufted carpets in which the latex is used to hold the backings together, to hold the fibers tighter and to bond the fiber to the backing. Jute is the backing on the carpet used for heavy berber carpets and for some wool carpets. A latex compound is applied to the tufted primary backing to anchor strongly to the base. As the tufted primary backing passes under a puddle of the same latex compound, a doctor's-blade forces the latex to penetrate into and around the yarn bundles. This process is followed by the secondary backing to the carpet to improve the stability.
- **Spraying**: The adhesive property of latex has been utilized in the spraying process for bonding paper, cloth, fiber, textile *etc*. The latex compounds are typically applied at 7–20% solids. Natural fiber (*e.g.* coconut) can be bonded by spraying the loose fibers with a suitable latex compound after the fibers have been formed into their configuration in the finished fabric. The process consists of spraying, drying the product, compressing the dried mass in a mould to obtain a desired shape and curing it in an air oven for the permanence of shape.

Vulcanizing involves applying heat at a given temperature for a given time to cure the product so it takes up the desired shape. Curing involves the chemical reactions occurring in the rubber mixture to produce crosslinking. Heating can be done in the mould using steam or in autoclaves, ovens, by using hot gases, *etc.*

During vulcanization, the latex film is heated, and a combination of sulfur, activator, accelerator, and heat cause crosslinking of the rubber, giving strength and elasticity to the film. There are two ways of producing a vulcanized latex film – one is by curing the latex in the liquid state. The material is then called pre-vulcanized latex. Pre-vulcanized latices are easy to handle and have a long storage life. Having deposited a film of latex rubber by the various methods available, only drying is required to obtain a suitable product. The alternative method is to incorporate the vulcanizing ingredients into the latex and to heat and thereby cure the latex film after it has been deposited and dried. This type of compound latex is called vulcanizing latex; it has a limited shelf life, but it is cheaper to produce than pre-vulcanized latex and has other important advantages such as better filler acceptance, improved tensile strength and tear resistance.

Oven interiors should be made of non-corrosive materials and should be lagged to conserve heat. The temperature range should be between 0 °C and 80 °C for pre-vulcanized latices or 0 °C and 140 °C if vulcanizing latices are to be used. In both drying and vulcanizing ovens, warm air circulation is important in keeping moisture build-up and curing or drying times to a minimum; and so exhaust fans should be fitted to the ovens.

For drying rubber by a vacuum process it is heated in a closed vessel by steam from which the air and moisture are removed by a vacuum pump. The vacuum dryer is used where fast drying of rubber is required. While a vacuum dryer is an expensive piece of equipment, the cost and upkeep of buildings and saving of time must be set against its prime cost and working expenses.

5.3 Solution Based Methods

Non-vulcanized (gum) rubber in an organic solvent is used as an adhesive to join two pieces of material together, which involves a chemical cohesion process. Practically all types of raw rubber may be used to produce rubber cements. Depending on the temperature at which the adhesion bond is formed during vulcanization, a distinction is made between hot-setting rubber cements, which form at high temperatures, usually above 100 °C, and low-setting rubber cements, which form at low temperatures; the latter is also called self-vulcanizing rubber adhesives. Rubber adhesives consist of solid rubber dissolved in such organic solvents as gasoline, ethyl acetate, toluene, naphtha or trichloroethane. The solvent used will depend on the drying and flammability considerations in the application. Alternatively, the solid rubber can be compounded with other additives and the mix dissolved in the solvent. Rubber adhesive is used in the assembly and for the repair of rubber and rubberized fabric items and the production of rubberized fabrics; it is also used to attach rubber to metal, wood, cement or glass.

5.3.1 Solution Manufacturing Processes

The process used to make rubber adhesive is relatively simple. Milled raw rubber can be shredded and agitated in the solvent until a clear solution is obtained. After the rubber is broken down into smaller pieces, it is mixed with the organic solvent, other additives are added and mixed uniformly, and then filled in containers. Alternatively, the solid rubber can be compounded with other additives and the mix dissolved in the solvent.

5.3.2 Solution Mixing

Rubber adhesives are prepared by mixing the components in a special apparatus called glue mixers. First, the blocks of rubber are broken into smaller pieces using two large rollers. The rubber may be reduced in large high-speed mixers equipped with sharp blades, which pulverize the rubber into a size similar to sawdust. The solvent ingredients are mixed in tanks equipped with paddles. The rubber is added slowly until it is wetted by the solvent and is suspended or dissolved in the solution.

Solution adhesives are applied at 7–20% total solids content by spray, roller, or a doctor's-blade and then dried at room temperature or in air ovens used to evaporate some hazardous solvents. The toughness, abrasion, and long term durability may be improved by using a vulcanizing rubber solution. Self-vulcanizing rubber solutions are supplied to the consumer in the form of two solutions, which are mixed before the adhesive is used; one solution contains raw rubber with several other ingredients, and the other solution contains a highly active vulcanizing system. The major characteristics of adhesion bonds, which depend mainly on the type of raw rubber used, include resistance to separation, breaking, corrosive mediums, moisture, heat and frost.

5.4 Solid Natural Rubber Based Methods

Natural rubber blends are difficult materials to process, because in both the raw and compounded state they have viscoelastic properties. The production sequence is mixing, forming and vulcanizing. Simple mixing ensures that the mixture has a uniform composition throughout its bulk. The solid rubber and the other additives (similar to those emulsified or dispersed in the latex process) have to be mixed. This is done with two basic machines; a two-roll mill in which the material is passed between two heavy metal rollers mounted horizontally, and an internal batch mixer (a Banbury mixer or an internal mixer) in which the materials are sheared between the internal rollers and the inside of the chamber. The machines are driven by powerful electric rotors and water cooling is necessary because heat is produced by the mixing of the materials. In the case of solid blending, the particle size need not change, but the distribution of particles throughout the mixture approaches a random distribution. Thus, the quality of the rubber with a formulation is determined as a consequence of the combined effect of the unit processes of mixing. The unit processes playing major roles in mixing filled compounds would be dispersion of the additives, chain scission of the polymer, and bound rubber formation. Difficulties in subsequent processes and inadequate end product properties result if either distributive or dispersive mixing is incomplete.

Early compounding took place either in single rotor machines such as Hancock's Pickle, or on two-roll mills. The first use of the two-roll mill was in the 1830s in the USA. Hancock's Pickle was patented in 1837,[5] although models had actually been in use from the early 1820s. However, a two-roll mill used for mastication and compounding of rubber is described by Edwin M. Chaffee[6] of the Roxbury India Rubber Company. Henry Goodyear, a brother of Charles Goodyear, who developed the steam-heated two-roll mill in 1844.[7] The first reference to non-intermeshing counter-rotating wheels with paddles and beaters appeared in 1865; the Quartz mill of Nathaniel Goodwin[8] of Newburyport, Massachusetts, although there is some doubt whether this machine would have been suitable for rubber due to its apparent lack of strength. A new internal mixer design that appears suitable for rubbers was a twin rotor design patented by Paul Pfleiderer in 1878–79 from London.[9] In 1916 Mr Fernley H. Banbury improved on an 'internal mixing machine' built by Werner & Pfleiderer[10–12] by designing a Banbury mixer which processed a ram holding ingredients in a mixing chamber and having rotors. The mixing chamber had a door at the bottom. Today preparation of the compound is accomplished predominantly in internal mixers, and the mill is used mostly to sheet out and cool the rubber compound coming from the internal batch mixer. In some cases the final mixing process, such as the incorporation of sulfur and vulcanization accelerators as well as other minor components, is also done in the mill.

5.4.1 Two-Roll Mills

Rubber mixing mills create uniform blends of natural and synthetic rubber with other raw materials. Most rubber mixers are electronically driven and the output depends on the size and specifications of the machine and the material being processed. Different types of machines have different purposes, such as the processing of rubber materials, mixing rubber, sheeting rubber, plasticizing and warming rubber and uniform mixing and blending of raw materials.

Mills were first used for compounding after 1835, but when internal mixers became available the mills were relegated to a simple sheeting use in the mill room. However for specific types of compound, particularly roller covering materials, the mill is still very often the preferred mixing unit. It is also widely used in developing parts of the world for normal compounding, mainly to prepare coloured, tacky or very hard compounds. The two-roll mill is predominantly used in small factories or for small batch sizes. Mills are also important as follow-up equipment after internal mixers or as breakdown and warm-up equipment in front of calenders or extruders.

A mill consists of two hollow metal cylinders rotating towards each other (see Figure 5.5). The roll axes were horizontal and parallel. A pair of hollow rollers can either be heated internally by steam, or cooled by passing cold water through the cavity where it is important that the wall thickness is uniform in order to obtain an even surface temperature. The distance between the cylinders (mill rolls) can be varied, typically between 0.25 and 2.0 cm, by having the bearing blocks of the front roll resisting against adjusting screws. The gap between the rolls, called a nip, allows the rubber to pass through to achieve mixing caused by the shearing action in the nip (Figure 5.6).

Mixing is always preceded by a masticating process, introduced in order to soften the rubber so that compounding ingredients can be added. The long polymer chains must be partially broken by mastication, mechanical shearing forces applied by repeatedly passing the dried rubber into the gap between the two mill rolls (the mill nip). The two horizontal parallel rolls are made from hard castings supported through strong bearings in the mill frame made from steel casting. The rolls run at different surface speeds against each other, and the distance between them is adjustable. The rolls are rotated at constant speeds with the back roll running a little faster. The difference in speed between the two rolls is called the friction ratio and allows a shearing action (friction) at the nip to disperse the ingredients and to force the compound to stay on one roll, preferably the front one. Conventionally mills have had a friction ratio of between 1:1.05 and 1:1.25 for the front-to-back roll to maintain the compound on one roller. For mixing some of the synthetic rubbers, near-even speeds is best, or even an inverse friction ratio, *i.e.* less than 1.0.[14] The rollers are gradually brought closer together until the soft rubber adheres to the slower moving roller and passes round with it, being subjected to a kneading process

Figure 5.5 Two-roll mills.
(Reproduced with the kind permission of the Rubber Research Institute of Thailand.)

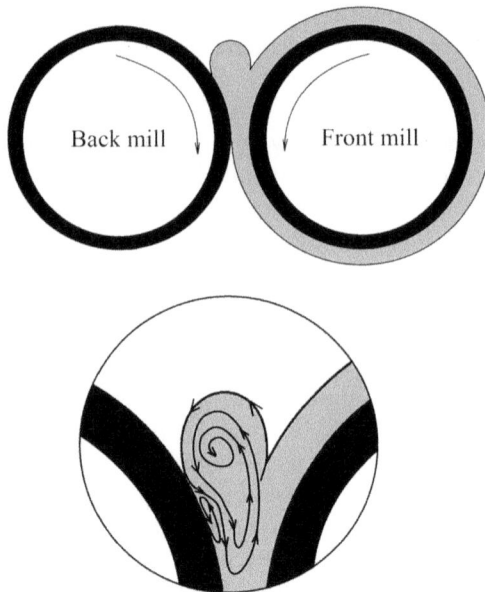

Figure 5.6 Conceptual view of rubber mill rolls.

as it passes the second roller, which revolves more rapidly. The rolls may be heated or cooled with a liquid circulating inside. At the beginning of the process the rollers are warmed, but later on friction may give rise to so much heat, generated by the very high viscosity of the rubber compounds during mixing, that it is necessary to pass cooling water into the inside of the roll because of the heat generated due to viscous flow that is generated continuously inside the rubber in the roll nip. The entire mixing operation is performed through repeated charging and discharging of rubber. The additives are charged while the rubber is bound onto a roll as shown in Figure 5.6. Modern mills are often fitted with hydraulic motors on each roll, dispensing with a gearbox, and hence become a very sophisticated tool in the mill room on which compound quality can be controlled.

5.4.2 Internal Batch Mixers

In large rubber factories, especially tire factories, the internal batch mixer has practically replaced the two-roll mill for the preparation of compounds. The mixing process in the batch mixer is accomplished inside a closed chamber by rotating kneading rotors. Traditional kneaders have two counter-rotating rotors, with each mixing rotor having two wings affixed on it. The two wing rotors typically rotate at different speeds through connecting gears. With internal batch mixers, an operator only performs charging and discharging whereas with the two-roll mill, the whole mixing operation is carried out by the operator. However, the variable nip opening on a mill, plus immediate visual feedback of the state of the mix, allows a good mill operator a high degree of control and consequently dispersion. The movement of the rotors achieves the mixing process either between the rotors or between the rotors and the chamber wall. In contrast to the mill, more uniform compounds and large batches can generally be prepared by the internal batch mixer. Internal batch mixers all have certain common components. There are:

- A chamber with a feed hopper and discharge door.
- Two relative rotations of the rotors producing both good dispersive and good distributive mixing.
- A floating weight (ram) which exerts pressure to push the batch down into the rotors and mixing chamber.
- A heating/cooling system, which controls the temperature of the chamber walls.

Figure 5.7 shows, schematically, the principal components of an internal batch mixer. Ingredients are dropped through a feeding hopper into the mixing chamber, which is jacketed for water-cooling. The chamber and rotors may be cored for heating or cooling to control the batch temperature. The feeding hopper can be closed by a pneumatically operated vertical ram, pressing onto the rubber materials. The ram would move up, to allow addition of ingredients to the nip, and it would move down to force the compound ingredients into the nip. Mixing can occur between the rotor to rotor or between the rotors to chamber wall. The distance between the rotors can be varied. The rubber compound can be discharged from the mixer, either by rotation of the whole mixing chamber or through a door at the bottom of the mixer (underneath the rotors).

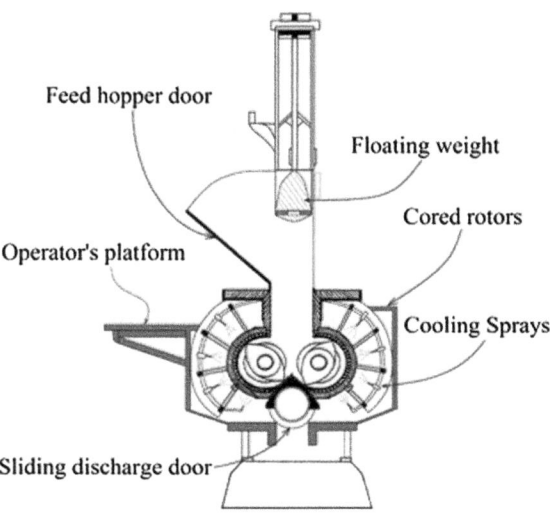

Figure 5.7 Internal batch mixer.

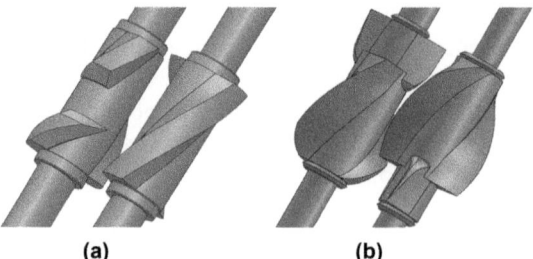

Figure 5.8 Basic rotor design: (a) intermeshing and (b) four-wing tangential.

The internal batch mixer is faster, eliminates dust and fumes hazards, uses less floor space, and is probably less operator sensitive. It has thus displaced the mill for most compounding operations. In modern, large-scale production facilities the material flow is extensively automated using microprocessor equipment which controls and monitors the whole mixing room. The internal batch mixers are made in two basic types of rotors, one is an intermeshing type (the Internal mixer) and another is non-intermeshing also called tangential type (the Banbury) as shown in Figure 5.8. Intermeshing rotors have to rotate at the same speed whereas tangential rotors, like the rolls of a mill, can rotate at different speeds. Generally, intermeshing rotors are bigger and achieve a better dispersion over a given mixing time. For equal chamber size, the tangential rotors accept the ingredients more easily and discharge them faster.

5.4.2.1 The Banbury Mixer

The Banbury mixer is a brand of internal batch mixer whose trademark is owned by the Farrel Corporation. Internal batch mixers such as the Banbury mixer are used for mixing or compounding rubber and plastics. As the first truly successful internal batch mixer, its name has become a general term for all tangential internal mixers. The material is obtained from the mouth of the feeding hopper into a tapering nip between the rotor wings and chamber sidewall by the rotor clipped to provide the compression and shear and the transfer of material around the mixing chamber from one rotor to the other in order to provide dispersion as shown in Figure 5.9. The mass transfer from one rotor to the other occurs due to the interlocking nature of the rotors using different rotor speeds. Various rotor shapes and different rotor speeds (1 : 1.1) are available for different mixing materials to make a good quality homogeneous mixture. With a sudden surface spiral rotor edge, the rotor edges are shaped in form of two-, four-, six-wing, *etc.* The two-wing rotors do not have as high productivity, hence four-wing rotors were initially developed for larger mixers.

5.4.2.2 Internal Mixers

The concept for the Intermix® was developed in the UK during the early 1930s by an engineer from the ITS Rubber Company. The construction and detailed

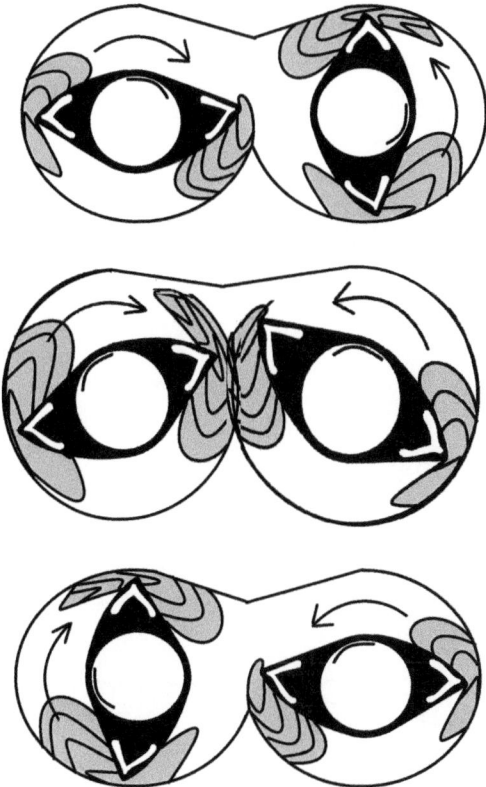

Figure 5.9 Pathlines of non-intermeshing rotors.

design of the Intermix® was contracted to Francis Shaw and Company of Manchester, who eventually acquired and patented the design. The Intermix® took a different approach to the problems of rubber mixing; the emphasis being given to the transfer of material along its length and in the opposite direction to the other rotor. Transfer from rotor to rotor occurs due to the interlocking nature of the rotors. Intermeshing rotors rotate at the same speed and the mixing effect is achieved as in a tangential system, but with this system mixing also takes place in the nip between the rotors. Compared to the tangential intermeshing system this system has more effective temperature control, drive power is higher but optimum fill levels are lower because of the narrow intermeshing zone. The use of intermeshing rotors has increased considerably in the technical rubber industry.

The internal mixer offers rapid mixing capabilities, from around 2 to 10 minutes, and thus requires an efficient cooling system. This is provided by drilled channels in the walls of the mixing chamber, through which water passes to control the mix temperature. The two rotors typically rotate at different speeds through connecting gears. The rotors and discharge door can also be water cooled (Figure 5.10). The temperature of the compound being mixed is

Figure 5.10 Internal mixer.
(Reproduced with the kind permission of Rubber Research Institute of Thailand.)

measured by a thermocouple in the side of the mixing chamber. Other parameters that can be measured and controlled during the mixing process are electrical power and time.

5.4.3 Continuous Mixers

Because of the numerous compounding ingredients used in the rubber industry and their varying physical form, it is difficult to achieve an economic and sufficiently accurate proportioning of the compounds in a continuously operating machine. Rubber extruders are machines that force material through a die or nozzle to give a profile strip. They fall into two types: those in which the pressure is produced by a ram and those in which it is produced by a screw. The screw extruder is the type most often used in the rubber industry; the ram extruder is a more specialized machine for short runs. In general, a screw extruder (see Figure 5.11) comprises:

- A feed hopper equipped with a level control feeder to receive the compound and pass it down into the flights of the screw.
- A screw to move the material forward.

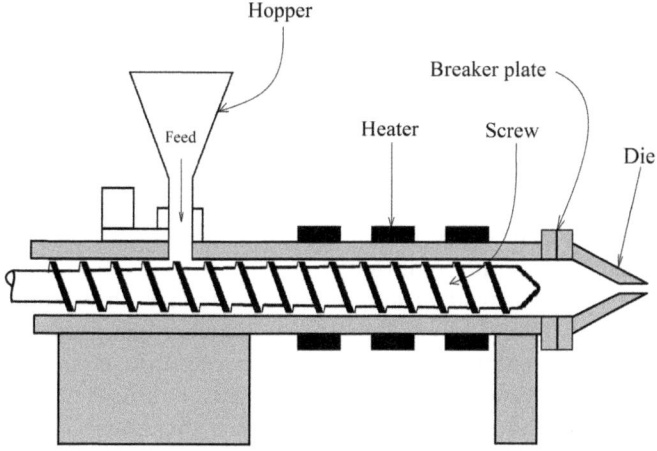

Figure 5.11 Diagram of a screw extruder.

- A barrel around the screw to contain the material, help it move, and provide part of the temperature control and to mix the rubber and other ingredients.
- A breaker plate to hold a die, to equalize the pressure from the screw and barrel, and to transport the compound to the die.
- A die, the hottest part of the machine, to form the softened rubber into the desired shape.
- A cooling mechanism (air ring, water bath, *etc.*) to freeze the molten polymer into the final shape.
- A puller to pull the polymer out of the die and through the cooling operation.
- Venting can also be installed in the barrel of an extruder to remove volatile matter from the compound.

5.4.3.1 Single-Screw Extrusion

Single-screw extrusion is one of the core operations in polymer processing and a key component in many other processing operations. The basic mixing element in the single-screw extruder is the fully flighted screw element which rotates in a stationary barrel and which melts, mixes, pressurizes and devolatilizes the ingredients of the formulation. The foremost goal of a single-screw extrusion process is to build pressure in a polymer melt so that it can be extruded through a die or injected into a mould. The screw has to lower the material's viscosity so that it will conform to the required shape, and the screw must also generate pressure to force the material into that required shape. The rubber screws have very short L/D and, in order to compress the air out of the rubber, the screws have varying pitches. The varying pitch is what gives the screw its compression. The screw design has an increased lead in the feed section and gradually decreases in pitch as it goes toward the discharge end of the screw. It can be divided into three parts: a feed section where the material enters through the feed hopper; a mixing section; and a discharge section. The feed section of the mixer is designed to optimize ingestion and delivery of feed material into the mixing section. Upon entering the mixing section, the feed material is softened and then continuously mixed, pumped, back-mixed and exited through the discharge section. The compounding extruder should have relatively shallow flights. A compression ratio (the channel depth at the end of this zone divided by the channel depth in the feed zone) of 2.5:1 is preferred and no additional mixing devices are generally required.

The earliest rubber extruders were hot feed extruders in which rubber compounds were first preheated by mill mastication before being fed as a strip to an extruder. The screw lengths were generally short and the channel depths relatively deep. Generally, most screw length–diameter ratios (L/D) for hot feed extruders were about 4 (see Figure 5.12), with a maximum ratio of about 5. Room temperature strips of rubber compounds were fed to cold feed rubber extruders but the quality of the extrudate produced was poor compared to hot feed extruders. The L/D of screws was in the range 15 to 20, and the screw channels were much shallower.[13] To enhance mixing in single-screw extruders,

Figure 5.12 Hot feed single-screw extruder.
(Reproduced with the kind permission of Rubber Research Institute of Thailand.)

special 'mixing devices' such as pins and vanes, are usually incorporated into the design.[14]

5.4.3.2 Twin-Screw Extrusion

The compounding of additives can be achieved with a twin-screw extruder, which can mix a rubber compound and convert it to a useful form simultaneously. Twin-screw extruders are designed to allow flexibility in how components are introduced, in addition to the ability to change screw configuration. The extruder comes configured with multiple feed ports, allowing the polymer components to be added as specified by the compounder. For example, the resin is generally added at the beginning of the extruder to ensure melting, mixing and uniformity of melt temperature, while fillers/reinforcing agents or liquid additives are added further downstream. Several studies have focused on the polymer–clay nanocomposites prepared by twin-screw extrusion using an original water exfoliation strategy.[15–17] In the reported method, the clay was suspended in water and this suspension was pumped at mid-extruder into the molten polymer. The water was later removed in the last third of the extrusion process through devolatilization under vacuum to provide a water-free material at the end of the compounding step.

Twin-screw extruders make use of starve feeding systems as opposed to a flood feeding system utilized with single-screw extruders. The flexibility of twin screw extrusion equipment allows this operation to be designed specifically for the formulation being processed. The twin-screw extruders, which can involve co-rotating or counter-rotating (see Figure 5.13) and intermeshing or non-intermeshing, are generally built as modular mixers. The type of mixing elements and their sequence are chosen to achieve particular mixing characteristics. In addition, the configurations of the screws may be varied using forward conveying elements, reverse conveying elements, kneading blocks, fully intermeshing co-rotating twin-screws providing the advantage of being self-wiping, fully flighted screws used in melting, and kneading discs (see Figure 5.14[1]) used especially for dispersive mixing.

Initially, the raw rubber (natural or synthetic) is mixed with several additives, which are chosen based upon the desired properties of the final product. The continuous mixer only mixes a small quantity of compound at any one time. The materials are processed in channels bounded by screw flights and barrel walls.

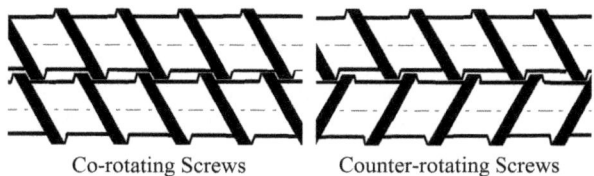

Figure 5.13 Conceptual view of the co-rotating and counter-rotating screws.

Figure 5.14 Conceptual view of the kneading discs.

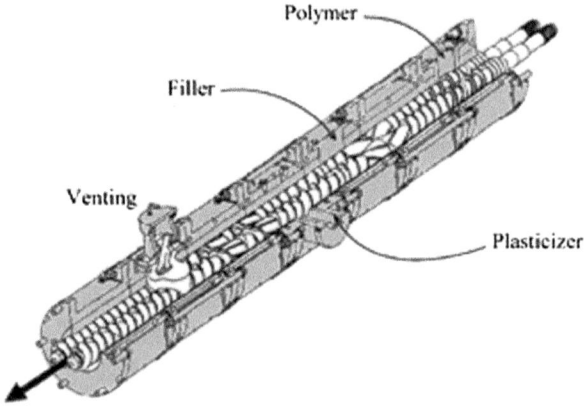

Figure 5.15 Diagram of a twin-screw extruder.

Barrels are also modular and utilize liquid cooling. The motor inputs energy into the process through rotating screws. Solids conveying and melting occurs in the first part of the process section. The polymers can be introduced into downstream barrel sections to facilitate sequential feeding. Side feeders for filler or plasticizer addition are also employed. Entrapped air, moisture, and volatiles are also removed by venting accomplished through the back end of the feed screw during the extrusion process. Discharge materials finally build and stabilize pressure to the die or front-end device as shown in Figure 5.15.

5.4.4 Solid Rubber Curing Processes

Modern rubber materials consist of approximately 60% polymers. The other part consists of vulcanization agents, softeners, accelerators, anti-aging agents and other chemicals. These additions are necessary to achieve the desired properties of the final product. A variety of methods exist for vulcanization.

5.4.4.1 Moulding Vulcanization

A vulcanizing mould consists essentially of two or more plates that can be brought together and separated by hydraulic pressure. The plates are usually heated by steam or electricity. The rubber compounds are vulcanized in various moulds between the heated plates under pressure. The principal moulding processes for rubber are compression moulding, transfer moulding and injection moulding.

Compression moulding involves placing a properly shaped blank from the unvulcanized rubber in each cavity of the mould. The mould is then closed and placed in a hydraulic press fitted with clamps. Under the applied pressure and heat, the rubber will flow and completely fill the mould cavity, which is then removed from the press. Compression moulding is the most important method using high pressure and temperature and is often used for bulky components requiring long low-temperature cures.

Transfer moulding differs from compression moulding in that the rubber is transferred through a hole into the mould cavity. In its simplest form a transfer mould consists of a ram in a cylinder fed with preheated slugs, and a mould cavity. A piece of the unvulcanized rubber is placed in the cylinder and covered by the ram. When the press is closed the ram forces the rubber through a hole into the actual mould cavity. This allows shorter cure times because of the heat generated as the rubber is forced to flow through the hole.

Injection moulding is similar to transfer moulding in that the rubber stack is forced into a closed mould cavity through a nozzle. A strip of the rubber is fed into a heated cylinder and masticated by a screw, which then moves forward like a ram and forces this preheated rubber through a nozzle into the mould cavities. As a result, rubber items can be vulcanized in very short times, there is better dimensional control, and less scrap.

5.4.4.2 Autoclave Vulcanization

Vulcanization can take place in hot air or in steam. Vulcanization in hot air ovens is not very efficient because of the poor heat transfer of hot air. Consequently, longer vulcanization times at lower temperatures are necessary to prevent aging caused by oxygen. Unlike hot air, saturated steam has better heat transfer and acts as an inert gas. Steam vulcanization takes place in autoclaves with a heated jacket and a closed chamber in which the articles are placed and the steam is introduced. All air in the autoclave needs to be displaced by steam before the cure time is started. This produces the best possible properties and most uniform cure.

5.4.4.3 Continuous Vulcanization

The compound is shaped and cured in a single line operation which produces very long lengths of tubes or cables. While there are several methods of continuous vulcanization, they employ pressureless systems and are based on the same principle: the shaped uncured product is transferred along a curing medium, for example liquid, hot air, steam, microwaves and infrared and high energy radiation.

5.5 Advantages and Disadvantages of Each Technique

The rubber industry is and has been notoriously conservative and reluctant to adapt to new concepts in materials, methods and machinery. Regarding the latter, this conservatism is sometimes justified because of the relatively high cost of machinery and its durability through many years of service and multiple refurbishments. The equipment for mixing rubber and chemical additives must fulfil the following requirements: provide steady-state conditions; give reproducibility of processing conditions; show versatility to adapt to the mixing and blending of all kinds of formulation; easy to operate, maintain and clean. To achieve optimum product quality, the equipment for compounding of fillers into a rubber must be capable of performing some of the following process tasks:

- Mixing efficiency providing incorporation and homogenization of fillers without exceeding degradation temperatures and uniform shear stress to each filler particle at any heat history and generating sufficiently high internal shear stresses to facilitate good dispersion.
- Easy operation to reduce labour intensity.
- Mix quality providing precise temperature control over the process to ensure narrow temperature distribution throughout the process and at discharge, again to regulate and minimize heat history.
- The performance and longevity of production.
- Providing short and uniform residence time distribution to minimize heat history.
- Low energy consumption.
- High productivity.

Using the above discussed factors which need to be considered in the choice of machinery has to be dealt with; the advantages and disadvantages of the mixing equipment are summarized in Tables 5.1 to 5.5.

Table 5.1 Advantages and disadvantages of latex mixing.

Type	Advantages	Disadvantages
Latex mixing	• Very much stronger and elastic as the elastomer chains have not been degraded by the mechanical work. • Cost-effective conventional vulcanization since even post-vulcanization takes place at a relatively low temperature. • Relatively inexpensive equipment.	• Restricts product thickness. • High consumption of energy for drying processes.

Table 5.2 Advantages and disadvantages of two-roll mills.

Type	Advantages	Disadvantages
Two roll mills	• Accepts all feed forms • Very versatile • Broad range of shear capability • Robust machines	• Difficult to automate • Batch-to-batch variation dependent on accurate feeding and process control • Varying power demand • Dirty operation • Difficult to incorporate fillers and other additives – gives long life cycle • Labour intensive • Low output • Difficult to achieve a uniform product

Table 5.3 Advantages and disadvantages of internal batch mixers.

Type	Advantages	Disadvantages
Internal batch mixers	• Accepts all feed forms • A wide variety of mixing operations • Relatively simple to operate • Good for short production runs • Broad range of shear • High output • Can be automated • Good for short production runs • Long life machines	• High capital cost • Varying power demand • Batch-to-batch variation dependent on accurate feeding and process control • Post mixer variable product heat history • Rapid temperature rise needs good control • Need post mixer forming • Can be labour intensive

Table 5.4 Advantages and disadvantages of single-screw extruders.

Type	Advantages	Disadvantages
Single-screw extruders	• Mechanically fairly simple • Relatively easy to operate • Easily automated • Capable of high pressure generation • High output • Energy efficient • Ease of process optimization • Uniform product shear and heat history	• High capital cost • Need free-flowing feed (particulate rubber) • Lack of positive conveying characteristics • Limited compounding and homogenizing capabilities • Large machines with long L/D • Need post mixer forming

Table 5.5 Advantages and disadvantages of twin-screw extruders.

Type	Advantages	Disadvantages
Twin-screw extruders	• Newer technology • Specialty applications • More complete melting allows machines to be shorter and gives better mixing • Self-wiping characteristic avoids any product hang-up in the machines • Easily automated • Ease of process optimization • Uniform product shear and heat history • Energy efficient • Geometry optimized for use	• High capital cost • Mechanically more complex with difficulties of fitting adequate thrust bearing to two closely positioned shafts

5.6 Conclusions

This chapter summarizes the current understanding of the rubber mixing process, which is to produce a product that has the ingredients dispersed and distributed sufficiently thoroughly that it will result in good cure in the following processes and give the required properties for end use applications. Mixing consists of mixing the raw rubber with several chemical additives. Compounding of rubber with fillers basically involves: the initial incorporation, wetting, followed by breaking down of agglomerates and finally distribution of the fillers within the polymer matrix. If the base compound is inadequately mixed, problems cascade down through the subsequent processes of shaping and curing into the end product. For latex mixing, the compounding ingredients and the latex are mixed and then the latex film is heated and cured after it has been deposited and dried. Solid rubber mixing involves mixing dry solid rubber and other compounding ingredients in mixing equipment. Solid rubber mixing is usually carried out in internal batch mixers, although some mixing is still carried out in open mills and nowadays using single- and twin-screw extruders. Extruders are being increasingly adopted as a means of giving a consistent product in a readily usable form, *i.e.* tube, cable or sheet.

References

1. K. F. Gazeley, A. D. T. Gorton and T. D. Pendle, in *Natural Rubber Science and Technology*, ed. E. D. Roberts, Oxford University Press, New York, 1998, p. 99.
2. N. Nakajima, *Rubber Chem. Technol.*, 1985, **58**, 1088.
3. N. Nakajima, *Rubber Chem. Technol.*, 1982, **55**, 931.
4. N. Nakajima, *Polym. Eng. Sci.*, 1979, **19**, 215.
5. T. Hancock, *Br. Pat.*, 7344, 1837.
6. E. M. Chaffee, *US Pat.*, 16, 1836.
7. E. H. Johnson, *India Rubber World*, 1950, **123**, 315.

8. N. Goodwin, *US Pat.*, 50, 115, 1865.
9. P. Pfleiderer, *German Pat.*, 10, 164, 1880.
10. P. Pfleiderer, *US Pat.*, 254,042, 1882.
11. D. C. Bogue, *IEC Fund.*, 1966, **5**, 253.
12. R. B. Bird and P. J. Carreau, *Chem. Eng. Sci.*, 1966, **23**, 427.
13. B. G. Crowther and H. M. Edmonson, in *Rubber Technology and Manufacture*, ed. C. M. Blow, Butterworth & Co. Ltd, London, 1971, p. 262.
14. A. Lawal and D. M. Kalyon, *Chem. Eng. Sci.*, 1995, **35**, 1325.
15. N. Hasegawa, H. Okamoto, M. Kato, A. Usuki and N. Sato, *Polymer*, 2003, **44**, 2933.
16. Z.-Z. Yu, G.-H. Hu, J. Varlet, A. Dasari and Y.-W. Mai, *J. Polym. Sci., Part B: Polym. Phys.*, 2005, **43**, 1100.
17. M. Kato, M. Matsushita and K. Fukumori, *Polym. Eng. Sci.*, 2004, **44**, 1205.

CHAPTER 6

Filler Migration in Natural Rubber Blends During the Mixing Process

HAI HONG LE,*[a,c] SYBILL ILISCH,[a,b] GERT HEINRICH[c,d] AND HANS-JOACHIM RADUSCH[a]

[a] Center of Engineering Sciences, Martin Luther University Halle-Wittenberg, D-06099 Halle (Saale), Germany; [b] Now: Styron Deutschland GmbH, Germany; [c] Leibniz-Institut für Polymerforschung Dresden e.V. (IPF), D-01069 Dresden, Germany; [d] Institut für Werkstoffwissenschaft, Technische Universität Dresden, D-01069 Dresden, Germany
*Email: hai.le.hong@iw.uni-halle.de

6.1 Introduction

Tire tread compounds have been prepared from various rubber blends of two or more rubbers, such as styrene-butadiene rubber (SBR), natural rubber (NR) and polybutadiene rubber (BR) and fillers like carbon black (CB) and/or silica. NR is an essential blend partner in passenger car tyre sidewall mixtures. The presence of NR is usually desirable, in particular for truck tyres, to obtain high mechanical performance, low heat build-up, good wear resistance, high flexibility at low temperature and high adhesion to steel cord. It is recognized that the desirable tyre properties are not only dependent on the characteristics of the rubber matrix and the crosslinking agent but also on the type and loading of filler used, as well as its dispersion and localization state in rubber blends. Assuming that filler transfer during masterbatch mixing is of minor importance, Hess et al.[1–3] demonstrated very well the effect of filler localization on hysteresis, tensile strength, fatigue and wear behaviour of NR/BR and SBR/BR tread compounds. Waddell[4] and Massie et al.[5] reported that the crack growth resistance of NR/BR blends with CB mainly dispersed in the NR phase has been significantly improved. Dynamic mechanical properties, hardness and fracture were found to be dependent on the filler localization in blends as well.[6-8] For the characterization of filler localization in low-filled blends microscopy methods like transmission electron microscopy (TEM)[1,9] and atomic force microscopy (AFM)[10-13] have been used. These methods are less suitable for highly filled rubber blends. Several methods based on electrical resistivity measurements,[14-17] nuclear magnetic resonance (NMR),[18,19] differential scanning calorimetry (DSC),[20,21] thermogravimetric analysis (TGA)[22,23] and dynamic mechanical analysis (DMA),[24-26] as well as the wetting concept,[7,27-30] have been developed for experimental determination of filler localization in highly filled rubber blends. Different concepts relating to the mechanisms of localization of filler in polymer blends have also been described in the literature. Some authors attributed the observed particle arrangements to melt viscosity effects,[31-33] but it is usually accepted that the phase selective localization of filler in rubber blends is governed by the polymer–filler interaction, which is in turn influenced by the nature of rubber components and filler.[34-38] In our previous works[7,27-30] it was found that the kinetics of filler localization are determined by the wetting process of filler by the matrix components in the early mixing stages and in the subsequent period the rubber–filler interaction is essential for the relocalization of filler. Compared to other rubbers like SBR, BR or EPDM, NR can wet filler much faster, *i.e.* more filler will be localized in the NR phase in the early state of mixing. However, the weak interaction between filler and NR surface, due to its non-polar nature, makes NR not favourable as a host for filler localization. The competition between high wetting speed and low rubber–filler interaction leads to a complex behaviour of filler localization in the rubber blends based on NR. In the present work the wetting concept was further developed to better understand the wetting mechanism of filler by different rubber types. It was used to experimentally determine the kinetics of filler localization in different rubber blends. In order to discuss the details of the filler transfer process during mixing the filler localization at an equilibrium state was predicted by means of the so-called Z-model developed previously.[39] The influence of processing and material parameters on the wetting kinetics and localization behaviour of different fillers in binary and ternary rubber blends has been the main focus of the present work.

6.2 Theoretical Prediction of Filler Localization in Rubber Blends at an Equilibrium State using the Z-Model

The thermodynamic criterion for miscibility of polymer blends is a negative Gibbs free energy of mixing. The Gibbs free energy of mixing is given by Equation (6.1):

$$\Delta G_m = \Delta H_m - T\Delta S_m \tag{6.1}$$

where ΔH_m is the change in enthalpy of mixing, and ΔS_m the change in entropy of mixing and T is the absolute temperature. The change in enthalpy during mixing is the result of dipole interactions, van der Waals interactions, acid–base interactions, Coulombic interactions and the interaction energy between the different components arising from hydrogen bonds. Due to the constraints of segmental mobility of polymer chains the change in entropy is usually too small to compensate for the change in mixing enthalpy. Thus, the Gibbs free energy of mixing is similar to the change in enthalpy of mixing:

$$\Delta G_m \approx \Delta H_m \qquad (6.2)$$

When a filler F is mixed into a binary rubber blend AB it will be wetted and bonded by the molecules A and B. Because polymer molecules can only come into contact with the outer layer of the filler particle, the effective particle volume is considered to be very small compared to the ineffective particle volume as well as the volume of the blend phase A and B, as schematically illustrated in Figure 6.1.

For the contact between polymer and filler, the change in enthalpy of mixing ΔH_m^{AF} of the phases A and F, and ΔH_m^{BF} of the phases B and F, can be described according to Hildebrand and Scott by Equations (6.3) and (6.4):[40]

$$\Delta H_m^{AF} = \frac{\Phi_A}{\Phi_A + \Phi_F^A} \frac{\Phi_F^A}{\Phi_A + \Phi_F^A} (\gamma_{AF})^2 \qquad (6.3)$$

and

$$\frac{\Delta H_m^{BF}}{V_{BF}} = \frac{\Phi_B}{\Phi_B + \Phi_F^B} \frac{\Phi_F^B}{\Phi_B + \Phi_F^B} (\gamma_{BF})^2 \qquad (6.4)$$

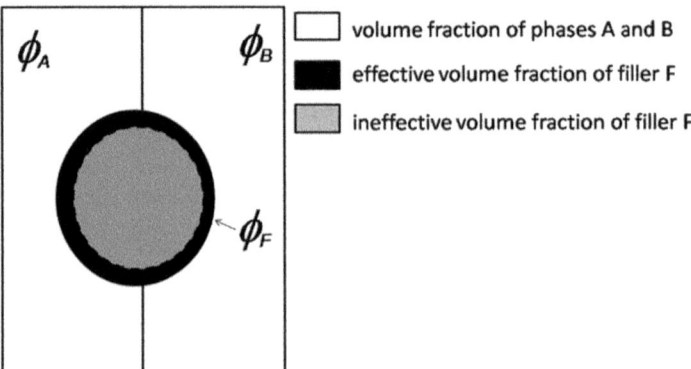

Figure 6.1 Effective and ineffective volumes of filler particles in a binary polymer blend.

with

$$\Phi_A + \Phi_B + \Phi_F = 1 \tag{6.5}$$

and

$$\Phi_F + \Phi_F^A + \Phi_F^B \tag{6.6}$$

where V_{AF} and V_{BF} are the average molar volumes of the two corresponding components. Φ_A and Φ_B are the volume fractions of the components A and B, respectively. Φ_F is the fraction of the effective volume of the filler F in the ternary system AFB. Φ_F^A and Φ_F^B are the effective volume fractions of F in A and B, respectively. The inner region (grey area, Figure 6.1) considered as the ineffective volume of F is not taken into consideration in Equations (6.3) and (6.4). K is a constant and takes a value of 1 according to Scatchard.[41,42] From Equations (6.3) and (6.4) it can be seen that in blends with high interfacial tension γ_{AB} requires more energy for dispersion. If the interfacial tension is low enough, a potential for molecular miscibility might exist.[43] At the interface of AB the component F migrates between A and B in order to minimize the Gibbs free energy of mixing ΔG_m^{AF} and ΔG_m^{BF} of the binary system AF and BF, respectively. The filler transfer process goes on until the thermodynamic equilibrium state is reached. Taking Equation (6.2) into account a thermodynamic equilibrium state is reached if:

$$\Delta G_m^{AF} = \Delta G_m^{BF} \Rightarrow \frac{\Delta H_m^{AF}}{V_{AF}} = \frac{\Delta H_m^{BF}}{V_{BF}} \tag{6.7}$$

Setting Equations (6.3) and (6.4) into Equation (6.7) we get:

$$\frac{\Phi_A}{\Phi_A + \Phi_F^A} \frac{\Phi_F^A}{\Phi_A + \Phi_F^A} (\gamma_{AF})^2 = \frac{\Phi_B}{\Phi_B + \Phi_F^B} \frac{\Phi_F^B}{\Phi_B + \Phi_F^B} (\gamma_{BF})^2 \tag{6.8}$$

Equation (6.8) can be rewritten as:

$$\frac{\Phi_F^A}{\Phi_F^B} = \frac{\Phi_B}{\Phi_A} \left(\frac{\Phi_A + \Phi_F^A}{\Phi_B + \Phi_F^B}\right)^2 \left(\frac{\gamma_{BF}}{\gamma_{AF}}\right)^2 \tag{6.9}$$

Because the effective volume fractions Φ_F^A and Φ_F^B are very small compared to the total filler volume and Φ_A as well as Φ_B of the host blend phases A and B, we can ignore Φ_F^A and Φ_F^B from the term $(\Phi_A + \Phi_F^A)^2 / (\Phi_B + \Phi_F^B)^2$ of Equation (6.9) in order to simplify. Equation (6.9) can now be expressed as follows:

$$\frac{\Phi_F^A}{\Phi_F^B} \approx \frac{\Phi_B}{\Phi_A} \left(\frac{\Phi_B}{\Phi_A}\right)^2 \left(\frac{\gamma_{BF}}{\gamma_{AF}}\right)^2 \Rightarrow \frac{\Phi_F^A}{\Phi_F^B} \approx \frac{\Phi_A}{\Phi_B} \left(\frac{\gamma_{BF}}{\gamma_{AF}}\right)^2 \tag{6.10}$$

According to our wetting concept,[27,28] describing the relationship between the wetted filler surface and the phase-specific filler localization, we can calculate the filler fraction in each blend phase A and B as follows:

$$\frac{\varphi_F^A}{\varphi_F^B} = \frac{\Phi_F^A}{\Phi_F^B} \approx \frac{\Phi_A}{\Phi_B}\left(\frac{\gamma_{BF}}{\gamma_{AF}}\right)^2 = n_{A/B}\left(\frac{\gamma_{BF}}{\gamma_{AF}}\right)^2 \qquad (6.11)$$

φ_F^A and φ_F^B are the weight fractions of the filler F in the phases A and B, respectively, with $\varphi_F^A + \varphi_F^B = 1$. $n_{A/B}$ is the blend ratio A to B. Using the Girifalco–Good equation,[44] describing the relationship between surface tension and interfacial tension, Equation (6.11) can be written as follows:

$$\frac{\varphi_F^A}{\varphi_F^B} = n_{A/B}\left(\frac{\gamma_B + \gamma_F - 2\sqrt{\gamma_B\gamma_F}}{\gamma_A + \gamma_F - 2\sqrt{\gamma_A\gamma_F}}\right)^2 \qquad (6.12)$$

γ_A, γ_B and γ_F are the surface tension values of the phases A, B and F, respectively. Setting $\varphi_F^B = 1 - \varphi_F^A$ into Equation (6.12) the weight fraction φ_F^A can be calculated using Equations (6.13) and (6.14):

$$\varphi_F^A = \frac{n_{A/B}\omega}{n_{A/B}\omega + 1} \qquad (6.13)$$

with

$$\omega = \left(\frac{\gamma_B + \gamma_F - 2\sqrt{\gamma_B\gamma_F}}{\gamma_A + \gamma_F - 2\sqrt{\gamma_A\gamma_F}}\right)^2 \qquad (6.14)$$

The validity of the Z-model was proved for different filled thermoplastic blends reported in the literature.[45] It was also extended to ternary rubber blends in our previous work.[30] For example, the filler localization in SBR/NBR/NR blends can be determined according to Equations (6.15) to (6.17).

$$\frac{\varphi_F^{NBR}}{\varphi_F^{SBR}} = n_{NBR/SBR}\left(\frac{\gamma_{SBR} + \gamma_F - 2\sqrt{\gamma_{SBR}\gamma_F}}{\gamma_{NBR} + \gamma_F - 2\sqrt{\gamma_{NBR}\gamma_F}}\right)^2 \qquad (6.15)$$

$$\frac{\varphi_F^{NBR}}{\varphi_F^{NR}} = n_{NBR/NR}\left(\frac{\gamma_{NR} + \gamma_F - 2\sqrt{\gamma_{NR}\gamma_F}}{\gamma_{NBR} + \gamma_F - 2\sqrt{\gamma_{NBR}\gamma_F}}\right)^2 \qquad (6.16)$$

$$\varphi_F^{SBR} + \varphi_F^{NBR} + \varphi_F^{NR} = 1 \qquad (6.17)$$

$n_{NBR/SBR}$ and $n_{NBR/NR}$ are the blend ratios of NBR to SBR and NBR to NR, respectively. γ_{SBR}, γ_{NBR} and γ_{NR} are the values of surface tension for the SBR, NBR and NR phases, respectively. By taking into consideration Equation

(6.17), the filler fractions φ_F^{SBR}, φ_F^{NBR} and φ_F^{NR} localized in the SBR, NBR and NR phases, respectively, can be calculated using Equations (6.18) to (6.22).

$$\varphi_F^{NBR} = \frac{1}{1 + \dfrac{1}{n_{NBR/SBR}\omega_{NBR/SBR}} + \dfrac{1}{n_{NBR/NR}\omega_{NBR/NR}}} \quad (6.18)$$

$$\varphi_F^{SBR} = \frac{\varphi_F^{NBR}}{n_{NBR/SBR}\omega_{NBR/SBR}} \quad (6.19)$$

$$\varphi_F^{NR} = \frac{\varphi_F^{NBR}}{n_{NBR/NR}\omega_{NBR/NR}} \quad (6.20)$$

with

$$\omega_{NBR/SBR} = \left(\frac{\gamma_{SBR} + \gamma_F - 2\sqrt{\gamma_{SBR}\gamma_F}}{\gamma_{NBR} + \gamma_F - 2\sqrt{\gamma_{NBR}\gamma_F}}\right)^2 \quad (6.21)$$

and

$$\omega_{NBR/NR} = \left(\frac{\gamma_{NR} + \gamma_F - 2\sqrt{\gamma_{NR}\gamma_F}}{\gamma_{NBR} + \gamma_F - 2\sqrt{\gamma_{NBR}\gamma_F}}\right)^2 \quad (6.22)$$

The filler loading in the blend phase, $S^{B(SBR)}$, $S^{B(NBR)}$ and $S^{B(NR)}$, respectively, can be calculated from the total filler loading added to S^B using Equations (6.23) to (6.25).

$$S^{B(SBR)} = \varphi_F^{SBR} \times S^B \quad (6.23)$$

$$S^{B(NBR)} = \varphi_F^{NBR} \times S^B \quad (6.24)$$

$$S^{B(NR)} = \varphi_F^{NR} \times S^B \quad (6.25)$$

6.3 The Wetting Concept for Experimental Determination of Filler Localization in Rubber Blends

The filler localization experimentally determined by the use of our wetting concept is correlated to the wetting behaviour of rubber as follows: the more the filler surface is wetted by a rubber phase, the more the filler is included and distributed in this phase.[27] When filler is mixed with a single rubber, for instance SBR, a part of the rubber chains will be bonded to the active centres, which are available on the filler surface. If the rubber is extracted from the filled uncured rubber compound with a suitable solvent, the rubber molecules bound on the filler surface remain in the rubber-filler gel. In our concept, we are interested in the fraction of rubber in this rubber–filler gel. The wetting behaviour is described by the so-called rubber layer $L(t)$, which can be calculated from the bonded rubber part R_b according to Equation (6.26):

$$L(t) = \frac{R_b(t)}{m_2(t)} = \frac{m_2(t) - m_1 \cdot c_s}{m_2(t)} \tag{6.26}$$

The mass m_1 corresponds to the rubber compound before extracting, and it is the sum of the mass of the bonded rubber part R_b, the mass of the soluble rubber part and filler. m_2 is the mass of the rubber–filler gel, which is the sum of the bonded rubber part and filler. c_s is the mass concentration of filler in the single rubber mixture or binary blends. t is the mixing time in an internal mixer. The difference between the terms rubber layer L and bound rubber, which has been commonly used in rubber technology, was discussed in our previous work.[27]

To understand the physical background of the rubber layer L, we correlated the wetting behaviour determined from our extraction experiment with the infiltration behaviour reported in the literature.[46–48] In a capillary flow of a Newtonian fluid in a small gap between two parallel plates, the position x of the liquid–air interface at time t is given by Equation (6.27):

$$x = \left(\frac{S\gamma_{lv}\cos\theta}{3\eta_1}\right)^{1/2} t^{1/2} \tag{6.27}$$

where S is the gap height, γ_{lv} is the surface tension of the liquid, $\cos\theta$ is the contact angle and η_1 is the liquid viscosity. Equation (6.27) is known as the Washburn model and has been widely applied in many studies. By taking into consideration the pressure difference between the inlet and the outlet ΔP, Lin et al.[49] modified Equation (6.27) to:

$$x = \left(\frac{S\gamma_{lv}\cos\theta}{3\eta_1} + \frac{S^2 \Delta P}{6\eta_1}\right)^{1/2} t^{1/2} \tag{6.28}$$

According to the infiltration experiments carried out by Bohin et al.[46] and Astruct et al.[48] and to simplify the calculations we assume that a silica agglomerate has a cylindrical shape with a length D_0, a cross-section A and a porosity ε, as well as density ρ_F (Figure 6.2(a)). A polymer melt with a density ρ_R infiltrates the agglomerate from the bottom as a result of capillary and viscous forces (Figure 6.2(b)). l^* is the infiltration length received after an infiltration time t. The infiltrated rubber part R_i can be calculated as follows:

$$R_i(t) = l^* A \rho_R \varepsilon \tag{6.29}$$

After extraction experiments the bonded rubber part R_b as illustrated in Figure 6.2(c) can be calculated from the infiltrated rubber part R_i and a factor h according to Equation (6.30):

The rubber–filler gel m_2 is calculated as follows:

$$R_b(t) = R_i(t) \times h \tag{6.30}$$

$$m_2(t) = l^* A \rho_R \varepsilon h + D_0 A \rho_F (1 - \varepsilon) \tag{6.31}$$

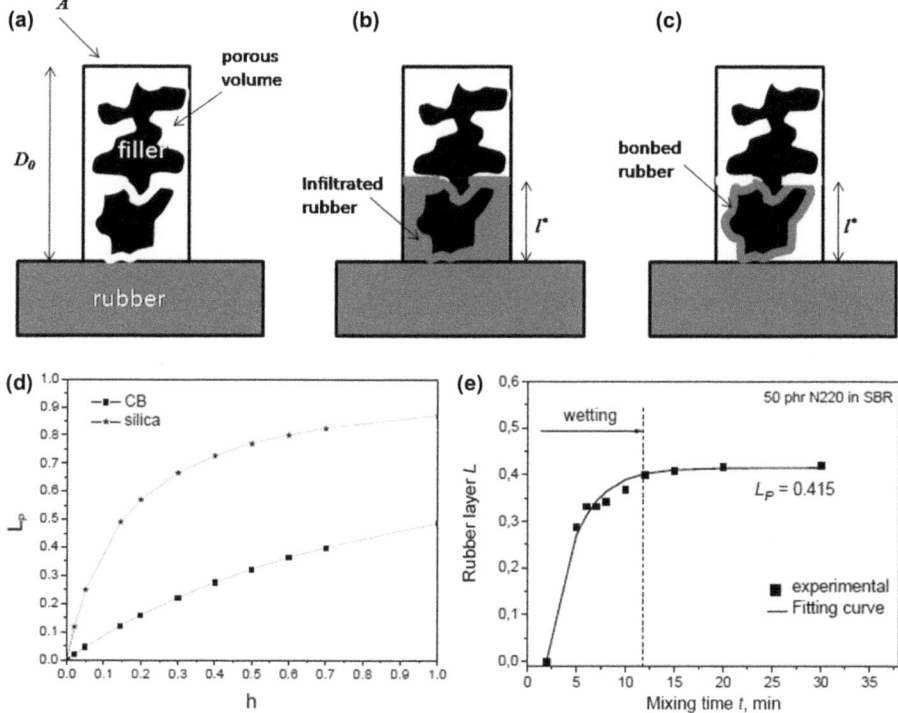

Figure 6.2 a) and (b) Infiltration test according to the work of Astruct. (c) Schematic illustration of the formation of the bonded rubber layer on the filler surface after the extraction experiment. (d) Correlation between the plateau value L_P and factor h. (e) Rubber layer L of a CB-filled SBR compound in relation to mixing time.[46]

Incorporating Equations (6.30) and (6.31) into Equation (6.26) the rubber layer L is calculated according to Equation (6.32):

$$L(t) = \frac{R_i(t) \times h}{m_2(t)} = \frac{l^* A \rho_R \varepsilon h}{l^* A \rho_R \varepsilon h + D_0 A \rho_F (1-\varepsilon)} = \frac{\varepsilon \dfrac{l^*}{D_0} h}{\varepsilon \dfrac{l^*}{D_0} h + \dfrac{\rho_F}{\rho_R}(1-\varepsilon)} \quad (6.32)$$

By assuming $x = l^*$ and substituting Equation (6.28) into Equation (6.32) we observe:

$$L(t) = \frac{hbt^{1/2}}{hbt^{1/2} + \dfrac{\rho_F}{\rho_R}(1-\varepsilon)} \quad (6.33)$$

where b is the wetting speed and can be described by Equation (6.34):

$$b = \frac{\varepsilon}{D_0}\left(\frac{S\gamma_{lv}\cos\theta}{3\eta_l} + \frac{S^2 \Delta P}{6\eta_l}\right)^{1/2} \quad (6.34)$$

In our model S is the pore size of the silica, ΔP is the pressure in the mixing chamber. When the filler is completely infiltrated by rubber, then $l^* = D_0$ and $L = L_P$. By setting $l^* = D_0$ into Equation (6.32) the plateau value L_P is determined by the factor h using Equation (6.35):

$$L_P = \frac{\varepsilon h}{\varepsilon h + \frac{\rho_F}{\rho_R}(1-\varepsilon)} \tag{6.35}$$

L_P can be experimentally determined using Equation (6.26) from extraction experiments of a sample with a sufficiently long mixing time.[7,27] Setting $\rho_R = 0.94$ g cm^{-3} and $\rho_F = 2.0$ g cm^{-3}, $\varepsilon = 0.93$ for silica as well as $\rho_F = 1.8$ g cm^{-3}, $\varepsilon = 0.63$ for CB into Equation (6.35), the correlation between L_P and h can be graphically described in Figure 6.2(d).

As an example, the rubber part in the gel of the single SBR mixture, L, is represented in Figure 6.2(e) as a function of mixing time of a batch mixing. It is obvious that in the first mixing period, L increases strongly with mixing time, since with the progressing mixing time permanently new filler surface is generated through the dispersion processes. In the second mixing period, in this case after reaching a mixing time of 12 minutes, the SBR part in the gel reaches a plateau value L_P that indicates the end of the wetting process. At this time the entire filler surface is wetted by SBR and no change of L is observed any more.

When filler is mixed into a ternary blend, for example a 33/33/34 SBR/NBR/NR blend, three blend components compete with each other to occupy the active centres on the filler surface to generate the rubber layer $L^{B(SBR/NBR/NR)}$ of the blend, which consists of the rubber layer $L^{B(SBR)}$, $L^{B(NBR)}$ and $L^{B(NR)}$ of the blend components SBR, NBR and NR, respectively, according to Equation (6.36).

$$L^{B(SBR/NBR/NR)}(t) = L^{B(SBR)}(t) + L^{B(NBR)}(t) + L^{B(NR)}(t) \tag{6.36}$$

The phase selective filler localization in a ternary SBR/NBR/NR blend can be determined according to Equations (6.37) to (6.39):

$$\frac{S^{B(SBR)}(t)}{S^{B(NBR)}(t)} = \frac{L_P^{NBR}}{L_P^{SBR}} \cdot \frac{L^{B(SBR)}(t)}{L^{B(NBR)}(t)} \tag{6.37}$$

$$\frac{S^{B(NBR)}(t)}{S^{B(NR)}(t)} = \frac{L_P^{NR}}{L_P^{NBR}} \cdot \frac{L^{B(NBR)}(t)}{L^{B(NR)}(t)} \tag{6.38}$$

$$S^B = S^{B(SBR)}(t) + S^{B(NBR)}(t) + S^{B(NR)}(t) \tag{6.39}$$

S^B is the total filler amount added to the ternary blend. $S^{B(SBR)}$, $S^{B(NBR)}$ and $S^{B(NR)}$ are the filler amounts localized in the SBR, NBR and NR blend phase at the mixing time t, respectively. The plateau values L_P^{SBR}, L_P^{NBR} and L_P^{NR} are the rubber layer at the end of the wetting process of filler by SBR, NBR and NR in single compounds.

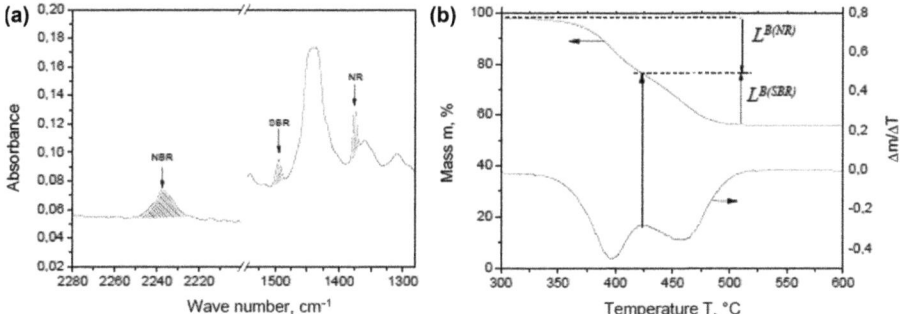

Figure 6.3 a) FTIR spectrum of the rubber–filler gel of a ternary 33/33/34 SBR/NBR/NR blend filled with 50 phr silica. (b) Principle of determination of the fraction of rubber components, $L^{B(NR)}$ and $L^{B(SBR)}$ from the TGA curve of a rubber–filler gel of a binary 50/50 SBR/NR blend.[30]

Fourier transform infrared (FTIR) analysis of the rubber–filler gel of blends can be used for the experimental determination of $L^{B(SBR)}$, $L^{B(NBR)}$ and $L^{B(NR)}$ from $L^{B(SBR/NBR/NR)}$.[30] A FTIR spectrum of a 33/33/34 SBR/NBR/NR blend is presented in Figure 6.3(a) with three characteristic peaks of the blend components. A^{SBR} is the area of a characteristic peak of the SBR phase at 1494 cm^{-1} (stretching vibration of the phenyl ring) and it was calculated from 1490 cm^{-1} to 1498 cm^{-1}. Correspondingly, A^{NBR} is the area of a characteristic peak of the NBR phase at 2238 cm^{-1} (stretching vibration of –C≡N) and calculated from 2223 cm^{-1} to 2254 cm^{-1}. A^{NR} is the area of a characteristic peak of the NR phase at 1376 cm^{-1} (bending vibration of –CH) and calculated from 1370 cm^{-1} to 1383 cm^{-1}.

The ratio $L^{B(SBR)}/L^{B(NBR)}$ and $L^{B(NBR)}/L^{B(NR)}$ can be determined from the ratio A^{SBR}/A^{NBR} and A^{NBR}/A^{NR} by means of a calibration curve. The procedure for creating the calibration curve was described in detail in our previous work.[30] By setting $L^{B(SBR)} = 0$ in Equations (6.37) and (6.39), the silica localization in an NBR/NR binary blend can be determined.

Due to the total adsorption of FTIR beams by CB in gel, FTIR cannot be used for analysis of the rubber–filler gel of CB-filled rubber blends. We proposed to use TGA to determine $L^{B(SBR)}$ and $L^{B(NR)}$ from $L^{B(SBR/NR)}$ of a binary CB-filled SBR/NR blend.[7,27] The contribution of $L^{B(SBR)}$ and $L^{B(NR)}$ was determined by means of differential TGA according to Figure 6.3(b).

6.4 Equipment and Experimental Methods

6.4.1 Preparation of Blends

To prepare filled rubber compounds and blends, rubbers were mixed with filler using an internal mixer. During the mixing time samples were taken out for further investigation. Used materials, formulations as well as mixing conditions will be given in details in each part later.

6.4.2 Characterization

6.4.2.1 Optical Microscopy

Optical microscopy has been used to characterize the CB dispersion. This method was described by Leigh-Dugmore[50] and modified by us. We produced gloss cuts by cutting stretched samples with dimensions of $1 \times 5 \times 20$ mm using a razor blade at room temperature and examined the cut surfaces by optical microscope with incident light. If the surface of the cut contains CB agglomerates or aggregates, the light scatters at this place and its area appears dark. With an image analysis program one can calculate the area of visible CB regions according to ASTM 2663. The degree of the macrodispersion was assessed as the amount of the non-dispersed agglomerates with an average diameter larger than 5 µm. From every sample six pictures were taken and from every picture six image analyses were carried out.

6.4.2.2 Transmission Electron Microscopy (TEM)

Ultra-thin sections approximately 35 nm thick were cut from compression-moulded plates with a diamond knife (35° cut angle, DIATOME, Switzerland) at $-140\,°C$ on a cryo-microtome and used for transmission electron microscopy (TEM) analysis. The slices were collected on a copper grid with a carbon-hole-foil. The specimens were investigated on a Zeiss Libra® 200MC (Zeiss, field emission cathode, point resolution 0.2 nm) with an accelerating voltage of 200 kV.

6.4.2.3 Scanning Electron Microscopy (SEM)

Scanning electron microscopy (SEM) investigations were performed on an Ultra Plus microscope (Zeiss, field emission cathode) operated at 2 kV accelerating voltage. To get a plain surface the samples were prepared by cutting a cryo-section using an ultra-microtome (Leica Ultracut UC7) with a diamond knife at $-140\,°C$. The cryo-sections were examined without any additional coating to avoid masks over the nanotubes and to allow a charge contrast imaging between the conductive CNT network and the rubber matrix.

6.4.2.4 Energy Dispersive X-Ray Analysis (EDX)

Energy dispersive X-ray analysis (EDX) (Voyager 1100, Fa. Noran Instruments) was used for a qualitative characterization of the composition of the rubber layer L of the composites.

6.4.2.5 Atomic Force Microscopy (AFM)

Morphological investigations were carried out using an atomic force microscope (Q-Scope 250, Quesant), operated in intermittent mode with a scan-head of 40 μm. Samples were produced by cutting in a cryo-chamber CN 30 of a rotary microtome HM 360 (Microm) with a diamond knife at –110 °C.

6.4.2.6 Equipment for Measuring Online Conductance

A conductivity sensor system was installed in the chamber of the internal mixer to measure the electrical signal of the conductive mixtures between the sensor and the chamber wall. The construction and position of the conductance sensors has been described in our previous works.[51–53]

6.4.2.7 Measuring Offline Conductivity

Measurement of electrical conductivity of uncured and cured samples was carried out at room temperature by means of a multimeter 2750 (Keithley). The shape of the conductive test specimens was a rectangular strip, whose ends were coated by silver paste in order to receive a good contact with the electrodes.

6.4.2.8 Extraction Experiment

To investigate the rubber–filler gel, 0.1 g of each raw mixture was stored for 7 days in 100 ml solvent. After 4 days the solvent was completely renewed. The solution was cast from the flask and the rubber–filler gel was taken out and dried to a constant mass.

6.4.2.9 Fourier Transform Infrared (FTIR) Analysis

Analysis of the rubber–filler gel was carried out using a Fourier transform infrared (FTIR) spectrometer S2000 (Perkin Elmer) equipped with a diamond single Golden Gate ATR cell (Specac).

6.4.2.10 Thermogravimetric Analysis (TGA)

Thermogravimetric analysis (TGA) of the gel was carried out using a thermobalance (Mettler Toledo) in the temperature range between 30 °C and 800 °C with a heating rate of 20 K/min.

6.4.2.11 Determination of Surface Energies

Wetting experiments (modified Wilhelmy method) were performed, using the dynamic contact angle meter and tensiometer DCAT 21 (DataPhysics Instruments GmbH, Filderstadt, Germany). For the Wilhelmy measurements, the filler particles were placed on a shallow plate. In the filler powder a 2×1 cm^2 piece of double-sided adhesive tape (TESA 55733, Beiersdorf, Hamburg, Germany), was immersed and gently moved, until the tape was uniformly coated by filler particles. The pellets of the granulated filler were pulverized finely in a mortar, before they were attached to the adhesive tape. Surplus particles that did not stick to the adhesive tape were blown away by a stream of nitrogen. The filler particle covered tape was used for Wilhelmy contact angle measurements without further modification.

Sessile drop contact angle measurements on a sheet of uncured rubber were conducted with the automatic contact angle meter OCA 40 Micro (DataPhysics Instruments GmbH, Filderstadt, Germany). The surface energies were calculated from the results of these wetting experiments. For this purpose a set of test liquids with different surface tensions (and polarities) was used: water (Millipore Milli-Q-Quality), formamide (Merck, Darmstadt, Germany), ethylene glycol (Fisher Scientific, Loughborough, UK), dodecane (Merck Schuchardt, Hohenbrunn, Germany), n-hexadecane (Merck, Darmstadt, Germany) and ethanol (Uvasol, Merck, Darmstadt, Germany). Surface energy calculations were performed by fitting the Fowkes equation.[54]

6.4.2.12 Small-Angle X-Ray Scattering (SAXS)

Small-angle X-ray scattering (SAXS) measurements were performed at room temperature using a rotating anode X-ray source RU-3HR (Rigaku) equipped with an X-ray optics device (Confocal Max-Flux, $\lambda = 0.154$ nm, Osmic Inc.) and a Bruker Hi-Star 2-D detector to detect the state of exfoliation. The generator voltage was 40 kV and generator current was 60 mA. The scattering vector q is defined by $q = 4\pi/\lambda \sin\theta$. All samples had a uniform thickness of 1.0 mm, *i.e.* the obtained peak area corresponds to the amount of ordered structures.

6.5 Results and Discussion

6.5.1 Silica Localization in Rubber Blends

6.5.1.1 Silica-Filled Blend Preparation

Solution styrene butadiene rubber (S-SBR) (Sprintan SLR-4601, Styron Deutschland GmbH) and acrylonitrile butadiene rubber (NBR) (Perbunan 3445F, Lanxess GmbH) with a nitrile content of 34% as well as NR (SMR 10, Standard Malaysian Rubber) were used as rubbers. The silica used was Ultrasil 7000 GR (Evonik Industries) with a specific surface area CTAB of 160 m^2/g and BET of 170 m^2/g. Stearic acid, zinc oxide (ZnO), N-cyclohexyl-benzothiazole-2-sulfenamide (CBS) and sulfur were used as processing and curing additives, respectively. The experimentally determined values of surface tension and Mooney viscosity of the materials used are given in Table 6.1.

Mixing experiments were performed by a Plasticorder PL 2000 internal mixer (Brabender). A rotor speed of 50 rpm and starting temperature of 50 °C were used for all tests. A fill factor of 0.7 was chosen for low-filled blends, while it was reduced to 0.6 for highly filled blends in order to keep the mass temperature below 100 °C. Two series of silica-filled blends, with and without additives, were prepared according to Tables 6.2 and 6.3 by variation of mixing time and silica loading.

Table 6.1 Surface tension and Mooney viscosity of materials used.

Materials	Surface tension[a] (mN/m)	Mooney viscosity MU ((ML 1+4) 100 °C)
SBR	24	45
NBR	27.2	45
NR	22	49
Ultrasil 7000 GR	73 ± 7^{52}	

[a]Values were experimentally obtained from the sessile drop contact angle measurements.

Table 6.2 Formulation and mixing conditions for preparation of filled NBR/NR and SBR/NR blends without any curing additives.

Mixing time (min)	Ingredient	NBR/NR (phr)	SBR/NR (phr)
0	NBR	50	
	SBR		50
	NR	50	50
5	7000 GR	10	50

Stopped and dumped at 7, 9, 12, 15 min.

Table 6.3 Formulation and mixing conditions for preparation of NBR/NR blends filled with different silica loading (with a constant loading of additives).

Mixing time (min)	Ingredient	Loading (phr)	(Stearic acid + ZnO + CBS)/silica[a] r
0	NBR	50	
	NR	50	
	Stearic acid	2	
	ZnO	3	
	CBS	2	
	Sulfur	1.5	
5	Silica	5 or	7/5
		7.5 or	7/7.5
		10 or	7/10
		20 or	7/20
		50 (highly filled)	7/50

Stopped and dumped at 15 min.
[a]r is the mass ratio of additives consisting of stearic acid, ZnO and CBS to silica.

6.5.1.2 Prediction of Silica Localization in Blends at an Equilibrium State

Using Equations (6.13) and (6.14) the filler fraction in the NR phase φ_F^{NR} in 50/50 NBR/NR blends can be calculated in dependence on the surface tension of the filler and presented in Figure 6.4.

The dependence of filler fraction in the NR phase φ_F^{NR} on the filler surface tension γ_F shows different behaviour in three ranges I, II and III. In range I, at a low filler surface tension far away from γ_{NR} an even localization of filler is

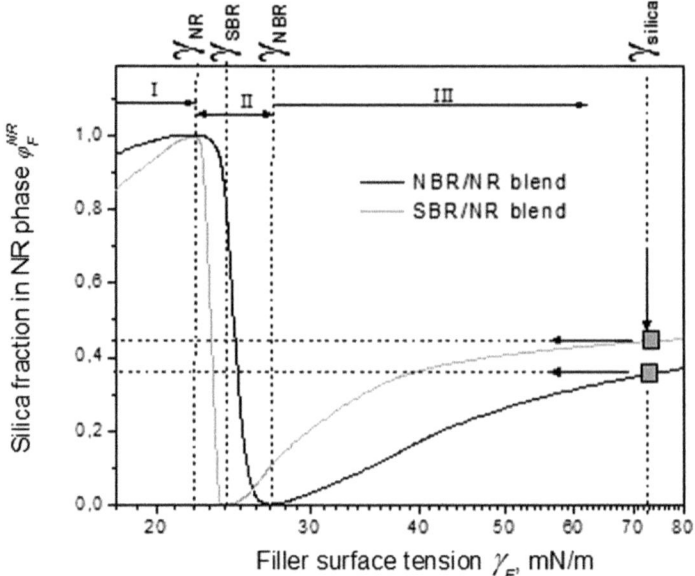

Figure 6.4 Master curve created by the Z-model presenting the filler fraction in the NR phase φ_F^{NR} in relation to the surface tension of filler for 50/50 NBR/NR and 50/50 SBR/NR blends.

nearly received because filler shows similar bad affinity to both blend phases. With increasing γ_F the filler fraction in the NR phase φ_F^{NR} increases and approaches a value of 1.0 when γ_F approaches γ_{NR}. In range II, passing γ_{NR} the filler fraction φ_F^{NR} decreases because the affinity of the filler to NR becomes worse and to NBR better. When γ_F approaches γ_{NBR}, a complete localization of filler in the NBR phase is nearly obtained. In range III, with increasing γ_F the filler loading φ_F^{NR} increases and approaches a value of 0.5 (even localization) at a high filler surface tension far away from that of both blend phases. From the master curve presented above it is easy to recognize some main features. First, an almost even localization of filler can be achieved when the filler surface tension varies greatly from those of both blend phases (similar poor affinity of filler to both rubber phases), or lies in between them (similar good affinity of filler to both rubber phases). Second, a very strong dependence of the filler localization on the filler surface tension is obtained in range II. A small change in filler surface tension in this range can lead to an extremely large change in filler localization. Third, according to the proposed model a localization of the filler at the interphase is not a thermodynamic equilibrium state. It is rather a result of an interplay between thermodynamic driving forces and rheological effects.[37,55–57] Elias et al.[37] stated that the presence of a large proportion of silica near the EVA/PP interface is due to the short mixing time. In other words, the nanoparticles do not have enough time to reach their preferred phase by Brownian motion. Furthermore, the formation of a packed layer at the interface by capillary interaction cannot be a relevant process for such high viscosity emulsions.

A weight fraction of silica in the NR phase $\varphi_F^{NR} = 0.36$ and 0.43 was predicted for 50/50 NBR/NR and 50/50 SBR/NR blends, respectively, by fitting $\gamma_F = 73$ mN/m[58] to the master curves as shown by two arrows in Figure 6.4.

6.5.1.3 Experimental Determination of Kinetics of Silica Localization in Rubber Blends

The silica fraction in the NBR and NR phase $S^{B(NBR)}$ and $S^{B(NR)}$ of NBR/NR blends prepared without curing additives according to Table 6.2 was experimentally determined by means of the wetting concept and is presented in Figure 6.5(a) as a function of mixing time. In the first mixing period (up to 8 minutes mixing time), the silica loading in both blend phases $S^{B(NBR)}$ and $S^{B(NR)}$ increases with mixing time. The first mixing period is attributed to the wetting process, in which the filler localization is mainly determined by the selective wetting behaviour of the filler by the blend components. The wetting speed $b^{NBR} = 0.5$ min$^{-1/2}$ and $b^{NR} = 0.45$ min$^{-1/2}$ determined in our previous work[39] indicate that NBR can infiltrate and wet silica only slightly faster than NR. However, the affinity of silica to NBR is better than to NR. Ziegler et al.[6] detected hydrogen bonding between the silanol groups of silica and C≡N groups of NBR by means of infrared spectroscopy (FTIR), while Kralevich et al.[59] stated that van der Waals forces are responsible for interaction between silica and non-polar NR. Ono et al.[60,61] applied high-resolution solid-state NMR for NR/silica composites and found no evidence of direct coupling between silanol groups and NR molecules at low processing temperatures. That is why in the first mixing period, higher silica loading was distributed into the NBR phase. After the wetting process is finished at 8 minutes, the filler loading in each blend phase remains unchanged with mixing time. The plateau value of 3.7 phr experimentally determined for $S^{B(NR)}$ corresponds very well with the predicted value of 0.36 of φ_F^{NR} determined from Figure 6.4.

The kinetics of silica localization in SBR/NR blends is presented in Figure 6.5(b). The silica loading in both phases $S^{B(SBR)}$ and $S^{B(NR)}$ increases in

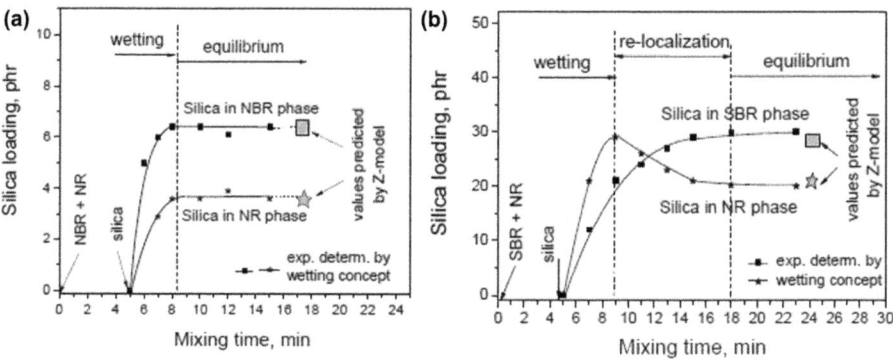

Figure 6.5 Kinetics of filler localization in (a) a 50/50 NBR/NR blend filled with 10 phr silica and (b) a 50/50 SBR/NR blend filled with 50 phr silica in the absence of curing additives and silane.

the first mixing period up to 9 minutes. In this period more silica was found in the NR phase. That is related to the fact that the wetting speed $b^{SBR} = 0.23$ min$^{-1/2}$ is much lower than that of NR.[28] $S^{B(NR)}$ reaches a maximum value at 9 minutes of mixing. Then, in the mixing period between 9 and 18 minutes $S^{B(NR)}$ decreases strongly, while $S^{B(SBR)}$ steadily increases. After 18 minutes both $S^{B(NR)}$ and $S^{B(SBR)}$ remain nearly constant. The strong decrease of $S^{B(NR)}$ between 9 and 18 minutes is related to the fact that the free SBR molecules replace the NR molecules bonded to the silica surface during the mixing process, because the weak-polar SBR has better affinity to the silica surface than the non-polar NR. Wang and Wolff[62] used inverse gas chromatography to investigate the adsorption of model substances to the silica surface. They found that the interaction of model substances to silica changes according to the order: NBR > SBR > NR > BR > EPR > IIR. Replacement of the bonded NR by SBR molecules is particularly favourable if NR exhibits more loosely bonded molecules. It is widely accepted that the bound rubber consists of a loosely bonded component and a tightly bonded one.[63] In contrast to the tightly bonded component, the loosely bonded component is much mobile and its binding energy to the silica surface is low. Serizawa et al.[64] investigated the formation of the loosely and tightly bonded components during the mixing time using NMR and found that at the first mixing stage, up to the black incorporation time (BIT), the amount of the loosely bonded component increases rapidly compared to the tightly bonded one. Above the BIT the tightly bonded component increases dominantly. Choi[65] quantified the amount of each component using extraction at 30°C and 90°C. At an extraction temperature of 30 °C both components remain on the silica surface. At an extraction temperature of 90 °C, the loosely bonded component is dissolved, and only the tightly bonded component remains in the bound rubber. Concerning the filler transfer taking place during preparation of polymer blends, Zaikin et al.[66,67] stated that the adsorption–desorption process or the replacement of the macromolecules of one polymer by those of another polymer is an important factor that determines the process of filler transfer within blend phases. Based on the experimental results with CB-filled polystyrene (PS)/polyethylene (PE) blends using different mixing technologies Gubbels et al.[68] concluded that the transfer of the filler from one phase (the less preferred one) to the second one is a kinetically controlled process under thermodynamic driving forces. The kinetics of this transfer process depends on the shear forces involved and the rheology of each polymer phase under processing conditions. Fenouillot et al.[38] stated in their review that movement of solid particles into the polymer–polymer interphase and their transfer process from one phase to the other is realized by three mechanisms: (i) the Brownian motion (self-diffusion) of the particles, which plays a significant role in blends of liquid emulsions; (ii) numerous collisions between solid particles and polymer droplets induced by shear during mixing; and (iii) trapping of filler particles in the inter-droplet zone during a collision between two dispersed droplets, where their coalescence may take place and lead to embedding the particles in a larger droplet. In the present work, we suppose that the loosely bonded component of $L^{B(NR)}$ is formed in the mixing period up to 9 minutes through a fast wetting process. Under thermodynamic driving forces the loosely bonded component of $L^{B(NR)}$ is easily replaced by the SBR molecules in the period between 9 and 18 minutes. In the period after 18 minutes no change in $S^{B(SBR)}$ and $S^{B(NR)}$ is observed. These plateau values of $S^{B(SBR)} = 30$ phr and $S^{B(NR)} = 20$ phr from the total filler loading of 50 phr correspond very well to the silica fractions $\varphi_F^{SBR} = 0.57$ and $\varphi_F^{NR} = 0.43$ predicted from Figure 6.4. An average transfer rate of 0.8 phr/min of silica from the NR to the SBR phase within the mixing period between 9 and 18 minutes was roughly determined from the curve presented in Figure 6.5(b).

6.5.1.4 Effect of Processing and Curing Additives on Silica Localization in Rubber Blends

Surface energies of silica have a low dispersive component and a high specific component when compared to CB with equivalent surface area and structure.[55] The higher specific component results in strong filler–filler interaction. The surface characteristics of silica can be changed by surface modification, for example by physical and chemical modification of silica. The specific component of the surface free energy (γ_s^{sp}) is significantly reduced, leading to improved interaction between silica and rubber for improved compatibility.[55] A reduction in filler–filler interaction results in better dispersion and reduced viscosity of the compound.

The use of stearic acid as a modifier for silica and other fillers like $CaCO_3$ and $Mg(OH)_2$ has been reported.[69–73] The authors found that the presence of adsorbed stearic acid on the filler surface reduces the hydrophilicity of the silica surface and enhances the compatibility between filler and matrix, which may lead to an improvement in filler dispersion and the related mechanical performance of composites. Kosmalska et al.[74] also investigated the adsorption of DPG, ZnO and sulfur on the silica surface and reported that the bonding of DPG/ZnO and ZnO to silica causes a reduction in the surface energy of silica from 66 mN/m to 28.75 mN/m and 35.49 mN/m, respectively. A similar effect of ZnO on the surface tension of silica was also found by Laning et al.[75] and Reuvekamp et al.[76] The adsorption of that additive and its impact on the scorch time and reduction of the crosslink density in silica-filled rubber compounds have been frequently characterized.[77]

In the present work the adsorption of stearic acid, ZnO, CBS and sulfur on the silica surface during the mixing process was characterized by means of measuring the online electrical conductance. As an example, the online conductance measured directly inside the mixing chamber during compounding of a NBR/NR blend is shown in Figure 6.6(a). The low conductance values of the NBR/NR blend are attributed to the presence of catalyst residues used in the polymerization processes. Upon addition of stearic acid, ZnO and CBS the online conductance increases significantly. This is related to the fact that in a polar medium like NBR additives are dissociated into ions, which give the compound higher ionic conductivity values. However, when additives are added to the silica/NBR/NR blend compound, no change in the online conductance is observed (Figure 6.6(b)). This phenomenon is obvious evidence for

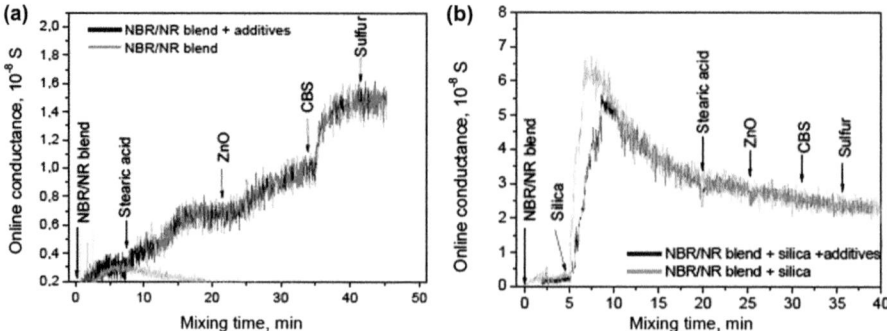

Figure 6.6 Online electrical conductance measured directly inside the mixing chamber during compounding of a (a) NBR/NR blend and (b) silica/NBR/NR blend without and with addition of additives.

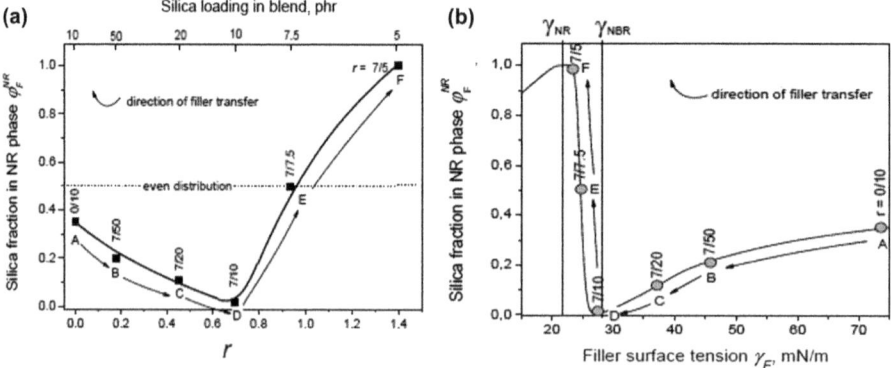

Figure 6.7 Silica fraction in the NR phase φ_F^{NR} of a 50/50 NBR/NR blend in relation to (a) the additives/silica ratio r and (b) the corresponding change of φ_F^{NR} along the master curve.[39]

the adsorption of additives to the silica surface during the mixing process. The decay of the online conductance is related to the evaporation of moisture from the compound. In all investigated cases sulfur did not affect the value of conductance. Thus, no information about its adsorption could be gained.

In order to understand the effect of processing and curing additives on the localization behaviour of silica, NBR/NR blends filled with different silica loadings were prepared by keeping constant the loading of additives according to Table 6.3. In Figure 6.7 the silica fraction in the NR phase φ_F^{NR} is presented in dependence on the curing additives/silica ratio r. Without curing additives $\varphi_F^{NR} = 0.36$ was determined as discussed above. With increasing ratio r, i.e. with decreasing silica loading, the silica fraction φ_F^{NR} decreases and reaches a value of zero at $r = 7/10$. Passing this value the silica fraction φ_F^{NR} strongly increases and reaches a value of 0.5 when $r = 7/7.5$ and 1.0 when $r = 7/5$. The inversion of filler localization at $r = 7/10$ is very interesting and will be explained later by means of the Z-model.

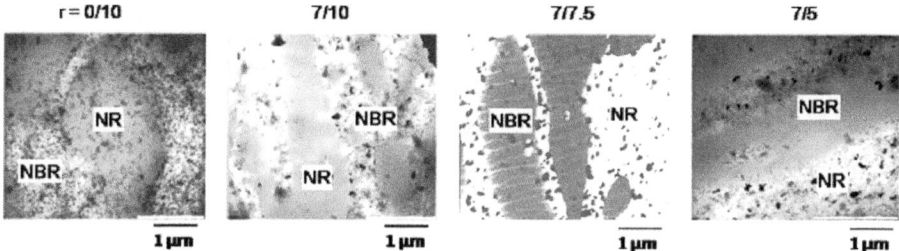

Figure 6.8 Morphology investigation of silica filled 50/50 NBR/NR blends with increasing ratio r.

The morphological investigation of NBR/NR filled with different silica loadings was carried out by AFM and TEM. The images of blends with different ratio r are presented in Figure 6.8 and support very well the silica localization shown in Figure 6.7(a).

The change in filler localization behaviour in rubber blends by variation of filler loading has also been frequently reported in the literature.[78–81] Phewphong et al.[80] observed a significant influence of silica loading on the filler localization in chlorinated polyethylene (CPE)/NR blends. He stated that the counter-balancing effects of relatively low viscosity of the NR phase and strong silica–CPE interaction, which change with increasing silica loading, are responsible for the change in silica localization. Maiti et al.[79] found with increasing silica content from 10 to 40 phr in NR/epoxidized natural rubber (ENR) blends, a decreasing weight fraction of silica in the ENR phase was observed. In that case, the authors explained that at the lower levels of filler loading, ENR accumulated more silica than NR. As the filler loading increased, ENR gradually became saturated and silica slowly migrated to the NR phase. In our opinion, this behaviour is related to the fact that the adsorption of additives like stearic acid, ZnO and CBS on the surface of silica will make it more hydrophobic. Regarding the effect on the filler localization with increasing silica loading by keeping constant the additive concentration, the surface of silica is differently modified, which leads to a change in the affinity of silica to the blend phases and consequently to a change in silica localization behaviour. The correlation between the change in silica surface tension and the corresponding silica loading in the NR phase φ_F^{NR} can be described well by means of the master curve shown in Figure 6.7(b). It is easy to see that when silica becomes more hydrophobic, i.e. γ_F approaches γ_{NBR}, silica will migrate from the non-polar NR to the polar NBR phase (A to D). When γ_F falls below γ_{NBR} and approaches γ_{NR}, silica becomes much more hydrophobic and migrates from the NBR to the NR phase (D to F).

6.5.1.5 Quantification of the Effect of Silane on Silica Localization

To quantify the effect of silane on the silica localization in rubber blends we prepared filled blends according to Table 6.4. Silane bistriethoxysilylpropyltetrasulfane Si69 (Evonik Industries AG) was used as coupling agent. It was

Table 6.4 Formulation and mixing conditions for preparation of silica-filled NBR/NR blends with different loadings of Si69.

Mixing time (min)	Ingredient	Loading (phr)	Initial temperature T_A (°C)
0	NBR	50	50 °C or 145 °C
	NR	50	
	Si69	1.5/3.5/6.0	
5	Silica	50	

Stopped and dumped at 15 min.

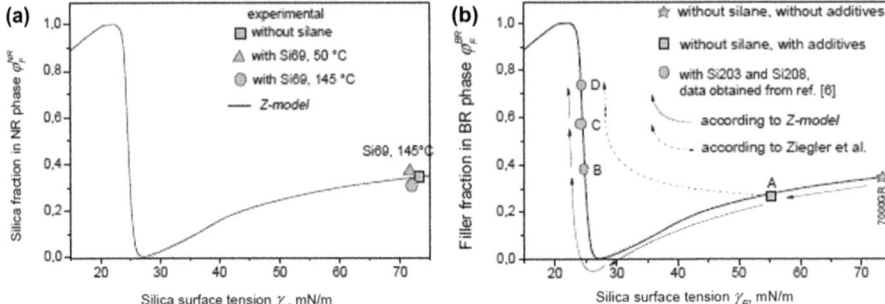

Figure 6.9 Effect of (a) Si69 and (b) Si203 on the silica localization in NBR/NR blends and NBR/BR blends, respectively.[39]

found that the use of Si69 did not affect the silica localization in NBR/NR blends regardless of mixing temperature, as seen in Figure 6.9(a). This may be related to the chemical structure of Si69, which has ethoxy groups available at both ends. When one end is chemically bonded to the silica surface, the OH groups of the other end may still remain free and keep silica hydrophilic.

In order to clarify the effect of the chemical structure of silane on silica localization we used the data presented in the work of Ziegler and Schuster.[6] Silica 7000 GR was pretreated by silane before mixing into 50/50 NBR/BR blends. Triethoxypropylsilane (Si_2O_3) and triethoxyoctylsilane (Si_2O_8) were used as coupling agent. After preparation of blends the silica localization was determined by means of a method based on data recorded from DMA. Both types of silane exhibit ethoxy groups only at one end of the molecule, which can be bonded to the silica surface. The other end of silane can shield the silica surface and make it more hydrophobic, which may lead to a significant change in silica localization. The sample name, the filler treated and the silica fraction in the BR phase φ_F^{BR} are given in Table 6.5.

According to the data presented in Table 6.5 Ziegler and Schuster[6] concluded that with increasing silane concentration, *i.e.* with decreasing surface tension of the silica surface γ_F, silica seems to migrate directly from the NBR to the BR phase (dotted arrow in Figure 6.9(b)). However, by means of the Z-model we recognize that the filler localization process is more complicated. The master curve of filler localization of a NBR/BR blend can be created using

Table 6.5 Silica treated with silane and its localization in NBR/BR blends.

Blend	Treated filler[a]	Silica fraction in BR phase[a] φ_F^{BR}	Silica surface tension[b] γ_F, mN/m
A	7000 GR + additives	0.27	57
B	7000 GR + additives + (2/3) Si$_2$O$_3$	0.38	24.7
C	7000 GR + additives + Si$_2$O$_3$	0.57	24.4
D	7000 GR + additives + Si$_2$O$_8$	0.74	24.1

[a]Data obtained from Ziegler and Schuster.[6]
[b]Data obtained from Figure 6.8(b).

$\gamma_{BR} = 22.2$ mN/m^{82} and is presented in Figure 6.9(b). The effect of silane treatment on silica localization can be quantified by fitting the filler fraction in the BR phase φ_F^{BR} from Table 6.5 to the master curve. Accordingly, when γ_F reduces and approaches γ_{NBR}, the loading of silica in BR φ_F^{BR} first reduces to zero and at the same time φ_F^{NBR} increases to 1.0. Silica starts to migrate back to the BR phase only when γ_F falls below γ_{NBR} (solid arrow).

6.5.1.6 Quantification of the Filler Surface Tension Change During Mixing

Silica surface tension, which was changed by adsorption of additives, can be determined by fitting the values of φ_F^{NR} observed from Figure 6.7(a) to the master curve presented in Figure 6.7(b). In Figure 6.10(a) the observed surface tension data is presented in dependence on the ratio r. With increasing ratio r the silica surface tension decays first strongly and then slowly approaches a constant value when the silica surface is saturated. The polar part of additives interacts with OH groups of silica and the non-polar part will shield the silica surface and determine the hydrophobicity of silica. Thus, the surface tension of the saturated silica surface is γ_{sat}, which is considered as the surface tension of the non-polar part of additives. The decrease of γ_F with r follows an exponential decay function and can be empirically described by Equation (6.40):

$$\gamma_F(r) = \gamma_{sat} + (\gamma_F - \gamma_{sat}) \cdot e^{-\frac{r}{a}} \quad (6.40)$$

where a is a factor describing the effectiveness of additives with respect to the reduction in the filler surface tension. Fitting Equation (6.40) to the data presented in Figure 6.10(a) by setting the surface tension value of silica $\gamma_F = 73$ mN/m, we get Equation (6.41):

$$\gamma_F(r) = 22.4 + 50.6 \cdot e^{-\frac{r}{0.28}} \quad (6.41)$$

If the silica surface is fully covered by additives a value of 22.4 mN/m is observed for γ_{sat}.

The effect of each additive with change in concentration on the silica localization was also investigated. By fitting the value of φ_F^{NR} into the master curve shown in Figure 6.7(b) the values of silica surface tension changed during the mixing process can be observed as a function of additive/silica ratio, as

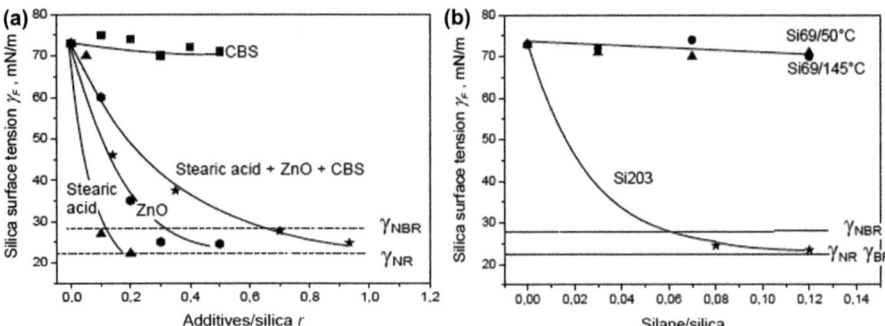

Figure 6.10 Silica surface tension changed during mixing by addition of (a) additives and (b) silane into 50/50 NBR/NR blends (surface tension values were obtained by fitting the data of φ_F^{NR} into Figures 6.6(b) and 6.8).[39]

presented in Figure 6.10(a). Stearic acid and ZnO show a strong effect compared to the mixture containing stearic acid, ZnO and CBS, while CBS does not show any effect. The moderate effect of the mixture of additives is related to the fact that stearic acid can react with ZnO and CBS rather than with the silica surface.

The silica surface tension by treatment with Si69 and Si203 is presented in Figure 6.10(b) in dependence on the silane/silica ratio. A clear difference between both silane types with regard to the impact on silica surface tension is observed. In a study by Castellano et al.,[83] the effect of the chemical structure of a triethoxysilane (TES), octadecyltriethoxysilane (ODTES) and bistriethoxysilylpropyltetrasulfane (TESPT) was investigated by inverse gas chromatography (IGC) at infinite dilution. Thermodynamic results indicate a higher polarity of the silica surface modified with TES as compared to that of the unmodified silica due to new OH groups formed by the hydrolysis of ethoxy groups on the silane. A grafting degree with 4.6 wt% of ODTES is enough to obtain a silica surface tension of 38.6 mN/m, similar to those of hexadecanol-modified silica (34 mN/m)[84] and of polyethylene (24–42 mN/m).[85,86] The corresponding value for TESPT (Si69) modified silica is 58 mN/m, which is not much different from the value 73 mN/m of the untreated silica; this suggests that the long alkyl chains of ODTES may form a shielding layer, leading to a low-polarity surface.

6.5.1.7 Ternary Blends of SBR/NBR/NR and SBR/BR/NR

Particularly synergistic properties can be obtained using ternary polymer blends that cannot be achieved with binary blends. Ternary blends based on NR, BR and EPDM have been used for tyre sidewalls and show excellent ultimate properties, better ozone resistance and fatigue resistance under dynamic load.[87] Specific morphology development, crosslinking behaviour and their correlation to mechanical properties of different ternary rubber blends like butyl rubber (IIR)/NR/SBR and NR/BR/SBR blends have been well characterized.[88–91] Addition of filler into ternary blends of acrylic rubber (ACM), fluorocarbon rubber (FPM) and multifunctional polyacrylate can change the phase morphology that leads to the change of end-use properties of these blends like stress and strain at break, damping properties, aging and swelling behavior.[92] In CB-filled ternary blends of PS, poly(methyl methacrylate) (PMMA) and styrene methyl methacrylate copolymer (SMMA), a layer of SMMA forms between the PS and PMMA phases.[93] A pre-addition of CB into the SMMA component imparts the ternary blends a high conductivity. However, CB migrates to the PS phase during the mixing process, and so the blends lose their conductivity after long mixing times. Blend morphology and silica distribution in BR/brominated isobutylene-co-p-methylstyrene (BIMS) and BR/BIMS/NR blends have been investigated by a Raman micro-imaging technique.[94] Depending on its concentration NR forms either a thin layer surrounding the dispersed BR phase or separate domains within the BR phase. Silica localizes unevenly in blends and tends to migrate to the boundary of the BIMS domains. No silica was found in the BR phase. Meier et al.[95] calculated the silica localization in the BR/SBR mix-phase and the NR phase of a ternary BR/SBR/NR blend by means of DMA measurements. The selective localization of silica within the BR/SBR mix-phase was not able to be determined, because the glass transition temperatures of both phases overlap. In the present work, we use our methods for prediction and determination of the filler localization in two ternary blends from SBR/NBR/NR and SBR/BR/NR. The formulation and mixing conditions for their preparation are given in Table 6.6.

Based on the surface tension data of the investigated rubbers and that of silica in the presence of additives, i.e. $\gamma_F = 45$ mN/m,[39] a silica loading in each phase of the prepared blends was predicted by fitting γ_F to the master curves shown in Figure 6.11. The results are summarized in Table 6.7.

The phase-specific localization of silica in ternary SBR/NBR/NR blends was determined by the wetting concept and is presented in Figure 6.12(a). The silica loading localized in the SBR, NBR and NR phase, respectively, increases immediately after addition of silica into the rubber blend. Due to the high wetting

Table 6.6 Mixing conditions and formulation of SBR/NBR/NR and SBR/BR/NR blends.

Mixing time (min)	Material	SBR/NBR/NR	SBR/BR/NR	Mixing conditions
0	SBR	33	33	
	NBR	33		
	BR		33	
	NR	34	34	
1	Stearic acid		2	$T_A = 50\,°C$
	ZnO		3	
	CBS		2	
	Sulfur		1.5	
2	Si69		5	
2	Silica		50	

Stopped at different mixing times.

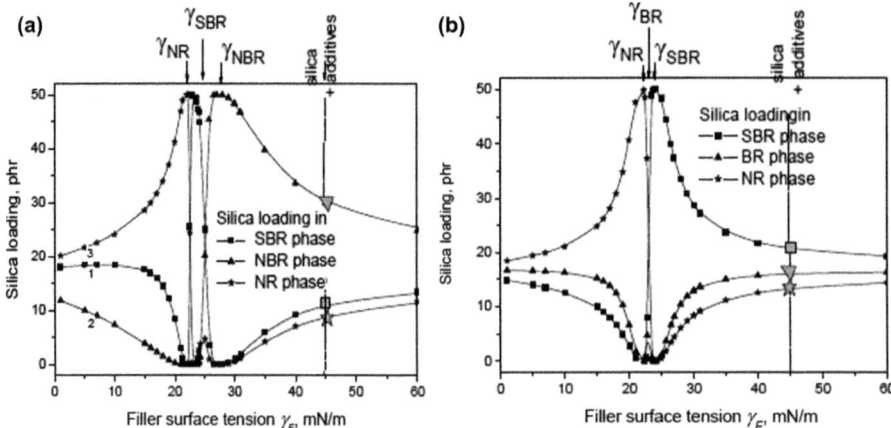

Figure 6.11 Master curves of filler localization of ternary (a) 33/33/34 SBR/NBR/NR and (b) 33/33/34 SBR/BR/NR blends created by the Z-model. (The value of silica surface tension in the presence of curing additives was taken from Figure 6.7(b)).[30,46]

Table 6.7 Filler loadings in different blend phases predicted by means of the Z-model.

Blend	Predicted filler loading in			
	SBR phase	NBR phase	BR phase	NR phase
33/33/34 SBR/NBR/NR blend	11	30		9
33/33/34 SBR/BR/NR blend	21		16	13

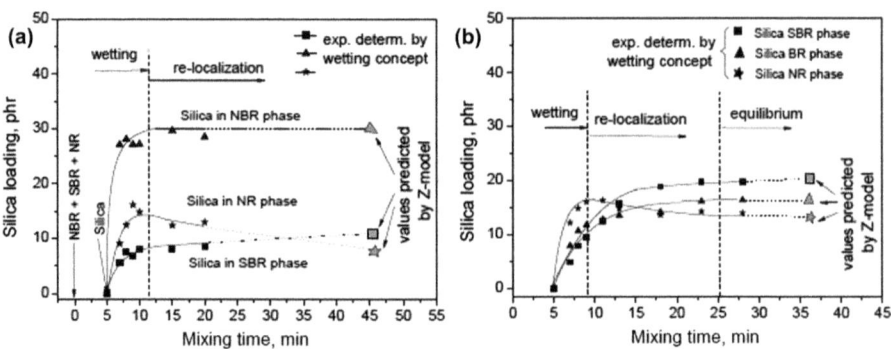

Figure 6.12 Kinetics of silica localization in 33/33/34 SBR/NBR/NR and 33/33/34 SBR/BR/NR blends filled with a silica loading of 50 phr in the presence of curing additives and silane.[30,46]

rate and better rubber–filler interaction, the NBR chains can wet more silica than NR and SBR in the first mixing stage (up to 10 minutes). After 10 minutes the loading of silica in the NBR phase reaches a plateau value of 30 phr, which corresponds very well to the predicted value. After that NBR seems not to take part in the filler distribution process any more. NR wets silica faster than SBR in the first mixing period. After 10 minutes, the replacement process of NR by SBR was also observed, which was observed above for binary SBR/NR blends. As a result of the replacement process, silica seems to migrate from the NR to the SBR phase. After 20 minutes the silica loadings determined for the NR and SBR phase are still far away from the predicted values. The lower filler transfer rate from the NR to the SBR phase in the ternary blend compared to that observed in the binary SBR/NR blend may be related to the mutual hindrance effect of the NBR phase. Linear extrapolation was used to detect the mixing time, which is needed for NR and SBR to reach the equilibrium state.

The kinetics of silica localization in SBR/BR/NR blends is presented in Figure 6.12(b). Due to the higher wetting rate, the NR chains can wet more silica than SBR and BR in the first mixing stage (up to 9 minutes). However, due to the very fast wetting process of silica a number of loosely bonded NR is formed, which is replaced by the SBR in the subsequent mixing state. The replacement process of NR by SBR on the silica surface took place also in this ternary blend. Thus, silica migrates from the NR phase to the SBR phase, while silica loading in the BR phase constantly increases. After 22 minutes of mixing the silica localization reaches its equilibrium state, which corresponds very well with the prediction.

6.5.2 Carbon Black Localization in Rubber Blends

6.5.2.1 Effect of Polarity and Viscosity of NR on CB Localization in Blends with SBR

The rubbers used were S-SBR (Sprintan SLR-4601, Styron Deutschland GmbH), NR (SMR 5, Nordmann & Rassmann) and epoxidized natural rubber (ENR 25 and ENR 50, Weber & Schaer GmbH & Co. KG, Germany) with 25 mol% and 50 mol% epoxy groups, respectively. Carbon black (Corax N220, Evonik Industries) was used as filler. To prepare filled rubber blends, rubbers were mixed with 50 phr filler in an internal mixer. The initial chamber wall temperature T_A was kept constant at 50 °C. The rotor speed was 50 rpm and the fill factor was 0.6.

The master curve of CB localization in 50/50 SBR/NR is shown in Figure 6.13(a). By fitting the CB surface tension γ_{CB} of 31 mN/m into the master curve a CB fraction of 0.8 is found in the SBR phase at the equilibrium state. The kinetics of CB localization in SBR/NR was experimentally determined by the wetting concept and is presented in Figure 6.13(b), which is similar to that of silica presented in Figure 6.5(b). Both wetting and relocalization processes are observed for CB filled SBR/NR blends. However, CB transfer takes place much slower than that of silica. A mixing time of more than 1000 minutes is needed for CB to reach the equilibrium state predicted by the Z-model. An average transfer speed of 0.02 phr/min can be roughly estimated for CB from Figure 6.13(b).

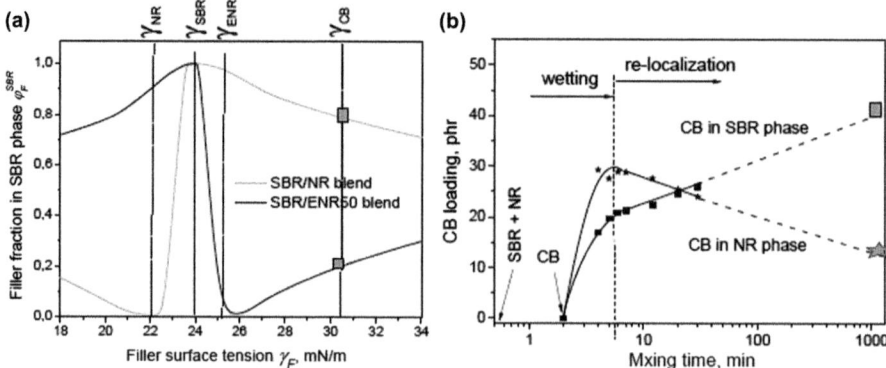

Figure 6.13 a) Master curves of filler localization of 50/50 SBR/NR and 50/50 SBR/ENR50 blends. (b) Kinetics of CB localization in SBR/NR blends.

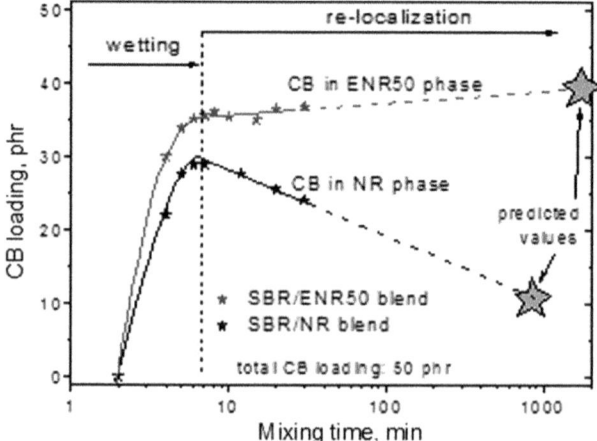

Figure 6.14 CB loading in the NR and ENR50 phase of 50/50 SBR/NR and 50/50 SBR/ENR50 blends, respectively, in relation to mixing time.

To characterize the effect of the functionalization of the NR phase on CB localization, a 50/50 SBR/ENR 50 blend was prepared and investigated, and compared with the behaviour of the SBR/NR blend. It was found that the wetting speed of the investigated rubbers increases according to the order $b^{SBR} < b^{NR} < b^{ENR25} < b^{ENR50}$.[27] With a higher value of b^{ENR50}, in the first mixing stage ENR 50 can wet more CB in SBR/ENR 50 compared to the NR phase in the SBR/NR blend according to Figure 6.14. Beyond the wetting process the CB loading in the ENR 50 phase still does not reach the predicted equilibrium, and so ENR 50 molecules continuously replace the SBR molecules on the CB surface. In contrast, NR is replaced by the SBR molecules as discussed above.

The effect of matrix viscosity on CB localization was characterized by varying the viscosity of the NR phase. To produce rubber with different

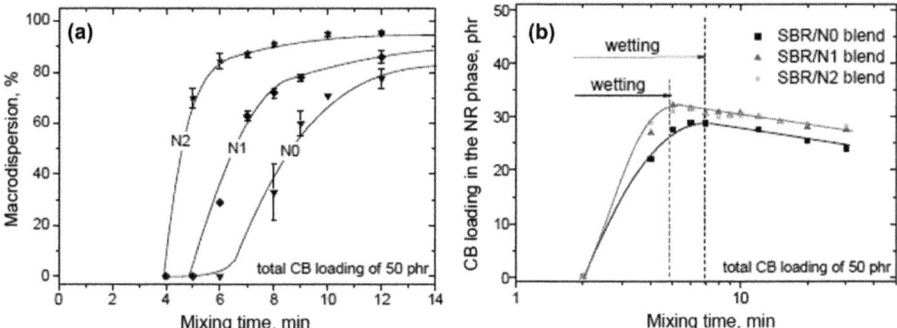

Figure 6.15 a) Effect of matrix viscosity on the development of a CB macrodispersion in NR single mixtures. (b) CB loading in the NR phase of 50/50 SBR/NR blends.

molecular weight by keeping the same polarity, NR was masticated in a two-roll mill for different periods of time. During the mastication process at room temperature, NR undergoes a chain scission, which results in a decrease in the molecular weight and viscosity. The designation of NR undergoing different mastication times is N0, N1 and N2. The filler–polymer interaction of these NR samples is the same. With decreasing molecular weight, the BIT determined by online measured electrical conductance shifts to shorter mixing times. As found in our previous work,[96] the shorter the BIT, the faster CB is dispersed. In Figure 6.15(a), the CB macrodispersion of N0, N1 and N2 is illustrated depending on mixing time. The development of CB dispersion shifts to the shorter mixing time with decreasing viscosity of NR.

The effect of the molecular weight of NR (N0, N1 and N2) on the phase-specific CB localization in the SBR/NR blend is shown in Figure 6.15(b). In the first mixing process, the NR phase with lower viscosity can wet CB faster than that with higher viscosity. This behaviour is in agreement with Equation (6.34), which shows that the wetting speed b increases with decreasing matrix viscosity. In the subsequent period, the viscosity seems not to affect the CB relocalization.

6.5.2.2 *Preparation of Rubber Blends with Defined Filler Localization Using Masterbatch Technology*

In rubber processing, masterbatch technology has been a common way to produce a defined distribution of filler in certain blend phases.[1–23,7,95] In all the works reported the filler transfer during the preparation of blends from masterbatches was considered to be negligible. In the present work we produced two blends filled with CB and silica using masterbatch technology in order to characterize the filler transfer within blend phases. SBR (Spintan SLR-4601, Styron Deutschland GmbH) and NR (SMR10, Standard Malaysian Rubber) as well as BR (Buna-cis 132, Styron Deutschland GmbH) were used as rubbers. CB (Corax N220, Evonik Industries AG) and silica (Ultrasil 7000 GR, Evonik Industries AG) were used as fillers. Masterbatches containing rubber and a given filler loading were prepared in an internal mixer for 20 minutes in order to obtain a good dispersion. The blend series A and B were prepared by mixing masterbatches for 10 minutes in an internal mixer with each other in order to produce blends with defined filler localization. The blend ratio SBR/NR of 50/50 and SBR/BR of 50/50 were chosen for series A and B, respectively. The total filler loading in the blends was 50 phr. The initial chamber temperature of 50 °C and rotor speed of 50 rpm were kept constant. The composition of the two blend series is given in Tables 6.8 and 6.9. It is worth noting that the surface tension of NR is similar to that of BR.

The mechanical properties, elastic modulus and tan δ at 60°C of CB filled SBR/NR blends are presented in Figure 6.16(a) as a function of the CB loading pre-mixed in the SBR phase. With increasing CB loading in the SBR phase the elastic modulus increases and passes a maximum value, when about 25 phr CB is localized in the SBR phase, and then decreases. It is well known that the elastic modulus of a blend is significantly dependent on the modulus of the matrix. By taking into consideration the viscosity and volume ratio of filled SBR and NR phases, which are dependent on the filler loading pre-mixed, the morphology of the filled blends can be predicted in Figure 6.16(a). It is clear that a co-continuous morphology is expected when CB is distributed evenly in both phases. A blend phase becomes the matrix if less CB is distributed in it. As

Table 6.8 Composition of 50/50 SBR/NR blends (series A).

	A1	A2	A3	A4	A5	A6
Given CB loading pre-distributed in SBR phase (phr)	0	12.5	25	30	37.5	50
Given CB loading pre-distributed in NR phase (phr)	50	37.5	25	20	12.5	0

Table 6.9 Composition of 50/50 SBR/BR blends (series B).

	B1	B2	B3	B4	B5
Given silica loading pre-distributed in SBR phase (phr)	10	20	30	40	50
Given silica loading pre-distributed in BR phase (phr)	40	30	20	10	0

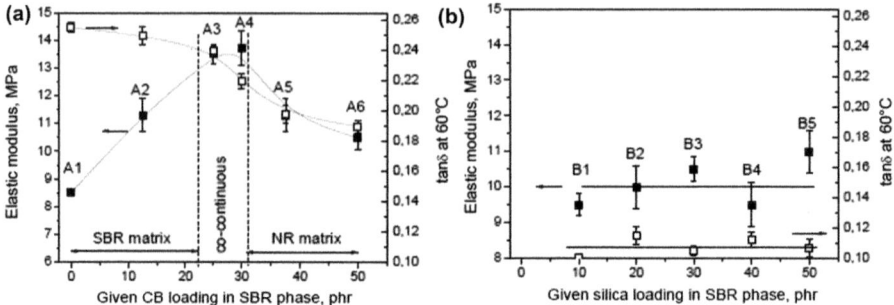

Figure 6.16 Elastic modulus and tan δ at 60 °C of (a) CB-filled SBR/NR blends (series A) and (b) silica-filled SBR/BR blends (series B) in dependence on the filler loading pre-mixed in the SBR phase.

a result the under-filled matrix determines the modulus of the blends. Only the blends with a co-continuous morphology and even filler distribution possess the highest modulus values. The value of tan δ at 60 °C, which is a measure of the rolling resistance of a tyre, decreases with increasing CB loading pre-mixed in the SBR phase.

In contrast to the strong dependence of mechanical properties on the CB localization shown in Figure 6.16(a), the mechanical properties of silica filled SBR/BR blends seem not to be influenced by the given silica loading pre-mixed in the SBR phase as seen in Figure 6.16(b).

In order to clarify the difference in mechanical behaviour between two blend series found in Figure 6.15 the filler loading in the SBR phase of the blends, which were prepared after blending two masterbatches together for 10 minutes, was experimentally determined by means of the wetting concept and is presented in Figure 6.17 as a function of the given filler loading pre-mixed in the SBR phase. In Figure 6.17(a) the CB loading in the SBR phase after 10 minutes of blending differs slightly from the given loading. The grey bar presents the filler loading at the thermodynamic equilibrium, which was determined by means of the Z-model. It is obvious that CB loading in the SBR phase tends to reach the predicted value indicated by the arrows. The low transfer speed of CB of 0.02 phr/min found from Figure 6.13 is the reason for the negligible transferred loading of CB in the blending process. Thus, the blends with defined CB localization could be prepared using masterbatch technology. In contrast, in Figure 6.17(b) the transfer of silica takes place strongly during the blending process. Compared to CB silica possesses an average transfer speed of about 0.8 phr/min, as determined in Figure 6.5(b). Thus, after 10 minutes of blending the silica loading in the SBR phase of all blends reaches a plateau value, which corresponds well to the one predicted. Because the silica localization in all blends prepared is similar, their mechanical properties are no different from each other.

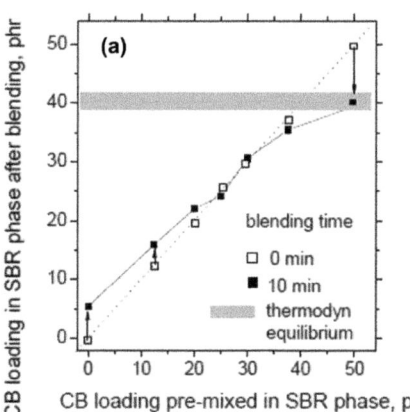 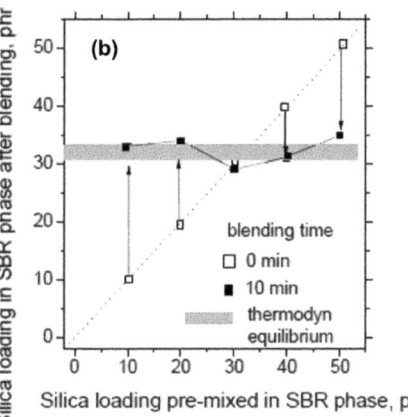

Figure 6.17 Filler loading in the SBR phase after 10 min of blending masterbatches determined by the wetting concept in relation to the given filler loading in the SBR phase of (a) SBR/NR and (b) SBR/BR blends.

6.5.3 Carbon Nanotube (CNT) Localization in Rubber Blends

6.5.3.1 CNT Dispersion and Localization in SBR/NR Blends

In order to characterize the kinetics of CNT localization in rubber blends, 20 phr of CNTs (Baytubes C150 HP, Bayer Material Science) was added into a 50/50 SBR/NR blend. The CNT fraction distributed in the SBR phase can be predicted by means of the Z-model as shown in Figure 6.18(a). A CNT fraction of 0.8 was found in the SBR at an equilibrium state.

To quantify and discuss the distribution kinetics of CNTs in the blend phases the wetting behaviour of the CNTs by blend components was investigated.[97] SBR wets the surface of CNT more slowly than NR, because the bulky styrene groups restrict its mobility and infiltration into the filler aggregates. The kinetics of CNT distribution in the SBR/NR blend was experimentally characterized by the wetting concept and is presented in Figure 6.18(b). Generally, CNTs exhibit a localization kinetics similar to that of CB shown in Figure 6.13(b). The CNT loading distributed in the SBR and NR phases increases rapidly in the wetting period up to 10 minutes. More CNTs are distributed in NR than in SBR because of the faster wetting rate of NR compared to SBR. After 10 minutes of mixing the amount of CNTs in NR starts to decrease, while the amount of CNTs in SBR continuously increases. The increase of CNT loading in SBR is attributed to the CNT transfer process from the NR phase to the SBR phase as a result of the replacement of NR on the CNT surface by SBR. Extrapolation to the predicted values indicates that CNTs need about 1200 minutes to reach the equilibrium state.

6.5.3.2 CNT Dispersion in Polychloroprene (CR) in Presence of Ionic Liquid

Obtaining a fine dispersion of CNTs in rubber has always been the most challenging task for their practical application. To overcome these problems, as

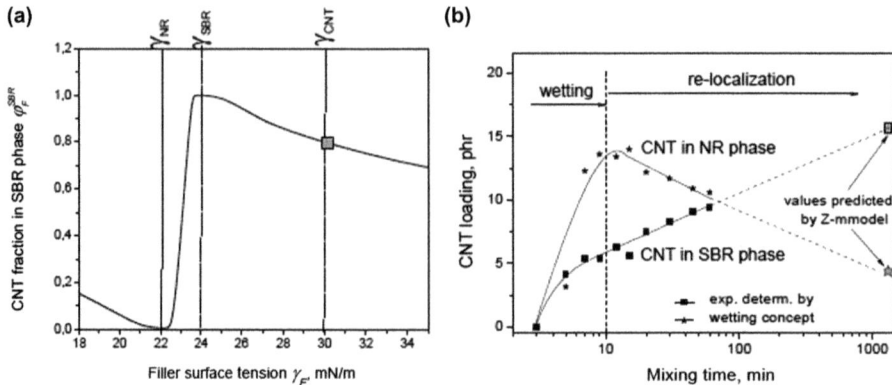

Figure 6.18 a) Prediction of CNT localization and (b) kinetics of CNT localization in SBR/NR blends.

well developing new preparation technologies there has been much progress in the functionalization of CNT surfaces with various organic molecules for better compatibility and dispersion of CNTs in rubber. For example, the processability and mechanical performance of rubber composites could be improved by introducing carboxylic acid groups[98,99] or multifunctional silane[100–102] onto the CNT surface. Recently, functionalization of CNTs with ionic liquids, a kind of molten salt with nearly zero vapour pressure and high thermal stability, is an interesting topic, because ionic liquids could provide a facile and promising method to control the surface properties of CNTs by means of cation–π interaction.[103–105] A CNT/ionic liquid mixture was mixed into silicone elastomers by Sekitani et al.[106] and hydrogenated nitrile rubber (HNBR) by Likozar et al.[107] to produce conductive rubber-like stretchable composites. Das et al.[108,109] used a series of ionic liquids as surfactants for blends of SBR/BR filled with CNTs in order to determine the coupling activity of ionic liquids between diene elastomers and CNTs. Recently, Subramaniam et al.[110,111] described a new simple route to disperse CNTs in polychloroprene (CR) using an ionic liquid, 1-butyl-3-methyl imidazolium bis(trifluoromethylsulphonyl) imide (BMI) and found that the use of BMI and a low concentration (5 phr) of CNTs in CR exhibited an electrical conductivity of 0.1 S/cm with a stretchability of $> 500\%$. In this regard, the present work focuses on the characterization of the selective wetting kinetics of CNTs in CR nanocomposites and its correlation to the filler dispersion. The effectiveness of BMI as a surfactant for CNTs will be characterized by means of online measurement of electrical conductance and discussed by taking into consideration the selective wetting behaviour of CNTs by BMI and rubber.

For experimental work, polychloroprene (CR) (Baypren 611, Lanxess GmbH) with Mooney viscosity MU (ML 1 + 4) 100 °C of 43 ± 6 was used as the rubber matrix. Ionic liquid BMI, which is basically made of an asymmetric heterocyclic cation 1-butyl-3-methyl imidazolium (BMI^+) and an anion bis(trifluoromethylsulphonyl)imide (BMI^-) (Sigma-Aldrich, Germany) was used as surfactant. According to the chemical structure of BMI a mass ratio BMI^+/BMI^- of 139/280 was calculated. CNT (Baytubes C150HP, Bayer Material Science) was used as filler. For convenient admixing of CNTs into the mixing chamber 5 phr CNTs were softly ground with 10 phr BMI until a black BMI/CNT paste was obtained. The composites were prepared in an internal mixer by keeping the following mixing conditions: initial chamber temperature T_A of 25°C, rotor speed of 70 rpm, fill factor of 0.72. The black BMI/CNT paste was admixed into the chamber after 3 minutes of mixing.

In Figure 6.19 the online conductance of the CNT composites without and with BMI, respectively, is presented as a function of mixing time. Upon addition of BMI the online conductance of nanocomposite with BMI increases and exhibits a typical conductance–time characteristic with $t_{onset} = 10$ min and $t_{Gmax} = 55$ min. At t_{onset} and t_{Gmax} the online conductance starts to rise and reaches the maximum value, respectively. According to our previous works[51,52] the macrodispersion of filler and the online conductance correlate closely. The largest change in the size of filler agglomerates, i.e. the dispersion of filler

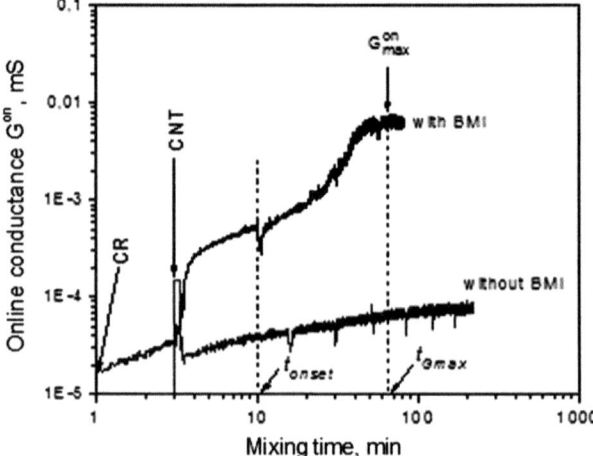

Figure 6.19 Online conductance of CNT-filled CR compounds without and with BMI.[113]

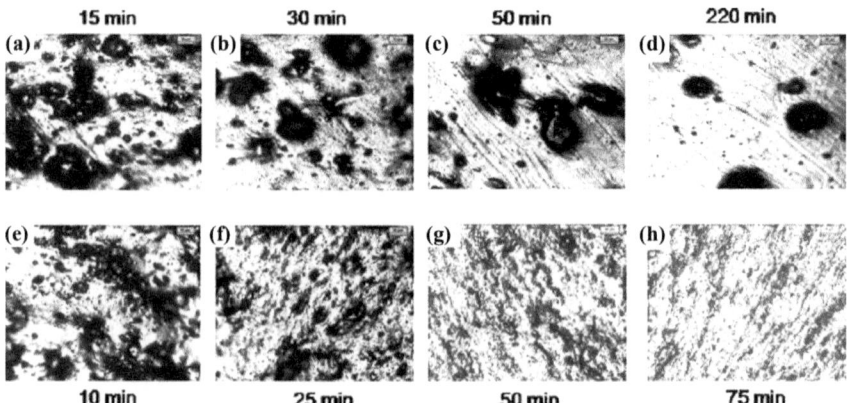

Figure 6.20 Optical microscopic images of CNT-CR composites without BMI (a)–(d), and with BMI (e)–(h) in relation to mixing time (image dimension 300 μm × 400 μm).

agglomerates into smaller aggregates or even individual tubes, is determined in the period between t_{onset} and t_{Gmax}. Thus, t_{onset} and t_{Gmax} have often been used as a way of characterizing the filler dispersion kinetics. Upon t_{Gmax} the online conductance decreases slightly, which is related to the better distribution of small aggregates throughout the matrix, as discussed previously.[51] For the composite without BMI the t_{onset} was not observed until a mixing time of up to 200 minutes.

The macrodispersion of filler in the CR matrix is studied by optical microscopy. In the images of composites without BMI shown in Figure 6.20(a)–(d), large agglomerates of Baytubes are still visible even after very long mixing times. The addition of BMI leads to a significant improvement in dispersion, as

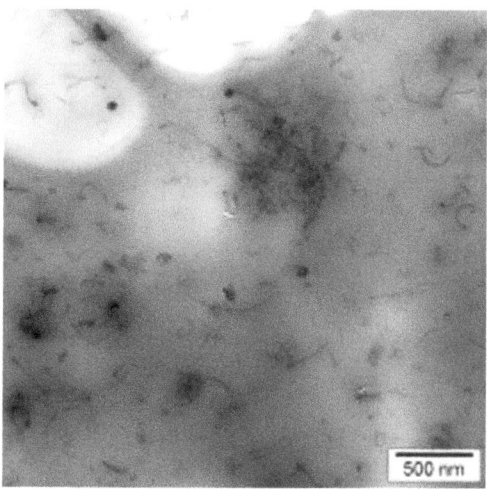

Figure 6.21 TEM image of CNT-CR composite with BMI.[113]

shown in Figure 6.20(e)–(h). The largest change in the size of CNT agglomerates is clearly determined in the range between $t_{onset} = 10$ min and $t_{Gmax} = 55$ min. At t_{Gmax} only some small agglomerates are observed. The better macrodispersion of Baytubes by addition of BMI may be attributed to the physical cation–π interaction between BMI$^+$ and the tubes and/or the perturbation of π–π stacking of multi-walls of the tubes as discussed in the literature[104,112] and proved by Raman investigation by Subramaniam et al.[110]

The microdispersion of CNTs is characterized by TEM images as shown in Figure 6.21. CNTs are well dispersed and form a continuous network throughout the CR matrix, which gives the composite a high conductivity.

6.5.3.3 Selective Wetting of CNTs

According to the wetting concept developed for silica filled rubber blends,[28] it is possible to quantify the composition of the rubber layer bonded to the silica surface by means of FTIR analysis of the silica–rubber gel. However for CNT filled compounds, due to the total absorption of infrared beams by CNTs existing in the gel, analysis of the gel by FTIR was not possible. As an alternative, we indirectly characterized the gel composition by investigating the extracted part using a FTIR spectrometer S2000 (Perkin Elmer) equipped with a diamond single Golden Gate ATR cell (Specac). The experimental procedure and the calculation were described in detail in our previous work.[113]

Substituting the surface tension values of CR and BMI into Equations (6.13) and (6.14), a master curve demonstrating the filler surface fraction wetted by the BMI molecules as a function of the filler surface tension can be created as seen in Figure 6.22. γ_{CR}, γ_{BMI} and γ_F are the surface tension values of CR and BMI as well as CNT, respectively. $n_{CR/BMI}$ is the ratio of CR to BMI. In the

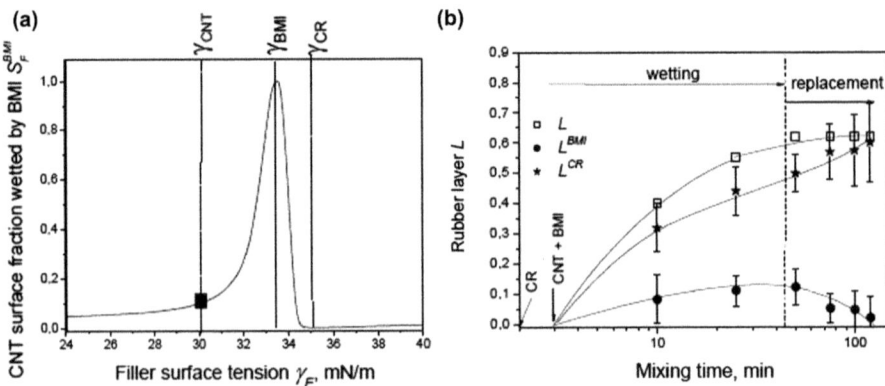

Figure 6.22 a) Master curve describing the CNT surface fraction wetted by BMI. (b) Kinetics of CNT wetting by CR and BMI.[113]

present work $\gamma_{CR} = 35$ mN/m, $\gamma_F = 30$ mN/m for CNTs were experimentally determined; $\gamma_{BMI} = 33.6$ mN/m was taken from the literature.[114]

Fitting the CNT surface tension into the master curve with the ratio $n_{CR/BMI} = 10$, a CNT surface fraction wetted by BMI S_F^{BMI} of 0.07 was found at a thermodynamic equilibrium state. The CNT surface fraction wetted by BMI is much smaller compared to that wetted by CR, although the affinity of CNTs to BMI is better than to CR. According to Equation (6.13) the CNT surface fraction wetted by BMI is determined by the thermodynamic driving force ($\gamma_{CR-F}/\gamma_{BMI-F}$) and the concentration compensation effect ($n_{CR/BMI}$). In the present work at a high value of $n_{CR/BMI}$ the concentration compensation effect dominates the thermodynamic effect and as a result, CR expectedly wets the large amount of CNT surface.

The experimental characterization of the selective wetting behaviour of CNT by CR and BMI was done by means of FTIR of the extracted parts. The rubber layer L and its contribution L^{BMI} and L^{CR} of the composites are presented in Figure 6.22(b) as a function of mixing time. It is obvious that in the first mixing period (up to 50 min) both L^{BMI} and L^{CR} increase. In this range BMI and CR concurrently infiltrate the CNT aggregates and wet the CNT surface. After the wetting process is complete, L^{BMI} decreases while L^{CR} continuously increases. Because the value of L is constant in the second period, it can be concluded that the free CR molecules replaced the bonded BMI on the CNT surface as a result of the concentration compensation effect. At 120 minutes mixing time BMI is completely replaced by CR, corresponding to the prediction made by the Z-model.

The EDX spectrogram of the filler–polymer gel of the composite with BMI after 120 minutes of mixing time is shown in Figure 6.23(a). No signals for fluorine at 0.67 keV and nitrogen at 0.39 keV of BMI^- and BMI^+, respectively, were found, while a strong peak of chlorine at 2.62 keV of CR can be observed. That corresponds very well to our FTIR analysis. The unbound part of BMI

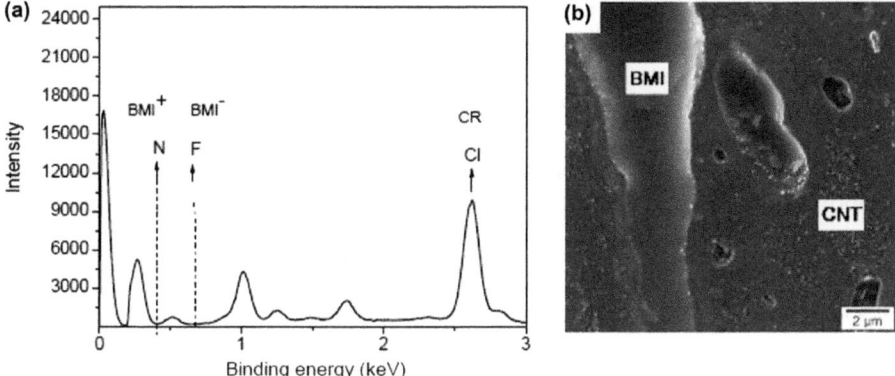

Figure 6.23 a) EDX spectrogram of CNT-CR gel. (b) SEM image of CR composite with BMI after 120 min mixing time.[113]

forms its own phase as seen in the SEM image (Figure 6.23(b)). A closer look at the BMI phase does not reveal any CNTs. Beside the BMI phase some non-dispersed CNT aggregates are still observed in submicron scale, which appear in the SEM micrograph as bright domains. The wetting BMI inside the non-dispersed aggregates could be replaced by the CR molecules, but this is difficult because CR needs time to infiltrate the aggregates. That is why the wetting and replacement process take place slowly in the composites, as discussed above.

Several recent studies have looked at the interaction between ionic liquids and CNTs as well as the molecular structure of their interphase. Likozar[115] investigated the adsorption kinetics of different ionic liquids into CNT/HNBR composites by immersing the cured composites in the ionic liquid/chloroform solvent. He observed a homogeneous distribution of anions in the composites by use of the fluorine signal detected by scanning electron microscopy/energy dispersive X-ray analysis (SEM/EDX). By means of fully atomistic molecular simulations Frolov et al.[116] studied the basic mechanisms of CNT interactions with several different room temperature ionic liquids in their mixtures with acetonitrile. It was found that two distinct layers of cations and anions are formed at the CNT surface. An increase in the length of the non-polar alkyl groups of cations increases the propensity of imidazolium-based cations to lay parallel to the CNT surface. Wang et al.[112] carried out Raman and IR measurements on mixtures of ionic liquids and single-walled carbon nanotubes (SWCNTs) and found that no strong interaction such as cation–π interaction exists between SWCNTs and imidazolium cations. It could be seen that the fluorine atoms of anions and the hydrogen atoms of the alkyl groups of cations are much closer to the SWCNTs than the nitrogen atoms and carbon atoms of the imidazolium rings. This indicates that the SWCNTs are surrounded by the polar parts of anions and non-polar parts of cations simultaneously. They proposed that the ionic liquids interact with SWCNTs through weak van der Waals interaction and the shielding effect of ionic liquids on the π–π stacking interaction among SWCNTs plays a key role in dispersing the SWCNTs. Their conclusion, however, is in contrast to that made by Fukusima et al.[103,104] and Ma et al.,[105] who stated that the specific interaction between the imidazolium ion component and the π-electronic nanotube surface is essential for the excellent dispersion of CNTs in ionic liquids. The time dependent presence of BMI^+ cations in the bound polymer regions as found in Figure 6.22(b) can be supported by the interaction mechanism proposed by Fukusima and Ma et al.[103–105]

6.5.4 Nanoclay Transfer in Rubber Blends

Regarding the complexity of the morphology development of nanoclay filled polymer blends the goal of the present work is an attempt to describe the dispersion and distribution kinetics of nanoclay in rubber blends. It is based on the development of a new *in situ* measurement technique to monitor the dispersion as well as fundamental studies to understand the relationship between various factors affecting the clay dispersion and distribution, respectively. In recent works,[117,118] an online measuring method has been introduced using the electrical conductance, directly measured during compounding, to monitor the intercalation and exfoliation process of nanoclay in a polymer matrix. In contrast to the conductance of CB or CNT filled rubber compounds, which is originated by the electron transport throughout the filler network, the online conductance of rubber–organoclay composites has an ionic nature due to the release of the quaternary ammonium cations, which are available in the galleries of the modified clays. The higher the degree of the nanoclay dispersion, the more cations are released and the higher the online conductance recorded. A close correlation between the online conductance chart and the kinetics of nanoclay dispersion has been observed. In the present work, the development of morphology of the clay filled HNBR/NR and HNBR/ENR blends is characterized by means of the methods we developed.[119]

HNBR (Zetpol 2030L, Zeon Deutschland) with acrylonitrile content of 36 wt% and NR (SMR10, Standard Malaysian Rubber), as well as epoxidized natural rubber (ENR 50, Weber & Schaer GmbH & Co. KG, Germany), with epoxidation grade of 50 wt%, were used as the rubber matrix. Organoclay (Nanofil 9, Süd-Chemie) modified by stearyl benzyl dimethyl ammonium chloride, with an average particle size of about 35 μm, and a weight loss on ignition of 35 wt%, was used as filler. Data from the clay provider, as well as our investigations, show that the organoclay has a basal spacing of 2.0 nm before compounding. Peroxide (Luperox 101, Atofina Chemicals) was used as a curing agent for the rubber compounds investigated. To prepare the blends, in the first mixing step NR and ENR were mixed with 10 phr nanoclay in an internal mixer at an initial chamber temperature of 50 °C and a rotor speed of 70 rpm in order to produce NR–clay and ENR–clay masterbatch, respectively. A mixing time of 10 min was used to produce both masterbatches. In the second step, pure HNBR was mixed with the masterbatch to obtain 50/50 HNBR/NR and HNBR/ENR blends, respectively, having a total clay

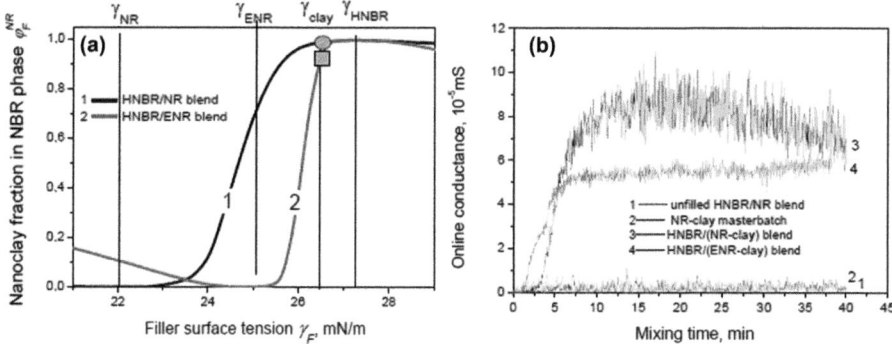

Figure 6.24 Master curves of (a) nanoclay localization and (b) development of online conductance of different blends.[119]

concentration of 5 phr. In the following they are called HNBR/(NR–clay masterbatch) and HNBR/(ENR–clay masterbatch) blends. For comparison purposes, an unfilled 50/50 HNBR/NR blend was also prepared.

Using data on the surface tension values of the rubbers and nanoclays used, the master curve of filler localization in 50/50 HNBR/NR and 50/50 HNBR/ENR blends was calculated and is presented in Figure 6.24(a). A clay fraction of about 0.9 and 1.0 localized in the HNBR phase of HNBR/ENR and HNBR/NR blends, respectively, was found, *i.e.* an almost complete localization of clay in the HNBR is predicted for both blends, when they reach a thermodynamic equilibrium state. Thus, at the beginning of the preparation of HNBR/(NR–clay masterbatch) and HNBR/(ENR–clay masterbatch) blends, nanoclay pre-mixed in the NR and ENR phase is far away from its equilibrium state. Nanoclay is expected to migrate to the HNBR phase during the blending process.

From the results of previous work[118] the rapid increase in the conductance signal just after adding nanoclay into HNBR could be attributed to ionic conduction. Compared to natural clay, the modified clay nanofiller is more compatible with organic polymers due to its organophilic nature with a lower surface energy. The polymer chains can intercalate within the galleries of the organoclay. It was shown that hydrogen bonds form between the ammonium ion of the intercalated amine and the nitrile group.[120] This seems to disassociate or weaken the attractive forces between the cationic head groups of the alkyl ammonium molecules and the negatively charged clay surface. During mixing the steady motion of rubber chains facilitates the transport of quaternary ammonium ions from the clay galleries towards the rubber matrix. The more contact area between organoclay and rubber is made along the mixing time through intercalation and exfoliation, the more the charge carriers primarily trapped inside the clay galleries are released, which imparts a higher conductance. In the present work, the online conductance charts of the unfilled HNBR/NR blend, HNBR/(NR–clay masterbatch) blend and HNBR/(ENR–clay masterbatch) blend are presented in Figure 6.24(b). The conductance curves of the unfilled 50/50 HNBR/NR blend (curve 1) and NR–clay masterbatch (curve 2) are very low. When HNBR was mixed with NR–clay masterbatch it was expected that the conductance curve of the HNBR/(NR–clay masterbatch) blend would lie at the same level as that of the unfilled HNBR/NR blend. However, the conductance curve of the HNBR/(NR–clay masterbatch) blend (curve 3) shows an unexpected chart with a high conductance compared to the unfilled HNBR/NR blend. In order to clarify the structural background of the conductance curve the blend morphology and nanoclay distribution in the blends were investigated.

The selective distribution of the clay in the different blend phases was determined by extraction experiments. The clay content located in the NR phase decreases from 5 phr at the start of the blending process to 1 phr after a mixing time of 40 minutes. Meanwhile, the clay loading in the HNBR phase increases from zero to 4 phr, which corresponds very well to the prediction made in Figure 6.24(a). The better interaction of the clay with the HNBR phase leads to the migration of nanoclay from the NR to the HNBR phase, which subsequently results in a strong increase in the online conductance (curve 3 in Figure 6.24(b)) in the first period (up to 20 minutes). The clay dislocation becomes obvious by taking a look at the AFM images presented in Figure 6.25(a)–(d). At 2 minutes a number of clay agglomerates (black dots) with a size of approx. 500 nm is visible in the NR phase (light area) and no clay in the HNBR phase (grey area). With increasing mixing time more and more clay is enriched in the HNBR phase. The AFM image of samples taken out at 40 minutes clearly shows a dominant localization of the clay tactoids in the HNBR domains. It is obvious that the morphology of the blend is changed as a function of mixing time. Up to 15 minutes the HNBR phase still appears as a co-continuous phase, which facilitates the ionic conduction of the blend. According to the rheological characterization, the viscosity of the NR phase is about half that of the HNBR phase. Therefore, in the unfilled 50/50 HNBR/NR an island matrix structure is expected. The highly viscous HNBR phase will be the dispersed phase in the not very viscous NR matrix. Indeed, after 15 minutes the continuous HNBR phase starts to be divided into domains that partly interrupt the motion of ionic species throughout the HNBR phase. As a result the online conductance subsequently decays in the period between 20 and 40 minutes. The extent of the conductance decay correlates with the reduction in the domain size of HNBR. In this period the diameter of HNBR domains reduces from about 2 μm to 0.7 μm and the corresponding online conductance from 8.26×10^{-5} mS to 6.86×10^{-5} mS.

The online conductance of the HNBR/(ENR–clay masterbatch) blend increases and reaches a plateau after a short time (curve 4, Figure 6.24(b)). On the basis of the development of the online conductance of the HNBR/(NR–clay masterbatch) blend discussed above, it is presumed that the increase in the online conductance in the first mixing period (up to 5 minutes) is due to migration of organoclay from the ENR phase to the HNBR phase. If the conductance chart of HNBR/(ENR–clay masterbatch) is compared with that of the HNBR/(NR–clay masterbatch) blend, an important difference is that the

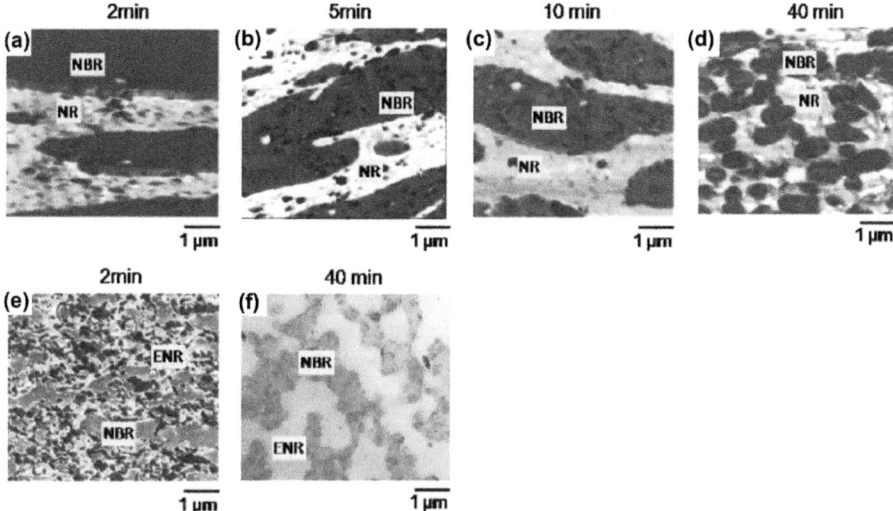

Figure 6.25 Morphology development of HNBR/NR–clay masterbatch blends (a)–(d)[119] and HNBR/ENR–clay masterbatch blends (e)–(f) in relation to mixing time.

former shows a plateau for longer mixing time, whereas for the latter the conductance starts decreasing after 15 minutes of mixing time. The development of morphology of the HNBR/(ENR–clay masterbatch) blend as a function of mixing time is shown in Figure 6.25(e)–(f). After 2 minutes of mixing time clay aggregates are observed in the ENR matrix surrounding the unfilled HNBR domains. As the mixing time increases, organoclay migrates to the HNBR phase. At the same time the island matrix morphology starts changing to a co-continuous morphology, as shown in Figure 6.25(f). Simultaneously, clay aggregates are broken down to smaller sizes. It can be clearly seen in Figure 6.25(f) that after 40 minutes of mixing time most of the clay platelets have transferred to the continuous HNBR phase, which is in line with our previous prediction. Formation of the continuous HNBR phase also explains the plateau in online conductance shown in Figure 6.24(b).

6.6 Conclusions

In the present work, a comprehensive description of filler localization kinetics and development of blend morphology in rubber blends containing NR could be done by combining the online measurement of electrical conductance and the wetting concept as well as the Z-model. The development of dispersion of different fillers with mixing time could be well characterized by the online conductance chart. The phase selective filler localization in binary and ternary rubber blends as a function of mixing time was experimentally determined by means of the wetting concept. For a quantitative prediction of filler localization in rubber blends a so-called Z-model based on thermodynamic data was proposed. Using this model, a master curve presenting a characteristic dependence of filler localization on the surface tension values of blend components was created. Compared to synthetic blend partners like SBR, BR or EPDM, NR exhibits a high filler wetting speed and low filler–rubber interaction, which influence the kinetics of filler localization in rubber blends containing NR in an opposite manner. During the first mixing stage, filler localization is strongly affected by the high wetting speed of NR and consequently more filler is localized in the NR phase. In the subsequent mixing state, after the wetting process is completed, the rubber–filler interaction dominantly influences the localization kinetics. As a result, a relocalization of filler within the blend phases takes place until the thermodynamic equilibrium state predicted by the Z-model is reached. It was found that the equilibrium state of silica filled blends can be reached within the experimental processing, while the filler localization of CB filled blends determined at the end of the mixing process is still far from the equilibrium state. This is related to the much higher transfer speed of silica compared to that of CB. Further investigations should be conducted to better understand this behaviour. The strong effect of curing and processing additives as well as silane on the silica localization is sufficiently explained by the Z-model by taking into consideration the deactivation of the silanol groups on the silica surface by adsorbed additives and silane. Using the master curve the surface tension of filler affected by addition of additives and silane can be estimated, which may be useful for evaluation and comparison of the effect of different coupling agents.

Acknowledgements

The authors would like to thank the German Research Foundation (DFG) for financial support, Dr Z. Ali for measurement of the online electrical conductance, Mrs K. Oßwald and Ms M. Keller for blend preparation and extraction experiments, Dr Klaus-Werner Stöckelhuber (Leibniz Institute of Polymer Research Dresden e.V. (IPF)) for surface tension measurement, Mrs Becker (Institute of Physics, Martin Luther University, Halle-Wittenberg, Germany) for EDX investigation, and Ms R. Boldt (Leibniz Institute of Polymer Research Dresden e.V. (IPF)) for TEM and SEM investigations.

References

1. W. M. Hess and V. E. Chirico, *Rubber Chem. Technol.*, 1977, **50**, 301.
2. W. M. Hess, C. R. Herd and P. C. Vegvari, *Rubber Chem. Technol.*, 1993, **66**, 329.
3. W. M. Hess, *Rubber Chem. Technol.*, 1991, **64**, 386.
4. H. W. Waddell, *Rubber Chem. Technol.*, 1998, **71**, 590.
5. J. Massie, R. Hirst and A. Halasa, *Rubber Chem. Technol.*, 1993, **66**, 276.
6. J. Ziegler and R. H. Schuster, *Kautsch. Gummi Kunstst.*, 2003, **56**, 159.
7. H. H. Le, S. Ilisch, G. R. Kasaliwal and H.-J. Radusch, *Kautsch. Gummi Kunstst.*, 2007, **60**, 241.
8. G. Meier, M. Klüppel, H. Geisler and R. H. Schuster, *Kautsch. Gummi Kunstst.*, 2005, **58**, 587.

9. M. H. Walters and D. N. Keyte, *Rubber Chem. Technol.*, 1965, **38**, 62.
10. V. Herrmann, K. Unseld and H. B. Fuchs, *Kautsch. Gummi Kunstst.*, 2001, **54**, 453.
11. A. H. Tsou and W. H. Waddell, *Kautsch. Gummi Kunstst.*, 2002, **55**, 382.
12. I. H. Jeon, H. Kim and S. S. G. Kim, *Rubber Chem. Technol.*, 2003, **76**, 1.
13. C. C. Wang, J. B. Donnet and T. K. Wang, *Rubber Chem. Technol.*, 2005, **78**, 17.
14. B. G. Soares, F. Gubbels and R. Jerome, *Rubber Chem. Technol.*, 1997, **70**, 60.
15. B. G. Soares, F. Gubbels, R. Jerome, E. Vanlathem, R. Deltour, S. Blacher and F. Brouers, *Chem. Mater.*, 1998, **10**, 1227.
16. F. Gubbels, R. Jerome, P. H. Teyssib, E. Vanlathem, R. Deltour, A. Calderone, V. Parentb and J. L. Bredas, *Macromolecules*, 1994, **27**, 1972.
17. H. H. Le, Z. Qamer, S. Ilisch and H.-J. Radusch, *Rubber Chem. Technol.*, 2006, **79**, 621.
18. W. Hu, M. D. Ellul, A. H. Tsou and S. Datta, *Rubber Chem. Technol.*, 2007, **80**, 1.
19. M. Kotani, H. Dohi, H. Kimura, K. Muraoka and H. Kaji, *Macromolecules*, 2007, **40**, 9451.
20. A. Sircar and T. G. Lamond, *Rubber Chem. Technol.*, 1973, **46**, 178.
21. J. M. Massie, R. C. Hirst and A. F. Halasa, *Rubber Chem. Technol.*, 1992, **66**, 276.
22. G. R. Cotton and L. J. Murphy, *Kautsch. Gummi Kunstst.*, 1988, **41**, 54.
23. C. D. Woolard and B. J. McFadzean, Proceedings of the 28th Annual Conference on Thermal Analysis and Application, Orlando, 2000.
24. S. Maiti, S. K. De and A. K. Bhowmick, *Rubber Chem. Technol.*, 1992, **65**, 293.
25. M. Klüppel, R. H. Schuster and J. Schaper, *Gummi Fasern Kunstst.*, 1998, **51**, 508.
26. M. Klüppel, R. H. Schuster and J. Schaper, *Rubber Chem. Technol.*, 1999, **72**, 91.
27. H. H. Le, S. Ilisch and H.-J. Radusch, *Rubber Chem. Technol.*, 2008, **81**, 767.
28. H. H. Le, S. Ilisch, D. Heidenreich, A. Wutzler and H.-J. Radusch, *Polym. Comp.*, 2010, **31**, 1701.
29. H. H. Le, K. Osswald, S. Ilisch, D. Heidenreich and H.-J. Radusch, *Rubber Chem. Technol.*, 2011, **84**, 41.
30. H. H. Le, K. Oßwald, S. Ilisch, T. Pham, K. W. Stöckelhuber, G. Heinrich and H.-J. Radusch, *Macromol. Mater. Eng.*, 2012, **297**, 464.
31. J. Y. Feng, C. M. Chan and J. X. Li, *Polym. Eng. Sci.*, 2003, **43**, 1058.
32. A. L. Persson and H. Bertilsson, *Polymer*, 1998, **39**, 5633.
33. R. Ibarra-Gomez, A. Marquez, L. Valle and O. S. Rodriguez-Fernandez, *Rubber Chem. Technol.*, 2003, **76**, 969.
34. M. Sumita, K. Sakata, S. Asai, K. Miyasaka and H. Nakagawa, *Polym. Bull.*, 1991, **25**, 265.
35. M. Sumita, K. Sakata, Y. Hayakawa, S. Asai, K. Miyasaka and M. Tanemura, *Colloid Polym. Sci.*, 1992, **270**, 134.
36. S. Wu, *Polymer Interface and Adhesion*, Marcel Decker, New York, 1982.

37. L. Elias, F. Fenouillot, J. T. Majeste, G. Martin and G. Cassagnau, *J. Polym. Sci., Part B: Polym. Phys.*, 2008, **46**, 1976.
38. F. Fenouillot, P. Cassagnau and J. C. Majeste, *Polymer*, 2009, **50**, 1333.
39. H. H. Le, K. Oßwald, S. Ilisch, X. T. Hoang, G. Heinrich and H.-J. Radusch, *J. Mater. Sci.*, 2012, **7**, 4270.
40. J. H. Hildebrand and R. L. Scott, in *The Solubility of Non-Electrolytes*, Dover Publications, New York, 1964.
41. G. Scatchard, *Chem. Rev.*, 1931, **8**, 321.
42. G. Scatchard, *Chem. Rev.*, 1949, **44**, 7.
43. D. R. Paul and S. Newman, *Polymer Blends*, Academic Press, New York, 1978.
44. L. A. Girifalco and R. J. Good, *J. Phys. Chem.*, 1957, **61**, 904.
45. H. H. Le, K. Oßwald, M. Keller, S. Ilisch, T. Pham, H.-J. Radusch, Theoretical prediction of filler localization in polymer blends: Validity proved by different systems, *14. Problemseminar: Polymerblends*, 14–15 September 2011, Halle, Proceedings and CD, p. 48.
46. H. H. Le, M. Keller, M. Hristov, S. Ilisch, T. H. Xuan, Q. K. Do, T. Pham, K.-W. Stöckelhuber, G. Heinrich and H.-J. Radusch, *Macromol. Mater. Eng.*, 2012.
47. F. Bohin, I. Manas-Zloczower and D. L. Feke, *Rubber Chem. Technol.*, 1994, **67**, 602.
48. M. Astruc, V. Collin, S. Rusch, P. Navard and E. Peuvrel-Disdier, *J. Appl. Polym. Sci.*, 2004, **91**, 3292.
49. C. M. Lin, W. J. Chang and T. H. Fang, *J. Electron. Packaging*, 2007, **129**, 48.
50. C. H. Leigh-Dugmore, *Rubber Chem. Technol.*, 1956, **29**, 1303.
51. H. H. Le, S. Ilisch, H. Steinberger and H.-J. Radusch, *Plast., Rubber Compos.: Macromol. Eng.*, 2008, **37**, 367.
52. H. H. Le, I. Prodanova, S. Ilisch and H.-J. Radusch, *Rubber Chem. Technol.*, 2004, **77**, 815.
53. H. H. Le, M. Tiwari, S. Ilisch and H.-J. Radusch, *Rubber Chem. Technol.*, 2006, **79**, 610.
54. F. M. Fowkes, *J. Phys. Chem.*, 1963, **67**, 2538.
55. M. J. Wang and S. Wolff, *Rubber Chem. Technol.*, 1992, **65**, 715.
56. Q. Zhang, H. Yang and Q. Fu, *Polymer*, 2001, **45**, 1913.
57. H. Yang, X. Zhang, C. Qu, B. Li, L. Zhang, Q. Zhang and Q. Fu, *Polymer*, 2007, **48**, 860.
58. U. Jönsson, M. Malmqvist and I. Ronberg, *Biochem. J.*, 1985, **227**, 363.
59. M. L. Kralevich and J. L. Koening, *Rubber Chem. Technol.*, 1998, **71**, 300.
60. S. Ono, M. Ito, H. Tokumitsu and K. Seki, *J. Appl. Polym. Sci.*, 1999, **74**, 2529.
61. S. Ono, Y. Kiuchi, J. Sawanobori and M. Ito, *Polym. Int.*, 1999, **48**, 1035.
62. M. J. Wang, S. Wolff and J. B. Donnet, *Rubber Chem. Technol.*, 1991, **64**, 714.
63. J. O'Brien, E. Cashell, G. E. Wardell and V. J. McBrietry, *Macromolecules*, 1976, **24**, 653.
64. H. Serizawa, M. Ito, T. Kanamoto, K. Tanaka and A. Nomura, *Polym. J.*, 1982, **14**, 149.

65. S. Choi, *Polym. Adv. Technol.*, 2002, **13**, 466.
66. A. E. Zaikin, R. R. Karimov and V. P. Arkhireev, *Colloid J.*, 2001, **63**, 53.
67. A. E. Zaikin, E. A. Zharinova and R. S. Bikmullin, *Polym. Sci. Ser. A*, 2007, **49**, 328.
68. F. Gubbels, R. Jerome, E. Vanlathem, R. Deltour, S. Bacher and F. Brouers, *Chem. Mater.*, 1998, **10**, 1227.
69. S. H. Ahn, S. H. Kim and S. G. Lee, *J. Appl. Polym. Sci.*, 2004, **94**, 812.
70. G. S. Deshmukh, S. U. Pathak, D. R. Peshwe and J. D. Ekhe, *Bull. Mater. Sci.*, 2010, **33**, 277.
71. A. O. Maged, A. Ayman and W. S. Ulrich, *Polymer*, 2004, **45**, 1177.
72. F. Erika and P. Bela, *J. Colloid Interface Sci.*, 1997, **194**, 269.
73. H. Huang, M. Tian, J. Yang, H. Li, W. Liang, L. Zhang and X. Li, *J. Appl. Polym. Sci.*, 2008, **107**, 3325.
74. A. Kosmalska, M. Zaborski and L. Slusarski, *Macromol. Symp.*, 2003, **194**, 269.
75. S. H. Laning, M. P. Wagner and J. W. Sellers, *J. Appl. Polym. Sci.*, 1959, **2**, 225.
76. L. A. E. M. Reuvekamp, S. C. Debnath, J. W. Ten Brinke, P. J. Van Swaaij and J. W. M. Noordermeer, *Rubber Chem. Technol.*, 2009, **76**, 34.
77. J. M. Pena, N. S. Allen, M. Edge, C. M. Liauw, O. Noiset and B. Valange, *J. Mater. Sci.*, 2001, **36**, 4419.
78. M. Kotani, H. Dohi, H. Kimura, K. Muraoka and H. Kaji, *Macromolecules*, 2007, **40**, 9451.
79. S. Maiti, S. K. De and A. K. Bhowmick, *Rubber Chem. Technol.*, 1992, **65**, 293.
80. P. Phewphong, P. Saeoui and Ch. Sirisinha, *Polym. Test.*, 2008, **27**, 873.
81. J. Wootthikanokkhan and N. Rattanathamwat, *J. Appl. Polym. Sci.*, 2006, **102**, 248.
82. K. W. Stöckelhuber, A. Das, R. Jurk and G. Heinrich, *Polymer*, 2010, **51**, 1954.
83. M. Castellano, L. Conzatti, A. Turturro, G. Costa and G. Busca, *J. Phys. Chem. B*, 2007, **111**, 4495.
84. A. Vidal, E. Papirer, M. J. Wang and J. B. Donnet, *Chromatographia*, 1987, **23**, 121.
85. S. Wu, *J. Colloid Interface Sci.*, 1969, **31**, 153.
86. Y. Tamai, *Prog. Colloid Polym. Sci.*, 1976, **61**, 93.
87. K. Sahakaro, R. N. Datta, J. Baaij and J. W. M. Noordermeer, *J. Appl. Polym. Sci.*, 2007, **103**, 2555.
88. S. Walheim, M. Ramstein and U. Steiner, *Langmuir*, 1999, **15**, 4828.
89. I. Luzinov, C. Pagnoulle and R. Jerome, *Polymer*, 2000, **41**, 3381.
90. T. Kojima, Y. Kikuchi and T. Inoue, *Polym. Eng. Sci.*, 1992, **32**, 1863.
91. V. K. Kaushik and Y. N. Sharma, *Polym. Bull.*, 1985, **13**, 373.
92. M. A. Kader and A. K. Bhowmick, *J. Appl. Polym. Sci.*, 2003, **90**, 278.
93. M. S. Lee, M. G. Ha, H. J. Ko, K. S. Yang, W. J. Lee and M. Park, *Fiber Polym.*, 2000, **1**, 32.
94. T. W. Zerda, G. Song and W. H. Waddeli, *Rubber Chem. Technol.*, 2003, **76**, 769.

95. G. Meier, M. Klüppel, H. Geisler and R. H. Schuster, *Kautsch. Gummi Kunstst.*, 2005, **58**, 587.
96. H. H. Le, M. Tiwari, S. Ilisch and H.-J. Radusch, *Kautsch. Gummi Kunstst.*, 2005, **58**, 575.
97. H. H. Le, G. Kasaliwal, S. Ilisch and H.-J. Radusch, *Kautsch. Gummi Kunstst.*, 2009, **62**, 326.
98. G. Sui, W. H. Zhong, X. P. Yang, Y. H. Yu and S. H. Zhao, *Polym. Adv. Technol.*, 2008, **19**, 1543.
99. A. A. Abdullateef, S. P. Thomas, M. A. Al Harthi, S. K. De, S. Bandyopadhyay and A. A. Basfar, *J. Appl. Polym. Sci.*, 2011.
100. M. J. Jiang, Z. M. Dang, S. H. Yao and J. Bai, *Chem. Phys. Lett.*, 2008, **457**, 352.
101. A. M. Shanmugharaj, J. H. Bae, K. Y. Lee, W. H. Noh, S. H. Lee and S. H. Ryu, *Compos. Sci. Technol.*, 2007, **67**, 1813.
102. H. X. Jiang, Q. Q. Ni and T. Natsuki, *Polym. Compos.*, 2011, **32**, 236.
103. T. Fukushima, A. Kosaka, Y. Ishimura, T. Yamamoto, T. Takigawa and N. Ishii, *Science*, 2003, **300**, 2072.
104. T. Fukushima and T. Aida, *Chem. Eur. J*, 2007, **13**, 5048.
105. J. C. Ma and D. A. Dougherty, *Chem. Rev.*, 1997, **97**, 1303.
106. T. Sekitani, Y. Noguchi, K. Hata, T. Fukushima, T. Aida and T. Someya, *Science*, 2008, **321**, 1468.
107. B. Likozar and Z. Major, *Appl. Surf. Sci.*, 2009, **257**, 565.
108. A. Das, K. W. Stöckelhuber, R. Jurk, J. Fritzsche, M. Klüppel and G. Heinrich, *Carbon*, 2009, **47**, 3313.
109. A. Das, K. W. Stöckelhuber, R. Jurk, M. Saphiannikova, J. Fritzsche, H. Lorenz and G. Heinrich, *Polymer*, 2008, **49**, 5276.
110. K. Subramaniam, A. Das and G. Heinrich, *Compos. Sci. Technol.*, 2011, **71**, 1441.
111. K. Subramaniam, A. Das, D. Steinhauser, M. Klüppel and G. Heinrich, *Eur. Polym. J.*, 2011, **47**, 2234.
112. J. Y. Wang, H. B. Chu and Y. Li, *ACS Nano*, 2008, **2**, 2540.
113. H. H. Le, X. T. Hoang, A. Das, U. Gohs, K. W. Stöckelhuber, R. Boldt, G. Heinrich, R. Adhikari and H.-J. Radusch, *Carbon*, 2012, **50**, 4543.
114. M. G. Freire, P. J. Carvalho, A. M. Fernandes, I. M. Marrucho, A. J. Queimada and J. A. P. Coutinho, *J. Colloid Interface Sci.*, 2007, **314**, 621.
115. B. Likozar, *Trans. F: Nanotechnol.*, 2010, **17**, 35.
116. A. I. Frolov, K. Kirchner, T. Kirchner and M. V. Fedorov, *Faraday Discuss.*, 2012, **154**, 235.
117. Z. Ali, H. H. Le, S. Ilisch and H.-J. Radusch, *J. Mater. Sci.*, 2009, **44**, 6427.
118. Z. Ali, H. H. Le, S. Ilisch, K. Busse and H.-J. Radusch, *J. Appl. Polym. Sci.*, 2009, **113**, 667.
119. Z. Ali, H. H. Le, S. Ilisch, T. Thurn-Albrecht and H.-J. Radusch, *Polymer*, 2010, **51**, 4580.
120. S. Sadhu and A. K. Bhowmick, *J. Polym. Sci., Part B: Polym. Phys.*, 2004, **42**, 1573.

CHAPTER 7
NR Blends and IPNs: Miscibility and Immiscibility

WIWAT PICHAYAKORN,* JIRAPORNCHAI SUKSAEREE AND PRAPAPORN BOONME

Department of Pharmaceutical Technology, Faculty of Pharmaceutical Sciences, Prince of Songkla University, Songkhla 90112, Thailand
*Email: wiwat.p@psu.ac.th

7.1 Introduction

Polymer blends and interpenetrating polymer networks (IPNs), which are mixtures of at least two polymers or copolymers, have been developed for increasingly important roles in improving the cost/performance ratio of simple polymers. For example, blending may be used to reduce the cost of an expensive polymer, to improve durability at high temperatures or to be heat sensitive, and to increase impact resistance. However, the main purpose of blending with natural rubber (NR) is to develop new polymeric materials with specific properties that differ from those of the pure polymers and preferably have a lower cost. This is better than synthesizing new NR polymers, especially if it involves less investment of money, time and effort. There are many reasons for using NR blends, such as to extend engineering resin performance, to dilute with a lower cost polymer, to develop materials with various desired properties, to produce a high-performance blend from synergistically interacting polymers, to adjust the composition of the blends according to customer requirements, and to improve the recycling of industrial wastes by using a different polymer blended composition.

RSC Polymer Chemistry Series No. 7
Natural Rubber Materials, Volume 1: Blends and IPNs
Edited by Sabu Thomas, Chin Han Chan, Laly A. Pothen, Rajisha K. R., Hanna J. Maria
© The Royal Society of Chemistry 2014
Published by the Royal Society of Chemistry, www.rsc.org

In general terms, polymer blends are produced by mixing or blending homogeneously two or more polymers, both in a fluid state as a solution or molten material. However, many mixtures of different polymers may not produce a homogeneous mix naturally due to their different chain structures. Thus, it is unlikely that these polymer blends will form a true solid solution. When these blends are changed to a solid form, one polymer is distributed within the continuous matrix of the other polymer. This distribution must be consistent, otherwise it will cause defects in the product, and the resulting mechanical properties of the polymer blends will be lower than those of the original polymers. Therefore, the miscibility of the polymers being used in the combination must be carefully considered.

The miscibility or immiscibility of possible NR blends and IPNs is the first parameter to be investigated before beginning the mixing. Many polymers cannot be blended into the NR chains although they may produce several improvements in the NR properties. Thus, this chapter describes the blending behaviours of NR with other polymers such as plastics, biopolymers and synthetic rubbers. At present, the thermodynamics of the Gibbs free energy is used to predict the compatibility of NR blends and IPNs. Various compatibilization methods are suggested that can improve the compatibility of heterogeneous NR blends and IPNs. In addition, this chapter describes some of the different techniques used to analyse and confirm the miscible and immiscible properties of NR based blends and IPNs.

7.2 Definitions

Polymer blends: Macroscopically homogeneous mixtures of two or more different components in which the continuous phase is polymeric.[1] They are prepared by mixing the two or more different components in a fluid state, either in solution or in molten forms, in order to obtain a new material with properties that are different but may be somewhere between those of the initial polymers. Finally, they may be dried to a solid state or used as the original homogeneous liquid mixture.

IPNs: A type of polymer blend prepared to modify the properties of NR. They are composed of two or more polymers with at least one being polymerized/crosslinked in their networks both without and/or with covalent bonds between the chains of the same or different polymer types.[2] They can be categorized into two types: non-covalent IPNs and covalent semi-IPNs (Figure 7.1). For the covalent semi-IPNs, the crosslinked covalent bonds occur between the different polymer chains. In addition, non-covalent IPNs can be further categorized into full IPNs and semi-IPNs. Full IPNs are defined as a combination of NR and other polymers in a network in which each is synthesized by polymerization with a crosslinking agent. In semi-IPNs, in contrast, only one type of polymeric component is crosslinked. However, there are no covalent bonds between the chains of the different polymer types, but the chains of the polymer become inserted into the framework of the other polymers. Moreover, pseudo-IPNs are defined as a type of IPN in which one of the

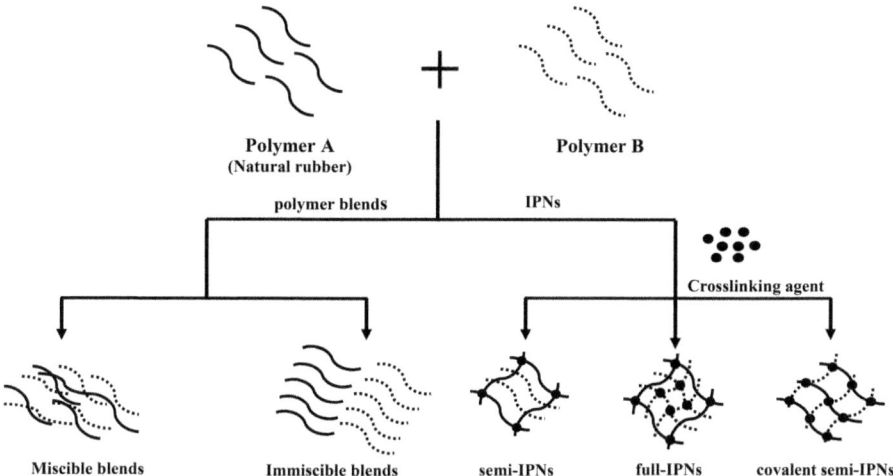

Figure 7.1 Structures of polymer blends and IPNs.
(Modified from Chen et al.,[3] Mathew et al.[13] and Tanaka and Araki.[22])

components has a linear structure.[2,3] They are similar to semi-IPNs in a pseudo or solution state.

Miscibility: The ability of a mixture to form a single phase, but perhaps at a particular temperature, pressure and composition.[1]

Immiscibility: The opposite of miscibility, where a mixture has more than one phase at a specific temperature, pressure and composition.[1]

Polymer blends can be classified as miscible or immiscible (Figure 7.1). These properties depend on the chemical structure, molar mass distributions and molecular architectures of the components. In principle, the constituents of these immiscible blends are separable by physical means. For miscible blends, however, the polymeric chains can be inserted into each of their own frameworks and exhibit a single glass transition temperature or have optical clarity, but only under specific conditions.

Partial miscibility: Both miscible and immiscible polymer blends that also exhibit macroscopically uniform physical properties usually caused by sufficiently strong interactions between the components of the mixture.[4,5] The partially miscible blends show either a homogeneous single phase or a heterogeneous phase, depending on certain conditions. This is called metastable miscibility. Sometimes, they may be only partially miscible yet may remain unchanged for an indefinite period. In some situations, however, phase separation of partially miscible systems may occur by activated nucleation.

Compatibility: The ability of the individual components in an immiscible polymer blend to exhibit interfacial adhesion under given conditions.[1] Special modification processes are required to improve the interfacial properties in an immiscible polymer blend, for example the addition of suitable compatibilizing agents, which can be classified into low molecular weight compounds, copolymers (block or graft copolymers) and crosslinking agents. These processes can stabilize and create a new polymer alloy from the immiscible polymer blends, and they have played an important role in the development of polymeric NR blends.[6,7] In these cases, the polymer blends are macroscopically uniform when observed at a scale larger than several times of visible wavelengths, due to the sufficiently strong interactions between the component polymers, but they may be immiscible at a microscopic scale and should be further investigated. These are called compatible polymer blends.

7.3 Miscibility and Immiscibility of NR Blends and IPNs

7.3.1 Techniques for Preparing NR Blends

Blends of NR with other polymers can be prepared by various mixing processes. Two main techniques, namely mechanical mixing and solution casting, have been widely used to produce NR polymer blends. The chemical bonding between NR and the blended polymers is not affected by the mixing method, but depends on their component types and concentrations.

Mechanical mixing, known as the melt mixing process, has been the most common method for producing polymer blends due to the ease of the scale-up process. Polymer blends can be obtained by exerting heat or force on two or more pure components in the molten state and then mixing them. This method reduces contamination and dilution problems, and ensures that the system does not change during moulding operations because the necessary heat maintains a low viscosity and low shear rates during the mixing. However, many other parameters should be considered and monitored such as viscosity, melting temperature, phase separation temperature and glass transition temperature. A two-roll mill is generally used for this method. This has two horizontal, parallel rotating rollers separated by a small gap, called the nip. The nip can be adjusted by moving the rollers. Blended material reaching the nip deforms by friction forces between the blended materials and the rollers, and flows through the nip in the direction of the roll motion. The front roller usually moves more slowly than the back roller. An internal mixer consists of a feed chamber, called a hopper, standing above a mixing chamber and a floating ram. The feed chamber forces the material into the mixing chamber, which can be set at different speeds. Dispersed mixing requires a high shear stress, and so it is important to remove heat efficiently to maintain a constant viscosity, by drilling the cooling system passages to the inner surface of a mixing chamber. An extruder is a mixing system that can exchange the material planes for a higher mixing efficacy. There are many approaches to improving the mixing in an extruder. The first approach is to use pins in the barrel to divert and divide the flow of the blended materials. The second approach is to use a cavity transfer mixer to promote the exchange of blended materials. Consequently, these two approaches can improve the mixing of the extruder by diversion and reorientation of the flow stream. The third approach is to use a twin screw to develop a shear force.

Solution casting is also a popular technique for NR blending. However, this method is appropriate only on a small scale, particularly a laboratory scale. Polymer blends can be prepared by dissolving or dispersing two or more pure components in an appropriate soluble solvent, and then mixing them. The solvent is finally eliminated by evaporation. Generally, three methods have been widely utilized to remove solvent. Firstly, the mixtures are transferred to a vacuum oven and the temperature is continuously increased to evaporate the solvent, and then cooled down to room temperature. Secondly, the mixtures are frozen and lyophilized under vacuum. Thirdly, when using high boiling point solvents, the mixtures are added into the non-solvent to precipitate the polymer blends, resulting in powder-formed products.[4,6,8–10]

7.3.2 Identification Parameters for Determination of Miscibility

The most popular and preferable parameter in identification of the miscibility of polymer blends is glass transition temperature. It is better than visual observation, and other methods such as light scattering, X-ray scattering and neutron scattering because it can identify at the molecular level and can be easily determined. Glass transition temperature is the temperature at which the polymeric chains have a combination of energy equal to the forces of attraction. Below this temperature, the polymer chains are locked into a random network with motion being restricted to vibrational, rotational and short order translational movements through small units of the polymer chains resulting in the glass morphology. Above this point, in contrast, translational movement of the entire chains is possible, and diffusion therefore occurs as might be required – this is called the rubbery stage. The observed glass transition temperature is affected by the frequency of testing in each sample, pressure, crystallinity and crosslinking. For miscible polymer blends, glass transition temperature does not relate to each copolymer because they are homogeneous materials at the molecular level, and result in a single peak of glass transition temperature. In contrast, immiscible blends exhibit multi-phase structures and have multiple glass transition temperatures of the original components for all conditions. If these blended systems are made of two or more polymers, they will have several peaks of glass transition temperature depending on the number of polymer components. Moreover, partially miscible blends show the glass transition temperature values of the blended components, which will shift from the pure blended components and toward each other for any blended components.

However, most NR blends and IPNs are heterogeneous due to the hydrophobic nature of NR. They may form mixtures that are nanometres in size when mixed with the other polymers rather than interpenetrating each other on a molecular scale. They are expected to have total or partial phase separation. In other words, these systems may be in a continuously homogeneous phase only on a macroscopic scale. Thus, the number of glass transition temperature peaks depends on the number and amount of the blended copolymers. The glass transition temperature peaks of each copolymer can be closely shifted up to the continuity of the components.

7.3.3 NR Blends and IPNs

NR and plastic blends have been reported for the preparation of thermoplastic NR blended systems. Many types of plastics have been blended with NR, such as polyethylene,[11] polypropylene,[12] polystyrene,[13,14] polyacrylates or polyacrylic acid,[15] and poly(methyl methacrylate).[9,16–20] These thermoplastic NRs provide new materials with different properties that range from a soft elastomer to a semi-rigid NR plastic for various industrial applications.[16,17,21,22] In addition, NR can improve the toughness of brittle thermoplastic materials. The NR and thermoplastic blends express greater ductility and impact strength at low temperatures. The requirements for certain mechanical and physical specifications can be achieved by optimizing the blended compositions and selecting the most suitable thermoplastics.

NR/poly(methyl methacrylate) blends have been the most widely used NR/plastic blends.[9,16–20] However, they have poor mechanical properties. Phase separation was found at the interface and indicated that these blends were immiscible.[9] The interfacial adhesion between the NR and poly(methyl methacrylate) surfaces could be increased by adding a graft copolymer of NR-*graft*-poly(methyl methacrylate) as a compatibilizer for improving the deformation nature of the polymer blends. Thus, these compatibilized NR/poly(methyl methacrylate) blends have improved mechanical properties compared to the uncompatibilized blends.

NR/poly(methyl methacrylate) based IPNs are also easily prepared in a pseudo-IPN solution by a sequential method using a high molecular weight NR and poly(methyl methacrylate) with divinyl benzene as a crosslinking agent and benzoyl peroxide as an initiator.[20] These IPNs clearly showed a compact and lower phase separation with a crosslinked density value of 142 g mol^{-1} that clearly produced a smooth morphology. The polymer chains of poly(methyl methacrylate) were dispersed in the continuous crosslinked NR phase. Only one glass transition temperature value was observed, indicating that the systems were compatible under these conditions.

Acrylic rubber was blended with NR at different ratios using a two-roll mill.[15] These blends were generally immiscible. However, the morphology of the NR/acrylic rubber blends was homogeneous when the blending contained 20% w/w NR content. Increasing the acrylic rubber content decreased the tensile properties, but increased the oil and heat resistances due to the specific properties of acrylic rubber.

NR/polystyrene IPNs were prepared using azobisisobutyronitrile, benzoyl peroxide or dicumyl peroxide as an initiator, and divinyl benzene as a crosslinking agent.[13,23] The crosslinking of polystyrene produced chemical entanglements that resulted in some intimate mixing of the phases and improved the homogeneity. When the blending had high levels of crosslinking agent, physical entanglements were generated and led to tighter and more compact blended networks. By increasing the crosslinked level of polystyrene with NR, the flexibility of the polymer chains became highly restricted, and the IPNs became brittle.

The naturally biodegradable polymers such as starch,[24,25] chitosan[26] and cellulose derived from natural sources[27,28] have produced a number of interesting NR blends and IPNs. These blended systems have an advantage in that they create fewer waste disposal problems compared to the petroleum based polymeric materials. The use of starch blends to enhance the biodegradability of conventional plastics has been reported by many researchers[24,29,30] in order to reduce the environmental impact of petroleum based plastic products and waste. The NR/maize starch blends exhibited a decrease in their mechanical strength due to the specific properties of starch.[24] However, the blended polymers showed a low interfacial interaction between the two phases due to the different polarity behaviour of the hydrophobic NR and the hydrophilic starch.

Much research has been conducted into producing new plastics based on thermoplastic starch and NR blends. Kahar and coworkers[30] reported the incorporation of tapioca starch into a thermoplastic NR/high density polyethylene blend at a ratio of 70/30 that could enhance its ability for environmental degradation due to the individual propensity of tapioca starch to absorb water. However, the Young's modulus, tensile strength and elongation at break decreased with increasing tapioca starch loading, due to the poor mechanical properties of this substance. The low viscosity of tapioca starch means that it can be easily dispersed in the thermoplastic NR phase, but this blend exhibited incompatibility. However, modifying the tapioca starch with citric acid increased the tensile properties and reduced the surface tension at the interphase between the tapioca starch and the thermoplastic NR system, due to the increasing hydrophilicity of tapioca starch. In addition, modified tapioca starch increased the dispersion level in the thermoplastic NR system.

Recently, Pichayakorn and coworkers reported the formation of mixtures of low protein NR and cellulose derivatives such as hydroxypropylmethyl cellulose,[27] sodium carboxymethyl cellulose[28] and methyl cellulose[28] by solution casting. These blended materials improved the mechanical properties of the films compared to those made from low protein NR itself. The Young's modulus and tensile strength increased, and the elongation at break values decreased with the cellulose derivative blending due to the individual poor mechanical properties of the cellulose derivatives. However, these properties increased with an increasing amount of cellulose derivatives. Nevertheless, the elasticity of the blended films was lower than that of the low protein NR film. This was due to the bulkiness of the anhydroglucose in the cellulose derivatives structure. The elasticity increased with increasing amounts of cellulose derivatives, which was directly related to the concentration of the methoxy groups in the cellulose derivative molecules.[31] However, they showed only one glass transition temperature, which was slightly changed from that of the pure NR.

This indicated that the NR/cellulose derivative blends were partially miscible only under experimental conditions.[27,28]

Polylactide is a synthetic biodegradable polymer of a thermoplastic polyester derived from biomass such as sugar, corn and beet. It possesses excellent physical and mechanical properties as well as being biocompatible and biodegradable. It has been restricted to packaging applications due to its high brittleness and poor crystallization behaviour. The blends of polylactide with NR were chosen to improve brittleness.[29,32,33] NR was blended with polylactide using simple melt blending in an internal mixer followed by a twin-screw extruder and compression moulding technique.[29] The rubber phase was uniformly dispersed in the continuous polylactide matrix with a droplet size in the range of 1–2 μm. The brittleness of polylactide was significantly improved. The elongation at break from 5% of pure polylactide increased to 200% by incorporation of 10% w/w NR as a blended material. Moreover, the NR molecules acted as a nucleating agent to improve the crystallization ability of polylactide. Although NR/polylactide blends can exhibit incompatible behaviour, it is possible to use them in the packaging sector because of their higher mechanical properties than pure polylactide. Therefore, these NR-blended materials are very promising for industrial applications.

NR is primarily utilized in the tyre industry because of its ideal properties. However, it has poor stability due to the existence of many double or unsaturated bonds. To improve the ambient stability of NR, synthetic rubber was chosen for blending. Zhang and coworkers[34] blended and vulcanized NR and chloroprene rubber at a weight ratio of 75/25 by using a two-roll mill. The NR/chloroprene rubber blends had improved mechanical properties in terms of their elongation strength and Shore A hardness. Moreover, the vulcanized chloroprene rubber blends had excellent oil resistance, thermal stability, self-extinguishing ability and ozone resistance.

Therefore, polymer blends have been widely used to improve properties, for product uniformity, processability, plant flexibility and high productivity of materials that can be widely used in further industrial applications. Table 7.1 summarizes some examples of the immiscibility of several NR/polymer blends in nature.

Table 7.1 Immiscibility of natural rubber blends in nature.

Immiscibility	References
Natural rubber/poly(methyl methacrylate)	18, 20
Natural rubber/polystyrene	13, 23
Natural rubber/polylactide	29, 32
Natural rubber/high-density polyethylene/tapioca starch	30
Natural rubber/maize starch	24
Natural rubber/chloroprene rubber	34
Natural rubber/ethylene vinyl acetate copolymer	32
Natural rubber/chlorosulfonated polyethylene	11
Natural rubber/acrylonitrile butadiene rubber	38, 39

7.4 Thermodynamics of NR Blends and IPNs

Assessment of the phase behaviour of blended mixtures is essential to determine the miscibility of polymer blends. One of the characteristics of miscible polymer blends is clarity at room temperature, but they may become turbid after raising or reducing the temperature. To predict the miscibility of polymer blends, many research workers have attempted to develop thermodynamic theories that can describe the phase behaviour of polymer blends.[7,9,18,20,24,35,36] These are not only used to design the specific properties of a mixture but also to minimize the operating cost. Due to the development of statistics, physics and fundamental thermodynamics, the knowledge of phase behaviour has been generally understood. According to the statistical thermodynamic methodology, there are two ways to formulate the theory of molecular liquid chains and their mixtures. Firstly, it depends on the ideas of predicting the probability of finding polymer segments in particular configurations such as by using the Ising model and the mean field model. Secondly, it is attributed to the assumption of there being a regular lattice structure of crystalline solids. Regarding the latter, it results in the so-called lattice model, free volume theory, or the hole theory.[4] However, the basic and very useful thermodynamic equation that will be able to describe the miscibility of two polymers is the Gibbs free energy (ΔG_m) equation:

$$\Delta G_m = \Delta H_m - T\Delta S_m \tag{7.1}$$

where H_m is enthalpy, S_m is entropy and T is the temperature in Kelvin. From the Gibbs free energy equation, a prediction can be made about whether a blend is miscible, immiscible or partially miscible by taking into account the sign of ΔG_m.

Miscible blends are homogeneous at the molecular level. They can be easy to form and this requires a negative sign for the Gibbs free energy of mixing. The domain size is comparable to the dimensions of the macromolecular statistical segment:

$$\Delta H_m - T\Delta S_m < 0 \tag{7.2}$$

From Equation (7.2), an exothermic mixture ($\Delta H_m < 0$) will spontaneously form. For endothermic mixtures, miscibility will occur only at a high temperature. In addition, most blends can be made miscible by increasing the pressure. These effects depend on the magnitude of the mixing heat ΔH_m. For $\Delta H_m < 0$, the miscibility is enhanced by compression. For $\Delta H_m > 0$, the miscibility is reduced. In contrast, with a positive sign for the ΔG_m, the blended system is immiscible. Some blended systems can exhibit a partial miscibility with a negative sign for ΔG_m. Thus, the second criterion of the sign of the second derivative of ΔG_m is required to specify the difference between partial and full miscibility.

For two-component blends, it is possible to construct a phase diagram that may exhibit a lower or upper critical solution temperature (Figure 7.2).

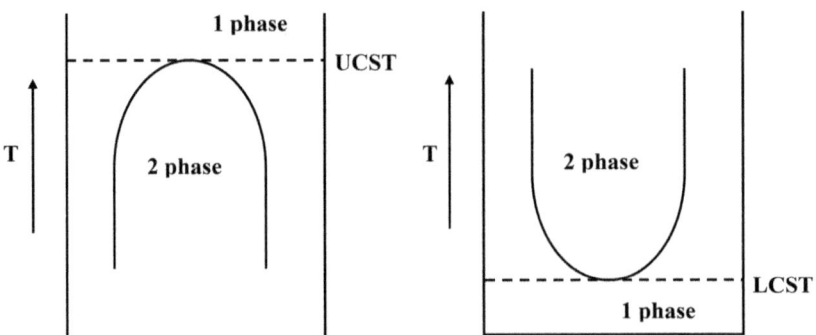

Figure 7.2 Phase diagrams of binary blends showing the upper and lower critical solution temperature behaviours (as UCST and LCST, respectively). (Modified from Strobl.[5])

Equation (7.1) can also demonstrate the phase diagram of polymer blends. Generally, polymer blends exhibit the upper critical solution temperature behaviour as the temperature increases and the $T\Delta S_m$ term should increase. On the other hand, when the ΔG_m term becomes positive and phase separation then takes place, this indicates that the lower critical solution temperature behaviour is more commonly seen. In this case, the enthalpy of mixing increases as the temperature increases, leading to the free energy being positive due to a phase separation phenomenon. Intermolecular attractive forces respond to the miscible behaviour and tend to disappear with the internal energy of the molecules. There are repulsive interactions and attractive forces between specific groups that can lead to incorporation of the two polymers. With increasing temperature, the fraction of the closed bonds decreases and the repulsive forces eventually dominate. This may induce the enthalpy of mixing to change from being negative at a low temperature to be positive at a higher one. Moreover, it is possibly related to the volume of shrinkage. As the temperature increases, the free volume for the local motion of monomers decreases. Hence, the mobility is reduced and leads to a lower entropy. For partially miscible blends, some homogeneous systems show a phase separation if the temperature increases and demonstrate lower critical solution temperature behaviour, while some change from being heterogeneous at room temperature to homogeneous at a higher temperature; this is called upper critical solution temperature behaviour.[5]

7.5 Phase Separation and Compatibilization

In most cases, NR blends are weak and brittle materials that mostly express incompatibility because the incorporation of a dispersed phase in a matrix phase leads to the presence of stress and a weak interface that affects the poor mechanical coupling between phases. The mechanical properties of a blend will be determined not only by the properties of its components, but also by the phase morphology and the interphase adhesion. This stress transfer in the blended substances is important for their applications. The phase morphology will normally be determined by factors that are process variables, such as the rheology of the blended components and the interfacial tension. This morphology is unlikely to be in thermodynamic equilibrium, but generally stabilizes against de-mixing by some methods. This usually means *via* quenching to below the glass transition temperature due to the occurrence of crystallinity in one or both phases, or occasionally by crosslinking. The compatibilization can be the principle interaction in complex ways to influence the final blended properties. One effect of compatibilizers is to reduce the interfacial tension and act as an emulsifying effect, and can lead to an extremely fine dispersion of one phase in another phase. Another effect is to increase the adhesion at the phase boundaries, giving an improved stress transfer. A third effect is to stabilize the dispersed phase against growth during annealing, by modifying the phase boundary interface.[7] Thus, the compatibilizers involve advantages such as by improving the dispersion of the interfacial energy, and enhancing the adhesion between the blending phases.

The technological definition of compatibilization, such as in the modification of NR blends, is the process that produces a desirable set of properties. These approaches can be defined and may assist the development of materials.[6,37] Table 7.2 shows the various types of compatibilization techniques for several NR/polymer blended applications.

7.5.1 Achievement of Thermodynamic Miscibility

Compatibilization by the achievement of thermodynamic miscibility is a concept that has been exploited in only a handful of situations to produce a commercial blend. The miscibility between polymers is determined by a balance of enthalpy and entropic contributions to the free energy of mixing. While the entropy of small molecules is high enough to ensure miscibility, the entropy of polymers is almost zero, which means that the entropy will determine the miscibility.[2,14]

7.5.2 Addition of Block or Graft Copolymers

The addition of block or graft copolymers represents the widest range of techniques to compatibilize blends. Block or graft copolymers containing some segments that are similar to the parts of the blended components are obvious choices as compatibilizers because they can increase the miscibility between the copolymer segments and the corresponding blended components. The appropriate structures and molecular weights of the copolymers are important factors that determine the distribution of these molecules between the blended interfaces.[8]

Poly(methyl methacrylate) is a hard and brittle material whose strength is greatly affected by stress concentrations.[16] It is usually chosen to disperse in the NR matrix, but the mixture forms as heterogeneous blends. A blend of NR/poly(methyl methacrylate) with a weight ratio of 60/40 was made compatible by the addition of a graft copolymer of NR-*graft*-poly(methyl methacrylate) as compatibilizer to enhance the interfacial adhesion between the poly(methyl

Table 7.2 Compatibilization techniques for natural rubber blends.

NR blends	Compatibilizer	Compatibilization techniques	References
Natural rubber/high-density polyethylene/tapioca starch	Modified tapioca starch with citric acid	Achievement of thermodynamic miscibility	28
Natural rubber/maize starch	Glycidyl methacrylate	Achievement of thermodynamic miscibility	22
Natural rubber-*graft*-poly(methyl methacrylate)/poly(methyl methacrylate)	Natural rubber-*graft*-poly(methyl methacrylate)	Addition of graft copolymers	9, 16
Natural rubber-*graft*-methyl methacrylic acid/poly(methyl methacrylate)	Natural rubber-*graft*-methyl methacrylic acid	Addition of graft copolymers	17
Natural rubber/ethylene vinyl acetate copolymer	Natural rubber-*graft*-polydimethyl(methacryloyloxymethyl) phosphonate	Addition of graft copolymers	33
Natural rubber/polylactide	Natural rubber-*graft*-glycidyl methacrylate	Addition of graft copolymers	30
Natural rubber/unsaturated polyester resin	Natural rubber-*graft*-polystyrene	Addition of graft copolymers	20
Natural rubber or maleated natural rubber/ethylene vinyl acetate	Phenolic-modified ethylene vinyl acetate	Addition of block copolymers	34
Natural rubber/chlorosulfonated polyethylene	Epoxidized natural rubber (Epoxyprene® 25)	Addition of block copolymers	11
Natural rubber/acrylonitrile butadiene rubber	Ethylene vinyl acetate copolymer	Addition of block copolymers	36, 37
Hydrogenated natural rubber/poly(methyl methacrylate-*co*-styrene)	Hydrogenated natural rubber and poly(methyl methacrylate-co-styrene)	Addition of functional or reactive polymers	38
Maleated natural rubber/poly(methyl methacrylate)	Maleated natural rubber	Addition of functional or reactive polymers	19
Maleated depolymerized natural rubber/epoxy resin	Maleated depolymerized natural rubber	Addition of functional or reactive polymers	39
Natural rubber/poly(methyl methacrylate)	Poly(methyl methacrylate) grafted on the backbone of natural rubber latex	*In situ* graft polymerization	18

methacrylate) and NR phases. The NR-*graft*-poly(methyl methacrylate) increased the tensile strength and hardness of the NR/poly(methyl methacrylate) blends by increasing by 0–5% the methyl methacrylic acid concentrations in the graft copolymerization process. A greater chemical interaction between the polar groups in the NR-*graft*-poly(methyl methacrylate) molecules and poly(methyl methacrylate) phases may be expected in compatible blends. However, addition of the NR-*graft*-poly(methyl methacrylate) with different degrees of graft copolymerization decreased the elongation at break.

NR/maize starch blends exhibit poor mechanical properties due to the characteristics of the starch and the low interfacial interactions between the two phases. The 1 part per hundred of rubber (phr) of glycidyl methacrylate was an appropriate compatibilizer to improve the compatibility and mechanical properties compared to the uncompatibilized blends because the epoxy group of the glycidyl methacrylate chemically interacted with the hydroxyl group of the maize starch and greatly decreased the cohesion energy and the crystallization of the starch. Therefore, starch molecules become better dispersed in the NR matrix and increase the interfacial interaction between the hydrophilic starch and the hydrophobic NR.[24]

Moreover, the 1% w/w of glycidyl methacrylate in the NR-*graft*-glycidyl methacrylate with 4.35% of grafting was also used as a compatibilizer for the NR/polylactide blends. This NR-*graft*-glycidyl methacrylate improved the compatibility between polylactide and NR phases. The epoxy groups in the NR-*graft*-glycidyl methacrylate reacted with the carboxyl groups of the polylactide chains in the blending process. However, this blended polymer increased the domain size of the NR particles in appearance and produced a poor distribution of the NR particles in the polylactide matrix.[32]

The addition of a block copolymer of 10 phr epoxidized NR into NR/chlorosulfonated polyethylene blends increased the tensile properties and tear strength of the blended system. The miscibility of the epoxidized NR was enhanced by an *in situ* grafting reaction of epoxidized NR onto the surface of the chlorosulfonated polyethylene particles. This graft copolymer reduced the interfacial tension between the chlorosulfonated polyethylene particles and the NR matrix, and the epoxidized NR acted as the load transferring agent between the NR and the chlorosulfonated polyethylene.[11]

Although an ethylene vinyl acetate copolymer was immiscible in NR blends,[35] addition of a 6 phr ethylene vinyl acetate block copolymer enabled compatibilization of heterogeneous NR/acrylonitrile butadiene rubber blends. These blends increased the tensile strength, the elongation at break and tear strength due to an increase in the interfacial adhesion between the blended components by increasing the rigidity of the matrix in the presence of the ethylene vinyl acetate copolymers.[38,39]

7.5.3 Addition of Functional or Reactive Polymers

The addition of functional or reactive polymers as compatibilizers has been reported. Usually, a small amount of the same polymer blends can be modified to add to the functional or reactive units that have an affinity to chemically react with the second blended components. Moreover, different molecular attractive forces are also possible, such as ionic bonding, hydrogen bonding, Van der Waals forces, dipole–dipole interaction, or hydrophobic interaction. The functional modification has to be achieved in a reactor or *via* a modified extrusion process.[7,8]

A modified NR can improve the blended systems that use it as an impact modifier, and a thermal and oxidative resistance improver for blending.[40] Thus, increasing the polar groups of NR may act to increase interfacial compatibility and to reduce the dimensions of the dispersed phase by interacting with the polar groups of the polymers.

Maleic anhydride has been grafted onto the backbone of a non-polar NR structure by the Diels–Alder and free radical reactions.[19,41] The free radicals generated at the C=C bond of maleic anhydride and the allylic carbon atoms of the NR backbone by heating and shearing then become one of the reactive sites for the grafting reaction. These grafting reactions with maleic anhydride into the polymer blends have been commercially popular and relatively inexpensive.

The low surface energy of NR has been improved so that it has a higher hydrophilicity and produces a higher adhesion of the NR blends with other polar polymers. When NR was used to prepare reactive blends with poly(methyl methacrylate) using a mixer, a compatible modified NR/poly(methyl methacrylate) blend was formed. When the sites for a high chemical interaction of the modified NR in the blended system between the different phases occurred, they exhibited a higher degree of thermal resistance and had an increased viscosity. Moreover, NR that had been modified by hydrogenation methods was one of the chemical modifications available to enhance the oxidative and thermal resistance of the diene-based NR before being blending with the poly(methyl methacrylate-*co*-styrene) copolymers.[40]

7.5.4 *In Situ* Graft Polymerization or Reactive Blending

A compatible thermoplastic blend can also be produced by reactive blending, which relies on the *in situ* formation of copolymers or interactive polymers. This differs from other compatibilization types in that the blended components themselves are either chosen or modified so that the inter-reaction occurs during the melt blending. Other compatibilizers are not added separately.[8] The continuous mixing processes such as using either a single- or twin-screw extruder are the most popular instruments for forming *in situ* graft polymers because in these processes the temperature can be controlled for the appropriate reaction.

The mechanisms required for the reaction mixture are (i) a graft or block copolymer formed by the chemical reaction between the chemical reactivities in the polymer chains, which can be initiated by the addition of an initiator to the mixing process; (ii) the formation of a block copolymer by an interchange reaction between the molecular chains of each type of the condensation-type polymer used in combination; (iii) the cutting and rebonding of each molecular polymer chain for a block or graft copolymer created under high shear conditions; and (iv) promoting the reaction by the addition of a catalyst.[5,6,10]

The methyl methacrylic acid monomers were converted to poly(methyl methacrylate), which was simultaneously grafted onto the backbone of the NR latex to form the graft copolymers by the use of radiation. These graft copolymers had an increased viscosity of the NR latex due to the increment in the total solid content. In addition, the graft copolymers had increased hardness as the degree of grafted poly(methyl methacrylate) onto the NR backbone increased, but they did not greatly influence the physical properties, *i.e.* tensile strength, modulus, elongation at break and thermal stability.[18]

7.6 Techniques for Measuring Miscibility and Immiscibility Properties

Numerous preparation techniques have now been developed for polymer blends. Basically, polymer blends can be classified into three categories according to their phased structure: miscible, immiscible and partially miscible blends. Miscible blends always show a stable homogeneous single-phase structure with one glass transition temperature, while the immiscible blends exhibit two or more phase structures together with two or more glass transition temperatures of the original components for all conditions. On the other hand, partially miscible blends can show either a homogeneous single phase or a heterogeneous phase, depending on certain conditions. However, a large number of commercial alloys and blends usually fall into this category.

It is clear that NR blends have different properties to their pure components. Among those blends, there can still be differences such as between the miscible and immiscible blends. A large number of methods have been introduced in order to study polymer blends more easier and more accurately. Four main groups with respect to the miscibility and immiscibility evaluations are briefly discussed, such as glass transition temperature, scattering, morphology, and infrared spectroscopy studies.[4]

7.6.1 Glass Transition Temperature Studies

It is generally known that miscible blends always exhibit one glass transition temperature, whereas immiscible blends show several original glass transition temperatures according to the number of components. In the case of partial miscibility, various glass transition temperatures can be obtained. However, it should be kept in mind that a single glass transition temperature does not always mean that there is complete miscibility. The amount of each component has a strong influence on the capability of equipment. Nowadays, there are several techniques for determining glass transition temperature values such as differential scanning calorimetry, differential thermal analysis, modulated differential scanning calorimetry, thermomechanical analysis, and dynamic mechanical analysis. Of these, the most popular is thermal analysis using differential scanning calorimetry. Another technique that should be mentioned is a mechanical technique using a dynamic mechanical thermal analyser. By subjecting polymers to small amplitude cyclic deformation, important information can be obtained concerning transitions that have occurred on the molecular scale. Moreover, other techniques can be used to determine glass transition temperature values such as the dielectric technique, dilatometric technique, radioluminescent spectroscopy, and thermogravimetric analysis.[42]

7.6.2 Scattering Studies

For normal amorphous systems, homogeneous mixtures are usually transparent, whereas heterogeneous mixtures are cloudy unless the components of the mixture have identical refractive indices. By variations in the temperature, pressure and composition of the mixture, the miscible blended appearance can change from being transparent to cloudy, or vice versa. The first appearance of cloudiness denotes the cloud point. Using a light scattering technique, one can investigate the phase separation phenomena, namely nucleation and growth, and spinodal decomposition. In order to use a scattering technique, a laser light is not only the source that can be used – X-ray and neutron analysis is also available. However, since the operating costs are very high and there is a limitation on the availability of such equipment, results from these methods are uncommon.[43]

7.6.3 Morphological Studies

High-powered microscopes can be used to reveal the internal structure, *e.g.* an interpenetrating phase, which is impossible to see with the naked eye. The normal technique used to observe phase boundaries under normal light is optical microscopy. However, as some blends have very tiny components, other more powerful techniques are required, *i.e.* transmission electron microscopy, scanning electron microscopy and atomic force microscopy. It should be noted that in order to see the structures clearly, preliminary treatments are sometimes necessary, for example etching or staining.[4,10]

7.6.4 Infrared Spectroscopy

A Fourier transform infrared spectrometer can be used to detect functional groups. This technique indirectly detects the miscibility by investigating some types of specific interactions in the blends, *e.g.* hydrogen bonding. Knowing about these molecular interactions can then be used to explain the shape of the phase diagrams. It is apparent that, in reality, there is no perfect technique to study all systems. The characteristics of different systems, the time available, operating costs, and the limitations of equipment need to be taken into consideration.[4]

7.7 Conclusions

Miscibility and immiscibility are important factors used to determine the properties of NR blends and IPNs. These parameters directly affect the completion or defects of new blended materials. Blends of NR with other polymers can be prepared by various mixing processes, such as mechanical mixing and solution casting. Normally, the NR is difficult to blend to produce an absolute miscibility with other polymers due to its hydrophobic property and immiscibility or partial miscibility are usually found. The thermodynamic Gibbs free energy and glass transition temperature can be used to predict miscible behaviours. Several techniques can also be used to improve the compatibility of NR blends with several polymers by achievement of thermodynamic miscibility, addition of block or graft copolymers, addition of functional or reactive polymers, and *in situ* graft polymerization or reactive blending. Several techniques can be used to determine the miscibility, however, observation of glass transition temperature is the most popular technique and can provide the most accurate results.

References

1. W. J. Work, K. Horie, M. Hess and R. F. T. Stepto, *Pure Appl. Chem.*, 2004, **76**, 1985.
2. Y. S. Lipatov and T. Alekseeva, *Phase-Separated Interpenetrating Polymer Networks*, ed. Y. S. Lipatov and T. Alekseeva, Springer Verlag, Berlin, 2007, pp. 1–227.
3. C. H. Chen, W. J. Chen, M. H. Chen and Y. M. Li, *Polymer*, 2000, **41**, 7961.
4. O. Olabisi, L. M. Robeson and M. T. Shaw, *Polymer-Polymer Miscibility*, ed. O. Olabisi, L. M. Robeson and M. T. Shaw, Academic Press, New York, 1979, pp. 19–193, 277–319.
5. G. R. Strobl, in *The Physics of Polymers: Concepts for Understanding Their Structures and Behavior*, ed. G. R. Strobl, Springer Verlag, Berlin, 2007, pp. 105–206.
6. G. G. Norman, in *Copolymers, Polyblends, and Composites*, ed. N. A. J. Platzer, American Chemical Society, Washington, 1975, pp. 76–84.
7. A. Ajji, in *Polymer Blends Handbook*, ed. L. A. Utracki, Kluwer Academic Publishers, the Netherlands, 2002, pp. 295–336.
8. S. B. Brown, in *Polymer Blends Handbook*, ed. L. A. Utracki, Kluwer Academic Publishers, the Netherlands, 2002, pp. 339–414.
9. Z. Oommen, G. Groeninckx and S. Thomas, *J. Polym. Sci., Part B: Polym. Phys.*, 2000, **38**, 525.
10. C. Paul, in *Blends of NR: Novel Techniques for Blending with Speciality Polymers*, ed. A. J. Tinker and K. P. Jones, Chapman & Hall, London, 1998, pp. 21–38.
11. V. Tanrattanakul and A. Petchkaew, *J. Appl. Polym. Sci.*, 2006, **99**, 127.
12. A. S. Hashim and S. K. Ong, *Polym. Int.*, 2002, **51**, 611.
13. A. P. Mathew, G. Groeninckx, G. H. Michler, H. J. Radusch and S. Thomas, *J. Polym. Sci., Part B: Polym. Phys.*, 2003, **41**, 1680.
14. S. H. Heidary, I. A. Amraei and A. Payami, *J. Appl. Polym. Sci.*, 2009, **113**, 2143.

15. J. Wootthikanokkhan and B. Tongrubbai, *J. Appl. Polym. Sci.*, 2002, **86**, 1532.
16. Z. Oommen and S. Thomas, *J. Appl. Polym. Sci.*, 1997, **65**, 1245.
17. L. Thiraphattaraphun, S. Kiatkamjornwong, P. Prasassarakich and S. Damronglerd, *J. Appl. Polym. Sci.*, 2001, **81**, 428.
18. F. Sundardi and S. Kadariah, *J. Appl. Polym. Sci.*, 1984, **29**, 1515.
19. C. Nakason, S. Saiwaree, S. Tatun and A. Kaesaman, *Polym. Test.*, 2006, **25**, 656.
20. R. M. Alcântara, A. P. P. Rodrigues and G. G. D. Barros, *Polymer*, 1999, **40**, 1651.
21. S. Chuayjuljit, P. Siridamrong and V. Pimpan, *J. Appl. Polym. Sci.*, 2004, **94**, 1496.
22. H. Tanaka and T. Araki, *Chem. Eng. Sci.*, 2006, **61**, 2108.
23. A. P. Mathew, S. Packirisamy, H. J. Radusch and S. Thomas, *Eur. Polym. J.*, 2001, **37**, 1921.
24. A. I. Khalaf and E. M. Sadek, *J. Appl. Polym. Sci.*, 2012, **125**, 959.
25. A. J. F. Carvalho, A. E. Job, N. Alves, A. A. S. Curvelo and A. Gandini, *Carbohydr. Polym.*, 2003, **53**, 95.
26. V. Rao and J. Johns, *J. Appl. Polym. Sci.*, 2008, **107**, 2217.
27. W. Pichayakorn, J. Suksaeree, P. Boonme, T. Amnuaikit, W. Taweepreda and G. C. Ritthidej, *Ind. Eng. Chem. Res.*, 2012, **51**, 8442.
28. W. Pichayakorn, J. Suksaeree, P. Boonme, T. Amnuaikit, W. Taweepreda and G. C. Ritthidej, *J. Membr. Sci.*, 2012, **411–412**, 81.
29. N. Bitinis, R. Verdejo, P. Cassagnau and M. A. Lopez-Manchado, *Mater. Chem. Phys.*, 2011, **129**, 823.
30. A. W. M. Kahar, H. Ismail and N. Othman, *J. Appl. Polym. Sci.*, 2012, **125**, 768.
31. T. S. H. Kwok, B. V. Sunderland and P. W. S. Heng, *Chem. Pharm. Bull.*, 2004, **52**, 790.
32. P. Juntuek, C. Ruksakulpiwat, P. Chumsamrong and Y. Ruksakulpiwat, *J. Appl. Polym. Sci.*, 2012, **125**, 745.
33. M. Kowalczyk and E. Piorkowska, *J. Appl. Polym. Sci.*, 2012, **124**, 4579.
34. P. Zhang, G. Huang and Z. Liu, *J. Appl. Polym. Sci.*, 2009, **111**, 673.
35. P. Intharapat, D. Derouet and C. Nakason, *Polym. Adv. Technol.*, 2010, **21**, 310.
36. W. Kaewsakul, A. Kaesaman and C. Nakason, *e-Polymers*, 2012, **5**, 1.
37. A. D. Litmanovich, N. A. Platé and Y. V. Kudryavtsev, *Prog. Polym. Sci.*, 2002, **27**, 915.
38. P. Kumari, C. K. Radhakrishnan, G. Unnikrishnan, S. Varghese and A. Sujith, *Chem. Eng. Tech.*, 2010, **33**, 97.
39. P. Kumari, C. Radhakrishnan, S. George and G. Unnikrishnan, *J. Polym. Res.*, 2008, **15**, 97.
40. S. Thawornwisit, K. Charmondusit, G. L. Rempel, N. Hinchiranan and P. Prasassarakich, *J. Elastomers Plast.*, 2010, **42**, 35.
41. K. D. Kumar and B. Kothandaraman, *eXPRESS Polym. Lett.*, 2008, **2**, 302.
42. D. Bikiaris, J. Prinos, M. Botev, C. Betchev and C. Panayiotou, *J. Appl. Polym. Sci.*, 2004, **93**, 726.
43. T. Hashimoto and H. Hasegawa, *Neutron News*, 1998, **9**, 32.

CHAPTER 8

Natural Rubber Based Non-Polar Synthetic Rubber Blends

SEIICHI KAWAHARA

Department of Materials Science and Technology, Faculty of Engineering, Nagaoka University of Technology, Nagaoka 940-2188, Japan
Email: kawahara@mst.nagaokaut.ac.jp

8.1 Introduction

The miscibility of natural rubber (NR) blends is one of the most important factors when designing NR products. For instance, when the NR is miscible with a dissimilar polymer on a molecular level, we may improve the properties of NR as a function of the composition of the polymer. This is significantly different from the design for immiscible NR blends, whose properties are greatly dependent upon the morphology of the blend but less so on the composition. In most cases, NR is immiscible with non-polar synthetic rubbers, *i.e.* NR/butadiene rubber (BR) with high *cis*-1,4-butadiene units,[1–4] NR/styrene-butadiene rubber (SBR),[5–10] NR/butyl rubber (IIR),[11,12] NR/silicone rubber (Q)[13,14] and NR/ethylene-propylene-diene rubber (EPDM).[15,16] This means it is important to find miscible NR blends and to control the morphology of the immiscible NR blends in a rational way. In this chapter, properties of NR blends are described from the viewpoint of miscibility, *i.e.* the miscible blend of NR/BR and the immiscible blend of NR/SBR.

8.2 Miscible NR Blends

8.2.1 Background

NR is miscible with BR having a large amount of 1,2-vinyl ethylene units,[17] but not with most polymers.[18] It is therefore quite important to investigate the miscibility of NR/BR blends. This has been carried out with high molecular weight synthetic isoprene rubber (IR)/BR blends as a model for NR/BR blends,[17,19,20] since NR contains about 6% w/w non-rubber components such as proteins, phospholipids, carbohydrates and so forth,[21] which may interfere in any interaction between NR and BR. Figure 8.1 shows a lower critical solution temperature (LCST) phase diagram for the IR/BR(32.3) blend,[19] in which a figure in parentheses represents the 1,2-unit content. The open circles represent the miscible state, whereas the closed circles represent the immiscible state. As for the IR/BR(32.3) (50/50) and (40/50) blends, the miscible state was accomplished at temperatures below 333 K. They entered an immiscible state at temperatures above 333 K. The locus of the phase boundary versus composition delineated a concave curve, characteristic of the LCST phase behaviour. The LCST of the IR/BR(32.3) blend was determined to be 333 K.

Figure 8.2 shows a temperature–copolymer composition phase diagram for the IR/BR blend, where the 1,2-unit content of BR is regarded as a copolymer composition.[19] In a plot of temperature versus 1,2-unit content of BR, we find a phase boundary between miscible and immiscible states. IR is immiscible with

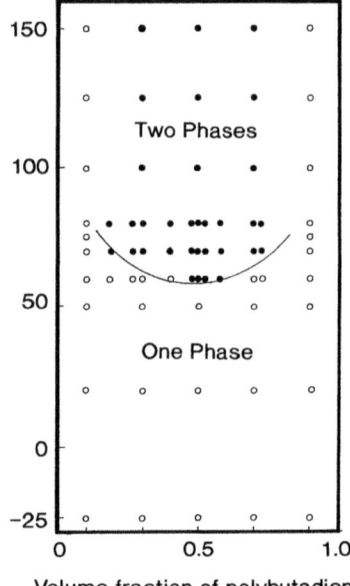

Figure 8.1 Lower critical solution temperature (LCST) phase diagram for the IR/BR(32.3) blend.

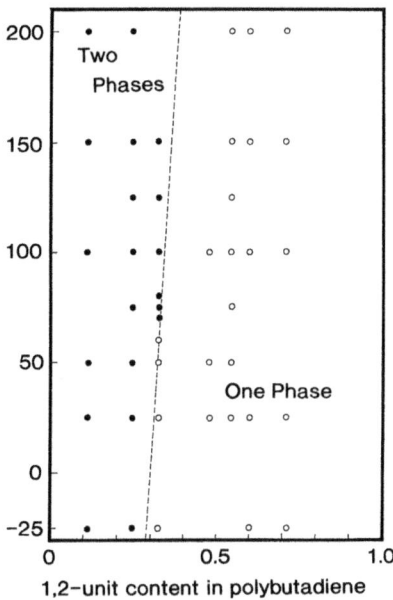

Figure 8.2 Temperature–copolymer composition phase diagram for the IR/BR(32.3) blend, where 1,2-unit content of BR(32.3) is regarded as a copolymer composition.

BR with low 1,2-unit content, whereas the reverse is found for BR with higher 1,2-unit content. The phase boundary between miscible and immiscible states appeared at 32.3% 1,2-unit content, where the LCST was observed to be 333 K for the blend.

This phase transition was confirmed to take place reversibly and the mass transfer was proportional to the thermodynamic driving force,[22] $T - T_s$, where T is the annealing temperature and T_s is the spinodal temperature. Moreover, the extremely large negative excess volume on mixing[23] and free volume fraction apart from additive rule[24] were observed as well as a conformational change on mixing estimated from spin lattice relaxation time measured by pulsed NMR spectroscopy.[25] These were proved positively by Roland and Trasc,[26] Hasegawa et al.,[27] Launger et al.,[28] Bahani et al.[29] and Roovers and Toporowski,[30] using various techniques.

However, the relationship between LCST and 1,2-unit content of BR was exclusively dependent upon the researchers. This was inferred to be due to the difference in molecular weight of the rubbers,[29] despite the fact that the IR and BR used in the previous works were high molecular weight polymers. To investigate the effect of molecular weight on miscibility and LCST phase behaviour, it is necessary to use NR and BR with 32.3% 1,2-unit content (BR(32.3)).

NR consists of an ω-terminal, two *trans*-1,4-isoprene units, long sequence *cis*-1,4-isoprene units (M_w about 2×10^6) and a α-terminal, which is linked to the phospholipid-containing fatty acid ester group, aligned in that order,[60] and it is rationally separated into a soluble fraction (NR-sol) and an insoluble fraction (NR-gel) in toluene.[31] NR-sol is branched polymer whose weight average molecular weight is about 2 million,[31,32] whereas NR-gel is a crosslinked polymer having three-dimensional network structures formed with physical crosslinking junctions at the ω-terminal due to the proteins and chemical crosslinking junctions at the α-terminal linking to the phospholipid.[33] Since the physical crosslinking junctions are held by aggregation of the proteins through hydrogen bonding formation, it may be cleaved by adding a polar solvent, *i.e.* alcohol, resulting in the dissolution of NR-gel into toluene.[34] If BR is mixed with the dissolved NR-gel, followed by the removal of the polar solvent to form the three-dimensional network structures, one may prepare a homogeneous crosslinked blend without segregation of the crosslinked polymers in the matrix. In this case, we may investigate the miscibility and phase behaviour of NR/BR blends, *i.e.* NR-sol/BR blends and NR-gel/BR blends, respectively.[35]

8.2.2 Characterization of NR-Sol and NR-Gel

Figure 8.3 shows a molecular weight distribution of NR-sol determined by size exclusion chromatography. The feature may be attributed to the bimodal distribution of molecular weight characteristic of NR.[36] The average molecular weight of the rubber was estimated using a calibration curve with standard polystyrene and is tabulated in Table 8.1. The weight average molecular weight, M_{wSEC}, and polydispersity, M_w/M_n, as well as the absolute weight average molecular weight, M_{wLALLS}, thus determined, were quite similar to those reported in the previous study.[37] NR-gel was characterized by measuring stress versus strain to prove the formation of the three-dimensional network structure of the polymer. To estimate the crosslink density, ν, the Mooney–Rivlin expression[38–40] was applied to the stress–strain relationship for NR-gel, as follows:

$$\sigma/(\lambda - 1/\lambda^2) = 2C_1 + 2C_2/\lambda \tag{8.1}$$

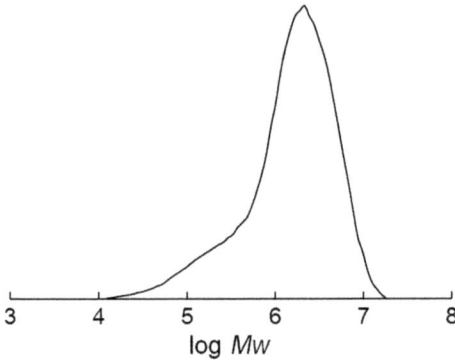

Figure 8.3 GPC curve for NR-sol.

Table 8.1 M_w, M_w/M_n and M_c of NR-sol and NR-gel.

Specimen	$M_w/10^5$ (g/mol) (LALLS)	$M_w/10^5$ (g/mol) (GPC)	M_w/M_n	ν from Mooney–Rivlin constant (10^{-6} mol/g)	ν from equilibrium swelling (10^{-6} mol/g)
NR-sol	23.2	26.8	4.62		
NR-gel				1.8	1.5
BR(32.3)		2.14	1.03		

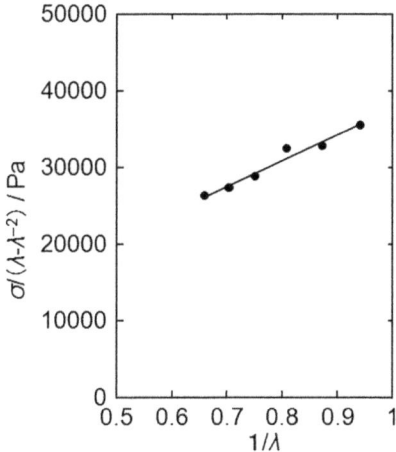

Figure 8.4 Mooney–Rivlin plot for NR-gel.

where σ is stress, λ is elongation, and C_1 and C_2 are the Mooney–Rivlin constants, respectively. Figure 8.4 shows the Mooney–Rivlin plot for NR-gel. At low strain, $\sigma/(\lambda - 1/\lambda^2)$ was proportional to $1/\lambda$, due to the linear elastic behaviour. As the strain increases, non-linear behaviour is observed. Thus, the least squares method was applied to the data points in the linear region to estimate the Mooney–Rivlin constants. Table 8.1 shows the crosslink density, ν, of NR-gel, estimated from the Mooney–Rivlin constants[41] as well as from equilibrium swelling measurement. The crosslink density estimated from the Mooney–Rivlin constants was quite similar to that estimated from the equilibrium swelling measurement. This is supporting evidence that NR-gel is crosslinked rubber having three-dimensional network structures formed with both the physical crosslinking junctions at the ω-terminal due to the proteins and the chemical crosslinking junctions at the α-terminal linking to the phospholipid, as reported in the previous paper.[36]

8.2.3 LCST Phase Behaviour

Figure 8.5 shows DSC thermograms for NR-sol, BR(32.3) and NR-sol/BR(32.3) (50/50) blends, respectively. The blend at 298 K, quenched by liquid nitrogen, showed one T_g between the T_gs of NR-sol and BR(32.3). This

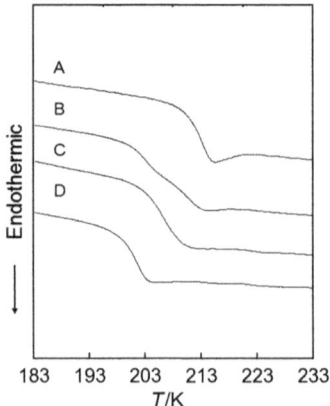

Figure 8.5 DSC thermogram of (a) NR-sol, (b) NR-sol/BR(32.3) blend annealed at 358 K, (c) NR-sol/BR(32.3) blend at 25 °C and (d) BR(32.3).

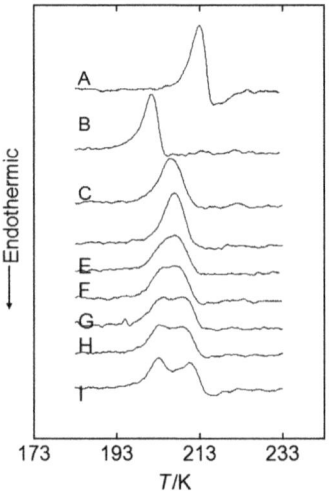

Figure 8.6 DSC curve of (a) NR, (b) BR(32.3) and NR-sol/BR(32.3) (50/50) blend annealed at (c) 298, (d) 323, (e) 328, (f) 331, (g) 333, (h) 343 and (i) 358 K for 2 h.

demonstrates that NR-sol is miscible with BR(32.3), despite the ultra-high M_w. In contrast, two T_gs were found for the blend annealed at 358 K for 2 h, suggesting that the phase separation occurred. Consequently, the LCST phase behaviour was found for NR-sol/BR(32.3) blend, in which a miscible–immiscible phase transition took place repeatedly. To draw the LCST phase diagram, the NR-sol/BR(32.3) blend was annealed at various temperatures between 233 K and 373 K. Figure 8.6 shows typical differential DSC thermograms for the blends, in which the differentiation was made to show T_g clearly because the T_gs of NR-sol and BR(32.3) were very close to each other. At 298 K, one T_g with a unimodal distribution was found, but it gradually changed to

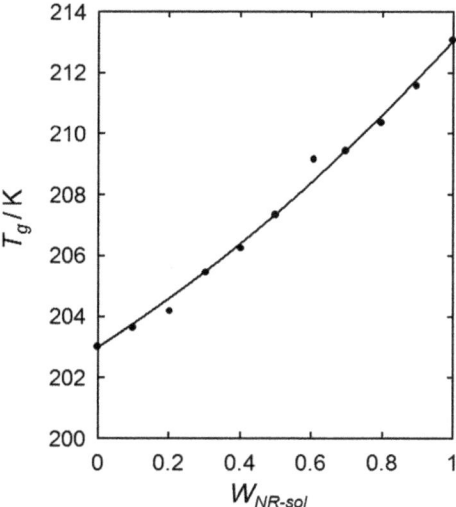

Figure 8.7 T_g of miscible NR-sol/BR(32.3) blend.

a bimodal distribution as the annealing temperature increased. The phase boundary temperature between the miscible and immiscible states was around 328 K for the 50/50 blends. Furthermore, two peak tops of the bimodal distribution of the thermograms were significantly dependent upon the annealing temperature. This may reflect coexistence compositions of two phases separated at the temperature. Since T_g depends upon the composition of the blend, a T_g–composition relationship was experimentally determined for the NR-sol/BR(32.3) blend. Figure 8.7 shows the T_g versus composition. The T_g was well approximated as a function of composition, as follows:

$$T_g = 2.6 W_{NR\text{-sol}^2} + 7.4 W_{NR\text{-sol}} + 203 \tag{8.2}$$

where $W_{NR\text{-sol}}$ is the weight fraction of NR-sol in the blend. This allows us to estimate a coexistence composition of each phase. To determine T_g for the phase-separated blend shown in Figure 8.6 correctly, the two overlapping peaks of the bimodal distribution were subjected to peak separation using a Gauss–Newton method, in which the distribution of onset of segmental motion was assumed to be Gaussian. The resulting coexistence curve is shown in Figure 8.8. The coexistence curve estimated using the (50/50) blend was consistent with those estimated using the (80/20), (70/30) and (60/40) blends. These corresponded to the phase boundary between the miscible and immiscible states determined by T_g measurements, expressed as open circles and closed circles, respectively. This demonstrates that the coexistence curve, determined by the method proposed in this work, is a bimodal curve in the equilibrium state for the blend of ultra-high molecular weight NR-sol with BR(32.3).

Figure 8.9 shows the differential DSC thermogram of the NR-gel/BR(32.3) (50/50) blend. The blend at 298 K, quenched with liquid nitrogen, showed one T_g, suggesting that the blend was miscible in the order of about 10 nm.[42]

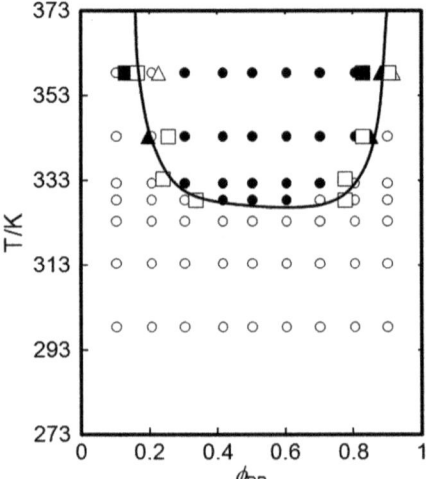

Figure 8.8 LCST phase diagram of NR-sol/BR(32.3) blend, obtained by coexistence curve estimated using (□) (50/50), (■) (60/40), (▲) (70/30), (△) (80/20) blends and by T_g measurements: (●), two T_gs; (○), one T_g.

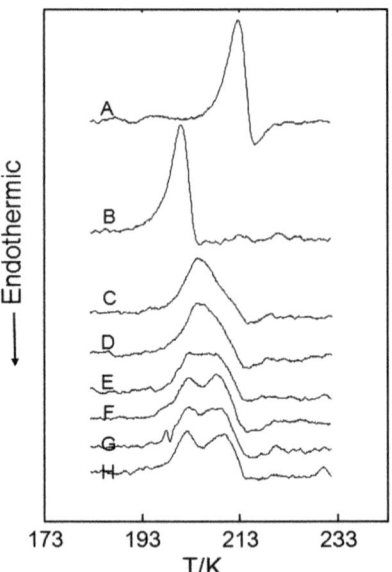

Figure 8.9 DSC curve for (a) BR(32.3), (b) miscible NR-gel/BR(32.3) (90/10) blend, (c) (80/20), (d) (70/30), (e) (60/40), (f) (50/50), (g) (40/60), (h) (30/70), (i) (20/80), (j) (10/90) and NR-sol.

However, the blend underwent phase separation at 333 K, 343 K and 358 K, as reflected by two T_gs. Hence, the LCST phase behaviour was also confirmed for the blend of NR-gel having three-dimensional network structures with BR(32.3).

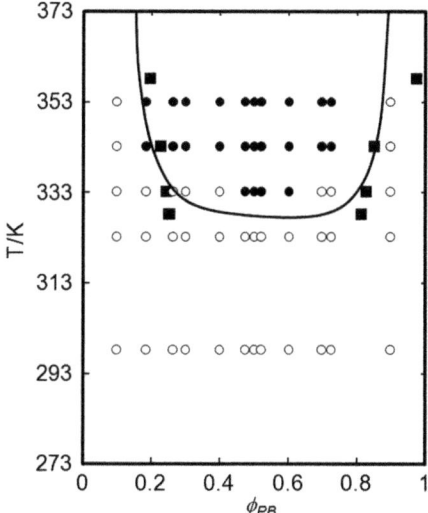

Figure 8.10 LCST phase diagram: (■) for NR-gel/BR(32.3) blend and solid line for NR-sol/BR(32.3) blend. The LCST phase diagram for IR/BR(32.3) blend is also shown: (●), two T_gs; (○), one T_g.

Figure 8.10 shows the LCST phase diagram for the NR-gel/BR(32.3) blend, in which the coexistence compositions were estimated from the T_g of the NR-gel/BR(32.3) (50/50) blend, using Equation (8.2). The coexistence curve for the NR-sol/BR(32.3) blend is also shown in the figure, together with the reported miscible and immiscible regions for the IR/BR(32.3)blend,[19] which were determined by DSC to be one T_g and two T_gs regions, respectively. The coexistence curve for the NR-gel/BR(32.3) blend appears at the same temperature–composition as that of the NR-sol/BR(32.3) blend. Furthermore, these coexistence curves corresponded to the reported phase boundary between the miscible and immiscible states for the IR/BR (32.3) blend.[19] Since the intermolecular attractive forces of these blends are expected to be the same, the only difference may be the molecular weight, i.e. $M_{wGPC} = 2.68 \times 10^6$ [g mol^{-1}] for NR-sol, 7.30×10^5 [g mol^{-1}] for IR, and $n = 1.5 \times 10^{-6}$ [mol cm^{-3}] for NR-gel. Therefore, it is concluded that the molecular weight of the polymer does not have a significant influence on the LCST of the NR-sol/BR(32.3), the NR-gel/BR(32.3) and the IR/BR(32.3) blends.

8.3 Immiscible NR Blends

8.3.1 Background

Many pairs of NR and polymers are immiscible, and this has been well summarized in the literature.[18] Among these pairs, the NR/SBR blend is worthy of note, since the crystallization of NR on an SBR matrix has been investigated.[43,44]

Rapid strain-induced crystallization has been associated with the outstanding mechanical properties characteristic of crosslinked NR compared to its synthetic analogue, *i.e.* crosslinked IR. This may be due in part to a reinforcing effect of strain-induced crystals on the properties of NR as a filler or physical crosslinking junctions. In fact, tensile and tear strengths of crosslinked NR are practically higher than those of crosslinked IR, at the high-speed limit of the tear test.[45] Therefore, it is important to control the rate of crystallization, in order to control the mechanical properties of not only NR itself, but also the NR blends.

The rate of crystallization of NR may be controlled by dispersing NR into SBR, in which heterogeneous nucleation would occur when the average diameter of the droplet was larger than about 1 μm.[46] Furthermore, the rate of crystallization was confirmed to depend on not only the diameter but also the dimension of the NR phase. Since isothermal crystallization of NR was promoted by fatty acids,[47–49] which were inherently present in NR and acted as a nucleating agent,[50] it may be possible to relate the rate of crystallization to distribution of the fatty acids among the NR domains.[51] Thus, the outstanding mechanical properties and the rapid strain-induced crystallization of NR may be represented to be a function of the amount of fatty acids, which is dependent upon the diameter and dimension of the NR domains.

In order to investigate both the outstanding mechanical properties and the rapid strain-induced crystallization of NR, the tear energy of the NR/SBR blend is useful, since the strain-induced crystallization of NR must play an important role in preventing a crack growth of the blend under large deformation.[43] The tear energy of non-crystalline polymers is well known to depend on the rate of tear and temperature,[52,53] and it is superposed according to the Williams–Landel–Ferry (WLF)[54] superposition relationship. The tear energy increases vertically, when the polymer is mixed with filler such as carbon black or silica.[43] Thus, the strain-induced crystallization of NR may be assessed by measuring the tear energy of the non-crystalline polymer containing NR dispersoid, due to the reinforcing effect of strain-induced crystals as a filler.

8.3.2 Characterization of NR/SBR Blends

Crosslink density, ν, of the samples used in the present study is tabulated in Table 8.2. The ν for SBR1 was 1.85×10^{-4} mol ml^{-1}, which was similar to that

Table 8.2 Crosslink density (ν; 10^{-4} mol/g) of the samples estimated from equilibrium swelling.

Sample	ν (10^{-4} mol/g) NR phase	SBR1 phase	SBR2 phase
NR/SBR1 (3/7)	0.742	2.10	
NR	0.842		
SBR1		1.85	
SBR2/SBR1 (3/7)		1.48	53.0

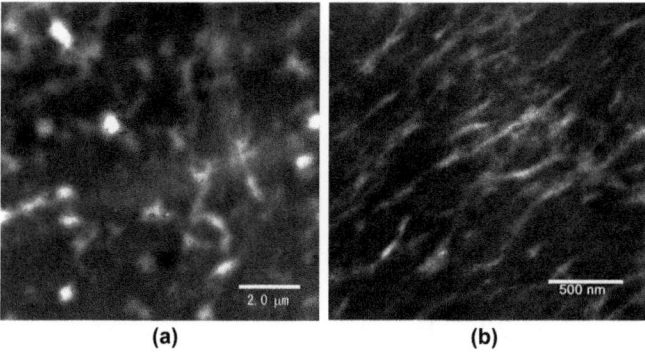

Figure 8.11 TEM photographs of (a) NR/SBR1 blend and (b) SBR2/SBR1 blend.

for the SBR1 phase in the NR/SBR1 and SBR2/SBR1 blends. On the other hand, the v for NR was identical to that for the NR phase in the NR/SBR1 blend. These may allow us to investigate the reinforcing effect of NR on the tear energy of SBR, since the crystallization of NR is known to be dependent upon both the crosslink density and the size of the dispersoid.

Figure 8.11(a) and (b) shows TEM photographs for the NR/SBR1 and SBR2/SBR1 blends, respectively, in which the dark phase is assigned to SBR1. In the photograph for the NR/SBR1 blend, NR was found to be a domain whose particle diameter ranged from 1 to 2 μm. In contrast, for the SBR2/SBR1 blend, SBR2 was dispersed in SBR1 as a fine ellipsoid of about 0.5 μm in major ellipse. Thus, the SBR1 was confirmed to be a matrix in the blends, whereas NR and SBR2 were domains.

8.3.3 Tear Energy

The tear energy of NR/SBR1 was compared with that of SBR1, in order to investigate the effect of the strain-induced crystallization of NR on the tear energy of SBR. The tear energy, G, for the NR/SBR1 blend measured at various cutting speeds, R, *i.e.* 0.5, 5.0, 50, 500 and 1000 mm min^{-1}, is shown in Figure 8.12. The G increased gradually as the R increased. At the definite R, the G depended on temperature, and increased abruptly at 273 K. In contrast, monotonic increase in the G of SBR1 is shown against both the R and temperature in Figure 8.13. The absolute value of the G of SBR1 was relatively lower than that of the NR/SBR1 blend. According to an earlier paper by Stager,[55] the difference in the G between SBR1 and the NR/SBR1 blend may be attributed to the effect of the strain-induced crystallization of NR.

Since the G of non-crystalline polymers is associated with an energy dissipation of the polymer, it is analysed in terms of the Williams-Landel-Ferry rate-temperature equivalence.[54] To investigate the reinforcing effect of NR to SBR1 in the present study, the rate-temperature equivalence was applied to the G of SBR1 and the NR/SBR1 blend, assuming that the energy dissipation was

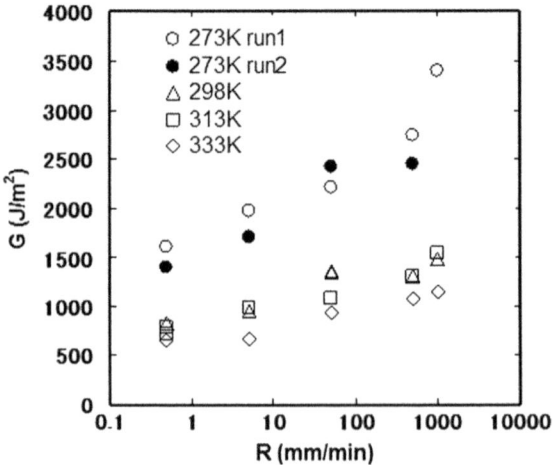

Figure 8.12 Temperature and cutting speed dependence of tear energy of NR/SBR1 blend.

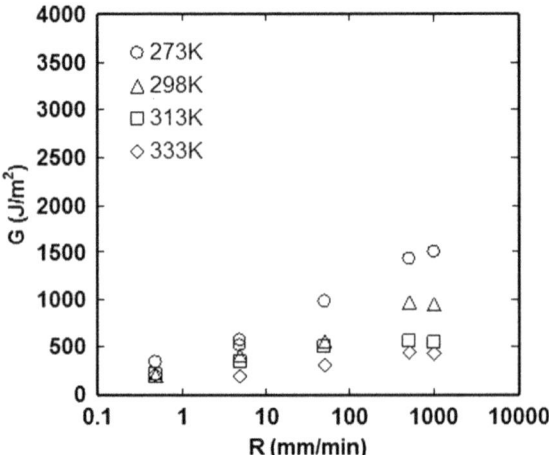

Figure 8.13 Temperature and cutting speed dependence of tear energy of SBR2/SBR1 blend.

independent of the crystalline domain, as in the case of the reinforcing effect of fillers such as carbon black and silica.[56,57]

The superposed G of the NR/SBR1 blend and SBR1 are shown in Figure 8.14, in which a rate–temperature shift factor, a_T, is defined as follows:

$$\log a_T = -8.86(T - T_s)/(101.6 + T - T_s) \tag{8.3}$$

where the reference temperature, T_s, was 268 K. T_s was 268 K. The G of SBR1 was completely superposed, as reported by Greensmith.[58] This suggests that the change in G was associated with an energy dissipation of SBR1. In contrast, the

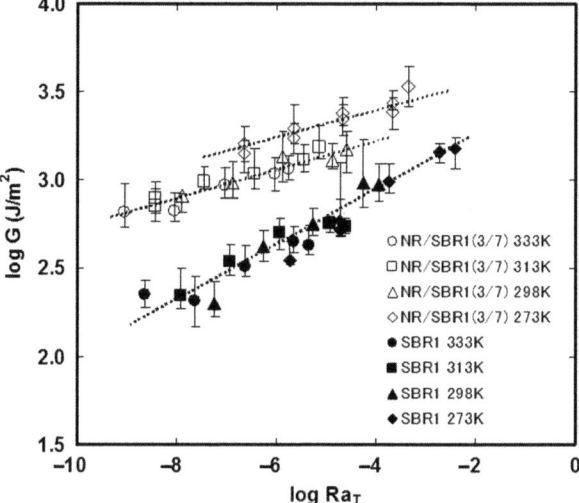

Figure 8.14 Master curves of tear energy for NR/SBR1 blend and SBR1.

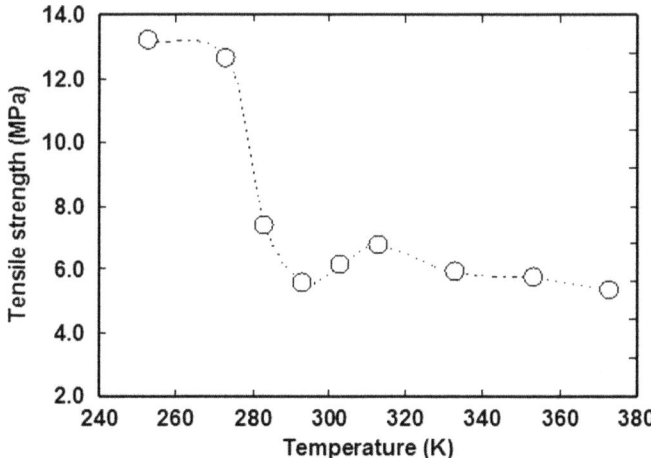

Figure 8.15 Temperature dependence of tensile strength for NR whose crosslink density was identical as NR phase in NR/SBR1 blend.

G of the NR/SBR1 blend was divided into two curves after superposition; one for the G measured at 273 K and the other at the temperatures ranging from 298 to 333 K. This may be expected to be due to either unsuitable a_T, used for the superposition of the G, or the strain-induced crystallization of NR.

To ascertain the effect of the strain-induced crystallization on the G of the NR/SBR1 blend, the tensile strength of NR was measured at various temperatures since it is well known to fall at the melting temperature of the strain-induced crystal.[45,59] Figure 8.15 shows the temperature dependence of the tensile strength of NR, whose crosslink density was the same as that of NR

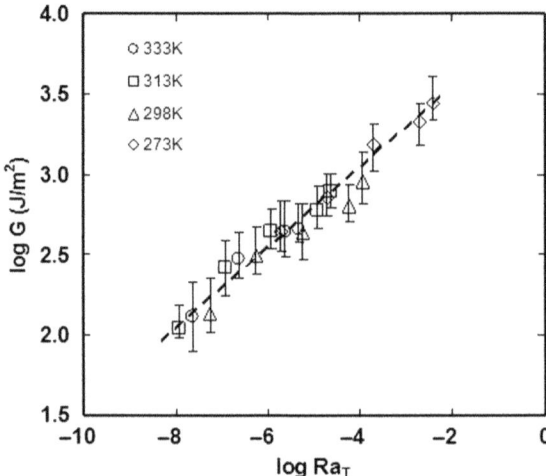

Figure 8.16 Master curves of tear energy for SBR2/SBR1 blend in which shift factor a_T of SBR1 was used.

in the NR/SBR1 blend. The tensile strength fell at about 293 K, at which melting of the strain-induced crystal of NR occurred. This may be supporting evidence that the abrupt drop in the G of the NR/SBR1 blend was due to melting of the strain-induced crystal at about 293 K.

Instead of dispersed NR, the effect of non-crystalline SBR2 on the G of SBR1 was also investigated. Figure 8.16 shows a logarithmic plot of the G versus Ra_T for the SBR2/SBR1 blend, in which the G of the SBR2/SBR1 blend was superposed with a_T of SBR1, as in the case of the NR/SBR1 blend. The logarithmic G of the SBR2/SBR1 blend was a function of logarithmic Ra_T and it was superposed to be a single curve. Furthermore, the value of the G of the SBR2/SBR1 blend was almost identical to that of SBR1, implying little reinforcing effect of SBR2. These are distinguished from the deviation of the G at 273 K from the master curve for the NR/SBR1 blend and the reinforcing effect of NR on the G. This demonstrates that the strain-induced crystallization of dispersed NR may play an important role in the enhancement of the G, which was dependent upon the temperature, crosslink density, strain, rate of strain and so forth.

Figure 8.17 shows the G for acetone-extracted NR/SBR1 (NR/SBR1-AE) blend, after superposition with a_T of SBR1, in which the fatty acids present in NR were removed by acetone extraction.[44] A single master curve was drawn for the NR/SBR1-AE after superposition, as in the case of the SBR2/SBR1 blend. This is quite different from the two curves drawn for the superposed G of the NR/SBR1 blend. The difference in the superposition may be due to the effect of fatty acid, since the fatty acids are mainly removed by acetone extraction. This may suggest that the fatty acids promote the strain-induced crystallization of NR.

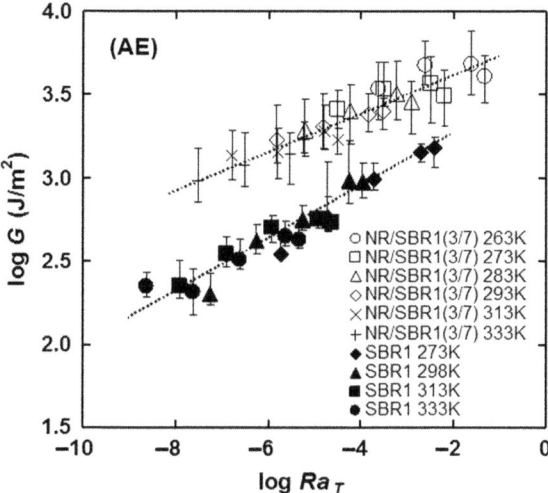

Figure 8.17 Master curve of G for the acetone-extracted strips of NR/SBR1 blend for which a_T of SBR1 was used.

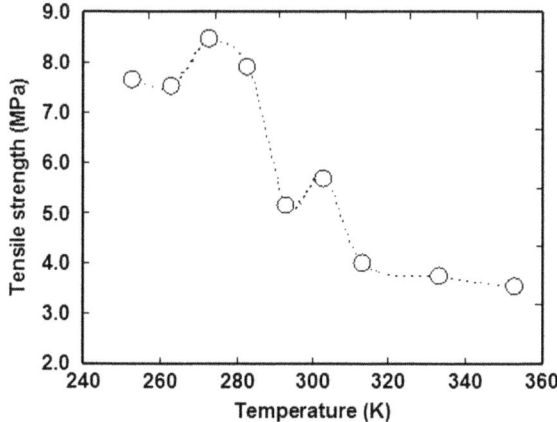

Figure 8.18 Temperature dependence of the tensile strength of NR-AE, the crosslink density of which was identical to that of the NR phase in the NR/SBR1-AE blend.

Figure 8.18 shows the temperature dependence of tensile strength for acetone-extracted NR (NR-AE) whose crosslink density was the same as that of the NR domain in the NR/SBR1-AE blend. The tensile strength of NR-AE fell at about 293 K, which was the same as the melting temperature of the strain-induced crystal of NR. This demonstrates that the melting temperature of the strain-induced crystal is not affected by the fatty acids. Consequently, the difference in the superposed G between the NR/SBR1 and NR/SBR-AE blends may be explained due to the difference in the rate of the strain-induced crystallization; that is, the rate of the strain-induced crystallization of the

NR/SBR1 blend may be higher than that of the NR/SBR-AE blend, as in the case of the isothermal crystallization of unstrained NR.[21,61]

8.4 Conclusions

NR-sol and NR-gel were miscible with BR(32.3) at 298 K, despite the ultra-high molecular weight and three-dimensional network structure that are present in the two rubbers, respectively. The LCST phase behaviour was found for the NR-sol/BR(32.3) and NR-gel/BR(32.3) blends by T_g measurements using DSC after quenching the annealed blends. The LCST of the NR-sol/BR(32.3) and NR-gel/BR(32.3) blends was 328 K, being identical to the LCST of the IR/BR(32.3) blend. It is concluded that the difference in molecular weight plays no significant role in the LCST of the NR-sol/BR(32.3), NR-gel/BR(32.3) and IR/BR(32.3) blends.

The tear energy of SBR1 increased after mixing with the strain-induced crystallizable NR but not with SBR2 having higher 1,2-unit content, in which crosslink density of SBR1 was similar to that of SBR1 in the NR/SBR1 blend. The tear energy of the NR/SBR1 blend at 273 K showed the deviation from the superposed curve drawn for the energy at 298, 313 and 333 K, which corresponded to increments of about 500 to 1000 J m^{-2}, while the tear energies of SBR1 and the SBR2/SBR1 blend draw a single curve by the superposition. After acetone extraction, the tear energy of the NR/SBR-AE blend was superposed to a single curve. The higher tear energy of the NR/SBR1 blend, *i.e.* the deviation, was attributed to the strain-induced crystallization of NR due to the effect of the fatty acids.

References

1. M. H. R. Ghoreishy, M. Alimardani, R. Z. Mehrabian and S. T. Gangali, *J. Appl. Polym. Sci.*, 2013, **126**, 1725.
2. Y. Inoue, M. Iwasa and H. Yoshida, *Netsu Sokutei*, 2012, **39**, 41.
3. N. Yan, H. S. Xia, Y. H. Zhan, G. X. Fei and C. Chen, *Plast., Rubber Compos.*, 2012, **41**, 365.
4. W. Zhang, X. Gong, S. Xuan and W. Jiang, *Ind. Eng. Chem. Res.*, 2011, **50**, 6704.
5. M. H. R. Ghoreishy, M. Bagheri-Jaghargh, G. Naderi and S. Soltani, *J. Appl. Polym. Sci.*, 2012, **125**, 3648.
6. M. A. Mansilla, L. Silva, W. Salgueiro, A. J. Marzocca and A. Somoza, *J. Appl. Polym. Sci.*, 2012, **125**, 992.
7. A. Boonmahitthisud and S. Chuayjuljit, *Polym.-Plast. Technol. Eng.*, 2012, **51**, 311.
8. S. P. Thomas, E. J. Mathew and C. V. Marykkutty, *J. Appl. Polym. Sci.*, 2011, **121**, 2257.
9. S. Chuayjuljit and W. Luecha, *J. Elastoplast.*, 2011, **43**, 407.
10. A. Aljaafari, *Mater. Des.*, 2010, **31**, 3207.

11. S. Fujinami, K. Nakajima and T. Nishi, *Nippon Gomu Kyokaishi*, 2011, **84**, 171.
12. M. Madani, *J. Polym. Res.*, 2010, **17**, 53.
13. T. T. N. Dang, J. K. Kim, S. H. Lee and K. J. Kim, *Compos. Interfaces*, 2011, **18**, 151.
14. T. T. N. Dang, J. K. Kim and K. J. Kim, *Int. Polym. Process.*, 2011, **26**, 368.
15. A. Alipour, G. Naderi, G. R. Bakhshandeh, H. Vali and S. Shokoohi, *Int. Polym. Process.*, 2011, **26**, 48.
16. K. Sahakaro, K. Sengloyluan and J. W. M. Noordermeer, *Gummi, Fasern, Kunstst.*, 2010, **63**, 775.
17. A. Yoshioka, K. Komuro, A. Ueda, H. Watanabe, S. Akita, T. Masuda and A. Nakajima, *Pure Appl. Chem.*, 1986, **58**, 1697.
18. A. J. Tinker and K. P. Jones, *Blends of Natural Rubber – Novel Techniques for Blending with Speciality Polymers*, Springer-Verlag, Berlin, 1998.
19. S. Kawahara, S. Akiyama and A. Ueda, *Polym. J.*, 1989, **21**, 221.
20. C. M. Roland, *Rubber Chem. Technol.*, 1989, **62**, 456.
21. Y. Tanaka, S. Kawahara and J. Tangpakdee, *Kautsch. Gummi Kunstst.*, **50**, 6.
22. S. Kawahara and S. Akiyama, *Polym. J.*, 1990, **22**, 361.
23. S. Kawahara and S. Akiyama, *Polym. J.*, 1991, **23**, 7.
24. S. Akiyama, S. Kawahara, I. Akiba, S. Iio, H.-H. Li and Y. Ujihira, *Polym. Bull.*, 2000, **45**, 275.
25. S. Kawahara, K. Sato and S. Akiyama, *J. Polym. Sci., Part B: Polym. Phys.*, 1994, **32**, 15.
26. C. M. Roland and C. A. Trasc, *Rubber Chem. Technol.*, 1989, **62**, 456.
27. H. Hasegawa, S. Sakurai, M. Takenaka and T. Hashimoto, *Macromolecules*, 1991, **24**, 1813.
28. J. Launger, R. Lay, S. Maas and W. Gronski, *Macromolecules*, 1995, **28**, 7010.
29. M. Bahani, F. Laupretre and L. Monnerie, *J. Polym. Sci., Part B: Polym. Phys. Ed.*, 1995, **33**, 167.
30. J. Roovers and P. M. Toporowski, *Macromolecules*, 1992, **25**, 3454.
31. A.-H. Eng, S. Ejiri, S. Kawahara and Y. Tanaka, *J. Appl. Polym. Sci.: Appl. Polym. Symp.*, 1994, **53**, 5.
32. J. T. Sakdapipanich, T. Kowitteerawut, K. Suchiva and Y. Tanaka, *Rubber Chem. Technol.*, 1999, **72**, 712.
33. Y. Tanaka, S. Kawahara and J. Tangpakdee, *Kautsch. Gummi Kunstst.*, 1997, **50**, 6.
34. S. Kawahara, T. Kakubo, M. Suzuki and Y. Tanaka, *Rubber Chem. Technol.*, 1999, **72**, 174.
35. S. Kawahara, Y. Asada, Y. Isono, K. Muraoka and Y. Minagawa, *Polym. J.*, 2002, **34**, 1.
36. A. Subramaniam, *Rubber Chem. Technol.*, 1972, **45**, 346.
37. S. Kawahara, T. Kakubo, N. Nishiyama, Y. Tanaka, Y. Isono and J. T. Sakdapipanich, *J. Appl. Polym. Sci.*, 2000, **78**, 1510.

38. M. Mooney, *J. Appl. Phys.*, 1940, **11**, 582.
39. M. Mooney, *J. Appl. Phys.*, 1948, **19**, 434.
40. R. S. Rivlin, *Philos. Trans. R. Soc., A*, 1948, **240**, 459.
41. L. Mullins, *J. Appl. Polym. Sci.*, 1959, **2**, 257.
42. K. C. Frish, D. Klempner and H. L. Frish, *Polym. Eng. Sci.*, 1982, **22**, 1143.
43. T. Kawazura, S. Kawahara and Y. Isono, *J. Appl. Polym. Sci.*, 2005, **98**, 613.
44. F. Noguchi, K. Akabori, Y. Yamamoto, T. Kawazura and S. Kawahara, *J. Phys.: Conf. Ser.*, 2009, **184**, 012020.
45. A. N. Gent, S. Kawahara and J. Zhao, *Rubber Chem. Technol.*, 1998, **71**, 668.
46. G. D. Wathen and A. N. Gent, *J. Nat. Rubber Res.*, 1990, **5**, 178.
47. A. N. Gent, *Trans. Faraday Soc.*, 1954, **50**, 521.
48. A. H. Eng, S. Ejiri, S. Kawahara and Y. Tanaka, *J. Appl. Polym. Sci.: Appl. Polym. Symp.*, 1994, **53**, 5.
49. N. Nishiyama, S. Kawahara, T. Kakubo, A. H. Eng and Y. Tanaka, *Rubber Chem. Technol.*, 1996, **69**, 608.
50. S. Kawahara, K. Takano, Y. Isono, M. Hikosaka, J. T. Sakdapipanich and Y. Tanaka, *Polym. J.*, 2004, **36**, 361.
51. T. Kawazura, S. Kawahara and Y. Isono, *Rubber Chem. Technol.*, 2003, **76**, 1164.
52. A. N. Gent, S. M. Lai, C. Nah and C. Wang, *Rubber Chem. Technol.*, 1994, **67**, 610.
53. A. N. Gent and S. M. Lai, *J. Polym. Sci., Part B: Polym. Phys.*, 1994, **32**, 1543.
54. M. L. Williams, R. F. Landel and J. D. Ferry, *J. Am. Chem. Soc.*, 1955, **77**, 3701.
55. R. G. Stager, E. D. von Meerwall and F. N. Kelley, *Rubber Chem. Technol.*, 1985, **58**, 913.
56. J. D. Ferry, *Viscoelastic Properties of Polymers*, Wiley, New York, 1961.
57. T. L. Smith, *Polym. Eng. Sci.*, 1965, **5**, 270.
58. H. W. Greensmith and A. G. Thomas, *J. Polym. Sci.*, 1955, **18**, 189.
59. A. G. Thomas and J. M. Whittle, *Rubber Chem. Technol.*, 1970, **43**, 222.
60. Y. Tanaka, *Rubber Chem. Technol.*, 2001, **74**, 355.
61. S. Kawahara, T. Kakubo, J. T. Sakdapipanich, Y. Isono and Y. Tanaka, *Polymer*, 2000, **41**, 7483.

CHAPTER 9
Natural Rubber Based Polar Synthetic Rubber Blends

KONSTANTINOS G. GATOS

Research and Development Center, Megaplast S. A., 38 Vasileos Konstantinou Avenue, 194 00, Athens, Greece
Email: kgatos@gmail.com

9.1 Introduction

Rubbers are classified as natural or synthetic, according to their origin. The first category relates to natural rubber obtained from plant sources while the latter is produced from monomers delivered by cracking and refining of petroleum.[1] According to the monomer selection, synthetic rubbers can be polar or non-polar. The polar class involves groups responsible for polarity like acrylonitrile (ACN), chlorine (Cl), epoxy (EP), carbonyl (CO), fluorine (F), *etc.* Important characteristics of polar rubbers include their resistance to swelling by oils and heat resistance.

Natural rubber (NR) is a very significant elastomer for the rubber industry due to its mechanical performance.[2] However, it fails to fulfil challenges under the bonnet and to meet the requirements of other demanding applications. Due to the unsaturated backbone it is susceptible to attack by oxygen, ozone and light. It is not oil resistant and it is swollen by aromatic, aliphatic and halogenated hydrocarbons. In general, it is unsuitable for use with organic liquids.[3] In an attempt to improve specific properties of NR, it is usually blended with synthetic polar rubbers. The grade selection and the relative amount of each elastomer are based on the set of final properties that are desired to be demonstrated by the blend. For example, blends of NR with nitrile rubber (NBR) are expected to enhance the resistance to swelling by oils[4] while mixing with chloroprene rubber (CR) improves thermal ageing and ozone attack.[5]

Although blends of NR with polar synthetic rubbers are usually immiscible, in practice miscibility is not a requirement for most rubber applications. Usually, micro-heterogeneity is dominant, while homogeneity is favoured by similarities in rubber viscosities and solubility parameters. Nevertheless, co-vulcanization is important for NR/polar synthetic rubber blends, wherein a single network structure including crosslinked macromolecules of both components is created. In that case, vulcanization should take place to similar levels in both parts, together with crosslinking across the interfaces.[6] A prerequisite for successful blending is the low interfacial tension, which enables on the one hand mixing at the interface and on the other hand small phase sizes.[7] In this respect, compatibilizers are beneficial for NR/polar synthetic rubber blending. As a result, not only is a smaller domain size obtained, but also less time is needed for mixing for the given preparation method.

9.2 Preparation Methods

The mixing procedure of NR with polar synthetic rubbers plays a significant role in the blending efficiency. These dissimilar elastomers can be either 'pre-blended' or 'phase mixed' with each other. In the first case the additives are added directly to the blend, while in the latter case the ingredients are initially added separately to each component of the blend followed by batch mixing.[8] The most widely used techniques include latex, solution and melt blending. In practice, the selection of each preparation method is based on the related vulcanizate application. For example, dipped goods and threads are produced by latex while tyres and profiles are made by melt compounding.[9]

9.2.1 Latex

Latex is considered to be an aqueous dispersion of fine rubber particles in the submicron to micron range.[10] NR latex is produced by the rubber tree while polar synthetic rubber latices are obtained by emulsion polymerization. Latex blending of NR with polar synthetic rubbers has been exploited extensively in the case of polyurethane rubber (PUR). Ricardo et al.[11] mixed NR/PUR in 75/25, 50/50 and 25/75 ratios by weight. Initially, NR latex with dry rubber content (DRC) of 60% was diluted with distilled water to obtain 40% of DRC and stabilized with aqueous sodium dodecyl sulfate (SDS). The pH of the latex was adjusted to 10 by adding 1% ammonia solution and the mixture was then centrifuged at 10 000 rpm for 1 h. In order to remove any SDS in this latex, centrifugation followed by washing with distilled water was repeated four times. In addition, PUR latex was prepared by a mixture of two molecular weights of poly(caprolactone) diol (PCL) and dimethylolpropionic acid (DMPA) linked with isophorone diisocyanate (IPDI). The DMPA was neutralized with triethylamine (TEA) and polyurethane was extended with ethylene diamine. The resultant PUR latex contained 30% solids. The related NR/PUR blends in the pre-described weight ratios were produced by casting the aqueous mixtures in PTFE moulds followed by water evaporation at room temperature for 9 days. Nevertheless, the resulting blends presented only a small degree of interpenetration of the two constituents as examined by NMR spectroscopy.

This limited phase interpenetration was best manifested in a NR/PUR blend wherein the NR latex was initially pre-vulcanized. Varghese et al.[12] prepared NR/PUR blends of 50/50 and 20/80 weight ratios. This sulfur pre-vulcanized NR latex was concentrated with 1% ammonia and presented 60% DRC. The NR latex was mixed with a polyester-based PUR latex of 50% DRC followed by stirring. Due to the different preparation routes of both latices, their particle size in the water dispersion differed significantly. The related mixture was cast in a mould built of glass plates and allowed to dry in air until transparent. It was then tempered for 30 min at 100 °C in an oven.

Latex blending of NR with polar synthetic rubbers presents several mixing advantages (e.g. eco-friendly dispersion medium, fine particle dispersion, etc.), but for other applications does not provide any cost advantage since coagulation, removal of water and drying are cost-intensive processes.[13]

9.2.2 Solution Mixing

The solution mixing of NR with synthetic rubbers is expected to give coarser particles because the low solution viscosity favours fast coagulation. Thus, macro-heterogeneity rather than micro-heterogeneity is promoted. However, solution blending with synthetic polar rubbers that are produced by solution polymerization is advantageous as the elastomer is produced *in situ*. This latter appoints compounding efficiency (*e.g.* better incorporation of additives).[14] In general, toluene has proved to be a suitable solvent for NR,[15,16] while for polar synthetic rubbers like NBR, chloroform[17] or methyl ethyl ketone (MEK)[18] has been used. The curative incorporation can take place either in solution or after the evaporation of the solvent.[19]

Gardiner[20] mixed toluene solutions of NR and CR at various weight ratios. The mixtures were shaken vigorously and then spread on a microscope slide. The blends were dried quickly and the outcome for all rubber percentages examined was immiscibility.

Ricardo et al.[11] used chloroform to mix NR with PUR. These rubbers had initially been prepared individually by the latex route. It was found that the rubber phase interpenetration in the solid film (after the total evaporation of the solvent) was very limited.

9.2.3 Melt Blending

By far the most common technique for mixing NR with polar synthetic elastomers is melt blending. Equipment such as open mills, internal mixers and extruders supports dispersive mixing. Also, the relatively high viscosity that appears after mixing prevents phase separation and promotes efficient blending. A significant point is the mastication of NR prior to mixing and the similarities of rubber viscosities at the compounding temperature.[14]

Murthy et al.[21] exploited an open mill to mix NR with NBR. The sulfur type curatives used, containing 2-mercaptobenzothiozole disulfide (MBTS), zinc oxide (ZnO) and stearic acid, along with napthenic oil, were added directly to the blend. In an attempt to minimize the possibility of the non-uniform dispersion of the curatives in such a blend, Sirisinha et al.[22] used dicumyl peroxide (DCP). Prior to mixing, NR was masticated in a water-cooled internal mixer for 15 min at 40 °C, setting the rotor speed at 55 rpm. In this way its Mooney viscosity value (ML1 + 4 at 100 °C) was reduced from 80 to 56. Then, NBR (35% ACN content) having a Mooney viscosity of 57 was added at a NR/NBR weight ratio of 20/80. DCP was charged in the compound after 11 min.

In the presence of compatibilizers, a similar process might be followed. Kumari et al.[23] mixed poly(ethylene-co-vinyl acetate) (EVA) with NR and NBR (34% ACN content). Blending took place in a two-roll mill at a nip gap of 1.3 mm and at a friction ratio of 1 : 1.4. The crosslink agent used was DCP at 2 phr and the EVA amount in the mixture did not exceed 6 phr. Kader et al.[24] used an internal mixer to compatibilize with *trans*-polyoctylene rubber (TOR) a Standard Malaysian Rubber with NBR (34% ACN content) at a 50/50 weight ratio. The TOR amount reached up to 40 wt% of the blend where the cure package was based on sulfur. ZnO and stearic acid were introduced in the internal mixer while sulfur and N-tert-butyl-2-benzothiazole sulfenamide (TBBS) were added on a two-roll mill.

Mixtures of NR with CR have been prepared previously either as 'pre-blended' or 'phase mixed'. Ismail and Leong[25] used a two-roll mill at 70 ± 5 °C to blend NR with CR at various weight ratios. Then ZnO, stearic acid, magnesium oxide (MgO), N-cyclohexyl-2-benzothiazyl sulfenamide (CBS), ethylene thiourea (ETU) and sulfur were incorporated in the batch. Although the 'pre-blending' route is more convenient, when a better dispersion of curatives is required in both rubbers, the 'phase-mixed' process is usually preferred. Zhang et al.[26] used a two-roll mill to mix individually NR and CR with their curatives. NR was compounded with ZnO, stearic acid, CBS, diphenyl guanidine (DPG) and sulfur. Similarly, CR was mixed with ZnO, stearic acid, MgO, CBS, tetramethylthiuram disulfide (TMTD) and sulfur. These two batches were blended at a NR/CR weight ratio of 75/25.

To assist melt blending of NR with CR, processing agents and aromatic oils are usually used.[27] In an attempt to replace the aromatic oils of petroleum origin, which are not considered to be eco-friendly, Kuriakose and Varghese[28] exploited rice bran oil and epoxidized rice bran oil. These cost-effective and non-toxic natural oils also worked as co-activators due to the considerable amount of free fatty acids they bear.

A 'pre-blended' technique has been followed by Tanrattanakul and Petchkaew[29] for mixing NR with chlorosulfonated polyethylene (CSM). NR was masticated on a two-roll mill prior to blending with CSM. Note that NR and CSM presented Mooney viscosity [ML (1 + 4) at 100 °C] of 66 and 63, respectively. The curing recipe mainly involved stearic acid, ZnO, MgO,

MBTS, TMTD and sulfur, which was incorporated in the batch on the open mill. A curing recipe substituting the previous accelerators with CBS has also been explored. The authors proposed ENR as a suitable compatibilizer for NR/CSM blends.

Several melt blending routes were tested for a NR/acrylic rubber (ACM) mixture by Wootthikanokkhan and Tunjongnawin.[30] The ACM used contained about 5 wt% of chlorine cure sites. Initially, NR was masticated for 15 min on a two-roll mill. Three different mixing protocols were performed. In the first place, the masticated NR was cross-blended with ACM on the open mill followed by incorporation of ZnO, stearic acid, 2-mercapto benzothiazole (MBT) and sodium stearate. The rubber sheet was left to cool down prior to the addition of sulfur for 3 min on the two-roll mill. Secondly, masticated NR was mixed with ZnO, stearic acid and MBT while ACM was added separately with sodium stearate. At this stage NR/ACM blending occurred and was completed after the incorporation of sulfur. In a third scheme, after the mixing of the above-mentioned chemicals in each separate rubber phase, sulfur was incorporated only in the ACM batch prior to rubber cross-blending. It was found that NR was less susceptible to mixing protocol changes than ACM, while the 'phase-mixing' route should be preferred for this system.

9.3 Blend Characteristics

The susceptibility of NR based polar synthetic rubber blends to preparation methods and protocols makes the use of testing procedures able to detect and monitor the characteristics of these blend structures imperative. Thus, a vast spectrum of test methods is used to measure properties of both the uncured stocks and the vulcanizates.[31]

9.3.1 Rheology

The rheological performance of rubber blends manifests the ease or difficulty of their fabrication.[32] Mooney viscosity measurement is usually used to determine the rheological behaviour of an uncured batch at a specific temperature. Other rheometers like plate-plate or capillary may be used for rheological characterization.

The Mooney viscosity of NR/NBR at a 20/80 weight ratio was studied by Sirisinha et al.[22] Blends that were prepared at several mixing times (15–40 min) and rotor speeds (40–60 rpm) in an internal mixer were evaluated. It was found that the Mooney viscosity (ML1 + 4 at 100 °C) of the blend was affected more by the mixing time than the rotor speed. The latter caused a more prominent temperature increase, which compensated for the shear stresses induced during blending. Thus, the efficiency of mechanical mastication was decreased. Moreover, by extending the mixing time up to 25 min the shear strain applied to the mixture was rather enhanced. Due to the effect of mastication, the viscosity of the blend was decreased. In another work, Sirisinha et al.[33] studied for the same system the effect of compatibilizers on Mooney viscosity.

Incorporation of carbon black (CB) increased the viscosity of the mixture while the smaller the CB particles the higher the Mooney viscosity value measured. This is related to the enhanced bound rubber content which followed the increase of the surface area of CB as the particle size was decreasing for the same filler content.

In a silica-filled NR/CR mix, Sae-oui et al.[5] studied the Mooney viscosity (ML1+4 at 100 °C) for various NR/CR weight ratios. As shown in Figure 9.1(a), CR presented the lowest viscosity value (i.e. 44) while by blending it with NR (having a Mooney viscosity value of 74), the batch became more viscous. The Mooney viscosity increment was almost linear up to a CR/NR weight ratio of 25/75, whereas it was saturated for more NR-rich compounds.

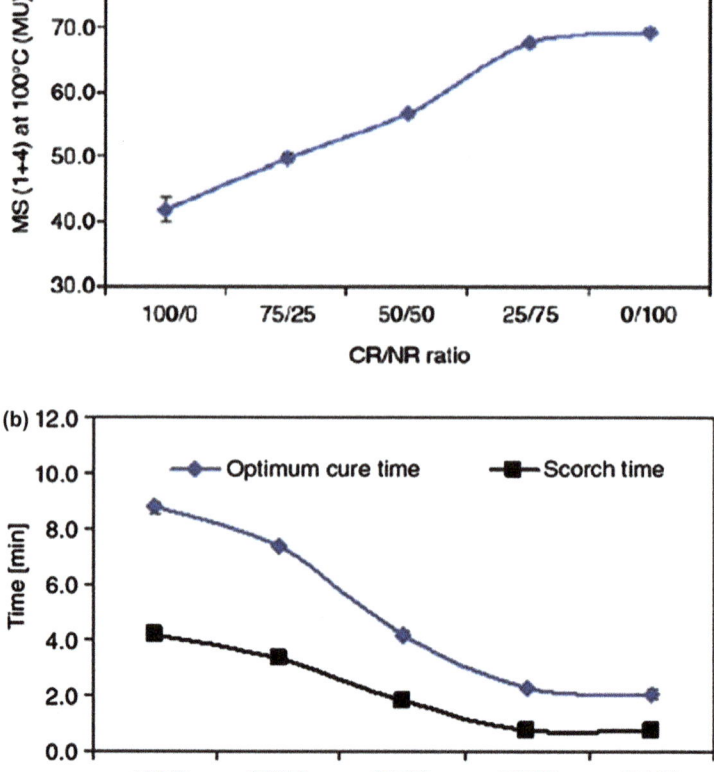

Figure 9.1 Effect of blend ratio on Mooney viscosity (a) and scorch time ($t_s 2$) along with optimum cure time ($t_c 90$) (b) of the compounds.
(Reproduced from Sae-oui et al.[5] with permission from Budapest University of Technology and Economics, Department of Polymer Engineering/BME-PT and GTE).

An opposite trend was presented by Kongvasana et al.,[34] where the minimum and maximum Mooney viscosity was obtained by NR/CR of 100/0 and 0/100 weight ratio, respectively. In this analysis the Mooney viscosity of the NR/CR blend was found to decrease with increasing NR content. Usually the Mooney viscosity of CR is lower than that of NR,[5,28] however for CR of relatively high molecular weight this scenario may change.

9.3.2 Curing

A prerequisite for delivering useful products made from NR based polar synthetic rubber blends is appropriate curing.[35,36] Vulcanization of mixtures comprising such dissimilar elastomers is affected by the interfacial area of the blend, the differential solubility and reactivity of the vulcanization agents, as well as the curative diffusion between the rubber zones.[20] There are several vulcanization formulations for such blends, but the correct selection of curing recipe is key in obtaining the desirable crosslink densities of the related rubber phases. In this respect, the essential crosslinking characteristic is the ability of the two elastomers to co-vulcanize.[37]

Lewan[4] studied the semi-efficient vulcanization of a blend of NR with NBR (41% ACN content) at a 50/50 weight ratio. NR and NBR were mixed separately with ZnO, stearic acid and antioxidant, followed by their cross-blending on an open mill with the sulfur curatives. Addition of CBS accelerator yielded a maldistribution of crosslinks in favour of NR when the mixture was cured at 150 °C. Incorporation of the rather polar tetramethylthiuram monosulfide (TMTM) secondary accelerator to this cure system changed the situation, promoting NBR crosslinking. However, curing at 180 °C of the same batches altered things significantly and both curing recipes delivered higher crosslink densities for NBR compared to NR. It has to be mentioned here that sulfur has been reported in general to be preferentially soluble in unsaturated elastomers.[38] For low temperature NR/NBR vulcanization, other cure recipes that gave a rather even distribution of crosslinks involve N-oxydiethylene-benzothiazole-2-sulphenamide (MBS) or TBBS as accelerators. Moreover, when MBS or TBBS were combined with MBTS or tetrabutylthiuram disulphide (TBTD) as secondary accelerators, the overall crosslink densities were increased without adversely affecting the crosslink distribution. Lewan[4] showed that when NR was mixed with NBR of lower ACN content (i.e. 34%), CBS accelerator failed to yield similar crosslink densities for the two rubber phases either at low (150 °C) or high (180 °C) vulcanization temperatures. The use of NBR with lower solubility parameter always created a maldistribution of crosslinks in favour of NBR for these curing systems. For a NR/NBR blend in which the NBR had 18% ACN content, Tinker[39] proposed replacing sulfur with bis-alkylphenol disulfide (BAPD), which is a sulfur donor. The accelerator used in the recipe was TMTM, and the curing temperature was set at 150 °C. It was found that BAPD resulted in a reduction of the preferential distribution of crosslinks.

Mathai et al.[40] examined blends of NR with NBR of 34% ACN content at various weight ratios under a conventional sulfur curing system containing CBS accelerator. The cure characteristics evaluated at 150 °C showed that the scorch time and the optimum cure time decreased as the weight ratio of NR increased in the blend. A more or less similar trend was presented by Ismail et al.[41] for a sulfur-cured recipe containing TBBS accelerator at 150 °C. The scorch and optimum cure time decrease was more pronounced when the blend involved NR having 50 mol% of epoxidation (ENR-50).

In a sulfur-cured NR/CR silica-filled blend containing TBBS as accelerator, Sae-oui et al.[5] showed that scorch and optimum cure time decreased as the weight ratio of NR increased in the mix (cf. Figure 9.1(b)). However, the crosslink density of the blend was reduced with increasing NR content. In order to counteract the reversion effect that can be observed in such blends (mainly due to hydrochloride diffusion into NR), Zhang et al.[26] involved organo-modified montmorillonite (OMMT) in the formulation of CR. The weight ratio of NR/CR was set at 75/25, and curing took place at 143 °C.

Wootthikanokkhan and Clythong[42] investigated the effect of accelerator type on curing in a NR/ACM blend. The masticated NR was mixed with ACM (containing 5 wt% of chlorine cure site) for 15 min on a two-roll mill. ZnO, stearic acid, accelerator and sodium stearate were then incorporated into the mixture. The batch was cooled before the sulfur was added. Three different accelerators were tested, namely zinc diethyldithiocarbamate (ZDEC), MBT and diphenyl guanidine (DPG). The blends vulcanized with ZDEC had a short cure time compared to those with MBT or DPG. The fast curing in the first case hindered the migration of ZDEC and sulfur to ACM and NR, respectively. The crosslink density did not change significantly when the vulcanization temperature decreased from 170 °C to 150 °C, but as expected, the optimum cure time was increased.

9.3.3 Swelling and Oil Resistance

Swelling is very significant for rubber blends. Differential swelling in solvents may be exploited to investigate the degree of interfacial adhesion of the rubber components in the blend.[39,43] Also, by using NMR spectroscopy and evaluating the peak broadening in the continuous wave ^{1}H-NMR spectra of swollen vulcanizates, the degree of crosslinking in the individual phases of such blends can be identified.[44] Nevertheless, one of the most common and valuable characteristics of NR/polar synthetic rubber blends is their resistance to swelling in oils.

Mathai et al.[40] cut circular specimens from sulfur-cured sheets of NR/NBR (34% ACN content) and immersed them in bottles of toluene at constant temperature until periodic weighing showed that equilibrium swelling had been attained. The toluene uptake of the rubber vulcanizate was increased by increasing the NR weight percentage in the blend (cf. Figure 9.2). This is in accordance with the inherent solvent resistance performance of NBR. The sorption behaviour of the same blend was tested for various types of penetrants

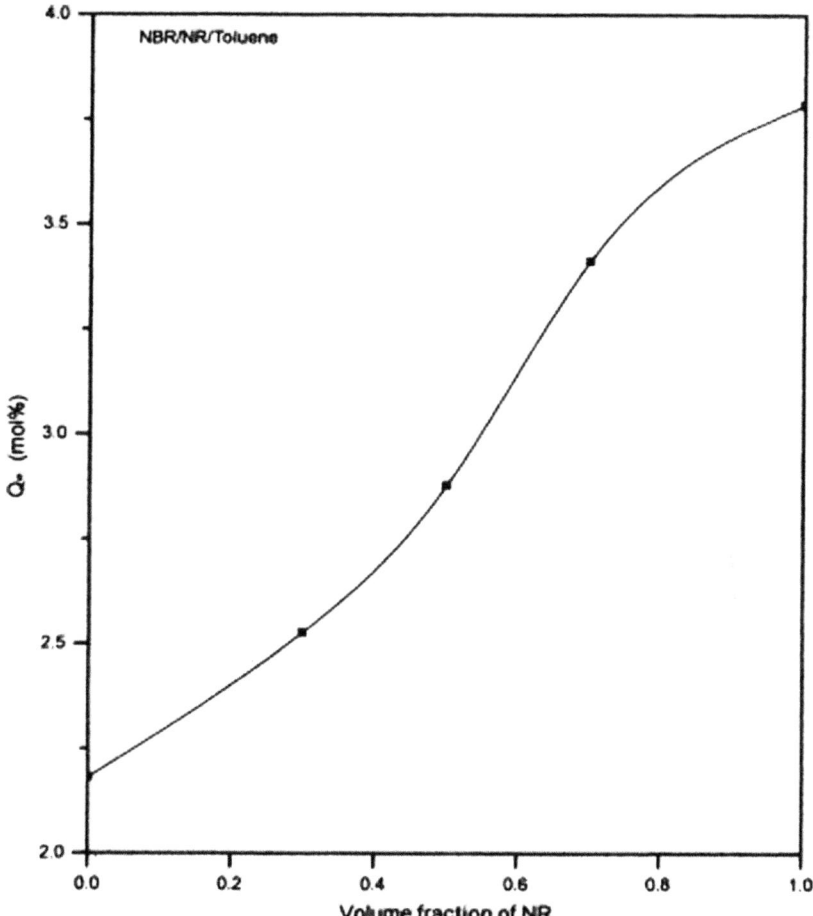

Figure 9.2 The variation in equilibrium toluene uptake ($Q\infty$) values with volume fraction of NR at 28 °C.
(Reproduced from Mathai et al.[40] with permission from Elsevier).

(*i.e.* benzene, toluene and *p*-xylene). A decrease in the molecular size of the penetrant (*i.e.* benzene < toluene < *p*-xylene) resulted in an increase in solvent uptake. Likewise, the presence of NBR in the blend was beneficial for solvent resistance. A similar but amplified trend was observed when the experiment took place at 70 °C. The increment of crosslink density which was detected as the NBR weight percentage in the blend increased and was advantageous to sorption hindrance behaviour of NR/NBR vulcanizate.

The presence of NBR in a sulfur-cured NR/NBR blend above 50/50 weight ratio significantly improves the oil resistance of the vulcanizate as measured after 70 h at 100 °C.[45] Moreover, for the same NBR weight percentage Karnika de Silva and Lewan[45] showed that the oil swelling can be further reduced when NR is partly substituted by methyl methacrylate grafted natural rubber. This graft copolymer is located at the NR/NBR interface, lowering interfacial tension and phase size. The enhanced oil resistance of NR/NBR blends can also be improved by adding fillers. Sirisinha et al.[33] mixed various CB and silica grades in a peroxide-cured NR/NBR blend of 20/80 weight ratio. A small CB particle size was beneficial for oil resistance while silica filler yielded a less pronounced effect. Nonetheless, crosslink density variations for these systems should also be considered.

For NR/XNBR blends, incorporation of thiophosphoryl compounds in the recipe has been proposed for reducing the swelling in a mixture of isooctane/toluene (70/30). That was found after immersing the vulcanizates in the solvent for 48 h at 30 °C.[46] Naskar et al.[47] credited this behaviour to crosslinking at the interface of NR with XNBR.

Sae-oui et al.[5] investigated the oil resistance of silica-filled sulfur-cured NR/CR blends. Samples were immersed in non-polar hydraulic oil at room temperature (i.e. 23 °C). The mass change of the blend after oil immersion was found to increase with increasing NR content, especially above a NR/CR weight ratio of 25/75. As expected, the lowest and the highest mass change were measured for neat CR and NR, respectively.

Wootthikanokkhan and Tunjongnawin[30] exploited differential swelling in a sulfur-cured NR/ACM (50/50) blend to measure the volume fraction of its rubber components. In this method each phase of the blend should be swollen in a solvent that does not swell the other rubber phase. After swelling to an equilibrium state, the volume fraction of rubber in the swollen gel (and crosslink density) of each rubber phase can be estimated. Accordingly, vulcanizates of NR/ACM were swollen to equilibrium for 7 days in the dark at 25 °C using cyclohexane and acetonitrile. The first solvent swells NR while the latter barely does. The opposite holds for ACM.

9.3.4 Morphology

Differences in viscosity values and solubility parameters between NR and polar synthetic rubbers usually produce immiscible blends. Table 9.1 shows the solubility parameters of some relevant rubber types.[7,48,49] The morphology of such blends is determined by the mixing procedure followed, the rheological properties and the degree of compatibility of the components involved. Investigating the morphology of the blend delivers significant information

Table 9.1 Solubility parameters of NR, NBR, CR and PUR.

	Solubility parameter ($MPa^{\frac{1}{2}}$)		
Type of rubber	Tinker[7]	Forrest[48]	Senichev and Tereshatov[49]
NR	16.7	16.2–17.4	16.5
NBR	up to 22	17.8 (18% ACN)–21.5 (40% ACN)	19.1
CR		16.6–19.2	18.1
PUR		20.0–21.1	18.4 (ether), 19.6 (ester)

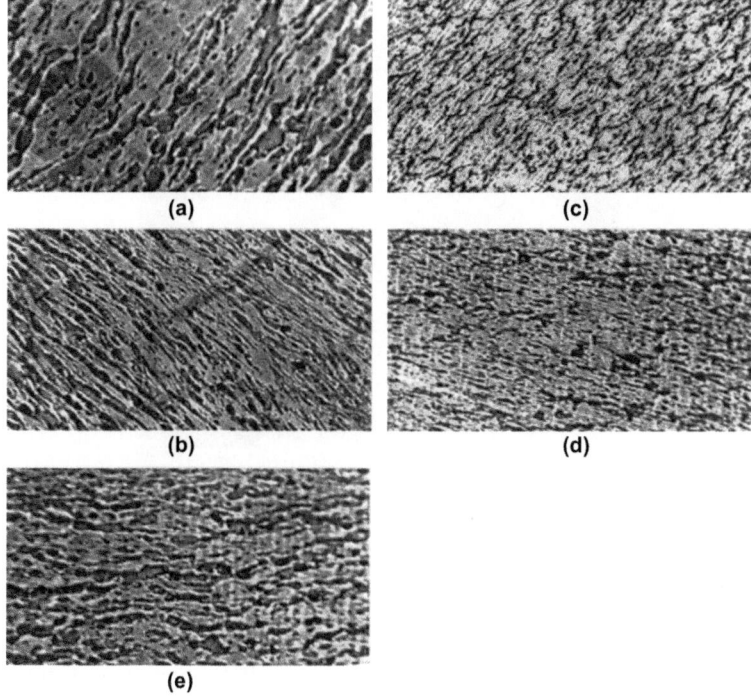

Figure 9.3 Micrographs (original magnification × 400) of blends prepared with blending times of (a) 15, (b) 20, (c) 25, (d) 30 and (e) 40 min.
(Reproduced from Sirisinha et al.[22] with permission from Wiley Periodicals Inc.).

regarding the network structure, the phase sizes and the distribution of additives, as well as indicating interface quality.

Sirisinha et al.[22] presented differences in the morphology of a NR/NBR blend (20/80 weight ratio) as a function of processing conditions (i.e. rotor speed and mixing time). The vulcanizates were cryogenically microtomed and the thin-sectioned specimens were observed by means of optical microscope. The size of the NR dispersed phase, which was almost independent of the rotor speed (not shown here), decreased with increasing blending time up to 25 min (cf. Figure 9.3). When the blending time increased further the dispersed phase was seen to coalesce. The phase size of NR was further reduced by adding liquid natural rubber (LNR) and epoxidized liquid natural rubber (ELNR) to the mix.[50] Besides, compatibilization of a NR/NBR blend (60/40 weight ratio) with poly(ethylene-co-vinyl acetate) at 6 phr produced more homogeneous fractured surfaces than the uncompatibilized one, as examined by scanning electron microscopy (SEM).[23]

Lewan[4] investigated the morphology of a NR/NBR vulcanizate of 45/55 weight ratio. The samples were stained with osmium tetroxide and examined by scanning transmission electron microscopy (STEM). As shown in Figure 9.4, the NBR phase, which appeared lighter than the NR phase, has a more discrete

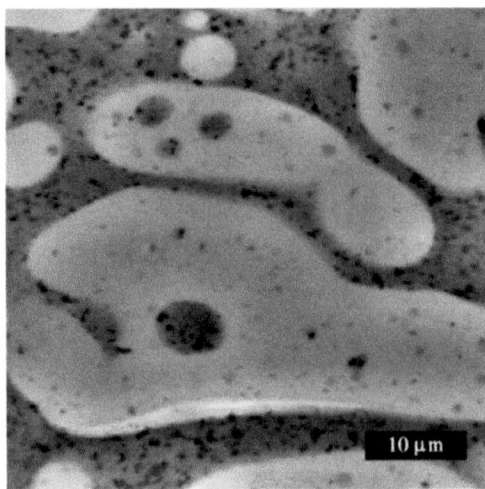

Figure 9.4 STEM micrograph of a 45:55 NR:NBR (41% ACN) gum blend vulcanizate.
(Reproduced from Lewan[4] with permission from Springer Science + Business Media).

character in the blend. At the same time the NBR phase was found to encapsulate some small NR regions. Moreover, ZnO (appearing as black spots) was found to be preferentially positioned in the NR phase. Information on interface adhesion between the two rubbers was obtained by exploiting equilibrium swelling of this vulcanizate to styrene. For this purpose, styrene was polymerized after equilibrium swelling. In this way, a mesh structure was created comprising strands of rubber in a polystyrene matrix. The mean size of the mesh cells is correlated with the molecular weight between the crosslinks. Ultrathin sections which were stained with osmium tetroxide were examined by transmission electron microscopy (TEM). Polystyrene (PS) was detected in the rubber phases (*e.g.* around ZnO particles), as well as along the interface. In the latter case, the band of PS was interrupted by strands of network material linking the rubber phases, indicating the existence of crosslinking across the interface. NR/NBR vulcanizates with better mechanical properties presented a thinner and more interrupted PS band. In order to reduce the resulting phase sizes in a vulcanizate of NR with NBR (34% ACN content), Kongsin and Lewan[51] incorporated CR. It was shown that the phase sizes reduced to about 10 μm (in a 50/5/45 NR/CR/NBR weight ratio) from more than 100 μm before compatibilization. In STEM micrographs CR was detected at the NR/NBR interface as a thin black line (due to its chlorine content). The same amount of CR (*i.e.* 5 phr) has also been used for the compatibilization of NR/NBR (20/80) vulcanizates reinforced with nanomodified fillers.[52]

Sae-oui *et al.*[5] analysed NR/CR vulcanizates by atomic force microscopy (AFM) of microtomed samples. CR appeared to be the continuous phase at weight ratios of 25/75 and 50/50 NR/CR. However, this situation is reversed

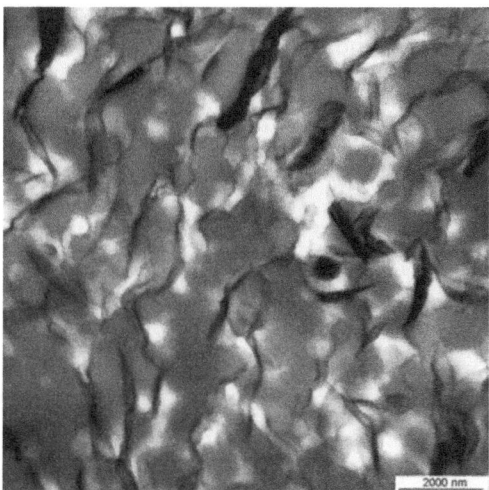

Figure 9.5 TEM picture taken from the film cast of the PUR/NR (1/1) latex blend containing 10 phr LS.
(Reproduced from Varghese et al.[12] with permission from Wiley Periodicals Inc.).

with further increases in the NR phase. In the AFM images the CR phase appeared darker than the NR phase.

The effect of curative type can be shown by the morphology of the blend. Wootthikanokkhan and Clythong[42] investigated by means of SEM the effect of changing the accelerator type on a blend of NR/ACM (50/50 weight ratio). The specimens for SEM examination were fractured after immersing them in liquid nitrogen followed by staining, washing and sputtering with gold. ACM appeared darker than the NR phase and the immiscibility was apparent in all SEM images. The structure developed was ACM rubber dispersed in a NR matrix, in which the dispersed phase was larger with MBT accelerator than with DPG. However, ZDEC gave rise to a rather fibre-like dispersed phase.

The immiscibility of a NR/PUR blend can be moderated by using filler. Varghese et al.[12] mixed sodium fluorohectorite with a 50/50 weight ratio of NR/PUR *via* the latex route. As shown through TEM investigation, this layered silicate (LS) is positioned at the interface of the pre-vulcanized NR particles with the PUR phase, creating a 'skeleton' structure (cf. Figure 9.5). Thus, the tendency for latex particles to coalescence is hindered and dispersed phase morphology prevails.[53]

9.3.5 Mechanical and Dynamic-Mechanical Behaviour

An important aspect defining the mechanical behaviour of NR based polar synthetic rubber blend vulcanizates is the mechanical response of the individual components and the interphase involved, influenced by the curative (and filler) distribution. The synergetic action of the blend phases escalates the ability of the developed network structure and morphology to withstand higher loads, increased elongations, demanding fatigue or dynamic stresses.

Within the NR phase the non-rubber components exist in a considerable percentage (*i.e.* about 6 wt%). These involve proteins, carbohydrates, phospholipids and metal ions developing multi-scaled structures especially important for the high green strength of NR.[54] The flexibility of the *cis*-1,4 structure appearing in NR polymer chains is altered by its partial transformation to a *trans*-1,4 structure during curing, in which the latter is more rigid. The various types of sulfur bridges (*i.e.* mono-, di-, poly-) along with the chain flexibility within network points are determined by the curing recipe and protocol followed. Besides, fillers (*e.g.* nanoclay) within the NR phase might induce a physical network during deformation that favours the alignment of NR chains, resulting in an increase in crystallization rate.[55] The eminent mechanical performance of NR is desirable to apply also in case of its blends with polar synthetic rubbers.

Ismail *et al.*[41] investigated sulfur-cured NR/NBR blends (of various weight ratios) in tensile testing. It was found that the modulus at 100% elongation increased along with the NBR content in the stock. This was related to the enhanced crosslinking induced in the vulcanizate with an excess of NBR. However, tensile strength followed an opposite trend. As the percentage of NBR increased above 25%, the strength of the vulcanizate reduced. For blends with NBR content above 50% the drop in tensile strength was far more evident, as schematically depicted in Figure 9.6. The latter was associated with the reduction in the crystallizable rubber component (*i.e.* NR) as NBR became the dominant and continuous phase. Likewise, elongation at break was reduced as

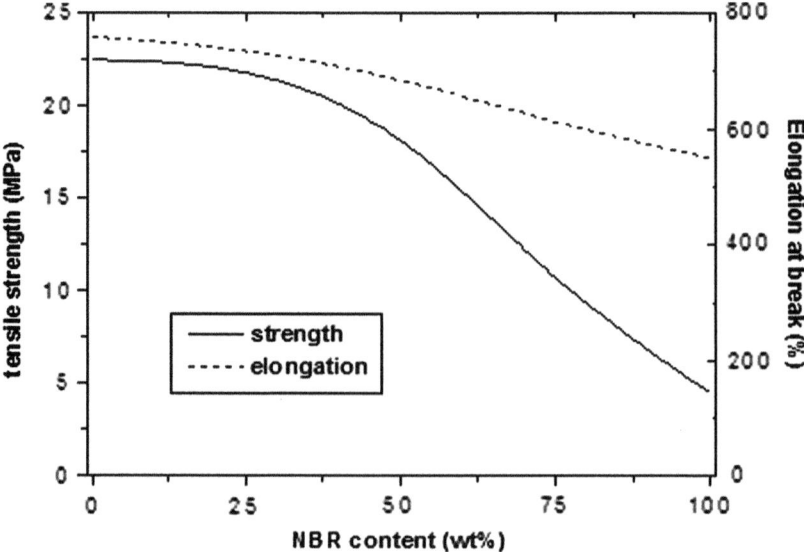

Figure 9.6 Tensile strength and elongation at break of NR/NBR blend as a function of NBR content.[41]

the weight ratio of NBR increased in the blend. In order to compensate for the polarity mismatch of these rubber components, epoxidized natural rubber with 50 mol% of epoxidation (ENR-50) fully substituted the NR.[41] Nevertheless, the related tensile properties followed a similar trend as in the case of NR but at relatively lower values.

Apart from the effect of the overall crosslink density of NR/NBR on the tensile strength of the blend, the crosslink distribution in separate phases is of great importance. Lewan[4] showed that for a 50/50 blend the higher tensile strength was evident for a NR/NBR crosslink density ratio of 1.0–1.2. Further tensile strength enhancement in a NR/NBR vulcanizate was obtained when methyl methacrylate grafted NR was added to the blend instead of NR.[45]

Accordingly, the action of CR has proved to be beneficial in NR/NBR mixtures especially when it is accompanied by a suitable curative package.[51] For a NR/NBR (20/80) blend, Kantala et al.[56] investigated the effect of ENR or CR as compatibilizer (at 5 phr). Both compatibilizers resulted in comparable tensile strength in the respective sulfur-cured vulcanizates. However, ENR performed better than CR when the same property was examined after immersing the compatibilized blend in hydraulic oil for 70 h at 125 °C.

For sulfur-cured NR/CR blends, Ismail and Leong[25] showed that the modulus at 100% elongation yielded a slight increase as NR approached 25%, but then decreased. This continued even below the initial value for pure CR as NR exceeded 75% in the vulcanizate. On the contrary, tensile strength showed a slight decrease from 0/100 to 25/75 NR/CR weight ratio, followed by an increase for higher NR percentages. The maximum tensile strength was maintained for 75% NR in the blend. By substituting NR with its polar version ENR-50, a similar property trend was observed, although attenuated (*i.e.* higher property values for all weight ratios). One reason for the enhanced performance of ENR/CR blends is the affinity between these rubbers, as has been proved by means of dynamic mechanical analysis (DMA).[57]

Wootthikanokkhan and Tongrubbai[58] examined the tensile properties of NR/ACM vulcanizates. It was found that increasing the NR content in the recipe improved the mechanical properties. However, that was not the case when the same specimens were heated for 24 h at 140 °C. After this thermal treatment, the vulcanizates with increased ACM weight ratio maintained to a greater extent their initial strain percentage compared to these with lower ACM content. To improve the mechanical response of NR/ACM sulfur-cured vulcanizates, incorporation of CB[59] or compatibilization with poly(isoprene-butyl acrylate) block copolymer (at 5 wt%) has been proposed.[60]

Varghese et al.[12] examined the dynamic mechanical properties of a NR/PUR (1/1) blend prepared by the latex route. The NR particles were pre-vulcanized with sulfur curatives prior to mixing with PUR latex; the latter phase remained practically uncured in the blend and so phase immiscibility was pronounced. As shown in Figure 9.7(a), the blend maintained the high storage modulus values at temperatures below the glass transition (T_g) of NR. At the same time, the presence of the PUR phase significantly improved the performance of the blend at the temperature range between the T_g of the two rubbers involved (*i.e.* from

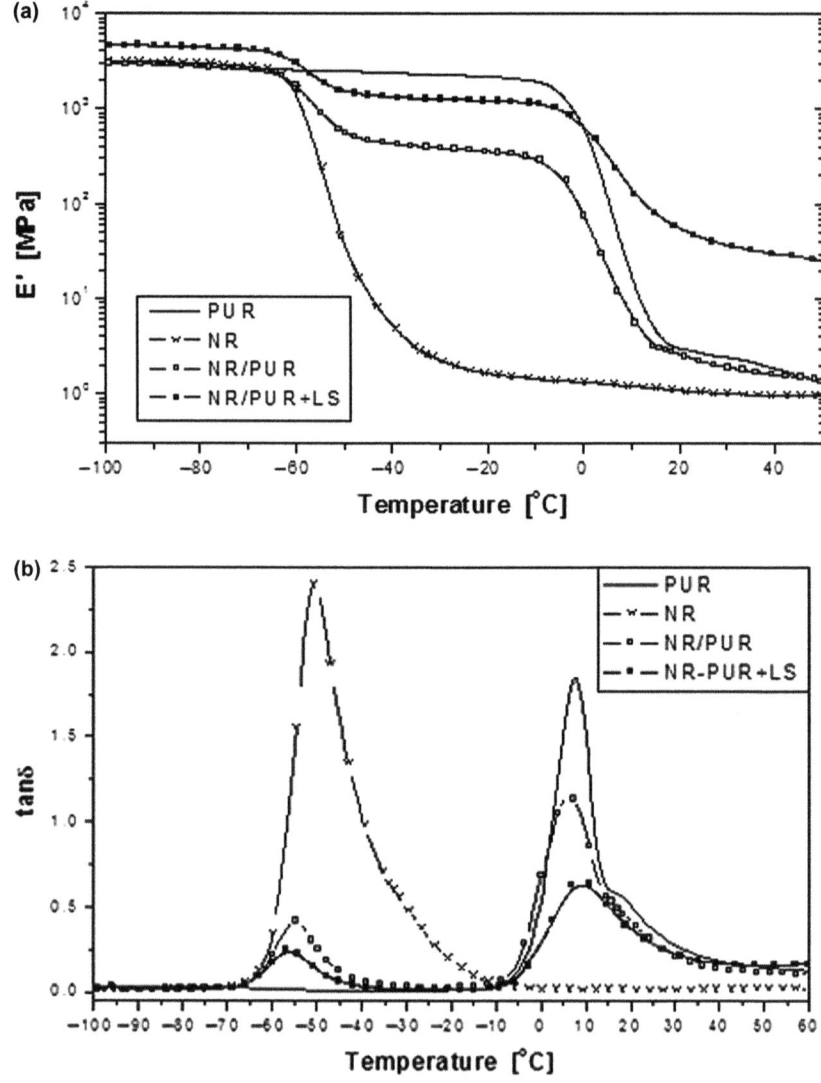

Figure 9.7 Storage modulus (a) and mechanical loss factor (b) as a function of temperature for NR, PUR, NR/PUR (1/1) and NR/PUR (1/1) compatibilized blend (in tension mode at 10 Hz frequency).[12]

about −60 up to 0 °C). The same trend holds also for the rubbery region, where both phases are above their T_g. The immiscibility of the NR and PUR phases in the blend is better manifested in Figure 9.7(b) by the existence of two clearly separated tan δ peaks.

Addition of LS as a compatibilizer reduced the damping performance of the NR/PUR blend, indicating reinforcing efficiency. For the NR phase, its tan δ peak was shifted slightly to lower temperatures, while for the PUR phase a slight shift to higher temperatures is seen. The first shift hints at limited interactions of the NR phase with the LS while the latter suggests the opposite for the PUR phase and the filler.[61] Addition of LS significantly improved the storage modulus of the blend over the whole temperature range (Figure 9.7(a)). This improvement was more pronounced in the rubbery region, mainly due to the morphology of the reinforced blend, in which the LS formed a skeleton-type structure around the pre-vulcanized NR particles (cf. Figure 9.5).[12]

9.3.6 Thermal Properties

The thermal properties of NR based polar synthetic rubber blends is expected to affect the application range of these vulcanizates.[62] Degradation, glass transition and crystallization temperatures determine the service conditions of each blend.[63] Additionally, the existence of single or multiple shifted or not glass transition temperatures of each blend supplies evidence for the miscibility of the components involved.[64,65] Nevertheless, in many cases further analysis might be required to determine miscibility.

Elizabeth et al.[66] investigated a sulfur-cured vulcanizate of NR mixed with hydrogenated nitrile rubber (HNBR) in a 50/50 weight ratio using differential scanning calorimetry (DSC). The T_gs of pure NR and HNBR vulcanizates were detected at about −63 and −25 °C, respectively. In the binary blend the position of the T_gs remained unaffected, indicating that the blend was immiscible. Addition of 3 or 7 phr of dichlorocarbene-modified NR (DCNR) of 15% chlorine content in the mix altered the T_g of the components marginally. Thus, the compatibilizing action of DCNR in the stock was characterized as rather poor.

Roychoudhury et al.[67] showed that the double T_g peak of a blend of ENR-50 with chlorosulphonated polyethylene (CSM) at 1:1 weight ratio was changed to a single one (at 2.7 °C) when carboxylated nitrile rubber (XNBR) was added in the stock at 50 wt% (i.e. ENR-50/CSM/XNBR; 25/25/50). Additionally, an exothermic peak was detected at 205 °C during a differential thermal analysis of a 33.3/33.3/33.3 mixture of this ternary blend. Taking into consideration that CSM, XNBR and ENR are soluble in chloroform whereas their blend is insoluble, the exothermic peak was attributed to self-crosslinking behaviour of this ternary compound (in the absence of curatives).

Similarly, for the ternary blend of ENR/CR/XNBR, Alex et al.[68] showed that the mixture can be miscible at 75 parts of ENR per 100 parts of CR/XNBR rubber. This was detected irrespective of the CR/XNBR ratio. The authors drew their conclusions about the miscibility by evaluating the T_g through DSC and DMA experiments. Also Mathai and Thomas,[69] despite the fact that their DMA findings on a sulfur-cured ENR/NBR blend exhibited a sharp single tan δ peak, continued their investigation by means of SEM. The morphological analysis revealed that the blend was immiscible (phase separated) and that the single peak was due to the initial close position of the tan δ peaks of ENR-50 and NBR with 34% ACN content. The polarity match of these rubber components probably brought closer their tan δ peak positions. However, the

Table 9.2 Glass transition temperatures and conductivity (σ) of NR, PUR, NR/PUR (50/50) blend and NR/PUR (50/50) blend compatibilized with 10 phr layered silicate (LS). Conductivity values correspond to the frequency of 10^{-3} Hz (Reproduced from Psarras et al.[72] with permission from Wiley Periodicals Inc.)

	Glass transition temperature (°C)	Conductivity (S.cm^{-1})
NR	−64.1	8.56×10^{-16}
PUR	−1.1	1.11×10^{-13}
NR/PUR (50/50)	−65.0/−6.4	3.56×10^{-13}
NR/PUR (50/50) + 10 phr LS	−65.4/−6.6	5.28×10^{-14}

authors concluded that miscibility cannot be simply judged by simple T_g measurements for peak positions closer than 20 °C.[69]

Pal et al.[70] showed that the thermal stability of CB-filled NR vulcanizate was enhanced by the addition of XNBR. The onset of degradation during thermogravimetric analysis in the presence of air was shifted from 258 °C for NR/XNBR vulcanizate of 80/20 weight ratio to 387 °C when this weight ratio was reversed. Similarly, the 10% degradation was shifted from 344 to 396 °C for the XNBR-rich blend. The NR/XNBR vulcanizate of 50/50 weight ratio presented onset and 10% degradation at 264 and 342 °C, respectively. This suggested that the performance of the blend was significantly improved when XNBR became the dominant phase.

Noriman and Ismail[71] showed that incorporation of 10 phr ENR-50 in a blend of a styrene butadiene rubber (SBR) with recycled NBR (at 50/50 weight ratio) reduced both the temperature at 5 and 50% weight loss during a thermogravimetric analysis in a nitrogen atmosphere. The first was reduced from 360 to 353 °C, the latter from 480 to 392 °C.

Psarras et al.[72] underlined the immiscibility of a latex route prepared blend of pre-vulcanized NR with PUR (50/50 weight ratio) by means of DSC. As given in Table 9.2, the T_gs of pure NR and PUR were detected at −64.1 and −1.1°C, respectively. For the NR/PUR blend, two clearly separated T_gs were evident, suggesting immiscibility. The position of the T_g assigned to NR remained unaffected in the blend while that related to PUR was shifted to lower temperature. Incorporation of LS did not change the T_g position significantly.

9.3.7 Dielectric Properties

Analysis of the dielectric response of NR is a subject that has recently attracted the attention of academia. The fact that NR is a multi-component biomaterial affects its dielectric relaxation behaviour mainly due to the naturally occurring network.[73] After sulfur vulcanization the developed network suppresses the fluctuations of the chain ends raising a slower process (i.e. α'-process) than the segmental one (i.e. α-process) due to the lipid.[74] Moreover, the strength of the first α'-process is highly affected by the presence of water compared to its negligent effect on the latter segmental one. Similarities in the apparent activation volumes for the dry and wet NR vulcanizates for both processes suggested an inter-relation of these processes due to the presence of stearic acid.[74]

The absence of polar side groups in the backbone of NR is responsible for the lack of any secondary relaxation process in its dielectric spectrum. Also, synthetic polar rubbers are dielectrically active materials. Apart from the glass to rubber transition (α-mode), local motions of polar side groups (β-mode) and rearrangement of small segments of polymer backbone (γ-mode) might arise in the related dielectric spectrum. Incorporation of additives might be responsible for the detection of interfacial polarization (IP) or the Maxwell–Wagner–Sillars (MWS) effect due to the accumulation of charges at the created interfaces.[61]

El-Nashar and Turky[75] presented the dielectric response of a peroxide-cured NR/NBR (1/1) vulcanizate for various mixing protocols. The highest values of permittivity in a permittivity versus frequency graph were obtained for the vulcanizate that had been prepared for 7 min at 140 °C. Other processing conditions, like those at 160 or 150 °C for 3 or 5 min, gave lower permittivity values at the examined frequency range. The NR/NBR vulcanizate with the highest permittivity values gave rise to the best mechanical properties.

Psarras *et al.* analysed extensively the dielectric response of a NR/PUR blend (50/50 weight ratio) at both ambient temperature[72] and a wide range of temperatures[76] by means of broad-band dielectric spectroscopy (BDS). The dielectric data were analysed through the electric modulus formalism, which corresponds actually to the inverse quantity of complex permittivity. This type of analysis minimizes the large variations in the permittivity and loss at low frequencies and neglects electrode polarization issues.[77] As shown in Figure 9.8, the high values for both the real (M') and imaginary part (M'') of the electric modulus for NR imply that it is a rubber of low dielectric permittivity and loss. In the NR spectrum of M' only one transition at the low frequency edge is detected (cf. Figure 9.8(a)). On the contrary, both PUR and NR/PUR blends yielded three distinct transitions in their M' spectra. All these transitions became more evident in Figure 9.8(b) on the imaginary part of electric modulus (M'') versus frequency. For NR the α-mode peak was not completely recorded due to the low value of the T_g. The detected peaks in the case of PUR and NR/PUR blends for scanning from the low frequency to its high end are assigned to IP, α-process and β-mode, respectively. The existence of IP for plain PUR is due to the interface created between its regions of hard and soft segments. The MWS effect for the NR/PUR blend is broader and it is amplified to higher frequencies. This phenomenon seems to be related to the heterogeneous morphology of this NR/PUR blend.[72] The relaxations of the α- and β-mode of PUR are affected marginally in the resulting blend.

The separate α-mode contribution of NR in the NR/PUR blend can be clearly observed in Figure 9.9 where the imaginary part of the electric modulus is depicted versus temperature and frequency. The temperature and frequency range was from -100 to 50 °C and 10^{-1} to 10^6 Hz, respectively.[61] Generally speaking, the frequency–temperature superposition shifts the loss peak position

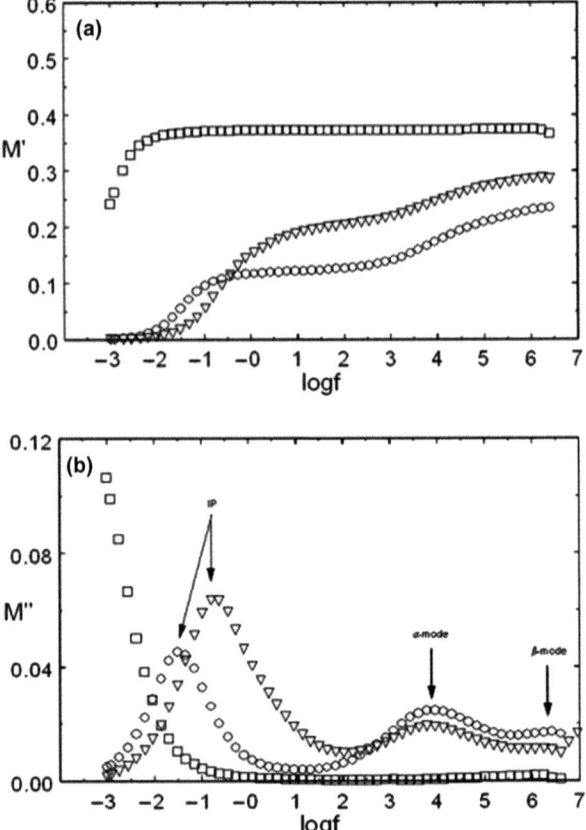

Figure 9.8 Real (a) and imaginary (b) part of electric modulus for the (□) NR, (○) PUR and (▽) PUR/NR systems.
(Reproduced from Psarras et al.[72] with permission from Wiley Periodicals Inc.).

of relaxation processes to higher frequencies with increasing temperature. The temperature dependence of each peak shift is a measure of the relaxation rate of the specific process.[78]

9.3.8 Infrared Absorbance

Fourier transform infrared spectroscopy (FTIR) has emerged as a valuable tool for rubber analysis. Siesler[79] monitored the onset, progress and decay of strain-induced crystallization of a sulfur-cured NR during a cyclic experiment. The infrared absorbance of NR at 1126 cm^{-1} assigned to C–CH$_3$ in-plane deformation vibration is a band sensitive to crystallinity. A thickness reference band was taken to be the 1662 cm^{-1} one assigned to ν(C=C) stretching. The ratio of the absorbance bands at 1126 cm^{-1} and 1662 cm^{-1} revealed the reversible nature of strain-induced crystallization of NR.

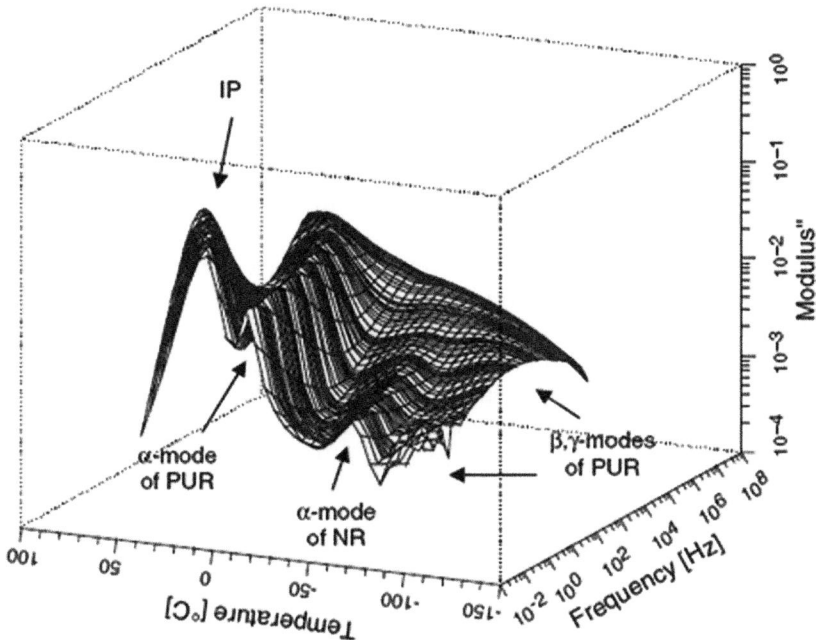

Figure 9.9 Imaginary part of electric modulus versus frequency and temperature for PUR/NR blend
(Reproduced from Psarras and Gatos[61] with permission from Springer Science + Business Media).

Temel et al.[80] evaluated the residual vulcanization accelerators in NR latex films by exploiting the peak area decline of characteristic IR bands of the related accelerators. Cook et al.[81] detected the oxidation of thermally treated NR at 150 °C in air through the IR absorption peak existence at 1720 cm^{-1} which is attributed to the C=O group. Narathichat et al.[82] noticed that the absorption attributed to the C=O group was increasing during oxidative degradation at the expense of the peak intensity at 835 cm^{-1}, which is attributed to =CH out-of-plane bending. The ratio of these bands was exploited to evaluate the thermo-oxidation during NR mastication at increased times and temperatures. The presence of grafted maleic anhydride (MA) on the NR backbone has been detected via the decrease in intensity of the peak at 1634 cm^{-1} assigned to C=C of NR and the increase of a band at 1708 cm^{-1} due to the C=O group of MA.[83]

FTIR can also be effectively applied in rubber blend analysis.[84] Several IR studies on vulcanized rubber products have been performed in connection with the pyrolysis method, in which the blend composition is analysed through the loss of characteristic IR bands.[6] Moreover, the existence of peak shift of characteristic bands in the case of NR blends before and after compatibilization has been used as an indication of preferable interactions of the compatibilizer with the blend components. This technique has been used in the case of NR/HNBR (50/50) blends compatibilized with dichlorocarbene-modified NR

(DCNR), in which the band at 2237 cm^{-1} assigned to C≡N stretching was shifted to 2240 cm^{-1} after compatibilization.[66] Nagode and Roland[85] reported miscibility of ENR-50 with CR (solvent-assisted mixing) at a 50/50 weight ratio as shown by means of FTIR and verified using DMA and DSC measurements. In this case, the subtraction of ENR and CR spectra from the spectrum of the blend revealed alterations in peak position and peak width of the FTIR bands. These were assigned to vibrations involving the chlorine atom of CR and the epoxy moiety of ENR in the blend.

Pre-vulcanized NR latex has been mixed with polyester-based PUR latex at a 1/1 weight ratio and IR spectra from the respective dried films have been recorded in the attenuation total reflection (ATR) mode. As shown in Figure 9.10 the FTIR spectra of PUR and NR show characteristic strong bands. In the case of NR such bands are at 834 cm^{-1} and 1375 cm^{-1} and are assigned to =CH out-of-plane bending and CH$_3$ asymmetric deformation, respectively.[86] In the case of PUR, characteristic peaks occur at 1716 cm^{-1} and 1264 cm^{-1} and are assigned to the stretching of urethane carbonyl groups C=O and the stretching of C–O in ester functionalities, respectively.[87,88] The bands due to NR retained their peak position in the spectrum of the NR/PUR blend without any peak shift (cf. Figure 9.10). On the contrary, the IR bands at 1716 cm^{-1} and 1264 cm^{-1} in the spectrum of pure PUR shift to 1719 cm^{-1} and 1275 cm^{-1} in case of the blend, respectively. Urethane groups are able to interact intermolecularly with themselves by hydrogen bonds forming microdomains.[89] Hydrogen bonded carbonyl shows a peak position at relatively lower

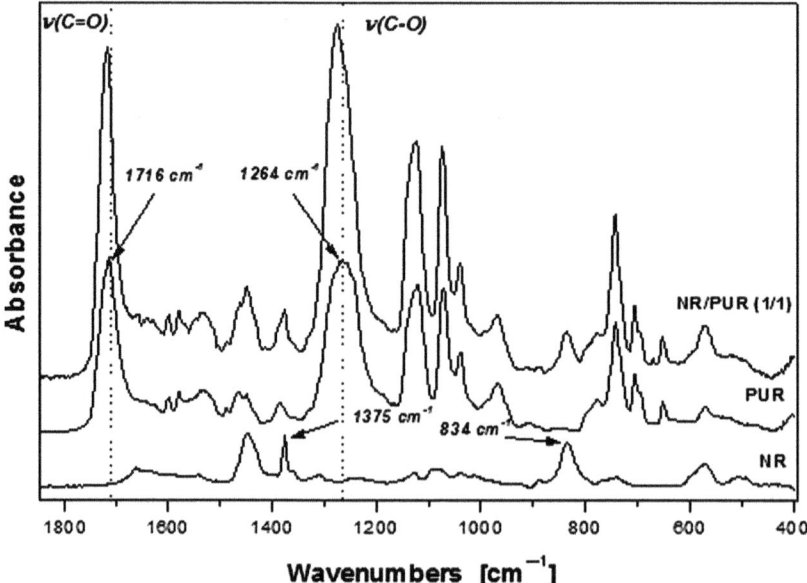

Figure 9.10 FTIR spectra of NR, PUR and NR/PUR (1/1) blend within the range of 400 to 1850 cm^{-1}.[12]

wavenumbers than free groups.[90–92] Thus, the observed peak shift of the 1716 cm^{-1} band to higher wavenumbers seems to be related to more weakly associated groups in the blend. However, this peak shift is rather marginal.[93]

9.3.9 Ageing and Other Properties

Blends of NR with polar synthetic rubbers have also been investigated for various other properties. Patcharaphun et al.[94] examined the quality of the welding line of a NR/NBR sulfur-cured specimen reinforced with CB. The welding line was created during compression moulding of two identical batches that were positioned at the opposite sides of a suitably designed mould. The tensile properties of the samples containing a weld line were lower than those without. Additionally, on increasing the NBR content in the mixture, the weld line strength was further reduced. However, this situation changed when the samples experienced thermal ageing and hot oil ageing. After such a treatment, the specimens with higher amounts of NBR maintained their strength with respect to the untreated vulcanizates.

Afifi and El-Wakil[83] measured the ultrasound velocity in NR/NBR (50/50) blends compatibilized with different weight percentages of NR-g-MA. It was found that at moderate concentrations of compatibilizer (2–6 phr) the ultrasonic velocity was increased. However, incorporating more NR-g-MA (up to 10 phr) decreased this property. This phenomenon was correlated with the degree of compatibility of the blend components. Increased ultrasound velocity relates to stronger interactions at the interfaces, thus better compatibility. The reason behind this is that the ultrasonic properties of the rubber blends are affected by the occurrence of phase inversion and variation of micro-voids within the blend.

Pal et al.[95] tested the abrasion resistance of NR/XNBR vulcanizate for various rubber weight ratios and CB grades. The weight ratio of 80/20 for NR/XNBR blends combined with a relatively low CB particle size was found to deliver the best abrasion resistance.

Elizabeth et al.[66] evaluated the heat build-up of a NR/HNBR sulfur-cured vulcanizate on changing the HNBR content. It was found that the lowest value was obtained for pure NR, whereas by increasing HNBR the heat build-up value was notably increased. For a NR/HNBR blend of 50/50 weight ratio, incorporation of 3 to 7 phr of dichlorocarbene-modified NR as compatibilizer decreased slightly the heat build-up value of this blend. The latter mix also yielded enhanced ozone resistance.

Ramesan et al.[96] investigated the ozone resistance of a NR/CR sulfur-cured vulcanizate as a function of the CR content. The specimens were exposed to 50 ppm ozone for 30 h at 40 °C. As schematically depicted in Figure 9.11, the time required for initial ozone cracking is greater as the CR content increases in the blend. When the CR phase becomes the dominant phase in the vulcanizate, significant property enhancement is evident.

A similar trend was shown for the flame resistance of the same NR/CR blend. Flame resistance was found to decrease on increasing the concentration

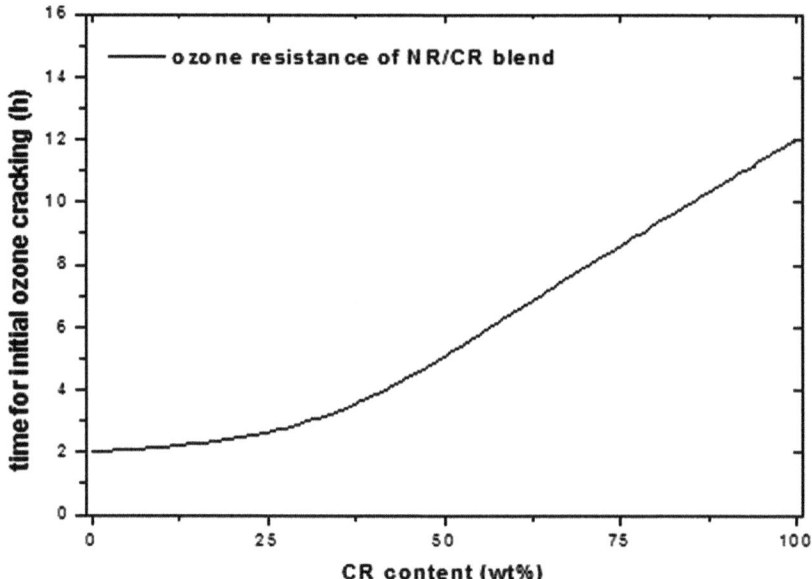

Figure 9.11 Ozone resistance of NR/CR blend as a function of the CR content.[96]

of NR. The enhanced performance was allowed by the char, which insulated the blend from the flame combustion.[96]

Psarras et al.[72] measured the conductivity of a NR/PUR (50/50) mixture. As presented in Table 9.2, the conductivity of NR was much lower than that of PUR. Nevertheless, the NR/PUR blend maintained the conductivity of PUR. Addition of LS in the blend decreased this value.

9.4 Applications

NR finds a plethora of applications in everyday life. It is mainly suitable for the manufacture of products that demand high mechanical and good dynamic characteristics, such as gloves, condoms, balloons, industrial hoses, tyres, adhesives, *etc.*[97]

In case of flexible inflations for milking machines, NR has been blended with NBR because the application requires good mechanical properties, and tear resistance along with enhanced resistance to swelling by butter oil. For optimum performance a NBR weight ratio of at least 50 wt% was selected for the blend.[98]

A mix of NR with NBR has been proposed for the manufacture of antistatic gloves.[99] Moreover, a NR/NBR blend with high NBR amount (*i.e.* 60–80 wt%) has been suggested for the composition of a sleeve for use in an impression packer.[100]

The solution for an oil-resistant elastomeric rubber composite particularly suitable for use as an elastic tape for garments has been found in a blend of NR,

ENR, NBR and CR. Among other properties (*i.e.* reduced oil swelling), the rubber tape made of this material combination is expected to yield good needle tear strength when sewn inside garments for direct skin contact. Such garment examples involve swimwear, headbands, shower caps, *etc.*[101]

In the manufacture of damping elements, both dynamic properties and resistance to thermo-oxidative environments are required. NR and CR appear to be suitable rubbers for such applications.[102]

NR blended with CR has been proposed for the blade material of a windscreen wiper. In this blend, incorporation of ENR (with epoxidation degree preferably from 15 to 25%) is suggested for enhanced performance of the vulcanizate.[103]

Blending of NR with PUR *via* the latex route has been claimed as beneficial for the manufacturing of condoms. Such a composition presents adequately high burst pressure for relevant applications.[104]

9.5 Outlook

NR is a material that has been investigated extensively by both industry and academia. Its superior mechanical properties have enabled its dominance in a wide range of applications. Nevertheless, its susceptibility to swelling by oils and thermal ageing, as well as its low ozone resistance, limit its application spectrum. Therefore, blending of NR with polar synthetic rubbers attempts to compensate for this weakness, but at the expense of mechanical properties. In an attempt to optimize the compromise, a more careful selection of curative package and mixing protocol should be specially designed for each specific NR/polar synthetic rubber blend. The maldistribution of crosslinks could be hindered, and so the overall behaviour of the blend is expected to be improved.

Additionally, suitable modification of NR[105–107] might be helpful for improving its blending efficiency and compatibility with polar synthetic rubbers. As widely accepted, incorporation of compatibilizers could supply a solution to this problem but as a consequence it increases the complexity of the compound.[108]

In any case, analytical techniques like NMR, BDS, FTIR and small-angle neutron scattering (SANS) will be much more involved in the analysis of relevant blends in the future.[109–111] Such investigations can give new insights and add to the knowledge already obtained by conventional characterization methods.

References

1. B. Rodgers, W. H. Waddell and W. Klingensmith, in *Encyclopedia of Polymer Science and Technology*, ed. H. F. Mark, Wiley, New York, 2004, p. 613.
2. N. M. Mathew, in *Rubber Technologist's Handbook*, ed. J. R. White and S. K. De, Rapra Technology, Exeter, 2001, p. 33.
3. R. B. Simpson, *Rubber Basics*, Rapra Technology, Shawbury, 2002, p. 77.
4. M. V. Lewan, in *Blends of Natural Rubber*, ed. A. J. Tinker and K. P. Jones, Chapman & Hall, London, 1998, p. 53.

5. P. Sae-oui, C. Sirisinha and K. Hatthapanit, *eXPRESS Polym. Lett.*, 2007, **1**, 8.
6. W. M. Hess, C. R. Herd and P. C. Vegvari, *Rubber Chem. Technol.*, 1993, **66**, 329.
7. A. J. Tinker, in *Blends of Natural Rubber*, ed. A. J. Tinker and K. P. Jones, Chapman & Hall, London, 1998, p. 1.
8. S. Datta, in *Science and Technology of Rubber*, ed. J. E. Mark, B. Erman and F. R. Eirich, Elsevier, New York, 2005, p. 538.
9. A. H. Eng and E. L. Ong, in *Handbook of Elastomers*, ed. A. K. Bhowmick and H. L. Stephens, Marcel Dekker, New York, 2001, p. 41.
10. K. G. Gatos and J. Karger-Kocsis, in *Rubber Clay Nanocomposites*, ed. M. Galimberti, Wiley, New Jersey, 2011, p. 371.
11. N. M. P. S. Ricardo, M. Lahtinen, C. Price and F. Heatley, *Polym. Int.*, 2002, **51**, 627.
12. S. Varghese, K. G. Gatos, A. A. Apostolov and J. Karger-Kocsis, *J. Appl. Polym. Sci.*, 2004, **92**, 543.
13. D. Mangaraj, *Rubber Chem. Technol.*, 2002, **75**, 365.
14. P. J. Corish and B. D. W. Powell, *Rubber Chem. Technol.*, 1974, **47**, 481.
15. L. Bokobza, *Kautsch. Gummi Kunstst.*, 2009, **62**, 23.
16. R. Magaraphan, W. Thaijaroen and R. Lim-Ochakun, *Rubber Chem. Technol.*, 2003, **76**, 406.
17. S. Sadhu and A. K. Bhowmick, *J. Polym. Sci., Part B: Polym. Phys.*, 2005, **43**, 1854.
18. W. G. Hwang, K. H. Wei and C. M. Wu, *Polym. Eng. Sci.*, 2004, **44**, 2117.
19. K. G. Gatos and J. Karger-Kocsis, in *Rubber Nanocomposites*, ed. S. Thomas and R. Stephen, Wiley, Singapore, 2010, p. 173.
20. J. B. Gardiner, *Rubber Chem. Technol.*, 1970, **43**, 370.
21. L. V. R. Murthy, S. Banerjee, B. Singh and R. S. Chauhan, *J. Appl. Polym. Sci.*, 1997, **65**, 731.
22. C. Sirisinha, S. Baulek-Limcharoen and J. Thunyarittikorn, *J. Appl. Polym. Sci.*, 2001, **82**, 1232.
23. P. Kumari, C. K. Radhakrishnan, G. P. Unnikrishnan, S. Varghese and A. Sujith, *Chem. Eng. Technol.*, 2010, **33**, 97.
24. M. A. Kader, W. D. Kim, S. Kaang and C. Nah, *Polym. Int.*, 2005, **54**, 120.
25. H. Ismail and H. C. Leong, *Polym. Test.*, 2001, **20**, 509.
26. P. Zhang, G. Huang and Z. Liu, *J. Appl. Polym. Sci.*, 2009, **111**, 673.
27. C. Fulin, C. Lan and L. Caihong, *Polym. Compos.*, 2007, **28**, 667.
28. A. P. Kuriakose and M. Varghese, *J. Appl. Polym. Sci.*, 2003, **90**, 4084.
29. V. Tanrattanakul and A. Petchkaew, *J. Appl. Polym. Sci.*, 2006, **99**, 127.
30. J. Wootthikanokkhan and P. Tunjongnawin, *Polym. Test.*, 2003, **22**, 305.
31. R. Brown, *Physical Testing of Rubber*, Springer, New York, 2006.
32. J. L. White, in *Science and Technology of Rubber*, ed. J. E. Mark, B. Erman and F. R. Eirich, Elsevier, New York, 2005, p. 237.
33. C. Sirisinha, S. Limcharoen and J. Thunyarittikorn, *J. Appl. Polym. Sci.*, 2003, **89**, 1156.

34. N. Kongvasana, A. Kositchaiyong, E. Wimolmala, C. Sirisinha and N. Sombatsompop, *Polym. Adv. Technol.*, 2011, **22**, 1014.
35. M. Akiba and A. S. Hashim, *Progr. Polym. Sci.*, 1997, **22**, 475.
36. A. S. Aprem, K. Joseph and S. Thomas, *Rubber Chem. Technol.*, 2005, **78**, 458.
37. V. A. Shershnev, *Rubber Chem. Technol.*, 1982, **55**, 537.
38. C. M. Roland, in *Handbook of Elastomers*, ed. A. K. Bhowmick and H. L. Stephens, Marcel Dekker, New York, 2001, p. 197.
39. A. J. Tinker, *Rubber Chem. Technol.*, 1990, **63**, 503.
40. A. E. Mathai, R. P. Singh and S. Thomas, *J. Membr. Sci.*, 2002, **202**, 35.
41. H. Ismail, S. Tan and B. T. Poh, *J. Elastomers Plast.*, 2001, **33**, 251.
42. J. Wootthikanokkhan and N. Clythong, *Rubber Chem. Technol.*, 2003, **76**, 1116.
43. R. L. Zapp, *Rubber Chem. Technol.*, 1973, **46**, 251.
44. M. J. R. Loadman and A. J. Tinker, *Rubber Chem. Technol.*, 1989, **62**, 234.
45. K. G. Karnika de Silva and M. V. Lewan, in *Blends of Natural Rubber*, ed. A. J. Tinker and K. P. Jones, Chapman & Hall, London, 1998, p. 68.
46. N. Naskar, S. C. Debnath and D. K. Basu, *J. Appl. Polym. Sci.*, 2002, **86**, 3286.
47. N. Naskar, S. C. Debnath and D. K. Basu, *J. Appl. Polym. Sci.*, 2001, **80**, 1725.
48. M. J. Forrest, *Rubber Analysis – Polymers, Compounds and Products*, Smithers Rapra Technology, Shawbury, 2001, p. 55.
49. V. Y. Senichev and V. V. Tereshatov, in *Handbook of Solvents*, ed. G. Wypych, Chem Tec Publishing, Toronto, 2001, p. 243.
50. C. Sirisinha, S. Limcharoen and J. Thunyarittikorn, *J. Appl. Polym. Sci.*, 2003, **87**, 83.
51. K. Kongsin and M. V. Lewan, in *Blends of Natural Rubber*, ed. A. J. Tinker and K. P. Jones, Chapman & Hall, London, 1998, p. 80.
52. S. P. Thomas, E. J. Mathew and C. V. Marykutty, *J. Appl. Polym. Sci.*, 2012, **124**, 4259.
53. S. H. Lee, M. Bailly and M. Kontopoulou, *Macromol. Mater. Eng.*, 2012, **297**, 95.
54. S. Toki, C. Burger, B. S. Hsiao, S. Amnuaypornsri, J. Sakdapipanich and Y. Tanaka, *J. Polym. Sci., Part B: Polym. Phys.*, 2008, **46**, 2456.
55. S. Toki and B. S. Hsiao, in *Recent Advances in Elastomeric Nanocomposites*, V. Mittal, J. K. Kim and K. Pal, Springer, Heidelberg, 2011, p. 135.
56. C. Kantala, E. Wimolmala, C. Sirisinha and N. Sombatsompop, *Polym. Adv. Technol.*, 2009, **20**, 448.
57. S. Bandyopadhyay, P. P. De, D. K. Tripathy and S. K. De, *Polymer*, 1995, **36**, 1979.
58. J. Wootthikanokkhan and B. Tongrubbai, *J. Appl. Polym. Sci.*, 2002, **86**, 1532.
59. J. Wootthikanokkhan and N. Rattanathamwat, *J. Appl. Polym. Sci.*, 2006, **102**, 248.
60. J. Wootthikanokkhan and B. Tongrubbai, *J. Appl. Polym. Sci.*, 2003, **88**, 921.

61. G. C. Psarras and K. G. Gatos, in *Recent Advances in Elastomeric Nanocomposites*, ed. V. Mittal, J. K. Kim and K. Pal, Springer, Heidelberg, 2011, p. 89.
62. R. S. Rajeev and P. P. De, in *Thermal Analysis of Rubbers and Rubbery Materials*, ed. N. R. Choudhury, P. P. De and N. K. Dutta, Smithers Rapra Technology, Shawbury, 2010, p. 381.
63. M. F. Bukhina and S. K. Kurlyand, *Low Temperature Behavior of Elastomers*, VSP, Leiden, 2007.
64. A. K. Sircar, M. L. Galaska, S. Rodrigues and R. P. Chartoff, *Rubber Chem. Technol.*, 1999, **72**, 513.
65. A. A. Yehia, A. A. Mansour and B. Stoll, *J. Therm. Anal.*, 1997, **48**, 1299.
66. K. I. Elizabeth, R. Alex, B. Kuriakose, S. Varghese and N. R. Peethambaran, *J. Appl. Polym. Sci.*, 2006, **101**, 4401.
67. A. Roychoudhury, P. P. De, A. K. Bhowmick and S. K. De, *Polymer*, 1992, **33**, 4737.
68. R. Alex, P. P. De and S. K. De, *Polymer*, 1991, **32**, 2345.
69. A. E. Mathai and S. Thomas, *J. Appl. Polym. Sci.*, 2005, **97**, 1561.
70. K. Pal, T. Das, S. K. Pal and C. K. Das, *Polym. Eng. Sci.*, 2008, **48**, 2410.
71. N. Z. Noriman and H. Ismail, *J. Appl. Polym. Sci.*, 2012, **123**, 779.
72. G. C. Psarras, K. G. Gatos and J. Karger-Kocsis, *J. Appl. Polym. Sci.*, 2007, **106**, 1405.
73. J. Carretero-González, T. A. Ezquerra, S. Amnuaypornsri, S. Toki, R. Verdejo, A. Sanz, J. Sakdapipanich, B. S. Hsiao and M. A. López-Manchado, *Soft Matter*, 2010, **6**, 3636.
74. P. Ortiz-Serna, R. Díaz-Calleja, M. J. Sanchis, G. Floudas, R. C. Nunes, A. F. Martins and L. L. Visconte, *Macromolecules*, 2010, **43**, 5094.
75. D. E. El-Nashar and G. Turky, *Polym.-Plast. Technol. Eng.*, 2003, **42**, 269.
76. G. C. Psarras, K. G. Gatos, P. K. Karahaliou, S. N. Georga, C. A. Krontiras and J. Karger-Kocsis, *eXPRESS Polym. Lett.*, 2007, **1**, 837.
77. G. M. Tsangaris, G. C. Psarras and N. Kouloumbi, *J. Mater. Sci.*, 1998, **33**, 2027.
78. A. Kalini, K. G. Gatos, P. K. Karahaliou, S. N. Georga, C. A. Krontiras and G. C. Psarras, *J. Polym. Sci., Part B: Polym. Phys.*, 2010, **48**, 2346.
79. H. W. Siesler, *Appl. Spectrosc.*, 1985, **39**, 761.
80. A. Temel, R. Schaller, M. Höchtl and W. Kern, *Rubber Chem. Technol.*, 2005, **78**, 28.
81. J. W. Cook, S. Edge, D. E. Packham and A. S. Thompson, *J. Appl. Polym. Sci.*, 1997, **65**, 1379.
82. M. Narathichat, K. Sahakaro and C. Nakason, *J. Appl. Polym. Sci.*, 2010, **115**, 1702.
83. H. Afifi and A. A. El-Wakil, *Polym.-Plast. Tech. Eng.*, 2008, **47**, 1032.
84. P. P. De, in *Spectroscopy of Rubbers and Rubbery Materials*, ed. V. M. Litvinov and P. P. De, Rapra Technology, Exeter, 2002, p. 77.
85. J. B. Nagode and C. M. Roland, *Polymer*, 1991, **32**, 505.
86. P. Nallasamy and S. Mohan, *Arab. J. Sci. Eng.*, 2004, **29**, 17.

87. A. Andersson, S. Lundmark, A. Magnusson and F. H. J. Maurer, *J. Cell. Plast.*, 2010, **46**, 73.
88. D. K. Graff, H. Wang, R. A. Palmer and J. R. Schoonover, *Macromolecules*, 1999, **32**, 7147.
89. G. A. Senich and W. J. MacKnight, *Macromolecules*, 1980, **13**, 106.
90. H. Oka, Y. Tokunaga, T. Masuda, H. Kiso and H. Yoshimura, *J. Cell. Plast.*, 2006, **42**, 307.
91. M. M. Coleman, D. J. Skrovanek, J. Hu and P. C. Painter, *Macromolecules*, 1988, **21**, 59.
92. N. B. Colthup, L. H. Daly and S. E. Wiberley, *Introduction to Infrared and Raman Spectroscopy*, Academic Press, San Diego, 1990, p. 289.
93. H. S. Lee, Y. K. Wang, W. J. Macknight and S. L. Hsu, *Macromolecules*, 1988, **21**, 270.
94. S. Patcharaphun, W. Chookaew and T. Tungkeunkunt, *Kasetsart J.: Nat. Sci.*, 2011, **45**, 909.
95. K. Pal, S. K. Pal, C. K. Das and J. K. Kim, *J. Appl. Polym. Sci.*, 2011, **120**, 710.
96. M. T. Ramesan, R. Alex and N. V. Khanh, *React. Funct. Polym.*, 2005, **62**, 41.
97. F. Sommer, in *Kautschuk Technologie*, ed. F. Röthemeyer and F. Sommer, Hanser, München, 2001, p. 13.
98. M. V. Lewan, in *Blends of Natural Rubber*, ed. A. J. Tinker and K. P. Jones, Chapman & Hall, London, 1998, p. 94.
99. C. S. Fong, W. W. Cheong and L. C. Yih, *Patent No. US 2 008 306 200*, 2008.
100. S. O. Hutchison, G. W. Anderson and G. L. Newby, *Patent No. US 3.983.906*, 1976.
101. R. M. Telang, *Patent No. US 6.756.106*, 2004.
102. T. Tetsuya, H. Katsuya and I. Hideyuki, *Patent No. US 5.984.283*, 1999.
103. E. Pieters, R. Lay and H. Parton, *Patent No. US 2010/0 205 764*, 2010.
104. D. M. Hill and C. Brodin, *Patent No. US 2010/0 229 873*, 2010.
105. S. Zhang, L. Cao, F. Shao, L. Chen, J. Jiao and W. Gao, *Polym. Adv. Technol.*, 2008, **19**, 54.
106. C. Nakason, A. Kaesaman, Z. Samoh, S. Homsin and S. Kiatkamjornwong, *Polym. Test.*, 2002, **21**, 449.
107. K. I. Elizabeth, R. Alex and S. Varghese, *Plast., Rubber Compos.*, 2008, **37**, 359.
108. A. Sujith, G. Unnikrishnan and C. K. Radhakrishnan, *J. Elastomers Plast.*, 2008, **40**, 17.
109. M. Klüppel, H. Menge, H. Schmidt, H. Schneider and R. H. Schuster, *Macromolecules*, 2001, **34**, 8107.
110. S. Gunasekaran, R. K. Natarajan and A. Kala, *Spectrochim. Acta, Part A*, 2007, **68**, 323.
111. Y. Ikeda, N. Higashitani, K. Hijikata, Y. Kokubo, Y. Morita, M. Shibayama, N. Osaka, T. Suzuki, H. Endo and S. Kohjiya, *Macromolecules*, 2009, **42**, 2741.

CHAPTER 10

Thermoplastic Elastomers from High-Density Polyethylene/Natural Rubber/Thermoplastic Tapioca Starch: Effects of Different Dynamic Vulcanization

MOHD KAHAR AB WAHAB,[a,b] NADRAS OTHMAN*[b] AND HANAFI ISMAIL[b]

[a] School of Materials Engineering, Universiti Malaysia Perlis, 02600 Jejawi, Perlis, Malaysia; [b] School of Materials and Mineral Resources Engineering, Engineering Campus, Universiti Sains Malaysia, Seri Ampangan 14300 Nibong Tebal, Seberang Perai Selatan, Penang, Malaysia
*Email: srnadras@eng.usm.my

10.1 Introduction

Thermoplastic elastomers (TPEs), which combine the characteristics of both thermoplastics and elastomers, can be classified into two major groups. The first is made up of block copolymers formed by polymerizing a thermoplastic monomer with an elastomer comonomer, as in the case of styrene block copolymers such as styrene-ethylene-butadiene-styrene (SEBS). Engineering TPEs such as thermoplastic urethanes (TPUs), copolyesters (COPEs) and polyether block amides (PEBAs) can be included in this first group. Other types of copolymer are from the polyolefin family, consisting of the polyolefin elastomers (POEs) and olefinic block copolymers (OBCs). The second category, commonly known as blended TPEs, include thermoplastic polyolefins (TPOs) which contain rubber as a dispersed phase in a thermoplastic matrix, typically a polyolefin. The blends usually come with the composition of a high thermoplastic–low rubber blend (*e.g.* a plastic/rubber ratio of 70:30). TPOs exhibit properties and functional performance similar to those of a conventional vulcanized rubber but can be processed in a molten state like a conventional thermoplastic. A further type of blended TPE is thermoplastic vulcanizates (TPVs), in which the elastomer is crosslinked during the melt blending process and contains plastic as a reinforcing or stiffening agent.

TPEs based on olefinic elastomer blends can be produced with a non-crosslinking elastomer (TPO) or crosslinked elastomer (TPV) phase, using a process known as dynamic vulcanization. In this process an elastomer is vulcanized during melt blending with a non-vulcanizing thermoplastic polymer. This process gives the particulate vulcanized elastomer a stable morphology in a continuous thermoplastic matrix. Dynamic vulcanization of the rubber phase in TPE blends also allow the crosslink rubber to transform the dispersed phase even though the percentage of rubber blend in the composition is more than 50%.[1] The increase in rubber viscosity through vulcanization affects the phase continuity and promotes the formation of the crosslinked rubber dispersed phase. The purpose of dynamic vulcanization is to produce materials with better mechanical properties, high thermal stability, hot oil resistance and thermoplastic fabricability.

One of the important families of blended TPEs is thermoplastic natural rubber (TPNR), based primarily on polyolefin and natural rubber (NR). TPNR blends have attracted much attention due to their ease of processability and a broad spectrum of properties available at a competitive price. The TPNR system consists of multi-phases of hard and soft domains from blending thermoplastics with NR. Generally, TPNR is prepared with various type of polyolefin, such as PP,[2,3] LDPE,[4] LLDPE[5] and HDPE.[6–9] TPNR is usually prepared by a melt mixing process using an internal mixer or twin-screw extruder. It presents a wide range of useful properties, excellent processing abilities, and can be produced at moderate cost.

Melt blending of NR and polyolefin is carried out at a temperature above the melting point in the case of thermoplastics, since its temperature is already above the NR melt temperature. For HDPE/NR and PP/NR blends, the recommended blending temperatures are about 150 °C and 180 °C, respectively.[10] Mixing conditions such as rotor speed and duration of mixing also influence the mechanical properties of the resultant blends. Consequently, the blend composition, temperature, mixing rate and processing time have direct bearing on the properties of tailor made TPNR. In such cases, the quality of the finished blends is normally judged by the physical and mechanical properties exhibited.

Polyolefins are considered to be the best choice for blending with NR on account of their high softening temperature and low glass transition temperature, which result in a versatile end product and are applicable over a wide range of temperature. Polyolefin-based NR blend compositions (HDPE/NR, PP/NR) can be varied to give materials of different mechanical properties ranging from soft elastomer to semi-rigid plastic. Soft grades of TPNR consist of high concentrations of rubber, while the harder grades can contain up to 30% of rubber. It is therefore possible to tailor TPNR to meet the desired mechanical and physical properties by optimizing the blend composition and selecting a suitable thermoplastic.

Studies of the dynamic vulcanization of polyolefins and NR have been reported in the literatures.[2,11–17] Polyolefin NR blends prepared by this selective crosslinking could replace the vulcanized rubber in end products where high resilience and strength is not essential but low-temperature performance is required.[18] They could also replace various flexible plastics such as plasticized polyvinyl chloride (PVC), ethylene vinyl acetate (EVA) and polypropylene comonomers.[19]

Previous studies have shown the potential effect of dynamic vulcanization on mechanical properties and morphology of TPE blends.[2,3,20–22] Ismail et al.[23,24] have reported the effect of dynamic vulcanization on the mechanical properties of NR/PP and NR/PP filled with rubber wood. The dynamic vulcanization system shows superior mechanical properties in terms of tensile strength, Young's modulus and elongation at break compared to direct TPNR blending.

Various crosslinking agents were used, such as sulfur systems, bismaleimide such as HVA-2 and peroxide.[17–19,23–28] George and coworkers[29] analysed the effects of amount of crosslinking agents on the rheological and mechanical properties of high-density polyethylene/acrylonitrile butadiene rubber (HDPE/NBR) blends. Punnarak et al.[30] investigated the various methods of vulcanization on reclaimed tyre rubber and high-density polyethylene (RTR/HDPE). They found that sulfur vulcanization methods created the highest degree of crosslink and filler–matrix interaction. One of the maleimides, m-phenylene bismaleimide, also known commercially as HVA-2, was found to be an effective crosslinking agent in the presence of a radical activator such as organic peroxide or MBTS.[10] Adding a free radical source such as dicumyl peroxide (DCP) was reported to accelerate the crosslinking of the NR phase and thus enhance the crosslink density of the blends.[28] However at a sufficiently high temperature, HVA-2 was able to vulcanize NR without the presence of free radical initiator.[25]

Unlike sulfur crosslinking, peroxide does not require an unsaturated bond to form crosslinks. Peroxide cross linking system only needs peroxide in the formulation to trigger the chemical reaction. Therefore care must be taken to avoid unwanted reaction with the peroxide. Some ingredients, which are not part of the cure system and usually common in sulfur systems, can interact with the peroxide and thus interfere with the cure. The main advantage of general peroxide-based crosslinking is that the blends possessed good high-temperature ageing resistance, good elastic behaviour and no discoloration of finished materials.[31] The density of rubber crosslinks depends on the decomposition of the peroxide selected. It was due to the initial and rate determining in the crosslinking step depends on the first order formation of free alkoxy radicals.[17] Therefore, an appropriate peroxide should be selected based on its decomposition temperature at its approximate processing temperature.

A more recent method of dynamic crosslinking, without using any curative agents, is through radiation crosslinking. The radiation crosslinking can be carried out by exposing the previously prepared materials with high-energy radiation such as X-rays, protons, electrons and neutron beams.[32] Upon irradiation, high local concentrations of free radicals are formed in the rubber chains. These free radicals can combine and form inter- or intramolecular crosslinks similar to peroxide crosslinks. The final properties of irradiated blends depend on the dosage, dose rate, irradiation conditions and added additives.[33] Zurina et al.[34] have reported that effective crosslinking can be achieved by irradiating the blends with electron beams. However in another comparative study of mechanical properties of EPDM-based blends by different methods of crosslinking, electron beam crosslinking was found to provide similar mechanical properties to peroxide crosslinking, but inferior to sulfur-based crosslinking.[35]

Recent concerns over the environmental impact of polymeric waste disposal has motivated the production of new biomaterial-based products. In the context of increased environmental problems, shortage of landfill and concerns about emissions of hazardous gas during incineration, the use of polysaccharide has become a popular alternative to conventional plastics. The combination of biodegradability, low cost and its availability in the form of a variety of plant sources has made it an attractive option as a partial replacement for conventional plastics. Considering their environmental benefits, the blending of polymeric materials with cheap natural biopolymers is now widely used in the manufacture of biodegradable or semi-biodegradable products. Since they are very cheap, the addition of this component can reduce further the overall cost of the product itself. This shows that the market for polysaccharide-filled TPNR is very promising, not only because of its biodegradability and the exciting properties that can be tailored but also from the point of view of economic advantages. Lignocellulosic fillers and starch are the two most frequently used polysaccharides as bio-based polymer products. The first has been intensively studied in the TPNR system; the latter is found only in thermoplastic or NR (mostly in latex form) matrices.

Two major methods have been developed to incorporate starch into commercial polymers. One is based on the use of granular starch as filler and the other is the use of gelatinized starch as one part of the polymer blend. The high viscosity and poor flow properties of starch make it difficult to process granular starch-based materials. Due to the processing difficulties and significant loss in tensile strength, blending granular starch with other polymers is limited to a starch content of 40% or less. However, the tensile strength and flow properties of starch-based blends can be significantly increased by converting native starch into a thermoplastic form. This conversion of starch into a thermoplastic is well known as a plasticization process and the transformed material is usually known as thermoplastic starch. This research work was carried out to study the possibility of producing a thermoplastic starch/TPNR blend, composed of

HDPE, NR and thermoplastic tapioca starch (TPS). Several previous studies have reported that blends with starch exhibit a marked decrease in tensile properties because the incorporation of starch interferes in the adhesion between the component phases.[21,36–41] Therefore, based on this research work, the mechanical properties were improved by NR vulcanization, which compensated for the deterioration in properties due to TPS loading.

10.2 Materials and Methodology

HDPE granulates with a density of 0.96 g cm^{-3} was supplied by Titan Chemicals, Malaysia. NR grade SMR L was obtained from Lembaga Getah Malaysia. The tapioca starch used was a food grade and obtained from Thye Huat Chan Sdn Bhd., Malaysia. Sulfur, zinc oxide, stearic acid and N-tert-butyl-2-benzothiazole sulfenamide (TBBS) were supplied by Bayer (M) Ltd, Malaysia. HVA-2 (Aldrich Chemical Company, Inc., USA) was used as a crosslinking agent. Reagent grade glycerol was obtained from Merck, Darmstadt, Germany and used as received.

10.2.1 Preparation of Thermoplastic Tapioca Starch (TPS)

Tapioca starch was dried at 80 °C in a vacuum oven for 24 h. It was first premixed with 35 wt% of glycerol using a kitchen blender (3000 rpm, 2 min) until a homogeneous mixture was obtained. Then, the mixture was stored overnight in a dry place. Later the mixture was processed using a heated two-roll mill at a temperature of 150 °C; the mixing time was 10 min.

10.2.2 Dynamic Vulcanization with HVA-2 and Sulfur Curative Agent

The blends with sulfur and HVA-2 crosslinker were prepared by melt blending of HDPE/NR/TPS in a Haake Rheomix 600 mixer equipped with roller rotors. The mixing was carried out at a temperature of 150°°C and the rotor speed was fixed at 55 rpm. Tables 10.1 and 10.2 show the weight proportions of HDPE/NR/TPS in the blends and the compounding recipes for the vulcanized blends. The ratio between HDPE and NR was fixed at 70/30 and TPS contents were varied from 5 wt% to 30 wt% relative to the overall weight of the blends.

10.2.3 Tensile Properties

Tensile properties were measured using an Instron Universal Testing Machine (model 3366) with crosshead speed 50 mm min^{-1} and tested at room temperature. Dumbbell samples (1 mm thick) with 50 mm gauge length were performed according to ASTM D 638. The mean value out of five samples was reported, with standard deviation to show the error range.

Table 10.1 The weight proportion of HDPE/NR/TPS blends.

Sample	Sample (weight proportion) (HDPE/NR)	TPS
HDPE/NR	100	0
Unvulcanized 5% TPS	95	5
Unvulcanized 10% TPS	90	10
Unvulcanized 20% TPS	80	20
Unvulcanized 30% TPS	70	30
Sulfur vulcanized 5% TPS	95	5
Sulfur vulcanized 10% TPS	90	10
Sulfur vulcanized 20% TPS	80	20
Sulfur vulcanized 30% TPS	70	30
HVA-2 vulcanized 5% TPS	95	5
HVA-2 vulcanized 10% TPS	90	10
HVA-2 vulcanized 20% TPS	80	20
HVA-2 vulcanized 30% TPS	70	30

Table 10.2 Compounding recipe for sulfur and HVA-2 vulcanized blends.

	Ingredients	Parts per 100 parts of blends
Sulfur curative agents	HDPE/NR/TPS	100
	Sulfur	1.2
	Zinc oxide	3
	N-tert-butyl-2-benzothiazole sulfenamide (TBBS)	0.3
	Stearic acid	1.5
HVA-2 crosslinker	HDPE/NR/TP	100
	N,N'-m-phenylenebismaleimide	0.5

10.2.4 Gel Content

The gel content of the samples was determined using a Soxhlet extraction technique. The samples were first extracted with boiling water (12 h) to remove the TPS phase and then they were extracted with hot xylene for 4 h and 8 h. The samples were placed in folded 120 mesh stainless steel cloth cages. The samples and cages were weighed before extraction. The extracted samples were then dried at 80 °C until there was no further weight loss. The gel content was calculated from the following equation:

$$\text{gel content} = 100 - \% \text{extract} \quad (10.1)$$

$$\% \text{extract} = \frac{\text{weight before extraction} - \text{weight after extraction}}{\text{original weight}} \quad (10.2)$$

10.2.5 Fourier Transform Infrared Spectroscopy (FTIR)

Fourier transform infrared spectroscopy (FTIR) of unvulcanized and vulcanized blends was measured using a Perkin Elmer System 2000 to characterize the structural changes during blend processing. The transmittance spectra regions were obtained between 4000 cm^{-1} and 800 cm^{-1} with a 4 cm^{-1} resolution.

10.2.6 Scanning Electron Microscopy (SEM)

Examination of the fracture surface of the samples was performed with a Leo Supra – 3SVP field emission scanning electron microscope at an acceleration rate of 20 kV. The fractured surfaces of the specimen were mounted on aluminium stubs and sputter coated with thin layer of gold to avoid electrostatic charge during examination.

10.2.7 Thermogravimetric Analysis (TGA)

Thermogravimetric analysis (TGA) of the blends was carried out using a Perkin Elmer Pyris 6 machine. Samples of the HDPE/NR/TPS blends were tested at a heating rate of 10 °C min^{-1} from ambient temperature to 600 °C. Two types of samples were studied: control and vulcanized HDPE/NR/TPS blends.

10.2.8 Dynamic Mechanical Thermal Analysis (DMTA)

A Mettler Toledo DMA861 was used to examine the dynamic mechanical properties of the HDPE/NR/TPS blends. Specimens of approximately 10.5 × 6.5 × 1 mm were tested in a tensile mode at a frequency of 1 Hz and oscillation amplitude was 20 μm. The analysis was run at 10 °C min^{-1} and started from −80 °C to 120 °C for the chosen blends. The temperature at the maximum point of the tan δ was taken as the T_g sample.

10.2.9 Differential Scanning Calorimetry (DSC)

Thermal analyses of unvulcanized and vulcanized HDPE/NR/TPS blends were performed using a Mettler Toledo differential scanning calorimeter TS0800GCL under a nitrogen atmosphere at a scan rate 10 °C min^{-1}. The samples were heated from ambient temperature to 190 °C and held at this temperature for 3 min to eliminate thermal history and destroy the polyethylene nuclei. Then the samples were slowly cooled at 10 °C min^{-1} to room temperature before the crystallized samples were subsequently heated to 230 °C at the same heating rate. The melting temperature (T_m) and the heat of fusion (ΔH_{exp}) were determined from the endothermic peak integration during the second heating.

10.3 Results and Discussion

10.3.1 Processing Characteristics

The melt processing characteristics of dynamic vulcanized HDPE/NR/TPS blends have been studied from the processing torque–time curve. Figures 10.1(a) and (b) show the processing torque of the melt mixed sulfur and

HVA-2 vulcanized HDPE/NR system with different TPS loadings. In the melt processing graph, the first and second peaks correspond to the shear torque before melting of HDPE and NR, respectively. The different peaks were obtained for all cases, which related to different amounts of HDPE and NR that were charged into the mixing chamber. The ratio between HDPE and NR was fixed at 70/30 and the weight percentages of HDPE and NR were reduced when TPS loading increased from 5 wt% to 30 wt%. The third peak appeared at around 7 min, which corresponded to the introduction of TPS. After 8 min, the processing torque stabilized, showing that a good level of mixing had been achieved.

The stabilization torque for mixing HDPE/NR with different TPS contents, as recorded at the end of mixing, can be seen in Figure 10.2. With the dynamic crosslinking occurring in the NR phase, the changes in stabilization torque

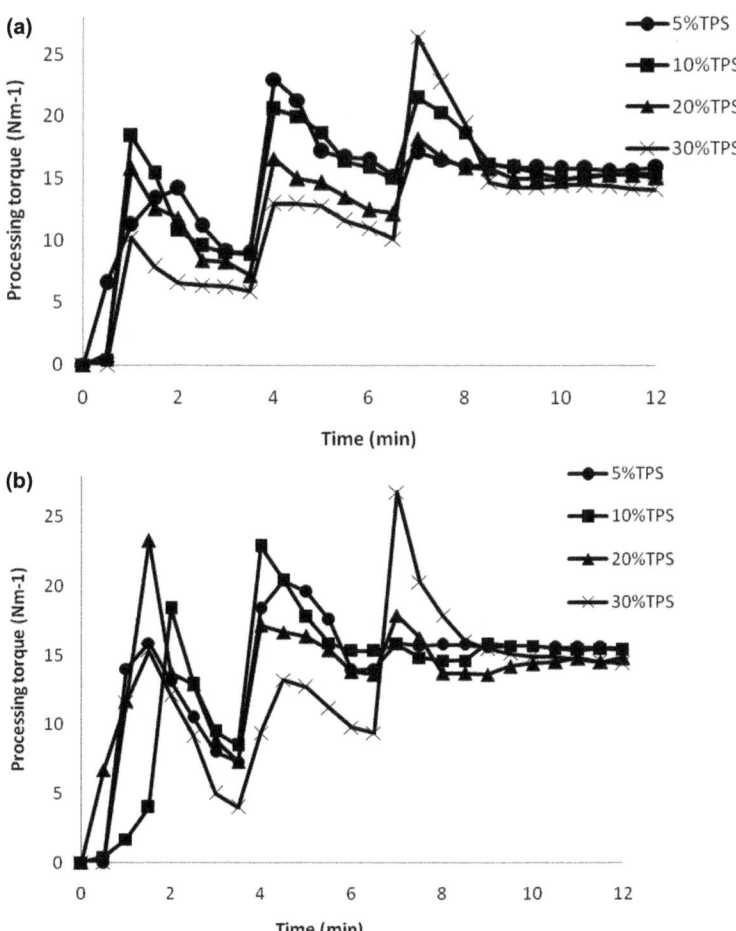

Figure 10.1 (a) The processing toque of sulphur-vulcanised HDPE/NR/TPS blends.[42]
(b) The processing toque of HVA-2-vulcanised HDPE/NR/TPS blends.[43]

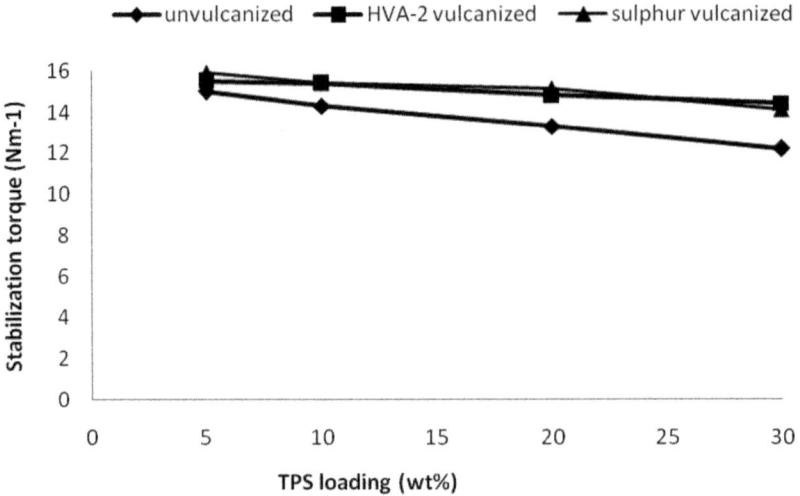

Figure 10.2 The effect of dynamic vulcanization on the stabilization torque of HDPE/NR/TPS blends as a function of TPS loading.

were affected only by TPS content. Therefore a comparison can be generally made to distinguish the effect of dynamic crosslinking for both HVA-2 vulcanized and unvulcanized blends. It clearly shows that the stabilization torque decreases as the TPS content increases. This can be attributed to the blends plasticized by glycerol in TPS and this plasticization effect was found to be stronger for the unvulcanized blends. The stabilization torque of sulphur and HVA-2 vulcanized blends which is reflected in the melt blends viscosity was observed to be linear even though TPS concentration has increased more than 30%. The increment in orthography can be correlated to the crosslinking formation in the NR phase.

10.3.2 Tensile Properties

Tensile strength, Young's modulus and elongation at break of 0–30% TPS content in unvulcanized and vulcanized HDPE/NR/TPS blends were compared and are shown in Figures 10.3 to 10.5. As TPS content increased, the tensile strength was found to decrease, as clearly shown in Figure 10.3. This was due to the poor interfacial adhesion and inability of TPS to support stress transfer from the HDPE/NR matrix phase. Effective stress transfer between phases was the major factor that determined the tensile properties of the blends. As shown in the SEM micrographs (Figures 10.9(c) and (d)), insufficient interfacial adhesion occurred and appeared as voids at the interphase. This poor adhesion cannot effectively transfer the stress and easily breaks when subjected to external loaded stress. On the other hand, the deterioration results were also due to the increase in particle size of TPS in the HDPE/NR blend. The addition of TPS at higher loading leads to particle agglomeration and this large TPS particle increases the stress concentration points in the blends. However, an

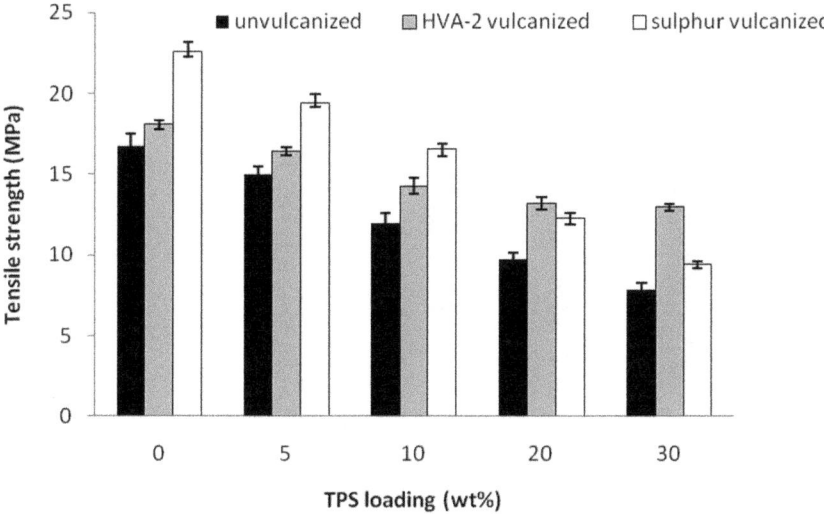

Figure 10.3 Tensile strength for unvulcanized, HVA-2 and sulfur curative system of the HDPE/NR/TPS blends.

interesting increase in tensile strength in sulfur and HVA-2 vulcanized HDPE/NR/TPS blends was observed in all studied blends. In the presence of a sulfur curative system, the tensile strength of the HDPE/NR matrix increased by ~5.6 MPa. At the lower TPS loadings (5% and 10%), tensile strength was increased by up to ~4.5 MPa when compared to unvulcanized counterparts. For HVA-2 vulcanized blends, tensile strength increased by 1.4 MPa for both HDPE/NR and the blend with 5% TPS content. At 10%, 20% and 30% TPS content, the tensile strength increased by 2.3, 3.4 and 5.1 MPa, respectively, when compared with unvulcanized counterparts. For all unvulcanized blends, tensile strength decreased marginally with the increase in TPS. On the other hand, the tensile strength of HVA-2 vulcanized blends decreased slightly for concentrations of TPS up to 10% and remained unaffected by further increases in TPS content up to 30%. The addition of HVA-2 and sulfur curative agent can promote the intra- and intermolecular linkages in NR phases and from the observations of tensile strength it is clear that the crosslink system can be used to improve the blend's properties. In other words, sulfur and HVA-2 vulcanization system have been applied in these blends as a way of compensating for the strength deterioration caused by incorporation of TPS.

Figure 10.4 shows the changes in Young's modulus before and after vulcanization using HVA-2 and sulfur as a function of different TPS content. Adding TPS to HDPE/NR blends can lead to a decrease in Young's modulus, especially at higher TPS content, due to the effect of incorporating soft TPS materials and the presence of glycerol in TPS. At low concentrations of TPS (≤ 10%), Young's modulus was observed to be almost the same for blends before and after vulcanization. A noticeable decrease can be observed at higher TPS content (up to 30%) for all the blend systems. For the case of the HVA-2

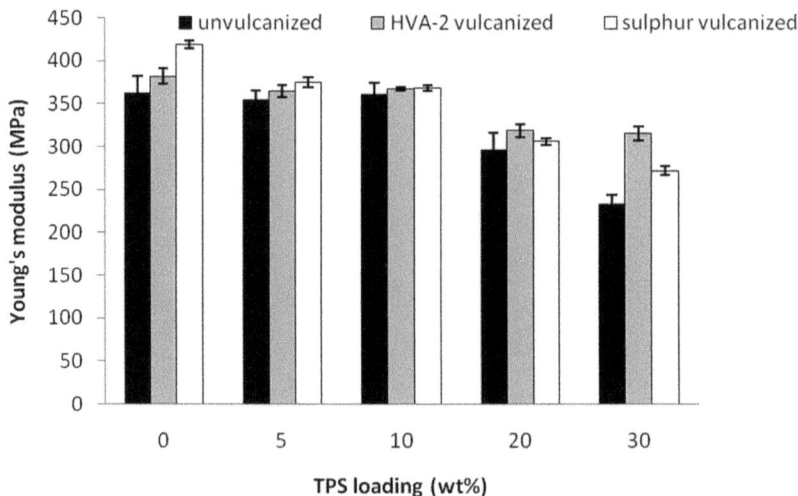

Figure 10.4 Young's modulus for unvulcanized, HVA-2 and sulfur curative system of the HDPE/NR/TPS blends.

system, a clear influence of vulcanization can be seen at 20% and 30% TPS content whereby the value of Young's modulus was maintained at around 319 MPa. On the other hand, it was found that the values significantly decrease to 297 MPa and 233 MPa for unvulcanized blends, respectively. The increment in Young's modulus was normally observed when the NR phase had been vulcanized, indicating the increase in blend stiffness. HVA-2 was actually responsible for forming a crosslink structure and increased the dimensional stability of the NR phase. Considering the fact that blends of HDPE/NR/TPS consist of low modulus components such as NR and TPS, increasing the formation of a crosslink structure within the NR phase can improve the blend's stiffness considerably. On top of that, the vulcanized NR phase can also prevent the polymer chains slipping and thereby result in a more rigid polymer blend.

Figure 10.5 shows that elongation at break for HDPE/NR/TPS blends decreased with the increase in TPS content, whilst sulfur and HVA-2 vulcanized HDPE/NR/TPS blends show better flexibility. The elongation at break demonstrated good agreement with the tensile strength trends. The exhibited increase in tensile strength value was in accordance with the increase in elongation at break. This phenomenon can be attributed to strain hardening. There was a slight decrease in elongation at break at 5 wt% and 10 wt% TPS loading for unvulcanized blends, whereas vulcanized blends showed an acceptable decrease up to 20 wt% TPS loading. With further increase in TPS content at 20 wt% and 30 wt%, unvulcanized blends showed a major decrease in elongation at break, possibly due to agglomeration of the TPS phase. For sulfur and HVA-2 vulcanized HDPE/NR/TPS blends, all tested compositions showed higher elongation at break when compared to their unvulcanized counterparts. At 20% and 30% TPS content, the elongation at break of sulfur

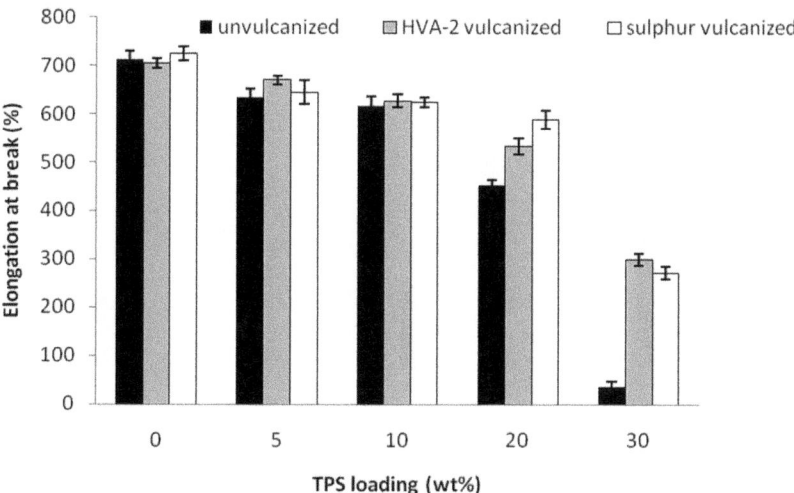

Figure 10.5 Elongation at break for unvulcanized, HVA-2 and sulfur curative system of the HDPE/NR/TPS blends.

vulcanized blends increased by 23.3% and 86.6%, and for the HVA-2 system the value increased by 15.1% and 88.2%, respectively. This indicates that better compatibility between the components and dynamic vulcanization implies better adhesion between NR and the HDPE/NR matrix.

10.3.3 Gel Content

Figures 10.6(a) and (b) show the remaining portion of the blends after extraction from water and xylene at 100 °C and 140 °C, respectively. Comparative studies were carried out for both unvulcanized and vulcanized blends after water (12 h) and xylene (4 h and 8 h) extraction. Water was used for TPS extraction, whereas xylene was used as an extracting solvent for HDPE and uncrosslinked NR. This method was generally used to determine the gel content related to the crosslink density. The degree of crosslink density was responsible for the blend characteristics and had a major influence on the tensile properties. After extraction with water (12 h), more than 90% of the blends were still remaining. This showed that water was not effective in extracting the TPS phase completely, which probably was due to TPS particles being entrapped in the matrix phase.

In the case of samples extracted in xylene for 4 h, the unvulcanized HDPE/NR blend was observed to have completely dissolved, whereas almost 51% and 36%, respectively. of sulfur and HVA-2 vulcanized HDPE/NR blends still remained undissolved. At 10% TPS loading, the gel content of sulfur and HVA-2 vulcanized blends was found to increase by 94.4% and 92.2% and by an increment of 36.4% and 49% for 30% TPS loading, respectively, when compared to the unvulcanized blends. Furthermore, as the extraction time was increased to 8 h, all the unvulcanized blends dissolved in hot xylene.

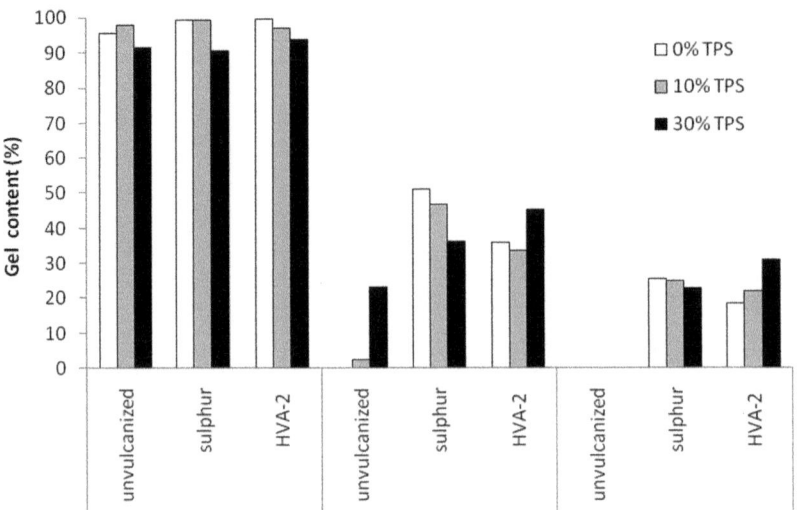

Figure 10.6 Effect of HVA-2 vulcanization on the gel content of the HDPE/NR/TPS blends.

The difference with sulfur vulcanized blends was almost 26%, 25% and 22.9% of the blends still remaining, with respect to 0%, 10% and 30% TPS content. A similar finding was observed in the HVA-2 system, where 18.4%, 22% and 31.1% of the blends still remained undissolved in xylene. Since all the unvulcanized samples were being dissolved, this study proposed that the suitable time for extraction of sulfur and HVA-2 vulcanized HDPE/NR/TPS was around 8 h. Furthermore, as can be seen at 30% TPS content, the gel content of HVA-2 vulcanized blends was observed to be higher than the NR component. This could be due to the vulcanizing fraction, which may hinder complete leaching of the TPS particles. The results obtained can be explained by the fact that a stable crosslink structure had been formed after the blends were subjected to HVA-2 crosslinker. The formation of a crosslink structure implies that the blends were resistant to chemical penetration and cannot be easily removed.

10.3.4 Structural Analysis

FTIR tests were carried out on unvulcanized HDPE/NR blends and the blends with 10% TPS loading and compared with their vulcanized counterparts. Figure 10.7 shows only minimal variations in vulcanized spectra from unvulcanized blends. The presence of crosslinker additive could not be detected and the interaction was expected to occur only between crosslinker and NR phase.

Xylene extraction of the vulcanized blends was further studied using FTIR. FTIR spectra for HDPE/NR, HDPE/NR-10% TPS and HDPE/NR-30% TPS blends before and after xylene extraction (12 h) is presented in Figures 10.8(a) and (b). The peaks of interest are the broad peaks at 3200–3400 cm^{-1}, 2915 cm^{-1}, 2847 cm^{-1}, 1472 cm^{-1}, 1462 cm^{-1}, 1112 cm^{-1}, 1041 cm^{-1}, 729 cm^{-1}

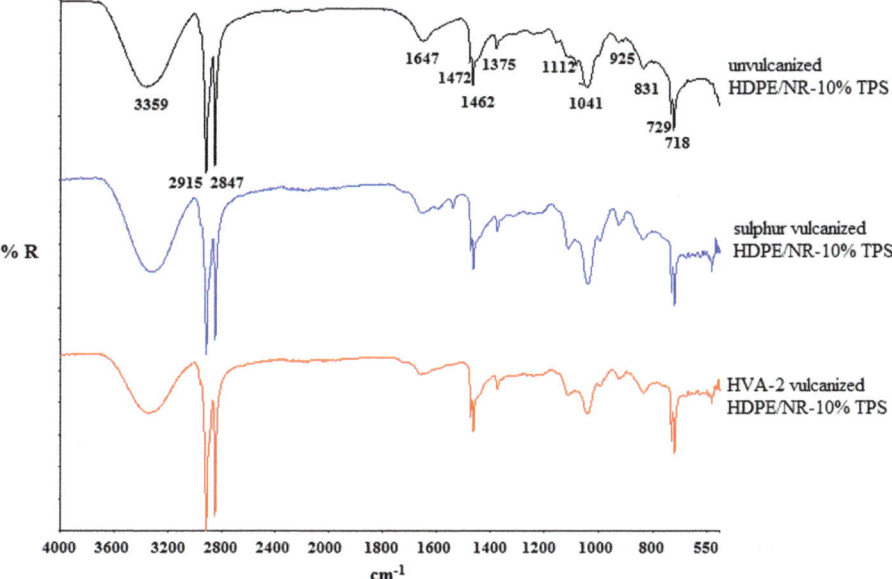

Figure 10.7 (a) HDPE/NR-10% TPS spectra for unvulcanized and vulcanized blends.

and 718 cm^{-1}. The band range occurred at 3200–3400 cm^{-1}, corresponding to the hydroxyl groups that were mainly from the TPS phase. The peaks at 2916 cm^{-1} and 2847 cm^{-1} represent the C–H asymmetry and symmetry stretching vibrations of CH$_2$ groups.[44] This peak was found to be of strong intensity in HVA-2 vulcanized blends and much weaker after xylene extraction. Other distinct peaks that featured HDPE were at 1472 cm^{-1} and 1462 cm^{-1}, which corresponded to the C–H bending vibrations of crystal and amorphous regions of HDPE.[45] In relation to these peaks, the doublet peak at 729 cm^{-1} and 718 cm^{-1} were also being used to refer to the crystallinity region of HDPE.[46] Both of these doublet peaks were found to be strong peaks for HVA-2 vulcanized spectra, but almost absent after xylene extraction, indicating most of the HDPE part had already been extracted from the blends. Xylene extraction spectra for both blends with 10% and 30% TPS also exhibited the absorption peak associated with the TPS phase. The absorption peak at 1041 cm^{-1} represented the vibration stretching in C–O–C and the 1112 cm^{-1} peak could be attributed to the vibration stretching of C–O in C–O–H groups in TPS.[47] As the blends underwent xylene extraction, the presence of these peaks confirmed that some of the TPS particles were still entrapped in the vulcanized NR phase.

10.3.5 Blend Morphology

Figure 10.9 shows the SEM micrographs of control, dynamically vulcanized (sulfur and HVA-2) HDPE/NR/TPS blends and the blends containing 10 wt%

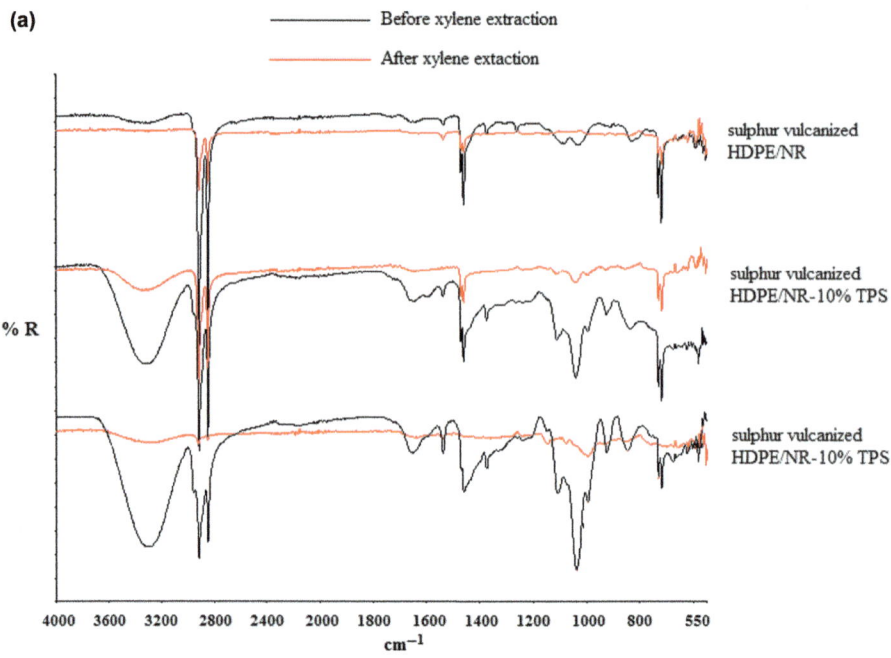
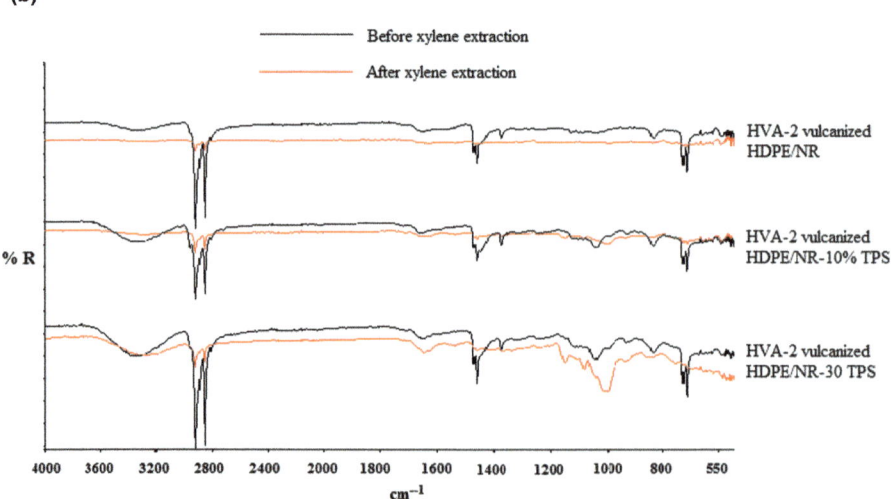

Figure 10.8 (a) Sulphur and (b) HVA-2 vulcanised HDPE/NR and HDPE/NR/TPS spectra before and after xylene extraction.[43]

and 30 wt% TPS. Figure 10.9(a) and (b) represent the fractured surface of unvulcanized and vulcanized HDPE/NR blends. SEM micrographs of unvulcanized and vulcanized HDPE/NR matrix have also been carried out in order to have an insight into the fracture behaviour and to compare with the HDPE/NR/TPS morphology. The fractured surface of the vulcanized blend

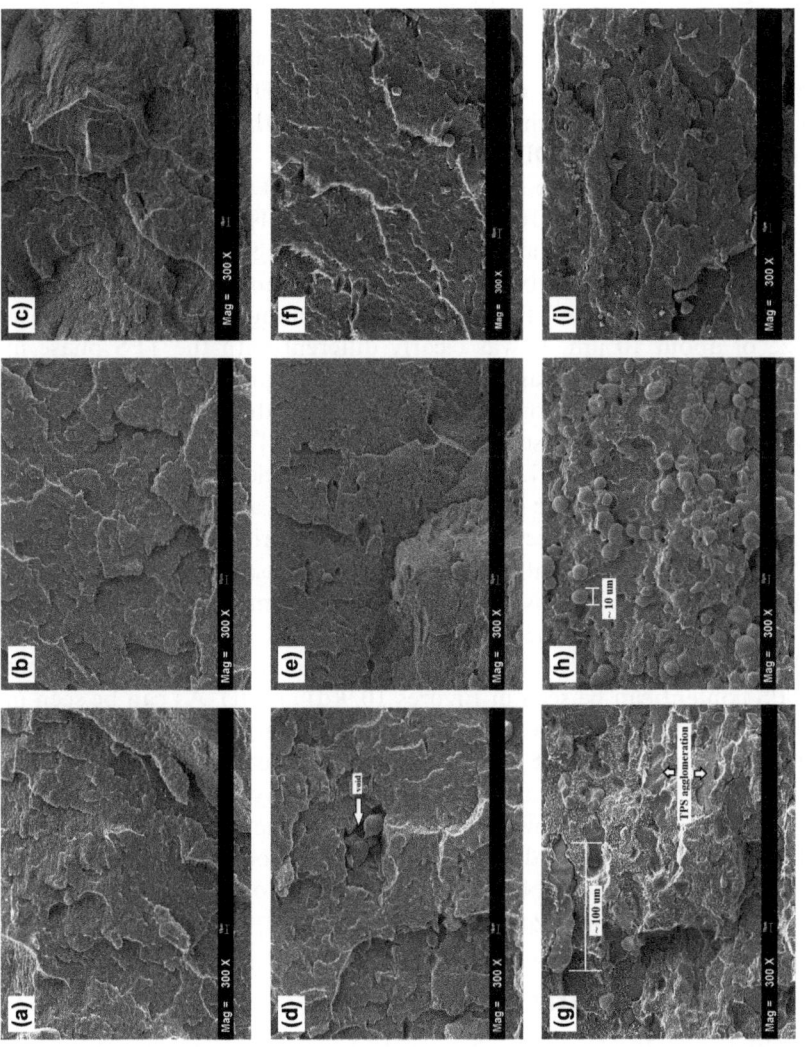

Figure 10.9 SEM micrographs of (a) unvulcanized HDPE/NR, (b) sulfur vulcanized HDPE/NR, (c) HVA-2 vulcanized HDPE/NR, (d) unvulcanized 10% TPS-HDPE/NR, (e) sulfur vulcanized 10% TPS-HDPE/NR, (f) HVA-2 vulcanized 10% TPS-HDPE/NR, (g) unvulcanized 30% TPS-HDPE/NR, (h) sulfur vulcanized 30% TPS-HDPE/NR and (i) HVA-2 vulcanized 30% TPSHDPE/NR.[42,43]

(Figure 10.9(b)) was coarser when compared to unvulcanized blends (Figure 10.9(a)), which could be associated with better fracture resistance to the external force applied on the blends. SEM images of unvulcanized blends (Figures 10.9(c) and (e)) show insufficient adhesion and there were some apparent voids present at the interphase between TPS and the HDPE/NR matrix. This is due to a compatibility problem between hydrophilic TPS and hydrophobic HDPE/NR. TPS particles did not tend to disperse homogeneously in the HDPE/NR matrix and their size ranged from ~ 10 μm to ~ 30 μm. In Figure 10.9(e), TPS particle size increased up to ~ 100 μm, indicating poor dispersion and a higher degree of phase separation. The TPS particles were clearly distinct from the adhering polymer. Besides, the crack fracture was clearly visible near the TPS-HDPE/NR interphase.

On the other hand, as can be seen in Figures 10.9(d) and (f), fine TPS particles were observed for both 10 wt% and 30 wt% TPS loading in sulfur-vulcanized HDPE/NR/TPS blends. With respect to the globular shape of TPS, this particle shape was only observed in sulfur-vulcanized HDPE/NR/TPS blends (Figure 10.9(f)). These TPS particles were homogeneously dispersed in smaller size of about 10 μm. It was clearly different with the TPS phase in unvulcanized blends, which was observed as a co-continuous agglomerate phase (Figure 10.9(e)). A network structure occurred during the NR vulcanization which increased the viscosity of the blends. This is resulted in the breaking-up of larger TPS particles into smaller particles which led to better dispersion of TPS phase. On the other hand, interfacial adhesion between TPS and vulcanized HDPE/NR phases improved and it contributed to a breakdown of the TPS particle into the smaller sized dispersed phase. As a result the TPS particle showed homogeneous dispersion and was well embedded in the HDPE/NR matrix even though at higher TPS loading.

For HVA-2 vulcanized blends, the fine TPS particles for both 10 wt% and 30 wt% TPS content can be seen in Figures 10.9(d) and (f). As can be seen in these figures, the smaller TPS particles, at around 10 μm, were homogeneously dispersed throughout the HDPE/NR matrix. In contrast to the TPS phase in unvulcanized blends, HVA-2 vulcanized blends displayed a discrete morphology with TPS and was well embedded in the HDPE/NR matrix (Figure 10.9(f)). This could be due to the fact that good interfacial adhesion exists between TPS and HDPE/NR, and it contributed to the breakdown of the TPS particle into the smaller sized dispersed phase.

10.3.6 Thermogravimetric Analysis

Figures 10.10(a) and 10.11(a) show the weight loss corresponding to unvulcanized and vulcanized HDPE/NR and HDPE/NR/TPS blends for different degradation temperatures. The thermograms showed unvulcanized and vulcanized HDPE/NR/TPS blends experienced a major thermal degradation in three steps. The first significant weight loss occurred at about 290–350 °C and could be attributed to the thermal degradation of amylose and amylopectin. The next stage started from 360 to 430 °C for all studied blends and was related to the

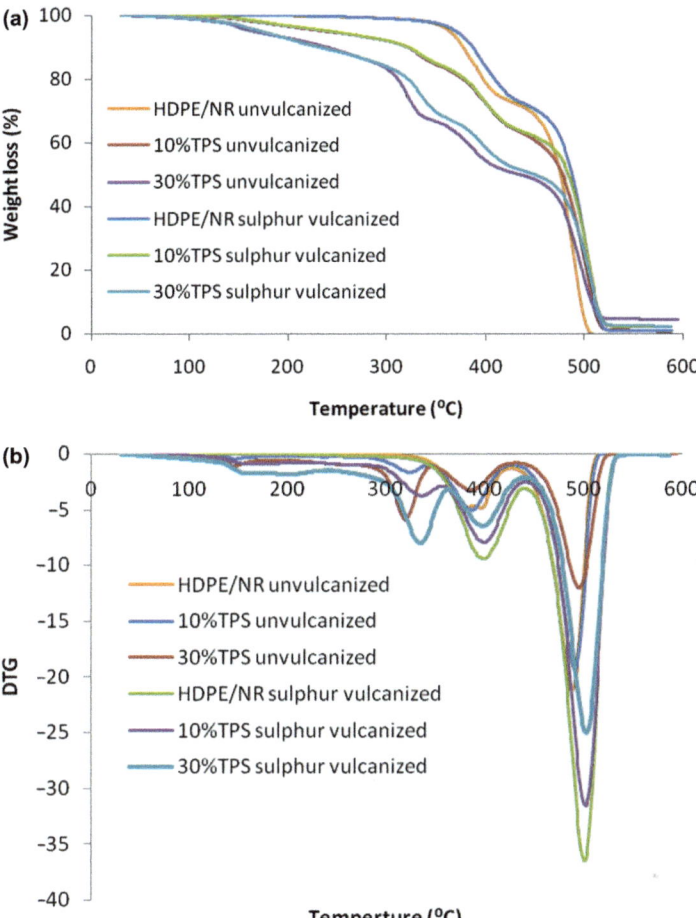

Figure 10.10 (a) Thermogravimetric analysis of unvulcanized and sulfur vulcanized HDPE/NR/TPS blends. (b) DTG of unvulcanized and sulfur vulcanized HDPE/NR/TPS blends.

decomposition temperature of the NR phase. A further weight loss at temperatures above 450 °C could be ascribed to the thermal decomposition of HDPE and an almost complete decomposition was observed to occur at 520 °C. The lower degradation temperatures of both unvulcanized and vulcanized HDPE/NR/TPS blends indicated that their thermal stability was less than that of HDPE/NR blends. The low thermal stability of TPS and poor interaction at the interface between TPS and the HDPE/NR matrix might lead to this behaviour.

The DTG peak measurements as observed from Figures 10.10(b) and 10.11(b) represent the maximum degradation rate of the blends. Therefore, as well as the TGA curves, DTG thermograms could also be used as a relative comparison for the thermal stability of HDPE/NR/TPS blends. The crosslink structure in vulcanized NR could affect the activation energy needed for

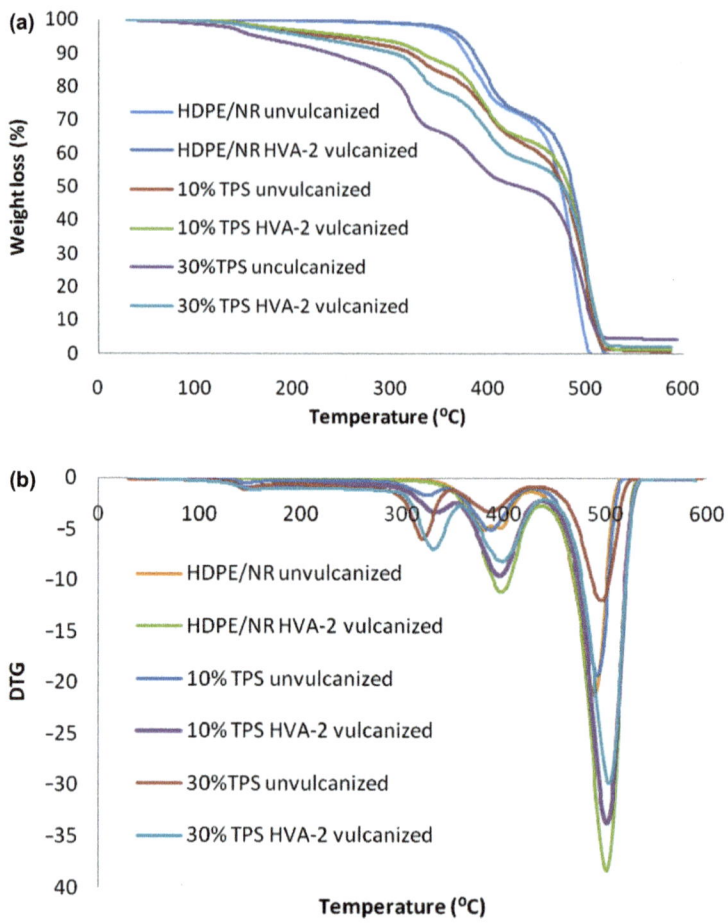

Figure 10.11 (a) Thermogravimetric analysis of unvulcanized and HVA-2 vulcanized HDPE/NR/TPS blends. (b) DTG of unvulcanized and HVA-2 vulcanized HDPE/NR/TPS blends.

thermal decomposition of the blends.[48,49] DTG thermograms of vulcanized blends showed the decomposition point for 10% and 30% TPS content was higher than those of unvulcanized counterparts. The degradation peaks of both vulcanized blends show a similar trend, which was around 10 °C higher than unvulcanized blends, corresponding to a significant increase in thermal stability. This behaviour occurred due to the crosslink structure in the NR phase, which could improve the barrier for diffusion of oxygen into materials, and thus could slow down the rate of degradation.

10.3.7 Differential Scanning Calorimetry

Thermal behaviour of unvulcanized and HVA-2 vulcanized HDPE/NR blends and the blends with 10% and 30% TPS for HVA-2 vulcanized were analysed using differential scanning calorimetry (DSC). Table 10.3 shows crystallization

Table 10.3 DSC data for unvulcanized, sulfur vulcanized and HVA-2 vulcanized blends.

	T_c onset (°C)	T_m onset (°C)	T_m peak (°C)	ΔH_m (J g^{-1})	X_c (%)	W_f
HDPE/NR unvulcanized	118.6	116.3	120.4	135.7	124.2	60.6	0.70
10% TPS unvulcanized	118.4	115.6	121.3	131.1	104.4	56.4	0.63
30% TPS unvulcanized	118.2	114.7	121.7	132.1	78.9	55.1	0.49
HDPE/NR sulfur vulcanized	118.5	115.6	121.0	131.2	82.5	57.7	0.70
10% TPS sulfur vulcanized	118.4	115.1	121.0	131.5	66.7	36.3	0.63
30% TPS sulfur vulcanized	118.6	116.5	121.3	130.5	43.8	30.6	0.49
HDPE/NR HVA-2 vulcanized	118.6	115.2	120.9	131.3	71.4	34.9	0.70
10% TPS HVA-2 vulcanized	118.2	114.7	121.3	131.9	32.1	17.4	0.63
30% TPS HVA-2 vulcanized	118.3	115.1	121.7	131.2	18.3	12.8	0.49

temperature (T_c), melting temperature (T_m), heat of fusion ΔH_m, the degree of crystallization (X_c) and weight fraction of HDPE (W_f) in the blends. DSC thermograms show single exothermic and endothermic peaks which reflect the crystal formation and melting of HDPE, since NR and starch usually do not have melting shape and start to decompose before the melting temperature. Unvulcanized HDPE/NR melts at a temperature peak of 135.7 °C and showed a noticeable decrease when the blends were subjected to the HVA-2 crosslinker. However, the melting temperature and crystallization temperature of HDPE were not much influenced by incorporation of TPS. Only a slight decrease was observed in T_c and T_m for HVA-2 vulcanized blends at 10% and 30% TPS but the degree of crystallization was greatly affected by TPS loading.

Crystallinity of the blends was calculated based on the following equation:

$$\text{Degree of crystallinity}, X_c(\%) = \frac{\Delta H_{\text{exp}}}{\Delta H_m \times Wf} \times 100 \qquad (10.3)$$

where ΔH_{exp} is the experimental data of heat fusion, ΔH_m is the heat of fusion of the 100% crystalline HDPE and W_f is the weight fraction of HDPE in the blend. The enthalpy of the fully crystalline HDPE was reported to be equal to 292 J g^{-1}.[50] As seen from Table 10.1, the X_c value tends to decrease with the incorporation of TPS, indicating that TPS particles could interfere with the mobility of the polymer and reduce its ability to form crystals. In the case of HVA-2 vulcanized blends, the degree of crystallinity was found to be affected by the network structure in the NR phase, and further decreased significantly with the incorporation of TPS. The X_c of the unvulcanized HDPE/NR blend was 60.6%, while for the HVA-2 vulcanized HDPE/NR HDPE/NR-10% TPS and HDPE/NR-30% TPS blends was 34.9%, 17.4% and 12.8%, respectively. Although the crosslinking process was taking place in the amorphous NR, it has been reported that the network structure formed near the surface of the chain fold would affect the nearby crystallinity.[21] The crosslinked structure had low mobility and would cause a disturbance in the crystalline formation of HDPE chains.

10.4 Conclusions

Based on this research work, it is demonstrated that dynamic vulcanization of HDPE/NR/TPS using sulfur and HVA-2 crosslinker was an effective way of improving the properties of the blends. The tensile properties increased with the addition of crosslinkers. All the tested samples displayed significant enhancement in tensile properties and better thermal stability, which could be attributed to crosslink formation in the NR phase. SEM studies of fractured surfaces showed that the vulcanized system had better dispersion of TPS particles and was well embedded in the HDPE/NR matrix. This shows that it could be effectively employed as one of the techniques to improve the homogeneity and tensile properties of HDPE/NR/TPS blends.

Acknowledgements

The authors would like to thank Universiti Sains Malaysia for the financial support provided by grant no: 1001.221.0.PBAHAN.8044003 and for the supply of raw materials. The authors also gratefully acknowledge the scholarship from Universiti Malaysia Perlis.

References

1. M. D. Edlul, A. H. Tsoub and W. Hu, *Polymer*, 2004, **56**, 185.
2. C. Nakason, K. Nuansomsri, A. Kaesamana and S. Kiatkamjornwong, *Polym. Test.*, 2006, **25**, 782.
3. C. Nakason, A. Worlee and S. Salaeh, *Polym. Test.*, 2008, **27**, 858.
4. A. K. Bhowmick and J. Heslop, *Polym. Degrad. Stab.*, 2001, **74**, 513.
5. H. M. Dahlan, M. D. Khairul Zaman and A. Ibrahim, *Radiat. Phys. Chem.*, 2002, **64**, 429.
6. S. Akhtar and S. S. Bhagawan, *Rubber Chem. Technol.*, 1987, **60**, 591.
7. P. Laokijcharoen and A. Y. Coran, *Rubber Chem. Technol.*, 1998, **71**, 966.
8. W. Pechurai, C. Nakason and K. Sahakaro, *Polym. Test.*, 2008, **27**, 621.
9. S. Pichaiyut, C. Nakason, A. Kaesaman and S. Kiatkamjornwong, *Polym. Test.*, 2008, **27**, 566.
10. A. Ibrahim and M. Dahlan, *Prog. Polym. Sci.*, 1998, **23**, 665.
11. D. J. Elliott and A. J. Tinker, *Ind. Gomma*, 1987, **31**, 30.
12. J. W. Teh, L. C. Tan, C. T. Chia, K. K. Tan and T. T. Teng, in *Proceeding of Composites Asia Pacific Conference*, Adelaide, 1989.

13. M. M. Sain, I. Simek, J. Beniska and P. Rosner, *J. Polym. Mater.*, 1990, **7**, 49.
14. K. G. Karmika, S. L. G. Rangith, S. S. Warnapura and W. P. M. Abeysekera, in *Proceedings of Conference of Natural Rubber: Current Developments in Product Manufacture and Applications*, Kuala Lumpur, 1993.
15. S. Cook, A. J. Tinker and J. Patel, *WO Patent Application*, 2003, 03/054078, A1.
16. B. Kuriakose and S. K. De, *Polym. Eng. Sci.*, 2004, **25**, 630.
17. A. Thitithammawong, C. Nakason, K. Sahakaro and J. Noordermeer, *Polym. Test.*, 2004, **26**, 537.
18. A. K. Tinker, *Polym. Commun.*, 1984, **25**, 325.
19. S. Varghese, R. Alex and B. Kuriakose, *J. Appl. Polym. Sci.*, 2004, **92**, 2063.
20. R. Asaletha, M. G. Kumaran and S. Thomas, *Eur. Polym. J.*, 1999, **35**, 253.
21. H. Huang, J. Yang, X. Liu and Y. Zhang, *Eur. Polym. J.*, 2002, **38**, 857.
22. Y. Yang, T. Chiba, H. Saito and T. Inoue, *Polymer*, 1998, **39**, 3365.
23. H. Ismail and Suryadiansyah, *Polym. Test.*, 2002, **21**, 389.
24. H. Ismail, Salmah and M. Nasir, *Polym. Test.*, 2001, **20**, 819.
25. A. Y. Coran, in *Encyclopedia of Polymer Science and Engineering*, 2nd edn, ed. H. F. Mark, N. M. Bikales, C. G. Overberger and G. Menges, John Wiley & Sons, New York, 1989, pp. 688–691.
26. A. S. Hashim and S. K. Wong, *Polym. Int.*, 2002, **51**, 611.
27. A. Hassan, M. U. Wahit and C. Y. Chee, *Polym. Test.*, 2003, **22**, 281.
28. M. Awang and H. Ismail, *Polym. Test.*, 2008, **27**, 321.
29. J. George, K. T. Varughese and S. Thomas, *Polymer*, 2000, **41**, 1507.
30. P. Punnarak, S. Tantayanon and V. Tangpasuthadol, *Polym. Degrad. Stabil.*, 2006, **91**, 3456.
31. P. R. Dluzneski, *Rubber World*, 2001, **34**, 201.
32. M. Akiba and A. S. Hashim, *Prog. Polym. Sci.*, 1997, **22**, 475.
33. A. Singh and J. Silverman (eds), *Radiation Processing of Polymers*, Hanser, Munich, 1992, pp. 1–14.
34. M. Zurina, H. Ismail and C. T. Ratnam, *Polym. Test.*, 2008, **27**, 480.
35. G. L. M. Vroomen, G. W. Visser and J. Gehring, *Rubber World*, 1991, **205**, 23.
36. R. Chandra and R. Rustgi, *Polym. Degrad. Stabil.*, 1997, **56**, 185.
37. D. Bikiaris, J. Prinos, K. Koutsopoulos, N. Vouroutzis, E. Pavlidou, N. Frangi and C. Panayiotou, *Polym. Degrad. Stabil.*, 1998, **59**, 287.
38. R. Shi, Z. Zhang, Q. Liu, Y. Han, L. Zhang, D. Chen and W. Tian, *Carbohydr. Polym.*, 2007, **69**, 748.
39. W. Ning, Y. Jiugao, M. Xiaofei and W. Ying, *Carbohydr. Polym.*, 2007, **67**, 446.
40. J. Raquez, Y. Nabar, M. Srinivasan, B. Shin, R. Narayan and P. Dubois, *Carbohydr. Polym.*, 2008, **74**, 159.
41. Z. Yang, M. Bhattacharya and U. R. Vaidyat, *Polymer*, 1996, **37**, 2137.
42. A. W. M. Kahar, H. Ismail and N. Othman, *J. Vinyl Addit. Technol.*, 2012, **18**, 192.

43. A. W. M. Kahar, H. Ismail and N. Othman, *J. Appl. Polym. Sci.*, 2012, **128**, 2479.
44. P. Pagès, F. Carrasco, J. Saurina and X. Colom, *J. Appl. Polym. Sci.*, 1996, **60**, 153.
45. G. Giancola, L. Richard, R. L. Lehman and J. D. Idol, *Powder Technol.*, 2012, **218**, 18.
46. X. Colom, J. Canavate, P. Pagès, J. Saurina and F. Carrasco, *J. Reinf. Plast. Compos.*, 2000, **19**, 818.
47. J. M. Fang, P. A. Fowler, J. Tomkinson and C. A. S. Hill, *Carbohydr. Polym.*, 2002, **47**, 245.
48. A. P. Mathew, S. Packirisamy and S. Thomas, *Polym. Degrad. Stabil.*, 2001, **72**, 423.
49. Q. Wang, F. Wang and K. Cheng, *Radiat. Phys. Chem.*, 2009, **78**, 1001.
50. J. Brandrup and E. H. Immergut, *Polymer Handbook*, 3rd edn, John Wiley & Sons Inc., New York, 1989. p. V-23.

CHAPTER 11

Natural Rubber/Engineering Thermoplastic Elastomer Blends

E. PURUSHOTHAMAN* AND MEHAR AL MINATH

Department of Chemistry, University of Calicut, Kerala, India
*Email: epurushot@yahoo.com

11.1 Introduction

The art of blending is supposed to be the way forward for plastics.[1] The blending of two or more polymers often yields another polymer with novel properties that cannot be achieved from any of the individual components. Scientific and commercial progress in the area of polymer blends to date has been quite significant and driven by the realization that blending can usually be implemented more rapidly and economically than the development of new polymers. New technology in this field makes it possible to modify polymers more easily and enhance their performance. A significant number of commercial polymer blends have become available, and continued efforts to create new materials with enhanced chemical or mechanical performance are on-going.[2] In addition, modifications in terms of processing characteristics, durability and cost can be achieved through polymer blending, which provides materials with unusual combinations of mechanical, thermal, chemical and morphological properties,[3,4] as the situation demands.

The introduction of thermoplastic elastomers (TPEs) during the late 1950s is one of the most important developments in the field of polymer science and technology. TPEs are a relatively new class of material, a blend of rubber and thermoplastic that combine a wide range of the physical properties of elastomers, such as elasticity, at room and service temperatures, and the excellent processing characteristics of thermoplastics at high temperatures. TPEs can be remoulded, display good processability and have many economic advantages.[5–7] Moreover, they provide immense potential for scrap and reject recycling, and ease of property manipulation through composition change.[8,9]

TPEs can be classified into three families:

1. Block copolymers or graft copolymers made up of soft and rigid polymer sequences. Styrene block copolymers like polystyrene (PS) blocks [styrene-butadiene-styrene (SBS), styrene-isoprene-styrene (SIS), and styrene-ethylene-butylene-styrene (SEBS)], and polyester TPEs belong to this family. Structurally, the thermoplastic blocks form physical network knots within the polydiene.
2. Thermoplastic elasto-ionomers, whose network knots are thermolabile ionic microdomains formed by associations between ionic groups (carboxylic or sulfonic acids) present in non-polar polymer chains.[10]
3. TPEs which are prepared by physical blending of an elastomer with a thermoplastic. Such TPEs are categorized into two types depending on rubber vulcanization: (i) thermoplastic olefins (TPOs) or thermoplastic elastomer polyolefins (TEOs); (ii) thermoplastic vulcanizates (TPVs).

The rubber phase of the TPO is unvulcanized. The plastics normally used are polyolefins such as polyethylene and polypropylene. In TPVs, the rubber phase is dynamically vulcanized.

Development of TPEs with various types of elastomers and polyolefins has been extensively reported by many researchers. Ethylene-propylene-diene monomer (EPDM) or its modified form is used as the elastomer in most polyolefin TPEs.[11] Natural rubber (NR) and thermoplastic blends have become an area of interest only recently. These materials are known as thermoplastic natural rubber (TPNR). The development of TPNR was principally based on the criteria set by EPDM blends with thermoplastics.[12] Two types are known, one belonging to the TPO class and the other belonging to the TPV class.

The development of TPNR is of great interest in the area of material science due to its unique properties and ecofriendly nature. The thermal resistance and ozone resistance of TPNR is superior to NR vulcanizates.[13]

Thermoplastics used to blend with NR include PS,[14] polyamide 6,[15] ethylene-vinyl acetate (EVA) copolymer,[16] poly(methyl methacrylate) (PMMA),[17,18] polypropylene (PP),[19–21] low-density polyethylene (LDPE),[22] linear low-density polyethylene (LLDPE)[23,24] and high-density polyethylene (HDPE).[25,26] To improve the properties of TPNR, modified NR is also used. ENR is the most frequently used modified NR. TPNR blends are prepared by blending NR and thermoplastics in various proportions. The role of rubber is to improve the impact strength and ductility of the plastic. Depending on the ratio, materials with a wide range of properties are obtained. The stiffness of the rubber is increased with the incorporation of plastic into the rubber matrix. The mechanical properties of TPNR again depend on the proportions of the rubber and thermoplastic components. The elastic properties of TPNR are considerably improved if the rubber phase is partially crosslinked during mixing, a process called dynamic vulcanization.[27]

11.2 Recent Developments in TPEs

There have been many significant developments in the area of TPEs over the past decade and extensive research has been conducted into TPNR. Apart from the development of blends, researchers are also looking at improving the properties of TPEs by reinforcing with suitable fillers. The development of biodegradable blends and composites of this class is also gaining momentum due to environmental concerns, and research is going on in the area of waste management by preparing TPNRs from recycled thermoplastics and NR.

Renju et al.[28] used expanded polystyrene (EPS), which makes up a considerable proportion by volume of thermoplastic waste in the environment, for blending with silica-reinforced NR. EPS and NR were initially grafted with maleic anhydride (MA) using dicumyl peroxide (DCP) to give a graft copolymer. They found that mechanical properties such as tensile strength, elongation at break, modulus, tear strength, compression set and hardness of the blend were either on a par with or better than that of virgin silica-filled NR compounds. The study also established the potential for using waste EPS. Guo et al.[29] prepared TPNR from scrap rubber powder (SRP) and LLDPE using a compatibilizer. The ENR/LLDPE-g-VM dual compatibilizer modified SRP/LLDPE showed good elasticity and comprehensive mechanical properties. It has potential applications in recycling SRP in large quantities and as an environmentally-friendly material. Similarly, Ismail and Suryadiansyah[30] developed polypropylene (PP)/recycled rubber (RR) blends. It was found that PP/RR blends have higher tensile strength and Young's modulus but lower elongation at break and stabilization torque than PP/NR blends.

Another research group led by Sukanya[31] blended RR with waste polyethylene (WPE) collected from municipal solid waste (MSW) in different proportions; composites were prepared with fly ash (FA) and characterized. Mechanical and dynamic mechanical properties of the blend and composites were studied in the presence as well as in the absence of a silane coupling agent (Si-69). It was found that the tensile strength, flexural strength, flexural modulus, impact strength and hardness properties of the FA composites improved in the presence of Si-69.

Natural fibre based TPNRs are also gaining importance as biodegradable blends. Ishak and coworkers developed composites of NR/HDPE blends reinforced with electron beam irradiated liquid natural rubber (LNR) coated rice husk (RH).[32] The composites filled with radiated RH showed the highest storage modulus (E') and low tan δ. Improved RH filler–matrix interfacial bond strength and adhesion to the matrix were achieved by coating the RH powder and curing the rubber coat by electron beam irradiation. TPE composites of EVA/NR blends filled with palm ash were prepared by Najib and coworkers.[33] An increase in palm ash loading in the composites resulted in an increase in stabilization torque, Young's modulus and swelling resistance of the composites, but a decrease in tensile strength and elongation at break. Kenaf bast fibres were used as reinforcing material in order to improve the mechanical properties of TPE composites of PP/NR and PP/EPDM blends by Anuar and Zuraida.[34]

New TPEs based on different plastics are being developed and characterized to obtain polymer materials of superior quality to suit specific applications. Narathichat et al.[35] prepared TPNR based on polyamide-12 (PA-12) and investigated the influence of blending techniques and types of NR on properties of the blends. It was found that simple blends with around 60 wt% rubber content exhibited a co-continuous phase structure, while the dynamically cured blends showed dispersed morphology. Furthermore, the ENR blends exhibited superior mechanical properties, stress relaxation behaviour, and fine grain morphology than those of the unmodified NR blends. Dynamic vulcanization led to enhancement of strength, hardness, stress relaxation properties and thermal resistance of the dynamically cured ENR/PA-12 blend. Suksawad et al.[36] prepared TPEs from deproteinized natural rubber (DPNR) by graft copolymerization of styrene. These blends exhibited outstanding mechanical properties even after processing due to the thermoplasticity of the DPNR-PS.

Panu and coworkers[37] found that the impact strength of reclaimed tyre rubber (RTR)/HDPE blends increased with increasing RTR up to 50% loading. This high impact strength is in the appropriate range for automotive applications.

11.3 Preparation of TPEs

Unlike conventional rubbers, TPEs are processed on thermoplastic machinery. Preparation requires no separate vulcanization stage, is easily done by internal mixers or extrusion and so productivity is high. The various methods of preparation of TPEs are discussed below.

11.3.1 Mixing

The mixing of rubber and thermoplastic to prepare TPEs is simplified as the number of additives necessary is considerably reduced. This saves time, energy, and capital costs of machinery. Varying the mixing temperatures near the melting points of the thermoplastic will naturally yield elastomer materials exhibiting different physical properties. The mixing rate and duration of mixing also influence the properties of the blend produced. Mixing is carried out using different machinery as shown below.

11.3.1.1 Brabender Plasticorder

One of the methods of preparing blends is using an intensive internal mixer known as a Brabender plasticorder. TPNR is normally prepared in the laboratory *via* mixing in a Banbury mixer or Brabender plasticorder attached to a mixer or a twin-screw compounder. The dried thermoplastic is first melted in the mixer and then masticated NR is added.[38,39] Additives such as compatibilizers or stabilizers are added before the addition of the elastomer.[40,41] The vulcanizing agent, such as DCP or sulfur, is added towards the end of mixing.[42,43] After mixing in the internal mixer, the blends are taken out and sheeted through a two-roll mill. This is then compression moulded in a hydraulic press.

11.3.1.2 Two-Roll Mixing Mill

When a two-roll mill is used for mixing, the components are not melted. The ingredients are compounded on an open two-roll mill and then compressively moulded in an electrically heated hydraulic press to the optimum cure; the moulding conditions are previously determined from torque data with a rheometer.[44,45]

11.3.1.3 Haake Rheocord

The Haake rheocord is an alternate procedure used to prepare TPEs. The component polymers are mixed thoroughly and then the curatives, if any, are added. Peroxides are added just before damping.[46] The material, while hot, is sheeted out on a mixing mill and then granulated. Aju et al. prepared NBR/PP and NR/PP blends by this method.[47]

11.3.1.4 Twin-Screw Extruder

Melt blending is also carried out using a twin-screw extruder. The screw speed is determined by the ease of processing and mechanical properties of the blends. The extrudate is quenched in a water bath at room temperature and pelletized. The blends are dried in a vacuum oven and re-extruded using the previous extrusion conditions. The blends are again dried in an oven and kept in a desiccator at room temperature. Tanrattanakul et al.[48] performed melt blending of nylon 6 and ENR using a twin-screw extruder.

11.3.2 Solution Casting

Solution casting is a very convenient method when thin films of the blends are required. The elastomer and the thermoplastic are dissolved in a suitable common solvent or a mixture of solvents and then cast on a glass plate.

It has been proved that the nature of solvent used for casting in the preparation of the blend influences the compatibility and related properties of PS-poly(vinyl methyl ether) (PVME) mixtures.[49] Clear films of PS-PVME mixtures were obtained on casting from solvents like benzene, toluene, etc., while visually incompatible films result upon casting from trichloroethylene and chloroform. Blends of PVC and LNR/epoxidized liquid natural rubber (ELNR) were prepared using a common solvent, 2-butanone.[50] Another research group prepared PVC/NR blends using THF as the solvent.[51]

11.4 Characterization of TPEs

Different characterization techniques are employed to understand the structure and properties of polymers and to establish the relationship between structure and different properties. The properties of polymers such as rheology, morphology, mechanical and thermal are also studied using different techniques.

11.4.1 Rheological Studies

Rheology is the science that studies the deformation and flow of materials in liquid, melt or solid form in terms of the material's elasticity and viscosity. This is accomplished by applying a precisely measured strain to the sample to deform it, and accurately measuring the resulting stress developed in the sample. The developed stresses are related to material properties through Hooke's and Newton's laws.

TPEs are processed using techniques that involve high shearing. It is therefore essential to examine the flow behaviour of these materials under various shearing conditions. The rheological parameters of the blend control its processability under operating conditions. The cure characteristics and thus the processability of the compounds are studied from the rheographs. These parameters are measured using a variable torque rheometer, a capillary rheometer or a moving die rheometer.

Here, the sample is subjected to a sinusoidal oscillating mechanical deformation, in the rheometer's preheated test chamber, which is opposed by a mechanical resistance, measured as torque. The time for 90% or 95% increase in torque for a rheograph is regarded as the optimum cure time. The rise in torque is considered to be proportional to increase in crosslink density and the crosslink density determines the physical properties of the compounded rubber.

Sujith et al.[52] carried out rheometric studies of intermediate super-abrasion furnace (ISAF) carbon black loaded NR/EVA composites of different blend compositions. A higher maximum torque value was found for high EVA loaded samples and this is attributed to the increase in crystallinity in them. EVA is semicrystalline, and so restricts the mobility of the macromolecular chains.

Riahi[53] and coworkers found that the values of torque obtained for NR/PP blends were in conformity with an increase in the crosslink density. Nakason et al.[54] showed that in EVA/PP blends, the mixed cure system has the highest final mixing torque, confirming maximum crosslinking in blends cured with a mixture of sulfur and DCP. Rheological studies of NR/LDPE blends and ground tyre rubber (GTR)/LDPE blends by Majid et al.[55] showed that all compositions exhibit a decrease in viscosity with increasing frequency, which is the characteristic of pseudoplastic behaviour of the blends. Investigations of the effect of compatibilizer on the rheological properties of NR/HDPE blends by Pechurai et al.[56] showed that at a given shear rate, the apparent shear stress and viscosity of the blends with phenolic modified polyethylene are higher than that of the blends without compatibilizer. This has been attributed to a chemical interaction between the NR and HDPE phases by the modified polyethylene compatibilizer. Also, at a given shear rate, the apparent shear viscosity increased with increasing loading levels of oil extended natural rubber (OENR) until 60 wt%, where the maximum value was observed.

11.4.2 Morphological Studies

Morphological studies are another important characterization technique for blends. The properties of the end product in blending depend on its morphology, which in turn is dependent on processing factors like type of mixer, rate and temperature of blending, component rheology, interfacial tension and crosslinking agent used. The morphology of the polymer blends can be understood by using different microscopic tools like optical microscopy, scanning electron microscopy (SEM), transmission electron microscopy (TEM) and atomic force microscopy (AFM). For polymer blends a minimum domain size of 1 μm can be examined by optical microscope. When the domain size is in the range of < 1 μm to 10 nm, SEM or TEM has been advantageously used.

For establishing morphology using SEM, samples are examined as freeze fractured surfaces or as microtome blocks of solid bulk samples. Contrast is achieved by one or more of the following methods:[57]

1. Solvent etching – where there is a large solubility difference in a particular solvent of the polymer being studied, e.g. PP/EP blends.
2. OsO_4 staining – when there is at least 5% unsaturation in the polymers being investigated, e.g. NR/EPDM, bromo isobutylene isoprene rubber/neoprene (BIIR/neoprene).
3. RuO_4 staining – this can be explored when there is no solubility difference or unsaturation.

Most of the morphological studies are carried out with the help of SEM analysis. Magida[58] and coworkers used SEM to visualize the microstructure of the blend of styrene-vinylacetate copolymer/natural rubber latex. Senna et al.[59] examined the morphology of the fractured surface of electron beam irradiated PP/ENR polymer blends by SEM to understand the relationship between blend morphology and mechanical properties. The photodegradation of TPE rubber/polyethylene blends was observed at different stages by Anil et al.[60] using SEM. The authors used SEM in their research work to follow the biodegadation behaviour of TPU/NR blends and their chitin-reinforced composites.[61] The SEM photographs of the sample before and after biodegradation are shown in Figure 11.1. Before biodegradation the samples exhibited a relatively smooth surface. However, after 90 days of soil burial in natural soil, the composites had many holes on their surface, indicating microbial attack. The SEM photographs indicate that initial biodegradation begins with surface erosion due to the action of microbes in the soil.

TEM is used whenever a more in-depth study of polymer phased morphologies (when domain sizes are < 1 μm) is required. TEM enables finer details to

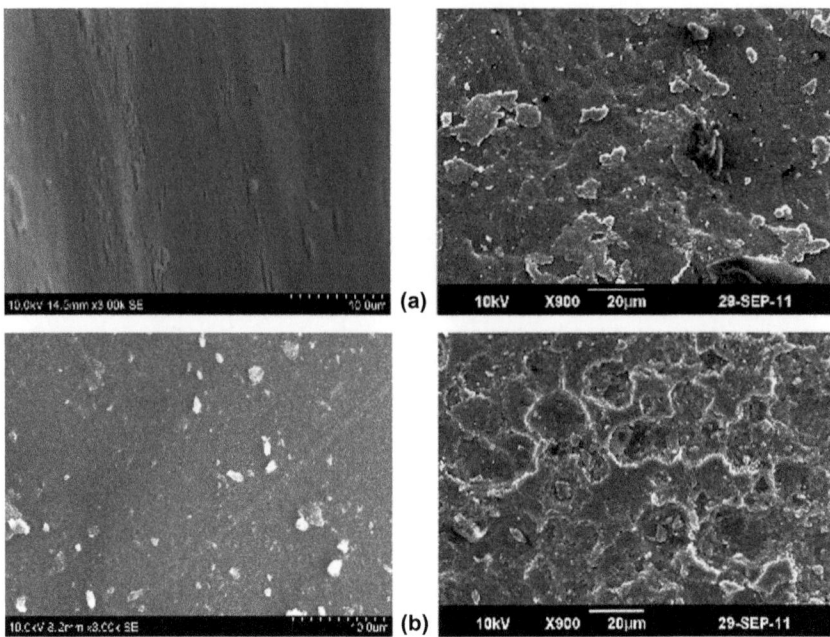

Figure 11.1 SEM images of the samples before and after the soil burial test. (a) TPU/NR blend. (b) TPU/NR chitin reinforced composite.

be examined – even as small as a single column of atoms, which is tens of thousands times smaller than the smallest resolvable object in a light microscope. Thin sections are required and it takes anywhere from 1 hour to 1 day per sample, depending on the nature of the sample. They must be ~100 nm thick and are usually prepared by microtoming with a diamond knife at near liquid nitrogen temperatures (−150 °C). The same contrasting media for SEM apply to TEM also. Indukuri et al.[62] carried out morphological characterization of poly(styrene-b-ethylene-co-butylene-b-styrene) TPEs and crosslinked NR using TEM studies. Sections of 40–100 nm thickness were obtained by cryomicrotoming bulk films at −90 °C. Contrasting was achieved by staining with ruthenium tetroxide for about 12 min. Another research group examined the morphology of NR-g-PMMA samples by TEM,[26] staining with ruthenium tetroxide. Honeker and Thomas[63] used TEM to assess the quality of orientation of poly(styrene-b-isoprene-b-styrene) (SIS) triblock copolymer with a PS cylinder by TEM of sections viewed parallel and perpendicular to the cylinder axis; this provided real space information on the deformed morphology.

AFM provides a 3D profile of the surface at nanoscale and does not require a conductive sample. The morphology of immiscible PP/NR blends was studied using AFM by Oh et al.[64] The AFM studies revealed the development of interfacial layers, interfacial roughening and improved interfacial adhesion between the PP and NR phases in the blends subjected to ultrasonic treatment.

At the same time weak adhesion and delamination at the interface were found in the untreated blends.

11.4.3 Scattering Analyses

Scattering experiments are widely used to study the structure of polymers. The crystallinity of polymer blends is determined through non-destructive analytical techniques like X-ray diffraction (XRD), small-angle X-ray scattering (SAXS) and ultra-small-angle X-ray scattering (USAXS). Based on the scattered intensity of an X-ray beam hitting a sample as a function of incident and scattered angle, polarization, and wavelength or energy, these techniques expose the crystal structure, chemical composition and physical properties of materials and thin films. The advantage of these techniques over crystallography is that a crystalline sample is not needed. Also, the wavelength of X-rays used for diffraction studies is of the order of 1 Å, which is of the same order as that of interatomic distances in solids. SAXS is a small-angle scattering (SAS) technique where the elastic scattering of X-rays (wavelength 0.1–0.2 nm) by a sample that is non-homogeneous in the nanometer range, is recorded at very low angles (typically 0.1–1^0). Even this angular range gives information about the shape and size of macromolecules, characteristic distances of partially ordered materials, pore sizes, and other data. SAXS is capable of delivering structural information of macromolecules between 5 and 25 nm, of repeat distances in partially ordered systems up to 150 nm. USAXS can resolve even larger dimensions.

Scattering of light is also widely used for structural analysis of polymers. The light scattering experiments provide information on structure much larger in scale than can be obtained by XRD, as the wavelength of light is almost three times greater than that of X-rays.

Neutron scattering experiments are also gaining importance recently for structural studies. The velocity of high-speed neutrons is decreased to the order of 1 Å before it is allowed to hit the sample.[65]

Anusree et al.[44] determined the degree of crystallinity of NR/EVA blend membranes from XRD patterns of the samples recorded with an X-ray diffractometer.

The structures of TPNR and their composites with carbon black were investigated by SAXS, and small-angle neutron scattering (SANS).[66] Thomas et al.[67] studied the deformation of cylindrical PS domains of a near single crystal styrene-isoprene-styrene (SIS) triblock copolymer by SAXS. The structural changes in the hexagonal lattice of cylinders on uniaxial tensile deformation were related to the macroscopic mechanical behaviour.

11.4.4 Mechanical Properties

An understanding of the mechanical behaviour of polymer blend materials becomes vital for creating innovative and economical designs for various components. The mechanical properties of polymer blends depend markedly on factors such as the nature of the polymer matrix, its crosslink density, miscibility of the components and polymer–polymer interactions. In rubber-modified thermoplastics, the role of rubber is to improve the impact strength and ductility of the plastic. The stiffness of the rubber is increased with incorporation of plastic into the rubber matrix.

The rubbery phase enhances the impact resistance by initiating highly dissipative deformation mechanisms.[68–71] Generally the phase size, cohesive strength of the rubber phase and the adhesion between phases influence the impact properties of polymer blends. In the case of blends of amorphous/semicrystalline polymers, the mechanical properties of the system are strongly related to the properties of the crystalline regions like crystallinity, crystalline morphology, and the size of the crystallites and their aggregates such as spherulites. The measurement of mechanical properties is concerned with load deformation or stress–strain relationships. Forces may be applied as tension, shear, torsion, and compression and bending. Mechanical properties such as tensile strength, tensile modulus, tear strength and Young's modulus are determined using a universal testing machine (UTM). Hardness is measured using a shore A durometer or shore D durometer, and abrasion resistance using a DIN abrader.

The macroscopic response of polymeric materials subjected to mechanical load is generally described by a stress–strain curve. The area under these curves is proportional to the total mechanical work that is available for conversion into other forms of energy. Various mechanical properties such as tensile strength, Young's modulus and toughness can be predicted from the stress–strain plot. The tensile strength is the stress needed to break a sample. Young's modulus, otherwise known as the modulus of elasticity or the tensile modulus, is the ratio of stress to strain. Stress–strain curves often are not straight-line plots, indicating that the modulus is changing with the amount of strain and hence, the initial slope is usually used as the modulus. Rigid materials, such as metals, have a high Young's modulus. Fibres also have high Young's modulus values, elastomers have low, and the values for plastics lie in between. The nature of the stress–strain curve indicates the type of polymeric material, as depicted in Figure 11.2.[72]

From the graph, it can be predicted that TPU is brittle and hard and, with the addition of NR, the brittleness and hardness of the blend decreases and the blend becomes more elastomeric in nature (Figure 11.3).

Mechanical studies of vulcanized TPNRs (vTPNRs) and unvulcanized TPNRs (uTPNRs) by Tanrattanakul et al.[38] showed that the uTPNRs had unclear yield points and high ductility, and a tensile behaviour similar to ductile thermoplastics, such as polyolefins. Its drawing strain was relatively constant compared to the tensile behaviour of vTPNRs. Yield strength and tensile strength, hardness and tension set of uTPNRs increased with increasing PP content. The tensile strength of uTPNRs was found to be in the same range as PP/EPDM TPEs. Majid and coworkers[55] found that introduction of 40 parts of NR into LLDPE decreased its mechanical properties, because of the poor compatibility between phases and formation of NR aggregates. The hardness

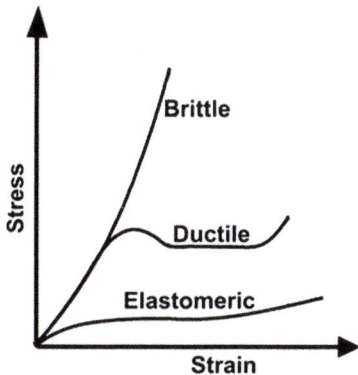

Figure 11.2 Representation of stress–strain curve of different types of polymeric materials subjected to mechanical load.

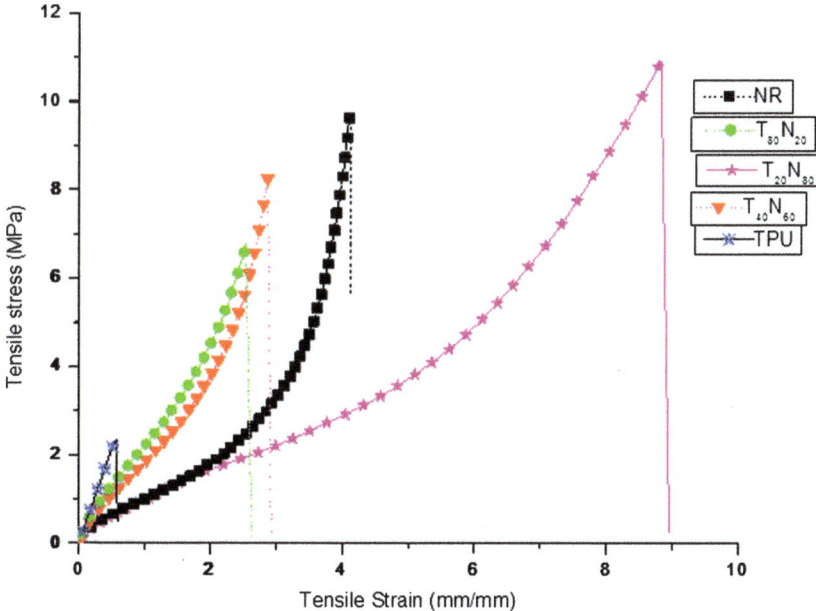

Figure 11.3 The stress–strain curve of TPU/NR blends.

of the blend was also lower than that of pure LLDPE. Substitution of 5–10 wt% of NR phase with GTR decreased the mechanical properties of the LLDPE/NR blend. However, modification of GTR by MA, and using DCP to provide vulcanization of rubber phases during reactive mixing of the blends, caused a dramatic increase in mechanical properties. Investigation of the mechanical properties of nitrile butadiene rubber (NBR)/PVC blends shows that the elastic modulus, tensile strength and degree of hardness increase with an increase in the amount of PVC. On the other hand, the elongation at break and abrasion resistance decrease on increasing the amount of PVC added to the rubber matrix.[45] The tensile strength, stress-at-peak, Young's modulus and flexural modulus showed a substantial enhancement upon vulcanization due to the presence of a crosslinked rubber phase in the case of rubber wood filled NR/PP blend composites.[42]

11.4.5 Thermal Analyses

Thermal analysis techniques are used to study the properties of polymers, blends and composites and to determine the kinetic parameters of their stability and degradation processes.[73,74] Here the property of a sample is continuously measured as the sample is programmed through a predetermined temperature profile. Among the most common techniques are thermogravimetry (TG) and differential scanning calorimetry (DSC). Dynamic mechanical analysis (DMA) and dielectric spectroscopy are essentially extensions of thermal analysis that can reveal more subtle transitions with temperature as they affect the complex modulus or the dielectric function of the material.

TG is a useful technique to determine the thermal stability of polymers and polymer blends. Polymers and their products slowly lose their useful properties because of polymer chain degradation under various environmental conditions. Heat is one of the degrading agents and its effect can be studied by thermal ageing. The thermal stability of individual polymers is greatly influenced by blending and it strongly depends on the interactions of the component polymers.[75] In TG the mass loss vs increasing temperature of the sample is recorded. TG results of TPU/ENR and TPU/NR blends and homopolymers are shown in Figures 11.4 and 11.5.

In TPU/NR and TPU/ENR blends, the thermal stability increases with increase in NR and ENR composition. Hence the thermal stability of TPUs can be improved by blending with ENR and NR; blending with ENR has enhanced stability.[61]

Subhra and coworkers reported the thermal behaviour of blends of poly(ethylene-*co*-acrylic acid)/ENR.[76] The thermal stability of styrene-vinyl acetate copolymer/NR blends was investigated by Magida and coworkers[58] using TG. The blends exhibited higher thermal stability than individual homopolymers. They also found that the irradiated blend was thermally more stable than the unirradiated blend. The thermal degradation behaviour of NR/PS semi- and full IPNs were studied as a function of blend ratio, crosslinker level and initiating system using TG by Aji and coworkers.[77] They found that the NR/PS semi- and full IPNs exhibited enhanced thermal stability compared to the homopolymers. Thermal analysis of ELNR by Radhakrishnan and coworkers[50] showed that epoxidation actually delayed the thermal degradation of LNR. They also carried out thermal analysis of PVC/LNR blends and found that the thermal degradation is independent of the components, indicating the immiscible nature of the blend.

In a DSC experiment the difference in energy input to a sample and a reference material is measured while the sample and reference are subjected to a

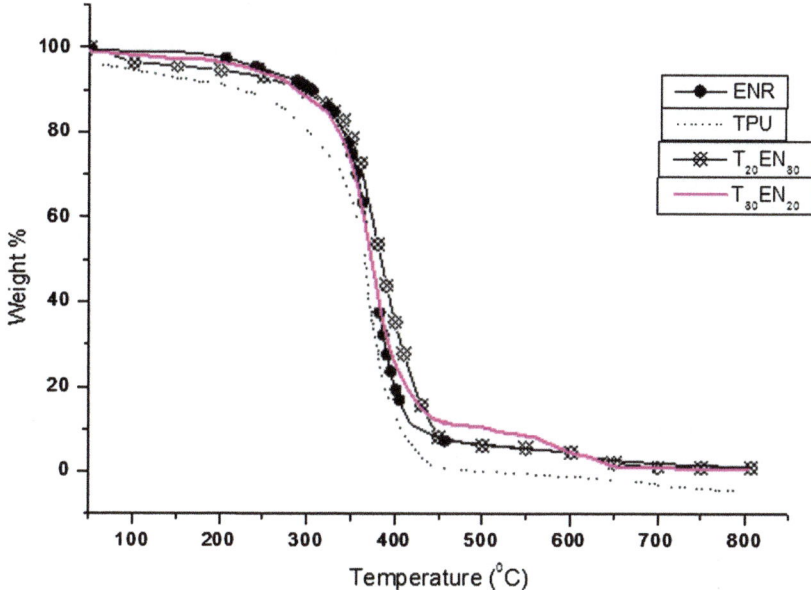

Figure 11.4 TG curves of TPU/ENR blends and homopolymers.

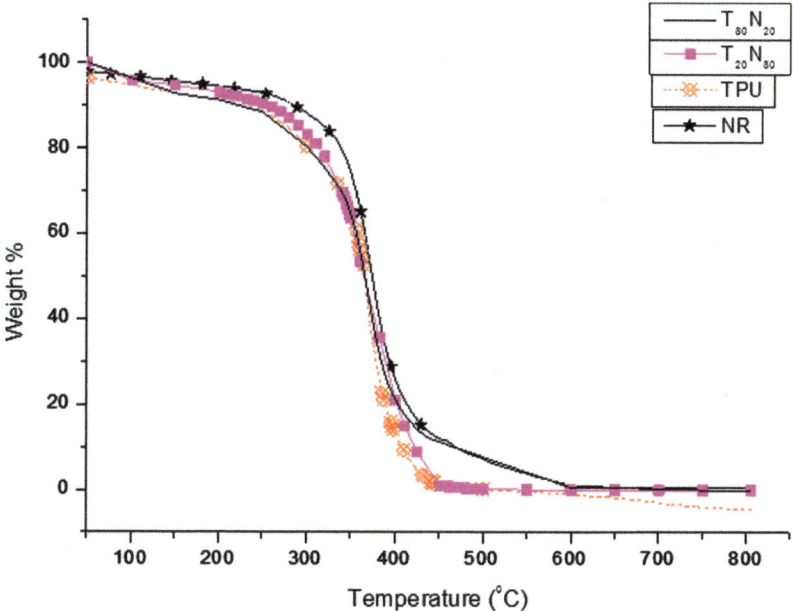

Figure 11.5 TG curves of TPU/NR blends and homopolymers.

controlled temperature programme.[57] DSC analyses are used to predict the miscibility of polymer systems. For an immiscible blend, the tan δ curves show the presence of two damping peaks corresponding to the glass transition temperatures (T_gs) of individual polymers. For a highly miscible blend, the curves show only a single peak in between the transition temperatures of the component polymers, whereas broadening of the transition occurs in the case of partially miscible systems. In the case of miscible or partially miscible blends the T_gs are shifted to higher or lower temperatures as a function of composition. A T_g outside the two component T_gs strongly indicates a plasticizing effect of the added component. However, in NR-polyolefin blends, the T_gs of NR (−67 °C) and polyolefin (−60 °C) are close together; the separation is less than 10 °C and as such the change in T_g in the blends will not be very indicative of the phase distribution.

DSC was used to follow the crystallization kinetics of PP and PP/NR blends by Joseph and coworkers.[47,78] They found that in non-isothermal crystallization, the energy released during the crystallization process appears to be a function of temperature rather than time, as in the case of isothermal crystallization. Also, 100% relative crystallinity was achieved by slow cooling in a narrow temperature region. The blend showed linear spherulite growth and the growth rate decreased with increasing T_c followed by a levelling off. The growth rates were unaltered by the blend ratio at high crystallization temperature, but with very low concentrations of NR phase, the growth rate showed a decrease at lower crystallization temperatures. In another study, PVC/ELNR blends were subjected to DSC analysis,[79] which showed T_g values corresponding to the component polymers at 20 mol% of epoxidation, indicating partial miscibility of a heterogeneous system, while a single transition at 50 mol% of epoxidation indicated phase merging for all compositions.

The blends prepared by the authors using TPU and NR, and TPU and ENR, show a single T_g (Figure 11.6), indicating the miscible nature of the blends.[61]

The DMA of aged PVC/ENR blends[80] revealed an increase in the T_g of the blends during thermal ageing; the highest T_g was for the PVC/ENR blend when aged at 140 °C for 2 days. It was also found that the presence of NBR in the ternary blend provided a resistance to this large increase in T_g. The thermal ageing under both conditions caused an increase in the tan δ peak widths. The intensity of the tan δ peak implies a drop in the molecular mobility in all three blends during ageing. The DMA results for an SBS/PS/SBR blend by Shih and coworkers[81] showed a shift of T_g and an increase in initial storage modulus.

11.4.6 Dielectric Properties

Electronic impedance spectroscopy (EIS) is used to measure the polymer dielectric properties and the changes in these properties with exposure time.[82] This is based on the interaction of an external field with the electric dipole moment of the sample, often expressed by permittivity. Dielectric thermal analysis (DETA) measures the permittivity, capacitance and dielectric loss of a polymer sample under an oscillating electric field as a function of temperature.[83] The dielectric properties of a blend system in general depend on structure, crystallinity, morphology and additives.[84]

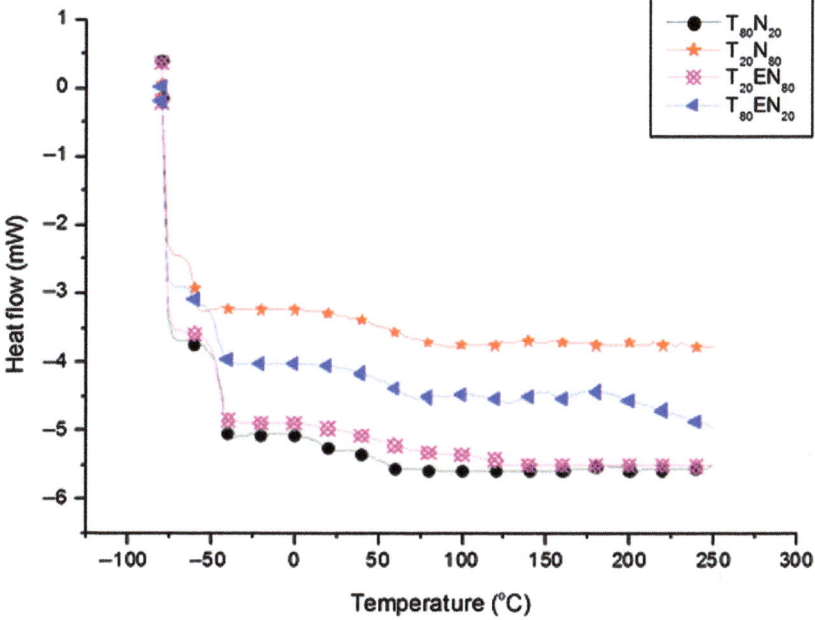

Figure 11.6 DSC thermogram of the TPU/NR and TPU/ENR blends.

The measurement of dielectric properties as a function of temperature was used as a way of monitoring the miscibility of two phased polymeric systems. Dielectric studies of NR/PS system by Aji P Mathew[85] showed that the resistivity decreases with increase in NR content at a given frequency. The volume resistivity was found to decrease with increase in PS crosslinking. But at high level of crosslinking the resistivity increases due to PS phase agglomeration. The dielectric constant, dielectric loss and ac conductivity of NR/PVA blend system were analysed by Jobish et al.[86] The dielectric constant and dielectric loss increased with the addition of PVA into NR.

11.5 Applications of TPEs

TPNRs possess excellent properties such as reduced permanent set, improved mechanical properties, greater resistance to attack by fluids, improved high temperature utility, etc.[87] and therefore provide very useful and attractive applications in different fields such as automotive parts, building materials and construction equipment, wire and cable insulation, and so on.[88] TPEs are generally used for making automobile parts where processes such as abrasion, flexing and tear are prominent factors leading to the fracture of the products.[89] TPNRs are currently finding markets in many applications where vulcanized rubbers have traditionally been used. They are also used in products that require better performance than can be obtained with general-purpose thermoplastics such as PVC and polyethylene (PE). The olefinic type TPNRs have potential uses in flexible automotive components such as bumpers and spoilers.[90] PP-based TPNRs are very important for a variety of applications in the automotive field. Blends of PP and NR combine the excellent processability of PP with excellent impact properties due to the presence of the elastomer phase.[47] TPNRs with a higher ratio of NR can replace vulcanized rubber and flexible plastics for applications in footwear, sports goods, seals and mountings, and a wide range of moulded and extruded goods, while TPNRs with a higher ratio of thermoplastic can be used for the manufacture of automotive bumpers and body protection strips. Magida et al.[58] developed a pressure-sensitive adhesive (PSA) by blending styrene-vinyl acetate copolymer latex (S-VAc) and natural rubber latex (NRL). The PSA was suggested to be used for adhering synthetic veneers to particle wood panels.

11.6 Conclusions

TPNR is a relatively new class of material obtained by blending NR and engineering thermoplastics such as PE, PS, EVA and PMMA. They possess many of the physical properties of elastomers, like elasticity at room and service temperatures, and the excellent processing characteristics of thermoplastics at elevated temperatures. Due to their excellent properties they have great potential for applications in engineering and consumer goods. TPNRs are prepared by the various methods used for blending elastomers and thermoplastic polymers. Different characterization techniques are used to determine the properties of newly prepared TPNRs. Recently, TPNRs have been prepared from waste thermoplastics, which helps to reduce the amount of plastic waste in the environment. Ecofriendly TPNR composites are also prepared using natural fibre reinforcing agents, which will definitely enhance the biodegradation capability of TPNRs.

References

1. P. Poomalai and Siddaramaiah, *J. Macromol. Sci., Part A: Pure Appl. Chem.*, 2005, **42**, 1399.
2. T. K. Kim, B. K. Kim, S. Y. Lee, Y. L. Cho, M. S. Kim and H. M. Jeong, *Macromol. Res.*, 2010, **18**, 177.
3. D. R. Paul and S. Newman, *Polymer Blends*, Academic Press, New York, 1978, p. 35.
4. J. G. Boner and P. S. Hope, *Polymer Blends and Alloys*, Chapman & Hall, Norwell, MA, 1993, p. 201.
5. J. G. Drobny, in *Handbook of Thermoplastic Elastomers*, William Andrew Publishing, New York, 2007, 142.
6. J. C. West and S. L. Cooper, in *Science and Technology of Rubber*, ed. F. R. Eirich, Academic Press, New York, 1978, p. 531.

7. A. Y. Coran, in *Handbook of Elastomers – New Developments and Technology*, ed. A. K. Bhowmick and H. L. Stephens, Marcel Dekker, New York, 1989, p. 249.
8. S. K. De and A. K. Bhowmick, *Thermoplastic Elastomers from Rubber-Plastic Blends*, Horwood, Chichester, 1990, p. 87.
9. M. A. Lopez-Manchado and M. Arroyo, *Polymer*, 2000, **41**, 7761.
10. D. Derouet, Q. N. Tran and J. L. Leblanc, *J. Appl. Polym. Sci.*, 2009, **112**, 788.
11. S. Abdou-Sabet and R. P. Patel, *Rubber Chem. Technol.*, 1991, **64**, 769.
12. A. Ibrahim and H. M. Dahlan, *Prog. Polym. Sci.*, 1998, **23**, 665.
13. D. J. Elliot, in *Developments in Rubber Technology*, ed. A. Whelan and K. S. Lee, Applied Science Publishers, London, 1982, pp. 1–44.
14. R. Asaletha, M. G. Kumaran and S. Thomas, *Eur. Polym. J.*, 1999, **35**, 253.
15. E. Carone Jr, U. Kopcak, M. C. Goncaves and S. P. Nunes, *Polymer*, 2000, **41**, 5929.
16. Z. Ghazali, A. F. Johnson and K. Z. Dahlan, *Radiat. Phys. Chem.*, 1999, **55**, 73.
17. M. F. Mina, F. Ania, F. J. Balta Calleja and T. Asano, *J. Appl. Polym. Sci.*, 2004, **91**, 205.
18. C. Nakason, W. Pechurai, K. Sahakaro and A. Kaesaman, *Polym. Adv. Technol.*, 2005, **16**, 592.
19. F. Riahi, D. Benachour and A. Douibi, *Int. J. Polym. Mater.*, 2004, **53**, 143.
20. O. H. Jeong Seok, A. I. Isayev and M. A. Rogunova, *Polymer*, 2003, **44**, 2337.
21. H. Azman, U. W. Mat and Y. C. Ching, *Polym. Test.*, 2003, **22**, 28.
22. A. K. Bhowmick, J. Heslop and J. R. White, *Polym. Degrad. Stab.*, 2001, **74**, 513.
23. H. M. Dahlan, M. D. K. Zaman and A. Ibrahim, *Polym. Test.*, 2002, **21**, 905.
24. H. M. Dahlan, M. D. K. Zaman and A. Ibrahim, *J. Appl. Polym. Sci.*, 2000, **78**, 1776.
25. C. Nakason, K. Nuansomsri, A. Kaesaman and S. Kiatkamjornwong, *Polym. Test.*, 2006, **25**, 782.
26. C. Nakason, S. Jamjinno, A. Kaesaman and S. Kiatkamjornwong, *Polym. Adv. Technol.*, 2008, **19**, 85.
27. N. M. Mathew, in *Rubber Technologist's Handbook, Volume 1*, ed. S. K. De and J. R. White, Rapra Technology Limited, Shawbury, 2001, p. 28.
28. R. V. Sekharan, T. A. Beena and T. T. Eby, *Mater. Des.*, 2012, **40**, 221.
29. B. Guo, Y. Cao, D. Jia and Q. Qiu, *Macromol. Mater. Eng.*, 2004, **289**, 360.
30. H. Ismail and Suryadiansyah, *Polym. Test.*, 2002, **21**, 389.
31. S. Satapathy, A. Nag and G. B. Nando, *Process Saf. Environ. Prot.*, 2010, **88**, 131.
32. I. Ahmad, C. E. Lane, H. M. Dahlan and I. Abdullah, *Compos. B: Eng.*, 2012, **43**, 3069.
33. N. N. Najib, H. Ismail and A. R. Azura, *Polym.-Plast. Technol. Eng.*, 2009, **48**, 1062.

34. H. Anuar and A. Zuraida, *Composites, Part B*, 2011, **42**, 462.
35. M. Narathichat, C. Kummerlöwe, N. Vennemann and C. Nakason, *J. Appl. Polym. Sci.*, 2011, **121**, 805.
36. P. Suksawad, Y. Yamamoto and S. Kawahara, *Eur. Polym. J.*, 2011, **47**, 330.
37. P. Panu, S. Tantayanon and V. Tangpasuthadol, *Polym. Degrad. Stab.*, 2006, **91**, 3456.
38. V. Tanrattanakul, K. Kosonmetee and P. Laokijcharoen, *J. Appl. Polym. Sci.*, 2009, **112**, 3267.
39. H. Ismail, L. Mega and H. P. S. A. Khalil, *Polym. Int.*, 2001, **50**, 606.
40. C. Nakason, M. Jarnthong, A. Kaesaman and S. Kiatkamjornwong, *J. Appl. Polym. Sci.*, 2008, **109**, 2694.
41. A. K. Bhowmick and J. R. White, *J. Mater. Sci.*, 2002, **37**, 5141.
42. H. Ismail, Salmah and M. Nasir, *Polym. Test.*, 2001, **20**, 819.
43. C. R. Kumar, I. Fuhrmann and J. Karger-Kocsis, *Polym. Degrad. Stab.*, 2002, **76**, 137.
44. S. Anusree, A. Sujith, C. K. Radhakrishnan and G. Unnikrishnan, *Polym. Eng. Sci.*, 2008, **48**, 198.
45. M. Abu-Abdeen and I. Elamer, *Mater. Des.*, 2010, **31**, 808.
46. S. Varghese, R. Alex and B. Kuriakose, *J. Appl. Polym. Sci.*, 2004, **92**, 2063.
47. A. Joseph, S. Lüftl, S. Seidler, S. Thomas and K. Joseph, *Polym. Eng. Sci.*, 2009, **49**, 1332.
48. V. Tanrattanakul, N. Sungthong and P. Raksa, *Polym. Test.*, 2008, **27**, 794.
49. M. Bank, J. Lengwell and C. Thies, *Macromolecules*, 1971, **4**, 43.
50. M. N. Radhakrishnan, G. V. Thomas and M. R. Gopinathan, *Polym. Degrad. Stab.*, 2007, **92**, 189.
51. M. N. Radhakrishnan and M. R. Gopinathan, *Polym. Bull.*, 2006, **56**, 619.
52. A. Sujith and G. Unnikrishnan, *J. Mater. Sci.*, 2005, **40**, 4625.
53. F. Riahi, D. Benachour and A. Douibi, *Int. J. Polym. Mater.*, 2004, **53**, 143.
54. C. Nakason, P. Wannavilai and A. Kaesaman, *Polym. Test.*, 2006, **25**, 34.
55. M. R. Abadchi, A. J. Arani and H. Nazockdast, *J. Appl. Polym. Sci.*, 2010, **115**, 2416.
56. W. Pechurai, C. Nakason and K. Sahakaro, *Polym. Test.*, 2008, **27**, 621.
57. N. P. Cheremisinoff, *Polymer Characterisation: Laboratory Techniques and Analysis*, Noyes Publications, New Jersey, 1996, p. 25.
58. M. M. Magida, Y. H. Gad and H. H. El-Nahas, *J. Appl. Polym. Sci.*, 2009, **114**, 157.
59. M. M. H. Senna, A. A. Abdel-Fattah and Y. K. Abdel-Monem, *Nucl. Instrum. Methods Phys. Res., Sect. B*, 2008, **266**, 2599.
60. A. K. Bhowmick, J. Heslop and J. R. White, *J. Appl. Polym. Sci.*, 2002, **86**, 2393.
61. M. A. Minnath, 'Development And Characterisation Of Thermoplastic Polyurethane-Natural Rubber/Epoxidised Natural Rubber (TPU-NR/ENR) Blends And Their Chitin Reinforced Composites', PhD thesis, Calicut University, 2013.
62. K. K. Indukuri and A. J. Lesser, *Polymer*, 2005, **46**, 7218.
63. C. C. Honeker and E. L. Thomas, *Macromolecules*, 2000, **33**, 9407.

64. J. S. Oh, A. I. Isayev and M. A. Rogunova, *Polymer*, 2003, **44**, 2337.
65. R.-J. Roe, in *Comprehensive Desk Reference of Polymer Characterization and Analysis*, ed. R. F. Brady Jr, Oxford University Press, New York, 2003, p. 338.
66. K. Yamauchi, S. Akasaka and H. Hasegawa, *Compos. A.*, 2005, **36**, 423.
67. C. C. Honeker, E. L. Thomas, R. J. Albalak, D. A. Hajduk, S. M. Gruner and M. C. Capel, *Macromolecules*, 2000, **33**, 9395.
68. I. Walker and A. A. Collyer, in *Plastics*, ed. A. A. Collyer, Chapman & Hall, London, 1994, p. 161.
69. C. B. Bucknall and R. J. Gaymans, in *Polymer Blends, Vol. 2*, eds. D. R. Paul and C. B. Bucknall, Wiley, New York, 2000, **6**, 177.
70. G. Christelle, *Adv. Polym. Sci.*, 2005, **43**, 188.
71. C. Nakason, S. Saiwari and A. Kaesaman, *Polym. Test.*, 2006, **25**, 413.
72. M. C. Mary, A. D. Douglas, S. Qilong, A. O. Susan, R. S. Nancy, R. W. Scott and S. M. Jeffrey, *Chem. Rev.*, 2009, **109**, 5755.
73. W. W. Sukowski, A. Danch, M. Moczyski, A. Rado, A. Sukowska and J. Borek, *J. Therm. Anal. Calorim.*, 2004, **78**, 905.
74. M. Ginic-Markovic, N. Roy Choundhry, J. G. Matisons and N. Dutta, *J. Therm. Anal. Calorim.*, 2001, **56**, 943.
75. H. Varghese, S. S. Bhagawan and S. Thomas, *J. Therm. Anal. Calorim.*, 2001, **63**, 749.
76. M. Subhra, P. G. Mukunda and G. B. Nando, *Polym. Degrad. Stab.*, 1995, **50**, 21.
77. P. M. Aji, S. Packirisamy and S. Thomas, *Polym. Degrad. Stab.*, 2001, **72**, 423.
78. A. Joseph, T. Koch, S. Seidler, S. Thomas and K. Joseph, *J. Appl. Polym. Sci.*, 2008, **109**, 1714.
79. M. N. Radhakrishnan, G. V. Thomas and M. R. Gopinathan, *Polym. Bull.*, 2007, **59**, 413.
80. M. C. S. Perera, U. S. Ishiaku and Z. A. M. Ishak, *Polym. Degrad. Stab.*, 2000, **68**, 393.
81. R.-S. Shih, *Polymer.*, 2011, **52**, 752.
82. W. S. Tait, in *Comprehensive Desk Reference of Polymer Characterization and Analysis*, ed. R. F. Brady Jr, Oxford University Press, New York, 2003, p. 669.
83. A. T. Riga, W-P. Pan, J. Cahoon. in *Comprehensive Desk Reference of Polymer Characterization and Analysis*, ed. R. F. Brady. Jr, Oxford University Press, New York, 2003, p. 325.
84. A. W. Birley, B. Hayworth and J. Batchelor, *Physics of Plastics: Processing, Properties and Materials Engineering*, Carl Hanser Publishers, Munich, 1992, p. 98.
85. A. P. Mathew, 'Interpenetrating Polymer Networks Based on Natural Rubber and Polystyrene', PhD thesis, M G University, 1999.
86. J. Jobish, C. Nakason and P. Praveen, *J. Non-Cryst. Solids*, 2012, **358**, 1113.
87. J. George, K. T. Varughese and S. Thomas, *Polymer*, 2000, **41**, 1507.
88. K. Chatterjee and K. Naskar, *eXPRESS Polym. Lett.*, 2007, **1**, 527.
89. Z. Oommen and S. Thomas, *J. Appl. Polym. Sci.*, 1997, **65**, 1245.
90. D. J. Elliott, *NR Technol.*, 1981, **12**, 59.

CHAPTER 12

Radiation Processing of Natural Rubber with Vinyl Plastics

CHANTARA THEVY RATNAM,*[a] ZURINA MOHAMAD[b] AND MOHAMMAD KHALID SIDDIQUI[c]

[a] Malaysian Nuclear Agency, Bangi, 43000 Kajang, Selangor DE, Malaysia; [b] Polymer Engineering Department, Faculty of Chemical Engineering, University Teknologi Malaysia, 81310 Skudai, Johor, Malaysia; [c] Faculty of Engineering, The University of Nottingham Malaysia Campus, Jalan Broga, 43500 Semenyih, Selangor DE, Malaysia
*Email: Chantara@nuclearmalaysia.gov.my

12.1 Introduction

Ionizing radiation has been found very useful in modifying the structure and properties of polymeric materials and it has been an area of enormous interest in the last few decades. Ionizing radiation has the unique ability to initiate polymerization and/or crosslinking reactions with the minimum of toxic chemicals, and so radiation processing is emerging as an excellent tool to produce polymeric materials with unique properties. Radiation processing is an alternative to conventional methods such as thermal and chemical processing in many industrial applications. Apart from polymerization and crosslinking, the irradiation-induced reactions in polymers may lead to degradation and/or grafting.

12.2 Radiation Effects on Polymers

When a polymer is subjected to high-energy radiation such as gamma rays and accelerated electron beams, it may undergo various effects, as shown in Table 12.1. These lead to grafting, crosslinking or degradation of the polymer, depending on its structure and irradiation conditions. Thus most polymers appear to fall into two distinct classes: those that crosslink and those that degrade.[1,2] Table 12.2 shows the classification of some polymers into the two groups when the polymer is irradiated at ambient temperature in a vacuum. Oxidative degradation may occur in the presence of oxygen. Polymers having a chain repeat structure of the type $-(CH_2CHR-)_n$ or $-(CH_2CH_2-)_n$ such as polyethylene, polystyrene (PS) and polyvinyl chloride (PVC) tend to crosslink upon exposure to irradiation. On the other hand, vinyl polymers in which there are two side chains attached to a single carbon, $-(CH_2CR_1R_2-)_n$, such as polyisobutylene, polymethacrylates and polyvinylidene chloride, tend to degrade upon irradiation.[3] All natural polymers, such as cellulose and other polysaccharides, are easily decomposed by radiation. Interestingly, natural rubber and its isomer gutta-percha are crosslinkable upon radiation.

12.3 Radiation Crosslinking of Polymers

The mechanism of irradiation crosslinking of polyolefins is similar to peroxide crosslinking. In both cases, the crosslinks are formed by a combination of alkyl radicals. The general view of radiation-induced reactions/oxidation in polymers can be found in the literature.[1,2] The advantages of radiation crosslinking as

Table 12.1 Effects of radiation on polymers.

Formation of free radicals
Formation of hydrogen and low molecular weight hydrocarbon
Formation of C–C bonds between molecules (crosslinking, grafting)
Cleavage of C–C bond (chain scission)
Increase in unsaturation
Breakdown of crystalline structure
Colouration
Oxidation

Table 12.2 Classification of some irradiated polymers.[1]

Crosslinking polymers	Degrading polymers
Polyethylene	Polyisobutylene
Polystyrene	Poly(α-methylstyrene)
Polyacrylate	Polymethacrylates
Polyacrylamide	Polymethacrylamide
Polyvinyl chloride	Poly(vinylidene chloride)
Polyamide	Cellulose and derivatives
Polyesters	Polytetrafluoroethylene
Polyvinylpyrrolidone	Polytrifluorochloroethylene
Rubbers	

compared to conventional chemical crosslinking are cost, speed, the ability to crosslink performed parts at or near room temperature, reduction in chemical ingredients and chemical residues for environmental or toxicological reasons and in many cases, superior material properties in the final product.[4] In a number of blend systems, irradiation has been used to induce crosslinking of one or more of the components, and/or the formation of crosslinks between the different phases, resulting in an improvement in physical properties. Crosslinking can also be used for fixing of non-equilibrium blend morphology.[4,5]

12.4 Radiation Sensitizers used as Crosslinking Agents

Radiation sensitizers are multifunctional vinyl monomers (MFA) that are highly reactive towards free radicals. Since all common MFAs contain terminal unsaturation, it can be expected that addition/polymerization is the principal mechanism by which they react in the polymer compound. These additives are used mainly to accelerate the radiation-induced crosslinking in the polymers. The addition of MFA to the polymer formulations suppresses the chain scission reactions and allows more crosslinking to occur.

12.5 Radiation Crosslinking of Natural Rubber (NR)

The structural formula of natural rubber (NR) is shown in Scheme 12.1. NR is sticky and non-elastic by nature. Crosslinking is a reaction of polymers to form a three-dimensional network. Crosslinking of rubber is called vulcanization. The crosslinking of NR makes it elastic. When NR is irradiated by high-energy radiation, hydrogen atoms of the trunk chain, mainly of methylene groups α to double bonds, are ejected and radical sites are formed (Scheme 12.2), and these radical sites are combined into C–C crosslinks.[6]

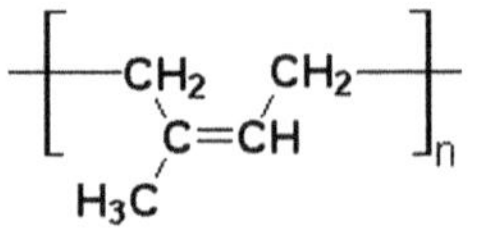

Scheme 12.1 Structural formulae of NR (*cis*-1,4-polyisoprene).

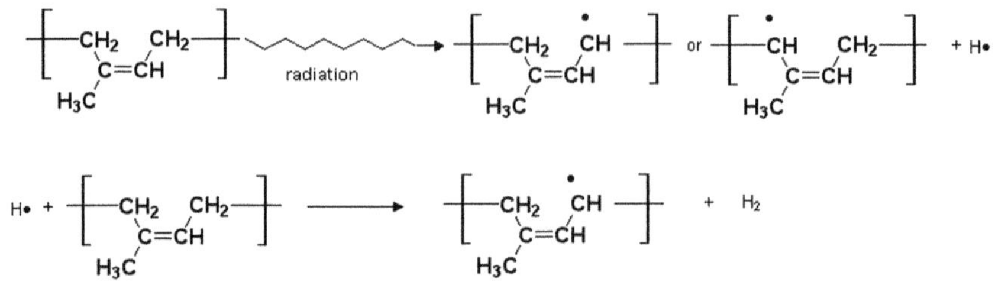

Scheme 12.2 NR radical formation by radiation.[7]

The addition of NR radicals to unsaturated C=C bonds forms crosslinks but the radiation crosslinking efficiency of NR is not high. This is believed to be due to loose packing of NR molecules with the *cis* structure and the presence of the methyl group. The structure of radiation-crosslinked NR shown in Scheme 12.3.[7]

12.5.1 The Properties of Radiation Crosslinked NR

In 1954, Charlesby[6] investigated the radiolysis of NR (smoked sheets) under reactor radiation. The distance between crosslinks (M_c) was found to decrease linearly with the radiation dose, indicating that the extent of crosslinking increased linearly with dose.

Crosslinking causes changes in the physical properties of NR, as summarized in Table 12.3.[7] Generally, the tensile strength and modulus increase on crosslinking, while the elongation at break decreases. Heat and solvent resistance are improved by crosslinking.

The properties of radiation vulcanized NR have been studied by a number of authors and compared with those of conventional vulcanizates.[7] It has been reported that radiation vulcanized rubber has inferior physical properties compared to sulfur vulcanizates. The maximum tensile strength achieved by the vulcanizates decreases in the order of sulfur vulcanizate > peroxide vulcanizate > radiation vulcanizate. Such differences in properties are attributed to the vulcanizate structure and bond energy. It has also been shown that radiation

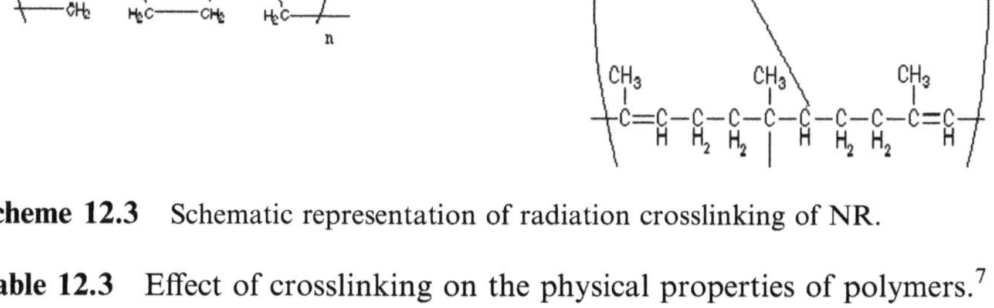

Scheme 12.3 Schematic representation of radiation crosslinking of NR.

Table 12.3 Effect of crosslinking on the physical properties of polymers.[7]

Increase	Decrease	No change
Tensile strength	Elongation at break	Electrical properties
Elastic modulus	Surface tackiness	
Melt viscosity		
Heat resistance		
Solvent resistance		

crosslinking isomerizes NR, which reduces its ability to crystallize. Lower crystallinity would result in lower tensile strength. The lower tensile strength of irradiated vulcanizate compared to sulfur vulcanized NR is also associated with the main-chain scissions of rubber molecules by radiation and ozone forced by irradiation in air. The addition of MFAs was found to be effective in preventing main-chain scissions and accelerating radiation-induced crosslinking. Upon addition of MFAs the tensile strength of radiation crosslinked NR was found to show a remarkable increase. Table 12.4 shows the physical properties of three vulcanization methods of NR containing 45 phr carbon black.[8]

It was found that sulfur, a classical vulcanizing agent, does not promote radiation crosslinking and even markedly inhibits the process, presumably as a result of free radical scavenging.[8]

On the other hand, carbon blacks are very efficient at promoting crosslinking of either NR, polybutadiene or nitrile rubber.[9]

12.6 Radiation Crosslinking of Epoxidized Natural Rubber (ENR)

The successful epoxidation of NR, by employing a hydrogen peroxide/formic acid system under controlled conditions, has ensured epoxidized rubber (ENR) its place in the elastomer market. Essentially, ENR of any desired level of epoxidation ranging from 1 to 90% can be prepared. Increasing the level of epoxidation systematically changes the properties of the rubber. The increase in T_g and Mooney viscosity of ENR with increase in epoxidation level is indicated in Table 12.5.[10]

Currently ENR 10, 25, 50 and 60 (10, 25, 50 and 60% mole epoxidation, respectively) can be found on the market. Scheme 12.4 shows the structural formulae of ENR.

The presence of an oxirane group in ENR was found to be effective in causing specific interaction with a second polymer. ENR exhibits improved oil resistance and decreased gas permeability, while retaining many of the properties of NR and also exhibiting some novel features.

Table 12.4 Tensile properties of NR vulcanized by sulfur, peroxide and radiation.[8]

Property	Vulcanization system used		
	Sulfur	Peroxide	Radiation
Tensile strength (Mpa)	24.5	20.0	24.0
Elongation at break (%)	510	450	430

Table 12.5 Typical properties of ENR and SMR L.[10]

Properties	SMR L	ENR 10	ENR 25	ENR 50
Glass transition temperature, T_g (°C)	−69	−60	−45	−25
Specific gravity	0.93	0.94	0.97	1.03
Mooney viscosity, $M_{L,\ 1+4}$ (100 °C)	50–70	90	110	140

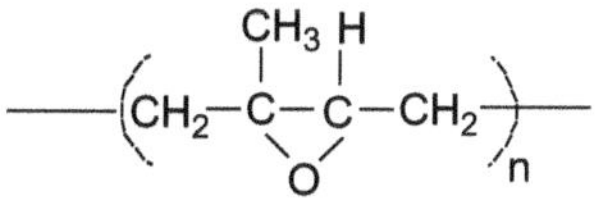

Scheme 12.4 Structural formulae of ENR.

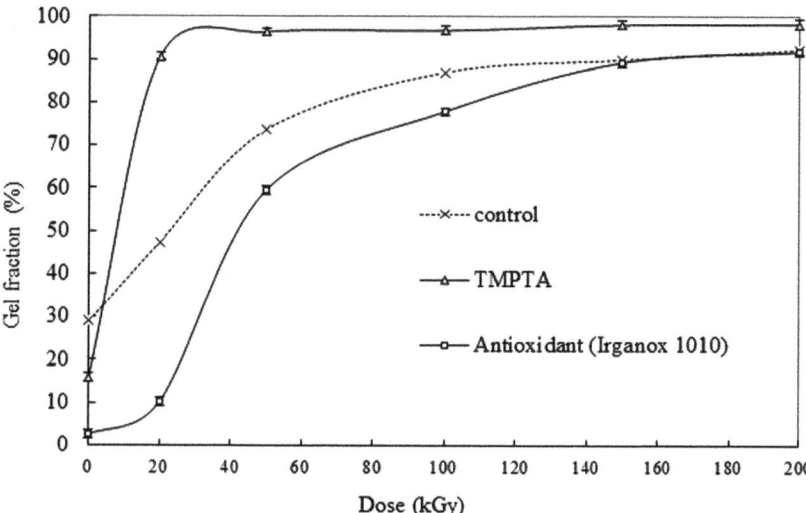

Figure 12.1 Effect of irradiation on the gel fraction of ENR50.

The effect of electron beam irradiation on ENR was reported by Ratnam et al.[11–13] Figure 12.1 demonstrates the effect of various additives on the gel fraction of ENR.[13] The effect of electron beam irradiation on the infrared spectra[10] and dynamic mechanical properties[11] of ENR is discussed in the literature.

12.7 Radiation Crosslinking of NR Based Blends

Studies over the past two decades have established that irradiation can be very useful in the processing of polymer blends.[14,15] Improvement in blend properties has been achieved through several different approaches, including crosslinking, scission, compatibilization, and morphology stabilization. Radiation crosslinking can be applied to a great number of polymer blends, which includes blends of NR and thermoplastics such as EVA, PVC, LLDPE and PP. Thus, in recent years, irradiation modification of blends has generated interest among researchers in the field. The enhancement in blend properties can be attributed to irradiation-induced crosslinking, along with irradiation-induced interactions between the polymers that can impart miscibility to incompatible or partially miscible blends. Studies of PVC/ENR, EVA/ENR and PVC/NR indicated that the enhancement in blend properties can be achieved through radiation-induced crosslinking.

12.7.1 Radiation Crosslinking of PVC/ENR Blends

The electron beam crosslinking of PVC/ENR blends by electron beam irradiation has been extensively reported by Ratnam and coworkers.[16–25] The influence of MFAs such as trimethylolpropane trimethacrylate (TMPTA), 1,6-hexaediol diacrylate (HDDA) and 2-ethylhexyl acrylate (EHA), as well as acrylated polyurethane oligomer on the 70/30 PVC/ENR blend, was investigated.[20] TMPTA was found to give the blend the best mechanical properties. Studies were also done on the effect of irradiation on various formulations of PVC/ENR blends with doses ranging from 50 to 200 kGy; these studies included the influence of various additives on the efficiency of the radiation crosslinking of the blends.

The variations in gel content with irradiation dose for 50/50 PVC/ENR containing various additives are shown in Figure 12.2. Apparently, the dose required to achieve 70% gel fraction increases in the following order: TMPTA < Irganox 1010 < TBLS.

Obviously, the gelation is induced at a lower dose when TMPTA is present in the compound. Such an observation is expected as the TMPTA employed in this study is well known as a reactive additive that forms crosslink bridges by an irradiation-induced free radical mechanism.[26,27] The increase in the dose to

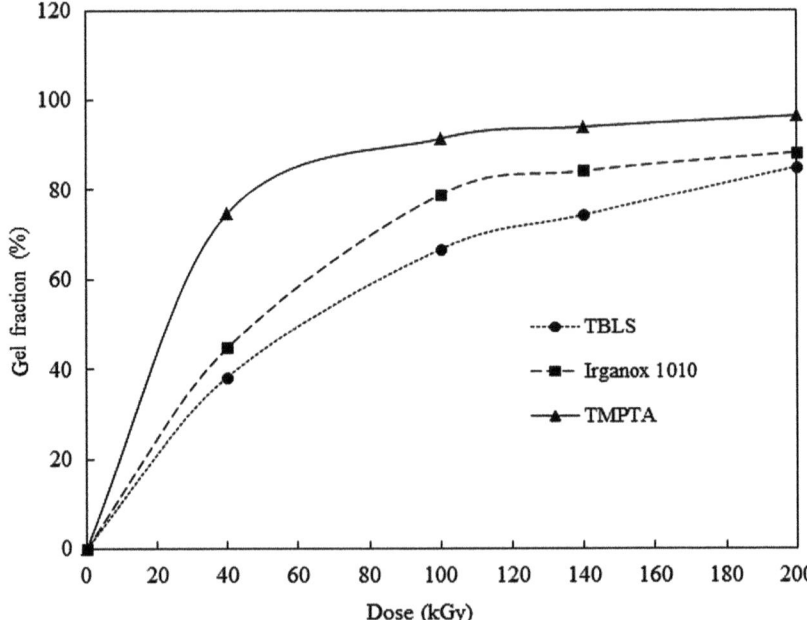

Figure 12.2 Effect of irradiation on the gel fraction of 50/50 PVC/ENR blend in the presence of additives.

achieve 70% gel fraction with the addition of Irganox 1010 and tribasic lead sulfate indicates the inhibition of irradiation-induced crosslinking in the blend. It has been reported that the most noticeable effect of the addition of antioxidants to polymers is the deactivation of the free radicals induced by the irradiation which otherwise could react to form crosslinks.[28]

Changes in the temperature dependence of tan δ and storage modulus of irradiated 50/50 PVC/ENR in the present of additives are shown in Figures 12.3 and 12.4. The temperature corresponding to the maximum in tan δ (tan $δ_{max}$) is taken as the T_g of the blend. At the dose studied, the 50/50 PVC/ENR blend exhibits only a single T_g corresponding to the cooperative segmental motions of the molecular chains. This suggests the single-phase morphology of the PVC/ENR blend is unchanged upon irradiation. Storage modulus rapidly decreases at the T_g zone due to the decrease in stiffness of the blend.

It is apparent from Figure 12.3 that the T_g of the blend shows the highest value with the addition of TMPTA and its lowest values in the presence of TBLS. This observation is in perfect agreement with the gel fraction results and is believed to be associated with the crosslinking induced by the irradiation. Reports on the relationship between the crosslink density and the T_g can be found elsewhere.[29] A similar trend in storage modulus with the addition of TMPTA, Irganox 1010 and TBLS can also be seen in Figure 12.4. The tan $δ_{max}$ (Figure 12.3) also exhibits its highest value for blends incorporated with TBLS and its lowest value for blends with added TMPTA. This is again attributed to the microstructural changes that have occurred through crosslink formation *via* the irradiation-induced reactions. The lower tan $δ_{max}$ is associated with higher crosslink density, as an elastic system should dissipate less energy as heat.

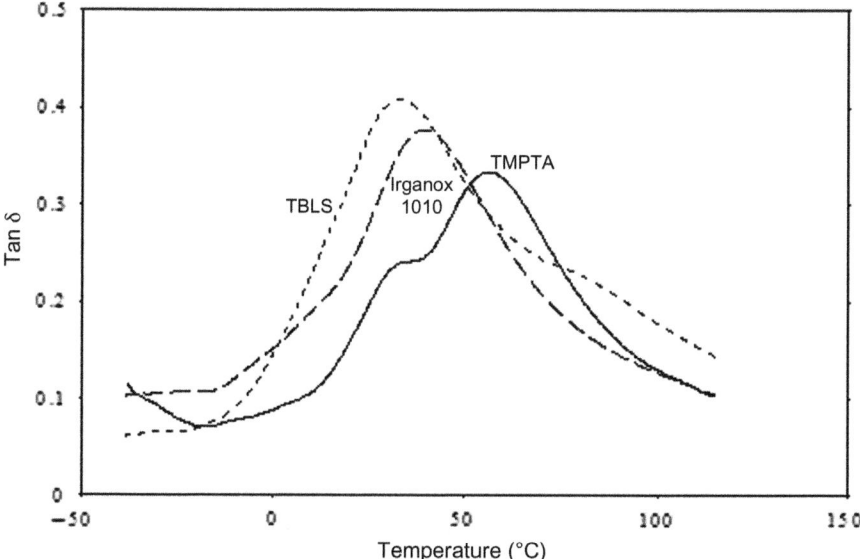

Figure 12.3 Temperature dependence of tan δ of 50/50 PVC/ENR blend irradiated at 100 kGy.

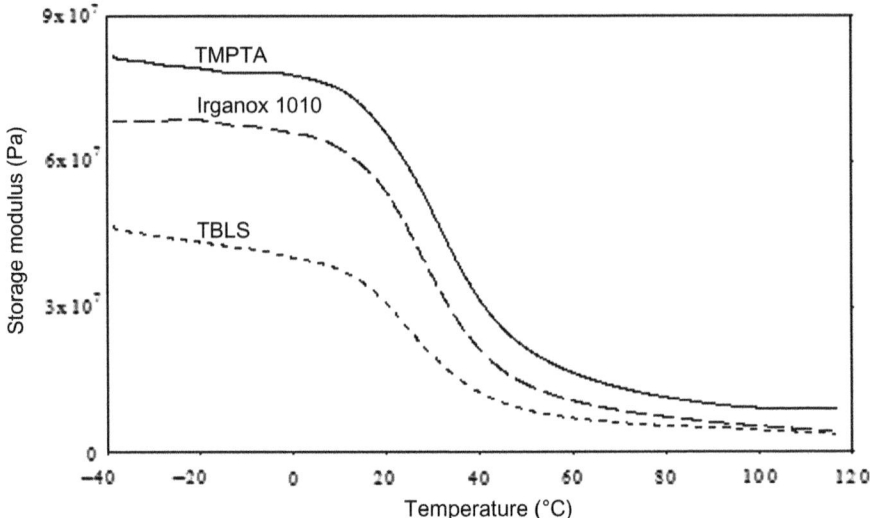

Figure 12.4 Temperature dependence of storage modulus of 50/50 PVC/ENR blend irradiated at 100 kGy.

Studies of the dynamic mechanical properties of PVC/ENR blends indicated that the compatibility of the system is improved upon radiation crosslinking.[24] Thus, akin to any other polymer system, the properties of NR based polymer blends can be improved by irradiation crosslinking. However, controlling the amount of scission and crosslinking in either of the two phases, as well as reactions at the interfaces, is of utmost importance for morphology fixation and property enhancement of the blend system. Generally, as a result of crosslinking the tensile strength, modulus and elasticity of polymers increases whilst the elongation at break decreases. On the other hand, chain scission leads to a decrease in the tensile strength, modulus and elasticity of polymers.

12.7.2 Radiation Crosslinking of EVA/ENR Blends

Ethylene vinyl acetate copolymers (EVA) are randomly structured polymers that offer excellent ozone resistance, weather resistance and excellent mechanical properties.[30] There has been growing interest in NR/EVA (polyethylene-co-vinyl acetate) in recent years. Koshy et al.[31] studied the influence of blend ratio and cure system on the degradation of NR/EVA blends and other studies[32] indicated that silica filled NR/EVA blends have higher storage and loss modulus. Jansen and co-workers[33] evaluated the mechanical and thermal properties of crosslinked blends of NR/EVA. Kannan et al.[34] investigated the structure–properties of EVA/ENR blends using FTIR and DMA techniques. They found that the blends showed high tensile strength and modulus due to higher levels of phase interactions. More recently, Zurina et al. reported observations on the properties on EVA/ENR blends that include the effect of blend ratio[35] and dynamic crosslinking.[36] Several attempts were made to investigate the effect of electron beam irradiation on the tensile properties, morphology and thermal properties of EVA/ENR blends.[37–40]

The variations in gel content with irradiation dose for 50/50 EVA/ENR blends and the blend containing the crosslinking agent TMPTA are shown in Figure 12.5.[37] The general rise in gel fraction with irradiation dose suggests that crosslinking may be the prevailing irradiation-induced reactions compared to chain scission. The acceleration of radiation-induced crosslinking in EVA/ENR blends upon addition of TMPTA, tripropyleneglycol diacrylate (TPGDA), HVA-2 and Surlyn ionomer has been reported.[37,40]

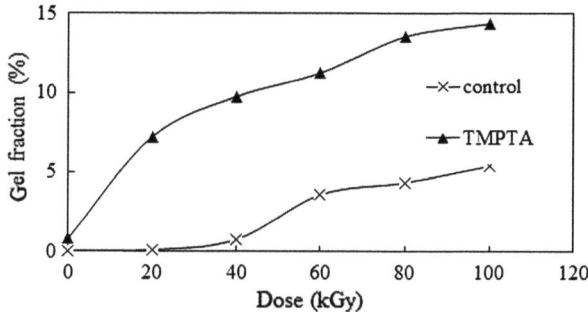

Figure 12.5 Effect of irradiation on the gel fraction of 50/50 EVA/ENR blend.

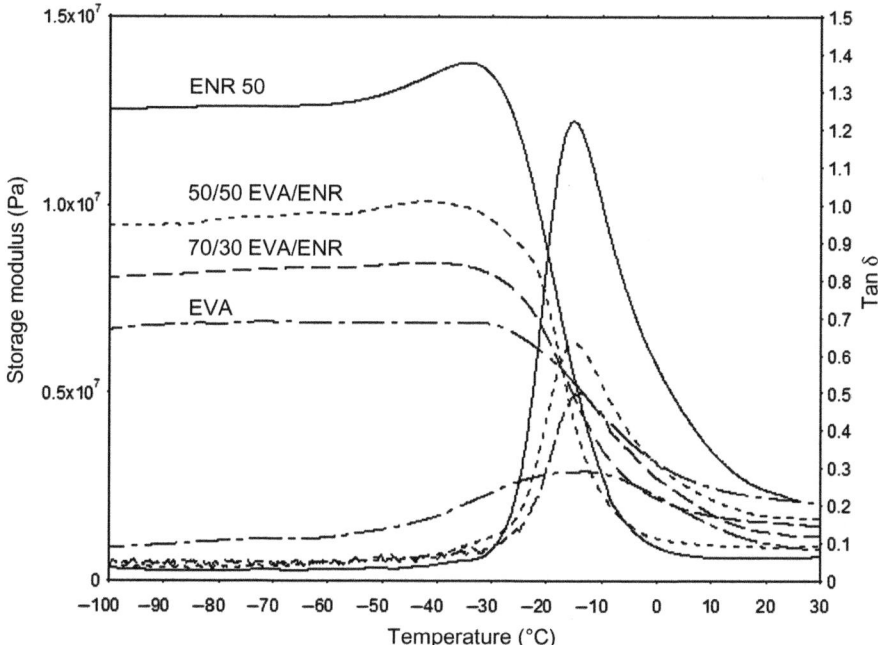

Figure 12.6 Temperature dependence of tan δ and storage modulus of EVA, ENR and EVA/ENR blends.

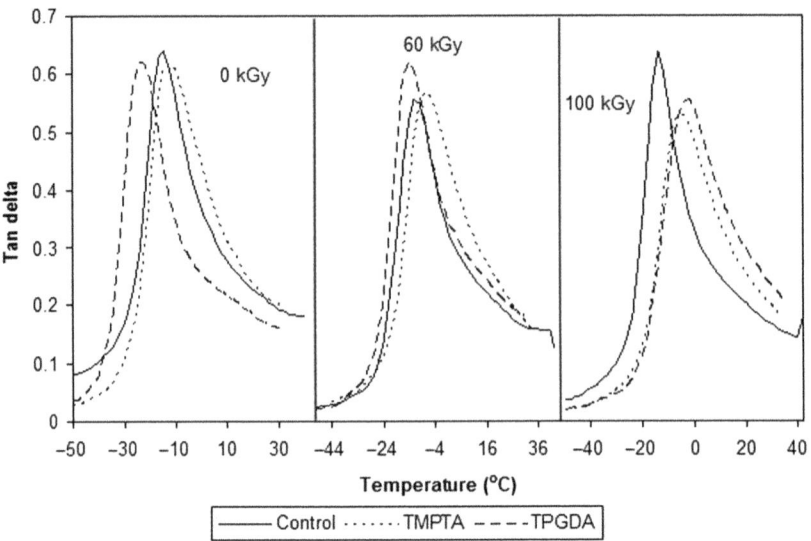

Figure 12.7 The effect of irradiation and addition of MFA on the temperature dependence of tan δ of 50/50 EVA/ENR blends.

The temperature dependence of tan δ and storage modulus of EVA, ENR and EVA/ENR blends is shown in Figure 12.6. A single T_g obtained for the 70/30 and 50/50 EVA/ENR blends indicates good compatibility between EVA and ENR.[37]

Figure 12.7 shows the effect of electron beam radiation and addition of MFA on the temperature dependence of tan δ of 50/50 EVA/ENR blends. The reduction in the T_g of unirradiated blends upon addition of TPGDA is due to its lubricant action; the addition of TMPTA increases the T_g of the unirradiated blend. This suggests that good interaction between TMPTA and the EVA/ENR matrix was achieved even before radiation. A shift in the T_g of EVA/ENR blends to high temperatures following electron beam irradiation was associated with the occurrence of radiation-induced crosslinking as well as improvement in the compatibility of the blend upon irradiation.[38] The elevation of T_g upon irradiation of the blends at 100 kGy in the presence of TMPTA and TPGDA was attributed to acceleration of the radiation-induced crosslinking in the EVA/ENR blend matrix by MFA (TMPTA and TPDGA).

Similar studies on ENR-50/EVA blends (50/50) containing N,N'-m-phenylene dimaleimide (HVA-2) indicated that it is effective in enhancing the radiation-induced crosslinking of the ENR-50 phase when compared to EVA. The irradiation-induced crosslinking in ENR/EVA blends in the presence of HVA-2 has also improved the compatibility of the blend.[39] A possible mechanism of crosslinking for ENR-50 and EVA with HVA-2 has been proposed.[39]

12.7.3 Radiation Crosslinking of PVC/NR Blends

Blends of NR and PVC were found to show separate T_gs, which indicates they are not compatible.[41] A more recent work by Radhakrishnan and

Gopinathan[42] revealed that the enhancement in compatibility between NR and PVC could be achieved by using NR/polyurethane block copolymer. Studies focusing on the effect of electron beam irradiation on the physical properties of PVC/ENR/NR ternary blends and PVC/NR blends were conducted with particular attention to irradiation-induced crosslinking. Figure 12.8 shows the effect of radiation on the gel fraction of ENR/NR/PVC ternary blends at various NR and ENR compositions. It is clear from Figure 12.8 that the gel fraction increases with the increase in radiation dose, indicating that the radiation-induced crosslinking has occurred in the ternary blend in the same way as in the binary PVC/ENR blend system.[16,20] Another salient point to be noted in Figure 12.8 is the drop in gel fraction with the increase in NR content. This trend is believed to be associated with the lower degree of crosslinking formed in the NR phase as compared to the ENR phase.

Figure 12.9 illustrates the effect of radiation on the T_s of an ENR/NR/PVC ternary blend containing 30% NR. It is clear from Figure 12.9 that an enhancement in the T_s of the ENR/NR/PVC of the ternary blend has occurred upon irradiation. Such enhancement in T_s is attributed to the formation of

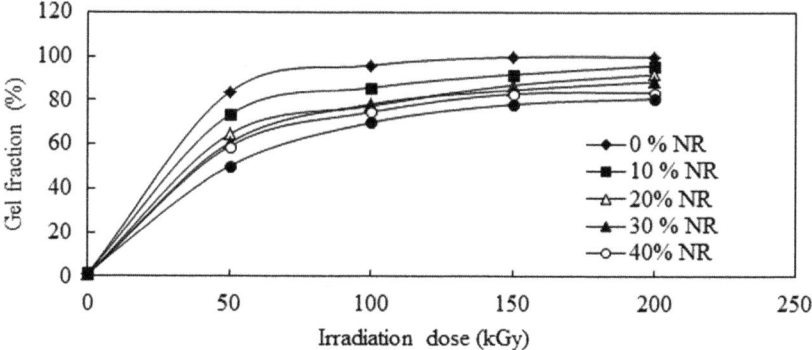

Figure 12.8 Effect of irradiation on the gel fraction of ENR/NR/PVC ternary blends.

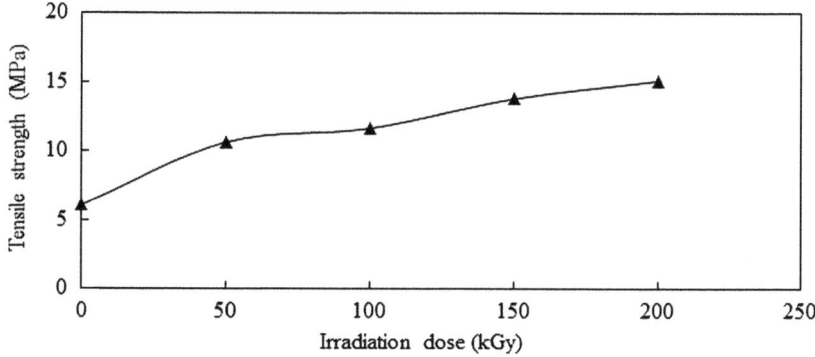

Figure 12.9 The effect of irradiation on the T_s of ENR/NR/PVC ternary blend containing 30% NR.

radiation-induced crosslinking in the blend, as evidenced from the gel fraction values.

The photomicrographs in Figure 12.10 show the morphological development that occurs as a consequence of irradiation for the ENR/NR/PVC ternary blend with 30% NR content. Prior to radiation the 30% NR blend showed the presence of fibrils on the fracture surface. The fracture surfaces of the irradiated samples appear rougher and more brittle compared to the unirradiated sample. The fracture paths are elongated and penetrate deeply into the material. No fibrils were observed after radiation. Such changes in fractured surfaces upon irradiation are attributed to the irradiation-induced crosslinking of the ternary blend.

Attempts were also made to study the acceleration of radiation-induced crosslinking in the PVC/NR blend system in the same way as PVC/ENR[22] and EVA/ENR.[37,40] It is evident from Figure 12.11 that addition of TMPTA significantly accelerates the irradiation-induced crosslinking in the PVC/NR blends, as observed for PVC/ENR[22] and EVA/ENR[37,40] blends. Figure 12.12 clearly shows that remarkable enhancement in the T_s has occurred upon irradiation of the NR/PVC blend in the presence of TMPTA. Such a trend is attributed to the formation of radiation-induced crosslinks and acceleration of the radiation-induced crosslinking by the TMPTA.

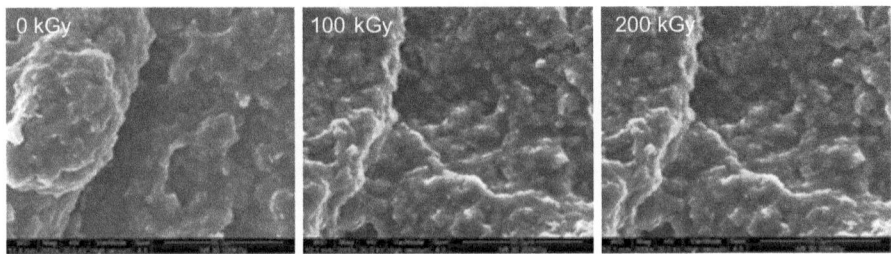

Figure 12.10 SEM micrographs of cryogenic fracture surfaces of ENR/NR/PVC ternary blend containing 30% NR (magnification 2000×).

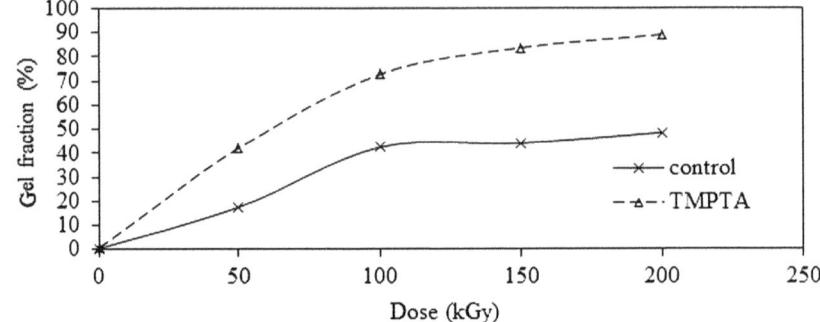

Figure 12.11 Effect of irradiation on the gel fraction of 50/50 NR/PVC blend.

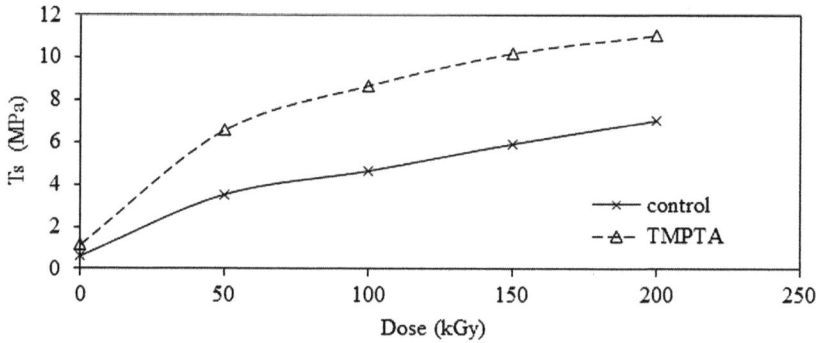

Figure 12.12 Effect of irradiation on the T_s of 50/50 NR/PVC blend.

12.8 Conclusions

Studies on the effect of electron beam irradiation on PVC/ENR, EVA/ENR and PVC/NR blends indicated that the blends undergo crosslinking upon irradiation. An acceleration of crosslinking occurs in the blends with the addition of MFA. Thus, akin to any other polymer system, the properties of NR based polymer blends can be improved by irradiation-induced crosslinking. However, control of the amount of scission and crosslinking in either of the two phases, as well as reactions at the interfaces, is of utmost importance for morphology fixation and property enhancement of the blend system.

Acknowledgements

The authors wish to thank the Ministry of Science, Technology and Innovation for sponsoring this work under an IRPA grant (IRPA 03-01-03-0013 EA001). We appreciate the cooperation given by ALURTRON staff of Nuclear Malaysia, Mr Wan Ali and Mr M. N. Falah.

References

1. A. Charlesby, *Atomic Radiation and Polymers*, Pergamon Press, London, 1960, pp. 134–158.
2. A. Chapiro, *Radiation Chemistry of Polymeric Systems*, Pergamon Press, Oxford, 1962, pp. 339–384.
3. A. A. Miller, E. J. Lawton and J. S. Balwit, *J. Polym. Sci.*, 1954, **14**, 503.
4. R. L. Clough, *Nucl. Instrum. Methods Phys. Res., Sect. B*, 2001, **185**, 8.
5. A. Singh and K. Bahari, in *Polymer Blend Handbook*, ed. L. A. Utracki, Kluwer Academic Publishers, London, 2002, pp. 757–854.
6. A. Charlesby, *Atomics*, 1954, **5**, 12.
7. K. Makuuchi, *An Introduction to RVNRL*, R.R.I Global Co. Ltd, Thailand, 2003, pp. 1–40.

8. A. A. Basfar, M. M. Abdel-Aziz and S. Mofti, *Radiat. Phys. Chem.*, 2002, **63**, 81.
9. A. Charlesby, J. Burrows and T. Bain, *The Rheology of Elastomers*, Pergamon Press, Welwyn Garden City, 1958, pp. 122–149.
10. H. Ismail, *Epoxidized Natural Rubber*, University Science Malaysia Publishers, Penang Malaysia, 2004, pp. 1–36.
11. C. T. Ratnam, M. Nasir, A. Baharin and K. Zaman, *Polym. Int.*, 2000, **49**, 1693.
12. C. T. Ratnam, M. Nasir, A. Baharin and K. Zaman, *Eur. Polym. J.*, 2001, **37**, 1667.
13. C. T. Ratnam, M. Nasir, A. Baharin and K. Zaman, *Nucl. Instrum. Methods Phys. Res., Sect. B*, 2000, **171**, 455.
14. L. Spanadel, *Radiat. Phys. Chem.*, 1979, **14**, 683.
15. S. Akhtar, P. P. De and S. K. De, *J. Appl. Polym. Sci.*, 1986, **32**, 4169.
16. C. T. Ratnam and K. Zaman, *Angew. Makromol. Chemie*, 1999, **269**, 42.
17. C. T. Ratnam and K. Zaman, *J. Sains Nukl. Malays.*, 1999, **17**, 37.
18. C. T. Ratnam and K. Zaman, *Polym. Degrad. Stab.*, 1999, **65**, 99.
19. C. T. Ratnam and K. Zaman, *Polym. Degrad. Stab.*, 1999, **65**, 481.
20. C. T. Ratnam and K. Zaman, *Nucl. Instrum. Methods Phys. Res., Sect. B*, 1999, **152**, 335.
21. C. T. Ratnam and K. Zaman, *J. Appl. Polym. Sci.*, 2001, **81**, 1914.
22. C. T. Ratnam, M. Nasir, A. Baharin and K. Zaman, *J. Appl. Polym. Sci.*, 2001, **81**, 1926.
23. C. T. Ratnam, M. Nasir, A. Baharin and K. Zaman, *Polym. Degrad. Stab.*, 2001, **72**, 147.
24. C. T. Ratnam, M. Nasir, A. Baharin and K. Zaman, *Polym. Int.*, 2001, **50**, 503.
25. C. T. Ratnam, M. Nasir, A. Baharin and K. Zaman, *Polym.-Plast. Technol. Eng.*, 2001, **40**, 561.
26. W. A. Salmon and L. D. Loan, *J. Appl. Polym. Sci.*, 1972, **16**, 671.
27. T. Czvikovszky, *Period. Polytech., Mech. Eng.*, 1994, **38**, 209.
28. W. W. Simons, *The Sadler Handbook of Infrared Spectra*, Sadler, Philadelphia, 1978.
29. T. G. Fox and S. Loshaek, *J. Polym. Sci.*, 1955, **15**, 371.
30. K. W. Doak, in *Encyclopedia of Polymer Science and Engineering*, 2nd edn., H. F. Mark, N. M. Bikales, C. G. Overbeger and G. Menges (eds.), Wiley, New York, 1985, Vol. 6, pp. 386–429.
31. A. T. Koshy, B. Kuriakose and S. Thomas, *Polym. Degrad. Stab.*, 1992, **36**, 137.
32. A. T. Koshy, B. Kuriakose, S. Thomas and S. Varughese, *Polym.-Plast. Technol. Eng.*, 1994, **33**, 149.
33. P. Jansen, A. S. Gomes and B. G. Soares, *J. Appl. Polym. Sci.*, 1996, **61**, 591.
34. S. Kannan, G. B. Nando, A. K. Bhowmick and N. M. Mathew, *J. Elastomers Plast.*, 1995, **27**, 268.

35. M. Zurina, H. Ismail and C. T. Ratnam, *J. Appl. Polym. Sci.*, 2006, **99**, 1504.
36. M. Zurina, H. Ismail and C. T. Ratnam, *Polym.-Plast. Technol. Eng.*, 2008, **47**, 1.
37. C. T. Ratnam, Z. Abdullah and H. Ismail, *Polym.-Plast. Technol. Eng.*, 2006, **45**, 555.
38. M. Zurina, H. Ismail and C. T. Ratnam, *Polym. Degrad. Stab.*, 2006, **91**, 2723.
39. M. Zurina, H. Ismail and C. T. Ratnam, *Polym. Test*, 2008, **27**, 480.
40. M. Zurina, H. Ismail and C. T. Ratnam, *J. Vinyl Addit. Technol.*, 2009, **15**, 47.
41. R. S. Popovic and R. G. Popovic, *Kautsch. Gummi Kunstst.*, 1999, **54**, 254.
42. M. N. Radhakrishnan Nair and M. R. Gopinathan Nair, *Polym. Bull.*, 2006, **56**, 619.

CHAPTER 13

Blends and IPNs of Natural Rubber with Acrylic Plastics

WIWAT PICHAYAKORN,* JIRAPORNCHAI SUKSAEREE AND PRAPAPORN BOONME

Department of Pharmaceutical Technology, Faculty of Pharmaceutical Sciences, Prince of Songkla University, Songkhla 90112, Thailand
*Email: wiwat.p@psu.ac.th

13.1 Introduction

Acrylate polymers, commonly known as acrylics or polyacrylates, are one group of polymers that can be referred to generally as plastics. They are synthetic compounds produced from the bonding of one or more acrylic acids or acrylic esters with other molecules. Polyacrylate and poly(methyl methacrylate) are well-known examples of these polymers. The typical monomers of acrylic acid and acrylic esters (Figure 13.1) are the simplest unsaturated carboxylic compounds, consisting of the carbonyl carbon directly connected to a vinyl group, which forms an acrylate polymer. They are easily formed because their double bonds are very reactive. Acrylate polymers are highly transparent, highly gas and chemical resistant, and are more resistant to impact than glass. They are characteristically superior in their oil and heat resistance, which has made them the materials of choice for such situations. They are available for use in a variety of sizes, shapes, colours and textures. They are used as dispersants, thickeners, adhesives, binders, and vehicles for paint. Moreover, they are produced industrially on a large scale as precursors to conversion esters. However, they can be easily scratched or cracked, and are more expensive than other polymers,

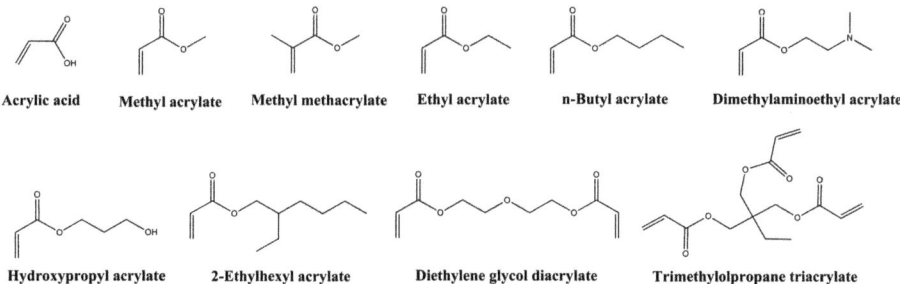

Figure 13.1 Typical monomers of acrylic acid and acrylic esters.

including natural rubber.[1] Thus, if their advantageous properties could be combined by blending to form polymer blends or interpenetrating polymer networks (IPNs) with natural rubber, they could be called thermoplastic natural rubbers, and would be more interesting materials for new applications.[2]

Natural rubber blends and IPNs are widely used materials in modern industries. They represent one of the most rapidly growing areas of research in polymer science.[3,4] Polymer blends are macroscopically homogeneous mixtures of two or more different components in which the continuous phase is polymeric. IPNs are a type of polymer blend in which at least one component is polymerized/crosslinked in their networks both without and/or with covalent bonds between the chains of the same or different polymer types. If only one type of polymeric component is crosslinked by itself, they are called semi-IPNs. If the crosslinked bonds occur between different types of polymer chains, they are called covalent semi-IPNs. In contrast, if each of the components is crosslinked by itself, they are called full IPNs.

The applications of natural rubber/acrylate blends and IPNs in many fields such as adhesion, colloidal stability, and the design of composite and biocompatible materials, require a fundamental understanding of their structures, phase phenomena and compositions of blends.[3,4] The thermodynamic, mechanical and rheological properties of these blends and IPNs are the subject of many investigations. Natural rubber/acrylate blends and IPNs can also be modified to produce better viscoelastic, mechanical, rheological and other important properties. They can be processed into a molten stage repeatedly to achieve a thermoplastic property, and they can exhibit the typical properties of rubber.[2] They can be used to create new and popular materials in material engineering for various applications.[4,5] Moreover, many of their advantageous properties have been widely developed for use in industrial applications such as cars, household appliances, medical devices, electrical cables and headphone cables.[6,7]

13.2 The History of Natural Rubber–Acrylate Blends and IPNs

One objective of developing polymer blends is to obtain new materials with different properties to those of the individual polymer components.[8] Natural rubber blends have been used extensively in many industries for many years. Generally, natural rubber/acrylate polymer blends themselves are highly incompatible due to the differences in their polarities. They exhibit poor physicomechanical properties and have poor physicochemical interactions across the phase boundaries.[6,9–11] These blends are characterized by having a coarse morphology, narrow interface, and very poor interaction between the phase boundaries.[11–14] Therefore, technological compatibilization of the immiscible pairs is necessary for natural rubber/acrylate blends. The main point of interest is to graft the acrylate copolymers into the natural rubber backbones to form special segments that can be chemically identified as the main reactive blends with other blended compositions or to act as blended compatibilizers. The polarity of natural rubber molecules can be modified by graft copolymerization to prepare natural rubber–acrylate blended products that have better properties than those of the unmodified natural rubber. The properties of these natural rubber graft copolymers can be improved, such as by having a higher hydrophilicity, or a higher chemical reactivity. This will lead to an increase in the inter-chain interactions between the natural rubber and acrylate molecules *via* the non-polar groups of natural rubber and the polar groups of the acrylate polymers, which have some better properties than those of the unmodified natural rubber.[6,9,10] Thus, they can be blended with the other polymers more easily than with the original natural rubber. These modified compounds can produce polymer blends with novel properties, due to the highly ambivalent nature of the rubber backbones and grafted side chains.[2]

13.3 Preparation Methods of Natural Rubber–Acrylate Blends and IPNs

13.3.1 Natural Rubber–Acrylate Blends

In industrial applications of thermoplastic natural rubbers, a raw natural rubber material can be modified by blending with acrylate polymers. These thermoplastic natural rubbers exhibit good damping properties and resistance to environmental conditions. The methyl methacrylate monomers, polyacrylate and poly(methyl methacrylate) are the most frequently used acrylic materials to blend with natural rubber. In general, the natural rubber and these acrylate polymers are reported to be heterogeneous blends due to their immiscible properties.[11] Several modifications of the natural rubber backbones have been used to improve the hydrophilicity of natural rubber and to provide more compatibility between natural rubber and the acrylate polymers, such as by catalysis to produce hydrogenated natural rubber, interaction with maleic anhydride to form maleated natural rubber, and grafting with methyl methacrylic acid, polyacrylate or poly(methyl methacrylate) before blending with another acrylic molecule (Figure 13.2).

Ho and coworkers[8] prepared stabilized colloidal latex blends of natural rubber and poly(methyl methacrylate). The monodispersed poly(methyl

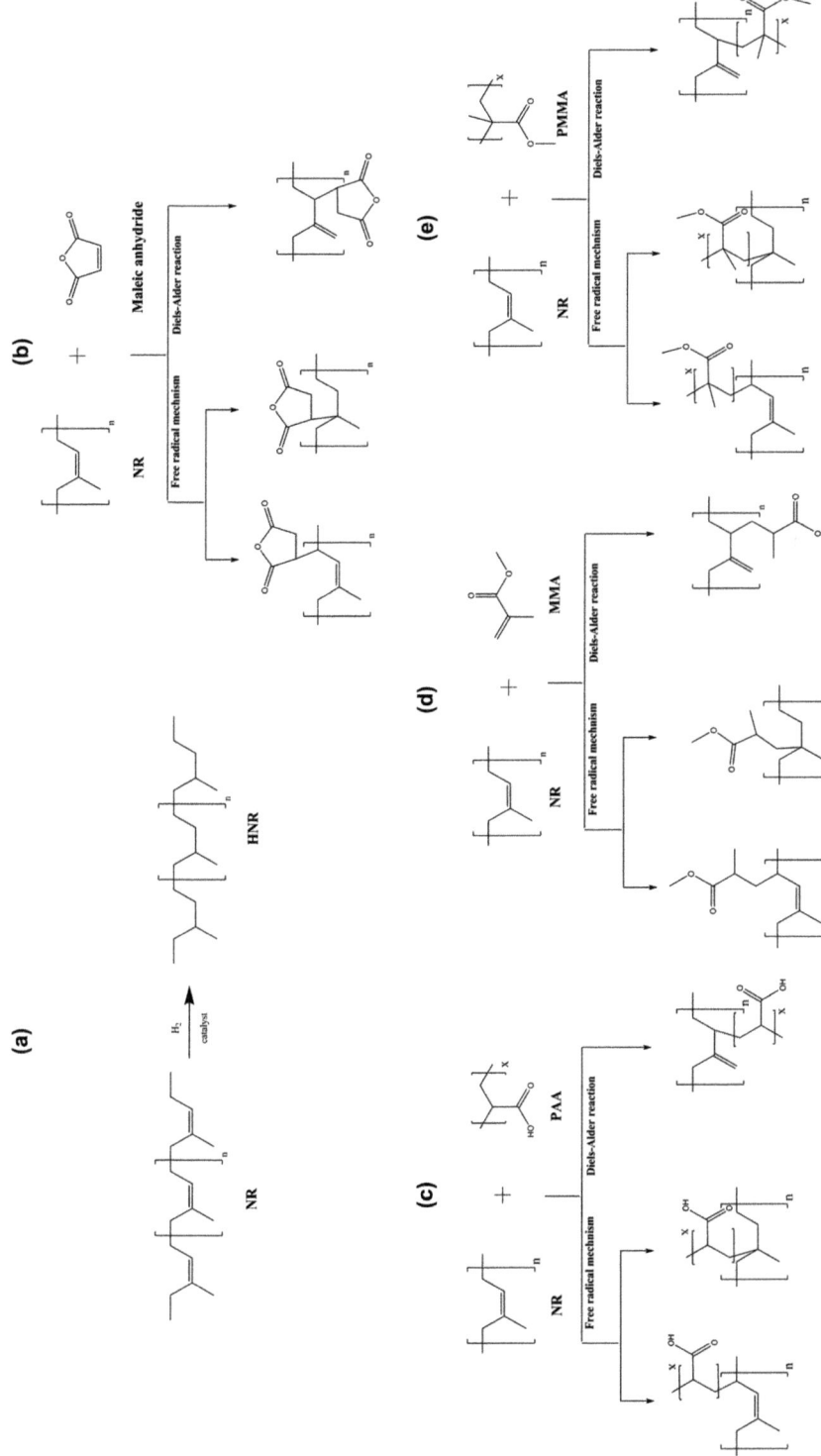

Figure 13.2 Possible modifications to the natural rubber backbone by (a) catalysis to hydrogenated natural rubber, (b) interacting with maleic anhydride, and grafting with (c) polyacrylate, (d) methyl methacrylic acid, and (e) poly(methyl methacrylate). (Modified from Nakason et al.[7].)

methacrylate) latex was prepared by modifying the film surface of the ammoniated natural rubber latex using sodium dodecyl sulfate as a surfactant. The particle size of the stabilized colloidal latex blends was 80 nm. Up to 50% w/w of poly(methyl methacrylate) could be blended into the natural rubber latex.

The preparation of natural rubber-*graft*-methyl methacrylic acid has been reported by Lenka and coworkers.[9,10,15] The vanadium ion was used as an initiator, which initiated the creation of free radicals on the backbone of natural rubber and this increased the interaction between the natural rubber and the methyl methacrylate surfaces.[9] The coordination complexes derived from the acetylacetonate of Mn(III) ions could also be used as an initiator to form the natural rubber-*graft*-methyl methacrylic acid.[10] Under different conditions, silver ions could be used as a catalyst to produce natural rubber-*graft*-methyl methacrylic acid with different concentrations of methyl methacrylic acid monomers, and potassium peroxydisulfate as an initiator.[15] Consequently, these methods were successful in the preparation of compatible blended natural rubber and methyl methacrylic acid by graft copolymerization. This compatibility was confirmed by nuclear magnetic resonance and infrared spectroscopy techniques. The interaction between natural rubber and methyl methacrylic acid was significantly increased and was useful for further blending with other polyacrylate molecules or different polymer types.

In addition, the compatibility between natural rubber and poly(methyl methacrylate) blends was also improved by using natural rubber-*graft*-methyl methacrylic acid as a compatibilizer.[2] The natural rubber macroradicals could react with methyl methacrylic acid by emulsion polymerization techniques to form graft copolymers using potassium persulfate as the initiator. The natural rubber-*graft*-methyl methacrylic acid/poly(methyl methacrylate) blends were also prepared using a melt-mixing system with different graft copolymer compositions in the natural rubber-*graft*-methyl methacrylic acid, and an appropriate blending ratio of these graft molecules and poly(methyl methacrylate). The poly(methyl methacrylate) pellets were melted in a two-roll mill and then the natural rubber-*graft*-methyl methacrylic acid was blended at 165°C until the surface of the blends became smooth. This preparation provided the graft copolymers that acted as an interfacial agent with an increased compatibility and adhesive force between the two phases of these blends.

In addition, the methyl methacrylic acid monomers could be polymerized to form poly(methyl methacrylate), which was grafted onto the backbone of the natural rubber latex to form natural rubber-*graft*-poly(methyl methacrylate) copolymers *via* a radiation technique.[16] Moreover, this graft technique also resulted in an increase in the viscosity of the natural rubber latex due to the increase in the total solid content of the latex. Increasing the proportion of grafted poly(methyl methacrylate) in the natural rubber-*graft*-poly(methyl methacrylate) copolymers increased the hardness of the polymer blends, but did not greatly influence their tensile strength, modulus, elongation at break, or thermal stability. In addition, Mina and coworkers[17] reported that natural rubber/poly(methyl methacrylate) blends could be prepared by the solution method. Concentrated natural rubber latex and methyl methacrylic acid solutions were mixed together in different proportions in a toluene solvent using a magnetic stirrer. Poly(methyl methacrylate) was produced *in situ* from the methyl methacrylic acid monomers by radiation polymerization using γ-rays from a Co^{60} source at 0.8 kGy. The natural rubber particles became entrapped into the poly(methyl methacrylate) networks as inclusion complexes to produce homogeneous blends that provided good morphology and mechanical properties. However, the hardness of these natural rubber/poly(methyl methacrylate) blends decreased in proportion to the decrease in the observed glass transition temperature.

The poly(methyl methacrylate) molecules were dispersed in the natural rubber matrix, or vice versa, to form spherical droplets, as observed by optical photographs or scanning electron microscopy.[13,17] The compatible natural rubber/poly(methyl methacrylate) blends had been made by the addition of the graft copolymer of natural rubber-*graft*-poly(methyl methacrylate) as the compatibilizing agent due to its ability to enhance the interfacial adhesion between the two homopolymers.[13] Moreover, Nakasorn and coworkers[6] reported that natural rubber-*graft*-poly(methyl methacrylate) could be blended with poly(methyl methacrylate) *via* a dynamic vulcanization technique with a conventional sulfur vulcanization system. The natural rubber-*graft*-poly(methyl methacrylate) was synthesized by a semi-batch emulsion polymerization technique *via* different bipolar redox initiation systems, *i.e.* cumene hydroperoxide and tetraethylene pentamine.[6]

Polyacrylate is a highly hydrophilic substance due to the presence of carboxylic acid groups. It provides water-attractive sites and can be exploited for its ability to strongly absorb water. Natural rubber is a hydrophobic polymer that limits the penetration of water molecules into its networks. Polyacrylate was chosen to blend with natural rubber to form swellable natural rubber/polyacrylate blended membranes. The mixed matrix membranes could swell and control the hydrophobic–hydrophilic polymer composites by blending with different ratios between the natural rubber and the polyacrylate. The concentrated natural rubber latex with 60% w/w dry rubber content was obtained by a centrifugation method to remove non-rubber components. The rubber was dispersed in 7% w/w ammonium hydroxide. Dry polyacrylate pellets were dissolved in 30% w/w ammonium hydroxide. Then, the natural rubber latex was mixed with the polyacrylate solution. Ethylene glycol was added as a crosslinking agent, and heated at 70 °C. Under these conditions, the polyacrylate was allowed to crosslink with natural rubber. The mixed matrix membrane was then formed by the casting method. The water sorption selectivity and swollen property of this mixed matrix membrane were found to increase when the polyacrylate content is increased.[18]

In addition, the compatibility of natural rubber and poly(methyl methacrylate) blends was improved by modification of the functional groups of the natural rubber molecules to be more hydrophilic. Nakasorn and coworkers[7] synthesized maleated natural rubber by using maleic anhydride. Grafting maleic anhydride onto the backbone of the non-polar natural rubber overcame the disadvantage of the low surface energy of natural rubber. This improved the hydrophilicity and adhesion of natural rubber. The maleated natural rubber had the appropriate reactive sites to blend with the poly(methyl methacrylate). The maleated natural rubber/poly(methyl methacrylate) blends were compatible due to the polar nature of the blended components. A high chemical interaction between the different phases might occur.

Furthermore, the C=C bonds in the natural rubber structure might induce poor thermal and oxidative resistance in the natural rubber blends. Thus, Thawornwisit and coworkers[19] proposed the preparation of hydrogenated natural rubber, which is one of the chemical modifications available to improve the oxidation and thermal resistance of diene-based natural rubber before blending with poly(methyl methacrylate-*co*-styrene). The poly(methyl methacrylate-*co*-styrene) was resistant to the outdoor environment and had excellent optical properties with a high refractive index, but it was extremely brittle and had low impact strength. Hydrogenated natural rubber could, however, be used as an impact modifier, as well as to improve its thermal and oxidative resistance for these acrylic plastics.

Therefore, these results strongly indicate that the natural rubber/poly(methyl methacrylate) blends are immiscible. However, this problem can be solved by using compatibilizers and reactive blends, such as by modifying the natural rubber backbone by grafting to the copolymers, adding polar groups, or by hydrogenation on the natural rubber structure before blending with poly(methyl methacrylate). The increasing hydrophilicity of natural rubber may act as an interfacial agent to increase the compatibility of blends with reducing dimensions of the dispersed phase.

13.3.2 Natural Rubber–Acrylate IPNs

Natural rubber blends have not been substantially limited to only reactive blends, such as those produced by grafting, epoxidation and halogenation studies. IPNs are one choice for modification of the natural rubber to increase the compatibility between natural rubber and acrylate polymers. Full IPNs of natural rubber and acrylate polymers are defined as a polymer blend having a crosslinked polymerization of each polymer type in their networks. Semi-IPNs, in contrast, are those in which only one type of polymeric component, mostly acrylate polymer, is crosslinked. The polymerization of IPNs would be a potential method for the modification of natural rubber.[20–26]

Amnuaypanich and coworkers[21] prepared semi-IPN pervaporation membranes from natural rubber and polyacrylate for the dehydration of water–ethanol mixtures using ethylene glycol as a crosslinking agent and Triton X-100 as a surfactant. The polyacrylate solution was mixed with natural rubber latex and the surfactant. Then, ethylene glycol was added into the dispersion to initiate the crosslinking of polyacrylate. The mixture was stirred at 70 °C for 3 hours in an atmosphere of nitrogen gas. Then, the dispersion was cast on a glass plate and dried at 70 °C for 24 hours in an oven, where polyacrylate became crosslinked in the immediate presence of natural rubber yielding the semi-IPN structure. This material was supposed to become compatible and more hydrophilic after blending the hydrophilic polymer of polyacrylate into the hydrophobic natural rubber. Increasing the polyacrylate content enhanced both the water permeation flux and the selectivity of the membrane.

The natural rubber-based IPNs obtained with a high molecular weight natural rubber and poly(methyl methacrylate) were easily prepared by a sequential method using divinyl benzene as the crosslinking agent and benzoyl peroxide as an initiator.[20] In addition, the natural rubber and poly(methyl methacrylate) blends were also easily prepared by polymerization with *tert*-butyl peroxy-2-ethylhexanoate as an initiator without a crosslinking agent.[22] They improved the phase adhesion, and both the natural rubber/poly(methyl methacrylate) blends and IPNs were compatible. Ethylene glycol dimethacrylate could also be used as a crosslinking agent and *tert*-butyl peroxy-2-ethylhexanoate was used as an initiator to form semi-IPNs based on natural rubber and poly(methyl methacrylate) with different levels of crosslinking agents.[23] The methyl methacrylic acid monomer, initiator and crosslinking agent were incubated with a swollen natural rubber at 5–8 °C for 24 hours, followed by heating at 80 °C for 22 hours. The natural rubber/poly(methyl methacrylate) semi-IPNs clearly showed two glass transition temperatures to indicate a phase separation of these materials. Nevertheless, the compatibility increased when the amount of crosslinking agent is increased. The synthesized IPNs led to an improvement in the compatibility and mechanical properties, which depended on the acrylate polymer content and the amount of crosslinking agent.[25,26] In addition, the full IPN based natural rubber and poly(methyl methacrylate) were also produced with different levels of ethylene glycol dimethacrylate as a crosslinking agent and azobisisobutyronitrile as an initiator in various concentrations of poly(methyl methacrylate).[24] The natural rubber sheets were immersed in a homogeneous mixture of methyl methacrylic acid, crosslinking agent and initiator. The swollen sheets were kept at 0 °C to achieve equilibrium of the distribution of the methyl methacrylic acid monomer in the matrix. These swollen networks were heated at 80 °C for 6 hours and at 100 °C for 2 hours to complete the polymerization and crosslinking of methyl methacrylic acid. When the concentration of *in situ* poly(methyl methacrylate) is increased, both phases tended to become continuous. When the crosslinking density of the natural rubber phase is increased, the poly(methyl methacrylate) domain size decreased, resulting in a nanometer sized morphology. Full IPNs showed better phase mixing than semi-IPNs due to the greater interpenetration resulting from the poly(methyl methacrylate) crosslinking.

13.4 Natural Rubber–Acrylate Blends and IPNs: Properties and Characterization Techniques

13.4.1 Morphological Properties

Normally, the morphological properties of natural rubber–acrylate blends and IPNs are firstly investigated to identify their compatibilities. Several techniques, such as optical microscopy, scanning electron microscopy, transmission electron microscopy and atomic force microscopy, are generally used for observation in both liquid and dried matrix specimens with different magnification levels depending on the size of the dispersed particles in these blends.

In the liquid phase, the poly(methyl methacrylate) mixed *in situ* with natural rubber in toluene solution formed a natural rubber/poly(methyl methacrylate) blend with a low natural rubber content, but it had an inhomogeneous appearance that was confirmed by optical microscopy (Figure 13.3). The inclusion complex of natural rubber in the poly(methyl methacrylate) resulted in non-uniformly dispersed, variably sized globules or spherical-like particles throughout the samples regardless of the concentrations. Furthermore, the natural rubber had the tendency to form a wider phase when its proportion in the solution was more than 3% w/w.[17]

The incompatibility of natural rubber/poly(methyl methacrylate) blends was also confirmed in solid form using scanning electron microscopy (Figure 13.4). For the solid natural rubber/poly(methyl methacrylate) blended films prepared by solution mixing and casting methods, the low level of the poly(methyl methacrylate) phase was found to be dispersed as domains in the continuous natural rubber matrix. The increasing amount of poly(methyl methacrylate)

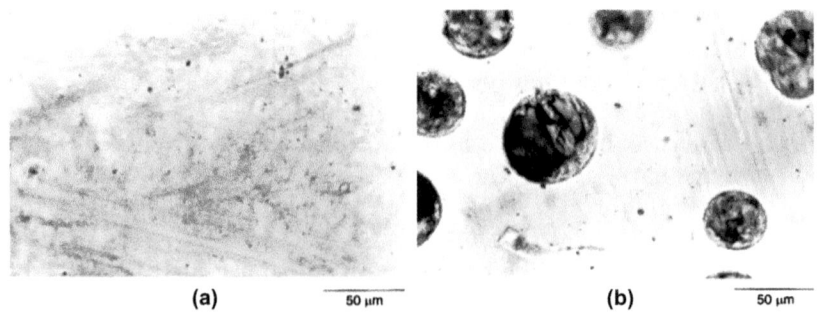

Figure 13.3 Morphologies of (a) poly(methyl methacrylate) and (b) 3% natural rubber blended in poly(methyl methacrylate) using an optical microscope. (Modified from Mina *et al.*[17]).

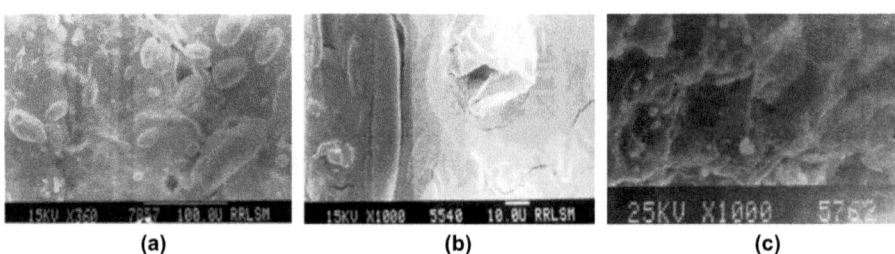

Figure 13.4 Scanning electron micrographs of natural rubber/poly(methyl methacrylate) blends at the ratios of (a) 70/30; (b) 50/50; (c) 30/70. (Modified from Oommen and Thomas[13]).

also increased in droplet size due to the agglomeration/coalescence of the dispersed particles. There was no evidence of interfacial adhesion seen in these blends, and the domain size, in this case, was very large. However, when there was a large amount of poly(methyl methacrylate) and a small amount of natural rubber, this dried film existed as a co-continuous phase. The appropriate interfacial adhesion in their blends resulted in a homogeneous form of natural rubber in the poly(methyl methacrylate) matrix.[13]

For the incompatible natural rubber/poly(methyl methacrylate) blends, the interfacial adhesion between the dispersed poly(methyl methacrylate) phase and the natural rubber matrix were also improved by the addition of a graft copolymer to form compatible blends. The domain size of the dispersed poly(methyl methacrylate) phase tended to decrease when the percentage of the graft copolymer in the polymer blends were increased. Thus, from this effect it could be concluded that the natural rubber/poly(methyl methacrylate) blends expressed compatible morphology in the presence of a graft copolymer that acted to induce interfacial adhesion between the poly(methyl methacrylate) droplets and the natural rubber matrix.[11,13] Moreover, the natural rubber-*graft*-poly(methyl methacrylate) prepared by using a semi-batch emulsion polymerization technique *via* a bipolar redox initiation system was blended with poly(methyl methacrylate) and vulcanized to form a thermoplastic vulcanizate.[6] Fracture morphology of the vulcanized 60/40 natural rubber-*graft*-poly(methyl methacrylate)/poly(methyl methacrylate) blends was carried out using cryogenic scanning electron microscopy. There was a higher graft efficacy of the natural rubber-*graft*-poly(methyl methacrylate) that provided more finely dispersed natural rubber domains and smaller cavity sizes in the natural rubber/poly(methyl methacrylate) matrix. This was due to the polar groups of the natural rubber-*graft*-poly(methyl methacrylate), which acted as an interfacial agent to increase the compatibility by reducing the dimensions of the dispersed phase. Moreover, the conventional vulcanization systems also produced smaller natural rubber particles that were dispersed in the poly(methyl methacrylate) matrix.

The thin sheets of natural rubber/poly(methyl methacrylate) blends prepared by a melt-mixing system with a two-roll mill and different grafting properties of the natural rubber-*graft*-methyl methacrylic acid monomer were also observed after their fracture under a scanning electron microscope (Figure 13.5).[2] It was found that the cavitation in the specimen decreased as the poly(methyl methacrylate) contents increased, together with an increase in the graft natural rubber content. In a similar way to the effect of the natural rubber-*graft*-poly(methyl methacrylate)as compatibilizer, at the same graft natural rubber and poly(methyl methacrylate) blended ratio, the fracture surface with the higher grafting amount produced a smoother surface than that of the lower grafting amount because of the higher percentage of the natural rubber–methyl methacrylic acid graft efficiency, which produced a higher interfacial adhesion between the two phases of the blended materials.

Blended matrices of maleated natural rubber and poly(methyl methacrylate) were prepared by mechanical mixing, and the fractured surface morphology of

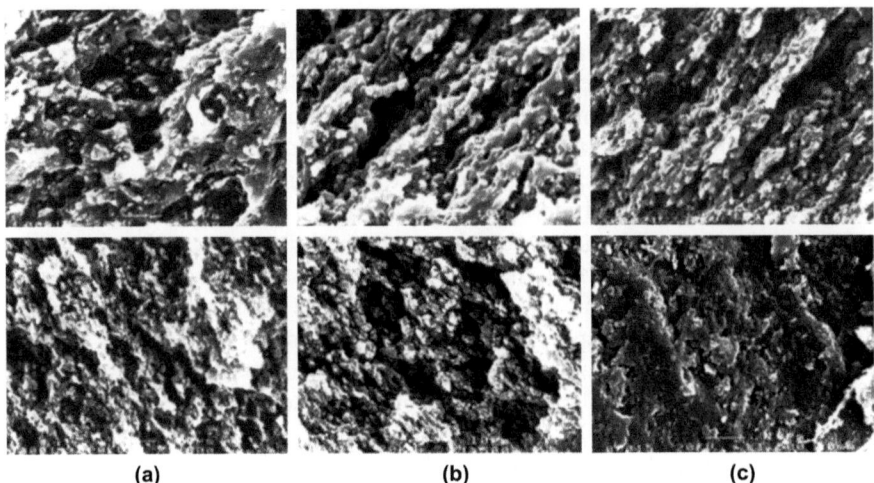

Figure 13.5 Fracture scanning electron micrographs of grafted natural rubber/poly(methyl methacrylate) blends at the ratios of (a) 70/30; (b) 60/40; (c) 50/50 in low (top) and high (bottom) levels of grafting amount. (Modified from Thiraphattaraphun et al.[2]).

these blends was also observed by scanning electron microscopy.[7] A higher amount of the graft of the maleated natural rubber copolymer resulted in a lower cavitation size in the maleated natural rubber/poly(methyl methacrylate) blended specimens. Furthermore, the poly(methyl methacrylate) minor phase was also more finely dispersed, with an increase in the graft efficacy of the maleated natural rubber. These results were similar to those using the natural rubber-*graft*-methyl methacrylic acid and natural rubber-*graft*-poly(methyl methacrylate) as compatibilizers. These graft copolymers could act as an interfacial agent by reducing the dimensions of the dispersed phase through interactions or chemical reactions with the polar groups in the blended poly(methyl methacrylate).

Modified hydrogenated natural rubber/poly(methyl methacrylate-*co*-styrene) blended sheets were investigated by scanning electron microscopy.[19] A rough surface with large holes on the fracture surface were found in the unmodified poly(methyl methacrylate-*co*-styrene) copolymer sheets due to their brittle properties. When the specimens were blended with hydrogenated natural rubber, however, their fracture surfaces were smoother, depending on the hydrogenated natural rubber content. It was concluded that hydrogenated natural rubber might act as an interfacial agent and provides better compatibility, and it acted as an impact modifier to improve the impact strength of the copolymer sheets. These factors increased the miscibility of the natural rubber and the acrylate polymers.

Atomic force microscopy has also been used to investigate the morphology of natural rubber blended films. The natural rubber and poly(methyl methacrylate) particles in the latex films can be easily distinguished and are clearly

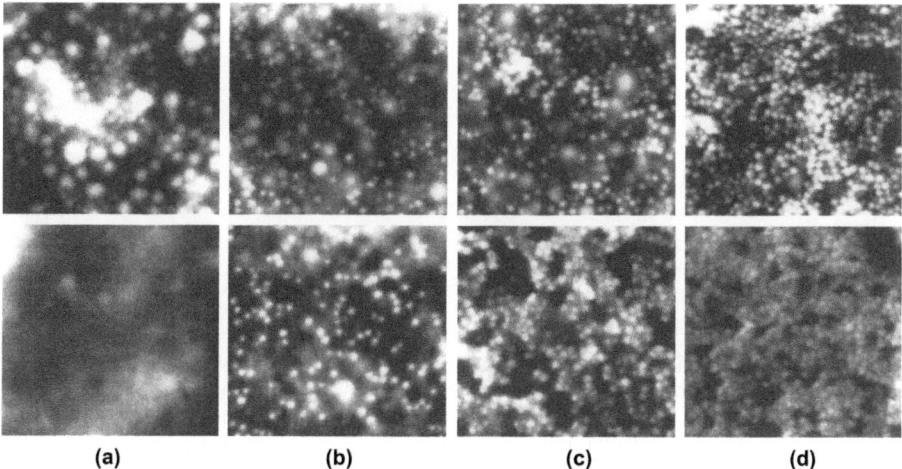

Figure 13.6 Atomic force microscopy top-view images of the air/polymer/film interfaces of natural rubber/poly(methyl methacrylate) blended films at the amounts of (a) 0%; (b) 10%; (c) 30% and (d) 50% poly(methyl methacrylate) with (top) unannealed and (bottom) annealed at 130°C for 1 hour. (Modified from Ho et al.[8]).

visible. Ho et al.[8] prepared a colloidal stable natural rubber latex containing up to 50% w/w poly(methyl methacrylate) blended by a polymerization reaction. The poly(methyl methacrylate) particles were spherical and highly monodispersed in the colloidal natural rubber. However, they exhibited many orderly packed regions separated by random disordered zones (Figure 13.6, top). In general, the natural rubber latex forms the tacky and translucent films, whereas the poly(methyl methacrylate) latex forms brittle and opaque films. It was found that the natural rubber/poly(methyl methacrylate) blended films became translucent at a 20% w/w poly(methyl methacrylate) blending. However, the clarity and tackiness of the natural rubber/poly(methyl methacrylate) blended films depended on the poly(methyl methacrylate) content, and became opaque and non-tacky when the poly(methyl methacrylate) blend was more than 30% w/w. After annealing, the natural rubber particles were continuously and completely observed in a flat and smooth film surface. In contrast, the poly(methyl methacrylate) particles did not change on the film surface. The morphology of these blended films showed an asymmetric distribution of the poly(methyl methacrylate) particles by atomic force microscopic investigations (Figure 13.6, bottom). The accumulation of the poly(methyl methacrylate) particles at the interface strongly depended on the size and concentration of the poly(methyl methacrylate) particles. They showed that the small poly(methyl methacrylate) particles on the surface still maintained their spherical shapes and were fairly well dispersed. However, some poly(methyl methacrylate) particles appeared to be completely coated by a soft molten natural rubber material. The size of the poly(methyl methacrylate) particles subjected to the annealing process were noticeably larger than the unannealed particles.

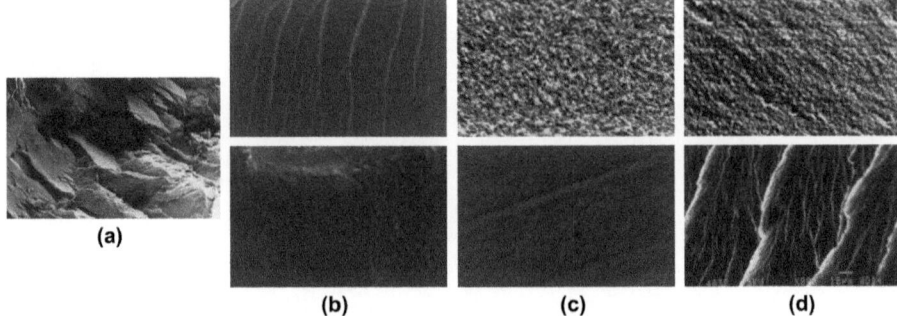

Figure 13.7 Scanning electron micrographs of (a) poly(methyl methacrylate) films and crosslinked natural rubber/poly(methyl methacrylate) pseudo-IPNs at the amounts of (b) 0%; (c) 10% and (d) 30% poly(methyl methacrylate) with (top) high and (bottom) low crosslinked density. (Modified from Alcântara et al.[20]).

In the IPNs of natural rubber with acrylate polymers, Alcântara and co-workers[20] synthesized pseudo-IPNs, a type of IPN in which one of the components has a linear structure, by a sequential method using a high molecular weight crosslinked natural rubber and linear poly(methyl methacrylate). These IPNs had a clear compact and smaller phase domain of poly(methyl methacrylate) by scanning electron microscopic observations (Figure 13.7). Normally, linear and crosslinked natural rubber have a smooth and wave-like morphology that is independent of the crosslinked density, and pure poly(methyl methacrylate) has a wave-like morphology with some fibrils. When the IPNs of natural rubber and poly(methyl methacrylate) with the highest crosslinked density were formed, however, a more compact and smaller phase domain was observed. Decreasing the crosslink density clearly revealed wave-like and smooth morphologies. At the lowest crosslink density, they showed the dispersed fibrils in the continuous crosslinked natural rubber phase. Moreover, these fibrils clearly increased their compactness and became more uniform in their distribution when the poly(methyl methacrylate) composition increased at a low crosslink density of natural rubber. Thus, the crosslinking reaction on the natural rubber affected the morphology of their IPNs.

Nanostructures of full IPNs based on natural rubber and poly(methyl methacrylate) prepared using different crosslinking levels of both the natural rubber and poly(methyl methacrylate) phases, and various poly(methyl methacrylate) contents were investigated by scanning electron microscopy and transmission electron microscopy, and similar results were obtained.[24] At the same crosslink density, a phase separation was observed with different morphologies in the poly(methyl methacrylate) blends with a range of contents of 20 to 60%. At lower poly(methyl methacrylate) contents, the poly(methyl methacrylate) domain had larger particles that were dispersed in the continuous natural rubber matrix phase. By increasing the poly(methyl methacrylate) content, the dimensions of these particles increased to form clusters of large domains surrounded by very small domains, and then changed to the dual phase and co-continuous morphologies, respectively, due to the deeper interpenetration of phases caused by the crosslinking effects of the second phase. Moreover, crosslinking of both phases facilitated deeper interpenetrations between both networks, and the degree of compatibility attained during formation of the IPNs. The full IPNs with an increasing crosslink density in both the natural rubber and poly(methyl methacrylate) were clearly observed in the scanning electron microscopic images. With a higher crosslinking of the natural rubber matrix, the phase distribution of poly(methyl methacrylate) in the natural rubber matrix became much smaller, ordered and compact. In addition, with a higher crosslinking of both phases a nanostructured morphology was clearly revealed, with the spherical domains of poly(methyl methacrylate) being nano-sized and finely dispersed. In this case, the crosslinking level at 2% natural rubber and 4% poly(methyl methacrylate) produced full IPNs having a fine structure and smooth morphology.

13.4.2 Mechanical Properties

The mechanical properties of polymer blends are generally measured as tensile strength in terms of Young's modulus, ultimate tensile strength, elongation at break, and adhesion properties in terms of peel strength and tack adhesion. The specimens of these polymer blends can be in the form of either films or another exact shape. These properties can normally be determined using a universal testing machine.

Generally, the natural rubber polymers have high elasticity caused by bond distortions. When force is applied, the bond lengths deviate from the equilibrium and any strain energy is stored electrostatically. However, natural rubber has a tendency to degrade rapidly due to the instability of the double bonds in its structure. Oxygen, ultraviolet light and ageing can break down the molecular structure of natural rubber. Processing techniques by the addition of modifiers and fillers can increase the usefulness of rubbers, and their durability for various applications. The double bonds in the natural rubber structure are available for chemical reactions. Modifications of natural rubber by graft copolymerization are used to produce new products that have better properties than those of the unmodified natural rubber, especially elastic properties.[2] The interaction between rubber and fillers can affect the mechanical properties of natural rubber. In addition, graft copolymers show novel properties due to the highly ambivalent nature of the rubber backbone and its grafted side chains. The graft copolymerization of acrylic substances, *i.e.* methyl methacrylic acid, polyacrylate and poly(methyl methacrylate), onto the natural rubber backbone with various initiator systems to give better mechanical properties has been studied by several workers.[6,7,9–12,26]

Other mechanical properties including modulus, tensile strength and elongation at break of the incompatible natural rubber/poly(methyl methacrylate) blends were very poor, especially in the values of stress at their break points.[12] Nevertheless, the modulus and tensile strength values of these polymer blends rapidly increased after the addition of natural rubber-*graft*-poly(methyl methacrylate) copolymer as compatibilizer. The improvement in mechanical properties was associated with strong interfacial interaction between the poly(methyl methacrylate) domains and the natural rubber matrix with the compatibilizer, which was in accord with the smoother morphology of the blends. However, their elongation at break values significantly decreased after addition of the graft copolymer.[11–13] The natural rubber-*graft*-poly(methyl methacrylate) with a different percentage molar ratio of natural rubber and methyl methacrylic acid in the graft polymerization process and different natural rubber-*graft*-poly(methyl methacrylate)/poly(methyl methacrylate) blended ratios were checked. Both the degrees of graft copolymerization and amounts of natural rubber blend affected the mechanical properties of acrylate blends.[12] The tensile strength, tear strength and hardness increased with an increasing poly(methyl methacrylate) content in the natural rubber-*graft*-poly(methyl methacrylate)/poly(methyl methacrylate) blends. They were also increased by decreasing the natural rubber and methyl methacrylic acid molar ratios in the graft polymerization due to the higher poly(methyl methacrylate) interacting with the natural rubber-grafted molecules that provided for higher polar groups, thus a higher poly(methyl methacrylate) content could be blended into the natural rubber-*graft*-poly(methyl methacrylate) copolymer. However, the elongation at break decreased with an increasing methyl methacrylic acid monomer concentration in the graft polymerization process. Poly(methyl methacrylate) is a hard and brittle thermoplastic material, the strength of which is greatly affected by the stress concentrations.[2] Likewise, when the natural rubber-*graft*-poly(methyl methacrylate)was blended with poly(methyl methacrylate) at various ratios of 75/25, 60/40, 50/50 and 40/60, the mechanical properties in terms of tensile strength, elongation at break and Shore A hardness were similar to those observed with increasing methyl methacrylic acid concentrations in the graft copolymerization due to the enhancing effect of the poly(methyl methacrylate) content on the thermoplastic natural rubber.[6]

In a similar way, Thiraphattaraphun and coworkers[2] reported the mechanical properties of natural rubber-*graft*-methyl methacrylic acid/poly(methyl methacrylate) blends prepared by the melt-mixing system. A trend was observed of increased tensile strength, tear strength and hardness properties but with a decrease in the level of elongation at break. This was caused by increasing the level of the methyl methacrylic acid concentrations in the natural rubber-*graft*-methyl methacrylic acid polymerization process and the poly(methyl methacrylate) contents in the natural rubber-*graft*-methyl methacrylic acid/poly(methyl methacrylate) blended compositions. The continuous phase of the natural rubber and poly(methyl methacrylate) produced inferior tear strength. This high tear strength of the compatibilized system was provided by the presence of a good adhesion between the two elastic and hard phase components. Moreover, the impact strength was the parameter essentially used to prevent the formation of cracks in the natural rubber-*graft*-methyl methacrylic acid/poly(methyl methacrylate) blended matrix. The impact energy and the interfacial adhesion between the two polymers of the natural rubber-*graft*-methyl methacrylic acid/poly(methyl methacrylate) blends increased with an increased natural rubber-*graft*-methyl methacrylic acid content. The effect of the high natural rubber-*graft*-methyl methacrylic acid content was to dissipate a large amount of energy in the blended materials. Thus, this blended matrix was difficult to crack and led to an increase in the impact strength of the blends.[27]

The semi-IPNs of natural rubber and poly(methyl methacrylate) also showed the effects of their compositions and levels of crosslinking in the poly(methyl methacrylate) phase on their mechanical properties. The modulus and tensile strength of these semi-IPNs increased as the composition and level of crosslinking in the poly(methyl methacrylate) phase was increased. These improved tensile properties were affected by the interpenetration and reinforcement of the crosslinked poly(methyl methacrylate) phase.[22,23] Likewise, the full IPNs showed a change from a rubbery to a plastic nature with increasing concentrations of poly(methyl methacrylate), and the morphology of the full IPNs changed to being compact.[24] The crosslinked poly(methyl methacrylate) phase enabled an improvement in the mechanical properties of the materials. The Shore A hardness values of the IPNs increased with the rise in poly(methyl methacrylate) content. The effect of the crosslink density of the natural rubber phase was clearly to influence the mechanical properties of these materials. However, the effect of the crosslink density of the poly(methyl methacrylate) phase in the full IPNs on their mechanical performances was different from those of the natural rubber phase.

Thus, these acrylate polymers are hard and brittle plastics that can have a direct effect on the mechanical properties in both the natural rubber blends and IPNs. They show an increase in modulus, tensile strength, tear strength, hardness and impact energy, but a decrease in elongation at break.

13.4.3 Thermal and Thermomechanical Properties

Thermal and thermomechanical analyses have also been used to study the thermal and mechanical behaviours of materials at different temperatures. Several methods are commonly used, such as dynamic mechanical thermal analysis, differential thermal analysis, dynamic mechanical analysis, modulated differential scanning calorimetry, thermogravimetric analysis and differential scanning calorimetry. These methods are generally interested in determining the glass transition temperature of the materials, *i.e.* the temperature at which there is a change from the rubbery to the glassy stage.

Basically, the glass transition temperature of natural rubber in given in various publications in the range of -50 to $-75\,^\circ\text{C}$.[2,7,20,22,23] Thiraphattaraphun and coworkers[2] also reported the thermal transition phenomena of the natural rubber-*graft*-methyl methacrylic acid obtained by the dynamic mechanical thermal analysis method. In the transition region, increasing the temperature resulted in a decrease in the storage modulus due to the increase in the chain stiffness of the polymer. Increasing the amount of methyl methacrylic acid also decreased the storage modulus of the graft copolymer, and so there was a reduction in the glass transition temperature value. This value of the natural rubber-*graft*-methyl methacrylic acid decreased from −52.15 °C to −56.81 °C after the natural rubber was reacted with increasing methyl methacrylic acid concentrations of 60 to 120 parts per hundred of rubber (phr) in the graft copolymerization process. This result indicated an improvement in mechanical properties by grafting the natural rubber with the methyl methacrylic acid monomer. The damping value or loss of the tangent (tan δ) peaks was also related to the glass transition temperature. The tan δ value showed a maximum around the transition region. The tan δ peaks decreased from 0.972 to 0.505 as the methyl methacrylic acid concentration increased from 60 to 120 phr, and their overall widths also decreased. This indicated that there was higher graft efficiency between the natural rubber and the methyl methacrylic acid units in the presence of a high amount of methyl methacrylic acid, which resulted in an increase in natural rubber chain immobilization.

Thermogravimetric analysis is generally the technique used to study the decomposition of substances. Amnuaypanich and coworkers[18] characterized the thermal behaviour of natural rubber–polyacrylate blends and their incorporations with zeolite 4A by this technique. The blended membranes showed a three-step loss of weight: first at 40–150 °C that could be attributed to the loss of absorbed water from the carboxylic groups in the polyacrylate molecules, second at 160–240 °C that was due to the disintegration of the crosslinked sites of the polyacrylate networks and ester linkages, and third at 280–400 °C that was due to destruction of the main chain polyacrylate and natural rubber leading to the major weight loss. The decomposition temperatures of the maleated natural rubber/poly(methyl methacrylate) blends were also studied by thermogravimetric analysis, and then confirmed by dynamic mechanical analysis.[7] Their decomposition temperatures decreased when the concentration of maleic anhydride graft copolymerization increased. In addition, they also decreased when the amount of poly(methyl methacrylate) increased in the maleated natural rubber/poly(methyl methacrylate) blends. Therefore, the maleated natural rubber/poly(methyl methacrylate) blends with a higher content of maleated natural rubber exhibited a higher degree of thermal resistance.

The differential scanning calorimetric thermograms of the pseudo semi-IPNs of crosslinked natural rubber and linear poly(methyl methacrylate) synthesized by Alcântara and coworkers[20] with different ratios between the crosslinked natural rubber and methyl methacrylic acid showed one glass transition temperature value in the range of −48.8 to −55.6 °C. The pseudo-IPNs with a higher crosslink density of natural rubber resulted in the glass transition temperature being closer to that of pure natural rubber, while the lower crosslink density showed a much higher value for the glass transition temperature. However, only one glass transition temperature value was observed and indicated that these systems were compatible. For the thermogravimetric analysis, the decomposition behaviour of the pseudo-IPNs with a high crosslink density revealed two broad decomposition steps as the decomposition of the pseudo-IPNs at 120–500 °C and the degradation of the small structural products of natural rubber at 505–750 °C. However, the lower crosslink density polymers showed only one decomposition step that started in the temperature range of 130–160 °C and was completed at 485–495 °C.

Jayasuriya and Houraton[22] prepared natural rubber/poly(methyl methacrylate) blends by polymerization of methyl methacrylic acid monomers with an initiator at a low temperature then heated them to a higher temperature. These natural rubber/poly(methyl methacrylate) blends formed uniformly distributed swollen natural rubber blends. The natural rubber/poly(methyl methacrylate) was compatibly blended at a low poly(methyl methacrylate) content and showed only one glass transition temperature in the dynamic mechanical thermal analysis characterization process. For those with more than 30% w/w poly(methyl methacrylate) content, these blends showed two glass transition temperatures by the dynamic mechanical thermal analysis method at around −45 to −48 °C and 154 to 158 °C, so it was assumed that these were the glass transition temperatures of the natural rubber and poly(methyl methacrylate), respectively. However, these glass transition temperatures of blended natural rubber had shifted from that of a pure natural rubber at −51 °C in the identical experiment and indicated some compatible mixing of the components. These results were also confirmed by the modulated differential scanning calorimetric data. Moreover, this similar synthetic method could also be used to prepare the natural rubber/poly(methyl methacrylate) semi-IPNs, but using ethylene glycol dimethacrylate as the crosslinking agent in the polymerizing reaction.[23] In these synthesized semi-IPNs, the dynamic mechanical thermal analysis results also possessed two glass transition temperatures to indicate the phase-separated nature of these materials at the higher poly(methyl methacrylate) contents, but with a slight shift of glass transition temperature of the natural rubber transition to a higher temperatures by 3–5 °C and the glass transition temperature of the poly(methyl methacrylate) transition was also shifted towards a lower temperature by about 5 °C to indicate that there had been a partial mixing of the components. The modulated differential scanning calorimetry also indicated that some enforced mixing probably occurred due to the interpenetration of the chain segments and the crosslinking of the poly(methyl methacrylate) phase that resulted in a partial mixing of the components in these semi-IPNs.[23]

The thermal behaviour of the full IPNs prepared from natural rubber and poly(methyl methacrylate) with ethylene glycol dimethacrylate as crosslinking agent was determined by both the thermogravimetric analysis and differential scanning calorimetry.[24] The presence of crosslinked poly(methyl methacrylate) in the full IPNs had no noticeable effect on the thermal behaviour of the materials which was confirmed by thermogravimetric analysis curves. All full IPNs with both low and high crosslink levels maintained the same thermal behaviour as the pure natural rubber matrix. Their degradation temperatures were found at around 350–420 °C. The differential scanning calorimetric thermograms showed two glass transition temperatures of natural rubber at below −50 °C and poly(methyl methacrylate) at higher than 130 °C for all full IPNs that was expressed by the phase separation of the multicomponents in these blending systems. At low poly(methyl methacrylate) contents, the β-relaxation of poly(methyl methacrylate) could be seen. The relaxation region of the higher poly(methyl methacrylate) contents produced a broader glass transition temperature and presented the major phase of the poly(methyl methacrylate) region in the full IPNs. A higher crosslinking level of poly(methyl methacrylate) at 8% led to the spreading of the relaxation, and there was no transition at a high temperature.

Thermal analysis could also be used to investigate the state of the water cluster in the structure of the polymer blends. Pervaporation membranes of the natural rubber/crosslinked polyacrylate semi-IPNs were prepared for the dehydration of water–ethanol mixtures. Their degradation was determined by thermogravimetric analysis. The thermogravimetric curve for the pervaporation of natural rubber/polyacrylate semi-IPN membranes showed a three-stage weight loss of the disintegration at the crosslinked sites of the polyacrylate networks at 170–230 °C followed by the decomposition of the natural rubber molecules at 350–400 °C, and finally, decomposition of the polyacrylate molecules at more than 400 °C. Increasing the amounts of polyacrylate in the semi-IPNs also increased the amount of weight loss at the crosslinked breakage temperature range. Moreover, they found some weight loss at 50–150 °C to indicate the loss of the absorbed water molecules in the semi-IPNs. The state of the absorbed water molecules in the semi-IPNs was further studied by differential scanning calorimetry. The water-swollen natural rubber/polyacrylate semi-IPNs showed a significant melting peak at 0 °C that was the transition peak of the free water in these semi-IPNs. There was no phase transition peak at a temperature below 0 to −50 °C, which indicated that there was no water bound by freezing in the polyacrylate molecules. However, the non-frozen bound water that possessed no detectable phase transition peak for the temperature lower than the freezing point of pure water could be calculated from the difference between the intensities of the free water and the total water. This non-frozen bound water arose from the strong interaction between the water molecules and the carboxylic groups of the polyacrylate polymer. The amount of non-frozen bound water increased with an increase in the amount of polyacrylate, while the free water showed the opposite trend. These results indicated that the functional groups of the polyacrylate molecules could strongly bind with water molecules and these appeared as non-frozen bound water.[21]

13.4.4 Rheological Properties

There are several techniques for studying rheological properties, but in this chapter we present the two techniques for rheological characterization that are used to determine the flow properties of the natural rubber reactive blending. First, the Mooney viscosity can determine the natural rubber that is related directly to the inherent property of its molecular weight. It is the most common instrument used to measure the bulk viscosity of rubber. It is defined as the shearing torque that can resist the rotation of a cylindrical metal disk/rotor embedded in the rubber within a cylindrical cavity.[6,7,14] Second, the shear flow property, likely to measure the shear rate and shear stress viscosity, can determine the flow behaviour of materials in liquid and soft solid forms. It can identify the plastic flow rather than the deformation of elasticity when the force is applied.[6,7]

Even though the natural rubber/poly(methyl methacrylate) blends are mostly immiscible, they could be improved to be more compatible by several methods. One such method is to add a third component called a compatibilizer to increase the interaction between the interfacial immiscible phases. The examples of a third component are block, graft or homopolymers that can interact with both the natural rubber and the poly(methyl methacrylate) phases. Generally, natural rubber exhibits a higher viscosity than the poly(methyl methacrylate). When the natural rubber contents are increased, the rheological viscosity of these blends also increases due to the natural rubber properties. For uncompatibilized natural rubber/poly(methyl methacrylate) blends, however, the viscosity of all blended compositions decreased with increasing shear stress and this was indicated by their pseudoplastic nature.[14] The pseudoplasticity is due to the random orientation and highly entangled state of the natural rubber molecules that become disentangled and oriented under a high shear stress, resulting in a reduction of the viscosity.[28] The decreasing viscosity at a higher shear rate might also be due to the shearing away of the dispersed phase of the incompatible natural rubber/poly(methyl methacrylate) blends. In general, poly(methyl methacrylate) acts as the dispersed phase. A strong interaction between the poly(methyl methacrylate) and natural rubber domains can be expected as they can be well packed together and form a wall structure. The incompatible natural rubber/poly(methyl methacrylate) blends have a sharp interface and the poor interaction between the two phases leads to an interlayer slip between the phases. For the compatibilized natural rubber/poly(methyl methacrylate) blends prepared with graft copolymers, the viscosity increased with increasing concentration of the graft copolymers. The increment in viscosity indicated that there was less slippage at the interface as a result of addition of the compatibilizer. The compatibilizer decreased the interfacial tension and hence the interaction between the natural rubber and the poly(methyl methacrylate) became stronger due to the increase in adhesion force between the two phases of these components.[14] Moreover, the shear viscosity of the compatible natural rubber-*graft*-poly(methyl methacrylate)/poly(methyl methacrylate) blends decreased with an increasing shear rate and this also indicated the shear-thinning pseudoplastic flow.[6] The different shear viscosity at a given shear rate of the natural rubber-*graft*-poly(methyl methacrylate)/poly(methyl methacrylate) blends was also attributed to the formation of crosslinking and different intermolecular networks in the rubber phase. In addition, the curing systems might play a significant role by controlling the particle size of the rubber domain dispersed in the thermoplastic matrix.

The Mooney viscosity value of the maleated natural rubbers/poly(methyl methacrylate) blends increased with an increased concentration of maleic anhydride used in the grafting copolymerization. The shear flow property of the pure natural rubber, maleated natural rubbers and poly(methyl methacrylate) was lower than that of the blends. The shear viscosity at a given shear rate of the blends increased with the increasing content of maleated natural rubber. This might be attributed to the increased levels of grafted maleates upon increasing the concentration of maleic anhydride monomers due to the higher chemical interaction between the polar groups in the maleated natural rubber molecules. The resistance to the flow behaviour of the polymer blends increased with an increase in the levels of the grafted maleates in the molecules and the amount of maleated natural rubber blends that produced a higher interactive force between the blended components. However, the balanced state between the maleated natural rubbers and poly(methyl methacrylate) was reached at their given blend ratio of 60/40. The increasing maleated natural rubber contents of higher than 60% resulted in a lowering of the chemical interaction between the maleated natural rubbers and the poly(methyl methacrylate) phases. Hence they showed a decreasing shear stress and viscosity. Compatible blends led to a positive deviation in the rheological properties.[7]

From these studies, it was concluded that several natural rubber/poly(methyl methacrylate) blends were compatible when the interactions between the immiscible phases such as graft copolymer as the compatibilizer was increased, and their viscosity behaviour also changed.

13.5 Applications of Natural Rubber–Acrylate Blends and IPNs

Acrylate polymers are often used in many applications that require good optical properties. However, they are unsuitable for use in the automotive industry because of their brittle characteristics. Thus, when natural rubber is blended with poly(methyl methacrylate), there is a big improvement in the elasticity of the brittle acrylate polymers. It is of interest that thermoplastic natural rubbers are relatively new products in the rubber industry and are fast-growing items in the polymer market. The acrylate polymers blended with natural rubber can improve various properties of both the natural rubber and the acrylate polymers such as elasticity, adhesion, processability properties, and transparency. These materials are known for their excellent processability, characteristic of acrylate polymers, and their elasticity property provided by natural rubber, thus they exhibit the typical properties of elastomeric materials and can be processed with thermoplastic processing equipment used to prepare acrylate polymers. Many of their interesting properties have been widely developed for several industrial applications such as in the automotive industry, household appliances, medical devices, electrical cables, and headphone cables.[6,7]

In different application fields, polymer electrolyte systems are special types of plastic materials that can be modified by the addition of additives such as plasticizers and fillers to improve the conductivity of solid polymer electrolytes and gelled polymer electrolyte systems. These systems can be used for a charge transport of the ionic type.[29] Several types of polymers such as poly(methyl methacrylate), poly(vinyl chloride) and poly(ethylene oxide) can be used as polymer hosts in these systems. However, they have not only polar substituents to transport ion salts but also require a low glass transition temperature to enhance mobility. The natural rubber-*graft*-poly(methyl methacrylate) polymerized matrices have been identified as possibilities for use in this application due to their polar characteristics. They can improve the mechanical properties of polymer electrolyte systems, increase adhesion to the electrode, increase solvent uptake, change solubility characteristics and enhance conductivity. Alias and coworkers[29] studied the gel-based polymer electrolyte systems with $LiCF_3SO_3$ consisting of a natural rubber-*graft*-poly(methyl methacrylate) with propylene carbonate as plasticizer. Plasticizers affected the polymer mobility and conductivity, which was dependent on the intrinsic nature of the plasticizer including viscosity, dielectric constant, polymer–plasticizer interaction, and ion–plasticizer coordination. The ionic conductivity of the natural rubber-*graft*-poly(methyl methacrylate) exhibited a maximum value of 1.85×10^{-4} S cm^{-1}. This identified the potential of natural rubber-*graft*-poly(methyl methacrylate) as a solid polymer electrolyte film in rechargeable battery systems by doping with Li^+ salts such as $LiClO_4$ and $LiBF_4$. The ionic conductivity increased with an increasing Li^+ salt loading until the optimum loading in the polymer host due to the increase in the number of conducting species in the electrolyte. At the maximum effective interaction between oxygen atoms and Li^+ ions in the electrolyte, the ionic conductivities obtained were $\sim 10^{-8}$ to 10^{-7} S cm^{-1}. The addition of 20% w/w $LiBF_4$ salts had a conductivity of 2.3×10^{-7} S cm^{-1}, while 15% w/w $LiClO_4$ salts had a conductivity of 4.0×10^{-8} S cm^{-1}.[30] The solid polymer electrolyte of natural rubber-*graft*-poly(methyl methacrylate) incorporated with the Li^+ salt of $LiCF_3SO_3$ at a weight ratio of 70/30 exhibited the highest ionic conductivity of 1.69×10^{-6} S cm^{-1} with an activation energy of 0.24 eV.[31]

13.6 Conclusions

Acrylate polymers or plastics are synthesized by bonding acrylic acid or acrylic esters with other molecules such as methyl methacrylic acid, polyacrylate and poly(methyl methacrylate. After blending and forming into IPNs with natural rubber *via* several techniques such as grafting and polymerization, these novel materials have the physical properties of both thermoplastics and rubbers, depending on the mixing ratios. These blends combine the typical advantages of both thermoplastic acrylate polymers and elastic rubber and are called thermoplastic natural rubbers. When heated to high temperatures during extrusion, injection and compression moulding, their individual chains slip to a plastic flow. When they are cooled, the chains are once again held firmly by the rubber. When subsequently heated, the chains can slip again to produce a plastic flow. Hence, the rubber can improve the impact strength and flexibility of the thermoplastic materials to the new blended substances that generally have a lower modulus and higher elasticity than pure acrylic plastics.

Furthermore, the rubber can improve other characteristics such as hydrophilicity and adhesion, leading to wider applications. Many techniques, *i.e.* morphological, mechanical, thermal, rheological and dielectric studies can be used to investigate the properties of natural rubber blends and IPNs with different types of acrylate polymers. Due to the interesting properties of natural rubber blends and IPNs with acrylate polymers, they can reach applications in many industrial fields such as automotive, household appliances, medical devices, electrical cables, and headphone cables.

References

1. J. Wootthikanokkhan and B. Tongrubbai, *J. Appl. Polym. Sci.*, 2003, **88**, 921.
2. L. Thiraphattaraphun, S. Kiatkamjornwong, P. Prasassarakich and S. Damronglerd, *J. Appl. Polym. Sci.*, 2001, **81**, 428.
3. Y. S. Lipatov, *Prog. Polym. Sci.*, 2002, **27**, 1721.
4. L. A. Utracki, in *Polymer Blends Handbook*, ed. L. A. Utracki, Kluwer Academic Publishers, the Netherlands, 2002, pp. 1–96.
5. L. H. Sperling and M. R. Hu, in *Polymer Blends Handbook*, ed. L. A. Utracki, Kluwer Academic Publishers, the Netherlands, 2002, pp. 417–443.
6. C. Nakason, W. Pechurai, K. Sahakaro and A. Kaesaman, *Polym. Adv. Technol.*, 2005, **16**, 592.
7. C. Nakason, S. Saiwaree, S. Tatun and A. Kaesaman, *Polym. Test.*, 2006, **25**, 656.
8. C. C. Ho, M. C. Khew and Y. F. Liew, *Surf. Interface Anal.*, 2001, **32**, 133.
9. S. Lenka, P. L. Nayak, A. P. Das and S. N. Mishra, *J. Appl. Polym. Sci.*, 1985, **30**, 429.
10. S. Lenka, P. L. Nayak, I. B. Mohanty and S. N. Mishra, *J. Appl. Polym. Sci.*, 1985, **30**, 2711.
11. Z. Oommen and S. Thomas, *Polym. Bull.*, 1993, **31**, 623.
12. Z. Oommen, M. R. Gopinathan Nair and S. Thomas, *Polym. Eng. Sci.*, 1996, **36**, 151.
13. Z. Oommen and S. Thomas, *J. Appl. Polym. Sci.*, 1997, **65**, 1245.
14. Z. Oommen, S. Thomas, C. K. Premalatha and B. Kuriakose, *Polymer*, 1997, **38**, 5611.
15. S. Lenka, P. L. Nayak and A. P. Das, *J. Appl. Polym. Sci.*, 1985, **30**, 2753.
16. F. Sundardi and S. Kadariah, *J. Appl. Polym. Sci.*, 1984, **29**, 1515.
17. M. F. Mina, F. Ania, F. J. Baltá Calleja and T. Asano, *J. Appl. Polym. Sci.*, 2004, **91**, 205.
18. S. Amnuaypanich, T. Naowanon, W. Wongthep and P. Phinyocheep, *J. Appl. Polym. Sci.*, 2012, **124**(Suppl 1), E319.
19. S. Thawornwisit, K. Charmondusit, G. L. Rempel, N. Hinchiranan and P. Prasassarakich, *J. Elastomers Plast.*, 2010, **42**, 35.
20. R. M. Alcântara, A. P. P. Rodrigues and G. G. D. Barros, *Polymer*, 1999, **40**, 1651.

21. S. Amnuaypanich and N. Kongchana, *J. Appl. Polym. Sci.*, 2009, **114**, 3501.
22. M. M. Jayasuriya and D. J. Hourston, *J. Appl. Polym. Sci.*, 2009, **112**, 3217.
23. M. M. Jayasuriya and D. J. Hourston, *J. Appl. Polym. Sci.*, 2012, **124**, 3558.
24. J. John, R. Suriyakala, S. Thomas, J. Mendez, A. Pius and S. Thomas, *J. Mater. Sci.*, 2010, **45**, 2892.
25. C. Vancaeyzeele, O. Fichet, B. Amana, S. Boileau and D. Teyssié, *Polymer*, 2006, **47**, 6048.
26. C. Vancaeyzeele, O. Fichet, S. Boileau and D. Teyssié, *Polymer*, 2005, **46**, 6888.
27. I. Walker and A. A. Collyer, in *Rubber Toughened Engineering Plastics*, ed. A. A. Collyer, Chapman & Hall, London, 1994, pp. 29–56.
28. L. A. Utracki and M. R. Kamal, in *Polymer Blends Handbook*, ed. L. A. Utracki, Kluwer Academic Publishers, the Netherlands, 2002, pp. 449–537.
29. Y. Alias, I. Ling and K. Kumutha, *Ionics*, 2005, **11**, 414.
30. M. Su'ait, A. Ahmad and M. Rahman, *Ionics*, 2009, **15**, 497.
31. K. S. Yap, L. P. Teo, L. N. Sim, S. R. Majid and A. K. Arof, *Physica B*, 2012, **407**, 2421.

CHAPTER 14

Photoreactive Nanomatrix Structures Formed by Graft Copolymerization of 1,9-Nonanediol Dimethacrylate onto Natural Rubber

ORAPHIN CHAIKUMPOLLERT,[a,b] NANTHAPORN PUKKATE[a] AND SEIICHI KAWAHARA*[a]

[a] Department of Materials Science and Technology, Faculty of Engineering, Nagaoka University of Technology, Nagaoka, Niigata 940-2188, Japan;
[b] National Metal and Materials Technology Center (MTEC), 114 Thailand Science Park, Paholyothin Road, Klong 1, Klong Luang, Pathumthani, 12120, Thailand
*Email: kawahara@mst.nagaokaut.ac.jp

14.1 Introduction

Photoreactive nanomatrix structures are novel functional nano-phase separated structures obtained by graft copolymerization, which consist of a dispersoid of major elastomer and a matrix of minor functional polymer having carbon–carbon double bonds as a photoreactive site. The photoreaction of the polymer in the nanomatrix may cement the nanomatrix structure with not only crosslinking junctions of the elastomer in the dispersoid and chemical linkages between the elastomer and the functional polymer but also crosslinking junctions of the functional polymer in the nanomatrix. In this regard, natural rubber, isolated from *Hevea brasiliensis*, is worthy of note as a component, because it is well known to be the best rubbery material, having outstanding mechanical properties; however it has poor oil resistance, weather resistance, ozone resistance, *etc.* To make up for these defects without losing its outstanding properties, the concept of nanomatrix-dispersed polymers may be significant, because natural rubber is improved with a small amount of functional polymer when the polymer forms a matrix.

In previous work, we found that the nanomatrix structure played an important role in the high proton conductivity of polymer electrolyte membranes[1] and the high elastic properties of rubbery materials,[2,3] and so forth. However, the resulting electrolyte membrane and the rubbery material were inferior in mechanical properties, such as lower tensile strength and tear energy. This has been explained by the lack of chemical linkages between the functional polymers in the nanomatrix. In order to improve the properties, it is necessary to introduce crosslinking junctions into the polymers in the nanomatrix, because the chemical linkages between the functional polymers prevent flow of the polymers in the nanomatrix. In this regard, the nanomatrix structure with photoreactive sites is expected to enhance the mechanical properties after UV irradiation, which forms a three-dimensional network structure in the nanomatrix.

In our previous work, the nanomatrix structure consisting of a natural rubber particle as the dispersoid of 1 μm in diameter and the functional polymer as the matrix of 15 nm in thickness was formed by coagulating the natural rubber particle covered with grafted functional polymer, such as polystyrene and polymethyl methacrylate.[4-7] Furthermore, the photoreactive rubber latex was prepared in our previous work by graft copolymerization of an inclusion complex of 1,9-nonanediol dimethacrylate (NDMA) with β-cyclodextrin (β-CD).[8] A methacryloyl group of poly(NDMA) grafted onto the natural rubber particle was confirmed to remain without any reaction, while the other was used for the graft copolymerization. Based on the results, the photoreactive nanomatrix structure may be formed by coagulating the latex of the photoreactive rubber.

However, the latex was not stabilized during the polymerization, and some of the reacted rubber particles were coagulated at 52% of conversion. The coagulation has been explained by the release of β-CD from the inclusion complex followed by crosslinking during graft copolymerization. In fact, the graft copolymerization was not completed, *i.e.* the coagulated particles were not covered with poly(NDMA). In order to cover all particles at high conversion of graft copolymerization, it is therefore necessary to make a stable inclusion complex. Recently, Satav and coworkers[9] demonstrated that a methacryloyl group of 1,4-butanediol dimethacrylate was protected with β-CD having a larger cavity of about 6.6 Å in diameter to form the stable inclusion complex. When we use β-CD to form the stable inclusion complex, all particles may be covered with poly(NDMA).

Here, we form the stable inclusion complex of β-CD and NDMA in water under various conditions (Figure 14.1). The resulting inclusion complex was subjected to graft copolymerization onto natural rubber particles with various radical initiators (Figure 14.2). Effects of the amount of inclusion complex and radical initiator on conversion of NDMA are investigated.

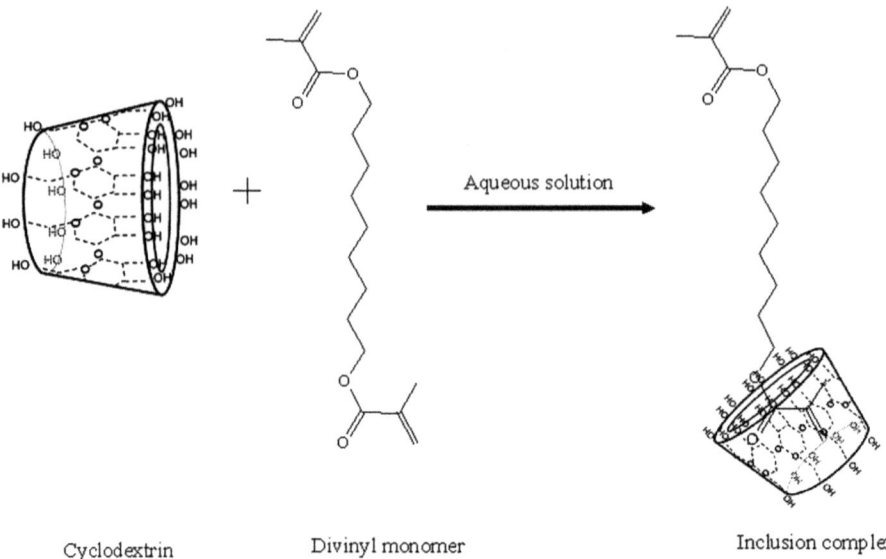

Figure 14.1 Schematic representation of inclusion complex formation of NDMA and β-CD before graft copolymerization.[9]

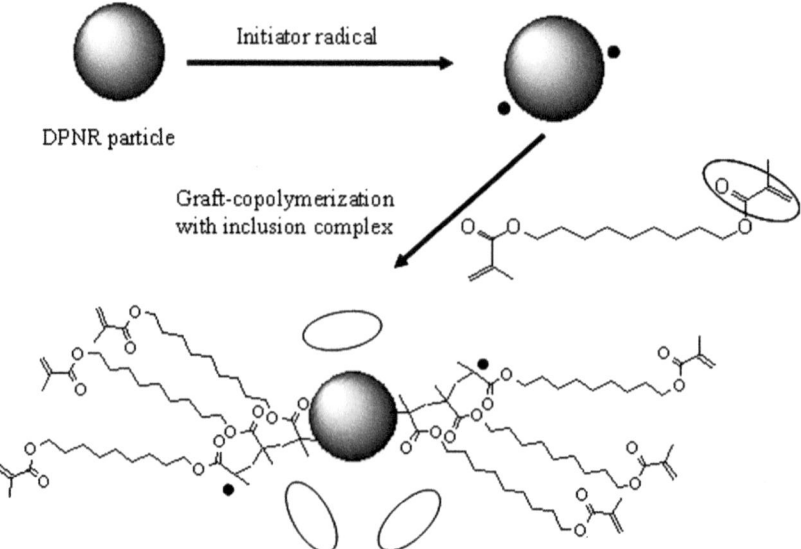

Figure 14.2 Schematic representation of graft copolymerization of inclusion complex onto DPNR in latex stage.

14.2 Inclusion Complex Formation of NDMA and β-CD

The inclusion complex was prepared by dissolving β-CD in ion-exchanged water. NDMA as a divinyl monomer was added in stoichiometric amount. Conditions of inclusion complex formation of NDMA and β-CD are tabulated in Table 14.1.

Figure 14.3 shows FT-IR spectra ranging from 1500 to 1900 cm^{-1} for NDMA, β-CD and the inclusion complex of NDMA with β-CD prepared at 363 K for 1–5 h, respectively. The spectrum of the inclusion complex showed a broad peak at about 1640 cm^{-1}, whereas the corresponding β-CD peak appeared at about 1649 cm^{-1}. This may be attributed to the interaction of

Table 14.1 Conditions for inclusion complex formation of NDMA and β-CD.

Feed of NDMA (mol)	0.044	0.044	0.044
Feed of β-CD (mol)	0.044	0.044	0.044
Ion-exchanged water (ml)	400	400	400
Reaction temperature (K)	303	323	363
Reaction time (h)	5	5	5
Feed of NDMA (mol)	0.044	0.044	0.044
Feed of β-CD (mol)	0.044	0.044	0.044
Ion-exchanged water (ml)	400	400	400
Reaction temperature (K)	303	323	363
Reaction time (h)	5	5	5

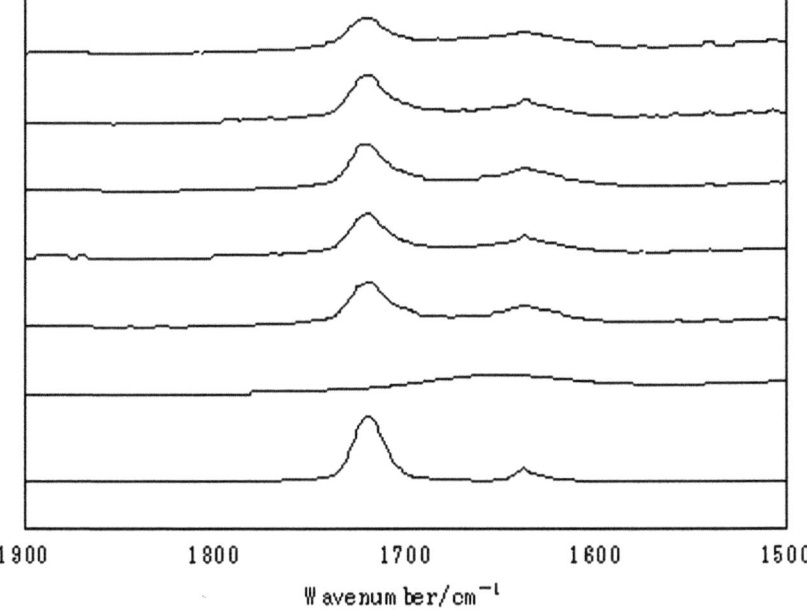

Figure 14.3 FTIR spectra for (a) NDMA, (b) β-CD, inclusion complex of β-CD and NDMA prepared at 363 K for (c) 1 h, (d) 2 h, (e) 3 h, (f) 4 h and (g) 5 h.

NDMA within the β-CD cavity.[10] As for NDMA, a stretching vibration peak of carbon–carbon double bonds and carbonyl groups appeared at about 1640 and 1719 cm^{-1}, respectively. After forming the inclusion complex of NDMA with β-CD, the peak height at 1719 cm^{-1} decreased with reaction time, while that at 1640 cm^{-1} did not change. The decrease in the peak height at 1719 cm^{-1} may be explained by the interaction between the carbonyl group of NDMA and β-CD at longer reaction times. The percentage of complex formation was estimated from a ratio of the peak height at 1719 cm^{-1} to that at 1640 cm^{-1}, as in the following equation:

$$\text{Complex formation}(\%) = \frac{(I_{C=O}/I_{C=C})_{Monomer} - (I_{C=O}/I_{C=C})_{Inclusion\ complex}}{(I_{C=O}/I_{C=C})_{Monomer}} \times 100$$

where $I_{C=O}$ and $I_{C=C}$ are peak height of a carbonyl group at 1719 cm^{-1} and a carbon–carbon double bond at 1640 cm^{-1}, respectively.

Figure 14.4 shows a plot of percentage of complex formation versus reaction time at various temperatures. The complex formation of NDMA with β-CD increased with increasing reaction time. Figure 14.4 also shows the effect of temperature on the complex formation. The ability to form the complex at 363 K was greater than those at 303 K and 323 K. This may be attributed to the higher solubility of NDMA in water at higher temperatures. The inclusion complex prepared from NDMA and β-CD at 363 K for 5 h was subjected to solid state ^{13}C-NMR, XRD and ^{1}H-NMR measurements.

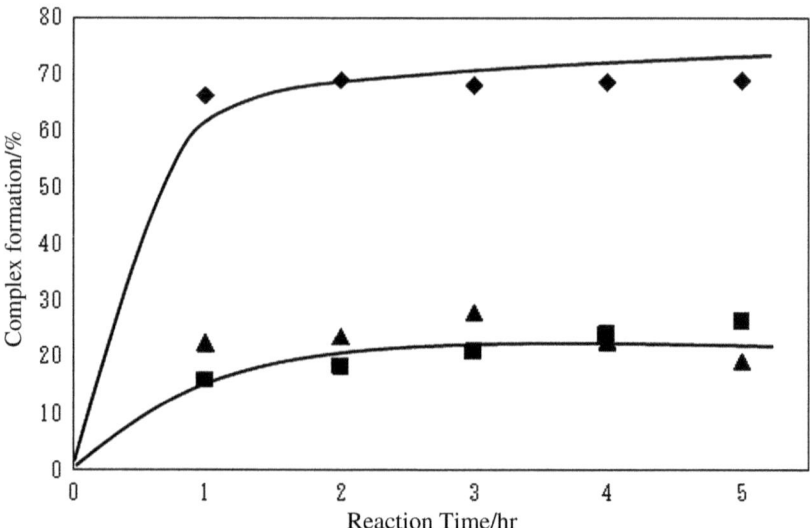

Figure 14.4 Plot of percentage of complex formation versus reaction time for the inclusion complex at various temperatures: (■) 302 K, (▲) 323 K and (◆) 363 K.

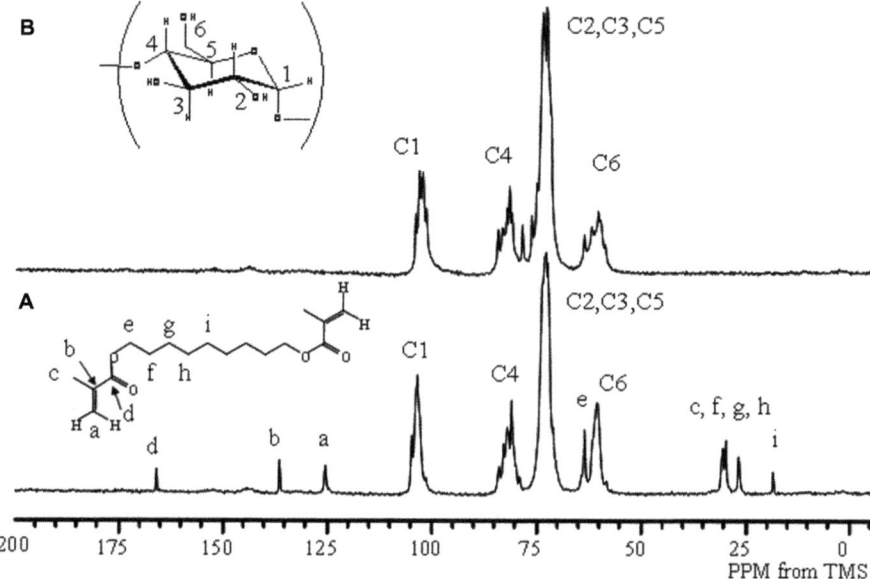

Figure 14.5 ^{13}C CP/MAS NMR spectra for (a) inclusion complex and (b) β-CD.

Solid state ^{13}C CP/MAS NMR spectra for the inclusion complex and β-CD are shown in Figure 14.5. In the spectrum of β-CD, four signals characteristic of C_6, C_2, C_3, C_5, C_4 and C_1 of β-CD appeared at 62.2, 70.0–75.0, 81.0 and 102.2 ppm, respectively. Each signal was observed as an overlapped signal due to each carbon of the glucose unit. After forming the inclusion complex, the spectrum showed apparent single signals for each carbon of the glucose unit accompanying signals of NDMA, which were assigned as shown in Figure 14.5. It is well known that β-CD retains less symmetrical cyclic conformation when it does not include any guest in the cavity.[11] In contrast, apparent single signals for each carbon of the glucose unit of β-CD in the inclusion complex indicate that a symmetrical cyclic conformation of β-CD is formed, *i.e.* the same environment for each glucose unit of β-CD due to formation of the inclusion complex.

Figure 14.6 shows XRD patterns for β-CD and the inclusion complex, respectively. The XRD pattern of β-CD was distinguished from that of the inclusion complex. The difference in XRD pattern may indicate a change in crystalline form due to the formation of the inclusion complex. The main peaks for β-CD appeared at $2\theta = 12.2°$, 14.3°, 21.8° and 22.7°, which indicate a cage-type packing, whereas the inclusion complex showed new peaks at $2\theta = 7.6°$, 11.3°, 12.8°, 16.8°, 20.1° and 33.16° accompanying the disappearance of the main peaks for native β-CD. As is well known, the peak at $2\theta = 7.6°$ and 20.1° in the inclusion complex is characteristic for the channel structure of β-CD including long guest molecules and polymers in particular.[9,10,12] Furthermore, the absence of uncomplexed β-CD was confirmed in the inclusion complex

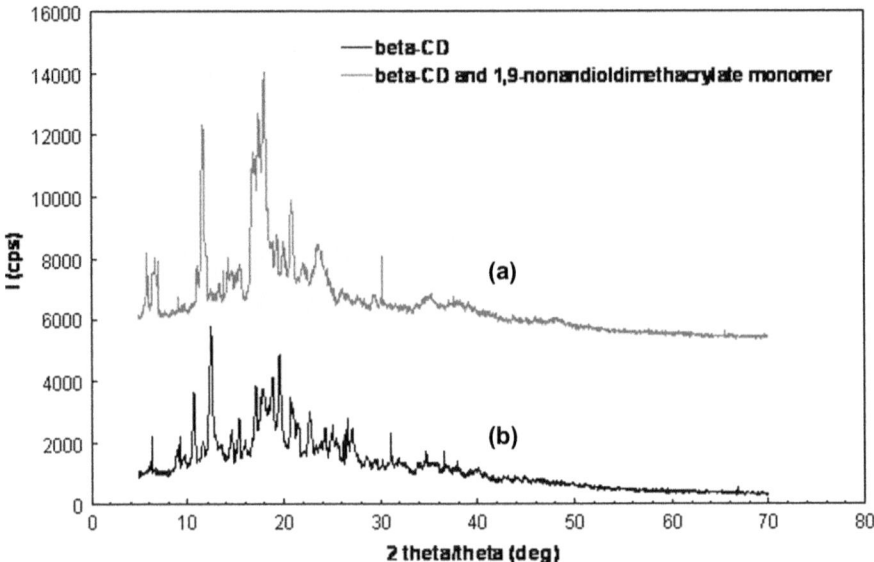

Figure 14.6 XRD diffractograms for (a) inclusion complex and (b) β-CD.

since there is no peak for β-CD. Therefore, it was demonstrated that the inclusion complex was formed with NDMA and β-CD.

In order to determine a ratio of NDMA and β-CD, ^{1}H-NMR spectra for NDMA, the inclusion complex and β-CD were measured in DMSO-d_6 as shown in Figure 14.7. Si signals for NDMA and five signals for β-CD appeared in the ^{1}H-NMR spectra, respectively. These signals were assigned as shown in Figure 14.7, according to previous reports.[8] After forming the inclusion complex, all signals in the ^{1}H-NMR spectrum were attributed to those of β-CD and NDMA. To estimate a ratio of NDMA and β-CD in the inclusion complex, integration of the peak at 4.1 ppm corresponding to four protons of the methylene group adjacent to the methacryloyloxy group of NDMA (-OCH$_2$-) was compared with the peak at 4.82 ppm corresponding to seven protons of the methine group in β-CD (H$_1$), although NDMA might be released from β-CD in DMSO-d_6. The integration showed a ratio of 1 : 1 of NDMA and β-CD in the inclusion complex. This suggests that the inclusion complex was formed in a ratio of 1 : 1.

14.3 Graft Copolymerization of Inclusion Complex onto DPNR Particles

Before graft copolymerization, natural rubber latex was purified to remove most of the protein, which is well known as a radical scavenger to disturb graft polymerization.[13] Figure 14.8 shows ^{1}H-NMR spectra for DPNR, NDMA and the graft copolymer (DPNR-*graft*-poly (NDMA)) prepared from

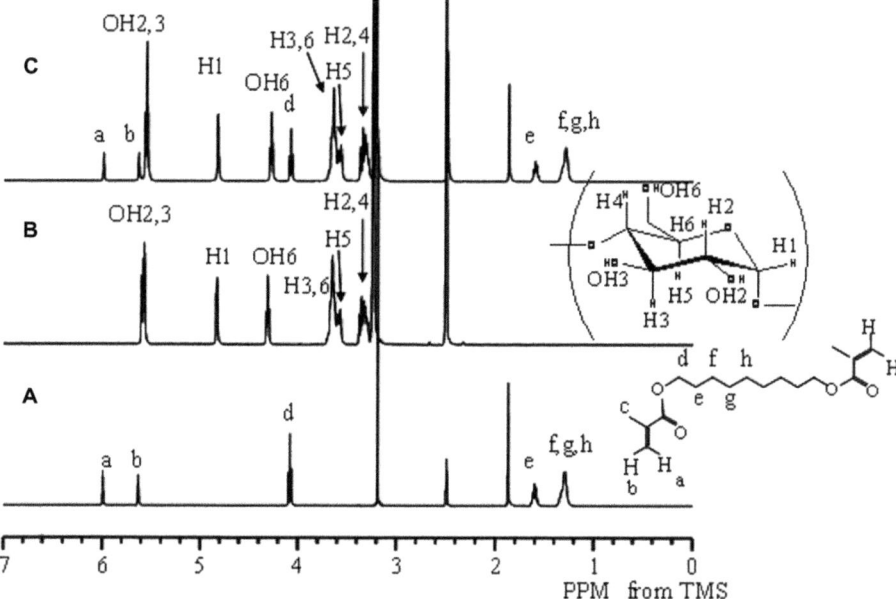

Figure 14.7 ^{1}H-NMR spectra for (a) NDMA monomer, (b) β-CD and (c) inclusion complex (solvent: DMSO-d_6).

20 w/w% DRC of DPNR latex and 150 g/kg rubber of the inclusion complex in the presence of 0.033 mol/kg rubber of TBHPO/TEPA at 303 K for 3 h. Signals in the spectra for DPNR and NDMA were assigned according to the literature.[8] After the graft copolymerization of the inclusion complex onto DPNR particles, two signals at 5.58 and 6.15 ppm appeared in the spectrum, which were assigned to unsaturated methylene protons of the methacryloyl group. Since the signals still appeared even after extraction of the graft copolymer with acetone for 24 h, the residual methacryloyl group is considered to link with natural rubber. Thus, the amount of residual methacryloyl group after graft copolymerization was estimated, as shown in the following equation:

$$\text{Amount of residual methacryloyl group}(\%) = \frac{I_{5.58}/(I_{4.10-4.20}/4)_{\text{graft copolymer}}}{I_{5.58}/(I_{4.10-4.20}/4)_{\text{monomer}}} \times 100$$

where I is the intensity of the signals and subscript numbers represent chemical shift (ppm) of the signals. The estimated amount of residual methacryloyl group is 45.3%. This suggests that formation of the inclusion complex maintained the methacryloyl group of NDMA after graft copolymerization.

Figure 14.9 shows FT-IR spectra ranging from 1600 to 1800 cm^{-1} for DPNR, DPNR-*graft*-poly(NDMA) prepared with TBHPO/TEPA, CHPO/TEPA, BPO or KPS as an initiator, respectively. A peak height was normalized with a height of reference peak at 1660 cm^{-1}, which was identified as a

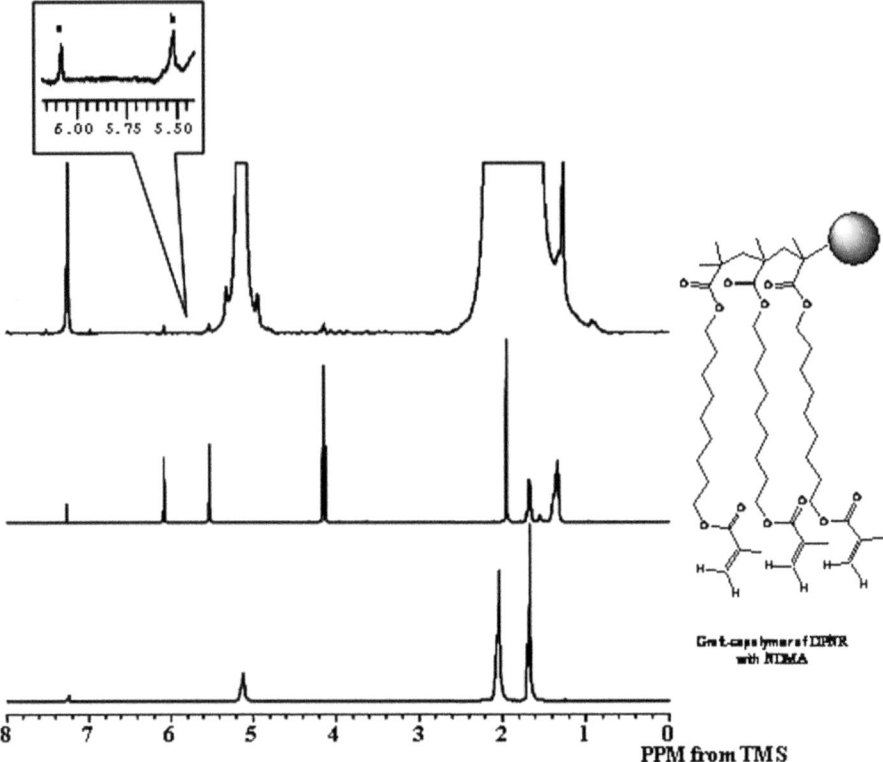

Figure 14.8 ¹H-NMR spectra for (a) DPNR, (b) NDMA and (c) DPNR-*graft*-NDMA (after acetone extraction) prepared from 20 w/w% DRC of DPNR, 15 w/w% inclusion complex for 3 h at 303 K (solvent: chloroform-d).

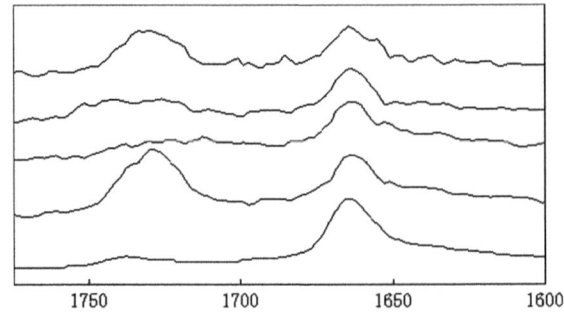

Figure 14.9 FT-IR spectra of (a) DPNR, DPNR-*graft*-NDMA copolymer using (b) TBHPO/TEPA, (c) CHPO/TEPA, (d) BPO and (e) KPS as a initiator.

stretching vibration of the carbon–carbon double bond of natural rubber. As for DPNR-*graft*-poly(NDMA), a stretching vibration peak of the carbonyl group of NDMA appeared at 1730 cm^{-1}. A conversion and content of NDMA in DPNR-*graft*-poly(NDMA) were estimated from a ratio of the peak height at 1730 cm^{-1} to that at 1660 cm^{-1} using a calibration curve.

$$\text{Content of NDMA}(\%) = \frac{\text{Weight of NDMA unit in the graft} - \text{copolymer}}{\text{Weight of graft} - \text{copolymer}} \times 100$$

$$\text{Conversion of NDMA}(\%) = \frac{\text{Weight of NDMA unit in the graft} - \text{coplymer}}{\text{Weight of feed of NDMA}} \times 100$$

Table 14.2 shows the estimated conversion and content of NDMA. The conversion of NDMA was dependent upon the initiator, in which the highest conversion, 69.0 w/w%, was accomplished by using 0.033 mol/kg rubber of TBHPO/TEPA. On the other hand, the conversion of NDMA increased with decreasing feed of the inclusion complex, while the content of NDMA was reached at a plateau at feed of 150 g/kg rubber of inclusion complex. It is quite noteworthy that no coagulation occurred during graft copolymerization even at higher conversion of more than 58.5%. This may be attributed to the formation of the stable inclusion complex between NDMA and β-CD. A suitable condition of the graft copolymerization was determined to be 150 g/kg rubber of inclusion complex and 0.033 mol/kg rubber of TBHPO/TEPA. The resulting product, DPNR-*graft*-poly(NDMA), prepared under suitable conditions, was subjected to TEM observation.

A TEM photograph for DPNR-*graft*-poly(NDMA) is shown in Figure 14. 10, in which a gloomy domain is natural rubber and a bright domain is poly(NDMA). As it is clearly seen, the natural rubber particle of about 1.0 μm in diameter was dispersed in a poly(NDMA) matrix of 10 nm in thickness to form a nanomatrix structure, while it did not contain poly(NDMA). Furthermore, a volume fraction of poly(NDMA) matrix was estimated by image analysis of the photograph to be about 3 w/w%, which corresponded to 1.81 w/w% estimated from the NDMA content shown in Table 14.2. These results indicate that the graft copolymerization occurs only on the surface of

Table 14.2 Various conditions of graft copolymerization of inclusion complex onto DPNR in latex stage.

No	Complex (g/kg-rubber)	Initiator (mol/kg rubber)		Reaction temp. (K)	NDMA content (w/w %)	NDMA conversion (w/w %)
1	150	KPS	0.033	353	1.38	44.4
2	150	CHPO/TEPA	0.033	313	0.06	2.1
3	150	BPO	0.033	353	0.25	8.1
4	150	TBHPO/TEPA	0.033	303	1.81	58.5
5	150	TBHPO/TEPA	0.01	303	0.85	27.3
6	150	TBHPO/TEPA	0.1	303	1.73	55.6

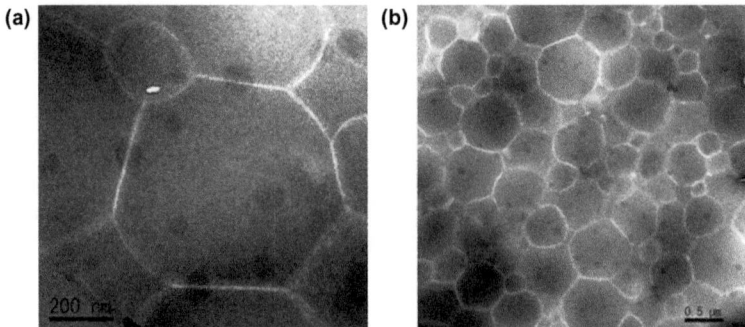

Figure 14.10 TEM photograph for DPNR-*graft*-poly(NDMA), (a) ×20,000, (b) ×5000.

natural rubber particles. This implies that formation of the inclusion complex changed the polarity of hydrophobic NDMA to inhibit penetrating NDMA into hydrophobic natural rubber particles. Consequently, the photoreactive nanomatrix structure was formed inspite of as low as 1.81 w/w% of NDMA content, which has never been seen. This novel photoreactive nanomatrix structure though graft copolymerization may be used for acrylic adhesive binders, clear coatings, *etc.*

14.4 Conclusions

The photoreactive nanomatrix structure was formed by graft copolymerization of the inclusion complex of NDMA with β-CD onto deproteinized natural rubber. The content, conversion and residual methacryloyl group of NDMA for the graft copolymerization were 1.81%, 58.5% and 45.3%, respectively, at NDMA feed of 150 g/kg rubber and initiator concentration of 0.033 mol/kg rubber. The TEM observation of the resulting graft copolymer showed that natural rubber particles of about 1.0 μm in diameter were dispersed in a poly(NDMA) matrix of 10 nm in thickness.

References

1. S. Kawahara, P. Suksawad, Y. Yamamoto and H. Kuroda, *Macromolecules*, 2009, **42**, 8557.
2. S. Kawahara, Y. Yamamoto, S. Fujii, Y. Isono, K. Niihara, H. Jinnai, H. Nishioka and A. Takaoka, *Macromolecules*, 2008, **41**, 4510.
3. K. Kosugi, R. Suthangkul, O. Chaikumpollert, Y. Yamamoto, J. Sakdapipanich, Y. Isono and S. Kawahara, *Colloid Polym. Sci.*, 2012, **290**, 1457.
4. N. Pukkate, Y. Yamamoto and S. Kawahara, *Colloid Polym. Sci.*, 2008, **286**, 411.
5. N. Pukkate, T. Kitai, Y. Yamamoto, T. Kawazura, J. Sakdapipanich and S. Kawahara, *Eur. Polym. J.*, 2007, **43**, 3208.

6. S. Kawahara, T. Kawazura, T. Sawada and Y. Isono, *Polymer*, 2003, **44**, 4527.
7. N. Hayatiyusof, S. Kawahara and M. Said, *J. Rubber Res.*, 2008, **11**, 97.
8. N. Pukkate, T. Horimai, O. Wakisaka, Y. Yamamoto and S. Kawahara, *J. Polym Sci., Part A: Polym. Chem*, 2009, **47**, 4111.
9. S. S. Satav, R. N. Karmalkar, M. G. Kulkarni, N. Mulpuri and G. N. Sastry, *J. Am. Chem. Soc.*, 2006, **128**, 7752.
10. S. Jambhekar, R. Casella and T. Maher, *Int. J. Pharm.*, 2004, **270**, 149.
11. A. Harada, J. Li and M. Kamachi, *Macromolecules*, 1993, **26**, 5698.
12. C. C. Rusa, T. A. Bullions, J. Fox, F. E. Porbeni, X. Wang and A. E. Tonelli, *Langmuir*, 2002, **18**, 10016.
13. S. Tuampoemsab and J. Sakdapipanich, *Kautsch. Gummi Kunstst*, 2007, **57**, 678.

CHAPTER 15

Blends and IPNs of Natural Rubber with Thermosetting Polymers

RAJU THOMAS,*[a] ISHAK AHMAD,[b] SAHRIM HJ. AHMAD[c] AND SHINU KOSHY[d]

[a] Department of Chemistry, Mar Thoma College, Tiruvalla-689 103, Kerala, India; [b] School of Chemical Sciences and Food Technology, Science and Technology, Universiti Kebangsaan Malaysia, 43600 Bangi, Selangor, Malaysia; [c] School of Applied Physics, Science and Technology, Universiti Kebangsaan Malaysia, 43600 Bangi, Selangor, Malaysia; [d] School of Chemical Sciences, Mahatma Gandhi University, Kottayam, Kerala, India
*Email: rajuthomastvla@gmail.com

15.1 Introduction

Thermoset resins occupy a significant role in the field of polymer technology as they have a wide range of applications in diverse fields of human endeavours. But when cured, they become brittle in nature. A considerable amount of work has been done to try and enhance their toughness, without decreasing the thermomechanical and other desirable properties of these materials. In order to improve their fracture resistance they are usually blended or reacted with various additives and modifiers, which generally form a second dispersed phase. Depending on the weakness of the host polymer matrix, incorporation of a dispersed phase can lead to a positive impact on properties such as optical clarity, toughness, elasticity, creep resistance and permeability. Depending on the end use, careful selection of additives such as fillers, rubbers and reinforcements is necessary.[1] The most frequently used modifiers are liquid rubbers.[2–5]

Epoxy resins are widely used as protective coatings and adhesives, for structural applications such as adhesives to bulk out structural components, with extensive use in the aerospace, dental and other medical fields. Moreover, epoxies are widely used as insulating and structural materials in manufacturing microelectronic devices and components such as computer chip packing and circuit boards, due to their excellent combination of chemical and corrosion resistance and good electrical properties. On the other hand, unsaturated polyester (UPE) resin, a resin prepared from a polyester synthesized by esterification of glycol, unsaturated acid and saturated acid dissolved in styrene monomers, is widely used in many applications, such as electronic equipment, containers, automobiles, and cultured marble because of its clarity, and excellent chemical and corrosion resistance.

Rubber toughening can be achieved by two distinct methods. One of the most common methods is to blend the epoxy/hardener mixture with functionalized liquid rubber having restricted solubility and limited compatibility. During cure the rubber undergoes phase separation, which leads to a two-phase microstructure. Rubber was added to the uncured thermoset resin and after the crosslinking reaction the rubber-modified polyester exhibited a two-phase microstructure consisting of relatively small rubbery particles dispersed in a matrix of unsaturated polyester. This microstructure resulted in the material possessing a higher toughness than the unmodified thermosets, with only a minimal reduction in other important properties, such as modulus. The morphology of the final modified thermoset can significantly affect the toughening mechanism and consequently its fracture toughness. It has been found that the dispersion of rubber particles within the glassy matrix can improve this property because the rubber particles can absorb and disperse the energy.[6] In this chapter, we are limiting our discussion to epoxy and unsaturated polyester resin based blends and IPNs.

15.2 Elastomer-Modified Epoxy Resin Systems

Epoxy monomers are highly viscous liquid molecules that contain two or more three-membered ring structures consisting of an oxygen atom bonded to two carbon atoms in a way that is formed, and is known as an epoxy, oxirane, ethoxyline or glycidyl group. But most epoxies are thermoplastic resins that have little value until they are cured with crosslinking agents. Several types of liquid natural rubbers (LNRs) with various functional groups are used successfully to increase the toughness of epoxy resins.

In an interesting study by Mathew et al.,[7] an attempt was made to toughen diglycidyl ether of bisphenol A (DGEBA) type epoxy resin with LNR possessing hydroxyl functionality (HTLNR); the epoxy monomer was cured using nadic methyl anhydride as hardener in the presence of N,N-dimethylbenzylamine as accelerator. Blends of epoxy resin/HTLNR containing 5, 10, 15 and 20 wt% elastomer were prepared using the melt mixing technique.

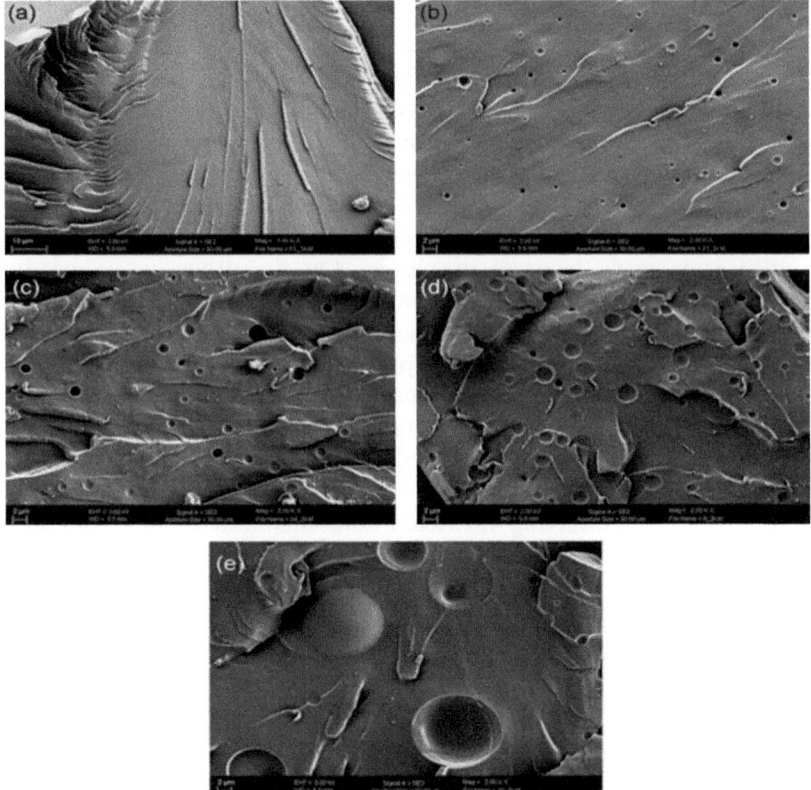

Figure 15.1 SEM micrographs[7] of (a) neat epoxy, (b) 5 wt%, (c) 10 wt%, (d) 15 wt %, (e) 20 wt%.

The unmodified epoxy system showed cracks at different planes on the smooth glassy surface, indicating brittle failure, while the fracture surfaces of the rubber-modified ones were rough, indicating massive shear deformation Figure 15.1. The morphology appeared as a phase-separated one with the dispersion of small rubber domains in the epoxy matrix. However, the particle size increased with increase in rubber content due to the coalescence process.

The viscoelastic properties of the blends were analysed using dynamic mechanical thermal analysis. The plot of tan δ versus temperature showed a single relaxation peak around 130 °C, corresponding to the T_g of the epoxy-rich phase for the unmodified epoxy network and for the blends. On further examination, a relaxation peak of very low amplitude at around −65 °C (called β-relaxation), for both modified and neat epoxy matrices, is found, which is attributed to the motions of glycidyl units in the network.[8] The T_g of the rubbery phase seems to be overlapped with β-relaxation and the T_g of the epoxy-rich phase slightly shifts toward the low-temperature side with the addition of the rubber. The decrease in T_g of the epoxy phase can be attributed to a dilution effect by the addition of the rubber phase, and may also result

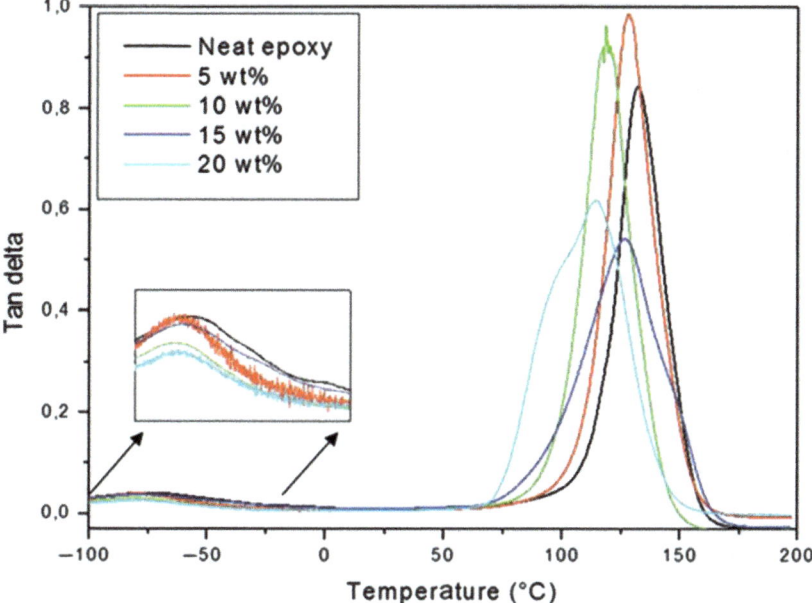

Figure 15.2 Variation of tan δ with respect to temperature for HTLNR modified epoxy.[7]

from the miscible rubber phase in the epoxy-rich phase. The presence of a secondary peak, at a lower temperature in the tan δ, suggests the presence of a fraction of IPN type of sponge type structure consisting of miscible epoxy/rubber phase with higher mobility due to a different degree of crosslinking. The authors believe that this anomalous behaviour could be due to the increase in the relative amount of dissolved rubber as rubber content increases and hence having a secondary peak at a lower T_g, as shown in Figure 15.2.

The fracture behaviour was found to increase with increase in rubber content, as shown in Figure 15.3. The fracture toughness of 0.85 MN/m$^{3/2}$ for the neat epoxy system is comparable with the literature values.[5] A maximum of around 100% increase was observed for 20 wt% a HTLNR-modified blend and the increase is due to cavitation and shear deformation of the rubber particles in the matrix.[9,10] The impact property was also enhanced for modified epoxy samples with an increase in HTLNR concentration up to 15 wt%, followed by a decrease.

The same group[11] has also observed enhanced impact and toughness behaviour of the blends due to the secondary phase separation effect on the same rubber-modified system. The effect is originated due to the combined effect of hydrodynamics, viscoelastic effects of the rubber phase, diffusion, surface tension, polymerisation reaction and phase separation. In a dynamic asymmetric system, the diffusion of the fast dynamic phase is prevented by the slow dynamic phase, and hence the growth of the fast dynamic phase is retarded by the slow dynamic phase. In the case of low viscosity blends the

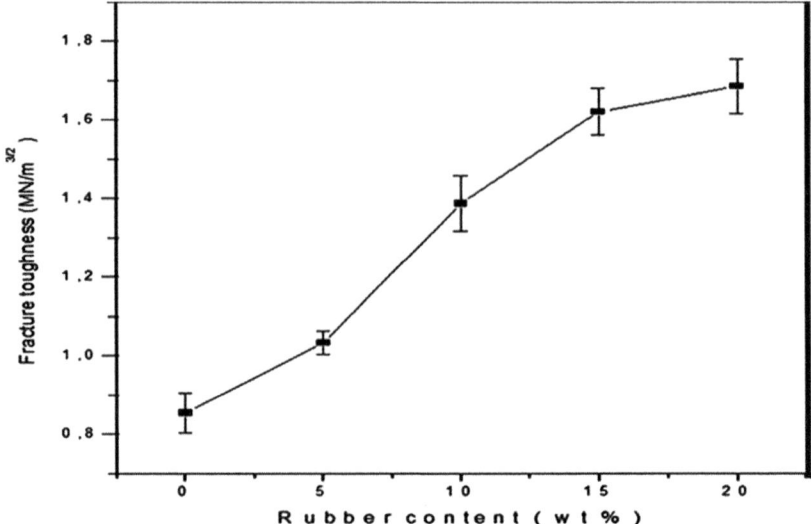

Figure 15.3 Variation of fracture toughness with rubber content.[7]

diffusion of the fast dynamic phase cannot follow geometrical growth and cannot establish local concentration equilibrium and hence double phase separation takes place.

The morphology of extracted and unextracted samples containing varying weights of elastomer are represented in Figure 15.4(a) to (i). The micrographs show crosslinked epoxy sub-inclusions in the rubber domains (particle in particle morphology). The etched surface for 10 wt% HLNR–epoxy as represented by Figure 15.4(d) shows holes uniformly distributed in the epoxy matrix and the unetched tensile fracture surface given by Figure 15.4(e) shows secondary phase separation of crosslinked epoxy in the dispersed rubber phase. The size of the domains reaches a critical value of 6 μm in the case of 15 wt% rubber-modified epoxy. The relatively dark region is the continuous epoxy matrix in which HLNR domains are dispersed.

The variation in the domain sizes showed that the size increases steadily (*i.e.* number average diameter varying from 1 to 9 μm) with increase in composition from 5 to 20 wt% rubber. It is known that rubber particles of 4–8 μm are responsible for toughening. Smaller particles of 1–3 μm toughen the system through shear banding, while larger particles toughen through cavitation and shear yielding.[12]

In a study by Hong *et al.*,[13] the effects of ENR on the curing behaviour and physical properties of an epoxy/dicyandiamide/2-methyl imidazole system were analysed. The participation of ENR in the curing reaction system changed the reaction to an exothermal one. The time to maximal curing rate, the glass transition temperature, the rate constant, and the reaction order of the epoxy system were also changed. The rubber is precipitated due to spinodal decomposition caused by the increase in molecular weight of the epoxy matrix.

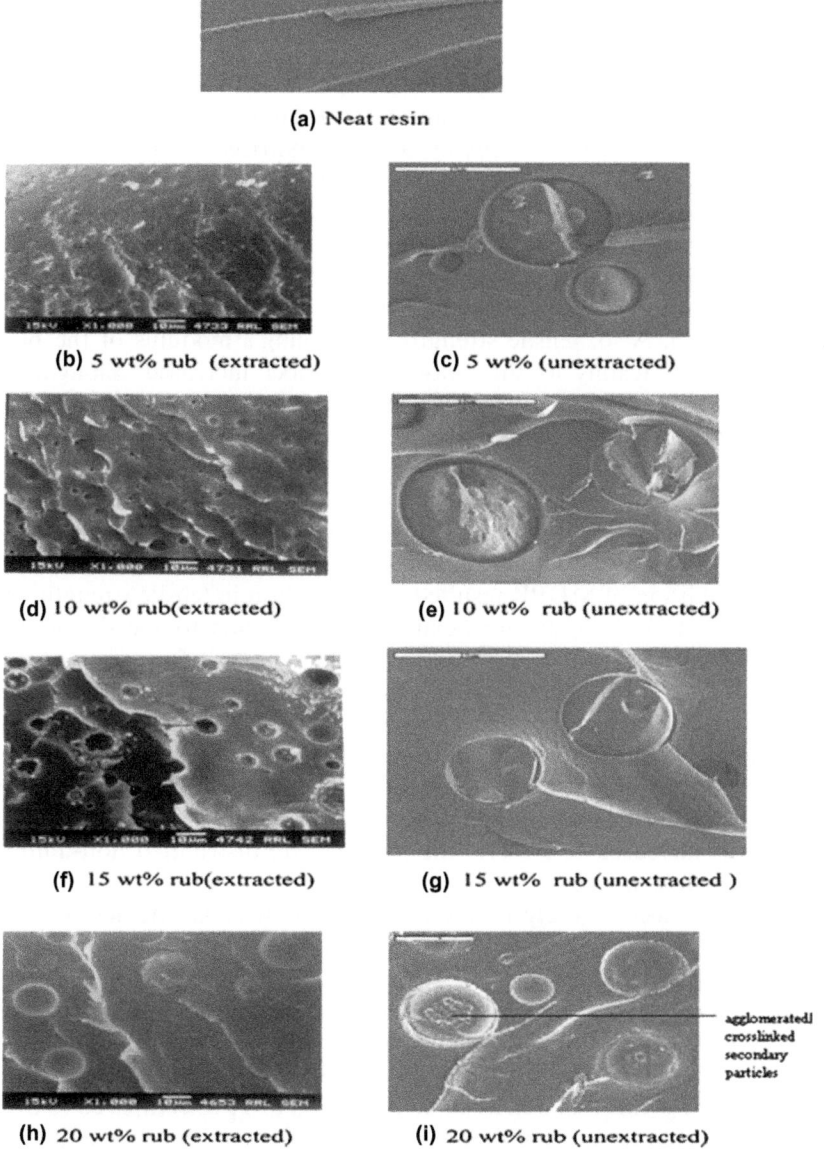

Figure 15.4 a) to (i) SEM of HLNR modified epoxy resin at the cure[11] (extracted and unextracted).

This has been reported in many other cases also.[2,14–16] The particle size of the rubber phase showed an increase with increase in the curing temperature and ENR content. Other studies also proved this.[17,18] The effect of ENR on the curing characteristics becomes more significant in the system with more elastomer. The lap shear strengths (LSS) obtained from different systems of epoxy resin modified with ENR increased with increase in the curing temperature due to a better cure at the higher temperature, but decreased with increasing ENR content, resulting from an inferior structure formed in the presence of ENR.

In another study by Chuayjuljit et al.,[19] epoxidized natural rubbers (ENRs) prepared from high ammonia concentrated NR latex via an in situ epoxidation method have been used to modify epoxy resin matrices. From TGA and DSC analysis, it was found that thermal decomposition temperature and glass transition temperatures of the modified epoxies were similar to epoxy resins. The impact strength increased up to only 5 phr modifier content and thereafter showed a decreasing trend. The SEM micrographs showed that rubber globular nodules were present at lower phr of rubber content, which resulted in higher impact strength. Also, tensile strength and Young's modulus of the blended resin decreased steadily as the amount of rubber increased. Elongation increased when the amount of rubber was not above 5 phr. Lower flexural strength and flexural moduli were obtained when the amount of rubber increased.

Kumar et al.[20] studied the effect of maleated depolymerized natural rubber (MDPR), functionalized by grafting depolymerized natural rubber (DPR) with maleic anhydride by a simple thermal reaction, as a modifier for the epoxy resin. The addition of MDPR resulted in a reduction in tensile strength, tensile modulus, flexural strength and flexural modulus, due to the dissolution of rubber in the epoxy matrix.[21,22] However, the ductile deformation of rubber before failure, the toughness property, was confirmed from the higher values of elongation at break and flexural strain to failure percentage values of the modified matrices. The fractured surfaces of the epoxy blend systems having 1 and 2 wt% of MDPR showed the distribution of uniformly distributed globular particles having dimensions in the range of 1–6 μm. However, when the rubber content was increased to 3 wt%, the particles are distributed non-uniformly, having a much larger size.

The FTIR spectral analysis has proved that there is interaction between MDPR and the liquid epoxy resin; the secondary alcoholic group of the epoxy resin opens the anhydride rings of maleated rubber to form ester and carboxyl groups. The carboxyl group interacts with the epoxide of the resin to generate new secondary alcohol groups. Thus there is interaction between the rubber and the resin. DSC traces of epoxy/MDPR blends showed no significant change in the T_g values as well as a narrow region of T_g as compared to the neat epoxy, suggesting limited compatibility between the components and the possibility of distinct phase separation.

Shear banding between the rubber particles clearly showed evidence of shear yielding and plastic deformation over large volumes which are the energy absorption mechanism operating in rubber-modified networks. The neat epoxy showed brittle fracture, suggesting the absence of a shear yielding mechanism. Also the poor fracture properties of the higher modified system (3 wt%) is due to the presence of large particles which are randomly distributed. The increase in the free volume of samples up to 2 phr loading of the modifier is in support of the elongation at break and toughness. The increase in the values can be expressed to provide more space for the chain movements, and hence toughness. This is lacking in higher weight content epoxy where the free volume value is less. The crosslinking of the resin by the modifier is evident from the slight decrease in the I_3 values up to 2 phr modifier addition as there is a chance for the reaction of each anhydride group with two epoxy rings at higher temperatures. Due to lower viscosity of the systems at higher temperatures, there are chances for the anhydride group to react with new epoxy rings in the other chains, resulting in a crosslinking network structure.

Recently a Malaysian group has studied[23] the toughening effect of different types of natural rubbers, LNR and liquid epoxidized natural rubber (LENR), on the epoxy matrix using dicyandiamide curing agent. The Izod impact test for all the compositions of LENR-toughened epoxy has shown that optimum impact strength was recorded at 5 phr rubber content in the matrix. However, no significant change in impact strength was reported for higher or lower weight content of LENR in the epoxy matrix. Meanwhile LNR-modified epoxy only reported an increase in the impact strength at 3 phr; excessive rubber in the matrix resulted in a decrease in impact strength. The reduction in the impact strength was explained by the particle size from the fracture surfaces. The size of the rubber particles in the LNR-modified epoxy composites varied from 0.5 μm to 30 μm as the concentration of rubber increased, with an optimal impact strength reported for 3 phr rubber. Meanwhile the size of rubber particles in the LENR-modified epoxy composites seems to be independent of the concentration of rubber inside the matrix.

The mechanical and morphology studies of LNR-modified epoxy have led to an explanation of the toughening phenomenon operating in these materials.[24] The morphology of neat epoxy fracture surface showed smooth, glassy, rivery fractured surface with ripples and no shear deformation line was found. It indicates that no significant plastic deformation had occurred and no energy dissipating mechanism was operating here. However, the fractured surface of LNR-toughened epoxy clearly showed two distinct phases – a continuous epoxy matrix and a dispersed rubber phase. Rubber particles will act as an energy dissipating centre in the epoxy matrix. As a result, rubber particles would respond to the triaxial stresses near the crack tip and cause localized shear yielding. This was followed by the rubber bridging mechanism in the crack tip zone in the epoxy matrix and plastic void growth initiated by cavitation or debonding of the rubbery particles from the surface.[25] In rubber-modified epoxies, deformation lines were observed to propagate through rubber particles and it proved the ductile fracture did occur. Increments of rubber particles size were observed when rubber composition was increasing. Similar observations were found by Lee et al.[26] According to Thomas et al.,[5] the effectiveness of rubber particles as stress concentrators and in dissipating energy can be influenced by rubber particle size. Crack energy propagation and dissipation mechanisms are suitable for particles with smaller sizes. Meanwhile, rubber particles will have less interfacial adhesion when particles became larger at higher compositions. It may cause rubber particles to be 'pulled out' from the matrix due to crack propagation through the interface between rubber particles and the matrix. Not much crack energy will be dissipated because the crack was not propagated through the rubber particles. Chikhi et al.[27] pointed out that the excessive liquid rubber composition led to agglomeration because they have not reacted with epoxy resin hence resulting in a poor interface with the matrix.

15.3 Elastomer-Modified Unsaturated Polyester Resin Systems

Unsaturated polyester resins mainly consist of alcohols and acids, called alkyd resin (UPA) with some carbon–carbon double bonds per molecule, and have the ability to react with a low molecular weight crosslinking agent, *i.e.* a vinyl monomer such as styrene, to produce thermoset polymers. The dense crosslinking and consequently the poor resistance against crack initiation and propagation and high volumetric shrinkage of UPA resin are the two most important drawbacks in its widespread applications. Many investigations have been carried out to toughen thermosetting resins by adding typical elastomers.

In an interesting study by Ahmad et al.,[28] unsaturated polyester (UPA) has been modified by LNR. The tensile strength and impact energy increased with increasing LNR concentration and the optimum concentration of LNR was found to be 10 phr. A finer dispersion of the rubber phase in UPA resin, which was caused by lower interfacial energy between the components, is attributed to good adhesion at phase boundaries in the 10% LNR/UPR blend. The strain at break increased with increasing LNR content. This was explained as due to the improvement of extensibility property with higher rubber content in the system. Apparently, the network of LNR in the UPR matrix plays a critical role in the enhancement of the tensile properties of blends. The rubber network provides a backbone for blends to resist cracking. The finer and more perfect network resulted in the resistance to cracking which accounted for higher tensile strength and elongation at break. However, further increase in LNR content resulted in a higher incompatibility of rubber and the hard segments of UPR. Further, it was reported that the UPA/LNR system, when blended with modified montmorillonite (MMT), improved the tensile strength compared to the unmodified MMT. This provided good interaction between the matrix and the fillers for the system with MMT and the fillers are homogeneously dispersed in the matrix which led to an increase in the tensile properties of the composite. The optimum composition of modified MMT in the composite was found at 4% filler loading.

In another study by Cherian et al.,[29] UPA was modified by blending with functionalized rubbers such as ENR, hydroxyl-terminated natural rubber, and maleated nitrile rubber. The elastomers bearing reactive functional groups showed better compatibility with the resin and improved the toughness and impact resistance of the cured resin substantially, compared to unmodified elastomers. Maleated nitrile rubber (NBR) was far superior to all other rubbers; the addition of 2.5% of the rubber resulted in maximum tensile strength. Addition of more rubber lowered the tensile strength. This was considered as due to the high degree of compatibility and the possible inter-component grafting of NBR and polyester chains. For all other elastomers, there was a simultaneous increase in tensile strength. A considerable enhancement of toughness and impact resistance has been recorded when the functional rubbers are used in place of unmodified rubbers. In the case of maleated NBR, the tensile strength of the cured UPR has been found to increase by as much as 98% simultaneously, improving the toughness by about 303%. The improvement in toughness and impact resistance has been achieved without seriously affecting other properties. Properties such as tensile strength have, in fact, shown considerable increase during the toughening process.

In another study,[30] UPA has been modified using graft copolymers of natural rubber and polystyrene, synthesized by free-radical grafting of styrene monomer onto natural rubber in latex form. The graft copolymers obtained and unsaturated polyester (UPE) resin were mixed and cast at room temperature using methyl ethyl ketone peroxide as an initiator and co-octoate as an accelerator. The samples prepared from ungrafted natural rubbers exhibited aggregation of the rubber component, whereas the samples prepared from grafted natural rubbers showed good dispersion of the rubber component in a glassy matrix of UPE resin. It was found that the amount of polystyrene grafted onto natural rubber and the graft copolymer content in the polymer blend significantly affected the mechanical properties (impact and flexural strength) of the blend samples. An increase in the amount of hard and brittle polystyrene in glassy matrix of UPE resin overshadowed the impact-absorbing ability of the rubber component, causing the impact strength of the blend samples to be lower than that of pure UPE resin. On the other hand, an increase in easily elongated uncrosslinked rubber molecules, as the graft copolymer content in blend samples increased, resulted in a decrease in their flexural strength.

In a study by Hisham et al.,[31] blends of unsaturated polyester resins prepared from recycled poly(ethylene terephthalate) (PET) bottle waste with LNR was carried out by varying the amount of LNR from 0 to 7.5 wt%. The properties of the blend samples were then compared to the optimum blend of LNR with commercial resin through the T_g, mechanical strength and morphology properties. The optimum tensile strength was obtained by the addition of 2.5% rubber, and thereafter showed a decreasing trend. This is attributed to the presence of hydroxyl (–OH), epoxy (C=O) and carbonyl (C–O)[32] functional groups present in the rubber which is prepared by a photochemical oxidation method. The presence of these functional groups allows the solubility of the elastomer in the uncured resin to form a homogeneous distribution of discrete incompatible elastomeric particles during crosslinking. Due to the resin shrinkage during the cure process, microvoids are generated. Also, small sizes of cracking zones represented smaller sizes of microvoids. On the other hand, samples having higher weight content of rubber exhibited large particles which lead to zones where cracking can pass through easily and reduced the value of the tensile stress.[33] The tensile modulus also showed the greatest value with the optimum concentration (2.5%) of rubber. This optimum concentration is believed to achieve good adhesion and homogeneous distribution of rubber particles in the blend system. During deformation, debonding occurred between internal cavitations of the rubber and the interface bonding of rubber particles with the UPA matrix interface. This results in the cavitation of the elastomeric particles[34–40] before the failure of the matrices, which ultimately leads to an energy dissipation mechanism and hence fracture toughness. Finer and more homogeneous distribution of the smaller particles was observed by morphological analysis with an improved adhesion at the phase boundary. The comparison of optimum blend samples showed that LNR acts as an effective modifier for the UPR based on recycled PET rather than for the commercial resin.

15.4 Conclusions

Successful attempts to toughen thermoset resins, especially epoxy and unsaturated polyester resins using functionalized liquid rubbers have been reported. In general, a well-dispersed and chemically bonded rubbery phase that, with a suitable volume fraction and particle size/distribution, will offer the best toughness increase to a specific thermoset system without significant loss of thermomechanical properties. The morphology developed during cure of the thermoset systems varied in particle size depending on the weight content of elastomers and the state of cure of systems. The major toughening mechanisms operating in these systems are shear yielding of the matrix and cavitation. Normally, the toughness enhancement operates only with low rubber content. More rubber inclusion will tend to flexibilize the matrix. The tensile strength, fracture toughness and impact strength of rubber-modified thermoset systems increases with an optimum content of elastomer. Unlike in thermoplastic-modified thermosets, the thermal properties are not so high because the elastomers are of low modulus nature.

References

1. A. Ibrahim and H. M. Dahlan, *Prog. Polym. Sci.*, 1998, **23**, 665.
2. C. W. Wise, W. D. Cook and A. A. Goodwin, *Polymer*, 2000, **41**, 4625.
3. V. D. Ramos, H. M. Costa, V. L. P. Soares and R. S. V. Nascimento, *Polym. Test.*, 2005, **24**, 387.
4. R. Thomas, J. Abraham, S. P. Thomas and S. Thomas, *J. Polym. Sci., Part B: Polym. Phys.*, 2004, **42**, 2531.
5. R. Thomas, Y. Ding, Y. He, L. Yang, P. Moldenaers, W. Yang, T. Czigany and S. Thomas, *Polymer*, 2008, **49**, 278.
6. F. Vazquez, M. Schneider, T. Pith and M. Lambla, *Polym. Int.*, 1999, **41**, 1.
7. V. S. Mathew, P. Jyotishkumar, S. C. George, P. Gopalakrishnan, L. Delbreilh, J. M. Saiter, P. J. Saikia and S. Thomas, *J. Appl. Polym. Sci.*, 2012, **125**, 804.

8. P. Jyotishkumar, J. Koetz, B. Tierisch, V. Strehmel, R. Haßler, C. Ozdilek, P. Moldenaers and S. Thomas, *J. Phys. Chem. B*, 2009, **113**, 5418.
9. A. F. Yee and R. A. Pearson, *J. Mater. Sci.*, 1986, **21**, 2475.
10. R. S. Raghava, *J. Polym. Sci., Part B: Polym. Phys.*, 1988, **26**, 65.
11. V. S. Mathew, C. Sinturel, S. C. George and S. Thomas, *J. Mater. Sci.*, 2010, **45**, 1769.
12. R. Bagheri and R. A. Pearson, *J. Mater. Sci.*, 1996, **31**, 3945.
13. S. G. Hong and C. K. Chan, *Thermochim. Acta*, 2004, **417**, 99.
14. K. Yamanaka, Y. Takagi and T. Inoue, *Polymer*, 1989, **60**, 1839.
15. P. K. Chan and A. D. Rey, *Macromolecules*, 1997, **30**, 2135.
16. T. Inoue, *Prog. Polym. Sci.*, 1995, **20**, 119.
17. S. C. Kunz, J. A. Sayre and R. A. Assink, *Polymer*, 1982, **23**, 1897.
18. P. Bussi and H. Ishida, *Polymer*, 1994, **35**, 956.
19. S. Chuayjuljit, N. Soatthiyanon and P. Potiyaraj, *J. Appl. Polym. Sci.*, 2006, **102**, 452.
20. K. Dinesh Kumar and B. Kothandaraman, *eXPRESS Polym. Lett.*, 2008, **2**, 302.
21. C. B. Bucknall, *Toughened Plastics*, Applied Science Publishers, London, 1977.
22. A. J. Kinloch, in *Rubber-Toughened Plastics*, ed. C. K. Riew, *Advances in Chemistry Series*, American Chemical Society, Washington DC, 1989, p. 67.
23. S. Y. E. Noum, S. H. Ahmad, R. Rasid, Y. C. Hock, L. Y. Seng, M. A. Tarawneh and S. Y. Yahya, in *Proceedings Malaysia Polymer International Conference*, 21–22 October 2009.
24. L. Y. Seng, S. H. Ahmad, R. Rasid, S. Y. E. Noum, Y. C. Hock and M. A. Tarawneh, *Sains Malays.*, 2011, **40**, 679.
25. Y. Huang, D. L. Hunston, A. J. Kinloch and C. K. Riew, in *Toughened Plastics 1*, ed. C. K. Riew and A. J. Kinloch, ACS Advances in Chemistry 233, American Chemical Society, Washington DC, 1993, p. 1.
26. W. H. Lee, K. A. Hodd and W. W. Wright, in *Rubber-Toughened Plastics*, ed. C. K. Riew, Advances in Chemistry Series, American Chemical Society, Washington DC, 1989, p. 263.
27. N. Chikhi, S. Fellahi and M. Bakar, *Eur. Polym. J.*, 2002, **38**, 251.
28. I. Ahmad and F. M. Hassan, *J. Reinf. Plast. Compos.*, 2010, **29**, 2834.
29. A. B. Cherian and E. T. Thachil, *J. Elastomers Plast.*, 2003, **35**, 367.
30. S. Chuayjuljit, P. Siridamrong and V. Pimpan, *J. Appl. Polym. Sci.*, 2004, **94**, 1496.
31. S. F. Hisham, I. Ahmad, R. Daik and A. Ramli, *Sains Malays.*, 2011, **40**, 729.
32. I. Abdullah and S. Ahmad, *Mater. Forum*, 1992, **16**, 353.
33. L. Suspene, Y. S. Yang and J. P. Pascault, in *Rubber-Toughened Plastics*, ed. C. K. Riew, Advances in Chemistry Series, American Chemical Society, Washington DC, 1989, p. 163.
34. W. D. Bascom and D. L. Hunston, in *Proceedings of the International Conference on Toughening of Plastics*, PRI, London, 1978, pp. 22, 978.

35. W. D. Bascom, R. L. Cottington, R. L. Jones and P. Peiper, *J. Appl. Polym. Sci.*, 1975, **19**, 2545.
36. W. D. Bascom and R. L. Cottington, *J. Adhes.*, 1976, **7**, 333.
37. W. D. Bascom, R. L. Cottington and C. O. Timmins, *Appl. Polym. Symp.*, 1977, **32**, 165.
38. W. D. Bascom, R. Y. Ting, R. J. Moulton, C. K. Riew and A. R. Siebert, *J. Mater. Sci.*, 1981, **16**, 2657.
39. H. Tamamoto, *Int. J. Fract.*, 1978, **14**, 347.
40. V. Tvergaard, *Int. J. Fract.*, 1981, **17**, 389.

CHAPTER 16

Natural Rubber Blends with Biopolymers

SILVIA MARIA MARTELLI,*[a] CAROL SZE KI LIN,[b] ZHENG SUN,[b] NATHALIE BEREZINA,[c] FARAYDE MATTA FAKHOURI[a,d] AND LUCIA HELENA INNOCENTINI-MEI[d]

[a] Faculty of Engineering, Federal University of Grande Dourados (UFGD), Dourados, Brazil; [b] School of Energy and Environment, City University of Hong Kong, Hong Kong, China; [c] Materia Nova, Rue des Foudriers 1, 7822 Ghislenghien, Belgium; [d] Department of Chemical Engineering, Campinas State University (UNICAMP), Campinas, Brazil
*Email: smmartelli@gmail.com

16.1 Introduction

An important change is taking place in the global vision of society and this change is being reflected in the policies of global polymer industries. The development of 'green' and 'environmentally friendly' bio-based polymers, polymers derived from renewable resources, is growing fast. The advantages of developing polymeric blends versus the synthesis of new polymeric materials have been well discussed in the literature and in this book. The use of blends allows the combination and optimization of properties from each of the components that form the blends. Biopolymers play a very important role in the rubber industry. Natural rubber (NR) was the first biopolymer used in the rubber industry and on the basis of this the whole rubber industry was built.[1]

NR is a renewable natural resource with many excellent properties, such as outstanding resilience, high strength and good processability.[2] Moreover, NR is widely used in industry and it is freely and naturally available at low cost. The unique mechanical properties of NR result from its highly stereoregular microstructure and the rotational freedom of α-methylene C–C bonds and from the entanglements that result from its high molecular weight, which contributes to its high elasticity.[3]

Different properties of NR-based materials can be obtained through blends with biopolymers. The rubber particles behave as stress concentrators, enhancing the fracture energy absorption of brittle polymers and ultimately resulting in a material with improved toughness. In order to impart toughness to polymers, the following criteria must be met:

- The rubber must be distributed as small domains in the polymer matrix.
- The rubber must have good interfacial adhesion to the polymer.
- The glass transition temperature of the rubber must be at least 20 °C lower than the use temperature.
- The molecular weight of the rubber must not be low.
- The rubber should not be miscible with the matrix polymer.
- The rubber must be thermally stable at the polymer processing temperatures.[4]

In the past few decades, several biopolymers have been examined for their uses as modifying components in rubber blends. In this chapter, the use of biopolymers such as lignin, polysaccharides, proteins and polyesters to produce blends with NR are discussed.

16.2 Natural Rubber/Lignin Blends

16.2.1 General Information

Lignin is a complex heteropolymer of cinnamyl alcohols, and it is the third most abundant natural polymer present in nature after cellulose and hemicellulose. It is estimated that the amount of lignin on earth is 300 billion metric tonnes, with an annual biosynthetic production rate of 20 billion metric tonnes.[5] Lignification has allowed the evolution of large arborescent land plants capable of survival in relatively arid environments.[6] It provides rigidity, internal transport of water and nutrients, and protection against attack by microorganisms.[7]

Lignins are amorphous polymers consisting of phenylpropane units, and their precursors are three aromatic alcohols (monolignols), namely p-coumaryl, coniferyl and sinapyl alcohols (Figure 16.1). Wood lignins mainly contain guaiacyl (G) and syringyl (S) units, whereas the lignins of herbaceous plants contain all three units (p-hydroxyphenyl (H), G, S) in significant amounts with different ratios.[8] The lignin matrix comprises a variety of functional groups, such as hydroxyl, methoxyl and carbonyl, which imparts a high polarity to the

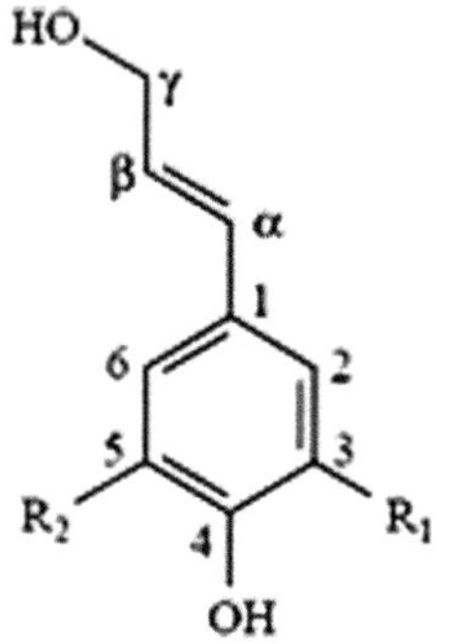

Coniferyl alcohol/guaiacyl: R_1 = OMe, R_2 = H
Sinapyl alcohol/syringyl: $R_1 = R_2$ = OMe
pCoumaryl alcohol: $R_1 = R_2$ = H

Figure 16.1 The three building blocks of lignin.[10]

lignin chains. Due to these properties, lignins can be used to produce new polymer blends.[9] Lignin exists in a wide range of molecular weights; the isolated lignins are in the range of 1000 to 12,000 depending upon the extent of degradation and condensation during isolation.

Lignins are a by-product obtained mainly from the pulp and paper industry. When lignin is isolated from wood (delignification), it undergoes significant chemical changes which depend on the method of isolation so that the lignins obtained under industrial conditions (technical lignins) are not identical in their structure to nature.[9a] Structurally, there are three main groups of lignins (i) softwoods (gymnosperms); (ii) hardwoods (angiosperms) and (iii) grasses (non-woody or herbaceous crops).

16.2.2 Blends and their Applications

Since the early stages in the technological developments of polymer science, the lignins have been used as reinforcement for rubber matrices.[10] Lignin was added to NR to improve rubber stability[11] and rubber barrier properties to aromatic hydrocarbons.[12] Among rubber-reinforcing agents, lignin is unique in its solubility in aqueous alkali, such solutions being compatible with the latex emulsions in all proportions. Dispersing agents commonly used with insoluble pigments are thus eliminated.[10]

The usual method for the preparation of the parent lignin–rubber mixture is the dissolution of lignin in an aqueous solution of alkali, mixing it with the rubber latex, heating the mixture, and the joint coagulation of the rubber and lignin by pouring the mixture into an acidic solution of acid while stirring. The reinforcing effect of lignin is determined by its particle size, which depends on a number of factors like prope, the precipitation conditions and several other factors. In the parent mixture lignin is not bound to rubber by chemical bonds. The reinforcing effect of lignin increases appreciably when lignin is precipitated in the presence of sodium silicate.[9a]

The first reported practical test of the use of kraft lignin in tyres was performed in 1964.[13] In this test, half the carbon black was substituted by kraft lignin, resulting in an improvement in the abrasion index of the rubber. Also, it has been reported that rubbers reinforced with kraft lignin are lighter than carbon black- or silica-filled rubber. The same authors also mentioned that lignin improves the ozone resistance of the rubber. Heat-treated lignin from softwood black liquor has been shown to stabilize rubbers against atmospheric degradation. More recently, lignin has been used in the production of vehicle tyres made of a material based upon a rubber containing particles of an unsulfonated lignin.[14]

It has been reported that the quality and strength of fibres made from the latex improves.[9a] Mixtures of rubber with kraft lignin are also suitable for use as rubber to glass adhesives. The lignin, after the addition of mineral acid, reduces spurting and evaporation. Kraft lignin also compares very favourably with other reinforcing agents such as carbon black because of its lower density.[9a]

Lignin-reinforced rubbers weigh appreciably less per unit volume owing to the low specific gravity (1.3) of lignin. Moreover, the brown colour of lignin permits a wide colour range, without sacrifice of mechanical properties, by blending with white pigments.[10] Lignin has also been recognized as an effective antioxidant in NR.[15]

A distinctive feature of the preparation of lignin-filled rubbers is that, in order to reinforce the latter, lignin is introduced at the latex stage because the introduction of powdered lignin into dry rubber does not lead to the 'reinforcement' effect. It is believed that the reason for this is the binding of lignin particles to one another by hydrogen bonds, which leads to condensation and growth of its particles and prevents dispersion of the lignin particles during the rolling of rubber.[16]

Gregorová et al. have studied a series of carbon black-filled NRs containing lignin from the viewpoint of their thermo-oxidative ageing. The results were compared with those from NR vulcanizates stabilized with the commercial rubber antioxidant N-phenyl-N-isopropyl-p-phenylene diamine (IPPD). The results obtained show that lignin exerts a stabilizing effect in carbon black-filled NR. Its effect is comparable with that of conventional synthetic antioxidant. Moreover, the addition of lignin increased the stabilizing effect of the IPPD.[17] Mixtures of rubber with kraft lignin are also suitable for use as adhesives.[9a] The mechanism of the reinforcement of raw rubbers by sulfate lignin can be accounted for mainly by its surface active properties and the coagulate is therefore precipitated in the form of fine particles. This leads to the formation of a material with low stress parameters for 300% elongation and a high hardness, so that the lignin-filled vulcanizates exhibit a lower resistance to wear than carbon black filled rubbers.[9a]

Various treatments of kraft lignin have been suggested for improving its rubber-reinforcing properties. By passing oxygen through an alkaline solution of the lignin, a superior reinforcing pigment for butadiene-styrene rubbers could be produced.[18] The mechanism of the reinforcement of rubbers by lignin is fairly complex and has not been completely investigated. Apart from the physical explanation, associated with the surface active properties of lignin, there is a chemical explanation of this phenomenon. The reinforcement of rubbers by alkali lignin can be accounted for by the interaction of the active groups of lignin with the double bonds or active groups of the rubbers, *i.e.* the formation of hydrogen bonds between the hydroxyl groups of lignin and the π electrons of the double bonds of the rubbers. When isocyanates are introduced, the high reactivity of the isocyanate groups promotes the rupture of the hydrogen bonds, which are unable to reinforce the dry lignin and the latter is thus converted into an active series.

Modified NR latex can be also blended with lignin to produce a paperboard barrier coating in order to replace unrecyclable wax coating materials. The blends containing lignin showed a reduction of water vapour permeability in coated paper when compared with coatings without lignin.[19]

16.3 Natural Rubber/Protein Blends

16.3.1 General Information

16.3.1.1 Biosynthesis and Biodiversity of Proteins

Proteins are natural polymers composed of amino acids linked by peptide bonds. They can be distinguished from other natural polymers such as fats and carbohydrates by the presence of the nitrogen atom. Other heteroatoms such as sulfur and, more rarely, phosphorus may also be present.

Two main biosynthetic pathways for the synthesis of proteins occur in nature: the ribosomal and the non-ribosomal – multienzyme – paths.[20] Significant differences between the proteins were obtained. The proteins obtained by the ribosomal pathway do not have any size limitation (up to 26,000 amino acid units), they are also mostly composed of combinations of 22 amino acids with only rare presence of D-amino acids (usually not more than one). Whereas the proteins, obtained by the multienzyme pathway, are much smaller, they count only 2 to 50 amino acid units. On the other hand the diversity of the amino acid units composing these proteins is much bigger: 2-, 3- and 4-amino compounds are commonly present as well as several D-amino acid units.

From the polymer point of view, proteins obtained by the ribosomal pathway, are mostly linear copolyamides with low dispersity. The proteins obtained by the multienzyme pathway are mostly peptides (oligomers). Among this latter group three naturally occurring proteins are poly(amino acid)s with important size distribution:[21] cyanophycin, composed of aspartic acid–arginine dipeptides; ε-poly-L-lysine and poly-γ-glutamic acid (Figure 16.2(a)). Another, extensively studied, homopolypeptide is the poly-α,β-aspartic acid (Figure 16.2(b)), although this particular polypeptide as yet has not been found in nature.[21,22] Finally, the development of the molecular biology and other mutagenesis techniques allowed access to the biosynthesis of proteins containing non-natural amino acids,[23] whereas organic chemists encounter major difficulties in the efficient synthesis of these molecules.

Due to the important variety of proteins, their properties and applications are very wide. Thus, many proteins have food and therapeutic applications,

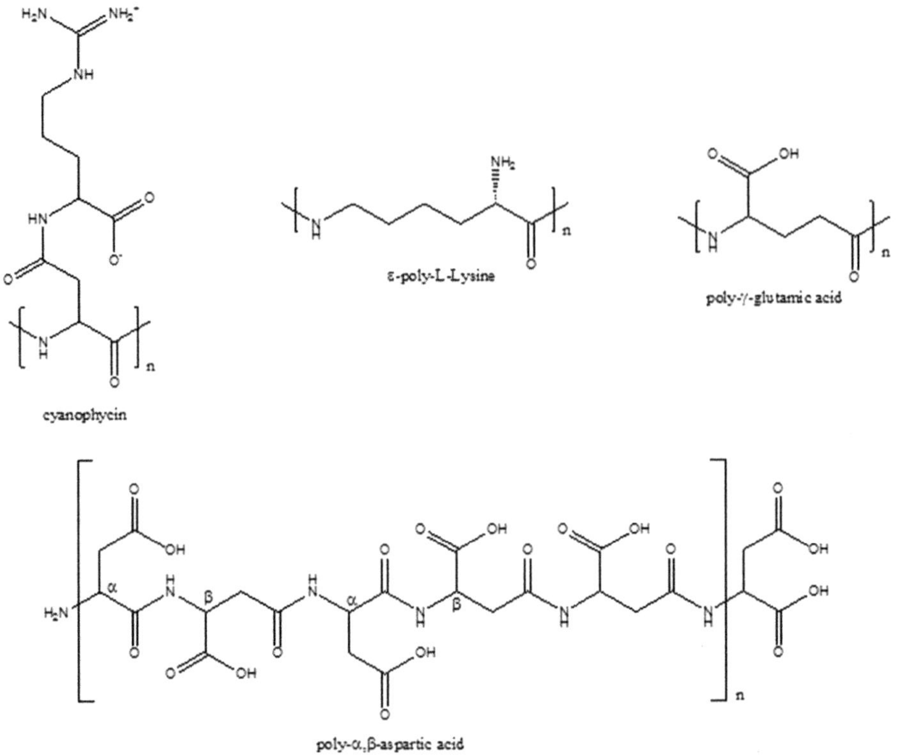

Figure 16.2 Molecular structures of poly(amino acid)s.

analytical enzymes and antibodies are used in medical diagnostics and as tools in genetic engineering technology.[24,25] The biocompatibility of proteins makes them interesting substrates for drug[25] and nutraceutical[26] delivery systems. ε-poly-L-lysine possesses remarkable antimicrobial properties and thus is widely used in different food applications, especially in Japan.[27] Poly-α,β-aspartic acid is mostly used as an anionic polydispersant in detergents.[23]

16.3.1.2 Materials Applications of Proteins

Several proteins have been extensively studied for their materials applications. Among them, soy protein is one of the most popular. Indeed, since the early 1930s it was used in phenol-formaldehyde blends for automotive applications.[28] However, soy protein is sensitive to moisture and exhibits relatively low strength properties. Thus stabilization by plasticization, compatibilization or crosslinkage is required to maintain long-term performance of soy protein-based plastic materials.[29] Also, several studies on soy protein-based blends with other natural polymers or their reinforcement by natural fibres have been performed. More recently studies on soy protein–nanoclay composites[30] and poly(butylene adipate-co-terephthalate) (PBAT) blends[31] were also performed.

This opens a novel route for soy protein-based plastics bearing improved mechanical properties.

Wheat gluten is another protein from plants, which was intensively studied for its material applications.[28] Wheat processing requires the presence of plasticizers such as glycerol or water to disrupt disulfide bonds. To control the formation of a crosslinked network, cysteine, formaldehyde or glutaraldehyde can be added. This contributes also to an improvement in moisture sensitivity and in elongation at break. On the other hand, the wheat gluten fibres have better properties compared to other protein-based bio products for biomedical applications.

Among the animal proteins, collagen (or its denatured form, gelatin) is the most abundant and it has been studied extensively.[32] As for other proteins collagen needs to be stabilized to improve its mechanical and moisture resistance properties. However, the main application of this protein and its blends are in the biomedical field, thus transglutaminases are more suitable crosslinking agents than aldehydes.[28] Also, the hydrolysed proteins (HP) from chrome-tanned leather waste, mainly composed of gelatine, were used to improve the biodegradability of synthetic polymers.[33]

Other proteins such as elastin, silk fibroin, keratin or albumin have been considered for materials applications very recently.[32] Due to the lack of availability or other purification and processing capacities, their cost remains relatively high. Consequently, up to now, their main material applications concern the high added value materials, mainly biomedical.

16.3.1.3 Indigenous Proteins of Natural Rubber

NR is the major component of latex, the main product of the rubber tree. Even if *Hevea brasiliensis* is widely recognized as the major commercial producer, latex can be produced by different plants, in which up to 12,000 species were found to be latex producers.[34] Nevertheless, important proteins or protein complexes occur in all species. Indeed, latex is synthesized from hydrophilic substrates by specific enzymes, mainly *cis*-prenyl transferases, which remain closely linked to the rubber particle interior.[35] Moreover, as NR latex is not a homogeneous fluid, the dispersion of proteins is not homogeneous either. Thus the purification of the NR remains low, and the final product often contains up to 6% of non-rubber components;[36] the overall protein content can reach 2% of latex fresh weight.[34]

The presence of indigenous proteins in NR used to be considered a significant drawback. These proteins are responsible for the 'latex allergy' affecting an average of 20 million people in the USA alone.[34,35] Several studies were thus initiated to circumvent this problem: biocompatible coatings based on poly(ethylene glycol) and hyaluronic acid,[37] combination of dilution/addition of water-soluble polymers (WSP)/centrifugation,[38] considering the 'self-cleaned' skim part of NR,[39] among others. Also, the negatively charged proteins present in the rubber phase of NR latex were described to influence the vulcanization of rubber by irradiation.[40]

This drawback can also be an advantage in some cases. Thus, the indigenous surfactant protein–lipid was successfully used to form complexes with poly(ethylene oxide) (PEO) during the formation of NR/polychloroprene composites.[41] The proteins present in NR were also found to be essential for the stabilization of silver nanoparticles in this material.[42]

16.3.2 Blends and their Applications

While relatively abundant literature is available on NR blends with other biopolymers, very few studies have been reported on protein/NR blends. In this section, we discuss aspects of the incorporation of proteins to synthetic rubbers, together with the analysis of protein/NR blends.

16.3.2.1 Natural Rubber and Proteins

The first blends of NR with proteins were reported in the early 1950s. The most commonly used protein at that time was casein,[43] although other proteins such as gelatine and soy protein were also mentioned.[44] These blends have found practical applications in adhesives[43a] or tyres.[43b] They have also been used to reinforce rubber products, especially towards the action of a kneader (mechanical stresses) and are capable of having, in addition, certain complementary mechanical properties.[44]

After a lack of interest in these blends for some 20 years, there were some new developments in this field. Thus, protein silver was incorporated in NR to produce antimicrobial latex compositions.[45] Proteins were also used to produce conductive rubber.[46] Collagen protein fibres were incorporated in NR to produce studless tyres exhibiting good properties at low temperatures.[47] Proteins were also cited as potential biofillers for accelerating the biodegradability of blends mainly composed of polyolefins and rubbers.[48]

More recently, it was established that non-water soluble amino acids can improve the anti-ageing properties of radiation vulcanized NR latex. Among these amino acids, (phenyl)alanine, asparagine and cystine are the most efficient. Cystine possesses an S–S linkage that can act as an antioxidant. Thus, fibrous structural protein from chicken, keratin, containing up to 44% of cystine, was tested for protecting NR against ageing. This study has, however, shown that the method of the extraction of keratin from chicken feathers is crucial. Indeed, extraction with a basic solution makes keratin pro-oxidant, whereas keratin obtained by a reductive extraction shows an antioxidant effect on NR.[49]

Also, a natural silk fibre (20%) reinforced polypropylene (PP) was blended with NR (up to 50% by weight) in order to increase the biodegradability and the overall sustainability of this composite.[50] The incorporation of NR contributed to the decrease in mechanical properties of the composite. Fortunately, this drawback can be compensated for by the application of gamma radiation.

16.3.2.2 Synthetic Rubbers and Proteins

Two proteins were particularly studied in blends with synthetic rubber: soy protein and sericin. Soy protein-styrene-butadiene composites have been given more specific attention. The reason for this interest is the fact that the dry soy protein aggregates have a very rigid nature. They were thus successfully used to significantly increase the shear elastic modulus of NR.[51] These results were further improved by using soy protein nanoparticle aggregates[52] and soy protein concentrates (SPC), obtained by coagulating and centrifuging an 11% defatted soy flour (DSF) dispersion.[53] Also soy spent flakes (SSF) were shown to be able to partially substitute carbon black (CB) co-fillers in rubber composites.[54]

Silk sericin is a natural protein derived from a silkworm *Bombyx mori*. Most of the sericin must be removed during silk production. Thus, sericin recovery and reuse increases the sustainability of silk production.[55] Moreover, the blending of sericin with rubber made the latter more biocompatible and gave it durable skincare properties.[56]

16.4 Natural Rubber/Polysaccharide Blends

16.4.1 General Information

Carbohydrates are found in the leaves, twigs, roots or seeds of plants. They comprise of a large group of compounds, all containing in their structure the fundamental chemical elements – carbon, hydrogen and oxygen. The existence of at least two functions in most organic carbohydrates allows them to undergo various chemical transformations, taking into account the differences in reactivity of various hydroxyl groups of the molecule: C=O and C–OH.[57] Carbohydrates are chemically represented by $(CH_2O)_n$ and may be divided into three main groups: monosaccharides, oligosaccharides and polysaccharides. Polysaccharides are formed by combining a large number of sugar units and between them stand out from starch and cellulose. Starch and cellulose are increasingly being used in other industrial applications to produce different materials such as rubber and biodegradable films.

16.4.1.1 Starch

Starch is the major storage form of carbohydrate in plants. It comprises two distinct structures: amylose and amylopectin. It can be derived from various sources such as corn, cassava, potato, sorghum bean, yam, peas, inter alia, various levels of amylose and amylopectin between them.[58]

Amylose is a linear molecule composed of glucose units with glucosidic α-1,4 linkages. It generally makes up 20–30% of the starch, and it is insoluble in water. Its linear structure can form helices to stabilize the molecule. The amylopectin is a branched molecule in which the starch fraction is approximately 70%, being composed of glucose units linked by α-1,4 linkages in the chain and α-1,6 interchain linkages.[58]

According to Bobbio (1995), the viscosity of the starch solution is greater than that of amylopectin. The approach of two or more amylose molecules is sufficient for the formation of H aggregates, which are numerous bridges between molecules, forming a crystalline area. In the nearest interchain amylopectin is partially, due to the hindrance of the chain branches, responsible for the low crystallinity of the molecule.[59] The amylose and amylopectin molecules are grouped to form granules, which have shape, size and crystalline areas differentiated and therefore identifying its origin. There are areas in the granule of starch with enhanced resistance to water penetration and hydrolysis due to the higher level of crystallinity, indicating a higher amylose content. The starch is practically insoluble in cold water and can absorb up to 30% of its weight without increasing the volume of the grains. However, it increases the amount of water absorbed upon heating, therefore the volume of the granules which come to occupy all the available volume.

16.4.1.2 Cellulose

Cellulose is the most abundant carbohydrate in nature. Its structure is similar to starch and glycogen. Cellulose is defined as a macromolecule, a non-branched chain of variable length of 1,4-linked β-D-anhydroglucopyranose units. It is a straight chain polymer composed of glucose units, and it is not absorbed by the human body in the absence of enzymes that are able to digest it.[60]

16.4.2 Blends and their Applications

Starch has been used as the main polymer in macromolecular compositions that can be processed as thermoplastics. In this case, the granular structure of starch is completely disrupted by the use of plasticizers under heating, giving rise to a continuous phase in the form of a viscous melt which can be processed following conventional plastic processing techniques such as injection moulding or extrusion. Since starch is hydrophilic, the drawback is that compounds prepared only with starch have poor barrier properties and solubility in water. In these cases, a good alternative is the blending of starch with hydrophobic materials.

NR has been blended with starch for a number of different applications. Arvanitoyannis *et al.* reported on biodegradable blends based on gelatinized starch and 1,4-transpolyisoprene (gutta-percha) for food packaging or biomedical applications.[61] Carvalho *et al.* studied thermoplastic starch/NR blends prepared directly using natural latex and corn starch. The blends were prepared in an intensive batch mixer at 150 °C, with NR content varying from 2.5 to 20%. The materials obtained were characterized by mechanical analysis (stress–strain) and by scanning electron microscopy. The dispersion of rubber in the thermoplastic starch matrix was homogeneous due to the presence of the aqueous medium, with rubber particles ranging from 2 to 8 mm. The process employed in this investigation made use of both starch and latex in their natural form, without any kind of purification. The authors also concluded the presence of the non-rubber constituents in the latex was responsible not only for ensuring the latex stability, but also for improving the compatibility between the thermoplastic starch and the NR phases.[62] The addition of rubber was, according to the authors, limited by phase separation as a function of glycerol content. In addition the glycerol used as a plasticizer seemed to contribute to both the plasticization of starch and to the improvement of the starch–rubber interface.[62]

Wang et al. studied blends of cassava starch modified by esterification and NR.[2] The samples were prepared by blending the modified starch with NR latex, and their morphology, thermal stability and mechanical properties were investigated. The authors concluded that the crystal structure of the starch particles becomes amorphous after esterification. The modified starch can be homogeneously dispersed in the NR matrix, with an average particle size around 200 nm. The loading of esterified starch leads to a significant improvement in thermal stability. According to the authors, when the esterified starch content is less than 20 parts per hundred (phr), the mechanical properties of the composites increase linearly.

Another study involving tapioca starch was carried out by Riyajan et al., with the aim of coating granular fertilizers to reduce their solubility in water, and for coating the fertilizer. The starch in this study was used to increase the hydrophilicity of NR by using potassium persulfate ($K_2S_2O_8$) as a catalyst. The modified starch was added to the NR latex in the presence of the non-ionic surfactant Terric16A16, at 60 °C for 3 h, and cast films on a glass plate were prepared to obtain NR-g-ST (natural rubber-grafted-starch). The authors concluded that the NR-g-ST was achieved by grafting with modified starch in the presence of $K_2S_2O_8$ as a catalyst. The structure of the modified NR was confirmed by ATR-FTIR.[63] In an aqueous medium the swelling rate of the modified NR decreased as a function of ST amount, due to its hydrophilic behaviour. In addition, the tensile strength of modified NR in the presence of ST, at 50 phr, was the highest compared to other samples. Also, the thermal stability of the modified NR-g-ST was higher than that for unmodified NR-g-ST and confirmed by TGA. After NR is grafted with ST, it can lead to intermolecular linkage between ST and NR molecules. In this way, the hydrophobic groups of NR are reduced to control the release of the coated urea fertilizer. Finally, the NR-g-ST membrane displayed a good barrier to controlling release of the urea fertilizer from the coated capsule, and is easily degraded in soil. This product, according to the authors, with good controlled release and water retention, could be especially useful in agricultural and horticultural applications.

Andrio et al. reported the effect of temperature and pressure on the permeability, diffusion and solubility coefficients of N_2, O_2 and CO_2 in NR reinforced with different amounts of cellulose. Rubber compounds were prepared by adding cellulose II, as 10% cellulose xanthate aqueous solution, to NR latex under stirring. Analysis of the results obtained by these authors showed that gas transport was severely hindered at the cellulose–rubber interface of the composites.[64] Barrer and Chio reported similar conclusions in studies with silicone rubber membranes, where the crystallization of the rubber caused a large decrease in permeability and diffusion coefficients and smaller decreases in solubility.[65]

Martins et al. also studied blends using cellulose. In their study, compositions of NR and cellulose II were prepared by co-coagulation of natural latex and cellulose xanthate mixtures. The influence of increasing amounts of cellulose II, varying from 0 to 30 phr, on processing, curing and mechanical properties were studied. The authors concluded that cellulose II has a large influence on cure characteristics and mechanical properties of the NR compositions studied. This filler facilitates the processing, participates in crosslinking and reinforces the NR compositions, indicating a good rubber–filler interaction. The best performance, according to the authors, was achieved by compositions with filler content of 15 phr.[66]

Kalf et al. studied the effect of grafting cellulose acetate and methylmethacrylate as compatibilizers on acrylonitrile butadiene rubber (NBR) and styrene-butadiene rubber (SBR) blends. Morphology studies of the samples show an improvement in interfacial adhesion between the NBR and SBR phases in the presence of the prepared compatibilizing agents. The authors also reported the samples with grafted compatibilizers showed superior crosslink density and thermal stability, as compared to the blends without graft copolymers.[67]

16.5 Natural Rubber/Polyester Blends

16.5.1 General Information

The blending of NR with two types of biodegradable plastics, namely polylactic acid (PLA) and polyhydroxyalkanoates (PHAs) are discussed in this section.

PLA (Figure 16.3(a)) is derived from lactic acid through the fermentation of corn starch, tapioca products or sugar cane. This aliphatic polyester is capable of degrading completely in both aerobic and anaerobic environments.[68] Another polyester family, the PHAs (Figure 16.3(b)), are produced by a wide variety of microorganisms under conditions of limiting nutritional elements (e.g. nitrogen, oxygen, magnesium and phosphorus) in the presence of excess carbon. They accumulate within cells as carbon and energy storage. PHAs represent a large group of polymers with distinct monomeric compositions and characteristics, of which poly-3-hydroxybutyrate (PHB) and its copolymer

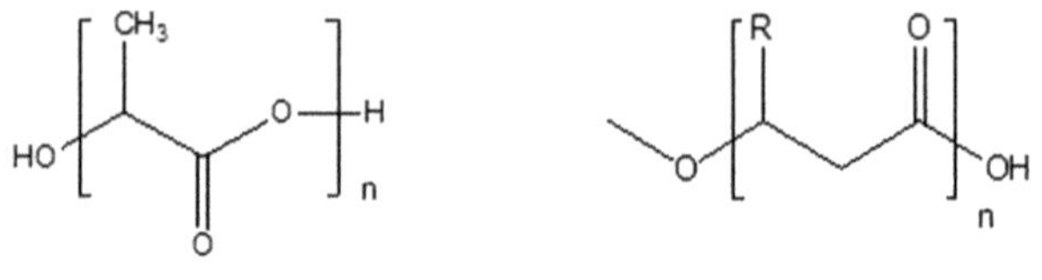

Figure 16.3 Structure of (a) PLA and (b) PHA.

poly(3-hydroxybutyrate-*co*-3-hydroxyvalerate) (PHBV) are the best known members.

Both PLA and PHAs possess excellent biodegradability and biocompatibility, and they have attracted significant commercial interest as promising candidates for renewable thermoplastics that can be used in a wide range of agricultural, industrial and biomedical applications. However, there are still major drawbacks that hinder their large-scale production. Taking PHB as an example, its nucleation density is low and so large spherulites with cracks and splits are formed during crystallization, resulting in high brittleness with poor impact resistance. The melting point is close to the thermal degradation temperature (170 °C), which causes a rather narrow window of processability. Finally, its production cost is much more higher than the conventional polymers (the price of 1 kg of polypropylene is US$1 but is US$2–8 for PHB). How to minimize these drawbacks? In addition to the optimization of processing technologies and conditions, blending could be a good option. Blending biodegradable plastics with other flexible polymers may produce novel materials with improved mechanical and thermal properties. NR has excellent properties such as high strength, resilience and elongation at break and so it has been successfully used as an impact modifier for various polymers.

16.5.2 Blends and their Applications

The mechanical properties of blends are highly dependent on the blending method used and sample preparation. The most commonly used technique for blending NR with PLA/PHA is melt blending, although solution blending and emulsion blending may also be used under certain conditions. Blends are generally prepared using an internal mixer or a twin-screw extruder. PLA should be dried at 70 °C for 4 h to eliminate moisture prior to mixing. Compression moulding and injection moulding can be used to prepare blend specimens. There are many factors affecting the effectiveness of blending, such as temperature, rotor speed and mixing time, and so optimal processing conditions need to be selected.

The appropriate particle diameter of NR is very important. Limiting it to ~1 μm and dispersing these small droplets into the matrix can effectively minimize crazing and crack progression, significantly increasing the toughness of blends. Bitinis *et al.* reported that in a PLA/NR blend, increasing the temperature from 160 °C to 200 °C led to an increase in the NR average droplet size from 1.15 to 2.30 μm.[4] In addition, mastication can be an effective way of generating the smaller size NR by decreasing the molecular weight. Jaratrotkamjorn *et al.* reported that the diameter of NR particles was reduced from 2.50 to 0.62 μm after mastication at 180 passes.[69] When blending with PLA, increasing the amount of NR mastication can significantly improve the tensile properties (optimal mastication number: 20–100) as well as the impact strength (optimal mastication number: 80–180).[69]

The rheological properties of the plastic phase (PLA/PHA) and the elastomer phase (NR) are vastly different, and so the blend is immiscible, leading to poor impact strength of the final products. To improve compatibility, compatibilizers are usually introduced as the third component. They can be block, homopolymer or graft copolymers, interacting with both phases to reduce the interfacial tension that is responsible for phase separation. Glycidyl methacrylate (GMA)-grafted polymers are often used as compatibilizers for PLA blends. The epoxy groups of GMA may react with carbonyl or hydroxyl groups located at the PLA chain ends. Previous studies have shown that glycidyl methacrylate-grafted NR (NR-g-GMA) can be employed as a compatibilizer for PLA/NR blends. As shown by the results of Fourier transform infrared (FTIR) spectroscopy, PLA has a band centred at 3500 cm^{-1} in the hydroxyl-stretching region due to the end carbonyl groups. The band subsequently disappears when NR-g-GMA is introduced, indicating reaction between epoxy groups and carbonyl groups during melt mixing. These results suggest there is good interaction between PLA and the compatibilizer (Figure 16.4). Juntuek et al. also reported that vetiver grass fibre can be used as the filler to cooperate with NR-g-GMA.[70] For PHAs, polybutadiene with maleic anhydride has been used as the compatibilizer system for the blending of PHB with epoxidized natural rubber.[71]

The blends can be characterized by several important aspects such as thermal, morphological and mechanical properties. These properties are reported in the following section.

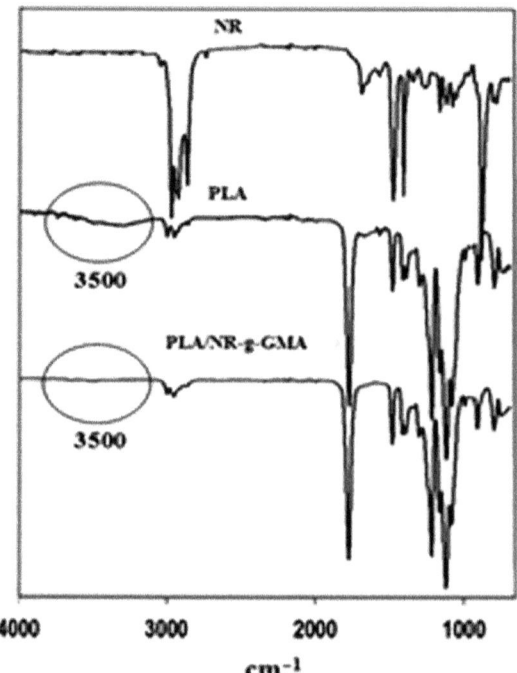

Figure 16.4 FT-IR spectra of NR, PLA and PLA/NR-g-GMA blend.[71]

16.5.2.1 Thermal Properties

The thermal degradation of PLA is a major limitation on its industrial applications. Results of gel permeation chromatography analysis show that after processing from compression or injection moulding, weight-average molecular weights ($\bar{M}_w$) and number molecular weights ($\bar{M}_n$) of PLA are decreased, indicating the degradation of PLA during processing (Table 16.1). The reason that injection moulding causes lower molecular weights than compression moulding is because the former has a higher shear rate, which creates more viscous heating, inducing the heat-sensitive PLA to be degraded faster.

Fortunately, the addition of NR can attenuate PLA degradation. The small molecules present in the NR, such as lipids or fatty acids, can migrate to the surface of the blend, acting as a lubricant to prevent the friction effects and local overheating, as well as the degradation of PLA. The incorporation of 10% of NR results in a significant reduction in torque and extrusion pressure. The onset degradation temperatures of distinct components can be determined by thermogravimetric analysis (TGA) (Table 16.2). It is clear that PLA/NR (90/10) has a slightly increased thermal stability, and the incorporation of compatibilizer further improves it. These data suggest a better interaction and dispersion of NR in the PLA matrix.[71]

The transition behaviour of the blend can be evaluated by differential scanning calorimetry (DSC). As shown in Figure 16.5, there is no significant change in glass transition temperature (T_g) of PLA with NR content (57.7–59.4 °C), which confirms the immiscible character of brittle and elastic polymers. However, a cold crystallization exothermic peak is observed. The cold crystallization temperatures (T_{cc}) are 127.6 °C for PLA/NR (90/10) and 123.5 °C for PLA/NR (80/20), respectively. The presence of cold crystallization suggests that the addition of NR may increase the crystallization rate of PLA, and the observed activities could result from the nucleating effect of NR.

Table 16.1 Molecular weights of PLA determined by gel permeation chromatography.

Samples	$\bar{M}_w$	$\bar{M}_n$
Pristine PLA	92,485	47,808
PLA (compression)	80,847	44,581
PLA (injection)	69,149	35,745

Table 16.2 Onset degradation temperatures determined by TGA.

Samples	Onset degradation temperatures (°C)
PLA	290.8
NR	305.3
PLA/NR (90/10)	294.2
PLA/NR/NR-g-GMA (90/9/1)	301.6

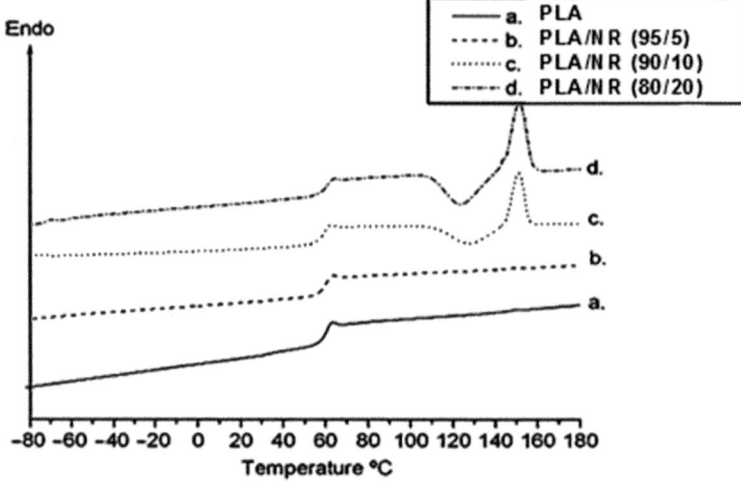

Figure 16.5 Melting curves of PLA/NR blends.[4]

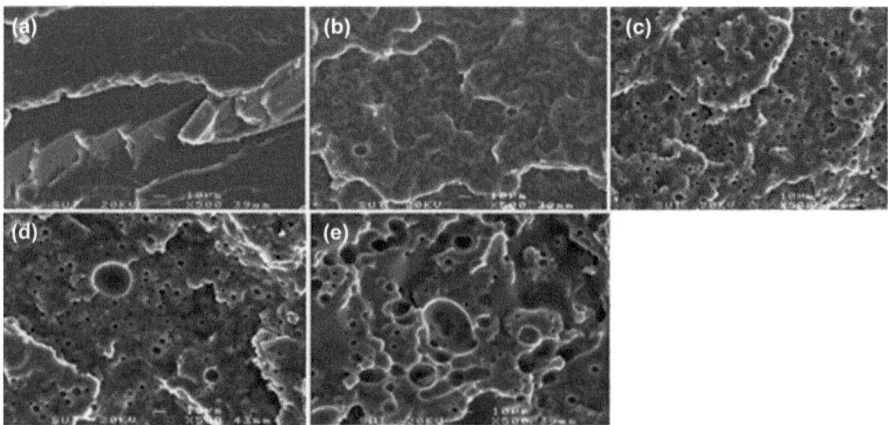

Figure 16.6 SEM micrographs of impact fractured surface.[71] (a) PLA; (b) PLA/NR (95/5); (c) PLA/NR (90/10); (d) PLA/NR (80/15); (e) PLA/NR (80/20).

16.5.2.2 Morphological and Mechanical Properties

Scanning electron microscopy (SEM) analysis can be carried out to examine the morphologies of the impact and tensile fractured surfaces of samples. As shown in the SEM micrographs of impact fractured surfaces (Figure 16.6(a)), unbonded rubber particles are separated from PLA in the blends, proving that PLA and NR are immiscible, and so a compatibilizer is needed. The average domain size of rubber particles in PLA/NR (90/10) is 5–10 μm (Figure 16.6(c)) and increases to over 10 μm in PLA/NR (80/20) (Figure 16.6(e)), which demonstrates the occurrence of coagulation effects. In general, in an immiscible blend system, polymers of the dispersed phase and continuous phase always

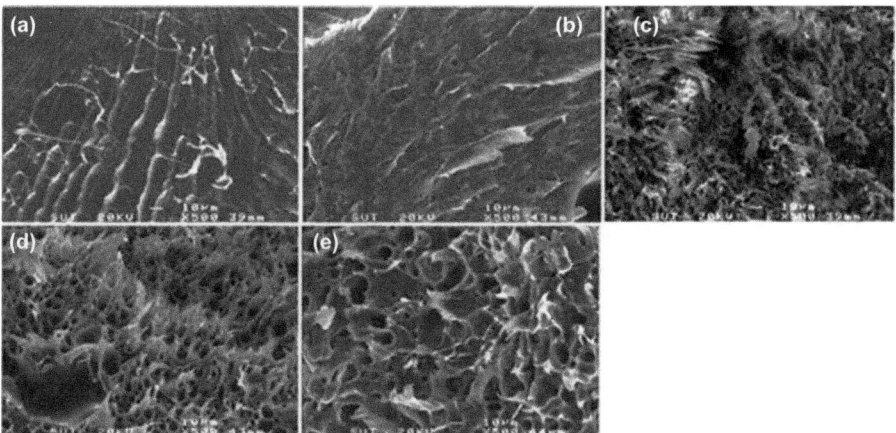

Figure 16.7 SEM micrographs of tensile fractured surface[71] (a) PLA; (b) PLA/NR (95/5); (c) PLA/NR (90/10); (d) PLA/NR (80/15); (e) PLA/NR (80/20).

Table 16.3 Mechanical properties of PLA/NR blends.

Samples	Young's modulus (GPa)	Tensile strength (MPa)	Elongation at break (%)
PLA	0.63 ± 0.04	52.99 ± 0.65	10.84 ± 1.64
PLA/NR (95/5)	0.62 ± 0.02	41.04 ± 1.60	25.75 ± 4.79
PLA/NR (90/10)	0.51 ± 0.04	29.66 ± 1.56	74.51 ± 9.06
PLA/NR (85/15)	0.50 ± 0.02	28.91 ± 0.89	23.39 ± 1.59
PLA/NR (85/15)	0.36 ± 0.02	18.23 ± 1.28	7.43 ± 0.30
PLA/NR/NR-g-GMA (90/9/1)	0.53 ± 0.02	27.46 ± 0.66	159.08 ± 21.27

coagulate separately when present in high concentrations. Figure 16.7 shows the SEM micrographs of tensile fractured surfaces. It is clear that longer fibrils are generated when NR is introduced at 5% and 10% (Figure 16.7(b), (c)), implying the change of PLA from brittle fracture to a ductile fracture. However, when increasing the concentration of NR to 15% or above, again NR starts to get debonded from PLA and forms large oval cavities (Figure 16.7(d), (e)). From SEM results, it can be concluded that in the blend system, the size of rubber particles and the interfacial adhesion between two phases are of crucial importance for the mechanical properties. It seems that 10% of NR (w/w) is an optimal content. These findings are consistent with the data given in Table 16.3. As a brittle polymer, the elongation at break of PLA is very low (10.8%). However, the addition of NR at 5% and 10% significantly increases it to 25.8% and 74.5%, respectively, and the incorporation of compatibilizer, PLA/NR/NR-g-GMA, further increases it to 159.1%. The ductility of PLA has been substantially improved. For Young's modulus and tensile strength, the addition of NR results in lowered values because of its rubbery nature. With respect to impact resistance, results from both Izod and Charpy modes show

Table 16.4 Impact strength of PLA/NR blends.

Samples	Izod impact strength (kJ/m^2)		Charpy impact strength (kJ/m^2)	
	Un-notched	Notched	Un-notched	Notched
PLA	19.55 ± 2.67	2.85 ± 0.66	19.24 ± 5.22	2.54 ± 0.55
PLA/NR (90/10)	Unbroken	6.36 ± 0.36	30.16 ± 5.90	4.29 ± 0.54

that 10% of NR markedly enhances the impact strength of PLA, on notched or un-notched samples (Table 16.4).

16.5.2.3 Biodegradable Properties

Bhatt *et al.* created blends of medium chain length PHA (mcl-PHA) with NR as well as other types of rubber, and they found that in addition to altering the thermal properties, the blending can also affect the biodegradable properties of PHA and NR. The blend films were effectively degraded by *Pseudomonas* sp. 202, the degradation ratio of which can be tailored by adjusting the rubber/PHA ratio.[72]

16.6 Conclusions and Outlook

The purpose of this chapter is to look at the most recent advances in NR/biopolymer blends. Since the beginning of NR technology, polymeric materials from renewable resources have been tested as additives for rubber blends and vulcanizates. Their influence on the vulcanization process, tensile strength, dynamic and mechanical properties have been discussed in this chapter. On the other hand, blending may improve the thermal and mechanical properties of biodegradable plastics, creating new materials with a good compromise between elastic modulus and impact resistance. Lignin was the first biopolymer to be used as an additive to NR, followed by starch, polyesters and proteins. It seems that the consideration of proteins for materials applications is only just emerging. Despite the fact that until now only little work has been done on the blending of proteins with NR, the diversity of proteins and of their applications will undoubtedly lead to more important investigations in the future. The combination of increasing economic necessity and industrial feasibility will secure the widespread usage of renewable raw materials. The protein/NR blends have an important role to play in this novel approach to materials science. The blending of NR can be used to create novel materials superior to the original PLA/PHA in terms of thermal and mechanical properties. As a result, these novel materials may have a broader application range with low cost advantage. In medicine, these upgraded biodegradable plastics can be used as surgically implanted devices, such as bioresorbable surgical sutures, biodegradable screws and plates for cartilage and bone fixation, biodegradable membranes for periodontal treatment, etc. Packaging could be another big potential market. Sustainable packaging is attracting increased consumer awareness and packaging applications made up over 50% of the global biodegradable plastics market in 2010. In pharmacology, the upgraded biodegradable plastics can be used as microscapsules and drug delivery systems.

The future for the application, development and commercialization of NR blends with biopolymers is assured if effort is rationalized and concentrated into critical mass in specific key areas. As supplemented with the relevant literature and in-depth discussion with the appropriate key players in the marketplace, this chapter further promotes the use of NR blends with biopolymers as a utilitarian and necessary feedstock for the chemical industries.

References

1. Z. Kramarova, P. Alexy, I. Chodak, E. Spirk, I. Hudec, B. Kosikova, A. Gregorova, P. Suri, J. Feranc, P. Bugaj and M. Duracka, *Polym. Adv. Technol.*, 2007, **18**, 135.
2. Z.-F. Wang, Z. Peng, S.-D. Li, H. Lin, K.-X. Zhang, X.-D. She and X. Fu, *Compos. Sci. Technol.*, 2009, **69**, 1797.
3. M. Valodkar and S. I. Thakore, *J. Appl. Polym. Sci.*, 2012, **124**, 3815.
4. N. Bitinis, R. Verdejo, P. Cassagnau and M. A. Lopez-Manchado, *Mater. Chem. Phys.*, 2011, **129**, 823.
5. A. U. Buranov and G. Mazza, *Ind. Crops Prod.*, 2008, **28**, 237.
6. L. A. Donaldson, *Phytochemistry*, 2001, **57**, 859.
7. R. Vanholme, K. Morreel, J. Ralph and W. Boerjan, *Curr. Opin. Plant Biol.*, 2008, **11**, 278.
8. N. G. Lewis and E. Yamamoto, *Annu. Rev. Plant Physiol. Plant Mol. Biol.*, 1990, **41**, 455.
9. (a) J. Wang, R. S. J. Manley and D. Feldman, *Prog. Polym. Sci.*, 1992, **17**, 611; (b) P. Alexy, B. Košíková and G. Podstránska, *Polymer*, 2000, **41**, 4901.
10. J. J. Keilen and A. Pollak, *Ind. Eng. Chem.*, 1947, **39**, 480.
11. L. T. Furlan, M. A. Rodrigues and M.-A. De Paoli, *Polym. Degrad. Stab.*, 1985, **13**, 337.
12. T. V. Mathew and S. Kuriakose, *Polym. Compos.*, 2007, **28**, 15.
13. West Virginia Pulp & Paper, *UK Patent*, GB976433, 1964.
14. C. Veas and E. Hotaling, *US Patent*, 20110073229, 2011.
15. M.-A. De Paoli and L. T. Furlan, *Polym. Degrad. Stab.*, 1985, **13**, 129.
16. E. G. Lyubeshkina, *Russ. Chem. Rev.*, 1983, **52**, 1196.
17. A. Gregorová, B. Košíková and R. Moravčík, *Polym. Degrad. Stab.*, 2006, **91**, 229.
18. Smith Paper Mills LTD Doward, U.S. Pat. 2.610.954, 1962.
19. H. Wang, A. J. Easteal and N. Edmonds, *Adv. Mater. Res.*, 2008, **47–50**, 93.
20. H. von Döhren, *Biopolymers: Polyamides and Complex Proteinaceous Materials I*, ed. S. R. Fahnestock and A. Steinbüchel, Wiley-VCH, Weinheim, 2003.

21. F. B. Oppermann-Sanio and A. Steinbüchel, in *Biopolymers: Polyamides and Complex Proteinaceous Materials I*, ed. S. R. Fahnestock and A. Steinbüchel, Wiley-VCH, Weinheim, 2003.
22. G. D'Alessio, *Prog. Biophys. Mol. Bio.*, 1999, **72**, 271–298.
23. W. Joengten, N. Müller, A. Mitschker and H. Schmidt, in *Biopolymers: Polyamides and Complex Proteinaceous Materials I*. ed. S. R. Fahnestock and A. Steinbüchel, Wiley-VCH, Weinheim, 2003.
24. (a) G. B. Kresse, *Basic Biotechnology*, Cambridge University Press, Cambridge, 2004; (b) N. M. Bergmann and N. A. Peppas, *Prog. Polym. Sci.*, 2008, **33**, 271.
25. A. S. Hoffman and P. S. Stayton, *Prog. Polym. Sci.*, 2007, **32**, 922.
26. H. Carrillo-Navas, J. Cruz-Olivares, V. Varela-Guerrero, L. Alamilla-Beltrán, E. J. Vernon-Carter and C. Pérez-Alonso, *Carbohydr. Polym.*, 2012, **87**, 1231.
27. T. Yoshida, J. Hiraki and T. Nagasawa, in *Biopolymers: Polyamides and Complex Proteinaceous Materials I*, ed. S. R. Fahnestock and A. Steinbüchel, Wiley-VCH, Weinheim, 2003.
28. J. M. Raquez, M. Deléglise, M. F. Lacrampe and P. Krawczak, *Prog. Polym. Sci.*, 2010, **35**, 487.
29. A. K. Mohanty, W. Liu, P. Tummula, L. T. Drzal, M. Manjusri and R. Narayan, in *Natural Fibers, Biopolymers, and Biocomposites*, ed. A. K. Mohanty, M. Misra and L. T. Drzal, Taylor & Francis, *Boca Raton*, 2005.
30. L. Yu, K. Dean and L. Li, *Prog. Polym. Sci.*, 2006, **31**, 576.
31. F. Chen and J. Zhang, *Polymer*, 2009, **50**, 3770.
32. A. Sionkowska, *Prog. Polym. Sci.*, 2011, **36**, 1254.
33. I. Tchmutin, N. Ryvkina, N. Saha and P. Saha, *Polym. Degrad. Stab.*, 2004, **86**, 411.
34. H. Y. Yeang, S. A. M. Arif, F. Yusof and E. Sunderasan, *Methods*, 2002, **27**, 32.
35. K. Cornish, *Phytochemistry*, 2001, **57**, 1123.
36. K. Nawamawat, J. T. Sakdapipanich, C. C. Ho, Y. Ma, J. Song and J. G. Vancso, *Colloids Surf., A.*, 2011, **390**, 157.
37. M. Yan, *React. Funct. Polym.*, 2000, **45**, 137.
38. D. F. Parra, C. F. P. Martins, H. D. C. Collantes and A. B. Lugao, *Nucl. Instrum. Methods Phys. Res., Sect. B*, 2005, **236**, 508.
39. M. M. Rippel, L. T. Lee, C. A. P. Leite and F. Galembeck, *J. Colloid Interface Sci.*, 2003, **268**, 330.
40. C. V. Chaudhari, Y. K. Bhardwaj, N. D. Patil, K. A. Dubey, V. Kumar and S. Sabharwal, *Radiat. Phys. Chem.*, 2005, **72**, 613.
41. K. Sanguansap, T. Suteewong, P. Saendee, U. Buranabunya and P. Tangboriboonrat, *Polymer*, 2005, **46**, 1373.
42. N. H. H. Abu Bakar, J. Ismail and M. Abu Bakar, *Mater. Chem. Phys.*, 2007, **104**, 276.
43. (a) US Rubber Company, *GB patent*, GB 685177, 1952; (b) Goodyear Tire & Rubber Company, *GB patent*, GB 801300, 1958.
44. R. L. Lehmann, B. J. Petussen, C. P. Pinazzi, *Nobel Bozel patent*, US 2931845, 1960.

45. Y. Umemura, Unitika LTD patent, EP 038258, 1988.
46. M. Bessho, K. Ikeda, Kanegafuchi Chemical Ind. patent, JP 62235342, 1987.
47. S. Iwama, Y. Yamaguchi, H. Takino, Toyo Tire & Rubber Company patent, JP 2219837, 1990.
48. (a) G. J. L. Griffin, Epron Ind. Ltd. patent, CA 1335005, 1995; (b) G. J. L. Griffin, Epron Ind. Ltd. patent, US 5212219, 1993.
49. L. V. Abad, L. S. Relleve, C. T. Aranilla, A. K. Aliganga, C. M. San Diego and A. M. dela Rosa, *Polym. Degrad. Stab.*, 2002, **76**, 275.
50. Q. T. H. Shubhra, A. K. M. M. Alam, M. A. Khan, M. Saha, D. Saha and M. A. Gafur, *Composites, Part A*, 2010, **41**, 1587.
51. L. Jong, *Composites, Part A*, 2005, **36**, 675.
52. L. Jong and S. C. Peterson, *Composites, Part A*, 2008, **39**, 1768.
53. L. Jong, *Composites, Part A*, 2006, **37**, 438.
54. L. Jong, *Composites, Part A*, 2007, **38**, 252.
55. Y.-Q. Zhang, *Biotechnol. Adv.*, 2002, **20**, 91.
56. K. Ueda, Seiren Co Ltd. patent, JP 2000169595, 2000.
57. O. R. Fennema, *Food Chemistry*, Fourth Edition. ed. S. Damodaran, K. L. Parkin and O. R. Fennema. Marcel Dekker Inc., 2007.
58. F. Xie, P. J. Halley and L. Avérous, *Prog. Polym. Sci.*, 2012, **37**, 595.
59. F. O. Bobbio and P. A. Bobbio, in *Introdução à química de alimentos*, ed. Varela, São Paulo, 1995.
60. H. P. S. Abdul Khalil, A. H. Bhat and A. F. Ireana Yusra, *Carbohydr. Polym.*, 2012, **87**, 963.
61. I. S. Arvanitoyannis, A. Nakayama and S.-i. Aiba, *Carbohydr. Polym.*, 1998, **37**, 371.
62. A. J. F. Carvalho, A. E. Job, N. Alves, A. A. S. Curvelo and A. Gandini, *Carbohydr. Polym.*, 2003, **53**, 95.
63. S.-A. Riyajan, Y. Sasithornsonti and P. Phinyocheep, *Carbohydr. Polym.*, 2012, **89**, 251.
64. A. Andrio, V. Compan, R. C. Reis-Nunes, M. L. Lopez and E. Riande, *J. Membr. Sci.*, 2000, **178**, 65.
65. R. M. Barrer and H. T. Chio, *J. Appl. Polym. Sci., Part C: Polym. Symp.*, 1965, 111.
66. A. F. Martins, L. L. Y. Visconte and R. C. R. Nunes, *Kautsch. Gummi Kunstst.*, 2002, **55**, 637.
67. A. I. Khalf, D. E. E. Nashar and N. A. Maziad, *Mater. Des.*, 2010, **31**, 2592.
68. R. Auras, B. Harte and S. Selke, *Macromol. Biosci.*, 2004, **4**, 835.
69. R. Jaratrotkamjorn, C. Khaokong and V. Tanrattanakul, *J. Appl. Polym. Sci.*, 2012, **124**, 5027.
70. P. Juntuek, C. Ruksakulpiwat, P. Chumsamrong and Y. Ruksakulpiwat, *J. Appl. Polym. Sci.*, 2011, **122**, 3152.
71. P. Juntuek, C. Ruksakulpiwat, P. Chumsamrong and Y. Ruksakulpiwat, *J. Appl. Polym. Sci.*, 2012, **125**, 745.
72. R. Bhatt, D. Shah, K. C. Patel and U. Trivedi, *Bioresour. Technol.*, 2008, **99**, 4615.

CHAPTER 17

Clay Reinforcement in Natural Rubber Based Blends: Micro and Nano Length Scales

YAMUNA MUNUSAMY,*[a] HANAFI ISMAIL[b] AND CHANTARA THEVY RATNAM[c]

[a] Petrochemical Engineering Department, Faculty of Engineering and Green Technology, Universiti Tunku Abdul Rahman, Jalan Universiti, Bandar Barat, 31900 Kampar, Perak, Malaysia; [b] Polymer Division, School of Materials and Mineral Resources Engineering, Universiti Sains Malaysia, 14300 Nibong Tebal, Penang, Malaysia; [c] Radiation Processing Technology Division, Malaysian Nuclear Agency, Bangi, 43000 Kajang, Selangor DE, Malaysia
*Email: yamunam@utar.edu.my

17.1 Introduction

A common practice to enhance the properties of rubber products is by loading large amounts of fillers that are either reinforcing or non-reinforcing. Traditionally carbon black, precipitated silica and calcium carbonate are used to reinforce the natural rubber matrix in bulk amounts, up to 80 phr in some cases. Addition of large quantities of fillers reduces the elasticity and processability and increases the weight of the natural rubber composites.

Thus, recent research into natural rubber composites has concentrated on the use of nanometric scale inorganic fillers, especially nanoclay, to produce nanocomposites. The nanoclay has increased specific surface area, which can form good interactions with polymeric chains. As a consequence a lower filler loading, normally less than 10 wt%, is needed to achieve required properties.[1]

Clay platelets are an interesting reinforcing filler because of their specific surface area, which is usually two orders of magnitude greater than conventional natural rubber fillers. Theoretically, therefore, a several fold increase in properties could be achieved at very low loadings of exfoliated clay platelets. In practice however, complete exfoliation is not easily obtained.[2]

To exfoliate clay, two considerations are important – polymer clay compatibility to ensure polymer chain penetration into clay galleries and stearic repulsion of the chains to push the platelets apart. Compatibility alone does not guarantee exfoliation. Sufficient shear forces during processing are also needed to delaminate the clay platelets apart from each other.[3–5]

Clay reinforcement has already been used in various polymer systems including polyamide, polypropylene, polystyrene and polycarbonate. However, the major challenges in using clay reinforcement in natural rubber are its nonpolar nature, the conventional processing methods used, such as a two-roll mill, and the high molecular weight of natural rubber polymeric chains.

Clay platelets can be more easily dispersed in polar polymers than in nonpolar polymers like natural rubber because of their hydrophilic nature. Natural rubber also consists of longer polymeric chains than most thermoplastics, and so the penetration of these long chains between the clay platelets becomes difficult. Moreover the conventional processing methods used do not increase the flowability of the natural rubber sufficiently to allow the delamination of clay platelets from each other.

17.2 Recent Developments

A common way of increasing the compatibility between a natural rubber matrix and clay nanofillers is by modifying the surface of the clay and by using compatibilizers. In order to render the hydrophilic layered silicates organophilic, the cation exchange capacity of clay is exploited. For example, pristine montmorillonite (MMT) clay usually contains hydrated Na^+ or K^+ ions, and so is only miscible with hydrophilic polymers such as poly(ethylene oxide) or poly(vinyl alcohol). In order to use MMT as a reinforcing filler in other polymer matrices, the hydrophilic silicate surface must be converted to an organophilic one. Generally, this can be done by ion exchange reaction with cationic surfactants including primary, secondary, tertiary and quaternary alkyl ammonium or alkyl phosphonium cations.

Alkyl ammonium or alkyl phosphonium cations in organosilicates lower the surface energy of the inorganic host and improve the wetting characteristics of the MMT in the polymer matrix. It also increases the gallery spacing.[6] Furthermore, the alkyl ammonium or alkyl phosphonium cations provide functional groups that can react with the polymer matrix to improve the strength of the interface between the MMT and the polymer matrix. Based on the cation exchange capacity of clay, the alkyl ammonium surfactant content in organic clay is usually over 30 wt%.[7–9]

To further improve the interaction between the modified clay and polymer matrix, a compatibilizer is used, which is compatible with the polymer matrix but at the same time more polar than the matrix. Epoxidized natural rubber[10] and maleated natural rubber[11] are among the common compatibilizers used in natural rubber–clay systems.

17.3 Preparation Methods

Determination of the right processing method and parameters are crucial in ensuring the exfoliation of clay platelets in any polymeric matrix. For economic reasons, many researchers have tried to adapt the conventional two-roll mill method to reinforce natural rubber matrices with clay. Liu *et al.* reported that kaolinite clay platelets were dispersed in a rubber matrix in a directional parallel arrangement, with 20–50 nm thickness, when the nanocomposites were prepared by a two-roll mill method.[12] This might be due to the shear force direction during preparation of the nanocomposites. Transmission electron microscopy (TEM) images and X-ray diffraction (XRD) results indicate intercalation of kaolinite clay in the natural rubber matrix rather than exfoliation. Similar results were obtained by Carli *et al.* in a natural rubber–Cloisite® clay system prepared by the two-roll mill method. Analysis by wide-angle X-ray diffraction (WAXD) showed that the Cloisite®-related diffraction peaks were displaced toward lower angles for all the nanocomposites in comparison to pristine Cloisite®, suggesting the intercalation of natural rubber chains in the Cloisite® structure. However, none of the peaks disappear, which indicates no exfoliation. TEM observation confirms this, where the images clearly show an intercalated structure, homogeneously dispersed into the rubber matrix.[13]

Natural rubber based clay reinforced nanocomposites can also be produced through melt intercalation techniques. In this technique, the clay is mixed with the polymer matrix in the molten state. The flowability of the polymeric chain is better, and so it can penetrate between the silicate layers of the clay. A common approach to ensuring exfoliation of clay during melt mixing is by adding a second polymer to the natural rubber. The clay will be either encapsulated earlier or selectively incorporated into the second polymer during the melt intercalation. In this ternary blend one must consider the polarity of each component, viscosity and the desired final properties when choosing the second host matrix for the clay. In practice, the production of ternary blends using natural rubber, a second polymer and clay is very rare. Many researchers still prefer to use compatibilizers to improve the interaction between clay and natural rubber. This might be due to a lack of understanding about the influence of the processing steps and sequences required to produce ternary blends with well dispersed clay layers.

Some may use the combined latex compounding and melt mixing approach to produce natural rubber nanocomposites. For instance, natural rubber/Ca-montmorillonite (Ca-MMT) nanocomposites with well exfoliated Ca-MMT were prepared by a combination of latex compounding and melt mixing. Firstly, a high Ca-MMT content masterbatch was co-coagulated by natural rubber latex and modified Ca-MMT aqueous suspension through latex compounding. The masterbatch was added into the system of styrene butadiene rubber and epoxidized natural rubber by subsequent melt mixing. The XRD and TEM results showed that intercalated and exfoliated nanocomposites were obtained by the masterbatch technique, but this two-step process is time-consuming and tedious.[14]

Solution intercalation can also be used to produce clay reinforced rubber based nanocomposites. Solution intercalation involves dissolving a polymer in a solvent, and then mixing with a clay–solvent dispersion, whereby the polymer will displace the solvent and intercalate within the interlayer. Finally the solvent is removed, yielding the nanocomposites.[15] This method was found to be very reliable, and the nanoclay remains dispersed after the solvent is removed. Solution intercalation consistently gives exfoliated nanocomposites, provided the clay organic treatment, solvent and blending conditions are considered.[16] For example in the EVA/natural rubber/organoclay nanocomposite system, complete exfoliation at 2 phr organoclay loading was achieved through solution intercalation compared to melt intercalation, as indicated in the XRD results in Figures 17.4 and 17.5.

The reason for better organoclay dispersion in solution blending can be explained by thermodynamic and kinetic factors.[17] When producing a nanocomposite, the polymer chains in the melt or solution have to penetrate between the silicate layers of the clay and therefore increase the interlayer spacing of the silicates. The structure formed is an intercalated nanostructure. Further penetration will cause the silicate layers to be delaminated from the initial ordered structure and exfoliation occurs. From a thermodynamic point of view molecules prefer to be in a disordered condition. Changes of molecules from an ordered state to a relatively disordered state will cause a gain in entropy and is more favourable. Thus, the penetration of polymer chains to the inside of silicate galleries is not favourable in melt intercalation[18] because the initial mobility of the polymer chain in the melt is higher compared to the mobility of the polymer chain in between the silicate layers. Intercalation of polymer chains in between the silicate layers will cause a loss of entropy.

In solution blending the organoclay is first stirred into a solvent for a prolonged period, during which time a large number of solvent molecules will be absorbed on the surface and in between the silicate layers. During polymer chain intercalation from solution to the space between the silicate layers, the solvent molecules will desorb from the silicate layers to make space for polymer chain penetration. The desorbed solvent molecules are freer to move and results in an entropic gain, which compensates for the decrease in entropy due to the demobilizing effect of polymeric chains between the silicate layers.[19] Therefore, compared to melt intercalation, intercalation of polymeric chains into the silicate layers is more thermodynamically preferred in solution blending. Moreover kinetically, the mobility of the polymer macromolecules is better in solution compared to the melt, due to the low viscosity of the solution. Therefore during mechanical stirring it is easier for the mobile polymer chain to penetrate between the individual silicate layers. However the solvent intercalation method is less popular compared to the two-roll mill and melt intercalation method for a number of reasons – it requires a large amount of solvent, is not environmentally friendly and is time consuming.

An alternative method of producing natural rubber based clay reinforced nanocomposites with outstanding properties is by using a spray drying technique. In this technique the silicate layers of clay will be well dispersed in an irradiated polymer latex and this mixture will be sprayed through hot air to produce micrometre-sized liquid droplets. When the solvent is fully evaporated, micrometre-sized polymer spheres with delaminated clay silicate layers on their surface are produced. These spheres can later be melt blended with natural rubber to produce ternary nanocomposites.[20] It is noteworthy that exfoliation of nanofillers can still be achieved without modification of the nanofiller surface, thus the expensive modification process can be eliminated.

17.3.1 Development of Ethylene Vinyl Acetate/Natural Rubber/ Organoclay Ternary Blends

Blends of ethylene vinyl acetate (EVA) with natural rubber at 50:50 composition exhibit improved electrical insulation properties and flexibility at low temperature (−55 °C) without the addition of plasticizer. Furthermore, the fully saturated EVA backbone imparts excellent heat, ozone and weather resistance, and the vinyl group provides the blend with good oil resistance properties.[21]

EVA/natural rubber/organoclay blends have been developed through melt intercalation techniques using different blending sequences to selectively incorporate organoclay into different host matrices. Formulations of the blends and the three different blending sequences adopted, B1, B2 and B3, are as shown in Table 17.1 and Figure 17.1, respectively. Samples with the same formulation were also prepared by solution blending to evaluate the effect of different preparation methods on the properties of the nanocomposites. The preparation flow of the solution blended samples is summarized in Figure 17.2.

To further improve the properties of these blends, sulfur and peroxide crosslinking agents were incorporated into the formulations, as shown in Table 17.2. These crosslinked nanocomposites were prepared using blending sequence B3. The clay nanofillers and the crosslinking agents were found to have a synergistic effect on the dispersion of the clay layers and also the crosslink density. This will be further discussed in Section 17.5. An alternative approach using electron beam irradiation was also used to introduce

Table 17.1 Composition of material to prepare EVA/natural rubber/ organoclay nanocomposites.

Material	Parts per hundred parts of resin (phr)					
SMR L	50	50	50	50	50	50
EVA	50	50	50	50	50	50
Organoclay	0	2	4	6	8	10

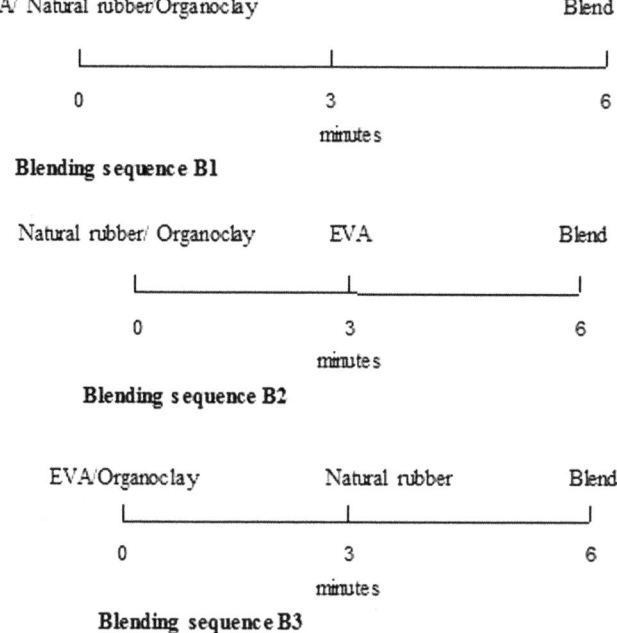

Figure 17.1 Blending sequences of nanocomposites in internal mixer.

crosslinking into the ternary blend system. An important advantage of irradiation crosslinking is the capability of introducing crosslinks into a polymer at low temperatures and after the product is formed.[22] The irradiation dosage needed to induce crosslinks in both of the matrices falls into the same range, 100–150 kGy, thus inhibiting the degradation of either one of the matrices during exposure to electron beam irradiation.[23,24]

17.4 Characterization of Nanocomposites

17.4.1 Morphology

Basically there are three types of structure formed by clay in any polymer matrix:

1. Tactoid/agglomerate – polymer chains encapsulate stacks of clay platelets.
2. Intercalate – polymer chains enter between clay parallel platelets and therefore increase the gallery spacing.
3. Delaminated or exfoliated – clay platelets are separated from one another in a plastic matrix. During exfoliation platelets at the outermost region of each packet cleave off, exposing more platelets for separation.

Many researchers have chosen X-ray diffraction (XRD) analysis and transmission electron microscopy (TEM) images to characterize the dispersion of

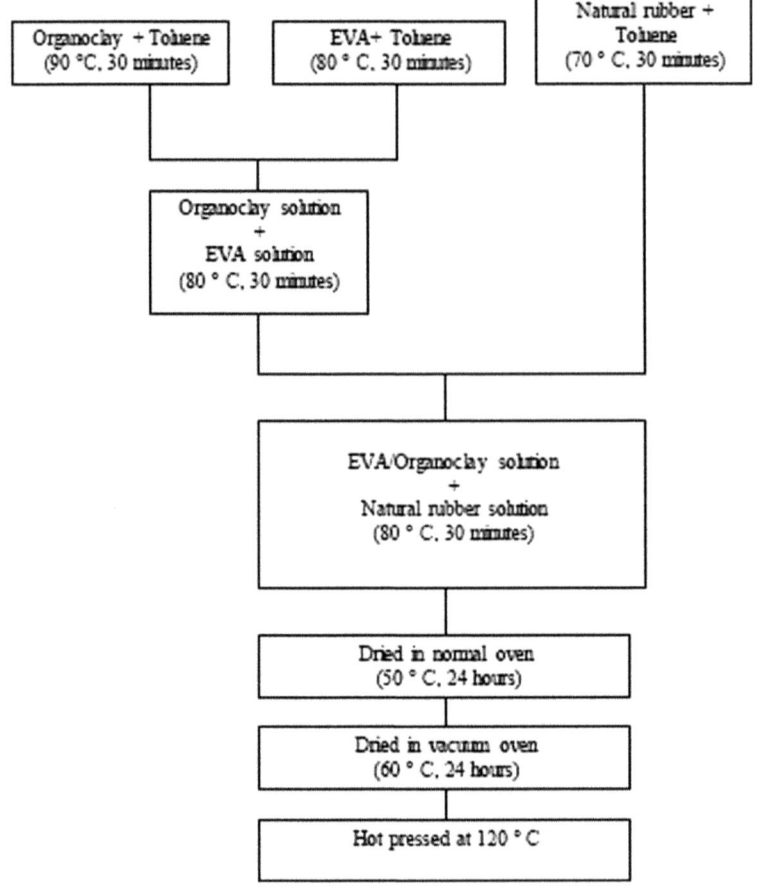

Figure 17.2 Preparation of solution blended nanocomposites.

Table 17.2 Crosslinking ingredient composition for sulfur and peroxide vulcanizates calculated to parts per hundred parts of rubber (phr).

	Parts per hundred parts of rubber (phr)						
System	NR	Sulfur	Zinc oxide	TMTD	CBS	Stearic acid	DCP
Sulfur	100.0	1.5	1.0	0.5	0.4	0.4	–
Peroxide	100.0	–	–	–	–	–	1.5

clay in polymer matrices. In XRD the intensity, width and position of the diffraction peak can be used to study the extent of clay dispersion. Since the XRD patterns are an average of the repetitive crystallographic planes of the clay, it is possible to identify an intercalated layer structure by this technique. However TEM is necessary to clearly define an exfoliated structure.[25] Expansion of the interlayer distance between the individual silicate layers will result in an intercalated ordered structure, which is reflected by formation of a new diffraction peak in XRD analysis. A decrease in the degree of coherent layer stacking would lead to peak broadening and intensity loss.[26]

Different blending sequences used to prepare the nanocomposites are expected to lead to different organoclay morphology in the polymer matrix and thus the extent of improvement in final properties will vary for each sample. In this case study, the mechanical properties and thermal stability were found to be higher when organoclay was selectively incorporated into the EVA phase through melt intercalation. TEM images, shown in Figure 17.3, indicate better dispersion of organoclay in the nanocomposites prepared through blending sequence B3.[27] An agglomerated structure can be observed for samples prepared through blending sequence B1 and B2 even at 2 phr organoclay loading. In blending sequence B3, the organoclay was selectively incorporated into the EVA phase. EVA has low melt viscosity and is more polar than natural rubber. During melt intercalation, the EVA chains enter between the organoclay silicate layers and increase the interlayer spacing. Initially each layer of silicate will have affinity towards each other. These affinities significantly decrease due to

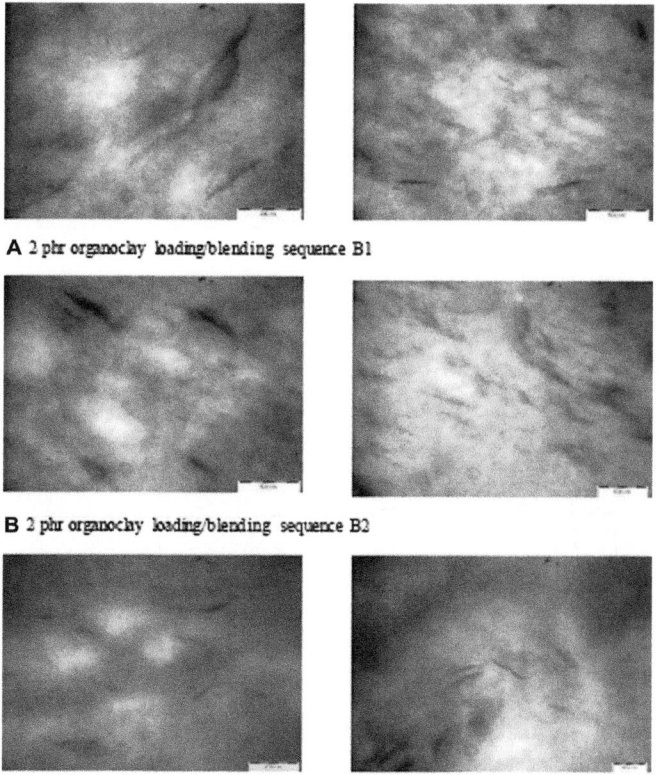

A 2 phr organoclay loading/blending sequence B1

B 2 phr organoclay loading/blending sequence B2

C 2 phr organoclay loading/blending sequence B3

Figure 17.3 TEM micrograph for EVA/NR nanocomposites with 2 phr organoclay loading.

the strong interfacial adhesion by hydrogen bonding between vinyl acetate groups of EVA and octadecylamine groups of organoclay. Therefore it will be much easier to destroy the layered structure of organoclay with sufficient shear force during melt intercalation.[28]

In blending sequence B2, the high viscosity of natural rubber and incompatibility between natural rubber and organoclay (because it does not have any polar groups in its backbone) caused poorer dispersion of the filler, which was first incorporated into the natural rubber phase. Thus agglomerates were formed.

A comparison of organoclay dispersion level between the melt intercalated and solution blended samples could be observed through the XRD pattern at 2 and 8 phr organoclay loading in Figures 17.4 and 17.5, respectively. All the

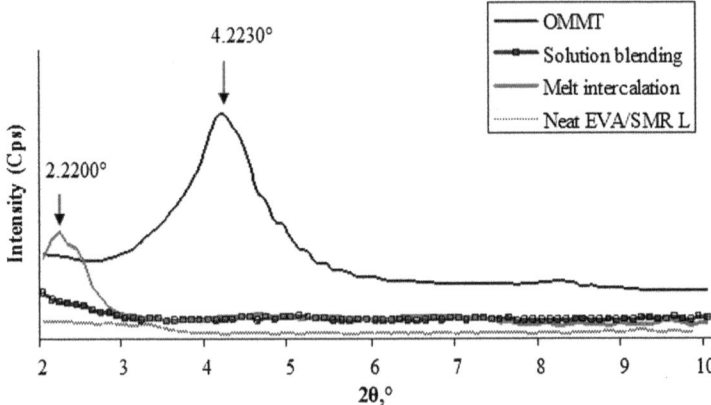

Figure 17.4 XRD peaks for organoclay, neat EVA/NR and its nanocomposites with 2 phr organoclay loading.

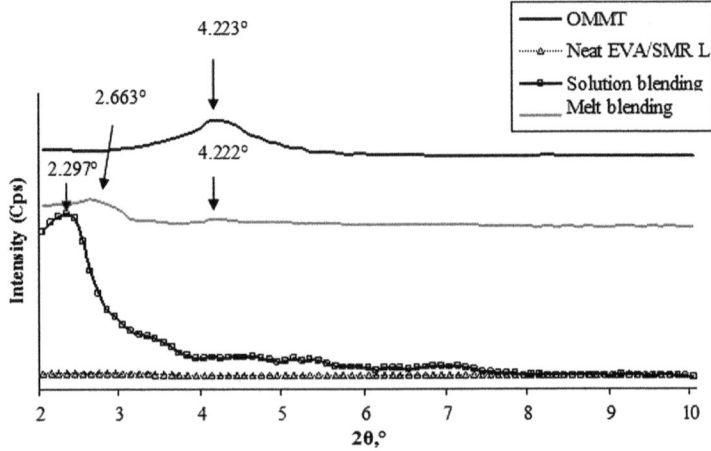

Figure 17.5 XRD patterns of pristine organoclay and EVA/NR nanocomposites with 8 phr organoclay loading.

Table 17.3 Comparison of diffraction angle and interlayer spacing for pristine organoclay and EVA/NR nanocomposites.

Sample	$2\theta,°$	d-value (Nm)
Organoclay	4.2230	2.1257
EVA/NR/organoclay 2 phr/melt blending	2.2220	4.1254
EVA/NR/organoclay 8 phr/melt blending	2.6630	3.3957
EVA/NR/organoclay 2 phr/solution blending	–	–
EVA/NR/organoclay 8 phr/solution blending	2.297	3.8420

corresponding diffraction values and interlayer distances are summarized in Table 17.3. The diffraction peak at $2\theta = 4.223°$ of the pristine organoclay is shifted to lower angles or diminished for all nanocomposites. This indicates penetration of the polymer matrix between the organoclay galleries, which will further expand the spacing between individual silicate layers. However, at 8 phr organoclay loading a second diffraction peak, $2\theta = 4.222°$, was observed for melt intercalated nanocomposites (see Figure 17.5). This peak indicates formation of agglomerated clay. When solution blending was used to prepare the nanocomposites, the d_{001} diffraction peak is completely absent at 2 phr organoclay loading (see Figure 17.4). This indicates that the ordered silicate layers were separated from each other and randomly dispersed in the polymer matrix.[29] Even at 8 phr organoclay loading, the diffraction angle was moved to a lower value compared to the melt intercalated samples and the second diffraction peak, which is associated with agglomerates, was not observed. Therefore the application of solution blending in preparing nanocomposites results in better dispersion of organoclay compared to melt intercalation.

High-resolution TEM only probes a very small volume, which may or may not be representative of the total composite. Moreover sample preparation for TEM analysis is difficult and time-consuming. Therefore, the method is too costly for routine characterization of nanocomposites. As a result researchers are now using melt rheology as a complementary method to analyse the dispersion/exfoliation in polymer–clay nanocomposites. Several authors found that the shear thinning index, n, gives a semi-quantitative measure of the degree of exfoliation.[30]

17.4.2 Mechanical Properties

Changes in the nanoscale properties of polymeric chains due to filler–matrix interactions will be reflected in macro-scale properties. In this case study, the highest tensile strength values were achieved at 2 phr organoclay loading for all nanocomposites prepared *via* blending sequences B1, B2 and B3, and these values decrease gradually at higher organoclay loadings, as shown in Figure 17.6. The clay layers are well dispersed at lower loading (Table 17.3). The reinforcing effect is reduced for nanocomposites with higher clay content owing to the poor dispersion of clay. The agglomerated clay has larger particle size and serves as stress concentration points and flaws for crack initiation,

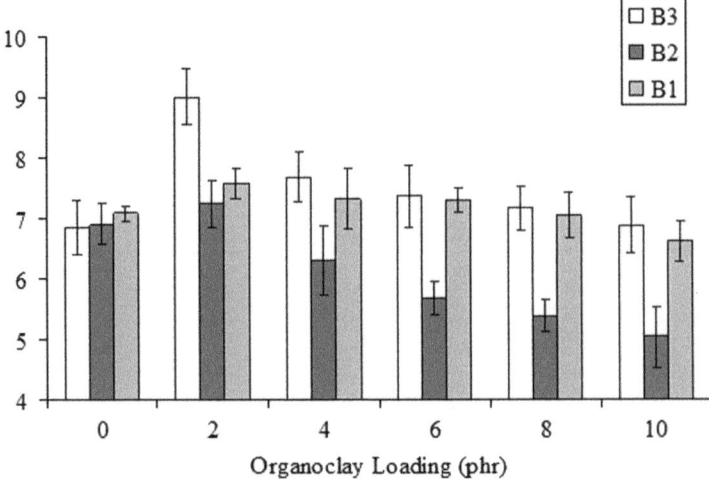

Figure 17.6 The effect of different blending sequences and organoclay loading on tensile strength of the EVA/SMR L nanocomposites.

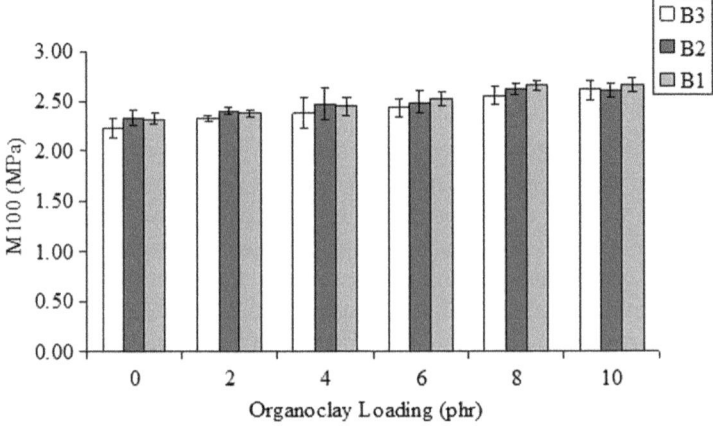

Figure 17.7 The effect of different blending sequences and organoclay loading on stress at 100% elongation (M100) of the EVA/NR nanocomposites.

resulting in premature failure.[5] Nanocomposites prepared *via* blending sequence B3 showed the highest improvement in tensile strength at 2 phr organoclay loading, 31.60% higher than pristine EVA/natural rubber blends. The high surface-to-volume ratio due to intercalation and exfoliation, and compatibility between organoclay and the EVA matrix,[31] contribute to the excellent reinforcing characteristics of the organoclay in nanocomposites prepared *via* blending sequence B3.

Figure 17.7 shows that with increasing loading of organoclay, there is an increase in stress at 100% elongation (M100) for all nanocomposites. Samples prepared *via* blending sequence B1 exhibit the highest improvement in M100.

M100 increased in the presence of organoclay due to the demobilizing effect it had on the matrix chains.³² In blending sequence B1, the organoclay was incorporated in both the EVA and natural rubber matrix. Therefore the demobilizing effect occurs in both the matrix giving rise to the M100 value compared to nanocomposites prepared *via* blending sequence B2 and B3.

The tensile fracture surface for neat EVA/natural rubber in Figure 17.8(a) is fairly smooth, with some tear lines and plastic flow. Generally the SEM micrographs for all the nanocomposites with 2 phr organoclay loading prepared *via* blending sequences B1, B2 and B3 in Figures 17.8(b) to (d) show a rough fracture surface compared to neat EVA/natural rubber. The tensile fracture surfaces for all the nanocomposites filled with 8 phr organoclay loading in Figures 17.8(e) to (g) exhibit no tear line or exhibit only one major fracture plane, which indicates no deflection was subjected to the propagation of crack growth during failure.

A 2 phr organoclay filled nanocomposite prepared *via* blending sequence B3 exhibits more matrix tearing with step-like appearances (indicated by white arrows) and lesser voids compared to nanocomposites prepared *via* blending

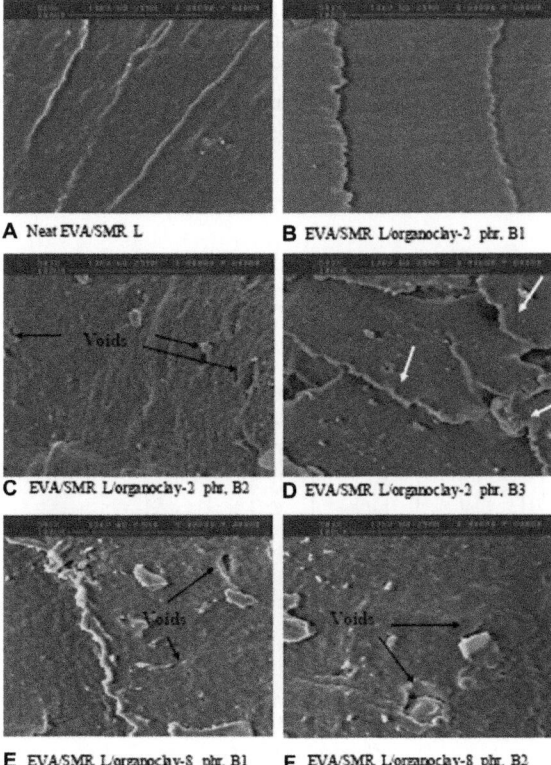

Figure 17.8 Tensile fracture surface for neat EVA/SMR L and its nanocomposites (2 and 8 phr) prepared *via* blending sequence B1, B2 and B3.

sequences B1 and B2. Well-dispersed silicates are very effective in deflecting the propagation of matrix tearing into tortuous paths, thus avoiding failure in one major plane. This result is in agreement with the highest increment in tensile strength for 2 phr organoclay filled nanocomposites prepared *via* blending sequence B3 compared to neat EVA/NR and other nanocomposites.

The comparison of tensile strength and M100 of the nanocomposites prepared through solution blending and melt intercalation (method B3) are presented in Figures 17.9 and 17.10, respectively. All the nanocomposites prepared *via* solution blending exhibit higher tensile strength and M100 value compared to melt intercalation samples. This is due to better dispersion of organoclay in the solution blending technique, which results in improved interaction between the organoclay particles and polymer matrix (proven XRD results in Figures 17.4 and 17.5).

Dynamic mechanical analysis (DMA) measures the responses of a given material to cyclic deformation as a function of temperature.[33] Figures 17.11 and 17.12 show the effect of organoclay loading and preparation method (solution blending and melt intercalation sequence B3) on the dynamic mechanical properties of EVA/natural rubber/organoclay nanocomposites at 2 phr loading. This result is in agreement with the tensile property results. The enhanced dispersion of organoclay in the solution blended nanocomposites caused the higher increment in storage modulus and glass transition (T_g) values compared to melt intercalated samples. These results suggest the existence of stronger adhesion between the filler and polymer matrix in solution blended samples compared to melt intercalated samples which will restrict the polymeric chain movement in nanocomposites.[34,35]

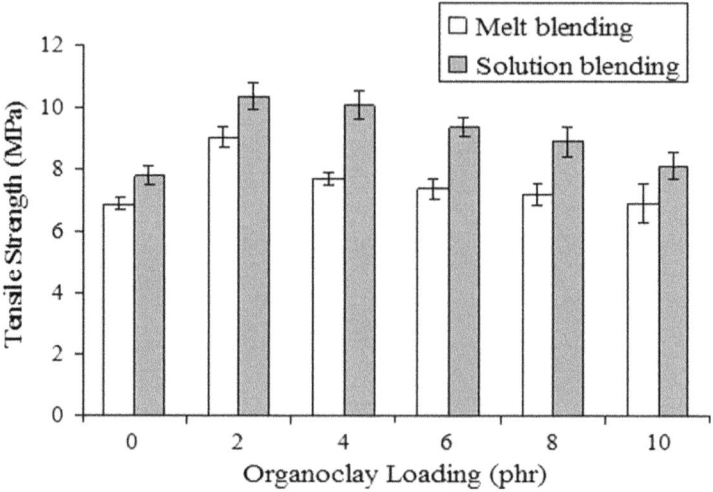

Figure 17.9 Tensile strength for neat EVA/NR and its nanocomposites prepared *via* solution blending and melt intercalation technique.

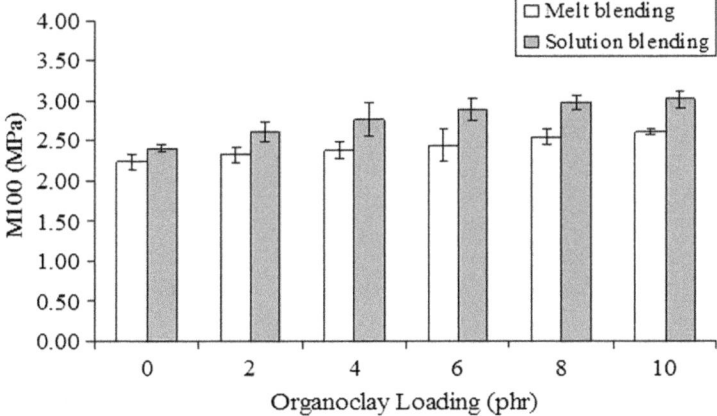

Figure 17.10 M100 for neat EVA/NR and its nanocomposites prepared *via* solution blending and melt intercalation technique.

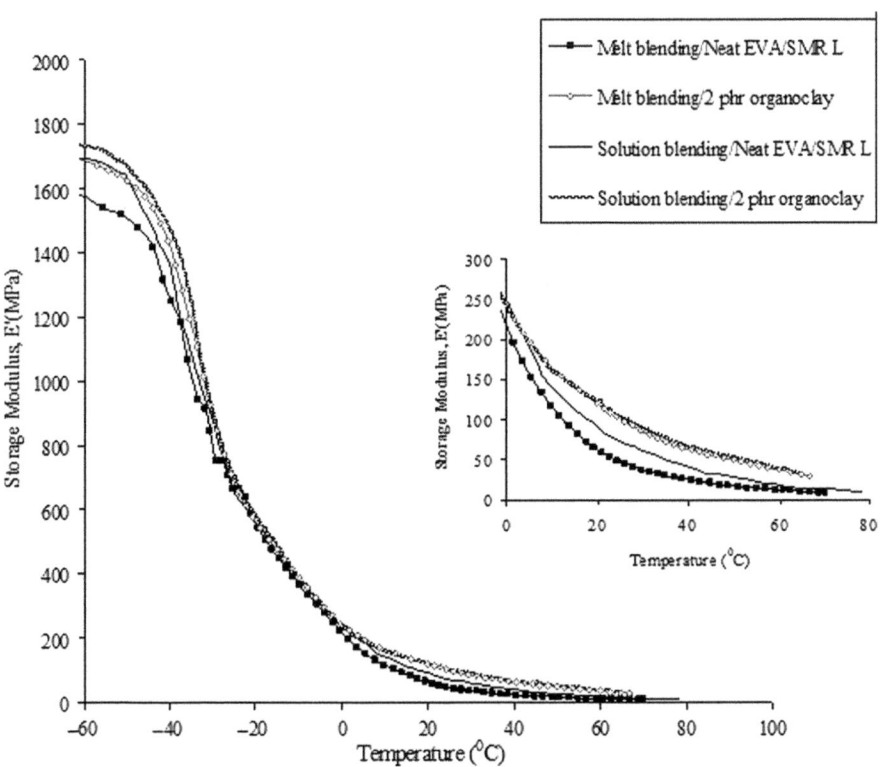

Figure 17.11 Storage modulus for neat EVA/NR and its nanocomposites with 2 phr organoclay loading.

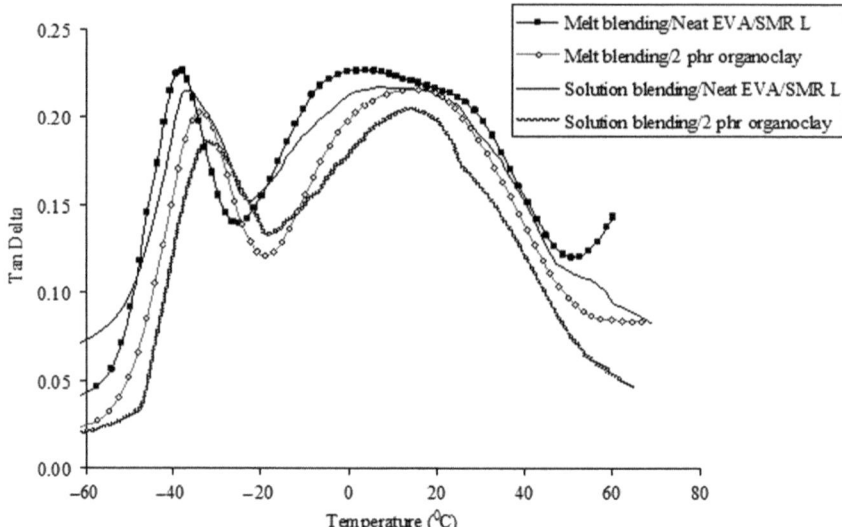

Figure 17.12 Tan δ for neat EVA/NR and its nanocomposites at 2 phr organoclay loading.

17.4.3 Thermal Properties

In some polymer nanocomposites the initial thermal decomposition will be accelerated due to the presence of alkyl ammonium surfactant on the surface of the clay. For instance in PVC the quaternary ammonium salts may accelerate the degradation of PVC.[36,37] However, this problem does not exist in clay reinforced natural rubber nanocomposites.

In the case of EVA/natural rubber/organoclay nanocomposites, the organoclay only influences the degree of nanocomposite decomposition but does not alter the mechanism by which it occurs, as shown in Figures 17.13 and 17.14. The results extracted from the thermogravimetric analysis graph are presented in Table 17.4. The decomposition temperature, T_{onset1} and T_{onset2} (temperature at 5% and 50% decomposition/temperature at 95% and 50% residual mass) improves and the weight loss decreases as the organoclay loading increases.

The improvement in thermal stability of the nanocomposites compared to the neat EVA/natural rubber is due to the barrier effect and insulating properties of organoclay. The well dispersed plate-like silicate layers form a tortuous path in the polymer matrix which gives a barrier effect and inhibits the diffusion of volatile degradation product from the inside of the polymer matrix.[38] Moreover the well-dispersed silicate layers restrict the movement of polymeric chains, hence reducing the free volume for diffusion of volatile degradation products.[39] Other researchers also confirm that organoclay tends to form a compact char-like residue on the surface of the nanocomposites when it is burnt. This char-like structure is incombustible and acts as an insulator which inhibits heat transfer to the inside of the nanocomposites.[40,41] At 8 phr

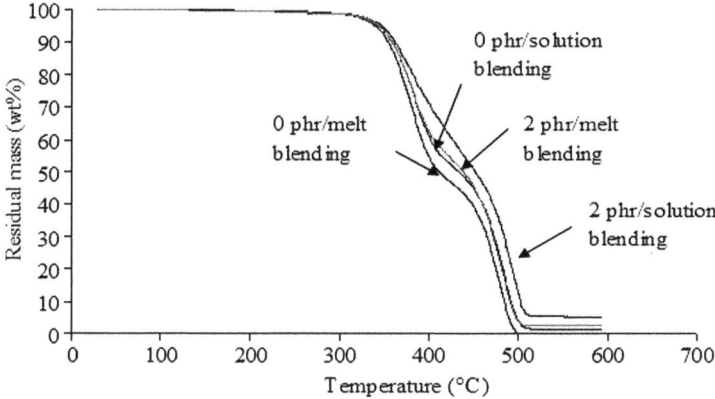

Figure 17.13 TGA curve for neat EVA/NR and its nanocomposites with 2 phr organoclay loading.

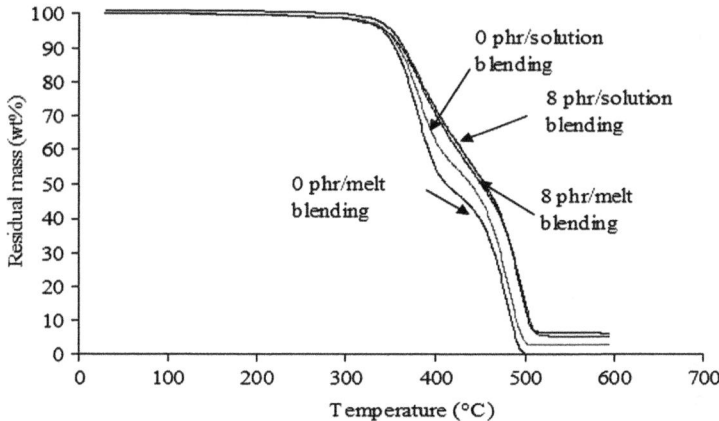

Figure 17.14 TGA curve for neat EVA/NR and its nanocomposites with 8 phr organoclay loading.

Table 17.4 The effect of different preparation methods and organoclay loadings on the degradation temperatures of neat EVA/NR and its nanocomposites.

Preparation method	Organoclay (phr)	$T_{5\%}$ (°C)	$T_{10\%}$ (°C)	$T_{max\ 1}$ (°C)	$T_{max\ 2}$ (°C)
Melt intercalation	0	343	445	379	481
	2	344	448	379	481
	8	344	450	381	484
Solution blending	0	343	425	386	483
	2	345	439	387	485
	8	352	455	389	496

organoclay loading more char will be formed compared to 2 phr organoclay loading and so the thermal stability of the nanocomposites improves proportionally to the increment in organoclay loading. The thermal stability of the nanocomposites prepared through solution blending is better because the dispersion of organoclay is better compared to melt intercalation.

17.4.4 Flammability

Good gas barrier properties of the organoclay in the nanocomposites and formation of char on the surface of the samples are the factors responsible for the increment in flame retardancy. These factors will impede oxygen diffusion into the nanocomposites and so reduce the fuel supply for burning.

In natural rubber blend nanocomposites, flame retardancy according to the UL 94 test method is only moderately reduced in the presence of organoclay. At 10 phr organoclay loading a stable char is formed and dripping could be eliminated. Formation of a stable char on the surface of 10 phr organoclay filled nanocomposites can be observed by SEM micrograph (Figure 17.15). At 4 phr organoclay loading the char is not stable and the matrix was burnt (Figure 17.15(a)), whereas at 10 phr loading a continuous char with few cracks was formed (Figure 17.15(b)). An improvement in flame retardancy with increasing amount of organoclay may also be attributed to the retarding combustion nature of silicate layers.

17.5 Crosslinking Techniques

17.5.1 Chemical Crosslinking

Chemical crosslinking of natural rubber using sulfur or peroxide as crosslinking agents is widely use to achieved the desired final properties. Thus it is crucial to understand the effect of clay on the crosslinking and vice versa.

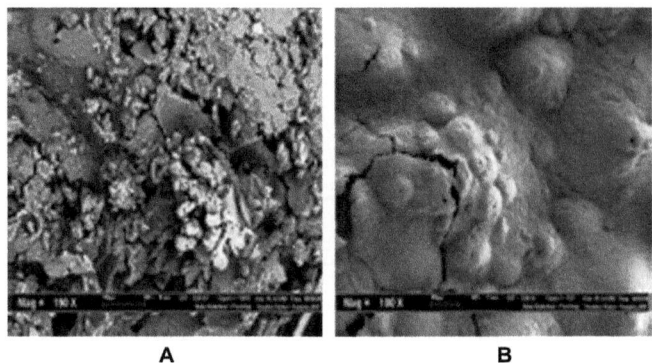

Figure 17.15 Burnt surface of the nanocomposites with (a) 4 phr organoclay loading, (b) 10 phr organoclay loading.

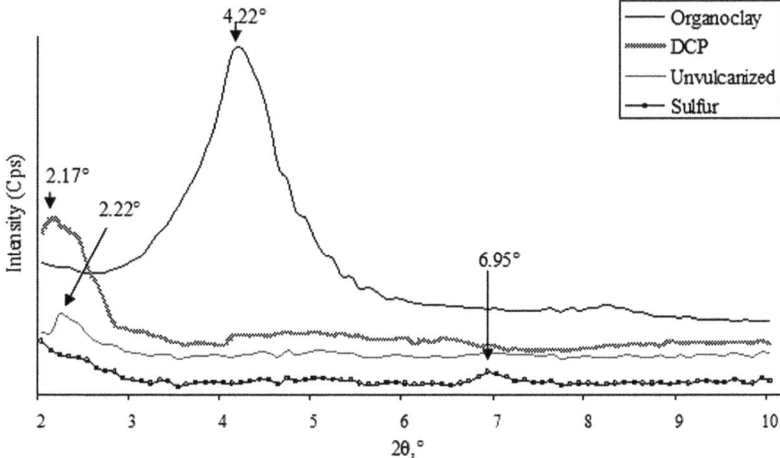

Figure 17.16 Effect of vulcanization on the XRD patterns of EVA/NR nanocomposites with 2 phr organoclay loading.

In the EVA/natural rubber/organoclay system, the dispersion of organoclay was influenced by these crosslinking agents. Figure 17.16 shows the XRD patterns of pristine organoclay and EVA/natural rubber nanocomposite vulcanizates, prepared with 2 phr organoclay loading. The diffraction values of sulfur and DCP vulcanizates were slightly lower than those for unvulcanized nanocomposites. This indicates further increment in interlayer distance and better levels of organoclay dispersion in the crosslinked matrix. The lower diffraction values for the vulcanizates can be associated with the presence of compounding ingredients.

Nanoparticles of ZnO in the sulfur compounding ingredients were found to intercalate in between kaolinite and montmorillonite layers.[42] Stearic acid, CBS and DCP are polar low molecular mass materials. The increment in individual silicate interlayer distance might be related to the adsorption of these curing ingredients on the clay layers. This phenomenon can be supported by the fact that the organoclay surface contains polar groups that can interact with the polar groups in these compounding ingredients. Adsorption of curing ingredients on the surface of filler has also been reported in silica filled polymers.[43]

In addition, some crosslinking ingredients can react with the organoclay modification agents. For example, octadecylamine can react with zinc oxide. Zinc curatives which are adsorbed on the surface of silicate layers can form complexes with octadecylamine organo-modification agent. As a result, the tethered octadecylamine chain will leave the clay surface and migrate into the rubber matrix to participate in sulfur vulcanization. These complexes will act as a co-curing agent, which accelerates sulfur vulcanization.[44]

The yield of crosslinking can be estimated from the gel fraction. The gel fraction values for EVA/natural rubber/organoclay nanocomposites are shown in Figure 17.17. An increment in organoclay loading does not show any clear influence in the gel content value of DCP vulcanizates. In the case of sulfur

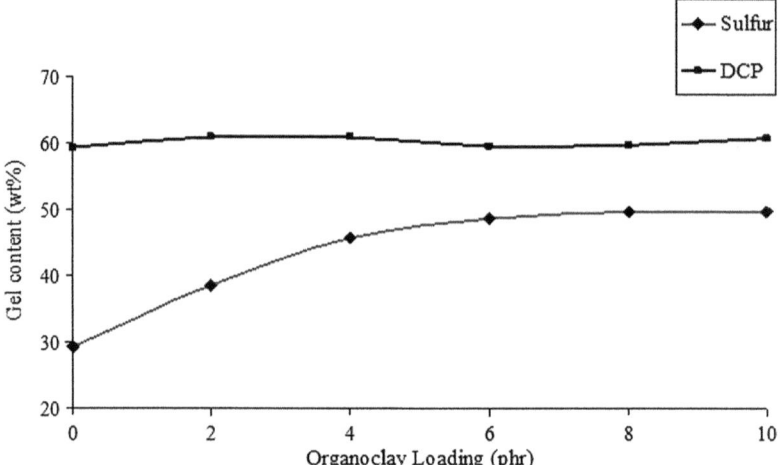

Figure 17.17 Effect of organoclay loading on the gel content value of sulfur and peroxide EVA/NR/organoclay nanocomposite vulcanizates.

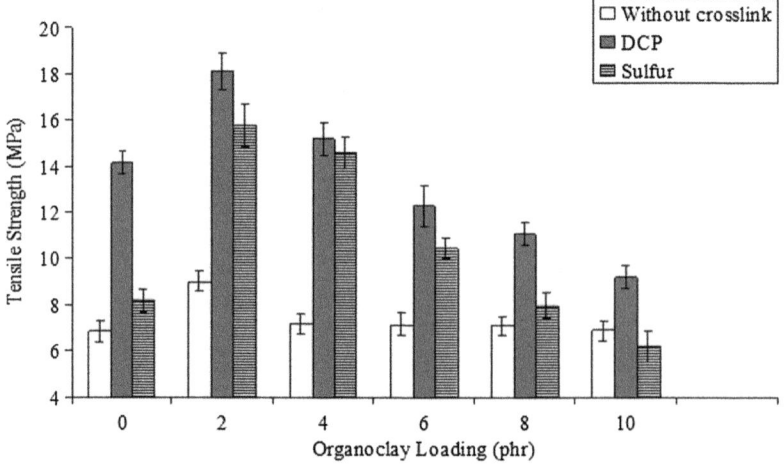

Figure 17.18 The effect of different crosslinking systems and organoclay loading on the tensile strength of EVA/NR/organoclay nanocomposites.

vulcanizates the gel content value increased linearly with organoclay loading. In previous research, organoclay was found to contribute to an obvious increment in crosslink density of natural rubber.[45] The increment in crosslink density in sulfur vulcanizates were a result of the presence of octadecylamine on the surface of organoclay. The amine intercalants of organoclay will act like an accelerator which participates in the formation of a zinc coordination complex and accelerates the crosslink process.

Figure 17.18 shows that the tensile strengths of sulfur and peroxide vulcanizates are higher than those for unvulcanized EVA/natural rubber nanocomposites at all organoclay loadings, except for EVA/natural rubber nanocomposite sulfur vulcanizate at 10 phr organoclay loading. The improvement in tensile strength is higher in peroxide vulcanizates compared to sulfur vulcanizates at similar organoclay loading. It is also obvious that the decrement in tensile strength as the organoclay loading increases is larger in sulfur vulcanizates compared to peroxide vulcanizates. The tensile strength decreased by 50.76% for nanocomposites vulcanized with peroxide when the organoclay loading increased from 2 to 10 phr, whereas, in the case of sulfur vulcanizates, a 61.03% decrement in tensile strength was observed when the organoclay loading increased from 2 to 10 phr. Crosslink density plays a crucial role in determining the mechanical properties of vulcanizates. However crosslink density is not linearly related to tensile strength.[46] In order to achieve a great improvement in the mechanical properties of vulcanizates, an optimum amount of crosslink density was needed (neither very low nor very high). Therefore, the lower tensile strength of nanocomposites prepared *via* sulfur crosslinking systems at higher organoclay loadings are mainly caused by the remarkable increase in the degree of crosslinking, as proven earlier *via* the gel content measurement. The excessive crosslinking of dispersed natural rubber phase makes it somewhat rigid and begin to act as stress concentrating points. Voids formed at the interface of excessively crosslinked rigid rubber particles and the continuous plastic phase. The formation of these voids is known as the cavitation phenomenon, and decrease the tensile strength of the sulfur vulcanizates at high organoclay loading.[47] At low crosslink density, the NR particles are not rigid and will still be attached to the continuous EVA phase. Therefore, the sulfur vulcanizates can withstand greater stress by undergoing increased extent of deformation before failure at low organoclay loading. DCP is capable of crosslinking both NR and EVA phases and therefore these vulcanizates can withstand greater stress. In addition, the more constant value of crosslink density at all organoclay loadings lead to better tensile properties compared to sulfur vulcanizates.

17.5.2 Irradiation Crosslinking

An important advantage of irradiation crosslinking is the capability of introducing crosslinks into polymers at low temperatures. Radiation crosslinking is a very useful technique to improve the thermal stability, stress crack resistance, solvent resistance and toughness of polymeric materials.[48] Previously the effect of irradiation crosslinking was studied in polyvinyl chloride (PVC)/epoxidized natural rubber (ENR 50) blends,[49] nylon 1010 and high impact polystyrene blends,[50] acrylonitrile butadiene copolymers (NBR)/PVC blends,[51] *etc*. When a polymer is exposed to electron beam irradiation, both crosslinking and degradation might occur depending on the dosage of irradiation, exposure time, structure of the polymer chain and the additive used.

The diffraction peak for the irradiated nanocomposites at 2 phr organoclay loading in Figure 17.19 is slightly moved to a lower angle compared to the non-irradiated sample. Thus compared to the non-irradiated nanocomposites, the

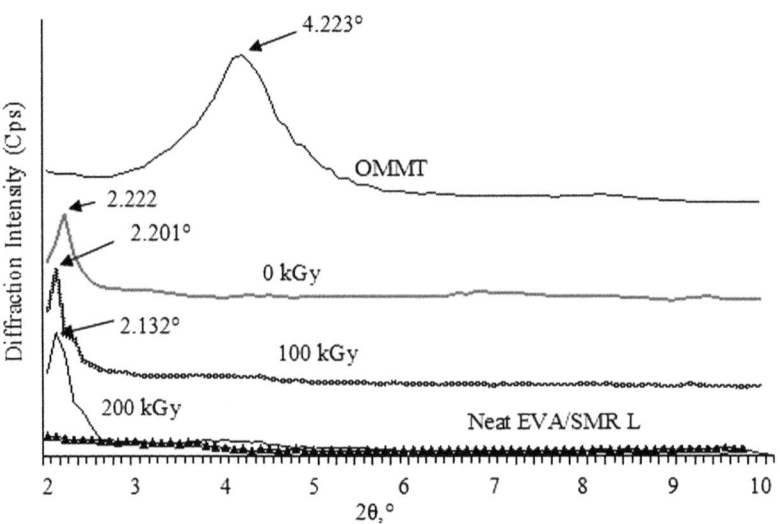

Figure 17.19 Effect of irradiation on the XRD patterns of EVA/NR nanocomposites with 2 phr organoclay loading.

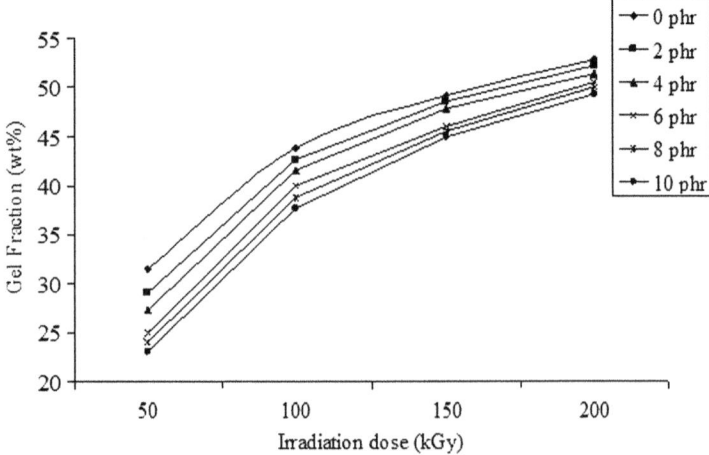

Figure 17.20 Effect of irradiation dose and organoclay loading on the gel fraction yield of EVA/NR blends and the nanocomposites.

interlayer spacing between the individual silicate layers have been further increased in the irradiated nanocomposite. This might be due to the formation of polar groups by electron beam irradiation, which will further enhance the intercalation of the organoclay in the matrix.

In the case of EVA/natural rubber/organoclay nanocomposites, at irradiation doses of 50, 100 and 150 kGy the gel fraction yield has been significantly reduced with the increment of organoclay loading, as shown in Figure 17.20.

The natural rubber and EVA radicals can be scavenged by organoclay. Thus radical–radical interaction to form crosslink networks will be hindered by organoclay. In addition, exfoliation and intercalation of organoclay nanoparticles also block the sites for crosslinking in the polymeric matrix. However, at 200 kGy the gel fraction yield was not much affected by organoclay loading.[52]

Basically there are three major factors which must be taken into consideration in order to understand the effect of irradiation on clay nanocomposites: Hoffmann degradation of organoclay modification agent, exfoliation and intercalation of organoclay and the barrier effect of organoclay.[53]

Alkylammonium can undergo Hoffman reaction, which will generate ammonium ions, acidic sites in the aluminosilicates and corresponding olefins. The acidic sites can accept a single electron from donor molecules with low ionization, leading to the formation of free radicals. This free radical can accelerate oxidation degradation or induce crosslinking of polymeric chains.[54] Oxidation degradation can be accelerated by oxygen from the atmosphere. Exfoliated and intercalated organoclay can form effective barrier layers which will inhibit the diffusion of oxygen from air into the polymeric matrix. Degradation *via* oxidation will be retarded.[55]

At the same time, crosslinking induced by free radicals will also be reduced due to the barrier effect. However, as the irradiation dose increases, more free radicals will be formed. The large amount of free radicals can overcome the barrier effect caused by well dispersed clay. This has been proved by the gel fraction yield. At lower irradiation doses the gel fraction is significantly reduced for the nanocomposites, whereas at 200 kGy the gel fraction for all the nanocomposites is almost the same as that for the pristine blend.

Thus in a blend which consists of two polymer components, when one of the components can only be crosslinked at higher irradiation dosage compared to the other, organoclay can be used to retard degradation.

17.6 Conclusions

The extent of clay dispersion and clay–polymer interaction is crucial in determining the formation of natural rubber based nanocomposites. Due to the low polarity and high viscosity of natural rubber, direct blending of clay nanoparticles into natural rubber will only yield micro-scale composites. Thus, it is more effective to blend clay nanoparticles into another polymer component before blending with natural rubber.

References

1. S. G. Prolongo, M. Campo, M. R. Gude, R. Chaos-Moran and A. Urena, *Compos. Sci. Technol.*, 2009, **69**, 349.
2. G. Mani, Q. Fan, S. C. Ugbolue and Y. Yang, *J. Appl. Polym. Sci.*, 1997, **97**, 218.
3. M. Alexandra and P. Dubois, *Mater. Sci. Eng.*, 2000, **28**, 1.

4. N. Hasegawa, H. Okamoto, M. Kato, A. Usuki and N. Sato, *Polymer*, 2003, **44**, 2933.
5. B. Yalcin and M. Cakmak, *Polymer*, 2004, **45**, 6623.
6. A. Dasari, Z. Z. Yu, Y. W. Mai, G. H. Hu and J. Varlet, and , *Compos. Sci. Technol.*, 2005, **66**, 2314.
7. J. R. Cho and D. R. Paul, *Polymer*, 2001, **42**, 1083.
8. S. W. Kim, W. H. Jo, M. S. Lee, M. B. Ko and J. Y. Jho, *Polymer*, 2002, **4**, 103.
9. T. D. Fornes, P. J. Yoon, D. L. Hunter, H. Keskkula and D. R. Paul, *Polymer*, 2001, **42**, 9929.
10. P. L. Teh, Z. A. Mohd Ishak, A. S. Hashim, J. Karger-Kocsis and U. S. Ishiaku, *Eur. Polym. J.*, 2004, **40**, 2513.
11. C. A. Rezende, F. C. Braganza, T. R. Doi, L.-T. Lee, F. Galembeck and F. Boue, *Polymer*, 2010, **51**, 1364.
12. Q. Liu, Y. Zhang and H. Xu, *Appl. Clay Sci.*, 2008, **42**, 232.
13. L. N. Carli, C. R. Roncato, A. Zanchet, R. S. Mauler, M. Giovanela, R. N. Brandalise and J. S. Crespo, *Appl. Clay Sci.*, 2011, **52**, 56.
14. J. Tan, X. Wang, Y. Luo and D. Jia, *Mater. Des.*, 2012, **34**, 825.
15. M. Frounchi, S. Dadbin, Z. Salehpour and M. Noferesti, *J. Membr. Sci.*, 2006, **282**, 142.
16. A. T. Koshy, B. Kuriakose and S. Thomas, *Polym. Degrad. Stab.*, 1992, **36**, 137.
17. S. Fillipi, E. Mameli, C. Marazzato and P. Magagnini, *Eur. Polym. J.*, 2007, **43**, 1645.
18. R. A. Vaia, K. D. Jandt, E. J. Kramer and P. G. Emmanuel, *Macromolecules*, 1995, **28**, 8080.
19. R. A. Vaia, H. Ishii and P. G. Emmanuel, *Chem. Mater.*, 1993, **5**, 1694.
20. S. I. Yun, D. Attard, V. Lo, J. Davis, H. Li, B. Latella, F. Tsvetkov, H. Noorman, S. Moricca, R. Knott, H. Hanley, M. Morcom, G. P. Simon and G. E. Gadd, *J. Appl. Polym. Sci.*, 2008, **108**, 1550.
21. M. Zurina, H. Ismail and C. T. Ratnam, *Polym. Degrad. Stab.*, 2006, **91**, 2723.
22. A. A. Basfar, *Polym. Degrad. Stab.*, 2002, **77**, 221.
23. H. M. Dahlan, M. D. K. Zaman and A. Ibrahim, *Radiat. Phys. Chem.*, 2002, **64**, 429.
24. M. Mateev and S. Karageorgiev, *Radiat. Phys. Chem.*, 1998, **51**, 205.
25. M. Valera-Zaragoza, E. R. Vargas, F. J. M. Rodríguez and B. M. Huerta-Martínez, *Polym. Degrad. Stab.*, 2006, **91**, 1319.
26. R. A. Vaia and E. P. Gianellis, *Macromolecules*, 2007, **30**, 8000.
27. H. Ismail, Y. Munusamy, M. Jaafar and C. T. Ratnam, *Polym.-Plast. Technol. Eng.*, 2008, **47**, 752.
28. H. A. Stretz, D. R. Paul, R. Li, H. Keskkula and P. E. Cassidy, *Polymer*, 2005, **46**, 2621.
29. A. D. Gianni, E. Amerio, O. Monyicelli and R. Bongivanni, *Appl. Clay Sci.*, 2008, **42**, 116.
30. R. Wagener and T. J. G. Reisinger, *Polymer*, 2003, **44**, 7513.

31. M. Aroyo, M. A. Lopez-Manchado, J. L. Valentin and J. Carretero, *Compos. Sci. Technol.*, 2007, **67**, 1330.
32. B. R. Guduri and A. S. Luyt, *J. Appl. Polym. Sci.*, 2007, **103**, 4095.
33. Y. Sun, Y. Luo and D. Jia, *J. Appl. Polym. Sci.*, 2008, **107**, 2786.
34. X. Wang, A. Huang, D. Jia and Y. Li, *Eur. Polym. J.*, 2008, **44**, 2184.
35. W. Zhong, X. Qiao, M. Cao, K. Sun and G. Zhang, *J. Appl. Polym. Sci.*, 2008, **107**, 1407.
36. F. Gong, M. Feng, C. Zhao, S. Zhang and M. Yang, *Polym. Degrad. Stab.*, 2004, **84**, 289.
37. A. Pozgav, I. Csapo, L. Szazdi and B. Pukanszky, *Mater. Res. Innovations*, 2004, **8**, 138.
38. Y. Wang, F. B. Chen, Y. C. Li and K. C. Wu, *Composites, Part B*, 2004, **35**, 111.
39. Y. Munusamy, H. Ismail, M. Mariatti and C. T. Ratnam, *J. Vinyl Addit. Technol.*, 2009, **15**, 244.
40. J. K. Pandey, K. R. Reddy, A. P. Kumar and R. P. Sing, *Polym. Degrad. Stab.*, 2005, **88**, 234.
41. A. Szep, A. Szabo, N. Toth, P. Anna and G. Marosi, *Polym. Degrad. Stab.*, 2006, **91**, 593.
42. K. G. Gatos and J. Karger-Kocsis, *Polymer*, 2005, **46**, 3069.
43. A. Usuki, A. Tukigase and M. Kato, *Polymer*, 2003, **43**, 2185.
44. Y. Munusamy, H. Ismail, M. Mariatti and C. T. Ratnam, *J. Reinf. Plast. Compos.*, 2008, **27**, 1925.
45. M. A. Lòpez-Manchado, M. Arroyo, B. Herrero and J. Biagiotti, *J. Appl. Polym. Sci.*, 2003, **89**, 1.
46. P. Boochathum and W. Prajudtake, *Eur. Polym. J.*, 2001, **37**, 417.
47. M. A. Soto-Oviedo and D. M. A. Paoli, *Polym. Bull.*, 2006, **56**, 75.
48. J. Sharif, K. Z. M. Dahlan and W. M. Z. Wan Yunus, *Radiat. Phys. Chem.*, 2006, **76**, 1698.
49. C. T. Ratnam and K. Zaman, *Polym. Degrad. Stab.*, 1999, **65**, 99.
50. W. Dong, G. Chen and W. Zhang, *Radiat. Phys. Chem.*, 2001, **60**, 629.
51. J. Li and C. Chan, *Polymer*, 2001, **42**, 6833.
52. Y. Munusamy, H. Ismail, M. Mariatti and C. T. Ratnam, *J. Vinyl Addit. Technol.*, 2009, **15**, 39.
53. Y. Munusamy, H. Ismail, M. Mariatti and C. T. Ratnam, *J. Appl. Polym. Sci.*, 2010, **117**, 865.
54. J. Sharif, K. Z. M. Dahlan and W. M. Z. Wan Yunus, *Radiat. Phys. Chem.*, 2006, **76**, 1698.
55. H. Lu, Y. Hu, Q. Kong, Z. Chen and W. Fan, *Polym. Adv. Technol.*, 2005, **16**, 688.

CHAPTER 18

Rheological Behaviour of Natural Rubber Based Blends

PLOENPIT BOOCHATHUM

Chemistry Department, Faculty of Science, King Mongkut's University of Technology Thonburi, 126 Pracha-utid Road, Bangmod, Tungkru, Bangkok, Thailand, 10140
Email: ploenpit.boo@kmutt.ac.th

18.1 Introduction

Under different deformation conditions, natural rubber (NR) may exhibit predominant viscous flow, elastic or viscoelastic behaviour. Thus, the time for the movement of the NR molecular chains, *i.e.* relaxation time, is vastly affected by those deformation rate and NR types. The variation of NR types such as smoked rubber sheet, rubber blocks such as skim block, STR 5L and STR 20, is another factor that influences rheological properties and processing of NR due to their different Mooney viscosity, molecular weight distribution and gel content. Types of NRs are based on the different production processes of NR in which rubber smoked sheets are produced from NR in the latex form whereas block types are produced from various types of NR sources such as from latex in the case of STR 5L, from rubber scrap in the case of STR 20 and from skim latex in the case of skim block.

In general, the huge molecular weight of Hevea NR reaching to 2.24×10^6 that is much higher than that of synthetic rubbers can cause the difficulty of processing.[1] Thus, high shearing force is necessary for high molecular weight NR to achieve the appropriate viscosity for ease of polymer dispersion in the case of blending process. Mastication is one of the important processes to reduce the viscosity of NR by breaking molecular chains of rubber molecules before adding chemicals or other polymers in the blending process apart from using processing aids such as peptizer, plasticizer or lubricant.

Not only the molecular weight and gel content difference, but also the non-rubber parts present in the NR molecules such as protein and fatty acid ester were found to play an important role in the rheological behaviour of NR. Recently, the rheological behaviour of low protein natural rubber (LPNR) was investigated using rubber processing analyser (RPA) in comparison with NR.[2] The results showed that complex viscosity (n^*) of LPNR was found to be lower than that of NR while the stress relaxation time of LPNR was faster than NR. On the other hand, the processing performance between LPNR and NR was found to show the significant discrepancy. This might be due to the lower molecular weight and lower torque of LPNR than those of NR. Recently, Boochathum and a coworker clarified the effect of protein and fatty acid ester present on NR molecules on the processability in the term of tan δ measured by using RPA.[3] SPNR; prepared from saponification reaction of NR latex, was found to consist both protein that measured in the term of nitrogen content and fatty acid ester content very much lower than DPNR; prepared from enzymatic deproteinization reaction of NR latex, and NR as summarized in Table 18.1.

It was remarkable that SPNR containing very low both protein and fatty acid content showed much easier processability, *i.e.* higher tan δ, than DPNR and NR while the processability of DPNR containing low proteins but relatively high fatty acid ester was not considerably different from that of NR as seen from comparative tan δ of the samples kept for 3 and 6 months measured at variation of frequency and strain shown in Figure 18.1.

As a result, it clarified that fatty acid ester was a great influential factor on the processability of NR. The relationship between non-rubber part content and the occurring gel content at the different storage intervals of 3 and 6 months showed that the protein and fatty acid ester present on the rubber molecules closely related to the occurring gel content as shown in Table 18.2. The processability might be assumed to closely relate to the gel content that was mainly caused by ester groups present on the NR molecules.

For the NR based blends, therefore, the rheological behaviour must be considered case by case such as blending with thermoplastics and blending with other rubbers. In the case of the chemically modified NR based-blends, it was not only the type of polymer used to blend with NR, but also the variation of the chemical modification of NR would control the different rheological property of the blends. Special-property NR based blends obtained from the

Table 18.1 Nitrogen content and ester content of rubbers.

Natural rubbers	N content (% w/w)	Ester content (mmol/kg rubber)
NR	0.20	34.30
DPNR	0.03	21.74
SPNR	0.02	4.78

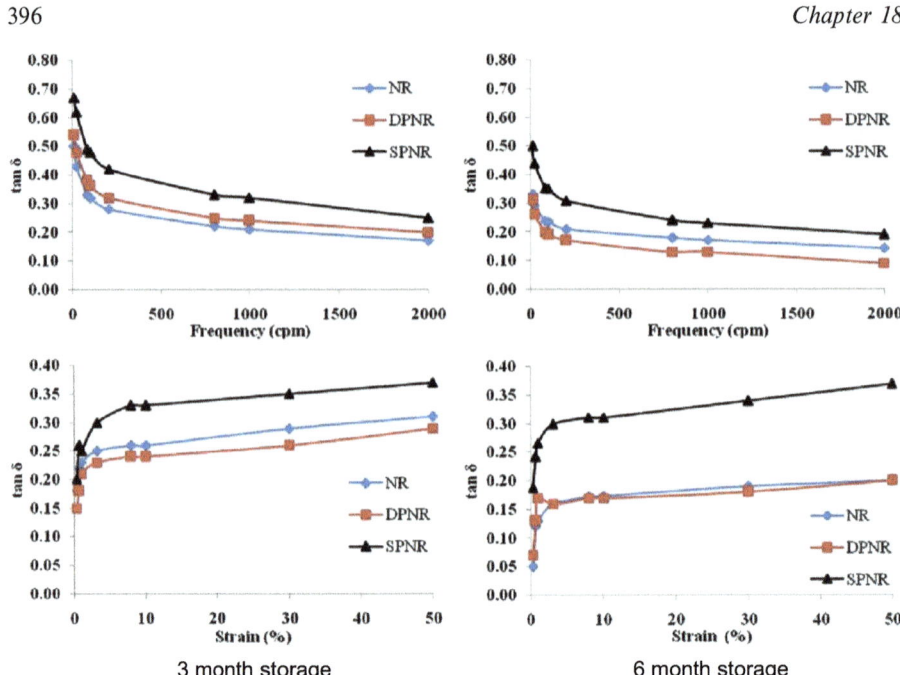

Figure 18.1 Tan δ of natural rubbers with different non-rubber content after 3- and 6-month storage measured at frequency- and strain-sweep modes.

Table 18.2 Gel content of rubbers at various storage periods.

Natural rubbers	Gel content (% w/w)		
	0 month	3 months	6 months
NR	43.97	45.66	47.50
DPNR	12.14	14.69	16.05
SPNR	0.00	0.00	0.83

chemical modification was believed to vastly affect the rheological behaviour of the blends and have been the most interesting research works up to present.

18.2 Rheological Behaviour

18.2.1 Natural Rubber–Thermoplastic Blends

It is widely known that the advantage of NR-based thermoplastic elastomer over the NR is the ease of processing, subsequently resulting in lower cost of production. Melt rheology of NR and ethylene-vinyl-acetate copolymer (EVA) blends were studied by Koshy et al.[4] The melt rheology of the blends between NR and EVA has been studied with reference to the effects of blend ratio, crosslinking systems, shear stress, and temperature. It was found that a positive deviation of viscosity of the blends between NR and EVA was observed at lower shear region. This rheological behaviour has been explained based on structural build-up of dispersed EVA domains in the continuous NR matrix. When EVA formed the dispersed phase, the viscosity of the blends was found to be a non-additive function of the viscosities of the component polymers at lower shear region, *i.e.* a positive deviation was observed.

Oommen *et al.* had studied melt rheological behaviour of the blends between NR and poly(methyl methacrylate) based on the effect of blend ratio, processing conditions and graft copolymer concentration as a function of shear stress and temperature.[5] It was clarified that the viscosity of the blends increased with the increase of the amount of NR. On the other hand, the flow behaviour of the blends was found to be influenced by dynamic vulcanization of the rubber phase.

Melt flow index and melt viscosity of NR and polypropylene blend in the ratio of 90:10 that was extruded in the presence of a peroxide [1,3-*bis*(*t*-butylperoxy)benzene] and coagent (trimethylolpropanetriacrylate, TMPTA) were investigated by Yoon *et al.*[6] The variation of melt viscosity that was characterized based on the crosslinking and chain scission of PP was studied. It was remarkable that at a constant content of the coagent, melt viscosity increased at a low and decreased at a high content of the peroxide, while, melt viscosity increased monotonically with the coagent concentration at constant peroxide content.

The NR in the form of air-dried sheet type and high-density polyethylene (HDPE) were blended using internal mixer at 180 °C.[7] Phenolic compatibilizers including dimethylol phenolic resin (SP-1045) and phenolic resin with active hydroxymethyl (methylol) groups (HRJ-10518) or liquid natural rubber (LNR) were applied as compatibilizers. At the mixing time of approximately 2 min, the NR was incorporated and the torque rose up due to the unmolten NR in the blends. When the NR was blended well with the HDPE, the torque decreased until it levelled off in approximately 5 min. Mixing torque was found to increase with an increased content of HDPE in the blend composition. Comparison of the mixing torque showed that the torque increased with an increase in the NR concentration higher than that increased with an increase in the HDPE concentration. This might be attributed to a higher shear viscosity of the pure NR to that of the pure HDPE as shown in the torque–time curves of various NR/HDPE blend ratios without the compatibilizers in Figure 18.2.

It was obvious that the mixing torque of the NR/HDPE blends corresponded to the apparent shear viscosity curve in which the mixing torque and apparent shear viscosity of the NR/HDPE blends can be arranged in the following sequence: $80/20 > 60/40 > 50/50 > 40/60 > 20/80$ as shown in Figure 18.3.

The mixing torque and apparent shear viscosity of the blends increased with increasing amounts of NR might be due to the natural characteristics of ADS that is produced from NR latex without shear cutting in the production process and it was dried at low temperature using the heat from sunshine. Therefore, ADS itself will have high Mooney viscosity in comparison with those other NR block types.

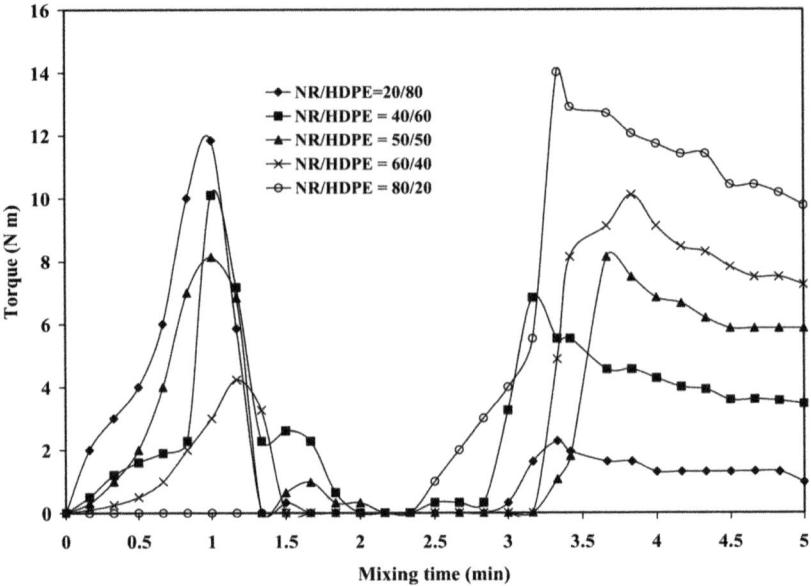

Figure 18.2 Mixing torque–time of NR/HDPE blends at various blend ratios without compatibilizers.

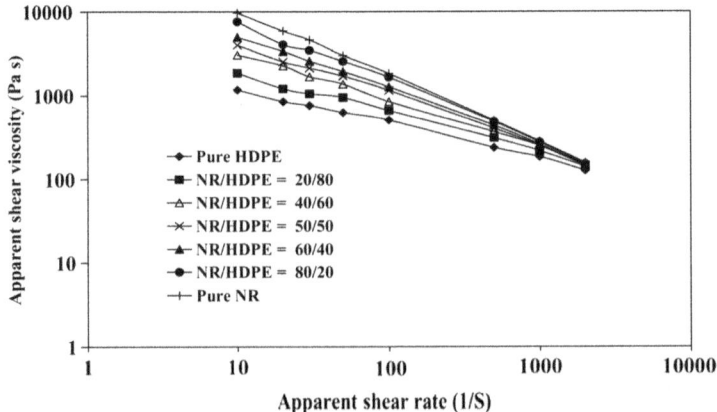

Figure 18.3 Relationship between apparent shear rate and shear viscosity of NR/HDPE at various blend ratios without compatibilizers.

Figure 18.4 showed the torque–time curves for various NR/HDPE blend ratios with a phenolic compatibilizer in the grade of HRJ-10518. A similar trend in the mixing torque to those NR/HDPE blends without using any compatibilizer was observed. That is, the mixing torque increased with increasing levels of NR in the blends. However, in the final mixing stage, at the mixing times of 3.5–5 min, the torque tended to be constant instead of decreasing as in the case of the blends without a compatibilizer. This might be

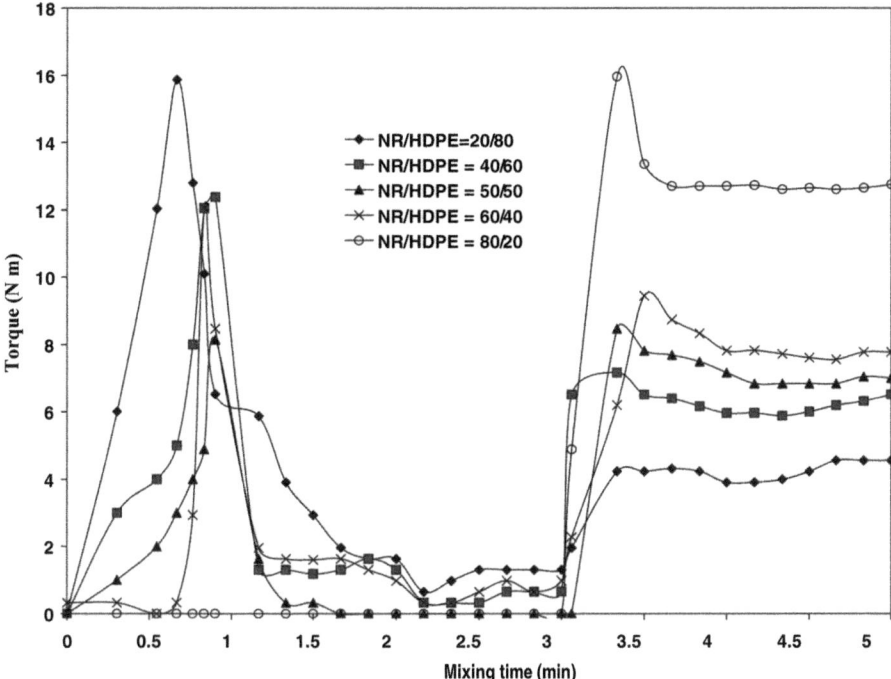

Figure 18.4 Mixing torque–time of NR/HDPE blends at various blend ratios and HRJ-10518 compatibilizer.

resulted by the compatibilizer that was able to bind the molecules between NR and HDPE molecules or bind NR molecules themselves. The phenolic compounds were capable of crosslinking the NR molecules to produce stronger molecular structures or causing interaction between the two different phases. Thus, a lower molecular disruption and a plateau in the torque–time curve at the final mixing stage would be obtained.

Figure 18.5 shows comparative torque–time curves for NR/HDPE blends at a fixed blend ratio of 80/20 with three types of blend compatibilizers added. It can be seen that the torque increased in the absence of a compatibilizer. Also, the phenolic resin with active hydroxymethyl (methylol) groups gave the highest mixing torque at the final mixing stage.

The processability of the blend among thermoplastic, synthetic rubber with special properties and NR (PP/EPDM/NR) compared to that of the blend among thermoplastic, synthetic rubber with special properties and 25 mol% epoxidized NR (PP/EPDM/ENR 25) was investigated by Ismail and co-workers.[8] The blends using PP with a MFI value of 14 g/10 min at 230 °C were prepared in the laboratory internal mixture at 180 °C with a rotor speed of 50 rpm followed by the compression moulding. Typical torque curves for PP/EPDM/NR and PP/EPDM/ENR 25 blends are shown in Figure 18.6. PP was charged into the mixing chamber for minutes. The torque increased due to the resistance exerted on the rotors by the unmelted PP. As the PP melted and

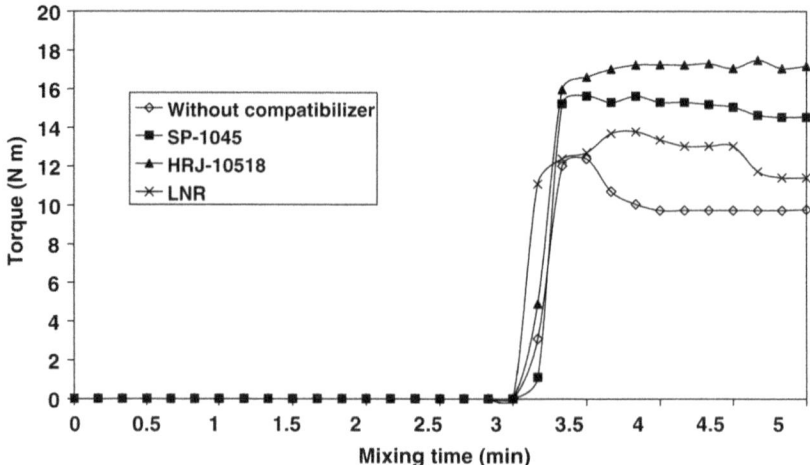

Figure 18.5 Comparison of torque–time curves of NR/HDPE blends at a blend ratio of 80/20 using compatibilizers.

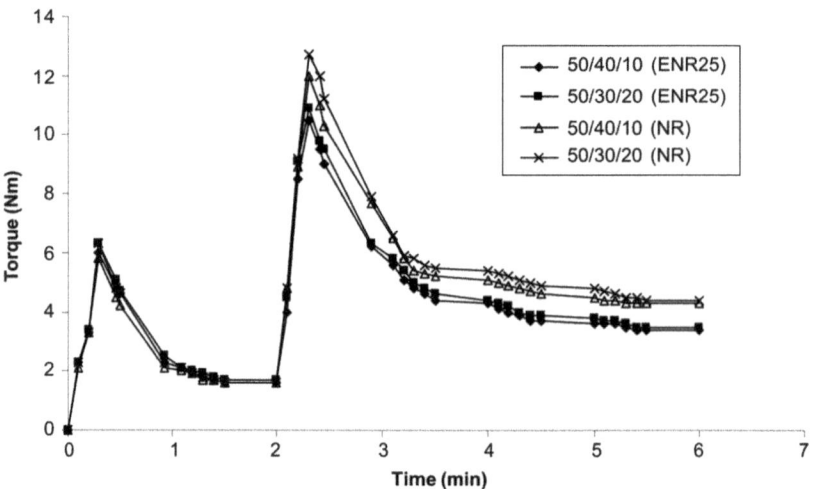

Figure 18.6 Torque–time curve obtained during blending of PP/EPDM/NR and PP/EPDM/ENR25.

subjected to mechanical shearing, the temperature inside the chamber increased resulting in a reduction of torque value (at 2 min). With the addition of rubber the torque increased sharply due to the increase in viscosity as rubber was added to the PP. As the mixing became homogeneous, the torque decreased. Upon completion of the mixing, the torque remained almost constant (at 6 min). In comparison between PP/EPDM/ENR 25 and PP/EPDM/NR blends, the torque values of blend containing NR were slightly higher than the blends containing ENR 25.

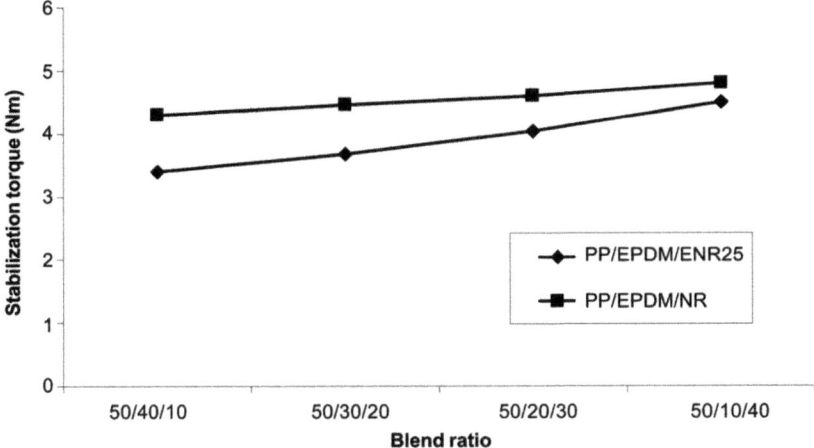

Figure 18.7 Effect of blend composition on stable torque of PP/EPDM/NR and PP/EPDM/ENR25 blends.

Figure 18.7 shows the stabilization torque values after 6 min of mixing in which the PP/EPDM/ENR 25 system had a slightly lower torque value than PP/EPDM/NR. It was indicated that the partial substitutions of EPDM with NR and ENR 25 resulted in more difficult processing because the torque values increased as NR and ENR 25 content increased. However, the PP/EPDM/ENR 25 blend was slightly easier to process than the PP/EPDM/NR blends, although morphological studies showed that PP/EPDM/NR blends exhibited more homogeneous phases than PP/EPDM/ENR 25.

Rheological properties of low grade NR (STR 20) blended with chlorinated polyethylene (CPE) with various blend composition ratios under oscillatory and steady shear flows were investigated by Wongwitthayakool et al.[9] Flow properties under capillary shear of uncured blends agree well with those under oscillatory shear in which the greater the NR content, the lower the apparent shear viscosity was observed. As mentioned above, the different sources of NR to produce STR 20 and STR 5L could affect their Mooney viscosity as well as gel content. Owing to this, the rheological properties of CPE/STR 20 was found to be different from that of CPE/STR 5L.

The rheological behaviour of NR and polystyrene (PS) blends had been carried out in the presence and absence of the NR-g-PS compatibilizer using a capillary rheometer and a melt flow indexer.[10]

Table 18.3 summarizes the blends with various compositions of polymers and compatibilizer.

Both in melt-mixed and solution-cast systems, the viscosity decreased with increase of shear stress, indicating pseudoplastic behaviour (Figure 18.8). In both cases, negative and positive deviations in viscosity could be seen at a high and low shear rate. However, as compared to solution-casted blends, in melt-mixed ones, degradation of NR and PS due to high temperature and shear was

Table 18.3 Composition of NR/PS blends.

Blend code	M_0	M_{30}	M_{50}	M_{70}	M_{100}	S_0	S_{30}	S_{50}	S_{70}	S_{100}	$SG_{(a)}$	$SG_{(b)}$	$SG_{(c)}$	$SG_{(d)}$
Wt % of NR	0	30	50	70	100	0	30	50	70	100	50	50	50	50
Wt % of PS	100	70	50	30	0	100	70	50	30	0	50	50	50	50
Wt % of graft	0	0	0	0	0	0	0	0	0	0	1.5	3.0	4.5	6.0

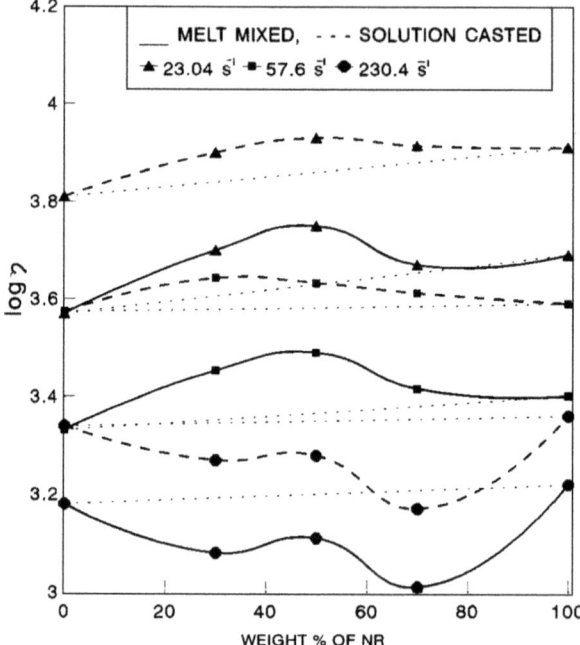

Figure 18.8 Variation of shear viscosity with weight per cent of NR at different shear rates (solution cast and melt mixed).

possible. In the absence of the compatibilizer, the blends showed a decrease in melt viscosity with an increase of shear stress, indicating pseudoplastic nature for both solution-cast blends using chloroform as the casting solvent and for the melt-mixed blends as shown in Figures 18.9 and 18.10, respectively.

It was explained that at closer to zero shear, the molecules were randomly oriented and highly entangled and therefore exhibited high viscosity. Under the application of shearing force, the polymer chains oriented, resulting in the reduction of shear viscosity and thus exhibited pseudoplastic behaviour. The reduction in viscosity of the blends at higher shear rate was also due to the decrease in the particle size of the dispersed domains. Solution-cast blends showed a higher viscosity as compared to melt-mixed blends. This is associated

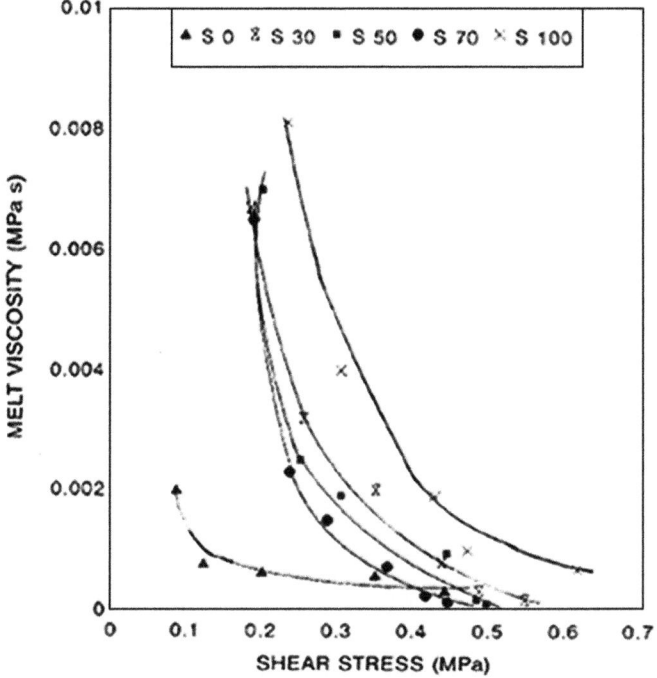

Figure 18.9 Variation of shear viscosity with shear stress of solution-cast blends.

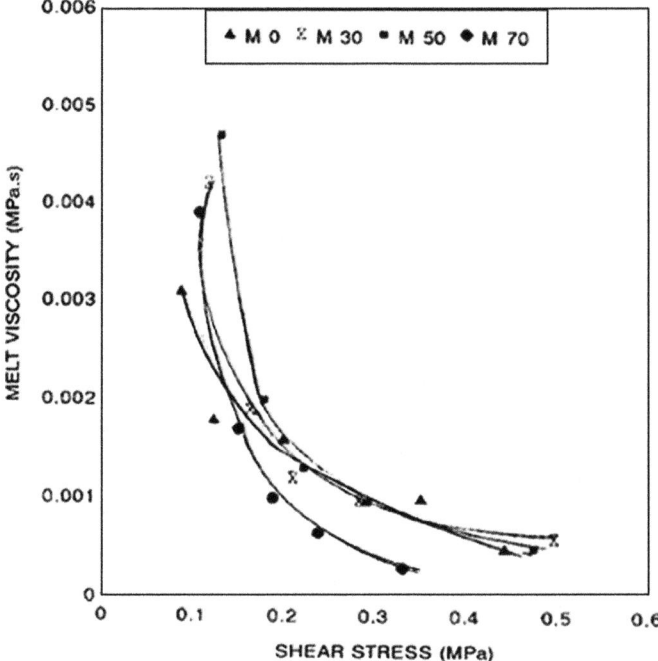

Figure 18.10 Variation of shear viscosity with shear stress of melt-mixed blends.

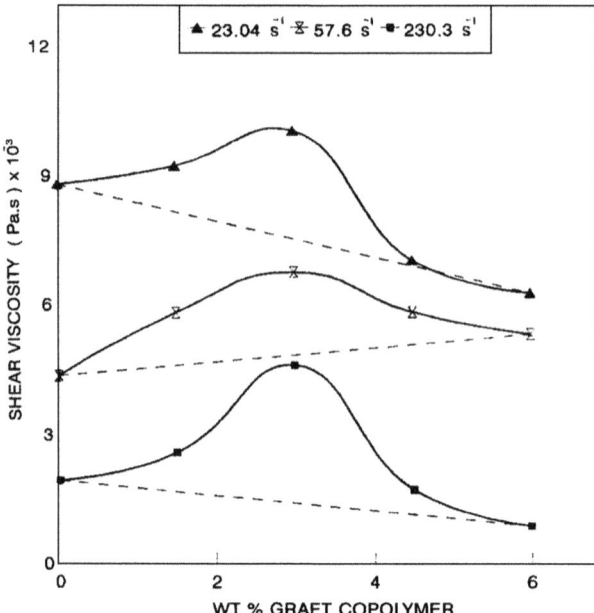

Figure 18.11 Variation of shear viscosity with the percentage of graft copolymer (50/50 NR–PS) solution-cast blends.

with the interlayer slip between the highly incompatible NR and PS phases. The effect of compatibilizer loading on the shear viscosity of the solution-cast 50/50 NR–PS blend at three different shear rates was given in Figure 18.11.

As the compatibilizer loading increased, the shear viscosity increased and then decreased at higher loading that showed a similar trend of their MFI. The initial increase in viscosity with copolymer loading indicated the higher interfacial interaction between the blend components at the interface. However, the decrease of the shear viscosity of the blends at higher loading of the compatibilizer might be associated with the formation of micelles which had a plasticizing action on the viscosity of the blends.

The rheological behaviour of NR blended with high density polyethylene (HDPE) was studied by Zhengping and coworkers.[11] It was clarified that the melt flow rate (MFR) of HDPE/NR blends at different weight ratio of NR in 100g HDPE initially increased with the increase of NR content reaching the maximum value at 10 g/100g HDPE and then decreased (Figure 18.12). Torque of all HDPE/NR blend ratios were found to be constant after 3 min of blending (Figure 18.13).

This might be due to a molecular chain scission of NR molecules caused by degradation during mastication taking place. According to the mastication degradation mechanism, long-chain molecular model and DSC results, it was revealed that partial NR chains were clamped by HDPE crystals and thus acted as tie molecules in the blend.

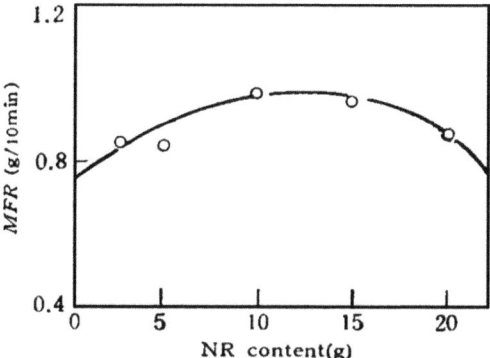

Figure 18.12 Melt flow rates of HDPE/NR blends.

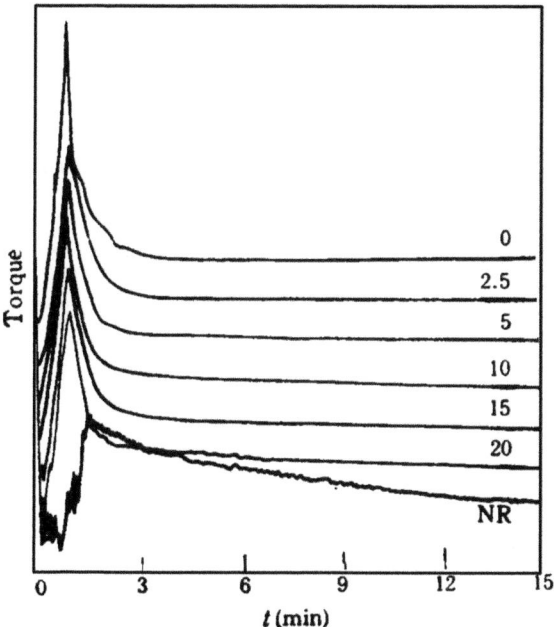

Figure 18.13 Rheological curves of HDPE, NR and their blends (data on the curves are NR contents in g/100 g HDPE).

18.2.2 Natural Rubber–Synthetic Rubber Blends

The blending between NR and synthetic rubbers such as nitrile rubber, butadiene rubber and ethylene propylene rubbers has been studying up to present. Though, these studies are not new, the blending process has to be still developed for achievement the best properties that are combined from both kinds of rubbers. Apart from the value added, properties including the mechanical and physical properties are expected to be obtained, however, rheology is one of the main issues to be considered for the NR and synthetic rubber blends.

In general, rubber–rubber blends could be classified as heterogeneous blends. As known, viscosity and shear stress differences depended on the quantity of the interface and specific interphase interactions, including entanglements in the rubber-rubber blends.

Vulcanizate of the blends of elastomeric chlorinated polyethylene (CPE) and NR (STR 20) with a blend composition ratio of 80/20 were prepared and recycled.[12] In the initial step, NR was masticated at 40 °C with a laboratory-size two-roll mill mixer to achieve a Mooney viscosity (ML1 + 4 at 100 °C) of approximately 60. For the blending process, CPE was melted on a front roll for 1 min at the set temperatures of 145 °C and 140 °C for the front and back rolls, respectively. Afterward, the masticated NR was charged and further mixed for 4 min. Compounding ingredients were then added and mixed for 5 min before discharging. Finally, the blend was compression-moulded into 1-mm-thick sheets under a clamping pressure of 15 MPa at 155 °C to achieve a stage of cure of 99%. The rheological study in this work was carried out under oscillatory shear flow with the use of a parallel-plate rheometer. All tests were performed at the test temperature of 170 °C. The results for rheological behaviour were recorded based on complex viscosity and damping factor (tan δ). The tan δ results, as shown in Figure 18.14, agreed very well with the G' results shown in Figure 18.15.

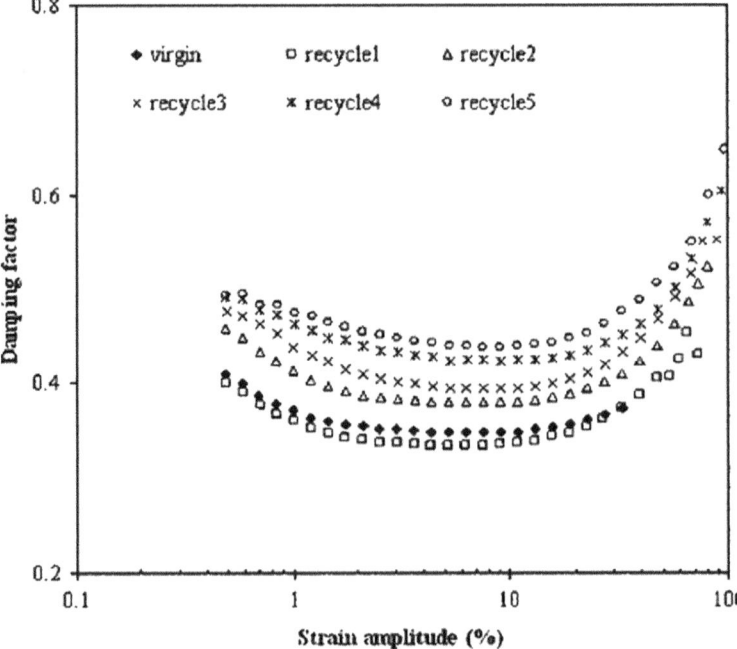

Figure 18.14 Tan δ as a function of the strain amplitude at 1 rad/s for CPE/NR blends with various recycling cycles.

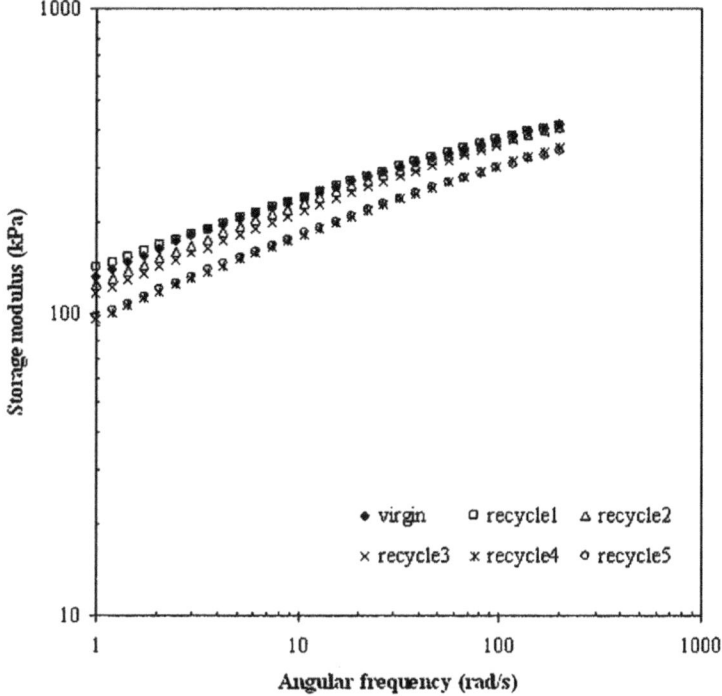

Figure 18.15 G' as a function of the angular frequency at 10% strain for CPE/NR blends with various recycling cycles.

The virgin and first recycled blends were classified as the group with the lowest tan δ values, which implied the highest magnitude of elastic behaviour *via* a high extent of molecular entanglement due to their good thermal stability. At a high strain of deformation, the upswing of tan δ of such blends was relatively small, and this supports the high degree of molecular entanglement, as discussed earlier. In contrast, the group subjected to a highly thermal history, that is, the fourth and fifth recycled blends, revealed the highest tan δ value because of the highest degree of thermal degradation *via* a molecular chain-scission mechanism occurring mainly in the NR phase.

The relationship between morphology and rheological properties of the blends between polybutadiene rubber (BR), poly(styrene-co-butadiene) rubber (SBR) and NR were studied.[13] Rheological properties were measured using rubber process analyser (RPA 2000). The BR used was composed of a high cis-1,4-polybutadiene content (98%) with Mooney viscosity of 52 (ML(1 + 4) at 100 °C). The SBR was composed of 23.5% styrene content with Mooney viscosity of 52 (ML(1 + 4) at 100 °C). The NR was Standard Malaysian Rubber with Mooney viscosity of 65 (ML(1 + 4) at 100 °C). For non-masticated NR the weight-average molecular weight was 1.25×10^6 and the polydispersity index was 5.1. After mastication the NR weight-average molecular weight was reduced to 8.2×10^5 and the polydispersity index was 3.7. The weight-average molecular weights of non-masticated BR and SBR were 6.0×10^5 and 5.7×10^5 and the corresponding polydispersity indexes were 3.1 and 3.3, respectively. After mastication, the BR and SBR average molecular weights did not change considerably. Blending of BR with NR and SBR with NR was carried out using Brabender plasticorder PLD 651 apparatus equipped with an internal mixer W 50. A total charge of 20 g was used. The rotor speed was 80 rpm and the temperature was 140 °C. For blending NR with SBR and BR, the SBR component was added first, while for the blend of NR with BR, the first component added was NR. The first component was masticated for 30 sec, and then the second component was admixed and blended for further 9 min. Rubber process analyser (RPA 2000) was used to measure the rheological properties of non-vulcanized blends at three selected temperatures: 80, 100 and 140 °C. All BR/NR blends had higher complex viscosity than pure components as shown in Figure 18.16.

The measured complex viscosity almost linearly increased with increase in weight fraction of BR. It might be assumed that the BR/NR blends behaved as heterogeneous blends with interphase interactions that were more pronounced for the blend with 75% BR content. Whereas the NR/SBR blends exhibited properties of blends with little interaction between the phases. In contrast, the observed increase in complex viscosity for BR/NR blends could worsen its processability.

The investigation of the effect of different percentages of nanoclay and matrix compositions on the rheological properties of NR/BR blend (75/25) prepared by open two-roll mill was recently carried out using RPA at temperature of 80 °C and frequency of 0.01–80 Hz by Alipour.[14] NR (SMR20); Mooney viscosity ML (1+4) at 100 °C, 55 and BR (97% cis-1,4); Mooney viscosity ML(1+4) at 100 °C, 58, were used in this research work. The nanoclay used was Cloisite 15 Å which was a natural montmorillonite modified with a dimethyl dehydrogenated tallow quaternary ammonium having a cation exchange capacity of 125 mequiv/100 g.ing ingredients. Nanocomposites were prepared by a laboratory open two-roll mill (Polymix 200-L) with the rotor speed of 80 rpm for 18 min at room temperature. NR and BR were blended and nanoclay that was pre-dried at temperature of 80 °C for 24 h was added to the blend. Complex viscosity (η^*) and elastic modulus (G') of the blends were in sensibly higher than that of pure NR/BR as shown in Figure 18.17. The falling trend of viscosity with frequency represented the pseudoplastic nature and shear-thinning behaviour of samples. At low and medium range of frequencies, both complex viscosity and elastic modulus significantly increased with clay loading. This was associated with interactions established between the matrix components and organoclay functional groups of which the nanocomposites structure consisted of intercalated structure as well as partially exfoliated structure. Thus, rheological behaviour of the NR/BR blend was markedly influenced by the clay loading.

The influence of dichlorocarbene modified natural rubber (DCNR) on the rheological behaviour of blends based on NR and hydrogenated nitrile rubber (HNBR) was studied.[15] The NR used in this study was ISNR 5 grade from India. The HNBR used was a Zetpol 2010 grade (Nippon Zeon, Japan) with

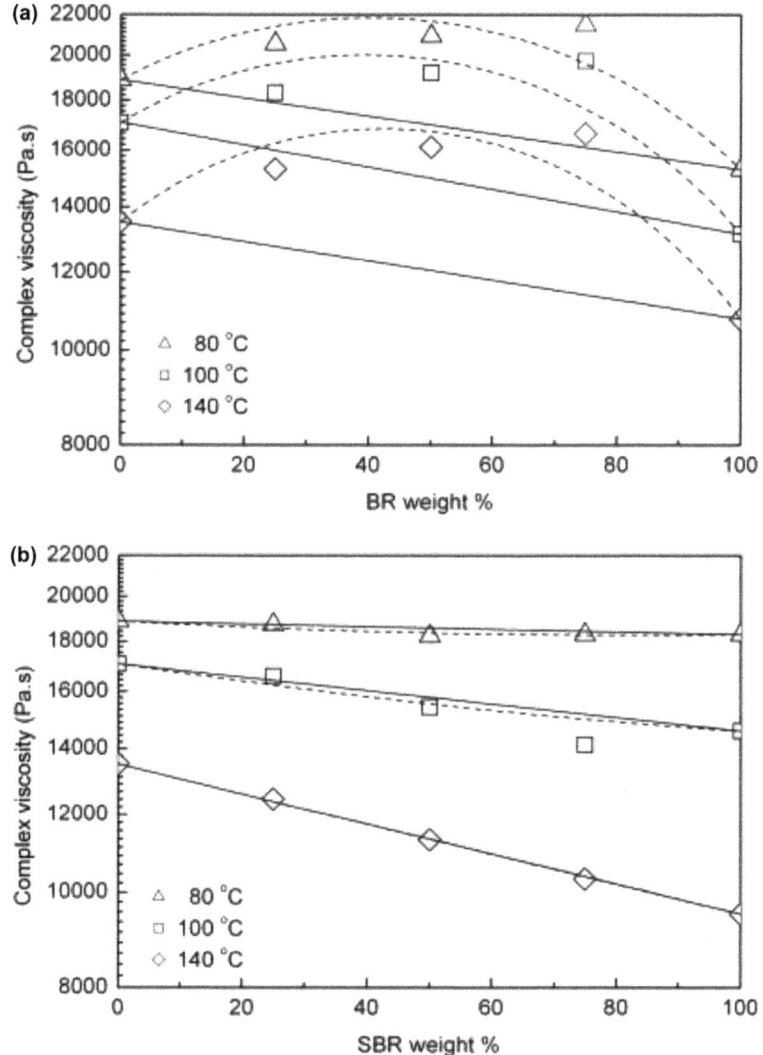

Figure 18.16 The experimental values of the complex viscosity (symbols) versus the composition of (a) BR/NR and (b) SBR/NR ($T = 80$, 100 and 140 °C; $\omega = 10$ rad/s). Solid curves (—) present log-additivity mixing rule predictions, dashed curves (---) present quadratic mixing rule predictions.

36% bound ACN content, an iodine value and a Mooney viscosity ML(1+4)100 °C were 11 g/100 g and 85, respectively. The blends were prepared on a two-roll mill. NR and HNBR were individually masticated for about 1 min and then blended together for a period of 4 min. For the preparation of the modified blend, DCNR was first masticated for 1 min incorporated with HNBR and then blended with NR. The filled mixes were prepared by adding the filler after mixing the rubbers. The total mixing time was 10 min. The formulation of NR/HNBR blends was shown in Table 18.4.

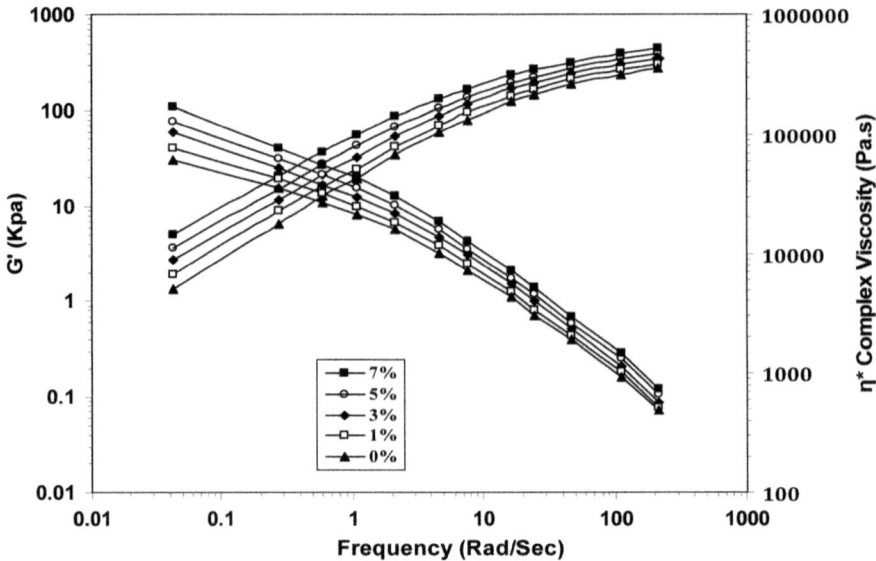

Figure 18.17 Rheological properties of NR/BR (75/25) nanocomposite samples.

Table 18.4 Formulation of NR/HNBR blends.

	N	H	C	NHa	NH	NHb	NHC3	NHC7	NHC10	NHB	NHC7B
NR	100			75	50	25	50	50	50	50	50
HNBR		100		25	50	75	50	50	50	50	50
DCNR			100				3	7	10		7
FEF										30	30

The rheological studies were carried out using a capillary rheometer, model 1474 as per ASTM D 5099-1933.

Figure 18.18 shows the shear viscosity as a function of shear rate at 110 °C for different blends of NR/HNBR. The viscosity of all mixes decreased with increasing shear rate showing a pseudoplastic flow behaviour. The molecules which were extensively entangled at normal conditions became disentangled and oriented in the direction of flow under the application of shear. As NR had a very highly entangled matrix, during application of load, the viscosity of NR reduced faster than for HNBR. HNBR showed a higher viscosity probably due to its polarity which was resulted from associative interactions whereas DCNR showed lower viscosity than HNBR at all shear rates, lower viscosity than NR at lower shear rate and higher viscosity than NR at higher shear rates. The incorporation of DCNR to the 50/50 blend increased its viscosity due to the interactions with the blend constituents. In the case of filled blends, DCNR increased the viscosity. The temperature dependence of the viscosity of both the gum and the filled 50/50 blend and the blend containing 7 parts of DCNR at three different shear rates are given in Figure 18.19. It was seen that viscosity decreased with temperature at all shear rates and this effect was more pronounced above 100 °C and at higher shear rates.

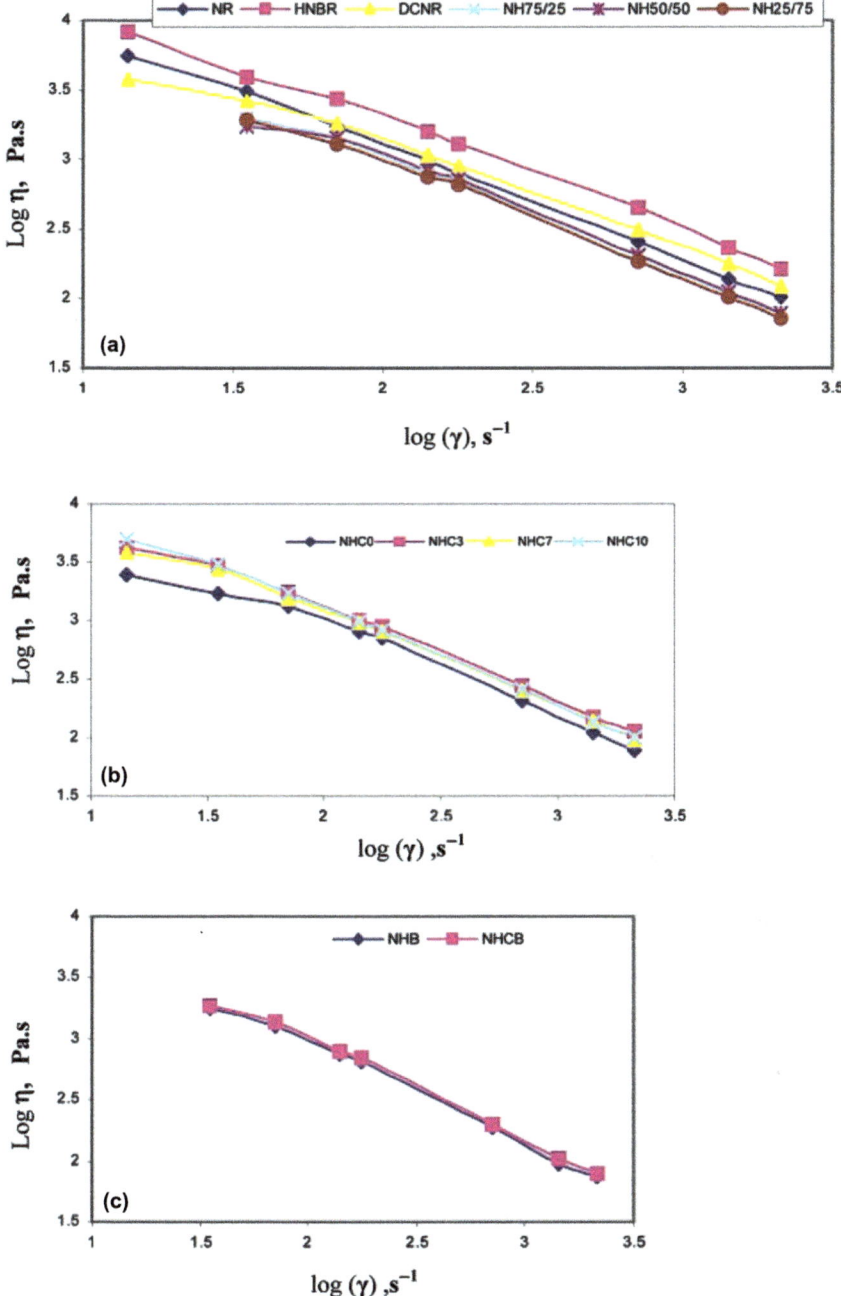

Figure 18.18 Variation of shear viscosity with shear rate of (a) NR, HNBR, DCNR and the blends (b) 50/50 blend containing different dosage of DCNR and (c) filled blends with and without DCNR.

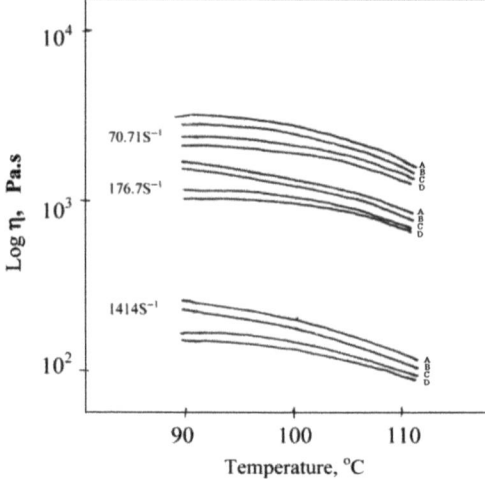

Figure 18.19 Variation of viscosity with temperature. (a) NHC, (b) NH, (c) NHCB and (d) NHB.

The rheological properties of the blends between the brominated EPDM (BEPDM) and NR (STR5L) at various compositions were investigated using a capillary extrusion in comparison with those of unmodified EPDM blended with STR5L.[16] Blending of BEPDM/STR 5L and EPDM/STR 5L were carried out in a laboratory-sized two-roll mill at a mixing temperature of 60 °C. At any blend composition, the higher amount rubber was masticated for 2 min after that the second rubber was then mixed and allowed to blend for another 6 min. Rheological properties in terms of shear stress and shear viscosity were studied using a Rosand single-bore capillary rheometer (model RH7, Rosand Precision Ltd., Stourbridge, West Midlands, England). A capillary die of diameter 2 mm, length 32 mm, and 180° entry angle with a length-to-radius (L/R) ratio of 32 was used as a long die. The apparent values of shear stress, shear rate and shear viscosity were calculated using the derivation of the Poiseuille law for capillary flow.[17]

$$\text{Apparent shear stress (Pa)}; \quad \tau = \frac{R \Delta P}{2L} \quad (18.1)$$

$$\text{Apparent shear rate (S}^{-1}); \quad \gamma'_{app} = \frac{4Q}{\pi R^3} \quad (18.2)$$

$$\text{Apparent shear viscosity (Pa s)}; \quad \eta_s = \frac{\tau}{\gamma'_{app}} \quad (18.3)$$

where ΔP is the pressure drop across the channel (in Pa), Q is the volumetric flow rate (in m^3 sec^{-1}), R is the capillary radius (in m), and L is the length of the capillary (in m). The plots of apparent shear stress versus apparent shear rate for STR5L/EPDM and STR5L/BEPDM blends with various blend compositions were shown in Figures 18.20 and Figure 18.21, respectively.

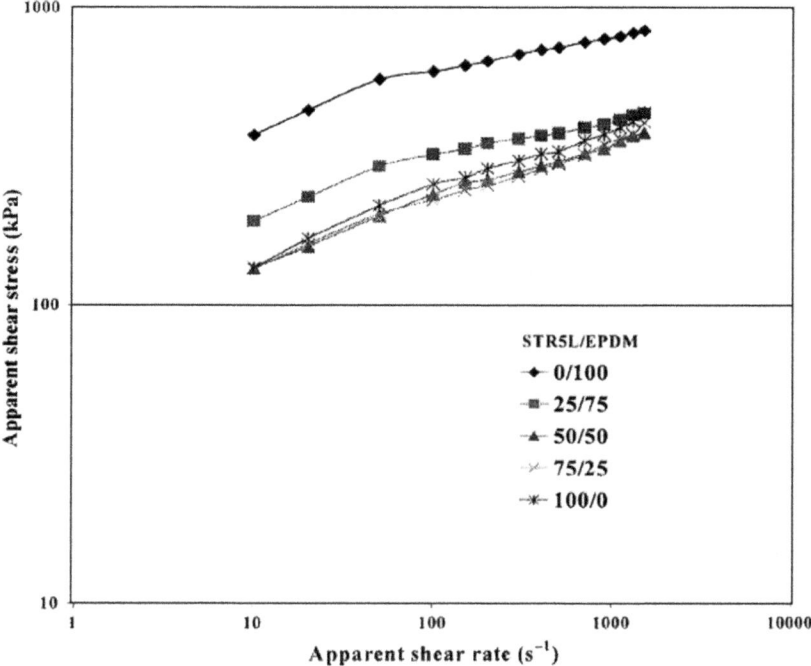

Figure 18.20 Effect of apparent shear rate on the apparent shear stress of STR5L/EPDM blends at various blend compositions.

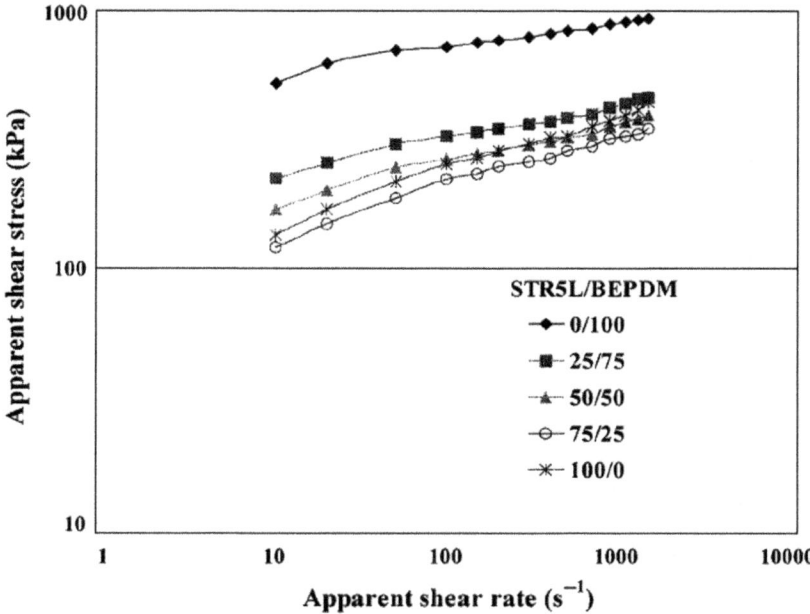

Figure 18.21 Effect of apparent shear rate on the apparent shear stress of STR5L/BEPDM blends at various blend compositions.

Table 18.5 The power law index (n) and the consistency of flow (K) for various blend compositions.

NR/EPDM blend	n	K (kPa)	NR/BEPDM blend	n	K (kPa)
0/100	0.14	293.0	0/100	0.10	444.6
25/75	0.15	149.9	25/75	0.14	169.0
50/50	0.20	88.7	50/50	0.16	124.5
75/25	0.21	85.1	75/25	0.20	81.8
100/0	0.22	86.8	100/0	0.22	86.8

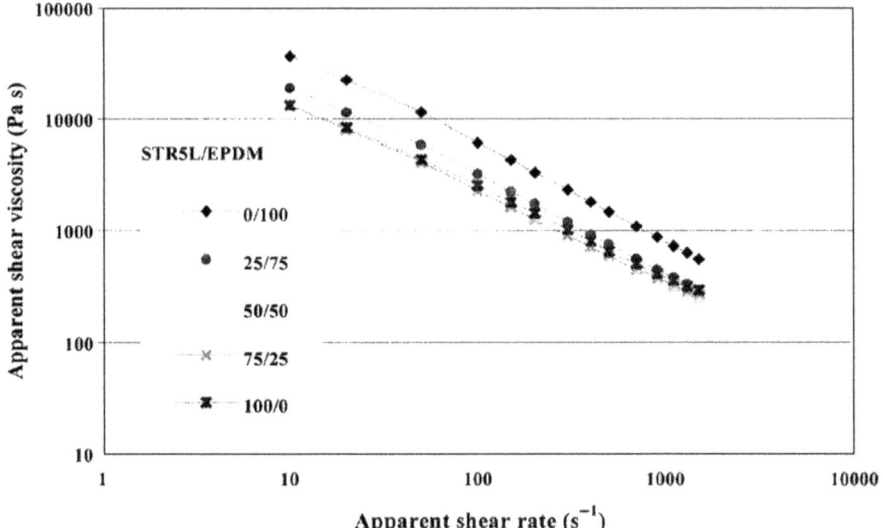

Figure 18.22 Effect of apparent shear rate on the apparent shear viscosity of STR5L/EPDM blends at various blend compositions.

Flow curves of all the blends showed reasonably straight lines, whose intercept K and slope n correspond to the power law equation (the Ostwald–de Waele equation):[18,19]

$$\tau = K(\gamma')^n \qquad (18.4)$$

where n is the power law index or the flow behaviour index, and K is the consistency of flow or viscosity coefficient index.

Table 18.5 showed the power law index and the consistency of flow of STR5L/EPDM and STR5L/BEPDM blends. Shear flow curves of the pure rubbers and their blends illustrated the pseudoplastic property as shear thinning behaviour with a power law index $n < 1$. Hence, the apparent shear viscosity of the two sets of blends decreased as the shear rate increased as shown in Figures 18.22 and 18.23 for STR5L/EPDM blend and STR5L/BEPDM blend, respectively.

Figure 18.24 compares the apparent shear viscosity with the level of EPDM or BEPDM in the blend composition at the apparent shear rates of 50, 150 and

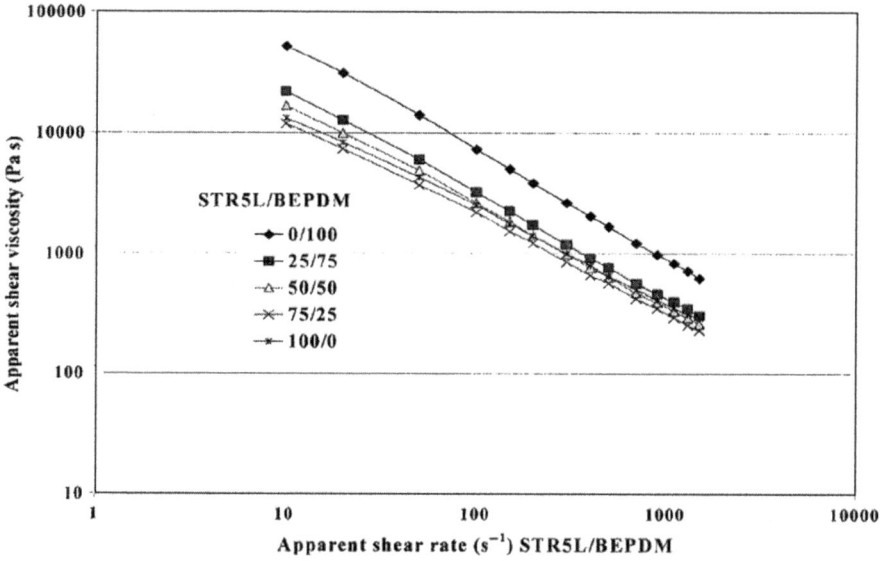

Figure 18.23 Effect of apparent shear rate on the apparent shear viscosity of STR5L/BEPDM blends at various blend compositions.

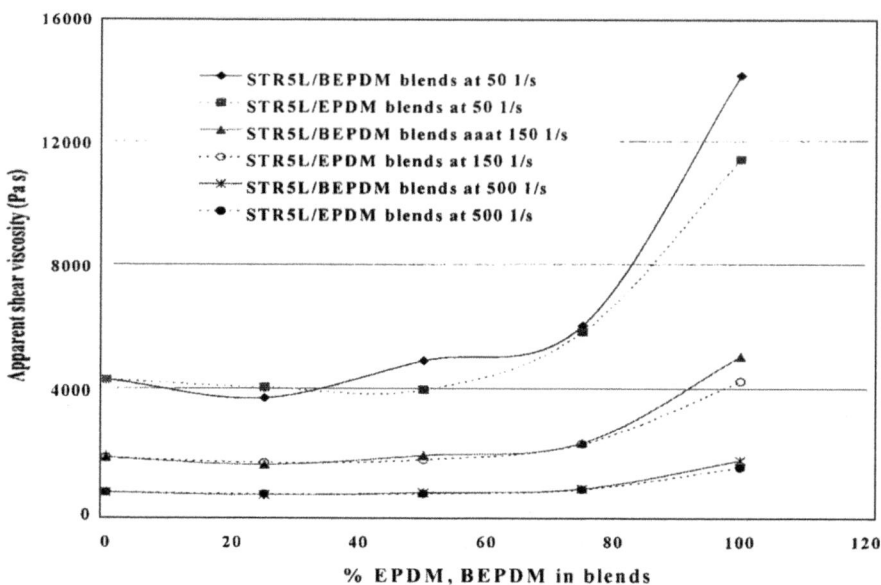

Figure 18.24 Comparison of apparent shear viscosity at apparent shear rates of 50, 150 and 500 s^{-1} for STR5L blended with various contents of EPDM and BEPDM.

500 s^{-1}. It was found that the apparent shear viscosity of the blends tended to increase with increasing levels of either EPDM or BEPDM due to the higher apparent shear viscosity of EPDM and BEPDM. However, at high apparent

shear rates, there was a remarkable difference in the apparent shear viscosity of the blends with increasing quantities of EPDM and BEPDM.

18.2.3 Chemically Modified Natural Rubber Blends

The rheological behaviour of various chemical modified NR that are blended with other types of polymer including thermoplastic, starch, synthetic rubber and NR itself are one of another great topics of interest. Maleated NR from air-dried rubber sheet (MNR) was blended with polypropylene (PP) in the weight ratio of 60/40 by a melt mixing process using two types of graft copolymer of polypropylene; maleic anhydride (PP-g-MA) and phenolic modified polypropylene (Ph-PP) as partly compatibilizers at various loading levels of 3, 5, 10, 15 and 20% w/w of PP.[20] Polypropylene was first incorporated into the mixing chamber of a Brabender Plasticorder and preheated for 6 min without rotation at 180 °C and then melted for 2 min at a rotor speed of 60 rpm. MNR was then added. The mixing was continued for 2 min. The compatibilizer (*i.e.* PP-g-MA or Ph-PP) was each incorporated into the blend and the mixing was continued for 3 min. A Rosand single bore capillary rheometer (model RH7, Rosand Precision Ltd, Gloucestershire, UK) was used to characterize shear flow properties in terms of relationship between apparent shear stress and shear viscosity with apparent shear rate. The test was carried out over a wide range of shear rates (100–1600 s^{-1}) at a test temperature of 180 °C. Dimensions of the capillary die used were 2 mm diameter, 32 mm length and 90° entry angle with aspect ratio (L/D) of 16/1. The material was first preheated in the rheometer barrel for 5 min under a pressure of approximately 3–5 MPa to get a compact mass. The apparent values of shear stress, shear rate and shear viscosity were calculated using the derivation of the Poiseuille law for capillary flow as described above in Equations (18.1) to (18.3).[17] A power law equation as described in Equation (18.4) was also applied to the relationship between apparent shear stress and shear rate (*i.e.* flow curves) for MNR/PP blends using various types and levels of compatibilizers.[18,19] The values of R and L used in work this were 1 mm and 32 mm, respectively. Plots of apparent shear stress versus shear rate (*i.e.* flow curves) of 60/40 MNR/PP blend using various concentrations of PP-g-MA and Ph-PP as a blend compatibilizer are shown in Figures 18.25 and 18.26, respectively.

Both cases showed that the shear stress increased with an increase of shear rate. At a given shear rate, the apparent shear stress first increased with an increasing level of PP-g-MA or Ph-PP compatibilizer until reaching the maximum value at a loading level of compatibilizer at 5% w/w. Increasing levels of a compatibilizer higher than 5% w/w caused a decreasing trend of the apparent shear stress as shown in Figures 18.27 and 18.28 for PP-*g*-MA compatibilizer and for Ph-PP compatibilizer, respectively.

This was claimed to be attributed to the maximum compatibilizing effect or chemical interaction between PP and MNR phases at a loading level of Ph-PP of 5% w/w. Plots of apparent shear viscosity versus shear rate of 60/40 MNR/PP blends with PP-g-MA and Ph-PP as a compatibilizer are shown in Figures 18.29 and 18.30, respectively.

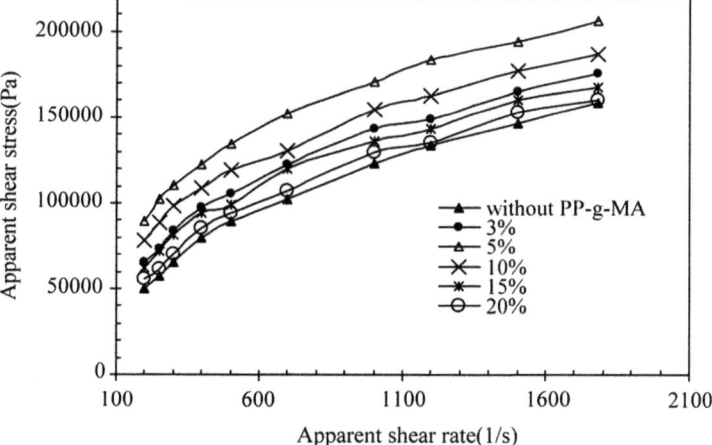

Figure 18.25 Relationship between apparent shear stress and shear rate for simple blends of 60/40 MNR/PP with various quantities of PP-g-MA.

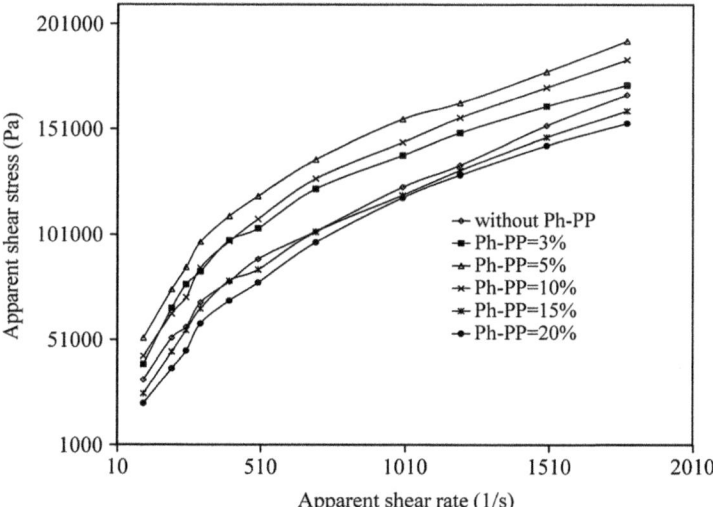

Figure 18.26 Relationship between apparent shear stress and shear rate for simple blends of 60/40 MNR/PP with various quantities of Ph-PP compatibilizer.

The maximum apparent shear viscosity was observed at a loading level of 5% w/w for both compatibilizers used. This trend of apparent shear viscosity was found to be the same as the apparent shear stress.

Two types of NR (ADS and STR 5L) were used to prepare maleated natural rubber (MNR). Melt rheological properties of MNRs (*i.e.* MNR-ADS and MNR-STR 5L) and their blends with various blend ratios were investigated.[21] Two sets of rubber blends including MNR-ADS and MNR-STR 5L were prepared using a two-roll mill at a mixing temperature of 60 °C. Mastication of

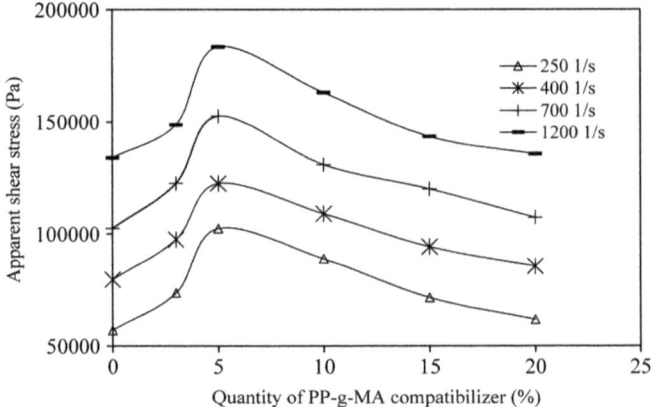

Figure 18.27 Apparent shear stress at constant shear rates for simple blends of 60/40 MNR/PP with various quantities of PP-g-MA compatibilizer.

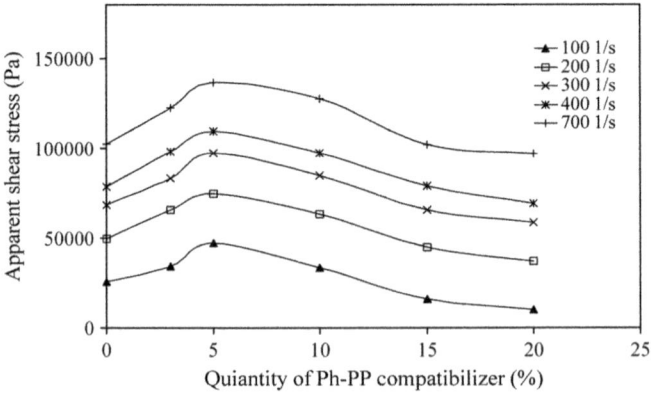

Figure 18.28 Apparent shear stress at constant shear rates for simple blends for 60/40 MNR/PP with various quantities of Ph-PP compatibilizer.

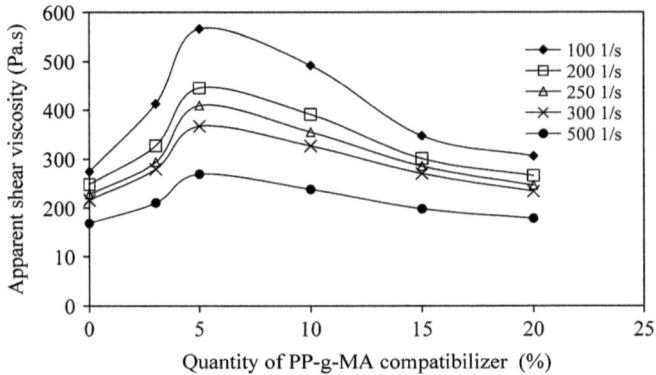

Figure 18.29 Apparent shear viscosity at constant shear rates for simple blends of 60/40 MNR/PP with various quantities of PP-g-MA compatibilizer.

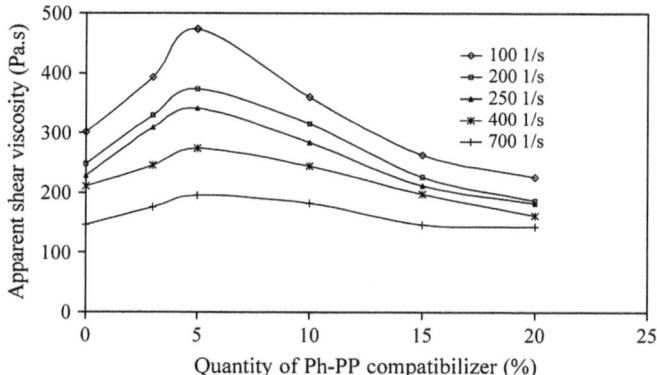

Figure 18.30 Apparent shear viscosity at constant shear rates for simple blends of 60/40 MNR/PP with various quantities of Ph-PP used as a blend compatibilizer.

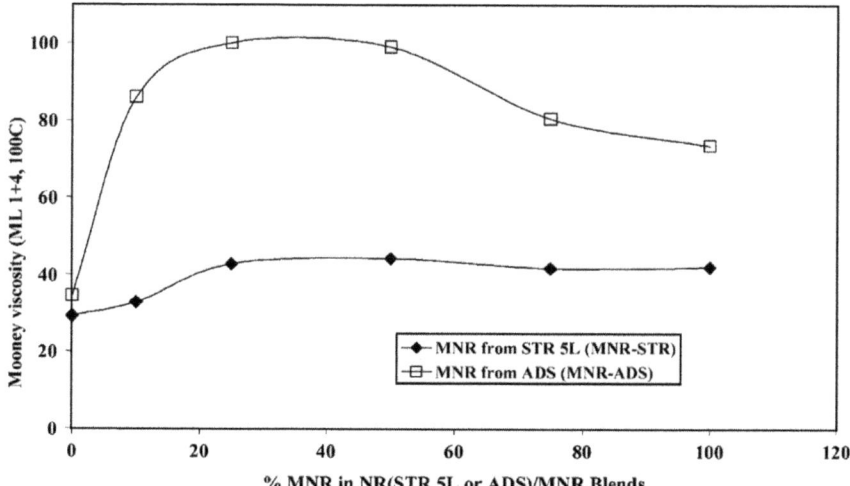

Figure 18.31 Mooney viscosity of rubber blends.

ADS or STR 5L was first performed for 5 min before incorporating the MNR. Mixing was continued for 5 min. The rheological behaviour of the blends were characterized using a SPRI Mooney viscometer (model AC/684/FD) and a Rosand single bore capillary rheometer, model RH7. The Mooney viscosities were tested at 100 °C using a large rotor, a preheating time of one min and a testing time of 4 min. The tests were performed according to ASTM D1646-89 (2000). The typical viscosity was reported as ML $(1+4, 100\ °C)$. It was found that the Mooney viscosity first increased with increasing level of MNR, and then decreased after a MNR level higher than 50% by weight. The maximum Mooney viscosity values were therefore observed in the range of MNR/NR ratios at 25/75 to 50/50 for both MNR-ADS/ADS and MNR-STR/STR 5L blends as shown in Figure 18.31.

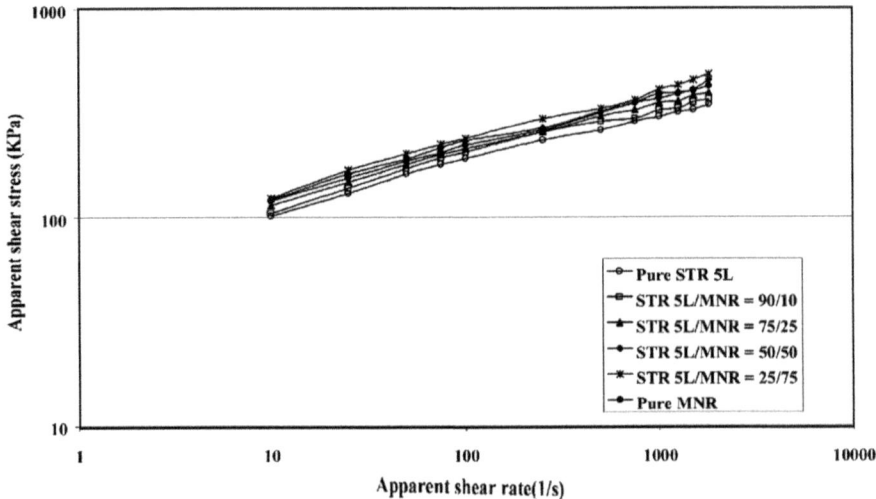

Figure 18.32 Plots of apparent shear stress versus apparent shear rate for MNR-STR blended with various quantities of STR 5L.

Mooney viscosities of MNR-ADS blends were found to be much higher than those of MNR-STR 5L blends. Rheological measurements were carried out the same method as described above.[16] The apparent values of shear stress, shear rate and shear viscosity were calculated using the derivation of the Poiseuille law for capillary flow as shown in Equations (18.1) to (18.3). Plots of the apparent shear stress versus apparent shear rate for various blend compositions of MNR-STR/STR 5L and MNR-ADS/ADS were shown in Figures 18.32 and 18.33, respectively. Straight lines of the flow curve were observed for all sets of the test. The results corresponded to the power law equation proposed by Ostwald as shown in Equation 18.4.

Figures 18.34 and 18.35 showed the apparent shear viscosity at various apparent shear rates for MNR-STR/STR 5L and MNR-ADS/ADS blends, respectively. The pseudoplastic (shear-thinning) behaviour in the flow of all types of blends was observed with the power law index, n, lower than 1. That is, the apparent shear viscosity decreased with an increase in the apparent shear rate. From the linear relation on a log–log scale, one can get the slope (n) and intercept (K), which are shown in the plots of the n and K values against the level of MNR in the blend compositions as shown in Figures 18.36 and 18.37, respectively.

The n value was found to be quite constant over the shear rate range studied. Moreover, for all blend compositions, the flow behaviour indices were less than 1, which indicated pseudoplastic (shear-thinning) behaviour. It was also found that the values of n for MNR-ADS/ADS blends were all higher than those of MNR-STR/STR 5L blends indicating the greater pseudoplasticity in the flow of melts of MNRSTR/STR 5L blends. The MNR/NR blends in this work exhibited very low n values (*i.e.*, $n < 0.25$). Therefore, the highly shear-thinning

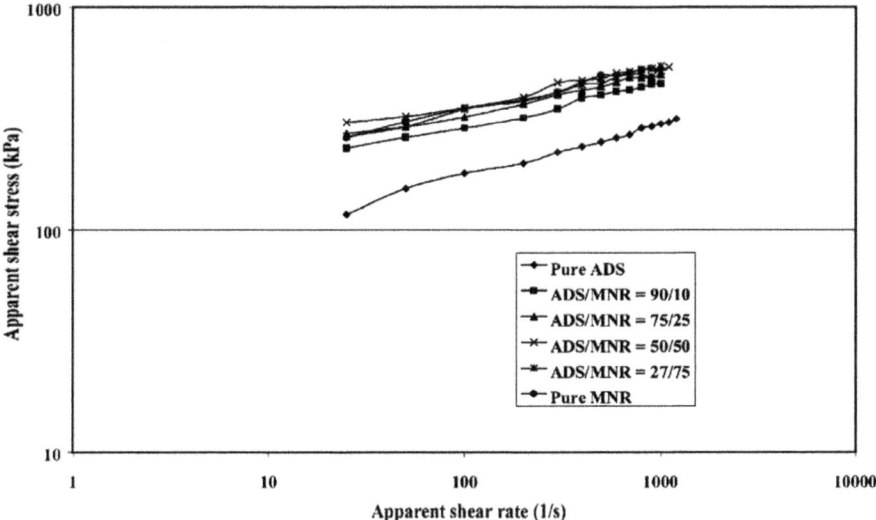

Figure 18.33 Plots of apparent shear stress versus apparent shear rate for MNR-ADS blended with various quantities of ADS.

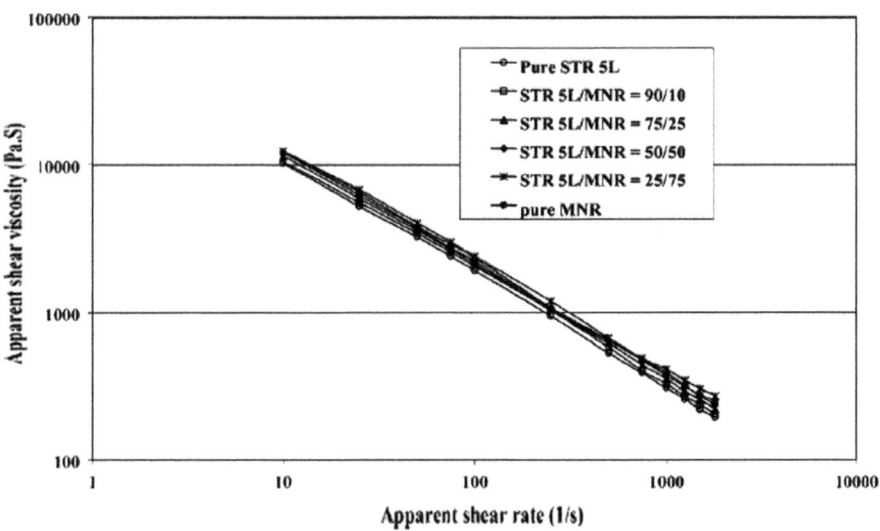

Figure 18.34 Plots of apparent shear viscosity versus apparent shear rate for MNR-STR blended with various quantities of STR 5L.

fluid flowed through the capillary almost as a plug moving at a uniform speed as the melt was sliding down against the channel wall. The consistency index, K, is a Newtonian viscosity if $n = 1$. By definition, the K value is related to the zero-shear viscosity (*i.e.*, shear viscosity at shear rate of zero) of the flowing rubber blends. The consistency index (K) of MNR-ADS/ADS blends that were

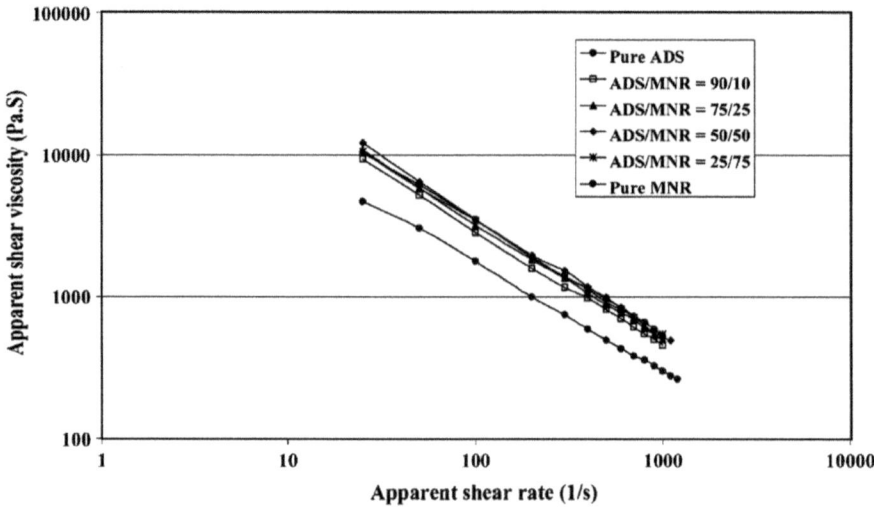

Figure 18.35 Plots of apparent shear viscosity versus apparent shear rate for MNR-ADS blended with various quantities of ADS.

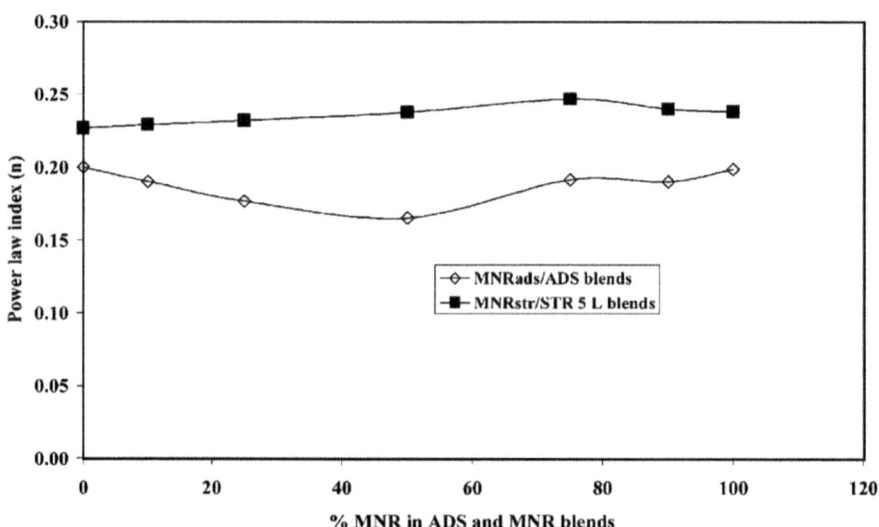

Figure 18.36 Power law index (n) for MNR-ADS and MNR-STR blended with various quantities of ADS and STR 5L, respectively.

plotted against various quantities of rubbers were found to be higher than those of MNR-STR/STR 5L blends similar to that of Mooney viscosity trend.

The melt rheological behaviour in terms of Mooney viscosity, apparent shear stress, and apparent shear viscosity at 100 °C of two types of blends that included a blend between maleated STR 5L (MNR) and cassava starch and a

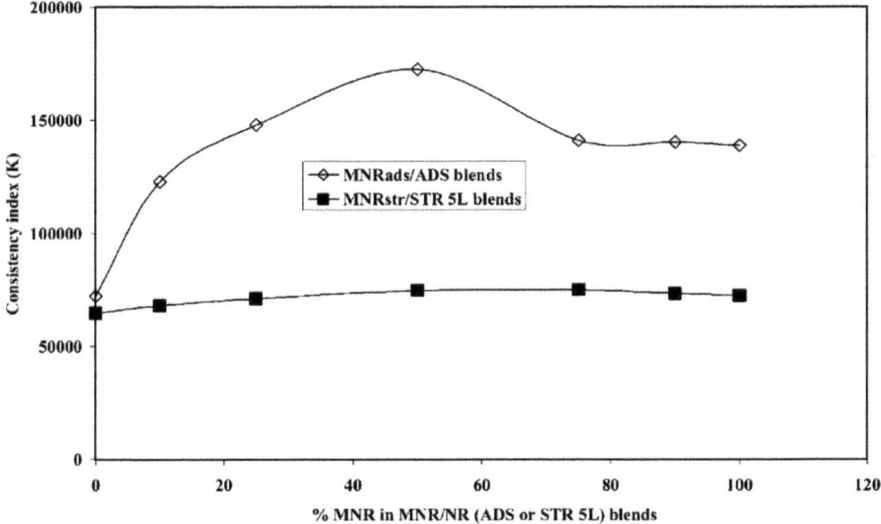

Figure 18.37 Consistency index (K) for MNR-ADS and MNR-STR blended with various quantities of ADS and STR 5L, respectively.

blend between STR 5L and cassava starch with and without MNR as a compatibilize was investigated by Kiatkamjornwong et al.[22] MNR and cassava starch (0, 20, 40 and 60 phr) were mixed using a two-roll mill at a mixing temperature of 60 °C. Mastication of 100 phr of MNR was first performed for 5 min before incorporating the cassava starch. Mixing was then continued for 5 min. The MNR–cassava starch compound was sheeted out and cut into small pieces. Two types of rheological technique were used to characterize the flow properties of rubber melts including SPRI Mooney viscometer (model AC/684/FD) and a Rosand single bore capillary rheometer model RH7. The measurement methods were the same as those described above mentioned.[16] The Mooney viscosity for all types of rubber blends was found to increase with increasing concentrations of cassava starch as shown in Figure 18.38.

Furthermore, the apparent shear stress and apparent shear viscosity of MNR/cassava starch blends were found to apparently increase and decrease with the apparent shear rate as shown in Figures 18.39 and 18.40, respectively, indicating a pseudoplastic behaviour of the blend. At a given shear rate, both the shear stress and the shear viscosity increased with increasing quantities of cassava starch added. That is, the higher flow curve was found when the higher quantity of cassava starch was incorporated into the blends. This implied that a higher pressure was needed during a melt flow in the capillary. The shear thinning behaviour of the flow was also observed. The increase in shear viscosity and shear stress obtained might be due to the chemical bonding between the anhydride group of MNR molecules and hydroxyl groups of cassava starch molecules. As a consequence, the polymer molecules could not be easily deformed under the shearing action for a capillary flow in the extruder.

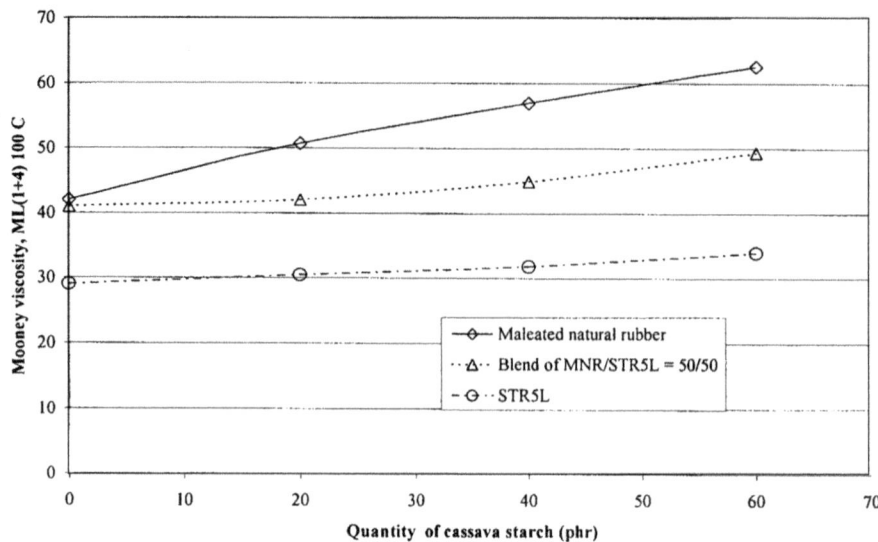

Figure 18.38 Mooney viscosity of rubber blends for various quantities of cassava starch.

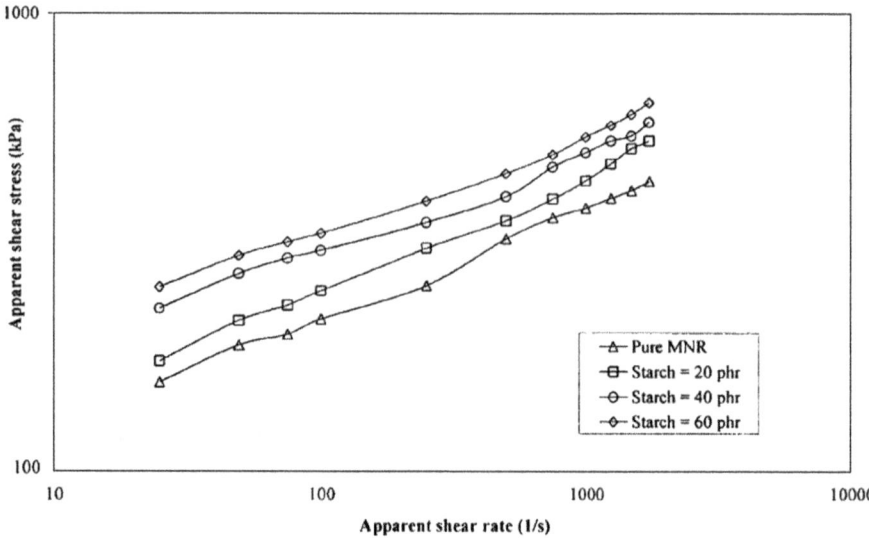

Figure 18.39 Relationship between apparent shear rate and apparent shear stress of MNR blended with various quantities of cassava starch.

Figure 18.41 showed the effect of apparent shear rate on apparent shear stress for the blends of MNR/STR 5L/40 phr cassava starch at various ratios of MNR to STR 5L. The flow behaviour according to the power law relationship was also observed. The blend of STR 5L exhibited the lowest flow curve, while, the blend of pure MNR gave the highest flow curve. It was believed that MNR

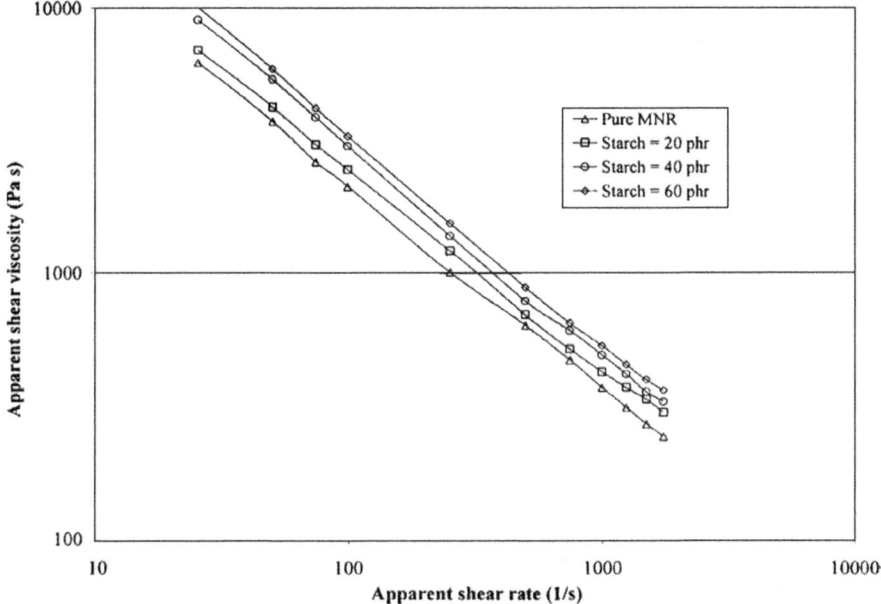

Figure 18.40 Relationship between apparent shear rate and apparent shear viscosity of MNR blended with various quantities of cassava starch.

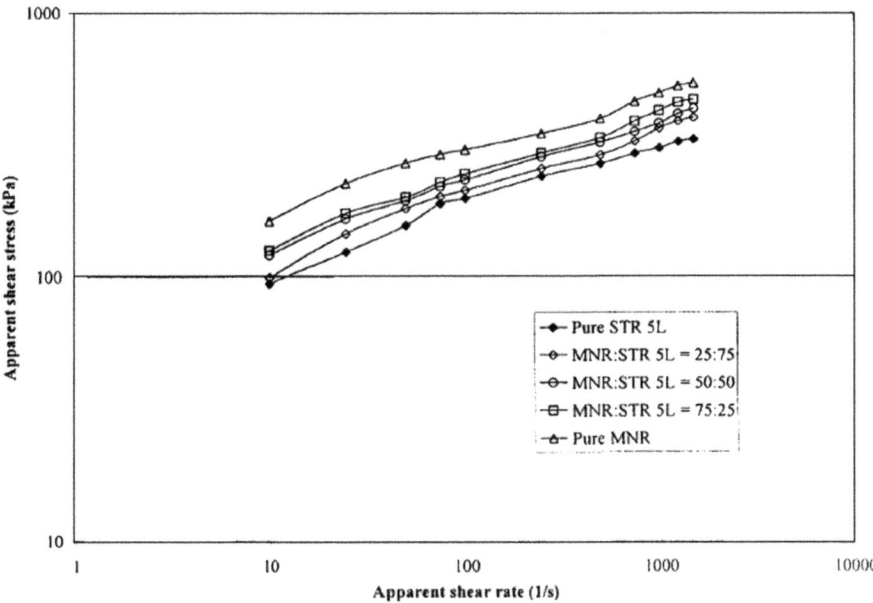

Figure 18.41 Relationship between apparent shear rate and apparent shear stress of rubber blends at the quantity of starch of 40 phr and different ratios of MNR to STR.

behaved as a compatibilizer of STR 5L/cassava starch blend resulting in higher flow curve with concentrations of MNR. It was clarified herein that that the chemical interaction between the polar groups in MNR and cassava molecules was responsible for the characteristics of the Mooney viscosity, apparent shear stress, and shear viscosity of the blends.

The rheology of the *in situ* formation of a graft copolymer between NR and polyamide 6 during processing was investigated.[23] Prior to all melt processing steps, the polyamide 6 was dried in a vacuum oven at 80 °C for at least 12 h. Blends of different compositions were prepared in a Haake torque rheometer fitted with a 50 ml mixing bowl and standard rotors at 240 °C and 60 rpm. Mixtures of NR and 3% of MA were prepared in a roll mill at room temperature and then mixed with polyamide 6 at different compositions in the mixing bowl. After mixing, blends were immediately quenched in cold water and dried in air and later compression moulded at 240 °C to obtain films of approximately 200 mm thickness. Melt behaviour of the blends was measured continuously during mixing using Haake torque rheometry.

Figure 18.42 showed characteristic torque values of binary polyamide 6/NR in comparison with polyamide 6/(NR with 3% w/w MA) blends obtained after 6 min of mixing. It was clear that MA-containing blends had much higher melt viscosity than polyamide 6/NR blends. It was assumed that these higher torque values were due to the occurrence of polyamide 6/NR grafting and also rubber crosslinking.

Rheological behaviour of the TPVs that were prepared from dynamic vulcanization of NR-*g*-PMMA and PMMA blends using various vulcanization systems including conventional vulcanization (CV) and efficient sulphur vulcanization (EV) systems were studied by Nakason *et al.*[24] The ingredient formulas of each vulcanization systems are summarized in Table 18.6.

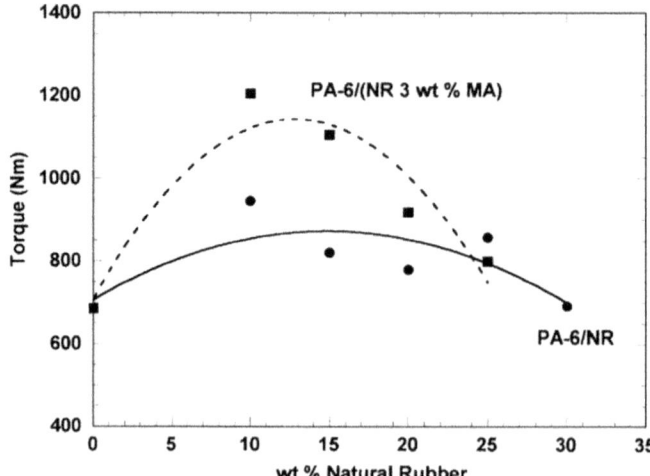

Figure 18.42 Haake torque of polyamide 6/NR and polyamide 6/(NR 3 wt% MA) as a function of NR content. Torque readings were taken after 6 min at 240 °C and 60 rpm.

Table 18.6 Compounding formulations used in the NR-g-PMMA/PMMA blends.

Ingredient	Quantity (phr)		
	CV^a	EV_1^b	EV_2^b
NR-g-MMA	60	60	60
PMMA	40	40	40
ZnO	3.6	3.6	3.6
Stearic acid	0.3	0.3	0.3
Wingstay L	0.6	0.6	0.6
TBBS	0.42	1.8	1.2
TMTD	–	–	0.6
Sulfur	2.1	0.18	0.18

^{a}CV is conventional vulcanization system.
^{b}EV$_1$ and EV$_2$ are efficient vulcanization systems.

Rosand single bore capillary rheometer (model RH7, Rosand Precision Ltd, Gloucestershire, UK) was used to characterize shear flow properties in terms of relationship between apparent shear stress and apparent shear viscosity with apparent shear rate. The test was carried out at a wide range of shear rates (5–1600 s^{-1}) at a test temperature of 200 °C. Dimensions of the capillary die used were 2 mm diameter, 32 mm length and 90° entry angle with the aspect ratio (length/diameter, L/D) of 16/1. The apparent values of shear stress, shear rate and shear viscosity were calculated using the derivation of the Poiseuille law for capillary flow as described in Equations (18.1) to (18.3). Plots of apparent shear stress versus apparent shear rate of TPVs that were prepared from 60/40 NR-g-PMMA/PMMA blend *via* dynamic vulcanization are shown in Figure 18.43.

It was found that the shear stress increased with an increase of shear rate. Furthermore, at a given shear rate, the neat PMMA and NR-g-PMMA exhibited lower shear stresses than those of the blends with dynamic vulcanization using various types of curing systems. This might be attributed to the difference phase morphology. That was, the neat NR-g-PMMA and PMMA were homogeneous while the TPVs were heterogeneous systems. It was found that the CV system provided polymer melt with lower shear stress at a given shear rate. This might be attributed to the CV system provided smaller rubber domains dispersed in the PMMA matrix. Therefore, processability of this type of TPV using injection moulding and plastic extrusion process might be more possible than those of its counterparts.

An apparent shear viscosity as shown in Figure 18.44 was found to decrease with the shear rate indicating that all types of the polymer melts were pseudoplastic (or shear-thinning behaviour) in nature. Shear viscosity at a given shear rate showed the same trend as the shear stress giving the order ranking as follows: EV2 > EV1 > CV > neat NR-g-PMMA > neat PMMA. This might be attributed to the formation of various types of crosslink and different intermolecular networks in the rubber phase and the particle size of the rubber domain dispersed in the thermoplastic matrix. To clarify the shear-thinning behaviour of the polymer melts, the power law equation was applied to the

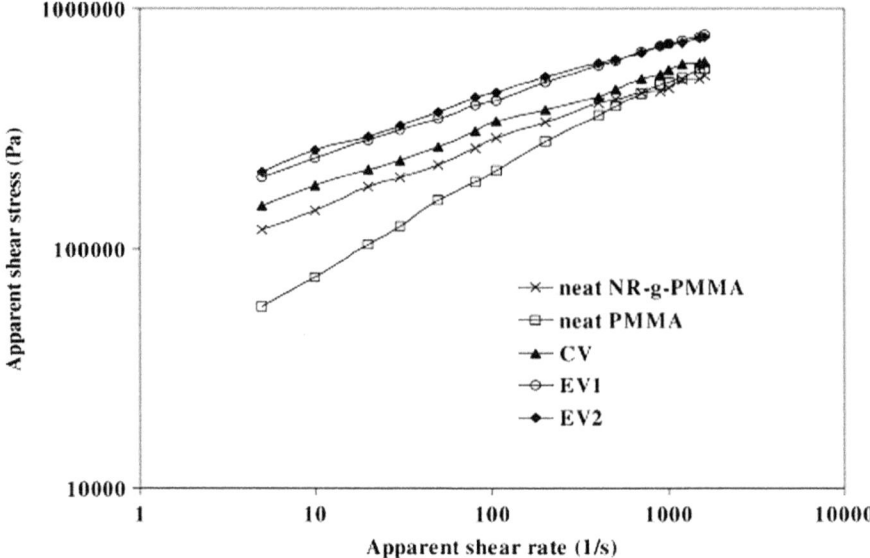

Figure 18.43 Relationship between apparent shear stress and shear rate of PMMA, NR-g-MMA and TPV of NR-g-PMMA/PMMA blends (60/40) at 200 °C.

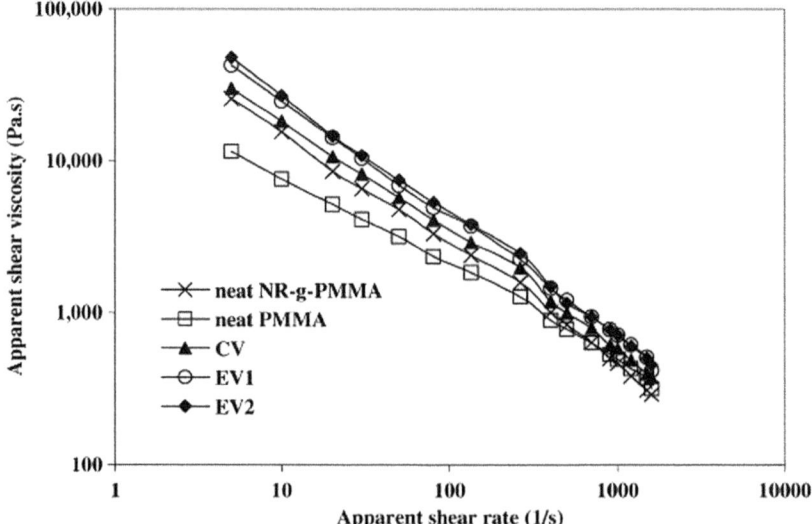

Figure 18.44 Relationship between apparent shear viscosity and shear rate of PMMA, NR-g-PMMA and TPVs prepared from NR-g-PMMA/PMMA blends (60/40) at 200 °C.

relationship between the shear rate and shear stress as shown in Equation (18.4). From the linear relationship of log-log scale of the apparent shear stress versus shear rate, it is possible to obtain the slope (n) and y-intercept (K).

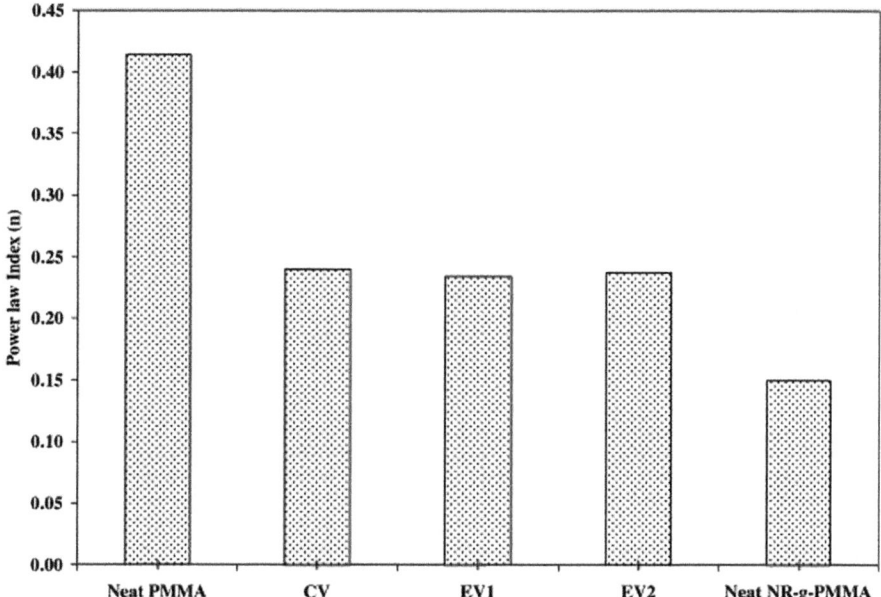

Figure 18.45 Power law index (n) of NR-g-PMMA, PMMA and TPVs prepared from NR-g-PMMA/PMMA blends using various vulcanization systems.

As from Figure 18.45, the value of n reflected the deviation of the flow profiles from uniform parabolic flow patterns (*i.e.* $n=1$ for Newtonian flow) to plug-like flow. The n values of the TPVs with various vulcanizing systems were similar and very low (*i.e.* $n \sim 0.24$). Furthermore, the neat NR-g-PMMA exhibited the lowest n value (*i.e.* $n=0.15$), while the neat PMMA showed the highest value (*i.e.* $n=0.41$). Therefore, the melt flow of the TPVs and the neat NR-g-PMMA exhibited highly pseudoplastic behaviour and the greater shear-thinning behaviour.

ENR prepared from chemical modification of NR latex has recently been commercialized. ENR was claimed to have oil resistance property, low gas permeability and more anti-oxidizing and damping than NR. Up to now, many research works have studied its applications and their blends with other polymers such as in this case, the blend between ENR and neoprene (CR) of which the effect of blending ratio on processability of ENR/CR blend was studied by Chiu *et al.*[25] Mastication of ENR (ENR-50, Malaysian Rubber Producers' Research Association), CR (ES-2-16K, Du Pont), vulcanizing agent, accelerator, and other related ingredients was carried out in a pressurized kneader (SYD-5, Star-King Enterprise Co., Taiwan) for 17 min under a rotation rate of 77 rpm. Each sample's composition was listed in Table 18.7.

The addition sequence of these ingredients was given in Table 18.8. The blended rubber was removed and cooled at room temperature for 30 min and then mixed using a two-roll mixer. In general, the processing of rubber compounds was performed under constant temperature and pressure. As known,

Table 18.7 Composition of ENR/CR blends.

Ingredient (phr)	EC1	EC2	EC3	EC4	EC5
ENR	100	75	50	25	0
CR	0	25	50	75	100
HAF	25	25	25	25	25
Stearic acid	0.66	0.66	0.66	0.66	0.66
$CaCO_3$	0.63	0.63	0.63	0.63	0.63
DM	0.7	0.7	0.7	0.7	0.7
M	0.2	0.2	0.2	0.2	0.2
NA-22	0	0.5	0.5	0.5	0.5
PBN	0	1	1	1	1
MgO	0	4	4	4	4
ZnO	5	5	5	5	5
S_8	2.5	2.5	2.5	2.5	2.5

phr, part per hundred parts of rubber; ENR, epoxidized natural rubber; CR, neoprene; HAF, carbon black (N330); $CaCO_3$, calcium carbonate; DM, 2-benzothiazole disulfide; M, 2-mercaptobenzothiazole; NA-22, ethylene thiourea (ETU); PBN, phenyl-naphthylamine; MgO, magnesium oxide; ZnO, zinc oxide; S_8, sulfur.

Table 18.8 Banbury mixing time of ENR/CR blends.

Addition sequence	Elapsed time, (min)	Holding time, (min)
Sequence a		
ENR-50	0–5	5
Filler (ZnO, $CaCO_3$, stearic acid, *etc.*)	5–6	1
HAF	6–9	3
DM, M, S_8	9–10	1
Blending	10–17	7
Total	17	17
Sequence b		
ENR-50/CR	0–5	5
Filler (MgO, $CaCO_3$, stearic acid, PBN, *etc.*)	5–6	1
HAF	6–9	3
DM, M, S_8, NA-22	9–10	1
Blending	10–17	7
Total	17	17
Sequence c		
CR	0–5	5
Filler (MgO, $CaCO_3$, stearic acid, PBN, *etc.*)	5–6	1
HAF	6–9	3
DM, M, S_8, NA-22	9–10	1
Blending	10–17	7
Total	17	17

the initial fluidity could be one of various factors that described the processability of rubber. Factors affecting the fluidity included the type of the rubber and the composition. One index for evaluating the initial fluidity was the Banbury processability index, which was ML1+4 at 100 °C. According to ASTM D1646, the Mooney viscosity was measured at 100 °C, preheated for 1 min, and tested for 4 min, at a rotational speed of 2 rpm, which gave an average shear rate of 1.6 s^{-1} by using a Mooney scorch tester (HI-8725, Hungta Co.,

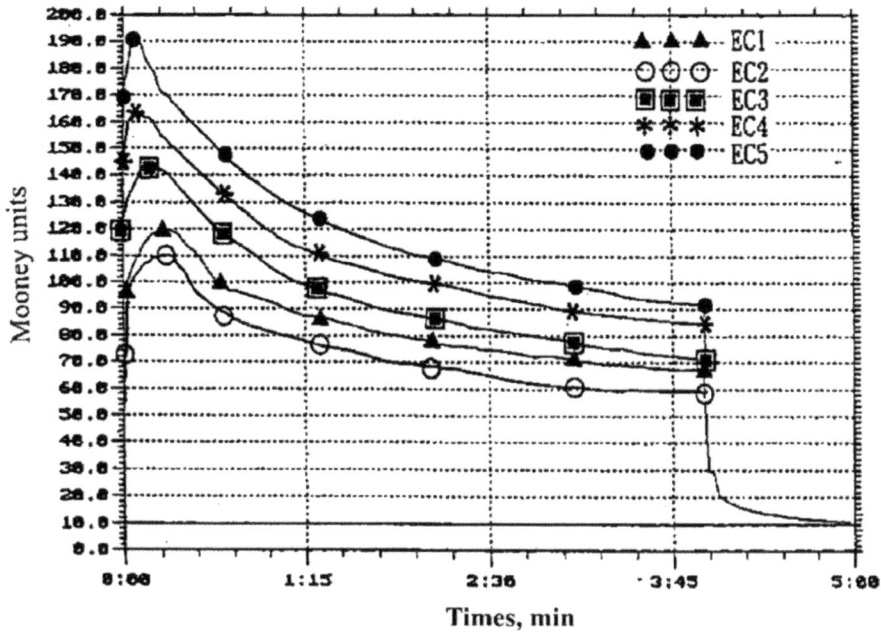

Figure 18.46 Mooney viscosity curves of ENR/CR blends at 100 °C.

Taiwan). It was found that the Mooney viscosity (ML1 + 4100 °C) of ENR was 72 (Mooney units) while that of CR was 93 (Mooney units) indicating the lower initial fluidity of CR than that of ENR (Figure 18.46).

Thus, the blending of ENR was able to lubricate the blending process resulting in the decrease of viscosity with ENR content (Figure 18.47). At an ENR/CR blending ratio of 75/25, the viscosity was the lowest, the fluidity was the highest. The effect of ENR on the plasticization of CR was the most pronounced. Therefore, the improvement in CR processability due to the blending of ENR was well established at this rubber ratio of 75/25.

The melt rheological behaviour of miscible blends of poly(vinyl chloride) (PVC) and ENR was studied by Varughese.[26,27] Blends of PVC and ENR were prepared in a Brabender plasticorder by the melt blending technique. The melt flow behaviour of these blends with respect to blend ratio and temperature had been examined using a melt flow indexer and capillary rheometer. They found that melt viscosity of plasticized PVC increased with ENR content at higher processing temperatures. On the other hand, some other research works in Ishiaku's group clarified the influence of ENR on shaping processes and the melt flow behaviour of PVC.[28,29] The melt flow behaviour of these blends with respect to blend ratio and temperature had been also examined using a melt flow indexer and capillary rheometer. The optimum mixing from rheometry by using the Brabender Plastogram as an indicator was found to be in very good agreement with the results obtained from capillary rheometry. It was found that ENR decreased the Brabender torque and the melt flow viscosity but

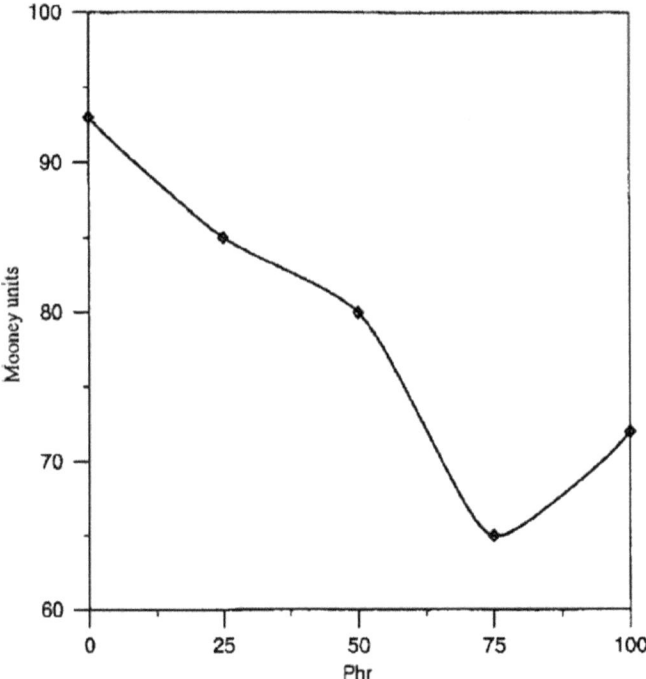

Figure 18.47 Effect of ENR content on Mooney viscosity of ENR/CR blends.

increased the melt flow index(MFI) of PVC in the blends. Moreover, the flow behaviour index (n') obtained from capillary rheometer data was found to be dependent on temperature and blend ratio.

ENRs with epoxide levels of 10, 20, 30, 40, and 50 mol% were blended with PMMA with various blend formulations and the shear-flow properties in terms of the shear stress and shear viscosity were characterized using A Rosand RH7 single-bore capillary rheometer (Gloucestershire, UK).[30] The tests were carried out over a wide range of shear rates (25–1600 s^{-1}) at a test temperature of 200 °C. The test was then carried out at a set of shear rates in a program *via* a microprocessor. The equations used to calculate the shear stress, shear viscosity, and shear rate are described above. The torque was observed to increase with the PMMA contents and the epoxide molar percentage as shown in Figures 18.48 and 18.49, respectively.

Furthermore, the shear stress (Figures 18.50 and 18.51) and shear viscosity (Figures 18.52 and 18.53) of the polymer blends in the molten state were found to increase as the ENR content and epoxide molar percentage increased. The reason for the increase in the torque, shear stress, and viscosity was claimed to be due to the chemical interactions between polar groups in the ENR and PMMA molecules of this partly miscible blends. On the other hand, all the ENR/PMMA blends exhibited shear-thinning behaviour. This was observed as a decrease in the shear viscosity with an increase in the shear rate.

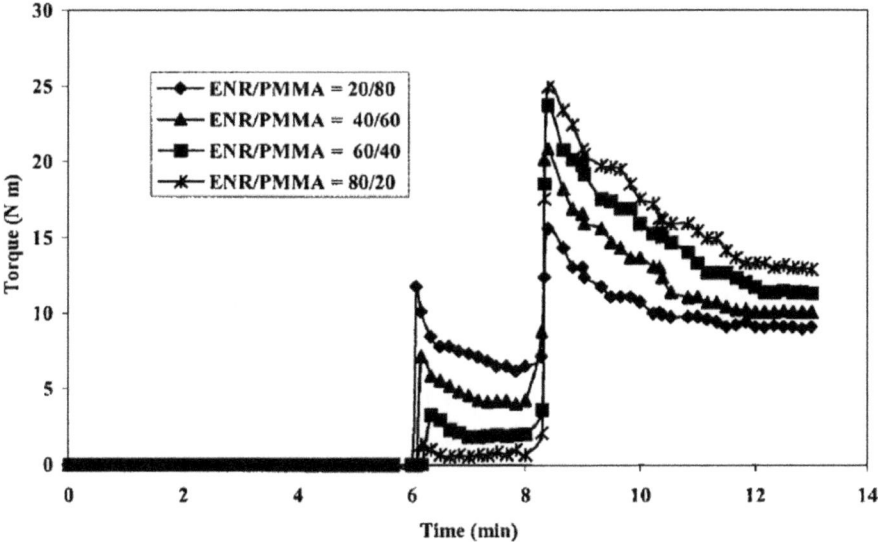

Figure 18.48 Torque–time curves for ENR-30/PMMA blends.

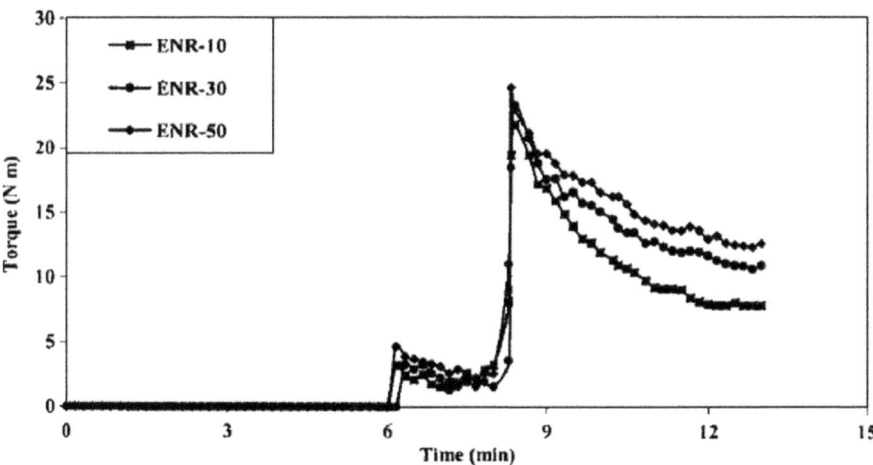

Figure 18.49 Torque–time curves for 60/40 ENR/PMMA blends with various ENRs.

Recently, blends of non-polar PP and polar ENR at 50:50 blend ratio were prepared by an in-line electron induced reactive processing technique.[31] PP and ENR were *in situ* compatibilized by using reactive modified PP which was prepared simultaneously by combination of polymer modification using triallyl cyanurate (TAC) with high-energy electrons. Brabender mixing chamber volume of 50 cm^3, with a rotor speed of 50 rpm at 180 °C in the presence of air was used for melt mixing process. The rheological measurements were carried out with an ARES rheometer (Rheometrics Scientific, New Castle, USA) with

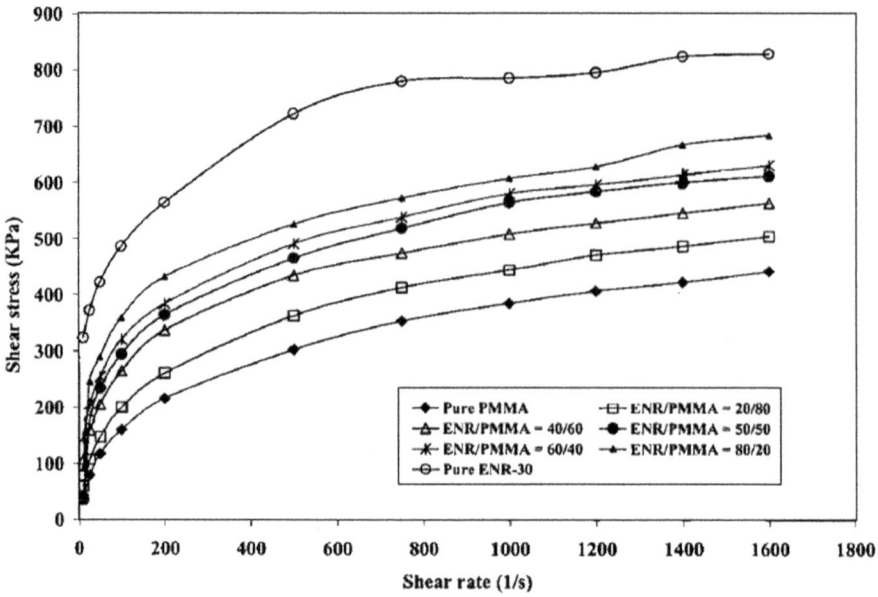

Figure 18.50 Relationship between the shear stress and shear rate of ENR-30/PMMA blends at 200 °C.

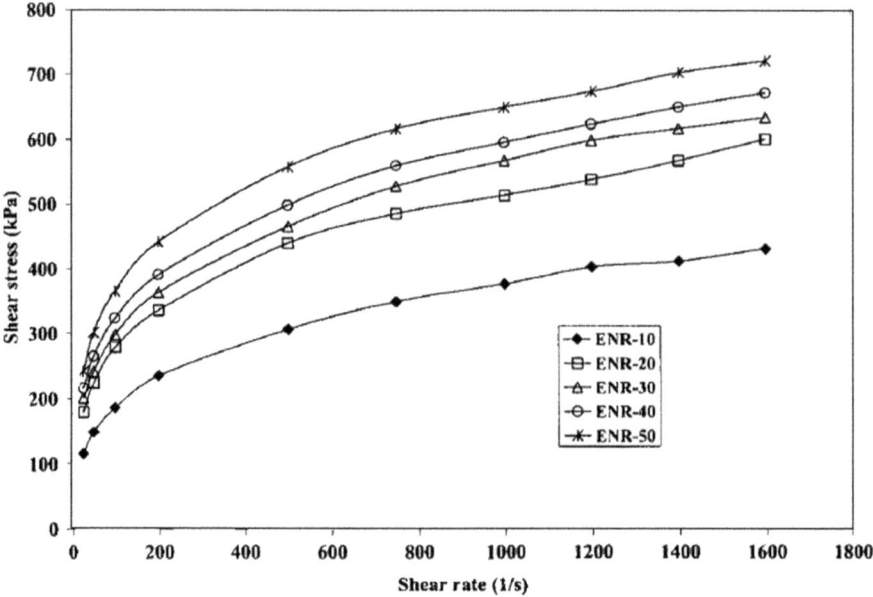

Figure 18.51 Relationship between the shear stress and shear rate of ENR/PMMA blends with various ENRs at 200 °C.

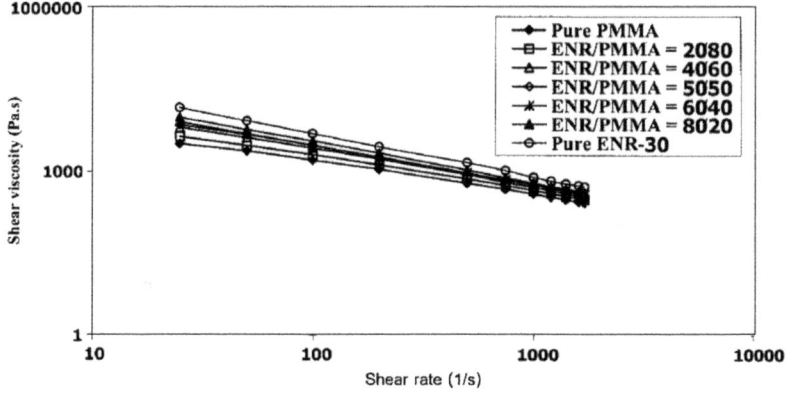

Figure 18.52 Effect of the shear rate on the melt viscosity of ENR-30/PMMA blends at various blend ratios.

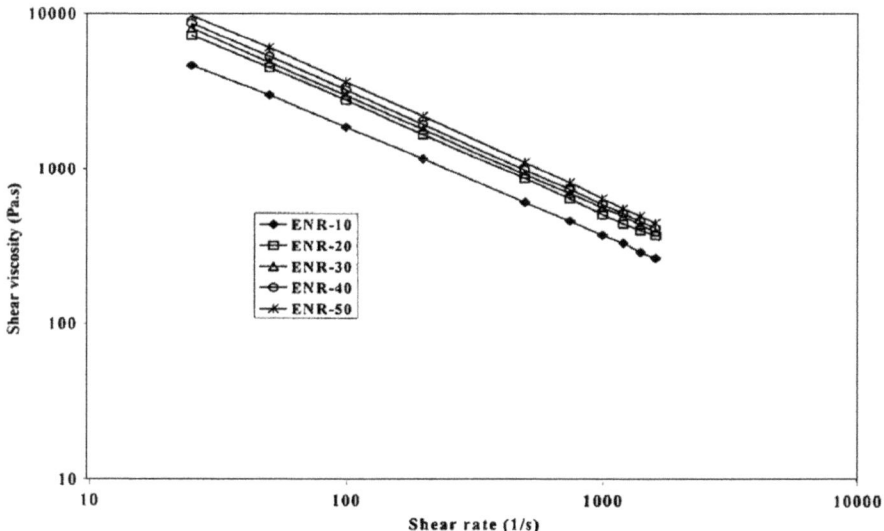

Figure 18.53 Effect of the shear rate on the melt viscosity of various ENRs in 60/40 ENR/PMMA.

torque transducers having a torque range from 0.02 to 2000 g cm. Parallel-plate geometry was applied to the formerly pressed plates with 1 mm thickness and 25 mm diameter. Frequency range was kept between 0.5 and 100 rad/sec and strain amplitude was kept under linear viscoelastic region. During each experiment, the temperature was maintained at 190 °C by constant heating of the sample under nitrogen atmosphere. Rheological studies revealed that samples modified with high-energy electrons had very high storage modulus, zero shear viscosity, and complex viscosity. Significant increment of zero shear viscosity as summarized in Table 18.9 could be attributed to the strong coupling between modified PP and ENR.

Table 18.9 Zero shear viscosities (η_0) of pure polymers and their blends.

Samples	Zero shear viscosity (η_0) (Pa s)
PP	3008
ENR	1260
PP/ENR	1850
PP/TAC/ENR	1620
PP/TAC/e/ENR	4040
PP/e/ENR	2000

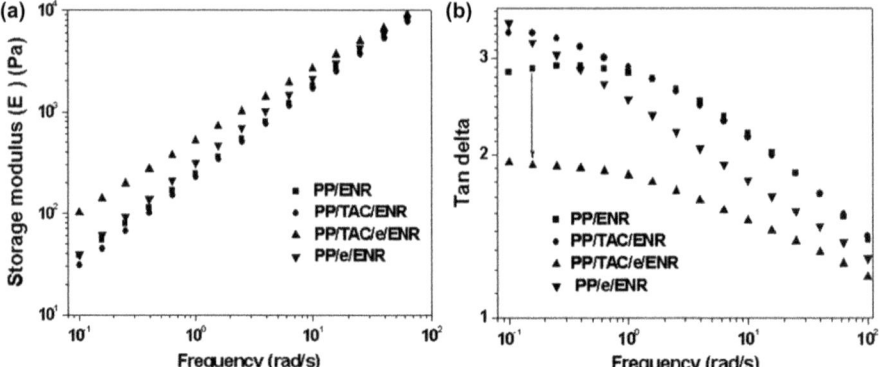

Figure 18.54 Rheological curves: (a) storage modulus and (b) tan δ for the respective blends.

The storage modulus (E) resulting from dynamic frequency scans was shown in Figure 18.54(a). The storage modulus parameter increased monotonically in the entire frequency range for PP/TAC/e/ENR, especially in the lower frequency region. The value of zero shear viscosity was also very high in case of PP/TAC/e/ENR which was due to strong coupling. In Figure 18.54(b), a sharp decrement in tan δ values could be seen in case of PP/TAC/e/ENR blend at low frequencies. On the other hand, observation of lowering in tan δ value also demonstrated the strong coupling between the modified PP and ENR. It is also observed that the complex viscosity of PP/ENR blends modified with electrons was higher compared to PP/ENR blends not modified by high-energy electrons. With increasing frequency, the complex viscosity decreased which showed shear thinning nature of the polymer blends.

The processability study of linear low density polyethylene (LLDPE) blended with the ozonolysed NR was investigated by Boochathum and a coworker.[32] The *in situ* ozonolysis reaction of NR latex was carried out to produce low molecular weight rubber with ketone and aldehyde at the end groups namely LA3; lower molecular weight compared to LA0; rubber without ozonolysis and LA7; lower molecular weight as compared to LA3.[33] The 60/40 LLDPE/NR blends were prepared by melt mixing at 150 °C in a Brabender Plasticorder with a mixing chamber capacity of 410 cm³, 60 rpm rotor speed. The blending methodology was as follows, firstly, LLDPE was added into the internal mixer, melted for 2 minutes, subsequently 1% w/w antioxidant (Flectol® H) and different loads of rubbers were added. The continuous mixing process was carried out until the torque was constant and then the blended samples were dumped and immediately laminated to thin sheets by two-roll mill. The sheets prepared were further compressed in a 12 metric ton hydraulic hot plate (Caver Model 3925) at 150 °C for 10 minutes. The processability of LLDPE/ozonolysed NR blends having different molecular weights (*i.e.*, functionalities) and different quantities of NR was investigated at various low strains using oscillatory rheological analyser, Rubber Processing Analyser (RPA2000, Alpha Technologies). The aim of using chemically modified NR in this research work was to maintain the processability of LLDPE while the elastic property of the blend was expected to be enhanced by NR. It was obvious that adding LA7 with the lowest molecular weights of $\overline{M}_w = 8.30 \times 10^5$ g/mol, $\overline{M}_n = 2.62 \times 10^5$ g/mol and MWD = 3.3 resulted in the lowest final mixing torque as shown in Figure 18.55. Moreover the addition of LA7 slightly reduced the processability of LLDPE since tan δ as a function of strain and temperature of LLLDPE/LA7 were found to mostly closer to those of LLDPE as shown in Figures 18.56 and 18.57, respectively.

Furthermore, the effect of LA7 loading on tan δ based on the strain sweep mode was studied (Figure 18.58). It was interesting that tan δ of LLDPE/LA7 (60/40) blend was much closer to that of LLDPE, indicating the similar processability with LLDPE at all range of strains. Therefore, blending of ozonolysed NR (LA7) with LLDPE rarely changed the processability behaviour of LLDPE and LA7 could be loaded up to 40% w/w LLDPE.

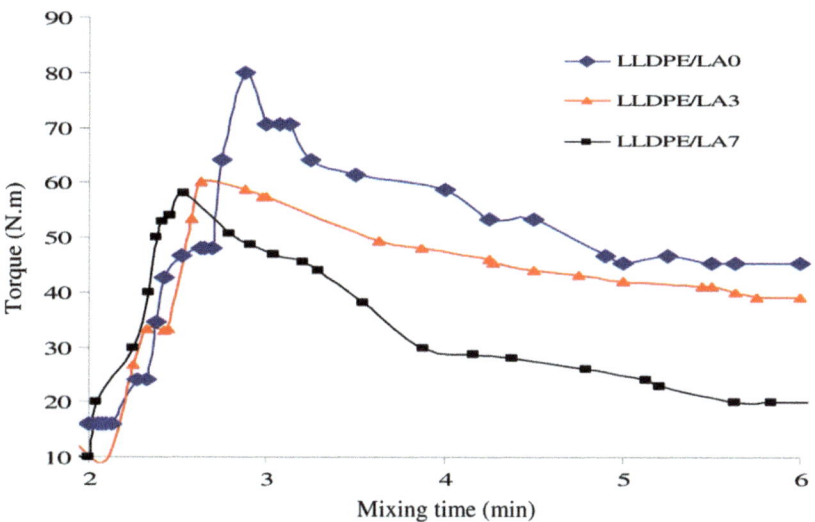

Figure 18.55 Effect of molecular weights of natural rubber on mixing torque of LLDPE/NR (60/40) blends.

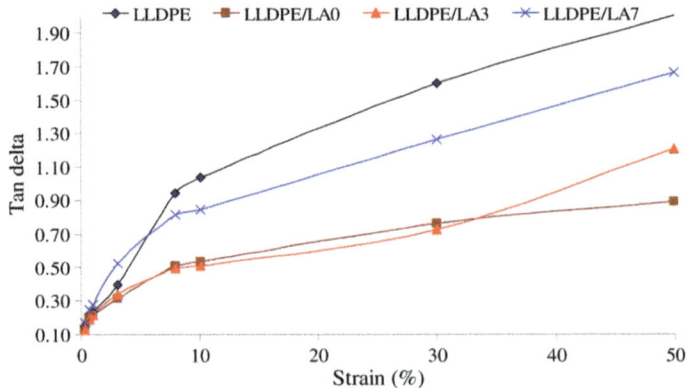

Figure 18.56 Tan δ as a function of strain sweep of various LLDPE/natural rubber blends in comparison with that of LLDPE.

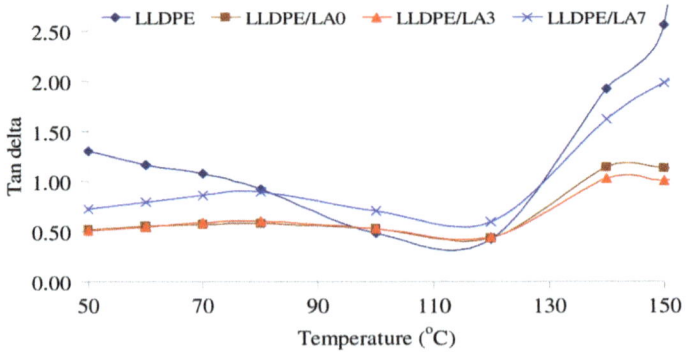

Figure 18.57 Tan δ as a function of temperature sweep of various LLDPE/natural rubber blends in comparison with that of LLDPE.

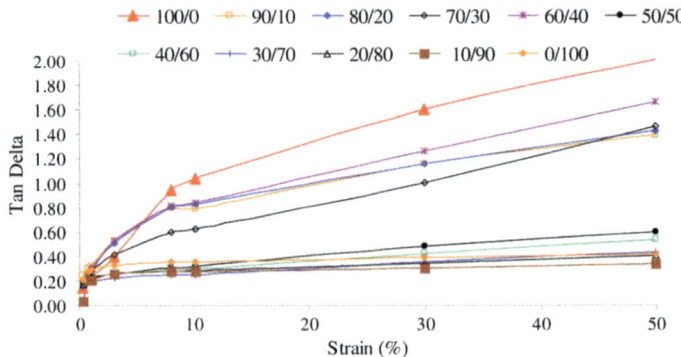

Figure 18.58 Tan δ as a function of strain sweep of various loadings of LA7 in LLDPE/LA7 blends.

References

1. K. N. G. Fuller and W. S. Fulton, *Polymer*, 1990, **31**, 609.
2. T. Gao, R. Xie, P. Li and M. Huang, *Adv. Mater. Res.*, 2011, **337**, 285.
3. S. Wiboolkul and P. Boochathum, in, *Proceedings of the 12th International Seminar on Elastomers*, National Metal and Materials Technology Center (MTEC), Thailand, 2010, pp. 83–85.
4. A. T. Koshy, B. Kuriakose, S. Thomas1, C. K. Premalatha and S. Varghese, *J. Appl. Polym. Sci.*, 1993, **49**, 901.
5. Z. Oommen, S. Thomas, C. K. Premalatha and B. Kuriakose, *Polymer*, 1997, **38**, 5611.
6. L. K. Yoon, C. H. Choi and B. K. Kim, *J. Appl. Polym. Sci.*, 1995, **56**, 239.
7. C. Nakason, S. Jamjinno, A. Kaesaman and S. Kiatkamjornwong, *Polym. Adv. Technol.*, 2008, **19**, 85.
8. Halimatuddahliana and H. Ismail, *Jurnal Teknologi*, 2003, **39**(A), 97.
9. P. Wongwitthayakool, P. Saeoui and C. Sirisinha, *Int. Polym. Process.*, 2009, **1**, 9.
10. R. Asaletha, G. Groeninckx, M. G. Kumaran and S. Thomas, *J. Appl. Polym. Sci.*, 1998, **69**, 2673.
11. F. Zhengping and X. Chengwei, *Acta Polym. Sin.*, 1995, **5**, 610.
12. P. Prasopnatra, P. Saeoui and C. Sirisinha, *J. Appl. Polym. Sci.*, 2009, **111**, 1051.
13. U. Šebenik, A. Z. Valant and M. Krajnc, *Polym. Eng. Sci.*, 46, 1649, **2006**.
14. A. Alipour, *2011 International Conference on Nanotechnology and Biosensors IPCBEE*, IACSIT Press, Singapore, 2011, **25**, pp. 44–48.
15. K. I. Elizabeth and A. Rosamma, *KAUTSCHUK UND GUMMI KUNSTSTOFFE*, 2005, **58**(12), 656.
16. C. Lewis, S. Bunyung and S. Kiatkamjornwong, *J. Appl. Polym. Sci.*, 2003, **89**, 837.
17. F. Cogswell, *Polymer Melt Rheology: A Guide for Industrial Practice*, Woodhead, Cambridge, 1981, p. 24.
18. J. A. Brydson, *Flow Properties of Polymer Melts*, Plastics Institute, London, 1970, p. 12.
19. C. D. Han, *Rheology in Polymer Processing*, Academic Press, New York, 1976.
20. C. Nakason, S. Saiwari and A. Kaesaman, *Polym. Test.*, 2006, **25**, 413.
21. C. Nakasona, A. Kaesamana, Z. Samoha, S. Homsina and S. Kiatkamjornwong, *Polym. Test.*, 2002, **21**, 449.
22. C. Nakason, A. Kaesman, S. Homsin and S. Kiatkamjornwong, *J. Appl. Polym. Sci.*, 2001, **81**, 2803.
23. E. Carone Jr, U. Kopcak, M. C. Goncalves and S. P. Nunes, *Polymer*, 2000, **41**, 5929.
24. C. Nakason, W. Pechurai, K. Sahakaro and A. Kaesaman, *Polym. Adv. Technol.*, 2005, **16**, 592.
25. H. T. Chiu, P. A. Tsai and T. C. Cheng, *J. Mater. Eng. Perform.*, 2006, **15**, 81.

26. K. T. Varughese, P. P. De, G. B. Nando and S. K. De, *J. Vinyl Technol.*, 1987, **9**, 161.
27. K. T. Varughese, *J. Appl. Polym. Sci.*, 1990, **39**, 205.
28. U. S. Ishiaku, M. Nasir and Z. A. Mohd Ishak, *J. Vinyl Technol.*, 1995, **1**, 42.
29. U. S. Ishiaku, M. Nasir and Z. A. Mohd Ishak, *J. Vinyl Technol.*, 1995, **1**, 66.
30. C. Nakason, Y. Panklieng and A. Kaesaman, *J. Appl. Polym. Sci.*, 2004, **92**, 3561.
31. S. Rooj, V. Thakur, U. Gohs, U. Wagenknecht, A. K. Bhowmick and G. Heinrich, *Polym. Adv. Technol.*, 2011, **22**, 2257.
32. S. Utara and P. Boochathum, *Process Material Properties, Processing Material for Properties (PMPIII)*, A publication of The Minerals, Metals & Materials Society (TMS), USA, 2009, pp. 409–414.
33. P. Boochathum and J. Sanchompu, *The 164th Meeting of the Rubber Division*, American Chemical Society (ACS), Cleveland, Lippincott & Peto, Inc. USA, 14–17 October 2003, p. 113.

CHAPTER 19

Spectroscopy: Natural Rubber Based Blends and IPNs

SA-AD RIYAJAN

Department of Materials Science and Technology, Faculty of Science, Prince of Songkla University, Thailand
Email: saadriyajan@hotmail.com

19.1 Introduction

Industrial products made from elastomer materials, especially natural rubber (NR) latex, have found extensive applications in tyres, tubing, surgical gloves, catheters, balloons and other products. But, the compatibility of devices made from NR is different when compared to those made of synthetic polymers due to rubber's elasticity, flexibility and resistance against splitting that provides NR products with more advantages than those of other polymers.[1] In addition, the advantage of NR is that it is a renewable resource. In order to improve their properties modification of NR polymers is essential. In previous work over the last 50 years, there have been three main approaches utilized to modify the properties of NR polymers including changes to its chemical microstructure, addition of different material reagents and blending with other polymers. Blending with other polymers is a technique to improve the properties of the polymer to get the new commercially useful materials having desired properties which cannot be derived from a single polymer. In addition, it is an easy, cheap and economical method and can produce new materials with novel properties. NR has been blended with many polymers such as PP, polyethylene (PE) and poly(vinyl alcohol) (PVA). Interpenetrating polymer networks (IPNs) is a one of the chemical blending and it consists of combination of two polymers, both

RSC Polymer Chemistry Series No. 7
Natural Rubber Materials, Volume 1: Blends and IPNs
Edited by Sabu Thomas, Chin Han Chan, Laly A. Pothen, Rajisha K. R., Hanna J. Maria

Published by the Royal Society of Chemistry, www.rsc.org

in network form, at least one of which is synthesized and/or crosslinked in the immediate presence of the other. The network cannot be separated unless chemical bonds are broken. The two or more networks can be envisioned to be entangled in such a way that they are concatenated and cannot be pulled apart, but not bonded to carbon–carbon chemical reaction. The IPN was divided into two types such as semi-IPNs and full IPNs. When one of the phases is crosslinked pseudo- or semi-IPNs will be formed. When both phases are crosslinked full IPNs are developed. The IPNs are generally prepared by sequential, simultaneous or latex polymerization techniques. The parameters affecting the properties of the IPN are dependent on phase behaviour, chemical additive agent, blend ratio and crosslinking levels. The aim of this chapter is to chemical structure and interaction of blend and IPN obtained from NR and the synthetic polymer as well as crosslinking agent. A complete chemical and physicochemical characterization of NR or their derivative based blends and IPNs is not possible without using spectroscopic techniques. This review focuses on the application of various spectroscopic methods including the FTIR, NMR, UV-vis, Raman spectroscopy and ESR for the chemical, physical interaction, degradation and blend composition of NR blends.

19.2 UV-Vis Spectroscopy

19.2.1 Introduction to UV-Vis Spectroscopy

Ultraviolet/visible (UV-vis) spectroscopy is useful as an analytical method for two reasons. Firstly, it can be used to identify certain functional groups in molecules, and secondly, it can be used for assaying. Unlike IR spectroscopy, UV-vis spectroscopy involves the absorption of electromagnetic radiation from the 200–800 nm range and the subsequent excitation of electrons to higher energy states. The absorption of ultraviolet/visible light by organic molecules is restricted to certain functional groups such as chromophores including carbonyl groups, carboxylic acid and hydroxyl groups that contain valence electrons of low excitation energy and the part of a molecule responsible for its colour.[2] The colour arises when a molecule absorbs certain wavelengths of visible light and transmits or reflects others. The UV-vis spectrum is complex and appears as a continuous absorption band because the superimposition of rotational and vibrational transitions on the electronic transitions gives a combination of overlapping lines.[2] Nowadays, the individual detection of electron transfers without superimposition by neighbouring vibrational bands can also be recorded. With UV-vis spectroscopy, it is possible to investigate the electron transfers between orbital or bands of atoms, ions and molecules existing in the gaseous, liquid and solid phase. Analysis of solutions and crystals usually takes place in transmission.

19.2.2 Sample Preparation and Typical Conditions for UV-Vis Spectroscopy Measurement

UV-vis spectra of NR and its derivative are usually recorded in organic solution and solid film. UV-vis spectroscopy analysis was carried out on a UV-vis spectrophotometer. The rubber samples of 1.0 g were heated in a hot air ageing chamber for different times.[3] The samples were dissolved in organic solvents such as tetrahydrofuran (THF) and chloroform at a concentration of 1.0 mg/ml, and then scanned from 200 to 600 nm on a UV spectrophotometer. A reference cell loaded with the THF was used to eliminate any signals due to the solvent itself.

19.2.3 Analysis of Polymer Blends

ENR /PEA blend[4]

The blends of ENR (50 mol%) and PEA (6 wt%) are demonstrated to be partially miscible up to 50% by weight of PEA and completely miscible beyond this proportion. For the 70:30 (ENR: PEA) blend, the T_gs shift toward an intermediate value but do not merge to form a single T_g, making the blend partially miscible. From UV results, the UV absorption band of its blend is found in range from 190 to 350 nm due to the electrons from one electronic level to the other in acid group of PEA. In ENR, the major peak shows at 247.5 nm, referring to n–σ* transition of the –O–C group of the secondary alcohol in ENR after mixing. The possible reaction of PEA and ENR is presented in Figure 19.1. The small absorption band of epoxy group in ENR appears in range of 195–196 nm. The absorption peak of ENR is shifted to a much lower value of 226 nm after blending with PEA because of an extensive etherification reaction between ENR and PEA. In addition, the peak of the ethylenic radical of ENR in the blend shows at 170–190 nm. This phenomenon is called the hypsochromic shift or the 'blue shift'. The miscibility has been assigned to the etherification reaction between –OH groups formed *in situ* during melt blending of ENR and –COOH groups of PEA. The occurrences of such reactions have been confirmed by UV and IR spectroscopic studies.

Chlorinated NR from latex[3]

A main absorption peak appears at around 245 nm on each spectrum as shown in Figure 19.2, which might be assigned to a conjugated polyene.[3] The intensity of the peak at 290 nm increased with increasing degrading time. This peak might be assigned to a conjugated polyene $-[C=C]_4-$.[3] Since this peak is very wide (this result is shown here), it is reasonable to presume that some other species which cannot be identified the time being are produced as well. This is different from the thermooxidative degradation of PVC in air. During the thermooxidative degradation of PVC, the conjugated polyene sequences $(-[C=C]-)_n$ with the ranging from 3 to 15 were produced from the elimination of HCl.[3] The possible reason is that the main backbone structures of PVC are of straight chains and the HCl can be easily eliminated through a 'zipper' dehydrochlorination reaction as shown in Figure 19.2.[3] But those of CNR from latex are of cycle and crosslinking structures[3] and the dehydrochlorination reaction might be retarded somehow.

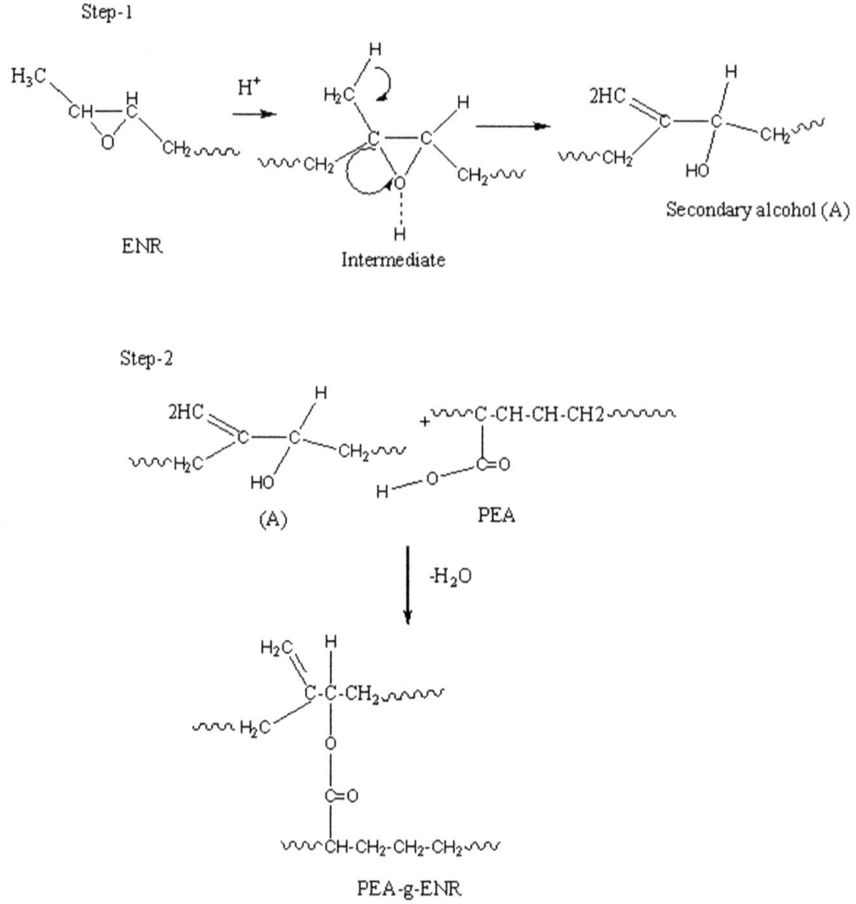

Figure 19.1 Possible reaction between PEA and ENR.

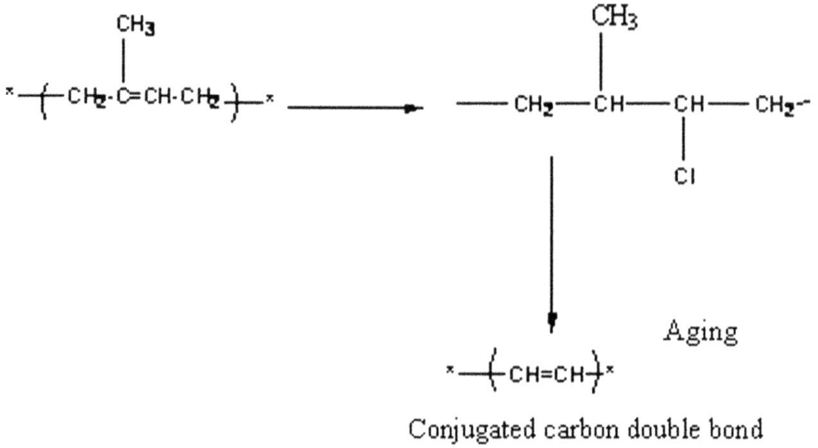

Figure 19.2 Possible degradation reaction of chlorinated NR under heating.

NR/polyaniline (PANI) blend[5]

Nowadays processing conducting polymers in form of blends and composites with commercial polymers is a well-established alternative as material for technological applications.[5] PANI and its derivate are one of the most interesting conducting polymers that have attracted considerable attention due to its good environmental stability, the control of the electronic and optical properties *via* the level of oxidation and protonation, low cost of raw material, and ease of synthesis. In this work conductive elastomeric films have been prepared by mixing a solution of NR dissolved in chloroform ($CHCl_3$) with a solution of poly(*o*-methoxyaniline) (POMA), a polyaniline derivative, dissolved also in $CHCl_3$, at different proportions. The final solution was casting onto Teflon mould placed in an oven with air circulation, at room temperature. After solvent evaporation the film obtained was removed from the Teflon mould by peeling them out. The primary doping of POMA was done by mixing it with dodecylbenzene sulfonic acid. Physical characterizations of the films were realized by UV-vis-NIR spectroscopy. The higher value of the electrical conductivity obtained was 10^{-5} S/cm, nine orders in magnitude higher compared with pure NR. From UV-vis-NIR and FTIR results, the POMA is responsible for the high electrical conductivity of the blend.

19.3 Fourier Transform Infrared Spectroscopy (FTIR)

19.3.1 Introduction to FTIR

Infrared spectroscopy is unarguably the most technique of characterizing materials of all forms. IR spectroscopy is one of the most vital and widely used analytical techniques available to scientists working on blend and IPN of NR. It is based on the vibrations of the atoms of a molecule. The infrared spectrum is commonly obtained by passing infrared electromagnetic radiation through a sample that possesses a permanent or induced dipole moment and determining what fraction of the incident radiation is absorbed at a particular energy. The energy of each peak in an absorption spectrum corresponds to the frequency of the vibration of a molecule part, thus allowing, qualitative identification of certain bond types in the sample. An IR spectrometer usually records the energy of the electromagnetic radiation that is transmitted through a sample as a function of the wave number or frequency. Nowadays, the total spectrum is analysed by an interference process and converted into the frequency or wave number range by means of a mathematical process known as the Fourier transform. FTIR has dramatically improved quality of infrared spectra and minimized the time required to obtain data. There are two general types of structural information that can be studied by IR spectroscopy: electronic structure (focused on valence and core electrons, which control the chemical and physical properties, among others) and geometric structure which gives information about the locations of all or a set of atoms in a molecule.

19.3.2 Sample Preparation and Typical Conditions for FTIR

There are three forms of sample including film, liquid and gas to analyse by FTIR.

Film form

The NR/PS blend sheet was dried in an oven at 50 °C for 24 h and kept in a desiccator before being further characterized. The chemical structure of the NR/PS blend sheet was analysed by an Attenuated Total Reflection Fourier Transform Infrared (ATR-FTIR) Bruker EQUINOK 55 measuring in the range of 4000–500 cm^{-1}.

Liquid form[6]

The $FeCl_3$ $6H_2O$ was added in polychloroprene (PCP)/NR blend in toluene. Then, this solution was poured into the tube and exposed to polychromatic light for either 24 h or one week under constant stirring. Some drops of the resulting solution were deposited on a polished NaCl crystal surface. After that, it was under reduced pressure and in the dark at room temperature to characterize the chemical structure by FTIR.

Gas form[7]

The output of the inert gas from the TGA was connected to a FTIR spectrometer through a heated line. The balance adapter, the transfer line and the FTIR gas cell can be heated until 523 K, thus avoiding the condensation of the less volatile compounds. On the other hand, the low volumes in the thermobalance microfurnace, transfer line and gas measurement cell permit low carrier gas flow rates to be used and allow a good detection of the gases evolved in the pyrolysis. In all the experiments, the transfer line and the gas measurement cell were maintained at 473 K.

Pyrolysis of sample[8]

A small quantity of polymer blend was introduced in a small tube and put in contact with the flame of a Bunsen burner. Volatile products, which condense at ambient temperature, were collected on a metallic spatula inserted in the tube at the moment of vapour evolution. The condensed liquid, collected along the time the entire polymer mass was reduced to inorganic residue, was carefully deposited on KBr windows in such a way to obtain acceptable homogeneity. This point was checked by means of the corresponding infrared spectra.

19.3.3 Analysis of Polymer Blends

Usually, FTIR is a powerful to detect the interaction from hydrogen bonding and chemical structure change of the rubber blend. Many works have studied the use of FTIR to detect for this purpose. The advantages of this technique are non-destructive and a precise measurement method which requires no external calibration, increase in speed, it collects a scan every second and increase in sensitivity – one second scans can be added together to ratio out random noise, allow greater optical throughput and be mechanically simple with only one moving part.

Considering the definition of compatibilizers,[6,7] it is a material which has been commonly used to enhance the compatibility of incompatible rubber blends, where the compatibilizers, often referred to as "interfacial agents" are able to improve interfacial adhesion between otherwise gross-phase-separated rubber pairs by reducing the interfacial energy between the phases as shown Figure 19.3. In addition, it improves process ability. In previous work, the block copolymer and grafting copolymer are applied in compatibilizers in the rubber blend. The interaction and hydrogen bonding from the compatibilizer and rubber phase are analyzed by spectroscopy. For example, the 50% mol epoxidation of ENR was used as a compatibilizer on the properties of SBR/NBRr blends.[7] FTIR analysis was performed to investigate the compatibility of SBR/NBRr blends containing ENR-50. The improvement of their properties upon compatibilization is due to an increase in crosslink density. FTIR analysis showed that ENR-50 is compatible with NBRr through the oxirane group and with SBR through the isoprene group. There are two types of hydrogen bonding that exist in this polymeric system, intermolecular and intramolecular hydrogen bonding amongst polymers observing the stronger hydrogen bonding at 3328 cm^{-1}, it also indicated that more intermolecular and intramolecular hydrogen bonds existed in the polymeric system. This was caused by the

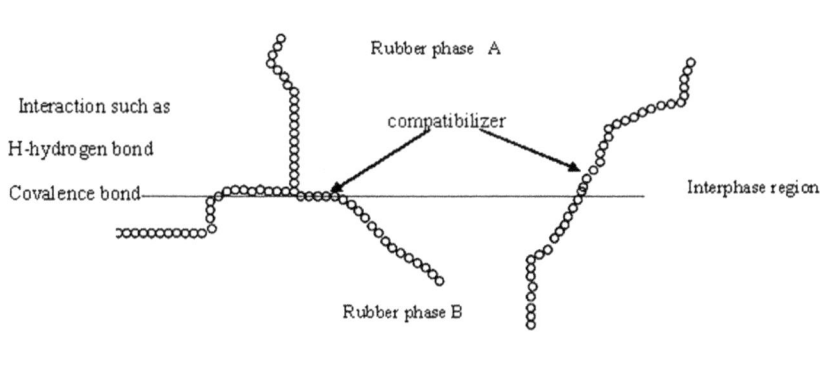

Figure 19.3 Possible model of compatibilizer use in rubber blends.

presence of ENR-50 in this compound, which provided more oxygen in the rubber blend through the oxirane group. The oxirane acts as the polar group and has a tendency to promote more hydrogen bonds with the nitrile group from NBRr, while the isoprene group from ENR-50 has good compatibility with carbon double bond from SBR. When NBRr increased from 5% to 50% in blend, ATR band of blend shifted the absorption at 1535 cm^{-1} to 1589 cm^{-1}. This strongly correlated with the rise of amine in the region from 1640 to 1560 cm^{-1} by the formation of crosslinks of NBRr with the oxirane ring opening of ENR-50 through sulfur linkage. The hydrogen bond absorption intensity of this sample at 3312 cm^{-1} also showed a slight increase. Since the amine itself has a tendency to form hydrogen bonding among oxygen and hydrogen of the polymeric structure. The total number of hydrogen bonds was also due to the formation of hydrogen bonds of hydroxyl from the oxirane ring opening reaction of ENR-50. These results was referred that ENR-50 is a compatibilizer for SBR/NBRr blend through compatible with rNBR through the oxirane group and with SBR through the isoprene group.

ENR/PEA blend[4]

FTIR was used to investigate the chemical structure of ENR and PEA and their blend and the main FTIR peak of these samples are summarized in Table 19.1. At high temperature, the epoxy group in ENR destroys into secondary alcohol, furan ring structures, aliphatic ethers and carbonyl groups. The asymmetric stretching (–C–O str) of the secondary alcohol group and intermolecular hydrogen bonding (R–O–H) are found at 1031 and 3489 cm^{-1}, respectively. The bands at 1065 and 1110 cm^{-1} are assigned the furanized ring structures and ether link, respectively. The carbonyl group in carboxylic acid group and its self-associated hydrogen bonding between two adjacent acid groups in the copolymer are located at 1074 and 3636 cm^{-1}, respectively. The ester group

Table 19.1 FTIR data of ENR and PEA.

Wave numbers (cm^{-1})	Functional group	Assignment of bands
Peaks of ENR-50		
3550-3220	Alcoholic –OH	Intermolecular hydrogen bonding R–O...H H...O–R
1140-11010	Aliphatic ether	Assym. C–O st.
1070-1060	Tetrahydrofuran rings	Ring vib.
1035-1030	Secondary alcohol	Assym. C–O st.
880-780	Cis-epoxide ring	Ring vib.
Peaks of PEA		
3650-3200	Acidic-OH	Intermolecular self-associated hydrogen bonding
1720-1700	Acidic C=O	C=O st.

from esterification reaction between ENR and PEA appeared in the region 1720–1750 cm^{-1}. It was found that the area of ester group in 30/70 ENR/PEA blend shows the highest value comparing to other samples. When the 30/70 ENR/PEA was heated at different temperatures from 120 to 160 °C exception at 150 °C, the area of the acid peak at 1704 cm^{-1} and the alcohol peak at 1031 cm^{-1} decreased as a function of temperature. But the area of ester group of esterification reaction from blends increase as a function of temperature. The area of epoxy group in blend decreased with increasing temperatures, exception at 140 °C.

NR/styrene butadiene rubber (SBR) blend[8]

FTIR have been employed extensively to monitor blend compositions from NR/SBR blend. The main bands for SBR and NR pyrolysis products are done. The main bands of NR/SBR blend corresponds to the out of plane bending vibrations of aromatic C–H and C=C groups of polystyrene at 750 and 700 cm^{-1}, respectively, and the out of plane bending vibrations of C–H of vinyl groups (990 and 910 cm^{-1}) and trans CH=CH at 960 cm^{-1} of butadiene. The thermal degradation of NR starts at temperatures between 300 and 500 °C and gives place to the formation of variable length oligomers as a consequence of polymeric chain random scissions. At high temperatures, there are many products including isoprene, dipentene and different unsaturated volatile products from thermal degradation of NR.[8] In case of PS, it degrades in such a way that monomer is considered the main degradation product together with about 30% of low molecular weight oligomers.[8] The degree of degradation of SBR is observed by FTIR. Thermal degradation of polybutadiene in SBR at temperatures close to 400 °C gives rise to the production of monomer and low molecular weight oligomer structures. Therefore, in SBR degradation in the temperature range studied, a mixture of polybutadiene and polystyrene oligomers as well as styrene and butadiene monomers will be collected. We have selected the following regions for the three components: styrene: region between 715 and 683 cm^{-1}, NR: region between 850 and 785 cm^{-1} and butadiene: region between 1010 and 855 cm^{-1}.

NR/PCR blend[9]

The effect of light and $FeCl_3 \cdot 6H_2O$ on PCR/NR blends in toluene solution were investigated to demonstrate the influence of each polymer on the degradation process. FTIR spectroscopy was used to characterize the degradation. The degradation process was also monitored by FTIR spectrum analysis of 50:50 PCP/NR blend films. The main bands of 50/50 PCP/NR blend shows at C=C band at 1660 cm^{-1}, multiple C–H and C–C bending, and stretch vibrations at 1445, 1431, 1303, 1202, 1118, and 1001 cm^{-1}, and also an asymmetric stretching and at 2960 cm^{-1} assigned to CH_3 groups. The C–Cl stretching and bending of PCP observed at 825 and 580 cm^{-1}.[10] The new bands of PCR/NR blend appear at ranges in 1510–1600 cm^{-1} and 3200–3640 cm^{-1}. This may be due to the chemical modification induced by the presence of $FeCl_3 6H_2O$ and the exposure of the parent solution to polychromatic light for 24 h. According to De Paoli and Rodrigues,[11] the broad band centred at 3300 cm^{-1} might be attributed to the formation of alcohol during the photodegradation of NR. Adam et al.[12] observed a broad band formed in the FTIR spectra of polyisoprene films irradiated with wavelength longer than 300 nm in the region of hydroxyl stretching vibrations just after light exposure. In addition, a new band at 835 cm^{-1} in the photodegradation of polyisoprene was assigned to epoxide groups. It has been reported that saturated and/or unsaturated acid chlorides are formed in the degradation of polychloroprene.[12,13] According to Delor et al.,[14] the bands at 1725 cm^{-1} could be assigned to ketone/carboxylic acid and to acid chloride at 1790 cm^{-1}. Adam et al.[12] considered that the formation of carbonyl products in the photodegradation of NR were much smaller when compared to the formation of alcohol. However, studying the photo-oxidation of polyisoprene, De Paoli et al.[11] verified the formation of an intense band at 1720 cm^{-1}. In the photodegradation of PCP by photo-Fenton process, the C_3H_5ClO moiety was characterized as a PCP degradation by-product.[9] The small band at 1800 cm^{-1} might be attributed to the formation of anhydride acid[12,15] The expected decrease in the C–Cl band (at 825 cm^{-1}) was not verified even for the film cast from the solution exposed to polychromatic light for one week, which is in opposition to our previous work.[16] This may be explained considering that NR groups (1,4 and 3,4 unsaturations) presented deformation in this region.

NR/PS foam (PSf) blend[17]

ATR-FTIR was then used for analysis of the chemical structures of the 50/50 NR/PSF blend in the presence of 10% MA and of the NR/PSf blended with 10% MA and 5% cellulose or polymer composite. Figure 19.4 shows the spectra of NR/PSf blended with 10% MA. The characteristic peaks at 3025 (C–H aromatic), 1610 (C–C aromatic), 2920 and 2849 (CH_2–CH_2), 1490 and 1450 (C_6H_5), 905 and 697 (C–H aromatic) cm^{-1} of PSf appeared in Figure 19.4(a) to (d). The bands at 1076 and 1664 cm^{-1} are attributed, respectively, to the symmetric C–S–C stretching vibrations and C=C of NR were also noticed. The grafted MA in the blend was deduced from the absorbance ratio of peaks at 1780–1784 cm^{-1} to 1835–1854 cm^{-1} (CH stretching on C=C of cis-1,4-polyisoprene) in Figure 19.4(b) (Zhang et al., 2010). The characteristic peaks at 3200–3400 cm^{-1} due to OH from cellulose fibre were observed in Figure 19.4(c) and (d). A shoulder at 1745 cm^{-1} confirmed that the small amount of ester groups remained in side chains of the polymer composite. After adding cellulose in NR/PSf blend, the intensity of peak at 3400 cm^{-1} of polymer composite decreased. In addition, the increase of bands at 1745 and 1270 cm^{-1} due to ester C=O and C–O–C of cellulose in polymer composite was detected. The peaks at 821, 896, 1046, 1382 and 2926 corresponding to the out of plane deformation of COOH, C–H deformation, C–O–C from β-1,4-glycosidic, OH bending and C–H stretching, respectively, also appeared.

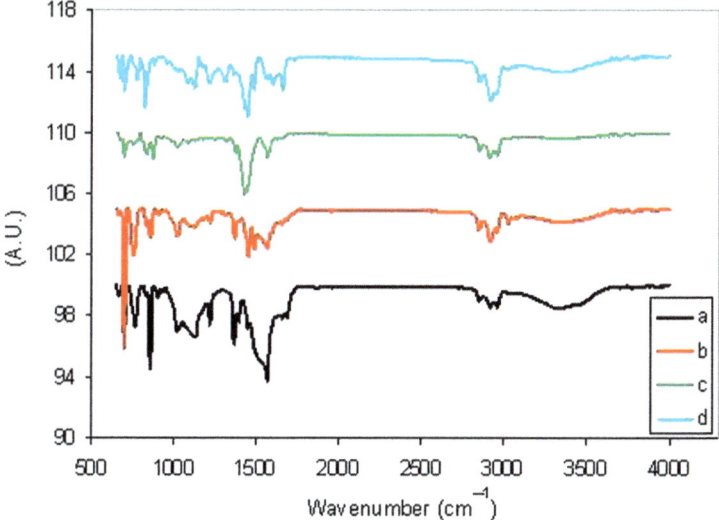

Figure 19.4 ATR-FTIR spectra of 50/50 NRG/PSF blended with 10% MA (a) before and (b) after immersion in toluene and polymer composite containing 50/50 NRG/PSF, 10% MA and 5% cellulose (c) before and (d) after immersion in toluene (our work).

NR/PS blend[18]

An intense characteristic band at 1778 cm^{-1} and a weak absorption band at 1854 cm^{-1} are observed. These bands can be assigned to grafted anhydride rings. They are due to symmetric (strong) and asymmetric (weak) carbonyl (C=O) stretching vibrations of succinic anhydride rings grafted on PS and NR. This proves the presence of grafted anhydride groups on the NR and PS chains and the possible chemical reaction between MA and PS as shown in Figure 19.5. The small shoulder at 1718 cm^{-1} (the carbonyl stretching frequency of the acid group) in the spectrum of the grafted blend arises from the acid peak resulting from MA modification. Therefore, the possible functionalization reaction of PS and NR with MA was expected. In general, the anhydride modification may result in anhydride or acid forms depending upon the addition or removal of water. Normal carbonyl group stretching frequency is 1760 cm^{-1}. According to literature,[18] the acid peak is indicated by 1714 cm^{-1} whereas the anhydride shows the absorption peaks at wave numbers of 1789 cm^{-1} (strong) and 1864 cm^{-1} (weak). We can also notice the peaks corresponding to absorbances at 1652 and 832 cm^{-1} due to the stretching of the C=C bond of the NR. The peak at 3026 cm^{-1} and 2868 cm^{-1} (not shown here) were assigned to the aromatic C–H stretching of polystyrene. These peaks corresponding to C–H stretching are slightly shifted and the intensities are reduced on addition of MA. The signals at 1375 cm^{-1} and 1446 cm^{-1} can be attributed to the aliphatic C–H stretching in natural rubber. The band at 1260 cm^{-1} corresponds to the C–O–C stretching in the ester moiety. In addition,

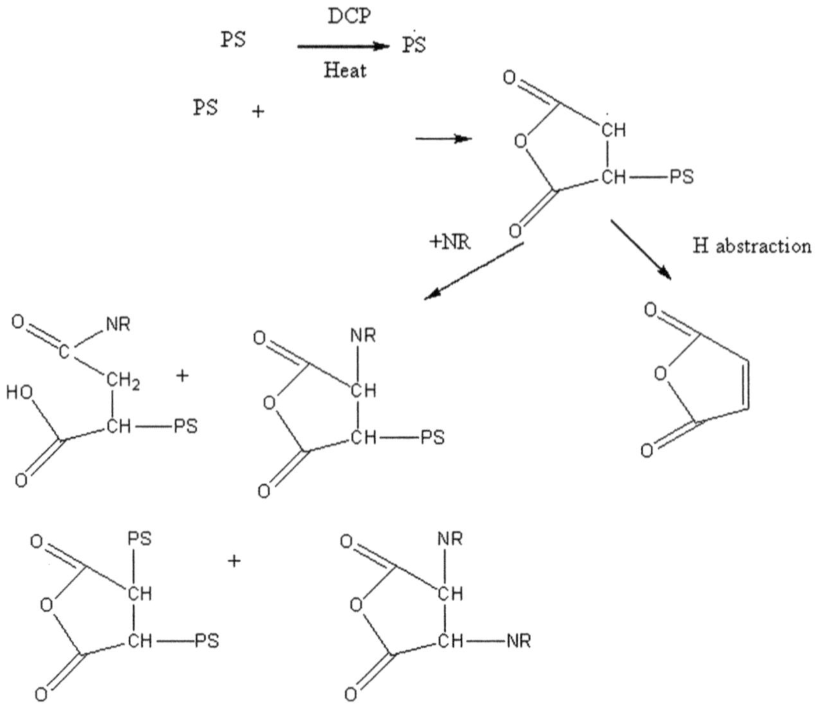

Figure 19.5 Possible grafting reactions of EPS and NR with MA (modified).[18]

the addition of silica in NR/PS blend was improved theirs mechanical.[18] The grafting between PS and NR initially grafted with MA using dicumyl peroxide (DCP) was confirmed by FTIR. The possible reaction between NR/PS and MA in presence of DCP and MA was presented in Figure 19.5. The DCP received heat and it give free radical form. At the same time, free radical DCP reacted with carbon– carbon of PS. Then, the PS free radical continued to react with NR molecular leading to the grafting copolymer through both anhydride linkage and opened ring linkage as shown in Figure 19.5. In addition, the good many other reactions arising from the presence of one/two NR/PS chains are occurred. Results show that an intense characteristic band at 1778 cm^{-1} and a weak absorption band at 1854 cm^{-1} are observed. These bands can be assigned to grafted anhydride rings due to symmetric (strong) and asymmetric (weak) carbonyl (C=O) stretching vibrations of succinic anhydride rings grafted on PS (polystyrene) and NR. This proves the presence of grafted anhydride groups on the NR and PS chains. The small shoulder at 1718 cm^{-1} (the carbonyl stretching frequency of the acid group) in the spectrum of the grafted blend arises from the acid peak resulting from MA modification. In general, the anhydride modification may result in anhydride or acid forms depending upon the addition or removal of water. Normal carbonyl group stretching frequency is 1760 cm^{-1}.[18] According to literature,[18] the acid peak is indicated by 1714 cm^{-1} whereas the anhydride shows the absorption peaks at wave numbers of

1789 cm^{-1} (strong) and 1864 cm^{-1} (weak).[19] We can also notice the peaks corresponding to absorbances at 1652 and 832 cm^{-1} due to the stretching of the C=C bond of the NR. The peak at 3026 cm^{-1} and 2868 cm^{-1} (not shown here) were assigned to the aromatic C–H stretching of polystyrene. These peaks corresponding to C–H stretching are slightly shifted and the intensities are reduced on addition of MA. The signals at 1375 cm^{-1} and 1446 cm^{-1} can be attributed to the aliphatic C–H stretching in natural rubber. The band at 1260 cm^{-1} corresponds to the C–O–C stretching in the ester moiety.

Chlorosulfonated rubber (CSM)/NR blend[20]

The compounds based on NR/CSM blend and butadiene acrylonitrile rubber (NBR)/CSM blend (50:50, w/w) with different loadings having 0, 20, 40, 50, 60, 80 and 100 phr of the filler with the average particle size of 40 nm were cured by sulfur under radiation doses (100, 200, 300 and 400 kGy) in the presence of oxygen. By using ATR-FTIR it was assessed that after exposure to doses of 100 kGy alcohols, ethers, lactones, anhydrides, esters and carboxylic acids are formed in materials. The formation of shorter polyene sequences and aromatic rings in aged samples are assumed on the basis of the obtained spectra. The FTIR spectra of aged samples confirmed the formation of various oxidation products, *i.e.* alcohols, ethers and small amounts of lactones, anhydrides, esters and carboxylic acids during radiation. Significant changes in spectra are induced by chain scission which was registered in the 1620–1450 cm^{-1} region. Gamma radiation leads to significant changes in the FTIR spectra of carbon black reinforced elastomers based on NBR/CSM and NR/CSM rubber blend. The broad increase in absorption can be found in C=O, O–H, and C–O stretching vibration domains for both blends. Significant changes appeared in the region of conjugated double bonds, too. Several oxidation products contribute to band at the hydroxyl region, *i.e.* alcohols (band at 1028–1075 cm^{-1}), ethers and small amounts of lactones, anhydrides, esters and carboxylic acids. Due to the formation of oxidation products, the bands related to CH_2, CH_3, =C–H and C=C vibrations decrease. CH_2 deformation band at 1458 cm^{-1}, CH_3 asymmetric deformation at 1372 cm^{-1}, and =C–H wagging at 831 cm^{-1} show a significant increase in a decrease in absorption at 1660 cm^{-1} is related to the loss of C=C bonds in 1,4-*cis* units. 1,4-*cis* units also absorb at 831 cm^{-1}. Besides the oxidation of C=C bonds in 1,4-*cis* units, *cis–trans* the isomerization can take place. The band at 2956 cm^{-1} related to methyl groups decreases, especially at the highest doses 200, 300 and 400 kGy. In thermally degraded NR/CSM and NBR/CSM rubber blends there have been identified di pentene, two isomeric forms of dimethylvinylcyclohexene, isoprene, and some aromatic compounds formed *via* the Diels–Alder reactions. Polyenes and polyenals show several bands in the 1500–1650 cm^{-1} range. Benzene C=C stretching vibrations bands are located between 1600–1450 cm^{-1}; hydrogen bond can cause the reduction of the stretching vibration frequency. The subtraction spectra reveal the occurrence of the broad band with a significant increase in absorption at 1470 cm^{-1}.

NR/cassava starch (St) blend[21]

First, the reduction in viscosity of St was subjected to the addition of $K_2S_2O_8$ at 60 °C for 45 min. $K_2S_2O_8$ was activated by heat and it changes into $K_2S_2O_8$ free radical. Then, the $K_2S_2O_8$ free radical reacted with St molecule. The degradation of St was occurred as shown in Figure 19.6. Finally, we acquired the short St chain. The viscosity of St dramatically decreased after addition of $K_2S_2O_8$ at 75 °C for 45 min to get the modified St. The free radicals from St attacked with carbon–carbon double bonds of NR which was activated by $K_2S_2O_8$ leading to the NR-g-St formation. At the same time, free radical $K_2S_2O_8$ reacted with carbon double of NR. Then, the St free radical continued to react with NR molecular leading to the NR-g-St formation as shown in Figure 19.6. The synthesized copolymers were characterized by the functional groups using ATR-FTIR technique. ATR-FTIR spectra of unmodified St and modified St are presented in Figure 19.7(a). The peak band at 1730 cm^{-1} corresponds to the carbonyl group vibration band and the absorbent band of hydroxyl group was observed at 3340 cm^{-1}. The absorption at 879 cm^{-1} was referred to a methane vibration and absorption at 1638 cm^{-1} is C–O vibrations in St. The vibrations of unsymmetry C–O–C bond and C–O bond in St are at 1154 cm^{-1} and 1045 cm^{-1}, respectively. After chemical modification of the St with $K_2S_2O_8$, the FTIR peak band was slightly altered as shown Figure 19.7(a). The FTIR spectra for St, NR-g-St before and after THF extractions for 48 h are presented in Figure 19.7(b). The results showed that the main FTIR band of St was observed at 3550–3200, 2933 cm^{-1} and the two band of 1155, and 1080 cm^{-1} for hydroxyl group, the C–H stretching and the C–O–C stretching from a triplet peak of St, respectively. The C–H stretching at 2965, 2928 cm^{-1}, the C=CH stretching at 1659 cm^{-1}, the –CH_3 deformation at 1452 cm^{-1}, the –CH_2 deformation at 1379 cm^{-1} and the C=CH bending at 843 cm^{-1} of NR-g-St appeared in the FTIR spectra (Figure 19.7(b)). The chemical structure of NR-g-St was confirmed again by both TGA and FTIR. After NR-g-St was heated from 50 to 330 °C by TGA technique, then its chemical structure was continually characterized by FTIR. It was clear that the notable difference between the FTIR spectra of St and NR-g-St was that the strongest peak for NR-g-St appeared at 1100 cm^{-1} (C–O–C) comparing to St. This data indicated that NR was grafted with St. In addition, the thermal stability of NR-g-St was the highest comparing other samples (see in thermal behaviour). They studied the preparation of thermoplastic starch foams from pristine cassava starch blended with NR latex by reactive blending and potassium persulfate as an initiator for graft copolymerization between the St and NR during baking. The chemical structure of NR-g-St was confirmed by ^{1}H-NMR and FTIR characterization. In case of FTIR, it was found a trace of an NR component in NR-g-St (this subtracted spectrum before and after Soxhlet extraction) as appeared at 2926–2854 cm^{-1}.

NR/chitosan (CST) blend

The synthesized copolymers from ENR and CTS were characterized by the functional groups using ATR-FTIR technique. The major characteristic peaks

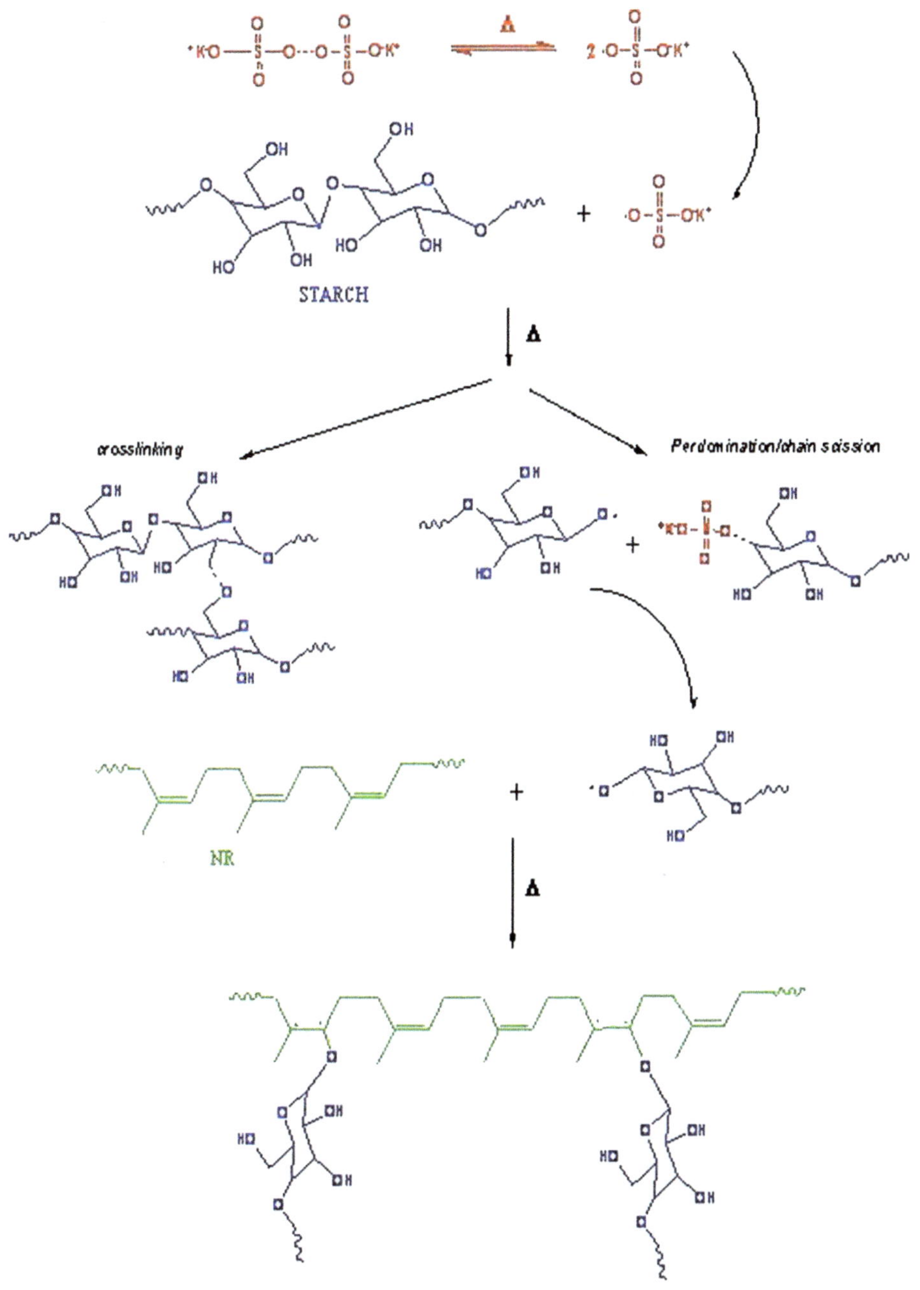

NR-g-STARCH

Figure 19.6 Possible reaction of NR grafting with starch in latex form (our work).

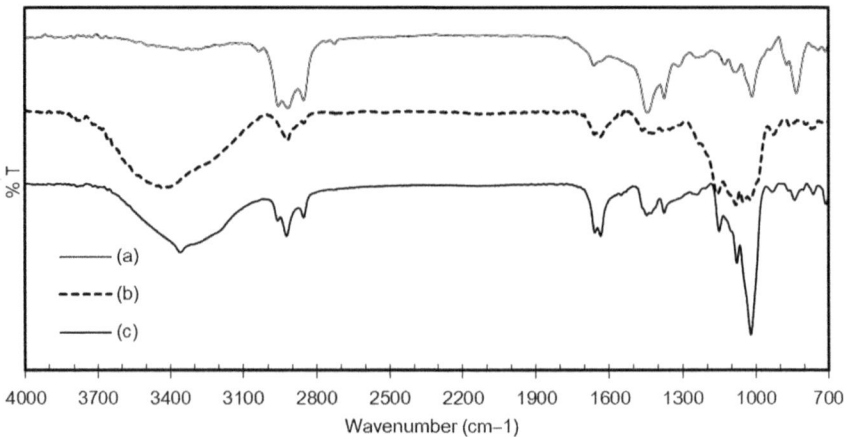

Figure 19.7 ATR-FTIR spectra of (a) NR, (b) St and (c) NR-g-CST (our work).

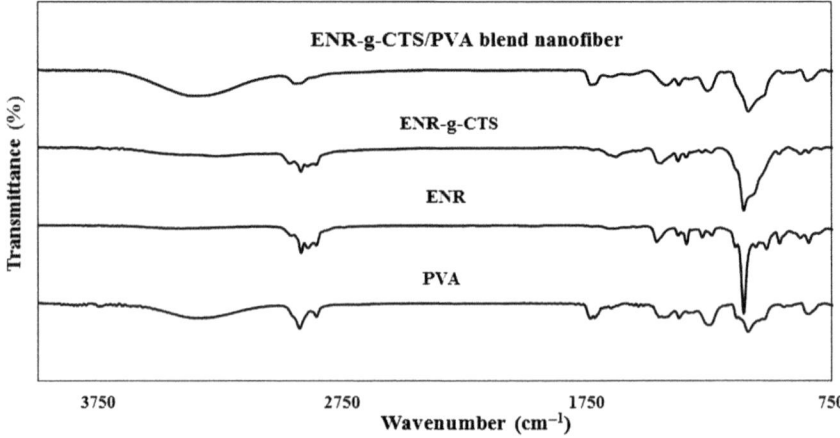

Figure 19.8 ATR-FTIR spectra of PVA, ENR, ENR-g-CTS and ENR-g-CTS/PVA blend (our work).

of CTS were found at 3450 cm^{-1} (O–H stretch), 2875 cm^{-1} (C–H stretch), 1650 cm^{-1} (C=O stretch of carbonyl group), 1375 cm^{-1} (C–H stretch of methyl group), 1155 cm^{-1} (bridge O stretch), and at 1092 cm^{-1} (C–O stretch) as shown in Figure 19.8. In case of ENR, the absorptions at 876 and 1252 cm^{-1} can be attributed to the epoxy group, and the weak absorption at 1735 cm^{-1} can be assigned to the carbonyl group of the ester. The wave number of carbon–carbon double bond (C=C) double bond was found at 1664 cm^{-1}. After adding CTS in ENR molecule, the chemical structure of ENR was changed. It was found that the new peak at 1154, 1089 and 3400 cm^{-1}. These new bands revealed that ENR was grafted successfully with CTS.

NR/PVA blend[22]

The possible reaction between PVA and MA was presented in Figure 19.9. The FTIR main bands of PVA were observed at 3200–3500 cm^{-1} and 2934–2904 cm^{-1} which were assigned to O–H stretching and C–H stretching as shown in Figure 19.9, respectively. The polyvinyl acetate residue in PVA was appeased at 1707 and 1096 cm^{-1}. In addition, the peaks at 1378 and 1328 are referred at –CH$_2$– wagging and –C–H– / –O–H– bending, respectively. The main peaks of PVA are presented in Table 19.2. After adding maleic acid in NR/PVA blend, the ester linkage of PVA crosslinked with maleic acid as shown in Figure 19.10 was confirmed at 1674 cm^{-1}.[22,23]

In addition, the C–O–C stretching of maleic acid ester in polymer blend was observed at 1152 cm^{-1}. The intensity of 3200–3500 cm^{-1} (O–H stretching of PVA) and 1293 cm^{-1} (O–H in-plane bending of PVA) dramatically decreased. This might be due to hydroxyl group from PVA crosslinked with maleic acid.

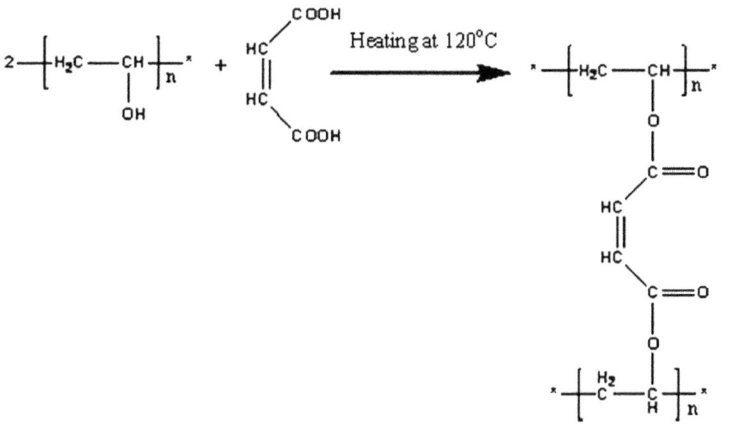

Figure 19.9 Possible crosslinking reaction between PVA and maleic acid.

Table 19.2 FTIR bands of PVA and PVA-M.[23]

Wave number (cm^{-1})		Peak assignments
PVA	PVA-M	
1293	–	O–H in-plane bending of PVA
1069	1064	C–O stretching and O–H in-plane bending vibrations coupled
997	998	C–O stretching and O–H in-plane bending vibrations coupled
3290	3295	O–H stretching of PVA
–	1152	C–O–C stretching of maleic acid ester
–	1663	C=C stretching of maleic acid ester
2967	2968	CH$_2$-symmetric stretching RCH2R'
2864	2864	C–H stretching (isolated) >C(OH)–H

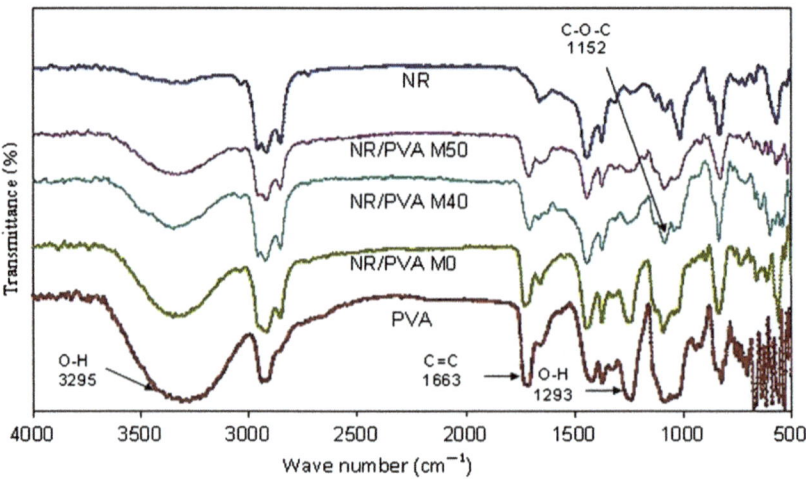

Figure 19.10 FTIR spectra of NR, PVA and NR/PVA blend with different maleic acid contents at 0, 40 and 50% (our work).

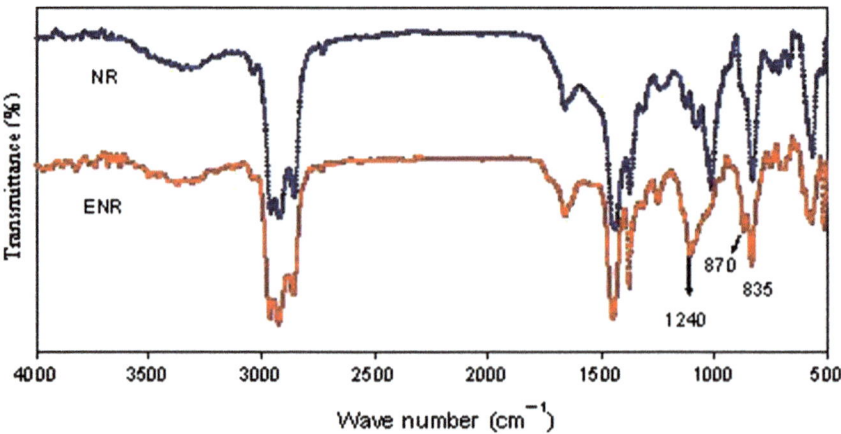

Figure 19.11 ATR-FTIR of NR and ENR (our work).

ENR/PVA blend

The semi-interpenetrating polymer network of NR and PVA containing maleic acid was prepared by the solution-latex method. The influence of the maleic acid content on the properties of the rubber blend was studied. The presence of a large number of hydroxyl groups in PVA resulting from the semi-interpenetration between NR and PVA and strong hydrogen bonding (perhaps both the intermolecular and intramolecular types), will affect the solubility of PVA in water. The FT-IR spectrum of PVA alone, ENR alone, the ENR/PVA blend and semi-IPN samples are shown in Figure 19.11. Maleic acid treatment of PVA produces intermediate heat stability. It is possible that in this

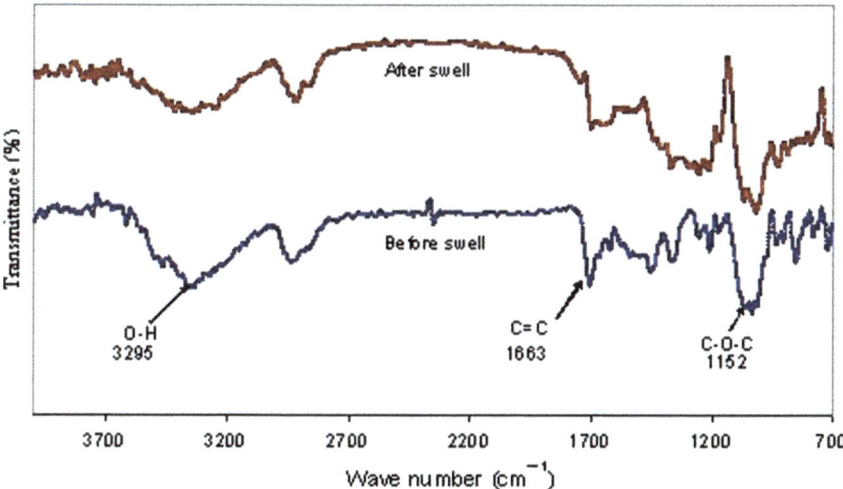

Figure 19.12 ATR-FTIR spectra of ENR/PVA blend in presence of maleic acid before and after swelling in water (our work).

temperature zone the total energy input is inadequate to break the double bond structure (–C=C–) of the heat-treated samples but it is sufficient to start the degradation of the –C–O– bond of the maleic acid treated PVA in the semi-IPN samples as shown in Figure 19.12. The ester linkage of the semi-interpenetrating sample is confirmed by FTIR. The absorption bands of the cis-1,4 polyisoprene in ENR is found at 2860 cm^{-1} (C–H stretching), 1665 cm^{-1} (C=C stretching), 1453 cm^{-1} (–CH_2– deformation), 1378 cm^{-1} and (methyl C–H deformation).

The absorptions at 876 cm^{-1} and 1252 cm^{-1} can be attributed to the epoxy group, and the weak absorption at 1735 cm^{-1} can be assigned to the carbonyl group of the ester. The main peaks of PVA alone shown at 1327 and 843, and 1087 cm^{-1} are attributed to C–H bending and C–O stretching, respectively. The changes of the characteristic spectra peaks reflect the chemical interactions when two or more substances are blended. In the typical spectrum of the semi-IPN sample sheet, the characteristic peak at 1729 cm^{-1} was shifted to 1730 cm^{-1}. This document indicates that there are hydrogen bonded interactions between the hydroxyl groups, carbonyl groups of PVA and carbonyl groups of maleic acid and the epoxy groups of ENR that helped them to be compatible.

ENR/gelatin blend

The main bands of ENR are C=C for cis-1,4 polyisoprene at 835 cm^{-1} and epoxy group at 875 cm^{-1} as shown in Figure 19.13. In addition, the band centred at 3384 cm^{-1} in ENR spectra plausibly from absorbed water in the sample. In case of gelatin, there were a strong absorption bands situated at about 3275, 1630 and 1540 cm^{-1}, referring to amide-III and free water, amide-II

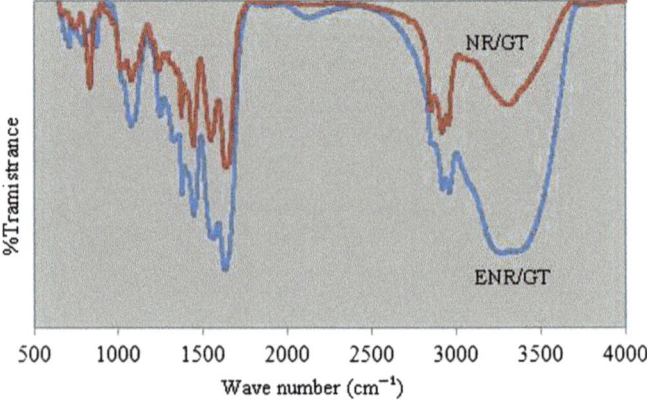

Figure 19.13 FTIR spectra of the NR/GT blend and ENR/GT blend (our work).

and amide-I, respectively.[24] The amide-I arises from stretching of C=O of amide in protein; the amide-II arises from bending vibration of N–H group and stretching vibration of C–N group while amide-III is related to the vibrations in plane of C–N and N–H groups of bound amide or vibrations of CH_2 group of glycine.[25] After adding ENR in gelatin, the FTIR band of gelatin shifts in the amide-I, amide-II and amide-III bands as well as the broader amide-III peak comparing gelatin alone. These results was referred that the presence of chemical interactions between gelatin and ENR molecules in the polymer matrix. The FTIR band of NR/GT blend was similar with ENR/GT blend. But the intensity of ENR/GT blend shows broader than NR/GT blend due to their chemical interaction between epoxy group from ENR and amide group of GT.

19.4 Nuclear Magnetic Resonance (NMR) Spectroscopy

19.4.1 Introduction to NMR Spectroscopy

NMR spectroscopy is one of the most powerful techniques for the structural and physicochemical study of organic compounds, both small molecules and polymers. It seems to be highly suitable for studying NR-based blends and IPNs. Different NMR techniques have been used to study NR-based blends and IPNs, including ^{13}C NMR, and ^{1}H NMR liquid state. However, only ^{13}C solid-state NMR can be used to study cured rubber, as this technique does not require the solubilization of the polymer.

19.4.2 Sample Preparation and Typical Conditions for NMR

Swollen state ^{1}H-NMR[26]

A specimen for the swollen state ^{1}H-NMR was carefully prepared by randomly sampling the specimen from different areas of the vulcanized sheet. In addition, sampling mass was controlled at 50 mg. The sample was allowed to swell to an equilibrium state in CDCl$_3$ for 5 days in the dark at 25 °C. A swollen state ^{1}H-NMR analysis was conducted by using an NMR spectroscopy.

Solid-state ^{13}C NMR spectroscopy[27]

The impact bar specimen of each sample was scrubbed into a powder form by using a flat file. The samples were then packed into the ZrO$_2$ rotor (4 mm in diameter). Solid-state ^{13}C NMR experiments were performed at 75.47 MHz at room temperature with a BRUCKER AVANCE 300 MHz wide bore spectrometer using a cross polarization (CP)/magic angle spinning (MAS) probe. Single pulse experiments combining MAS and high power proton decoupling were carried out using a pulse width of 90 (4 ms) with a repetition time of 4 s. A spectral width of 25 kHz and 16 K data points was used for data collection. The MAS frequency was 10 kHz. The decoupling radio frequency was 60 kHz. The CP/MAS experiments were done to observe the dynamics of the PET and PET blends from the point of view of ^{13}C atoms with variable contact times. For the present research, a rotor spinning frequency of 7 kHz was used. The ^{13}C CP magnetization curves versus contact time of each carbon of the samples were built up from the line intensities and was fitted using Origin 7.5. High resolution solid state NMR experiments were carried out at 100 MHz for C-13 using Varian Unity 400 Spectrometer. The instrument is equipped with a high-power amplifier for proton decoupling and a varian VT probe. NMR results were externally referenced to the aromatic peak of hexamethyl benzene (132.2 ppm). Samples were packed in silicon nitride rotors with Torlon end caps and were spun at a speed of 5000 Hz. Spinning speed and sample weight were carefully controlled to obtain reproducibility of the results. The sample temperature was maintained at room temperature. The ^{13}C-NMR spectrum of adamantine sodium sulfite is shown in Figure 19.14 and it gave the signal at 29.52 (methylene) and 38.54 (methane).

19.4.3 Analysis of Polymer Blends

NR/poly(ethylene terephthalate) (PET) blend[27]

An investigation of blend morphology as well as the molecular characteristic of the blend component would result in a better understanding of the compatibilization behaviours which affect the mechanical properties of the blends. In this study, solid-state CP/MAS ^{13}C NMR was used for this purpose. The PET and its blend are solid materials and are insoluble at room temperature in organic solvent. Then analysis of the molecular characteristic in solid form can be done by solid-state ^{13}C NMR spectroscopy. In this study, the molecular characteristics of PET and PET/NR blend were investigated by using a solid-state ^{13}C NMR with MAS and CP/MAS instrument. The solid-state CP/MAS ^{13}C NMR spectroscopy revealed that the interaction between the PET and the NR occurred though the observation of an increase in cross polarization time (TD) of the carbonyl carbon and a decrease of TH 1r relaxation of the carbonyl

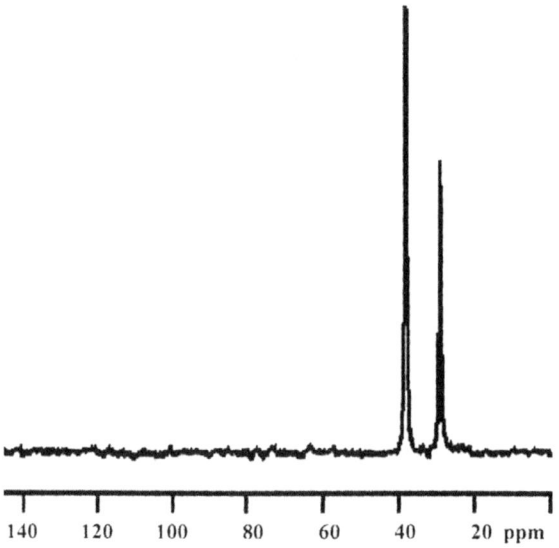

Figure 19.14 ^{13}C-solid state NMR spectrum of adamantine sodium sulfite (our work).

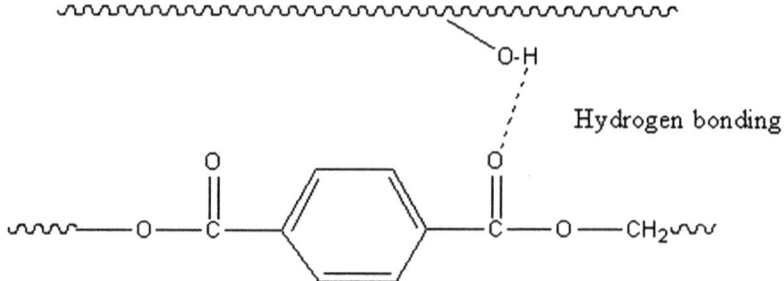

Figure 19.15 Possible interaction *via* hydrogen bonding between the carbonyl group of PET with abnormal groups such as hydroxyl function in NR.

group in the PET/NR blend. This should come from the interaction between the carbonyl group of PET with the abnormal groups such as hydroxyl function in NR, resulting in improving the compatibility of the studied blends, hence increasing the toughness as shown in Figure 19.15.

ENR/PVA blend

Figure 19.16 shows the spectrum of ^{13}C NMR of PVA, ENR/PVA blend with or without maleic acid. The carbon signal of PVA appeased at 45.0 ppm due to CH_2 in PVA. The ^{13}C signals at 63.0 70.0 and 76.0 ppm are assigned to CH–OH

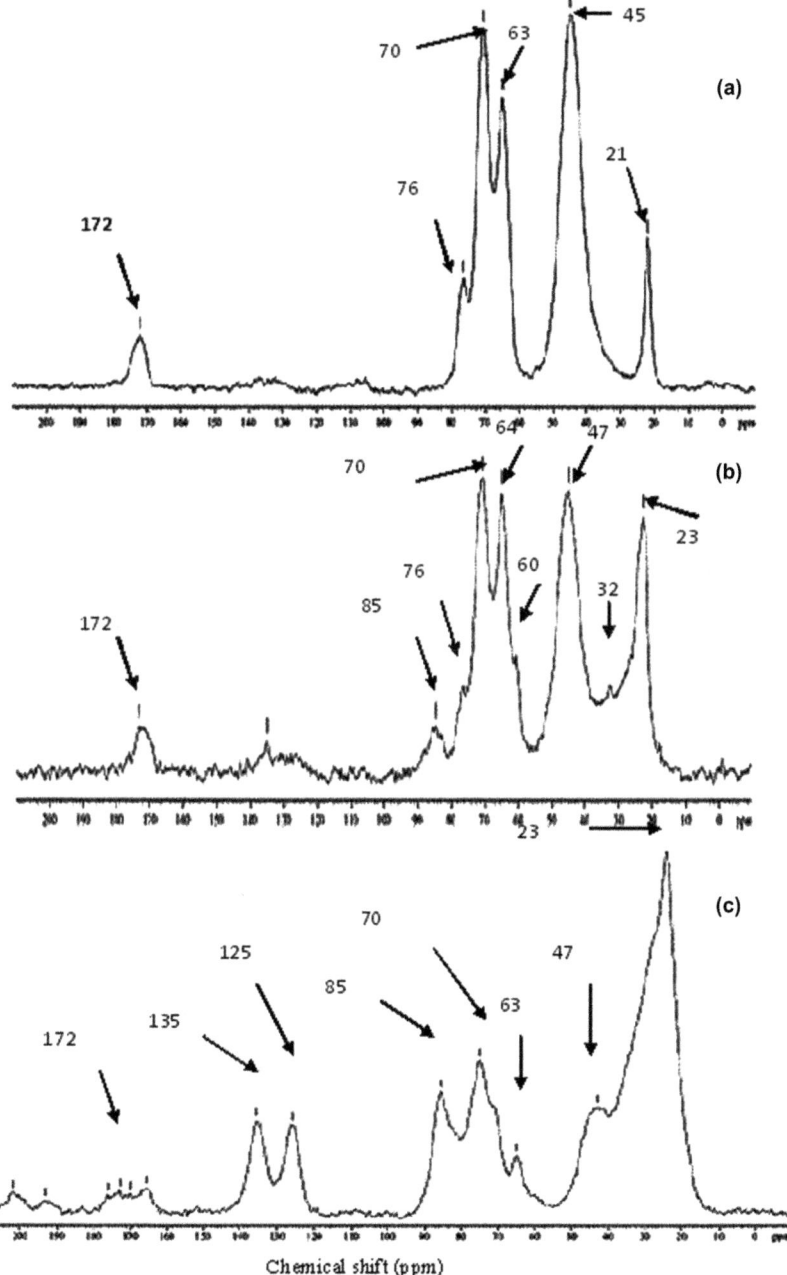

Figure 19.16 Solid-state ^{13}C NMR spectra of (a) PVA, (b) ENR/PVA blend and (c) ENR/PVA blend with maleic acid (our work).

and their hydrogel bonding (Saito et al., 2007). In addition, the carboxylic group in PVA was observed at 172.0 ppm. In case of ENR/PVA blend, the ^{13}C signal of ENR was found at 23.0 ppm referring to CH_3 and two signals such as 32.0 and 47.0 ppm assigned to CH_2. The ^{13}C signal of epoxy group in ENR shows at 60.0 and 64.0 ppm. The ^{13}C signal of PVA in blend was not change comparing to PVA alone. After adding MA in blend, the new functional group of blend was observed at 85 ppm. This may be due to ester linkage from PVA and MA. In addition, the intensity of ^{13}C-NMR signal at 63 and 70.0 ppm in PVA dramatically decreased due to chemical reaction between PVA and ENR. The main ^{13}C-NMR signals of PVA, ENR and maleic acid are summarized in Table 19.3.

Table 19.3 ^{13}C-NMR signal of PVA, ENR and maleic acid.

Component	Signal (ppm)	Assignment
PVA (Saito et al., 2007)	45	CH_2
	63, 70, 76	CH-OH Carbon, several peaks due to inter and/or intramolecular hydrogen bonding
	172	—O—C(=O)—CH_3
	23	CH_3 alkyl peak
	26.2, 32, 47	CH_2
ENR (IIczyszyn et al., 2003)	60	CH_3-C(-O-)-CH- (epoxide)
	64	CH_3-C(-O-)-CH- (epoxide)
	125	=CH-
	135	Quaternary carbon of natural rubber
	133	-CH=
	140	-CH=
	170	OH—C(=O)—
	173	OH—C(=O)—
Maleic acid (Jayaseksekara et al., 2004)	85	—O—C(=O)—CH_3
Eter-linkages		

NR/nitrile rubber (NBR) blend[28]

Lewan et al.[28] investigated the effects of accelerator types on crosslinking distribution in NR and NBR blends observing from a swollen state ^{1}H NMR method. It was found that the NR phase had a higher crosslink density than the NBR phase containing N-cyclohexyl-2-benzothiazole sulfonamide. But, by adding a secondary accelerator, crosslink density in the NBR phase became greater than that in the NR phase. In the case of using CBS alone, the type of formulation (efficient, semi-efficient or conventional), type of accelerator, extent of cure, and different concentration of ingredients (sulfur, activator) was revealed by solid state NMR.[29] Solid-state NMR was also used to evaluate different parameters in to identify elastomers in the tertiary blend of NR/SBR/BR.[29]

NR/St blend

The ^{13}C solid state NMR spectra of the St and NR/St blend are presented in Figure 19.17. In case of St, the main signals ^{13}C including glycosidic carbon C1, C4, C2, (72.2 and C6 in St show 101.2, 81.4, 72.2 and 61.9 ppm, respectively. The signal of C4 carbons is described to be characteristic for the amorphous glassy phase. This may be due to not detected in hydrated, highly crystalline St samples. In addition, the relative integrated intensity of the C4 signal at 81.4 ppm is smaller than the total integrated intensity of the whole spectrum. This result indicated that additional C4 signal is overlapping with the signal f C2, 3, 5 carbons around 72 ppm. In case of NR/St blend, ^{13}C solid state NMR spectra of NR/St blend are virtually the same as in respective native samples and no significant changes indicating the mutual interaction were found. The ^{13}C-NMR signals of carbon–carbon double bonds in isoprene of NR in blend shows 135 and 125 ppm. Three signal characteristics of methyl and two methylene carbons of saturated unit are detected at 23 and 26 and 32 ppm, respectively.

NR/polyacrylic acid (ACM) blend[26]

The mobility of the rubber molecules in NR/ACM blend was investigate by using ^{1}H-NMR observing from NMR peak width which increased with the crosslink density determined by another technique, NR vulcanizates and ACM vulcanizates with a variety of sulfur contents.[26] The swollen state ^{1}H-NMR technique may be used to investigate the crosslinking of each phase in the rubber blends from NR and ACM. In addition, the peak width of ^{1}H-NMR also changed with the amount of sampling mass and therefore, the optimum sampling mass was found to be 50 mg. Moreover, the polymer blend sample was randomly sampled from different parts of the sample sheet to enhance the accuracy of the data. Two characteristic peaks located at 5.15 and 4.15 ppm with good resolution can be seen. Two peaks represent olefinic protons in NR molecules and alkoxy protons (—OCH$_2$—) in ACM molecules,

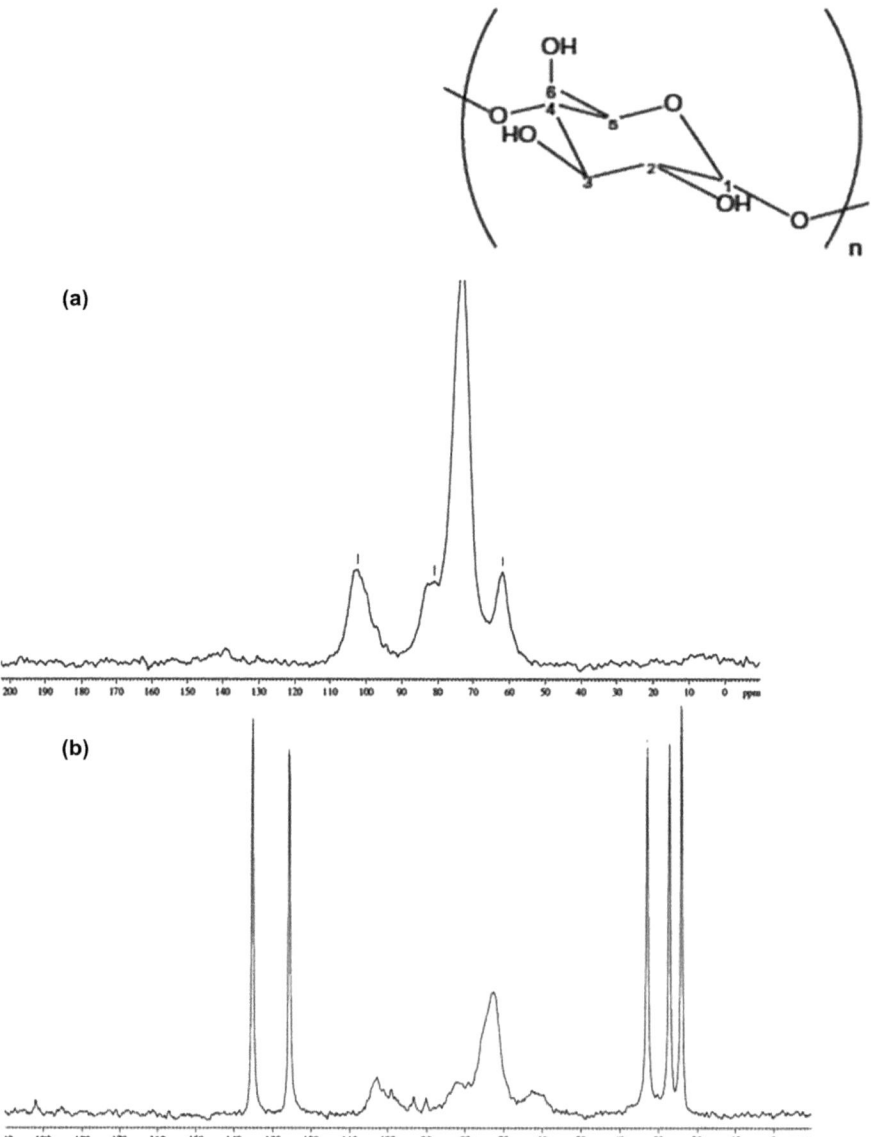

Figure 19.17 Solid-state ^{13}C NMR spectra of (a) St and (b) NR/St blend (our work).

respectively. The percentage of H in each phase in the rubber blends can be evaluated by using Equation (19.1), without conducting a deconvolution:

$$H(\%) = (b/a) \times 100 \qquad (19.1)$$

where b is the width of a particular peak was given as the signal strength at a reference point on the side of the peak, a is a percentage of the peak signal strength (a).

The offset of the reference line (b) was +0.15 ppm for the olefinic peak of NR and −0.15 ppm for the alkoxy peak of ACM. In addition, it is worth mentioning that the NMR experiment was repeated and differences in the H (%) values were found to be less than 1.0%. The swollen state ^{1}H-NMR experiment was repeated two or three times and the average values of H (%) were reported for each type of rubber blend. The width of the ^{1}H-NMR peak at 5.15 ppm in blend raised with the amount of sulfur (for a given sulfur/MBT ratio) was investigated. The width of the ^{1}H-NMR peak at 4.15 ppm in blend also increased as a function of sulfur content (for a given sulfur/soap ratio). The H (%) from ^{1}H-NMR peak at 5.15 ppm of sample containing sulfur content at 1.0, 1.5, 2.0, 2.5 and 3.0 was 8.82, 12.02, 14.28, 16.85 and 30.00, respectively. The trends observing from ^{1}H-NMR were also in good agreement with changes in torque analysing from moving die rheometer (MDR), increased. Two characteristic peaks located at 5.15 and 4.15 ppm with good resolution can be seen. The effect of mixing scheme on the distribution of crosslinks in the blend was investigated. Three different mixing schemes (1, 2 and 3) were described in detail as follows. In the case of scheme 1, the masticated NR and ACM were blended on a two-roll mill and then vulcanizing agents were added. The sample was then cooled prior to further mixing with sulfur for 3 min. The masticated NR and the ACM were premixed with the relevant chemicals (*i.e.* the ZnO, stearic acid, MBT for NR and the sodium stearate for ACM) prior to blending the two compounded rubbers. Sulfur was added at the final stage to avoid pre-vulcanization (scorching). In the final case, the sulfur was premixed with the ACM compound prior to blending with the NR compound. Again, the time and sequence for mixing other vulcanizing agent were carefully controlled with respect to that in the case of mixing scheme 2. All compounded blends were characterized by using the MDR, prior to vulcanization in the hydraulic hot press at 170 °C to their optimum cure time. Results show that value of H (%) of ACM phase in the blends from mixing schemes 2 and 3 were higher than that for the blend from mixing scheme 1. This means that by mixing the accelerator and activator with NR prior to blending, mobility of the ACM molecules decreased. In addition, the values of H (%) of the blend which experienced mixing schemes 2 and 3 were very close, suggesting that the mixing of sulfur with ACM compound prior to blending did not reduce the mobility of the ACM molecules. These results seem to be in a good agreement with those obtained from differential swelling and may be explained in the same fashion. It is interesting that the H (%) value of the NR phase significantly increased when the sulfur was mixed with the ACM compound prior to blending. Again, the NMR experiment was repeated and a similar result was observed. This result was not in good agreement with that obtained from the differential swelling which showed that the volume fraction of the NR phase did not significantly change with the mixing scheme. In general, the NR phase in the blend prepared by mixing scheme 3 should contain a lower amount of sulfur, taking into account the fact that sulfur was not directly compounded with NR. Therefore, crosslink density in the NR

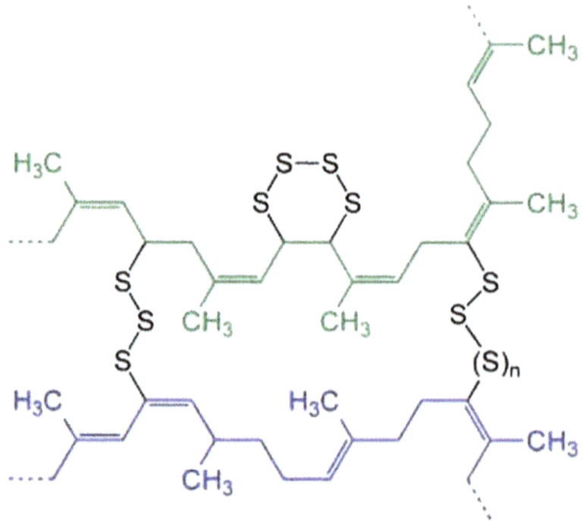

Figure 19.18 Possible crosslinking of NR with sulphur (http://en.wikipedia.org/wiki/Vulcanization, 2012).

phase should be decreased rather than increased as was implied by the NMR technique. But, the crosslinking characteristic was also altered in terms of relative amounts of mono-, di- and polysulfidic bonds in the blend. The possible sulfur crosslinking of NR is presented in Figure 19.18.

NR/polyurethane blend[30]

The solid state NMR was used to reveal the dynamics of NR/polyurethane (PU) blends. The NR and PU constituents largely occupy separated domains scale of >10 nm which was confirmed by the NMR spectra and relaxation properties. In addition, there is only a small degree of interpenetration of the two polymers analysed by NMR technique.

PVC/ENR50 blends[31]

The ^{13}C NMR is a technique to investigate the interaction between PVA and ENR50 and its degradation behaviour. ENR50 shows a peak at 60 ppm due to the methine and quaternary carbons in the epoxy group. The signals at 126 and 135 ppm correspond to the olefin double bond carbons in the unmodified isoprene units of ENR. After ENR was blended with PVC, the signals at 58 ± 60 ppm of PVC and ENR50 overlap in the spectrum of the blend. The ^{13}C-NMR spectra of the PVC/ENR blends after aging at 80 °C for 7 days changed comparing to the PVC/NBR blends before aging and it was found the new peaks at 70 ± 90 ppm. When temperature of aging increased from 80 to 140 °C, the broadened signals of new peak in blends were observed. In addition, PVC generates HCl at high temperatures. The new peaks appear in the region 70 ± 90 ppm due to furan ring carbons and ether linkage carbons and carbons attached to hydroxy groups.[32] In case of ENR, the epoxy group ring opens during degradation and HCl from PVC will react with the epoxy groups of the

ENR. PVC and ENR are crosslinked during high temperature moldings.[33] A proton solution state NMR of this sample shows only very small peaks at 3.4 ± 3.7 ppm corresponding to the furan rings.

PVC/ENR50/NBR blends[31]

The alkyl signals at 23 ppm (ENR carbons), 32 ppm (overlap of ENR and NBR carbons), 47 ppm (PVC methylene carbon) and 60 ppm (overlap of PVC methine carbon and ENR epoxy carbons) are observed in both unaged and aged blends. After aging at 80 °C for 7 days, the new peaks of PVC/NBR/ENR sample were found at between 70 ± 90 ppm. When the amount of ENR in the ternary blend is half that of PVC/ENR blend, the signal areas of ternary blend were found to be halved. Since the presence of NBR provides a protection against the ring opening reaction of the epoxy groups in the ENR molecules. When ageing at 140 °C of this blend, the colour of this sample show black due to the formation of conjugated unsaturation by dehydrochlorination from PVC. The proton decoupling field strength was approximately 60 kHz. During block decay (BD) experiments the delays between pulses were 30 s.[31] To enhance the signal to noise (S/N), the cross polarization (CP) technique was applied with a delay of 4 s between pulses. The BD/CP peak area ratio values calculated for the carbons at 47 ppm from PVC, 32 ppm from NBR and 23 ppm from ENR) in the degraded and undegraded blends are discussed. All samples were normalized to the methylene carbon (relatively rigid carbon) of the PVC at 47 ppm so that comparison between the spectra could be done.[28] All ENR50 and NBR carbons have BD/CP intensity ratio values higher than unity in unaged blends implying that ENR and NBR molecules are more mobile in these blends compared to PVC units. There is an increase in the BD/CP peak ratio of the NBR carbon when the PVC/NBR blend was aged at 80 °C for 7 days. This may be due to softening of the NBR molecules or increase in rigidity of PVC molecules during aging. Similar results have been observed in other blends.[34] Aging the blend PVC/NBR at 140 °C, have caused a drop in the BD/CP peak ratio of the NBR carbons. The mobility of the NBR carbons becomes similar to the PVC carbons. The BD/CP peak ratio of the PVC/ENR blend reduces to unity in the sample aged at 80 °C for 7 days due to the formation of the large membered rings in the ENR molecules which make these molecules more rigid. This result confirms that the presence of NBR provides a protection against the ring opening reactions in the ENR molecules. The increased peak width of the peaks in the CP/MAS spectra of the blends after aging was observed. In the PVC/NBR blend, there is no significant change in the PVC peak width during aging at either condition. In case of PVC/ENR blend, the widths of the peaks of both polymers increased. The increase in the PVC carbon peak is less but the rubber peaks have nearly doubled. This is due to a drop in the molecular mobility as a result of aging or more chemical shift dispersion as a result of more heterogeneous immediate environment.[35] In the ternary blend, the peaks at 23 and 47 ppm could be used to determine the relaxation times of ENR and PVC.

NR/PSf blend

Figure 19.19 shows the solid state ^{13}C NMR spectra of PSf containing maleic anhydride, NR and 50/50 NR/PSf blend containing MA, recorded at room temperature. The resonances of PS show at 146 and 128 ppm due to

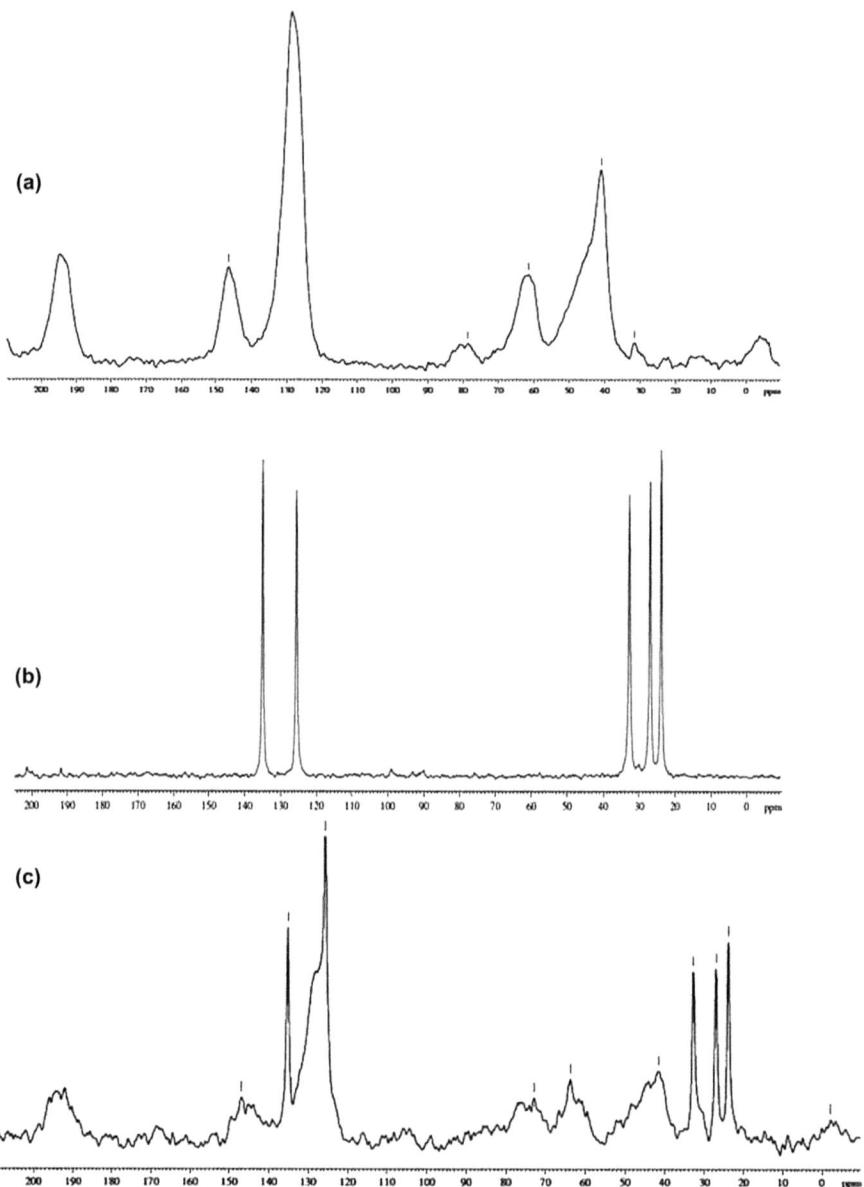

Figure 19.19 Solid-state ^{13}C NMR spectra of (a) PSf, (b) NR and (c) PSf/NR blend (our work).

non-protonated or quaternary ring (PS) and protonated aromatic carbons, respectively.

In addition, the methylene and methane carbon resonances of PSf were found at 41 and 46 ppm, respectively. In case of NR, five carbon signals were observed at 135, 125, 32, 26 and 22 ppm. After blending, there is detectable chemical shift difference or line shape change between PSf alone and NR alone. Therefore, it concluded that the ^{13}C chemical shift itself can provide direct information about the interaction between PSf and NR.

ENR/St blend

The solid state ^{13}C NMR spectra of (a) St and (b) ENR/St are displaying in Figure 19.20.

There are four ^{13}C-NMR signals in St. ^{13}C NMR Signals at 99–104, 81–84 ppm are assigned to C-1 and C-4 respectively in St. The signal at 59–62 ppm is assigned to hydroxymethyl carbon-6 in hexopyranoses. The large signal around 70–73 ppm is referenced with C-2, 3, 5 in hexopyranoses. After ENR was blended with St, the intensity of the signal at 84 ppm increased due to interaction between hydroxyl groups in ENR and St.

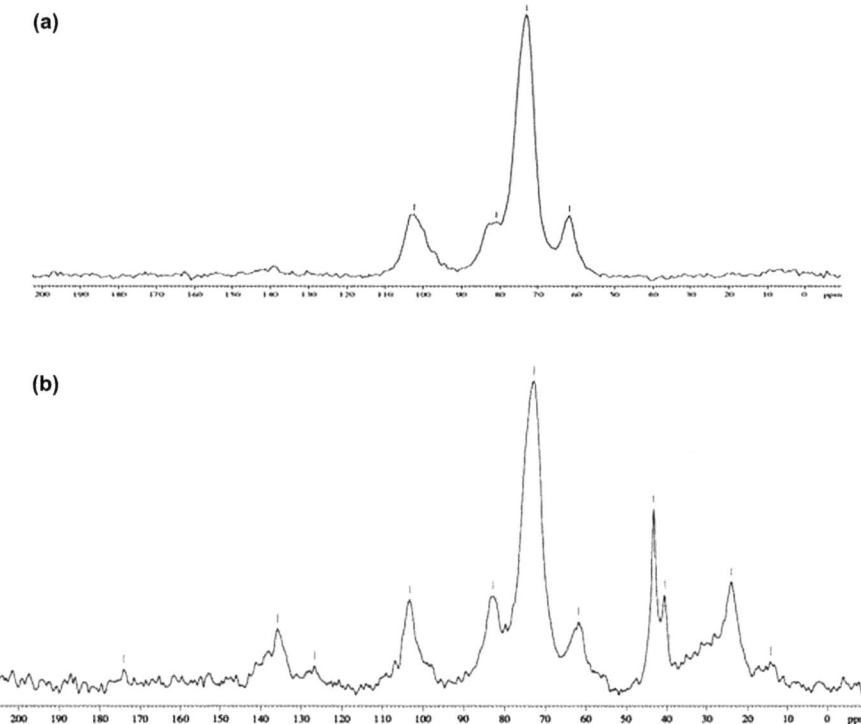

Figure 19.20 Solid-state ^{13}C NMR spectra of (a) St and (b) ENR/St blend (our work).

19.5 Raman Spectroscopy

19.5.1 Introduction to Raman Spectroscopy

The technique of Raman spectroscopy has rarely been used to analyse the chemical structure changes of elastomers due to problems of fluorescence. At present, the use of Fourier transforms collection techniques and spectral accumulations have applied to solve this problem having many of the barriers to the application of this technique within a routine laboratory environment. But the advantage of this work is a non-destructive analysis method as shown in Table 19.4.[36] In case of too much laser power, if the power is focused on a small point (in a microscope), sample can be destructive. Since FT-Raman systems employ higher power lasers. In previous, it has been used in a number of studies to examine the morphological changes in a variety of polymer. Recently, Raman spectroscopy has been developed to employ as an imaging instrumentation. Sample surface scanning in X- and Y-axis and sample depth (Z-axis) can be carried out by modifying the focus of the laser beam from the Raman microscope. Therefore, three-dimensional images can be thus built by using special software. The surface and bulk properties of immiscible rubber blend were investigated by Raman spectroscopy. The Raman spectrometry clearly elucidates the identification of phases between the dispersed phase and the matrix (continuous phase) of the immiscible rubber blends. It can be possible to obtain Raman spectra of NR only when near infrared light is used as an excitation source. From that point of view, FT-Raman spectroscopy is applied to investigate a structural change of NR, which does not provide microscopic information.

Table 19.4 Advantages and Disadvantages of Raman spectroscopy.[36]

Advantages	Drawback
• Can be used with solids and liquids • No sample preparation needed • Not interfered by water • Non-destructive • Highly specific like a chemical fingerprint of a material • Raman spectra are acquired quickly within seconds • Samples can be analyzed through glass or a polymer packaging • Laser light and Raman scattered light can be transmitted by optical fibers over long distances for remote analysis • Raman spectra can be collected from a very small volume (< 1 µm in diameter) Inorganic materials are normallys easier analyzed by Raman than by infrared spectroscopy	• Can not be used for metals or alloys. • The Raman effect is very weak. The detection needs a sensitive and highly optimized instrumentation • Fluorescence of impurities or of the sample itself can hide the Raman spectrum Sample heating through the intense laser radiation can destroy the sample or cover the Raman

Since the near infrared Raman system used here is equipped with a microscope, the rubber degradation can be analysed with a spatial resolution of 1 μm. This section describes some of the areas related to the analysis of its blend and its IPN NR and its blend which have been investigated using this technique.

19.5.2 Sample Preparation and Typical Conditions for Raman Spectroscopy

Usually, no Raman spectra can be derived from bulk solids, liquids, tablets, polymers, paper. Analysis can also be carried out through many containers. The Raman spectra were recorded with the 633 nm laser line (He-Ne laser, Spectra Physics, 4.5 mW maximum power) using a micro-Raman Renishaw 2000 equipped with an Leica microscope, a CCD detector, and a spectrograph with a 2400 g/mm grating (2 cm^{-1} spectral resolution). Of particular interest is the observation of main chain modifications during vulcanization and the ability to observe the conversion of insoluble to soluble sulfur under conditions appropriate to compounding and vulcanization. The influence of crystallization, both temperature and strain induced, on the FT-Raman spectrum of NR is also demonstrated. Near infrared Raman spectroscopy is applied to investigate the structural changes of the NR exposed to ozone and UV light. High quality microscopic Raman spectra of natural rubber are obtained and the depth profile of ozone and UV treated NR surface is analysed. Raman spectroscopy used in the microscope mode is very efficient for the characterization of interactions between the components in a polymer blends.

19.5.3 Analysis of Polymer Blends

NR[37]

The factors of degradation of the NR are oxygen, heat, ozone and UV. Here, the depth profile of the degradation of NR owing to ozone and UV treatment is investigated. The cross-section of freshly cut NR is used as a new, reference sample without any degradation. Then, the reference sample was exposed to ozone and UV/ozone for 6 h. Near infrared Raman spectra were recorded from the ozone and UV exposed surface to inner region, to the depth of 20 μm with 6 points line scan. There is no obvious change of Raman spectra in the depth profile of the new reference sample. The spectral changes of UV/ozone treated sample as a function of the depth from the surface. The band at 1670 cm^{-1} is assigned to the stretching mode of C=C double bond in NR.[4] The band intensity of 1670 cm^{-1} decreases and the broad band around 1600 cm^{-1} increased in intensity at the surface. This may be due to UV/ozone treatment results in the cleavage of C=C double bonds. The band around 1600 cm^{-1} is thought to result from a conjugated diene and the feature might come from the degradation compounds. To compare the spectral change quantitatively, the relative intensities of 1670 cm^{-1} band against 1450 cm^{-1} CH bending mode are calculated. Under both UV and ozone treatments, larger spectral changes of Raman spectra are observed.

PVDF and NR latex blend[38]

The interaction between NR and PVDF has been investigated by Raman spectroscopy. Because severe processing conditions are used to produce PVDF films, it is important to follow the molecular structure and crystalline phase of PVDF along this fabrication process. The Raman band of PVDF was observed at 795 cm^{-1} characteristic of the phase and assigned to CH$_2$ rocking. In terms of polarization, the incident laser was used either circularly or in other two directions perpendicular to each other. Upon compressing/annealing the PVDF powder with or without latex of NR, no significant anisotropy is induced in the polymer films. It can be seen that the phase is maintained until the melting point and it is reached again by cooling the film until room temperature.

NR/polyaniline blend[39]

Blends possessing the elastomeric properties of NR and the conducting properties of conducting polymer (polyaniline, PANI) were obtained, which are promising for further application in deformation sensors. Characterization was carried out by Raman spectroscopy. Evidence for chemical interaction between PANI and NR was observed, which allowed the conclusion that the NR latex itself is able partially to induce both the primary doping of PANI (by protonation) and the secondary doping of PANI (by changing the chain conformation). Moreover, the electrical conductivity reached in the blends was dependent on the doping conditions used, as observed by Raman scattering. The NR is responsible for these changes. For example, the Raman band at 1166 cm^{-1} (C–H bending of quinoid ring) dominates the powder spectrum whereas the band around 1601 cm^{-1} (C–C stretching benzenoid ring) is the strongest one for the blend sample. It was observed on exposing these films to m-cresol vapour during different time intervals that the bands at 1488 and 1520 cm^{-1} collapsed in a broad band around 1500 cm^{-1} and the bands at 1584 cm^{-1} (C–C stretching quinoid) and 1620 cm^{-1} (C–C stretching benzenoid) decreased in intensity. But the band around 1592 cm^{-1} (C–C stretching benzenoid) increases. It is important to point out that the bands around 1171, 1512 and 1601 cm^{-1} present the same FWHM with the different doping methods used here. These results referred that the NR rather than the HCl or corona discharge might be inducing in the PANI a doping effect similar to that induced by m-cresol (secondary doping). The main peaks of wave numbers (cm^{-1}, FWHM (cm^{-1}) in parentheses) and relative intensities from the Raman spectra for blend cast film containing 20% of emeraldine PANI in NR were summarized in Table 19.5.

Table 19.5 Wave numbers (cm^{-1}, FWHM (cm^{-1}) in parentheses) and relative intensities from the Raman spectra for blend cast film containing 20% of emeraldine PANI in NR and blend cast films containing 20% of emeraldine PANI in NR de-doped by NH$_4$OH or doped either by HCl or by corona discharge.[39]

PANI-NR NH$_4$OH	PANI-NR	PANI-NR HCl	PANI-NR corona	Assignment
1661 (17) 14	1661(20)46	1661 (22)80	1664 (11)31	NR
1638 (15) 11	1636(18)57	1637 (21)63	1646 (40)81	Cyclized strcture or tertiary N
1601 (38) 55	1604 (28) 100	1601 (34) 100	1601 (34)100	C–C benenoid stretching
	1568 (22)20	1568 (27)43	1569 (29)37	C=C stretching
1554 (16)6				C=C stretching
	1512(29)24	1507 (27)23	1512(26)28	N–H bending (in –plaine)
1465 (55)100				C=C and C=N
	1447 (48)32	1443 (42)25	1440 (57)26	
1414(23)19				
	1370 (17) 13	1370(26)51	1370(24)43	Cyclized structure or tertiary N
1327 (40)13	1332 (49) 62	1332(46)90	1332 (49)78	C–N stretching
1219 (25)41				
	1231 (33)8	1237 (62)13	1236 (32)6	C–N benzene diamine stretching
1163(29)83				C–H bending quinoid
	1171 (28)48	1172(31)75	1172 (29)54	C–H bending benzenoid
	1141 (32)20	1139 (38)43	1140 (29)30	
	1038 (13)8	1038 (15)12	1039 (24)12	C–H def.
995 (29)4	995 (32) 14	997 (36)26	999 (31)28	C–H wag
869 (32)7		869 (35)12	880 (22)5	Ben.ring def.
869 (30) 22				Quin.ring def.
	818(34) 12	819 (45) 30	819 (36)16	C–H (out-of-plane)
808 (40)18				C–H (out-of-plane)
778 (27) 32				Quin.ring def.
746 (15)21				
722 (29) 7	723 (26)6	722 (36)13	727 (32)9	Quin.ring bend.
645 (28)6	644 (35)8	646 (20)3	634 (41)13	C–C ring def. (out-of-plane)
601(35)8	601(35)17	603(22)22	603 (28)26	Amine def. (in-plaine)
573 (20)12	573 (17)30	576 (19)51	576 (18)73	Cyclized structure or tertiary N
524(39)30				Quinoid def.
	510(26)8	505(28)16	503 (45)11	C–N–C torsion
420 (32)29	417 (18)27	414(19)30	415 (20)25	Ring cef. or C–H wag

19.6 Electron Spin Resonance (ESR) Spectroscopy

19.6.1 Introduction to ESR

Electron spin resonance (ESR) spectroscopy is a method for studying the structure of materials concerning about unpaired electrons which is an electron that occupies an orbital of an atom singly. The principal of EPR is similar to those of nuclear magnetic resonance (NMR). However, EPR is electron spins that are excited instead of the spins of atomic nuclei and the most stable molecules have all their electrons paired. Therefore, the EPR technique is less widely used than NMR. The advantage of EPR is a great specificity, since ordinary chemical solvents.

19.6.2 Sample Preparation and Typical Conditions for ESR

The 50 mg of samples is measured at room temperature in the rectangular cavity at microwave power level far below saturation. The ESR spectra were recorded as first derivative curves and the free radical concentrations were calculated by numerical double integration of these spectra using the data system.

19.6.3 Analysis of Polymer Blends[40]

In previous, ESR was used to identify the interaction of PP/ENR blends. Spectroscopic analysis of electron beam irradiated PP/ENR polymer blends. PP degrades very rapidly when irradiated in air and the degradation occurs during irradiation as well as during post-irradiation. Also, the mechanism of degradation induced by irradiation in air is a free radical mechanism involving the formation of peroxides and hydroperoxides. Subsequently, main chain scission occurs by breakdown of these peroxides with the formation of oxidation products including carbonyl compounds. This degradation processes is related to the number of free radicals formed during irradiation as well as the decay rate of these radicals. The ESR spectra of electron beam irradiation at a dose of 50 kGy for PP and PP/ENR blends measured immediately after irradiation and after different periods of time. The initial ESR spectrum of PP/ENR blends exhibits the lines due to the alkyl and allyl radicals which are formed during electron irradiation. After PP and PP/ENR blends are exposed to oxygen, the peroxy radicals observed by ESR. The peroxy radicals are started to be a dominant after 3 h for PP/ENR blends due to the presence of the epoxy rings which form the following R and RO structures. The initial radical concentration increases with increasing ENR ratio and irradiation dose except in the case of the PP/ENR blend. The presence of 10% of ENR does not go systematically with blend composition, but the free radical concentration produced during irradiation increases as a function of irradiation dose. The results showed the drop in radical concentration proceeds for the blends containing 20% and the increases for blends with 30% ENR. This behaviour indicates that the change in polymer matrix is more effective on radical concentration than radical formation. The free radical concentration measured immediately after irradiation is the result of radical formation and radical decay (termination) during irradiation. In this regard, the radical formation will eventually increases with increasing the ratios of ENR (epoxy rings and double bonds). The ENR as an amorphous polymer will facilitate the mobility of radicals and consequently reduces the amount of radicals. This state of affairs will continue until the balance between radical formation and decay occurred for the blends with higher ratios of ENR. The decay of radicals of all the blends is almost the same within the studied range of time dependent on the ENR ratio. However, the radical decay of PP/ENR blends is faster than PP polymer within the first 5 h. This effect may be due to faster transformation of alkyl radicals to more stable, peroxy radicals in this blend ratio.

19.7 Applications

UV-vis, FTIR, NMR, Raman spectroscopy and ESR are the most powerful ways of analysing polymer blends based on NR and its derivatives. For example, NMR provides unique and important molecular motional and interaction profiles containing pivotal information on NR blend function. Some of the applications of spectroscopy are listed below:

Solution structure: the only technique for atomic-resolution structure of polymer blend based on NR in organic solutions confirmed by UV-vis spectroscopy, FTIR and NMR.
Molecular dynamics: the most powerful method for quantifying motional properties of polymer blends.
Crystalline behaviour: the most powerful tool for determining the structures of folding intermediates of rubber molecules in blend.
Ionization state: the most powerful tool for finding the chemical properties of functional groups in polymer blend, such as the ionization states of ionizable groups at the active sites of rubber molecules.
Hydrogen bonding: spectroscopy for the detection of hydrogen bonding interactions in polymer blend.
Cured properties of rubber blend: solid state NMR has the potential for determining atomic-resolution structures of rubber blend.
Chemical analysis: spectroscopy for chemical identification and conformational analysis of chemicals in rubber blend.
Degradation mechanism: spectroscopy to investigate the change in chemical microstructure after degradation process.
Dynamic properties: NMR spectroscopy to mobility of polymer molecule in rubber blend.

Based on NMR results, the effective internal correlation times have been calculated from the relaxation data to characterize intramolecular motions that are more rapid than overall rotational diffusion.

The applications of polymer blend made from NR are listed in Table 19.6.[41–45] The possible IPN from NR and PSf containing natural fibre

Table 19.6 Possible application of polymer blend based on NR.

Sample blend			
Polymer 1	Polymer 2	Application	References
SPNR	PSf	Packaging	17
Skim rubber	PVA	Adhesive	41
NR	St	Encapsulation of fertilizer	21
NR	Nitril rubber	Membrane for benzene	42
NR	PVA	Membrane for ethanol-water	43
NR	Chitosan	Electronic material	44

could be expected to use in artificial wood application, cup and picture frame. This is due to the preponderance of non-elastomeric substances in the material (Table 19.6).

19.8 Conclusions

NR is a biopolymer with unique structures and interesting properties such as mechanical properties; its non-toxic, have a wide range of applications and the raw material sources for their production are unlimited. Blending of NR with other polymer improve the properties. In order to acquire a deeper understanding of the mechanism of these properties, it is necessary to analyse the structure of its blend and IPN. The various spectroscopic techniques including UV-vis, FTIR, NMR, Raman and ESR spectroscopy are very useful and important in the structural analysis of blend and IPN from NR and their derivatives to understand the relationship between structure and properties. Knowledge of the microstructure of rubber blends is essential for an understanding of structure–property–activity relationships through chemical and physical interaction, degradation and blend composition as well as dynamic properties.

Acknowledgements

The authors are grateful for the fundamental research grant from The Thailand Research Fund/Commission on Higher Education: Grant No: (MRG5380190), (RTA 5480007) and Prince of Songkla University (SCI 540032S) that supported this work.

References

1. S. Riyajan and S. Chaiponban, *Kautsch. Gummi Kunstst.*, 2010, **63**, 70.
2. D. A. Skoog, F. J. Holler and S. R. Crouch, *Principles of Instrumental Analysis*, Belmont, CA, Thomson Brooks/Cole, 6th edn, 2007, pp. 169–173.
3. H. P. Yu, S. D. Li, J. P. Zhong and K. Xu, *Thermochim. Acta*, 2004, **410**, 119.
4. S. Mohanty, S. Roy, R. N. Santra and G. B. Nando, *J. Appl. Polym. Sci.*, 1995, **58**, 1947.

5. M. Depret and J. A. Malmonge, Project title: Conductive elastomeric blend of natural rubber and polyaniline derivative. Université de Valenciennes et du Hainaut, 2012, available from: http://www.deb.uminho.pt/tacts-meta/uep_project03.htm.
6. A. R. Freitas, A. F. Rubira and E. C. Muniz, *Polym. Degrad. Stab.*, 2008, **93**, 601.
7. A. Marcilla, A. Gomez and S. Menargues, *Polym. Degrad. Stab.*, 2005, **89**, 454e.
8. M. J. Fernández-Berridi, N. González, A. Mugica and C. Bernicot, *Thermochim. Acta*, 2006, **444**, 65.
9. A. R. Freitas, A. F. Rubira and E. C. Muniz, *Polym. Degrad. Stab.*, 2008, **93**, 601.
10. M. Celina, J. Wise, D. K. Ottesen, K. T. Gillen and R. L. Clough, *Polym. Degrad. Stab.*, 2000, **68**, 171.
11. M. A. De Paoli and M. A. Rodrigues, *Eur. Polym. J.*, 1985, **21**, 15.
12. C. Adam, J. Lacoste and J. Lemaire, *Polym. Degrad. Stab.*, 1991, **32**, 51.
13. W. Kaminsky, C. Mennerich, J. T. Andersson and S. Gotting, *Polym. Degrad. Stabil.*, 2001, **71**, 39.
14. F. Delor, J. Lacoste, J. Lemaire, N. Barrois-Oudin and C. Cardinet, *Polym. Degrad. Stab.*, 1996, **53**, 361.
15. J. B. Lambert, H. F. Shurvell, D. A. Lightner and R. G. Cooks, *Organic Structural Spectroscopy*, 1998, Prentice-Hall, Upper Saddle River, NJ.
16. A. R. Freitas, G. J. Vidotti, A. F. Rubira and E. C. Muniz, *Polym. Degrad. Stab.*, 2005, **87**, 425.
17. S. Riyajan, I. Intharit and P. Tangboriboonrat, *Ind. Crops Prod.*, 2012, **36**, 376.
18. R. V. Sekharan, B. T. Abraham and E. T. Thachil, *Mater. Des.*, 2012, **40**, 221.
19. T. Nampitch and P. Buakaew, *Kasetsart J. Nat. Sci.*, 2006, **40**, 7.
20. M. Marinovic-Cincovic, V. Jovanovic, S. S. Zija-Jovanovic and J. Budinski-Smendic, *Chem. Ind. Chem. Eng. Q.*, 2009, **15**, 291.
21. S. Riyajan, Y. Sasithornsonti and P. Phinyocheep, *Carbohyd. Polym.*, 2012, **89**, 251.
22. S. Riyajan, S. Chaiponban and K. Tanbumrung, *Chem. Eng. J*, 2009, **153**, 199.
23. J. M. Gohil, A. Bhattacharya and P. Ray, *J. Polym. Res.*, 2006, **13**, 161.
24. P. Bergo and P. J. A. Sobral, *Food Hydrocolloids*, 2007, **21**, 1285.
25. V. Schmidt, C. Giacomelli and V. Soldi, *Polym. Degrad. Stab.*, 2005, **87**, 25.
26. J. Wootthikanokkhan and P. Tunjongnawin, *Polym. Test.*, 2003, **22**, 305.
27. P. Phinyocheep, J. Saelao and J. Y. Buzaré, *Polymer*, 2007, **48**, 5702.
28. M. V. Lewan, in *Blends of Natural Rubber*, ed. A. J. Tinker and K. P. Jones, Chapman and Hall, London, 1998, p. 52.
29. J. T. Koenig, *Rubber Chem. Technol.*, 2000, **73**, 385.
30. N. M. P. S. Ricardo, M. Lahtinen, C. Price and F. B. Heatley, *Polym. Int.*, 2002, **51**, 627.

31. M. C. S. Perera, U. S. Ishiaku and Z. A. M. Ishak, *Polym. Degrad. Stab.*, 2000, **68**, 393.
32. M. C. S. Perera, J. A. Elix and J. H. Bradbury, *J. Polym. Sci., Part A: Polym. Chem.*, 1988, **26**, 637.
33. A. Ibrahim and M. Dahlan, *Prog. Polym. Sci.*, 1998, **23**, 665.
34. K. S. Jack and A. K. Whittaker, *Macromolecules*, 1997, **30**, 3560.
35. K. Takegoshi and K. Hikichi, *J. Chem. Phys.*, 1991, **94**, 3200.
36. Advantages and Disadvantages of Raman Spectroscopy, available from http://www.raman.de/htmlEN/home/advantageEn.html, 2012.
37. Raman Spectroscopy of Natural Rubber, available from http://www.publish.csiro.au/?act = view_file&file_id = SA0402274.pdf, 2012.
38. R. D. Simoes, A. E. Job, D. L. Chinaglia, V. Zucolotto, J. C. Camargo-Filho, N. Alves, J. A. Giacometti, O. N. Oliveira Jr and C. J. L. Constantino, *J. Raman Spectrosc.*, 2005, **36**, 1118.
39. A. E. Job, C. J. L. Constantino, T. S. G. Mendes, M. Y. Teruya, N. Alves and L. H. C. Mattoso, *J. Raman Spectrosc.*, 2003, **34**, 831.
40. M. M. H. Senna, A. A. Abdel-Fattah and Y. K. Abdel-Monem, *Nucl. Instrum. Methods Phys. Res., Sect. B*, 2008, **266**, 2599.
41. S. Riyajan and N. Pheweaw, *Rubber Chem. Technol.*, 2012, **85**, 547.
42. A. E. Mathai, R. P. Singh and S. Thoma, *J. Membr. Sci.*, 2002, **202**, 35.
43. S. Amnuaypanich, J. Patthana and P. Phinyocheep, *Chem. Eng. Sci.*, 2009, **64**, 4908.
44. J. Johns and C. Nakason, *J. Non-Cryst. Solids*, 2011, **357**, 1816.
45. S. Riyajan, S. Chusri and S. P. Voravuthikunchai, *Rubber Chem. Technol.*, 2012, **85**, 147.

CHAPTER 20

Mechanical and Viscoelastic Properties of Natural Rubber Based Blends and IPNs

WIWAT PICHAYAKORN,* JIRAPORNCHAI SUKSAEREE AND PRAPAPORN BOONME

Department of Pharmaceutical Technology, Faculty of Pharmaceutical Sciences, Prince of Songkla University, Songkhla 90112, Thailand
*Email: wiwat.p@psu.ac.th

20.1 Introduction

The mechanical and viscoelastic behaviours of natural rubber based blends and interpenetrating polymer networks (IPNs) are functions of their structures or morphologies. These properties of blended materials are generally not constant and depend on the chemical nature and type of the polymer blends, and also environmental factors involved with any measurements. Preparations of natural rubber blends and IPNs are well known as effective modification methods used to improve the original mechanical and viscoelastic properties of one or both of the components, or to obtain new natural rubber blended materials that exhibit widely variable properties.[1–4] The most common consideration for their mechanical properties include strength, ductility, hardness, impact resistance and fracture toughness, each of which can be deformed by tension, compression, shear, flexure, torsion and impact methods, or a combination of two or more methods.[5] Moreover, the viscoelasticity theory is a way to predict the behaviours of deformation of natural rubber blends and IPNs. The time and temperature sensitivities are determined from the viscoelasticities of the polymer blends. The natural rubber blended materials exhibit a combination of viscous and elastic behaviour when undergoing deformation. They have elements of both of these properties such as (i) resistance to the shear flow and strain linearly with time when a stress is applied, and (ii) showing instantaneous elasticity when being stretched and just as quickly returning to their original state once the stress is removed. In addition, when a stress is applied, the viscoelasticity can be a result of molecular rearrangement by changes to the position of some parts of the long chain polymer blends.[6] Generally, the natural rubber blends behave like an amorphous polymer with a glass transition temperature below the ambient temperature, and that may display nonlinear but recoverable deformation or even exhibit viscous flow. They cause an increase in terms of ductility, fracture toughness and elongation, while the strength property usually decreases. Their glass transition temperatures above room temperature probably cause an increase in the strength property which is expressed by brittle blended materials. The natural rubber blending may display a linear elastic behaviour, yield phenomenon, plastic deformation or cold drawing. The ductility property may increase or decrease with increasing temperature.[6] For traditional engineering practice, the mechanical and viscoelastic properties can determine the range of usefulness of the natural rubber blended materials and establish the lifetime that can be expected. In addition, they are also used to classify and identify the blended materials.[5,6] Therefore, this chapter discusses the mechanical and viscoelastic behaviours of natural rubber blends and IPNs with thermoplastics, thermosets, synthetic rubbers and biopolymers that find potential uses in industrial applications.

20.2 Instruments and Techniques for Mechanical and Viscoelastic Evaluations

20.2.1 Mechanical Properties

Mechanical properties are the parameters used to measure the forces able to deform the natural rubber blended materials such as elongation, compression, twist and breakage as a function of an applied load, time, temperature or other conditions by testing materials. Results of these tests depend on the size and shape of the specimens of the tested materials. Generally, the specimens are cut into a specific shape and their mechanical properties tested with an accurate load cell capacity and crosshead speed by a tensile machine such as an Instron testing machine or universal testing machine until they deform.[6–16]

The most important mechanical property of natural rubber blended materials is their tensile or stress–strain relationship as shown in Figure 20.1. Stress is the force applied to produce deformation of a unit area of a test specimen. The standard unit of this value presents in Pascals (Pa) or pounds per square inch (psi). Strain is the ratio of elongation or deformation to the gauge length of the test specimen per unit of the original length. It is expressed as a

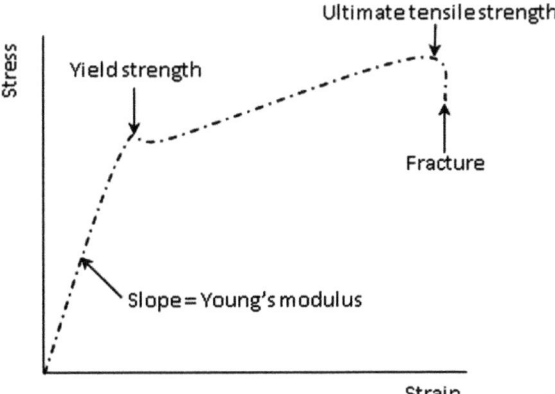

Figure 20.1 Tensile stress–strain curve of polymeric materials with the various stages of deformation.
(Modified from Ebewele[6]).

dimensionless number. The modulus of elasticity or Young's modulus is the linear slope of the stress–strain curve. Normally at low extensions or low strain, the stress–strain relationship of plastics shows a linear behaviour known as linear viscoelasticity. However, plastics generally show nonlinear behaviour of the stress–strain relationship. Therefore, modulus values for plastics are determined at very low extensions when the stress–strain curve is often a reasonably straight line. All elastic modulus values can be computed using the average initial cross-sectional area of the test specimens in the calculations. The result is expressed in Pa. The modulus of elasticity is applied to describe the stiffness or rigidity of plastics. Initially the modulus is high, until a point is reached to the plastic 'yields' or deforms. Prior to the yield point, the elongation is reversible. At the yield point, enough stress has been applied to cause the molecules to untangle and flow over one another, and further elongation is irreversible. Eventually, the sample breaks.[5,6,17–19]

The yield point is the first point on the stress–strain curves at which an increase in strain occurs without an increase in stress. After the yield point the specimen exhibits non-recoverable behaviour. Thus, this point would normally represent the limit of elasticity. The stress at the yield point is specified as the tensile strength at the yield, or yield stress. Tensile strength at the yield is an often quoted property, especially if it has a higher value than the ultimate tensile strength at break.

Elasticity can be determined as two values. The first value of elasticity, ultimate tensile strength, is the maximum tensile stress applied to break or fracture a test specimen. It is calculated by dividing the maximum load by the original minimum cross-section area of the specimen. The result is expressed in the force per unit area, usually megaPascals (MPa). The second value, elongation, is also used to measure the extension of materials before they lose their original shapes that are the deformation to the gauge length of the test materials. This value is calculated by dividing the maximum load by the

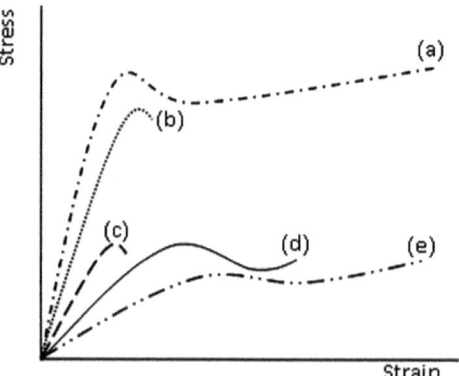

Figure 20.2 Stress–strain behaviour of various types for polymeric materials: (a) hard and tough, (b) hard and strong, (c) hard and brittle, (d) soft and weak, and (e) soft and tough.
(Modified from Ebewele[6]).

original minimum cross-sectional area of the specimen. Both ultimate tensile strength and elongation values can refer to the elastic property of materials. However, elasticity behaviour takes a time for the strain to return to zero. It is not immediately reversible after the applied force is stopped.[6] Examples of stress–strain behaviours of polymeric materials with various characteristics are illustrated in Figure 20.2. A hard and tough material such as polycarbonate is characterized by a high modulus, high yield stress, high elongation at break, and high ultimate strength. A hard and strong material such as polyacetal (polyoxymethylene) has a high modulus, high yield stress, usually a high ultimate strength, and low elongation. A hard and brittle material such as a general-purpose phenolic is characterized by a high modulus and low elongation. A soft and weak material is characterized by a low modulus, low yield stress and moderate elongation at break. A soft and tough material such as polyethylene shows a low yield stress, but very high elongation at break.[6]

Ductility is the ability of natural rubber blended materials to deform before breaking. However, ductility and elasticity are two mechanical properties that are usually confused, but they are opposite values. In some cases, the shear and torsion forces can deform the angle of perpendicular to twist which affects the elasticity of the tensile testing.[6] Although both ductility and elasticity properties can deal with stretching, ductility also accesses the ability of natural rubber blended materials to be bent or twisted. The natural rubber blended materials with a good ductility should not break or deform under these conditions. In addition, when stretching is applied for the ductility of blended materials, there is usually no possibility for the materials to return to their original forms.[13]

Toughness is the ability of natural rubber blended materials to absorb energy before fracturing or rupturing. It is a combination of the strength and ductility parameters that require a balance between them. The natural rubber blended materials having high strength and high ductility are strong and tough. They have more toughness than materials with a low strength and high ductility that are strong and brittle. Although toughness and strength may be related in mechanical properties, they are different. While strength focuses on the force needed to break the materials, toughness focuses on the amount of energy which the natural rubber blended materials can withstand. If an item can endure a high level of shock, it is considered to be tough. The toughness value expresses the slow absorption of energy of the blended materials. There are several names used to refer to toughness properties such as impact toughness, notch toughness and fracture toughness. In general, the measurement of toughness is calculated from the area under the stress–strain curve from a tensile testing. This value is simply called natural rubber blended toughness in terms of energy per volume of materials.[6,20]

Hardness is the resistance to localize the deformation of the permanent shape of natural rubber blended materials when a force is applied. It is generally determined by strong intermolecular bonds, but the behaviour of solid materials under force is complex. Therefore, there are different measurements of hardness such as the indentation, scratching, cutting or bending hardness. However, these values depend on ductility, elastic stiffness, strength, toughness and viscoelasticity.[6] A durometer is the device used to measure the hardness of a material and was first developed by Albert F. Shore in 1920. It is typically used to measure the hardness of polymers and rubbers. It can measure the depth of an indentation in the materials created by a given force on a standardized presser foot. The indentation depth is dependent on the hardness and viscoelastic properties of the materials, the shape of the presser foot, and the duration of the test. There are several scales of durometers used for materials with different properties. The two most common scales are the type A and type D (Shore A and D) that use slightly different measurement systems. The Shore A often measures softer plastics, while the Shore D is for harder materials.[6]

20.2.2 Viscoelastic Properties

Viscoelasticity is one of the important mechanical properties of blended materials. Viscoelastic behaviour is the intermediate character between liquid and solid states that combines the viscous and elastic responses under mechanical stress. When a force is applied to blended materials, they can flow in the same as being liquids. The natural rubber blended materials do not stretch, but they will only gradually return to their original shapes when the force is released. This property depends on temperature, pressure, time, chemical composition, molecular weight, distribution, branching, crystallinity, and the composite of blending conditions and systems.[6]

The viscoelasticity can be categorized as either linear or nonlinear, but only the linear viscoelasticity can be described theoretically with uncomplicated mathematics. The fundamental viscoelastic parameters of a linear viscoelastic system do not depend on the magnitude of the stress or strain.[21] Therefore, the linear viscoelastic regime is always used for studying the mechanical properties of viscoelastic blended materials. One of the accepted techniques for investigating the viscoelastic behaviours of natural rubber blended materials is the dynamic mechanical testing based on the fundamentally different responses of viscous and elastic elements to a varying stress or strain. This technique provides the information of sample structures without deformation.[22]

The viscoelastic behaviours can be determined by dynamic and transient tests. The transient test involves the imposition of a step change in stress/strain and the observation of the subsequent development at each time of the strain/stress. The dynamic test involves the application of a harmonically varying strain or stress. The oscillation test is one type of dynamic method for determining the rheological properties of the natural rubber blended materials in the ground state. It does not alter the static structure of the natural rubber blended materials. Dynamic mechanical parameters such as storage modulus (G'), loss modulus (G''), and the loss tangent (tan δ) of the polymers have been used to determine the glass transition region, relaxation modulus, degree of crystallinity, molecular orientation, crosslinking and phase separation.[4,23] The miscibility of the blends can easily be investigated from the shift in their glass transition temperatures. For the miscible natural rubber blends, only one glass transition temperature is obtained at a glass transition intermediate between the two or more components. When there is broadening of the transition peak or two or more peaks occur, it indicates the partial miscibility and/or immiscibility between two or more components.[4,22,24] It has been reported that the dynamic mechanical techniques can provide a meaningful correlation between the molecular structure and the mechanical properties.[23] The important role of dynamic mechanical analysis for the behavioural study of natural rubber blends is widely accepted.

The applied strain is the independent variable, so the stress vector (τ^*) is resolved into its in-phase (τ') and out-of-phase (τ'') components as shown in Equation (20.1).

$$\tau^* = \tau' + i\tau'' \qquad (20.1)$$

where i represents the out-of-phase unit vector.

A G' in-phase and a G'' out-of-phase are defined as Equations (20.2) and (20.3), respectively.

$$G' = \tau'/\tau^* \qquad (20.2)$$

$$G'' = \tau''/\tau^* \qquad (20.3)$$

The complex modulus (G^*) is the vector sum of the in-phase and out-of-phase moduli as shown in Equation (20.4):

$$G^* = G' + iG'' = \tau^*/\gamma^* = (\tau' + i\tau'')/\gamma^* \qquad (20.4)$$

where γ^* represents strain vector.

The tan δ is defined as in Equation (20.5):

$$\tan \delta = \tau''/\tau' = G''/G' \qquad (20.5)$$

The G' represents the energy stored elastically in the natural rubber blended materials during their straining. Therefore, G' is the 'storage modulus'. If the applied mechanical energy is not stored elastically, it must be 'lost' or converted to heat through molecular friction. It presents as a viscous dissipation within the natural rubber blended materials. It is represented by G'', which is known as the 'loss modulus'.[1,3,6,8]

According to the literature, the viscoelastic properties are determined by a dynamic mechanical thermal analyser (DMTA) or dynamic mechanical analyser (DMA) such as MKIIDMA, DMA Q 800, DMTA MK-II, DMTA 1V analysers, and RheoTech MDPT at different frequencies and temperature ranges with a constant heating rate. The tested samples are taken as rectangular bars.[3,4,9,12,16,23] A DMTA is a reliable tool to get unambiguous temperature values of the dynamic glass transition and rubber plateau. Many researchers have taken both theoretical and experimental approaches to investigating these properties of the blends.[1,3,4,16,23]

20.3 Mechanical and Viscoelastic Properties of Natural Rubber Blends and IPNs

The properties of a natural rubber can be extensively modified by blending with other polymers. These natural rubber blends provide an attractive way of obtaining new materials with unique combinations of properties. The blending of different polymers usually gives rise to new materials with properties that cannot be achieved from the individual components.[1] There are generally many good reasons for developing natural rubber based blends and IPNs that can produce new polymeric materials for industrial applications. Many attractive properties of the components can be synergistic. Natural rubber based blends and IPNs can be produced with many types of polymer, categorized as (i) thermoplastics such as polyethylene, polypropylene, polyacrylates, polylactide, polystyrene, polycarbonate, nylon, *etc.*, (ii) thermosets such as polyethylene resins, epoxy resins, urea formaldehyde, *etc.*, (iii) synthetic rubbers such as butadiene rubber, nitrile rubber, silicone rubber, styrene butadiene rubber, chloroprene/neoprene, *etc.*, and (iv) biopolymers such as cellulose, pectin, chitin, chitosan, *etc.* However, some natural rubber blends may be incompatible due to the differences in polarity according to their chemical structures. Thus, addition of a compatibilizer is one method used to improve the compatibility of immiscible blends that retain the chemical or physical properties of the individual homopolymers. The claim for using compatibilizers is to reduce the interfacial energy between the phases and permit a finer dispersion during mixing, provide a measure of stability against gross segregation, and result in an improved interfacial adhesion.[2,4,5,11,15,20,25,26] Moreover, many polymers including homopolymers or compatibilizers directly affect the mechanical and viscoelastic properties of natural rubber based blends and IPNs depending on the polymer types.

20.3.1 Natural Rubber/Thermoplastics

Thermoplastics or thermosoftening plastics are polymers that become homogenized liquid forms after being heated, and then returning to a rigid state after being cooled sufficiently. Moreover, when they are frozen, they become fragile glass-like materials. Thus, they can be remelted, remoulded, reshaped, and repeatedly frozen. These properties make thermoplastics that are recyclable and reversible materials. Most thermoplastics are high molecular weight polymers such as polyethylene, polypropylene, polyacrylates, polylactide, polystyrene, polycarbonate, and nylon with their chains associated through Van der Waals forces, dipole-dipole interactions, hydrogen bonding, or stacking of aromatic rings. These thermoplastic polymers are relatively easy to use in manufacturing processes. In contrast, natural rubber is a thermoset polymer that exhibits different characteristics such as heat resistance, flexibility, durability, tear resistance, abrasion resistance, melt resistance, chemical resistance, and impact resistance. Thus, many researchers focus on the enhanced properties of blended materials that are suitable for a specific application and produce better combinations of properties between natural rubber and thermoplastics. Thermoplastic natural rubbers are the blended materials of rubber and thermoplastics that combine the physical properties of conventional thermoset rubber and the excellent processing characteristics of thermoplastics. In some cases, either modifying the blended compounds or the use of a compatibilizer is used to overcome the incompatibility of these blends. To generate molten blending with a small sized dispersed phase and to reduce interfacial tension that will allow for stress transfer between phases, the use of compatibilizers becomes one of the attractive options. Therefore, blended compatibilizers are used to increase the blended compatibility.[1,3,9] Thus, thermoplastic natural rubbers can show the advantages that are typical of both the natural rubber and thermoplastic materials.

The variation of dynamic mechanical properties such as tan δ, G' and G'' versus temperature for the homopolymers and graft copolymers of natural rubber/poly(methyl methacrylate) blends with and without natural rubber-*graft*-poly(methyl methacrylate) as compatibilizer has confirmed the incompatibility between natural rubber and poly(methyl methacrylate). The tan δ of this blended polymer had two transitions that corresponded to natural rubber and poly(methyl methacrylate). This suggested the presence of a phase separation located at the interface and indicated an immiscible blend. The G's of this blended polymer were at the transition and plateau regions that also corresponded to natural rubber and poly(methyl methacrylate). Moreover, the incompatible natural rubber/poly(methyl methacrylate) blends had very low mechanical properties such as modulus, tensile strength, elongation at break, and tear strength. However, the problem of this incompatibility could be improved by the addition of a graft copolymer. A natural rubber-*graft*-poly(methyl methacrylate) compatibilizer improved the interfacial adhesion properties between the natural rubber and poly(methyl methacrylate) surfaces, reduced the size of the dispersed phase and resulted in an increase of the G'. Furthermore, the addition of a graft copolymer increased the mechanical properties such as the change of the deformation nature of the polymer blends. Thus, compatibilized blends have improved the mechanical properties of natural rubber/poly(methyl methacrylate) blends compared to the uncompatibilized blends.[4]

The natural rubber-*graft*-polydimethyl(methacryloyloxymethyl) phosphonate could be used as a compatibilizer to modify the blending process between natural rubber and ethylene vinyl acetate copolymer.[9] This blended polymer without compatibilizer exhibited poorer mechanical properties than the compatibilized blends. The optimum level of compatibilizer loading was 9% w/w. G' and the mechanical properties in terms of Young's modulus, tensile strength, and elongation at break increased with an increasing level of compatibilizer loading until it reached the maximum value at its optimum level. The highest interaction between natural rubber and ethylene vinyl acetate phases in the blended polymer produced the highest elasticity and complex viscosity, but the tan δ and tension decreased until it obtained its lowest value at an optimum level of compatibilizer. However, the Shore A hardness was not different between the uncompatibilized and compatibilized blends. This indicated that the compatibilizer molecules were forced to locate at the interface of the natural rubber and ethylene vinyl acetate. The segments of the grafted natural rubber were wetted by the compatibilizer component, while the polar functional groups in this graft copolymer interacted with the polar functional groups in the ethylene vinyl acetate molecules. The addition of compatibilizer in the range of 1–8% w/w greatly increased the G' due to gradually increasing the coverage of the interfacial area and hence producing the interfacial adhesion between the phases of the natural rubber/ethylene vinyl acetate blends. The compatibilizer at a concentration of 9% w/w reached a critical micelle concentration. When the loading level of the compatibilizer was higher than 9% w/w, the blending may be attributed to formation of micelles in the ethylene vinyl acetate matrix due to the excess amount of compatibilizer. This resulted in the decrease of the G' values in the polymer blends.

Kaewsakul and coworkers[1] prepared natural rubber/ethylene vinyl acetate blends, and detected the phase continuity and phase inversion phenomena by the dynamic properties in terms of the relationship between tan δ and the weight fraction of rubber in the blends. Different forms of natural rubber were exploited such as unmodified natural rubber or an air dried sheet, or modified as a maleated natural rubber. In addition, phenolic modified ethylene vinyl acetate was used as a compatibilizer. The incorporation of the two types of natural rubber exhibited decreasing tan δ values. In addition, when the natural rubber content in the blended system increased, this greatly reduced the tan δ values. This effect might be attributed to the dispersion of the natural rubber blends to change the continuous phase morphology due to a lower dissipated energy and the higher elasticity nature of the natural rubber. However, the maleated natural rubber/ethylene vinyl acetate blends exhibited the strongest interaction between the polar functional groups of the maleated natural rubber and ethylene vinyl acetate. This blending had low tan δ values compared to the unmodified natural rubber/ethylene vinyl acetate blends. Nevertheless, the tan δ values depended on the reactive blending of the polymer blends.

Natural rubber/polystyrene-based IPNs were synthesized using azobisisobutyronitrile, benzoyl peroxide, or dicumyl peroxide as an initiator, and divinyl benzene as a crosslinking agent.[3,27] For traditional natural rubber/polystyrene blends, the tan δ that calculated from G', and G'' had two distinct transitions that corresponded to the natural rubber and the polystyrene phases. This indicated that the system was immiscible at the molecular level. In both semi-IPNs and full IPNs, however, an increasing miscibility and intermixing between the two phases of natural rubber and polystyrene was found. The crosslinking of polystyrene produced chemical entanglements that resulted in some intimate mixing of the phases which improved the homogeneity. With high crosslinker levels in the polymer blends, the physical entanglements were generated and led to the tighter and more compact blended networks. In the semi-IPNs, the higher natural rubber/polystyrene crosslinking resulted in brittle IPNs, and the flexibility of the polymer chains became highly restricted. The damping property gradually decreased, whereas the G' increased. In contrast, the magnitude of the tan δ peak increased when the natural rubber content increased and this enhanced the elasticity. The different initiator systems had different viscoelastic behaviours. The dicumyl peroxide produced the maximum phase mixing and the highest G' due to the intimate and effective crosslinking of the phases and the strongest interaction. In the full IPNs, the natural rubber content in the composition increased the tan δ peak and the damping property, but the G' decreased. However, the semi-IPNs had better damping properties than the full IPNs because of the presence of the linear polystyrene chains.

Jayasuriya and Houraton[26] prepared natural rubber/poly(methyl methacrylate) blends by semi-IPNs methods using 0.5 and 1.5 mol% ethylene glycol dimethacrylate as a crosslinking agent. These semi-IPNs possessed two glass transition temperatures at higher poly(methyl methacrylate) contents to indicate a phase-separated nature of these materials. In contrast, the increment of the level of crosslinking in the poly(methyl methacrylate) phase slightly changed the glass transition temperatures of both the natural rubber and the poly(methyl methacrylate) in the semi-IPNs blends. It increased natural rubber's glass transition temperature and decreased poly(methyl methacrylate)'s glass transition temperature when determined by dynamic mechanical thermal analysis. This indicated that some grafting of the poly(methyl methacrylate) to the natural rubber backbones resulting in partial compatibility of these blends. In addition, the maximum tan δ of the natural rubber component decreased when the poly(methyl methacrylate) content increased. The G' value increased with an increasing content of poly(methyl methacrylate) from 30 to 50% w/w. This behaviour was caused by the reinforcement effect of the hard and glassy poly(methyl methacrylate) phases. The degree of component mixing increased with the increasing level of crosslinker in the poly(methyl methacrylate) phase that improved the compatibility of their blends. It was concluded that the natural rubber phase was the continuous phase in all these samples. Therefore, the blends behaved as reinforced elastomers. The mechanical properties such as modulus, tensile strength, and hysteresis behaviour increased with increasing poly(methyl methacrylate) contents in the semi-IPNs. Moreover, increasing the degree of crosslinking of the poly(methyl methacrylate) component also improved the tensile properties of the semi-IPNs. This might be attributed to the increase of interpenetration and reinforcement effects in the semi-IPNs that were disclosed by the hard and glassy crosslinked poly(methyl methacrylate) phases in continuous dispersions in the natural rubber matrix.

The carbon black-reinforced full natural rubber/polystyrene IPNs were synthesized using dicumyl peroxide as a crosslinking agent.[8] They presented the natural rubber as the dominant phase in the full IPNs networks. When the IPNs expressed a broad peak in the spectra of the temperature dependence of tan δ, they showed the enhanced mechanical properties. The tensile properties in terms of modulus, tensile strength, and elongation at break were improved when the polystyrene contents increased to the optimum level at 12% w/w. The increasing tensile strength was due to the interpenetration and interlocking of the natural rubber and polystyrene networks that resulted in the reinforcement of the natural rubber phase. In contrast, the decreasing tensile strength expressed weak to declining mechanical properties due to the phase separation of the natural rubber and polystyrene that presented two discrete tan δ peaks. For this effect, polystyrene acted as rigid filler, and its tensile strength declined. Thus, the dynamic and mechanical properties were related and the mechanical properties could be predicted from the dynamic properties. Moreover, the carbon black-reinforced full IPNs clearly decreased the miscibility of the natural rubber and polystyrene, and the resulting phase separation occurred with less polystyrene content.

20.3.2 Natural Rubber/Thermosets

Thermosetting polymers are often highly strong polymers crosslinked by covalent bonds; therefore, they are irreversible polymeric materials and difficult to recycle. These covalent bonds form during the initial moulding process that provides the material with a stable structure. When these polymers are heated, they do not melt, cannot be shaped or formed to any great extent, and will definitely not flow. They decompose at temperature before their melting points are reached. However, they are also better suited to high temperature applications up to their decomposition temperature. Thus, they are more likely to be used in situations where thermal stability is required. They tend to lack tensile strength and are more brittle.[2,12] Principal examples of thermosets include the epoxy resins, phenol-formaldehyde resins, and unsaturated polyester resins.

Epoxy resins are defined as thermosetting polymers containing at least one epoxide group. This group is also termed as an epoxy, oxirane or ethoxyline group. These polymers are widely used in many industrial applications as structural adhesives to form a matrix in high performance materials, such as surface coatings, engineering composites, and electrical laminates to sealant. They are amorphous states with highly crosslinked polymers that often exhibit high tensile strength, modulus, impact and fracture strengths. These properties cause poor peeling and shear strength of epoxy based adhesives with brittleness.

Moreover, they express excellent chemical and corrosion resistance, good dimensional stability and low shrinkage.[2] However, many publications have suggested that the blending between epoxy resins and natural rubber is further required to develop more the advanced blends.

Unsaturated polyester resins are prepared from polyesters synthesized by esterification of glycol, unsaturated and saturated acids. They have the disadvantages of poor impact strength. They are highly brittle polymers, but can be improved by the dispersion of natural rubber particles in the glassy matrix due to the energy absorption by these rubber particles that can dissipate and absorb the impact load. The possibility to blend these resins with natural rubber using sodium lauryl sulfate, toluene, or ammonia as dispersion aids has been reported.[14] These blended mixtures were cured at room temperature using methyl ethyl ketone peroxide and cobalt octoate as an initiator and accelerator, respectively. Unsaturated polyesters/natural rubber blends were easy to break when an impact load was applied. In addition, the impact strength of the blended polymers depended on the types of dispersion aid for which toluene had a greater effect compared to sodium lauryl sulfate and ammonia. These blends had decreased elastic modulus, tensile strength, and flexural strength with increasing natural rubber content in the blended systems compared to the pure resins. One characteristic of natural rubber is that it has a low elastic modulus. Thus, the unsaturated polyester resins/natural rubber blends would be expected to decrease in both their tensile and flexural strengths. These blends had no strong interaction between the unsaturated polyester resins and natural rubber as far as their hardness and brittleness were concerned. However, the unsaturated polyester resins that were blended with natural rubber or synthetic rubber were immiscible in nature due to the difference in their chemical structures.[28] One of the improvements to their compatibility was to use a graft copolymer between rubber and polystyrene by free radical grafting of styrene monomers onto the rubber chains.[25] The natural rubber-*graft*-polystyrene/unsaturated polyester blends had a significant effect on their mechanical properties with increasing hardness and brittleness in a glassy matrix of unsaturated polyester resins. Thus, these blended polymers had low impact and flexural strengths compared to pure unsaturated polyester resins. When the graft copolymer contents in the natural rubber-*graft*-polystyrene/unsaturated polyester blends increased, however, the elongation at break increased. Thus, the grafted polystyrene segments supported a good dispersion of the natural rubber molecules in the glassy matrix of unsaturated polyester resins and increased the miscibility of these blended systems.

The attempt to blend natural rubber with epoxy resins resulted from the abundance of natural rubber and that it was a renewable resource. Nevertheless, interfacial adhesion between natural rubber and epoxy resins was weak due to the hydrophobic nature of natural rubber. Thus, it was an interesting experiment to blend the toughened epoxy resins with synthetic reactive liquid rubber. In order to achieve an efficient stress transfer between rubber and the epoxy matrix, the rubber must have reactive functional groups.[2,29] Kumar and Kothandaraman[2] used the diglycidyl ether of bisphenol A epoxy resins as the major phase modified with maleated depolymerized natural rubber as the minor phase. The maleated depolymerized natural rubber is a graft copolymer that is synthesized from maleic anhydride onto depolymerized natural rubber. The mechanical properties in terms of tensile strength, tensile modulus, elongation at break, flexural strength, flexural modulus and flexural strain to failure were reported. An increasing content of the maleated depolymerized natural rubber affected to be highly soluble in the epoxy resin matrix. Thus, the tensile strength, tensile modulus, flexural strength, and flexural modulus decreased when the maleated depolymerized natural rubber was mixed in epoxy phase. However, the elongation at break and flexural strain of the epoxy/maleated depolymerized natural rubber blends increased after the maleated depolymerized natural rubber was blended, due to the increase of the ductile deformation. Therefore, the maleated depolymerized natural rubber significantly improved the toughness property of the epoxy resins. In addition, the epoxy resins had increasing impact strength properties after blending with the maleated depolymerized natural rubber up to 2% w/w. Nevertheless, the miscibility of epoxy/maleated depolymerized natural rubber blends decreased when the maleated depolymerized natural rubber concentration was more than 3% w/w because it produced a random phase separation in the epoxy matrix. Thus, this blended system provided good compatibility only when the maleated depolymerized natural rubber was mixed at a low concentration. Recently, the diglycidyl ether of bisphenol A epoxy resins were also blended with hydroxyl terminated liquid natural rubber.[12] The epoxy monomer was cured with nadic methyl anhydride, a hardener, in the presence of N,N dimethyl benzyl amine as an accelerator. The formation of the blended polymer became a phase-separated structure when the natural rubber was added into an anhydride hardener/epoxy monomer mixture. The domain size of the natural rubber in the blended structure increased with an increasing natural rubber contents due to the coalescence effect. Nevertheless, the glass transition temperature of the natural rubber phase overlapped with the β relaxation of the epoxy resins at −65 °C. The viscoelastic properties of the epoxy phase in the blended systems in terms of the glass transition temperature from the dynamic mechanical spectra, tan δ, log G' and G'' slightly decreased with the addition of natural rubber due to the dilution effect, and might also have resulted from the miscible rubber phase in the epoxy rich phase. However, tan δ often presented a single relaxation for the natural rubber/epoxy blends around 130 °C that corresponded to the glass transition temperature of the epoxy phase. The mechanical properties such as the impact and fracture toughness were found to be greater than the unmodified epoxy resins and increased with the increasing natural rubber content. The natural rubber droplets might act as stress concentrators leading to plastic deformations in the surrounding matrix. They might take up a significant amount of the applied stress that could increase the mechanical properties by increasing the cavitation and shear deformation in their matrix. Therefore, the heterogeneous natural rubber/epoxy blends exhibited good interfacial adhesion between the two phases that increased the mechanical properties in the rubber modified epoxy systems.

20.3.3 Natural Rubber/Synthetic Rubbers

The natural rubber does not generally exhibit all the desired properties for use in the rubber industry. Thus, it is possible to obtain better mechanical and physical properties at a lower cost by blending natural rubber with synthetic rubbers. Normally, natural rubber is deteriorated by ozone and thermal attacks due to its highly unsaturated backbone, and it also shows low oil and chemical resistances due to its non-polarity. However, these properties can be achieved by blending it with low unsaturated ethylene propylene diene monomer rubber, styrene butadiene rubber, carboxylate styrene butadiene rubber, nitrile butadiene rubber, chloroprene rubber, chlorosulfonated polyethylene rubber, and acrylonitrile butadiene rubber.[7,11,15,16,23]

Varkey and coworkers[16] studied the mechanical and viscoelastic properties of natural rubber/styrene butadiene rubber latex blends. The miscibility of these blends depended on the compositions, the viscosity of the individual components, and the processing history. The low viscosity components generally acted as the continuous phase. However, the amount of each phase was also the dominant factor to indicate the continuous action. Less than 50% w/w styrene butadiene rubber contents revealed a two phase structure in which styrene butadiene rubber domains were dispersed in the continuous natural rubber matrix. However, phase inversion occurred to form natural rubber dispersions in the styrene butadiene rubber phase when the styrene butadiene rubber content increased to more than 50% w/w. Nevertheless, these blended polymers exhibited immiscibility that had two glass transition temperatures of both natural rubber and styrene butadiene rubber from the dynamic mechanical analysis or viscoelastic data. Their glass transition temperatures shifted to closer values when the natural rubber content was increased in the blended systems due to the increased interaction of the individual components. The viscoelastic properties as the tan δ, G' and G'' of the blended polymers depended on the structural crystallinity and the extent of crosslinking of their blends. By increasing the natural rubber contents, these values of the blended polymers decreased. The mechanical properties such as tensile strength, modulus, elongation at break, and hardness also increased due to the increasing degree of the crosslinked density and the influential crystallization of natural rubber. They led to strengthening of the three-dimensional networks.

Natural rubber/chlorosulfonated polyethylene rubber blends also exhibited immiscibility.[15] Chlorosulfonated polyethylene rubber is the synthetic rubber used for applications in electric cables, hoses for liquid chemicals, waterproof cloths, floor tiles, and oil-resistant seals. It is chosen to blend with natural rubber to improve the resistance of natural rubber to ozone, oil, heat, flame and non-polar chemicals. This is due to the effect of the polarity of the chlorine groups in the chlorosulfonated polyethylene rubber. The tensile strength, elongation at break, and tear strength of these blends decreased with the increasing chlorosulfonated polyethylene rubber contents. In addition, the compatible natural rubber/chlorosulfonated polyethylene rubber blends were improved by adding the epoxidized natural rubber (Epoxyprene® 25) as a compatibilizer. The components were blended with natural rubber by the two-roll mill processing. Only 1–3 part per hundred of rubber (phr) of epoxidized natural rubber was sufficient to increase the compatibility of these blends. Moreover, it also increased the tensile strength but slightly decreased the elongation at break, and it did not improve the tear strength of these blends. The optimum compatibilizer concentration to improve the mechanical properties was 10 phr. However, its content higher than 10 phr decreased the tensile properties. These blended polymers exhibited two tan δ peaks belonging to natural rubber and chlorosulfonated polyethylene rubber. The addition of epoxidized natural rubber into these blends decreased the tan δ peak of chlorosulfonated polyethylene rubber, but it also revealed the compatibility of these blends. This miscibility was due to the *in situ* grafting reaction occurred by crosslinking between epoxidized natural rubber and chlorosulfonated polyethylene rubber on the surface of the rubber particles. This reduced the interfacial tension between the chlorosulfonated polyethylene rubber particles and the natural rubber matrix, and the epoxidized natural rubber acted as the load transferring agent between the natural rubber and the chlorosulfonated polyethylene rubber.

The natural rubber and acrylonitrile butadiene rubber were blended using ethylene vinyl acetate as a compatibilizer.[11] Although this compatibilizer did not improve the miscibility of the heterogeneous natural rubber/acrylonitrile butadiene rubber blends, the addition of ethylene vinyl acetate up to 6 phr increased the tensile strength, elongation at break, and tear strength due to the increasing interfacial adhesion between the blended components and then rigidity of the matrix. Furthermore, higher compatibilizer loading might affect the saturation of the interface adhesion that increased the interfacial tension, resulting in a reduction of the tensile strength, elongation at break, and tear strength. Thus, the optimum concentration of the ethylene vinyl acetate as compatibilizer for natural rubber/acrylonitrile butadiene rubber blends was 6 phr.

Carboxylate styrene butadiene rubber is a polar rubber that shows better adhesion to polar substrates such as textile fibres, paper fibres, and metals. It enhances the polymer tensile strength, and improves the resistance to hydrocarbon oils. It can easily modify the properties of natural rubber in their blends.[23] Natural rubber/carboxylate styrene butadiene rubber blends had better mechanical properties in terms of modulus, tensile strength, and elongation at break than their individual natural rubber components due to the strain-induced crystallization from the carboxylate styrene butadiene rubber and increased adhesion at their surfaces. The high strength of the carboxylate styrene butadiene rubber was attributed to the self-curing nature of the polymer blends. In addition, the strength of the cohesion in the carboxylate styrene butadiene rubber was much higher than that in natural rubber because of its high polarity. However, the blending ratio of carboxylate styrene butadiene rubber at 50% w/w exhibited maximum mechanical properties. The further addition for more than 50% w/w reduced the mechanical properties of the blends due to the increase of the carboxylate styrene butadiene rubber rigid phase. In addition, the tension values after failure decreased with the increasing contents of the carboxylate styrene butadiene rubber in the systems. Thus, the blended polymers were hard and brittle when the carboxylate styrene butadiene rubber contents increased. Nevertheless, from the tan δ, G' and G'' results, the polymer blends were immiscible showing two glass transition temperature values that corresponded to those of natural rubber and the carboxylate styrene butadiene rubber. The tan δ, G' and G'' values increased with the increasing content of carboxylate styrene butadiene rubber due to its higher intermolecular interaction in the blends that could better resist material deformations. These maximum values were also obtained when the carboxylate styrene butadiene rubber content was 50% w/w that was similar to the results of their mechanical properties. Although these polymer blends were thermodynamically immiscible, their miscible blends were formed in a latex dispersion without phase separation or flocculating. Moreover, these rubber blends could be vulcanized by the conventional sulfur-cured systems or gamma radiation.

Generally, natural rubber requires particular fillers to obtain their most desired properties as well as to make highly suitable rubber compounds for a variety of applications.[20] Recently, the polystyrene-encapsulated nanosilica (polystyrene-nSiO$_2$) was successfully used as a reinforcing filler in natural rubber/carboxylate styrene butadiene rubber blends to improve the mechanical properties in terms of modulus, tensile strength, and elongation at break.[7] The addition of the polystyrene-nSiO$_2$ reinforcing fillers increased the tensile strength and elongation at break due to the increasing strong interaction between the carboxyl groups of carboxylate styrene butadiene rubber and the hydroxyl groups of the nSiO$_2$ *via* hydrogen bonding. Moreover, the polystyrene shell also improved the adhesion at the interfaces between the nSiO$_2$ and the rubber matrix because the polystyrene molecules could prevent the aggregation of the nSiO$_2$ particles. This polystyrene shell improved the compatibility of the nanofiller and the rubber matrix. A slight shift of glass transition temperatures of the individual natural rubber and the carboxylate styrene butadiene rubber in their blends was observed that indicated the partial miscibility of the components. Polystyrene-nSiO$_2$ slightly increased the tan δ and G' compared to natural rubber/carboxylate styrene butadiene rubber blends without a reinforcing filler. Therefore, the polystyrene-nSiO$_2$ nanofillers were used to improve the reinforcement of the natural rubber/carboxylate styrene butadiene rubber blends for a variety of applications.

20.3.4 Natural Rubber/Biopolymers

Biopolymers are biodegradable and environmentally friendly polymers. They can be classified into four types, *i.e.* sugar, starch, cellulose, and synthetic biomaterials. The blending of natural rubber with biopolymers has been attractive and has created considerable interest in industry and academically for developing biodegradable materials of polymer blends at a reasonable cost and with appropriately combined properties of both the natural rubber and the other biopolymers. In addition, biopolymers possess generally useful properties such as biodegradability, biocompatibility, and non-toxicity leading to their extensive use over a wide range of applications.

Chitosan is produced commercially by deacetylation of chitin that is the main component of fungi cell walls, exoskeletons of arthropods such as crabs, lobsters, shrimps, insects, molluscs, including the squid. The addition of chitosan into the natural rubber phase increased the brittleness and hardness of the blended polymers that were related to the increase of the modulus, tensile strength, and Shore A hardness value, but a decrease of the elongation at break compared to pure natural rubber due to the flexibility of the fairly brittle chitosan.[30] Thus, the flexibility of the polymer chain in these blended systems was highly restricted. In addition, the mechanical properties of natural rubber/chitosan blends increased when the natural rubber/chitosan was aged at 55 °C for 10 days due to thermal crosslinking in the natural rubber phase. This increased the adhesion between natural rubber and chitosan phases in the blended systems.

Starch has various advantages compared to the other biopolymers. It is an inexpensive, abundant supply, renewable, environmental amity, and a fully biodegradable material. The cassava or tapioca starch was first selected to blend with natural rubber due to its high transparency, glistening sheen, and gloss. The composites of gelatinized cassava starch and natural rubber were polymeric materials displaying excellent physical and adhesive properties. They had the high degree of hardness and brittleness so that the tensile strength and elongation at break decreased with increasing cassava starch contents. In natural rubber blends with high cassava starch contents, however, the systems produced the agglomeration of starch particles. This problem could be solved by using cassava starch blended with an natural rubber-*graft*-poly(methyl methacrylate) that allowed a mix of cassava starch up to 60 phr.[31] Nevertheless, these blends did not improve the mechanical properties in terms of tensile strength, elongation at break, and tear strength that each continuously decreased with an increasing cassava starch content. This might be due to the high concentration of the cassava starch particles in the compound that inability to support stress transferred from the elastomeric phase. However, these blended polymers might increase the rigidity of cassava starch and natural rubber interfaces upon increasing the content of cassava starch. Hence, they became tight and hardened. Furthermore, the maize starch blended natural rubber also exhibited the same trend of mechanical properties as the natural rubber/cassava starch.[33] Increasing the maize starch contents decreased the mechanical properties due to the intrinsic properties of the individual starch. Although these blended polymers showed poor mechanical properties due to a low interfacial interaction between the two phases, they were improved by addition of a compatibilizer. The glycidyl methacrylate was an appropriated compatibilizer that improved the mechanical properties of natural rubber/maize starch blends. These properties increased compared to the uncompatibilized blends when the concentration of glycidyl methacrylate was increased up to 1 phr. The epoxy group of the glycidyl methacrylate interacted chemically with the hydroxyl group of maize starch that greatly decreased the cohesion energy and crystallization of starch. Therefore, starch molecules had a good dispersion in the natural rubber matrix that increased the interfacial interaction between the hydrophilicity of starch and the hydrophobicity of the natural rubber.

Cellulose derivatives are the synthetic biopolymers commonly used for blending with natural rubber in both liquid and solid forms. Hydroxypropylmethyl cellulose, methylcellulose, and sodium carboxymethylcellulose have been blended into the natural rubber latex.[33–35] Natural rubber blended films were produced, and their mechanical and viscoelastic properties were investigated. The glass transition temperature of natural rubber/cellulose derivative blends slightly shifted from that of the pure natural rubber, but the compatibility of these blends was confirmed. The natural rubber/cellulose derivative blends had increased modulus, ultimate tensile strength, elasticity, and adhesive properties. The tensile strength was significantly increased when the amount of cellulose derivatives in the blended systems increased while the adhesiveness did not always increase with different polymer blends. Natural rubber/methylcellulose blends produced films with the highest tensile property.

Polylactide, a derivative product from renewable resources such as corn starch, tapioca, or sugarcane, is a synthetic biodegradable polymer or bioplastic that can be degraded under certain conditions such as the presence of oxygen. It is stiff and brittle at room temperature. The toughness of the polylactide was improved by blending with natural rubber.[36] However, the natural rubber/polylactide blends exhibited immiscibility due to the difference in their polarities. The tan δ and G'' peak temperatures of the blended polymers clearly indicated two glass transition temperatures in the blends that confirmed the incompatibility between natural rubber and polylactide. Since the neat polylactide produced a brittle fracture, however, the natural rubber improved the tensile property, ductility, and toughness in their blends due to the reduction of the plastic deformational behaviour. The natural rubber particles could disperse in polylactide matrix and these polymer blends were highly biodegradable. Natural rubber/polylactide blends also provided soft and ductile materials. In addition, the compatibility between natural rubber and polylactide blends was further improved by addition of a compatibilizer, such as natural rubber-*graft*-glycidyl methacrylate.[10] It increased the interaction between the natural rubber and polylactide components because the epoxy group in natural rubber-*graft*-glycidyl methacrylate could react with the carboxyl groups of polylactide chains during blending, and resulted in a better dispersion of the natural rubber particles in the polylactide matrix. Furthermore, it significantly improved the mechanical properties such as tensile strength, elongation at break, and impact strength compared to the pure polylactide and natural rubber/polylactide blends without compatibilizer. These properties increased up to their maximum values when the content of the compatibilizer was increased to 1% w/w and the percentage of the grafting of natural rubber-*graft*-glycidyl methacrylate increased to 4.35%. Increasing the compatibilizer amounts higher than the optimum level to produce the blended polymers showed some large domain size of natural rubber particles and a poor distribution of natural rubber particles in the polylactide matrix.

20.4 Conclusions

Natural rubber based-blends and IPNs have been developed to improve the physical and chemical properties of conventional natural rubber for applications in many industrial products. They can provide different materials that express various improved properties by blending with several types of polymer such as thermoplastics, thermosets, synthetic rubbers, and biopolymers, and may also adding some compatibilizers. However, the level of these blends also directly affects their mechanical and viscoelastic properties. The mechanical properties of these polymer blended materials can be determined by several mechanical instruments such as tensile machine and Shore durometer. In addition, the viscoelastic properties can mostly be determined by some thermal analyser such as dynamic mechanical thermal analysis and dynamic mechanical analysis to provide the glass transition temperature values of polymer blends. For most of these natural rubber blends and IPNs, increasing the level of polymer and compatibilizer blends resulted in an increase of the mechanical properties until reached an optimum level, and then their values decreased. On the other hand, the viscoelastic behaviours mainly depended on the intermolecular forces of each material blend that can be used to investigate the miscibility of them. Therefore, the natural rubber blends and IPNs with different components should be specifically investigated in their mechanical and viscoelastic properties to obtain the optimum blended materials for use in several applications.

References

1. W. Kaewsakul, A. Kaesaman and C. Nakason, *e-Polymers*, 2012, **5**, 1.
2. K. D. Kumar and B. Kothandaraman, *eXPRESS Polym. Lett.*, 2008, **2**, 302.
3. A. P. Mathew, G. Groeninckx, G. H. Michler, H. J. Radusch and S. Thomas, *J. Polym. Sci., Part B: Polym. Phys.*, 2003, **41**, 1680.
4. Z. Oommen, G. Groeninckx and S. Thomas, *J. Polym. Sci., Part B: Polym. Phys.*, 2000, **38**, 525.
5. J.-M. Vergnaud and I.-D. Rosca, in *Rubber Curing and Properties*, ed. J.-M. Vergnaud and I.-D. Rosca, CRC Press, Florida, 2008, pp. 135–147.
6. R. O. Ebewele, in *Polymer Science and Technology*, ed. R. O. Ebewele, CRC Press, Florida, 2000, pp. 337–381.
7. S. Chuayjujit and W. Luecha, *J. Elastomers Plast.*, 2011, **43**, 407.
8. S. H. Heidary, I. A. Amraei and A. Payami, *J. Appl. Polym. Sci.*, 2009, **113**, 2143.
9. P. Intharapat, D. Derouet and C. Nakason, *Polym. Adv. Technol.*, 2010, **21**, 310.
10. P. Juntuek, C. Ruksakulpiwat, P. Chumsamrong and Y. Ruksakulpiwat, *J. Appl. Polym. Sci.*, 2012, **125**, 745.
11. P. Kumari, C. Radhakrishnan, S. George and G. Unnikrishnan, *J. Polym. Res.*, 2008, **15**, 97.
12. V. S. Mathew, P. Jyotishkumar, S. C. George, P. Gopalakrishnan, L. Delbreilh, J. M. Saiter, P. J. Saikia and S. Thomas, *J. Appl. Polym. Sci.*, 2012, **125**, 804.

13. T. G. Nieh, J. Wadsworth and O. D. Sherby, in *Superplasticity in Metals and Ceramics*, ed. T. G. Nieh, J. Wadsworth and O. D. Sherby, Cambridge University Press, Cambridge, 1997, pp. 189–207.
14. P. Pachpinyo, P. Lertprasertpong, S. Chuayjuljit, R. Sirisook and V. Pimpan, *J. Appl. Polym. Sci.*, 2006, **101**, 4238.
15. V. Tanrattanakul and A. Petchkaew, *J. Appl. Polym. Sci.*, 2006, **99**, 127.
16. J. T. Varkey, S. Augustine, G. Groeninckx, S. S. Bhagawan, S. S. Rao and S. Thomas, *J. Polym. Sci., Part B: Polym. Phys.*, 2000, **38**, 2189.
17. J. Bicerano, in *Prediction of Polymer Properties*, ed. J. Bicerano, CRC Press, Florida, 2002, pp. 222.
18. A. Kumar and R. K. Gupta, in *Fundamentals of Polymer Engineering*, ed. A. Kumar and R. K. Gupta, CRC Press, Florida, 2003, pp. 487–525.
19. P. A. Schweitzer, in *Metallic Materials*, ed. P. A. Schweitzer, CRC Press, Florida, 2003, pp. 58–97.
20. I. Walker and A. A. Collyer, in *Rubber Toughened Engineering Plastics*, ed. A. A. Collyer, Chapman & Hall, London, 1994, pp. 29–56.
21. H. F. Brinson and L. C. Brinson, in *Polymer Engineering Science and Viscoelasticity*, ed. H. F. Brinson and L. C. Brinson, Springer Science + Business Media, New York, 2008, pp. 221–364.
22. C. S. Brazel and S. L. Rosen, in *Fundamental Principles of Polymeric Materials*, ed. S. L. Rosen, John Wiley & Sons, New York, 1993, pp. 276–307.
23. R. Stephen, K. V. S. N. Raju, S. V. Nair, S. Varghese, Z. Oommen and S. Thomas, *J. Appl. Polym. Sci.*, 2003, **88**, 2639.
24. T. Murayama, in *Dynamic Mechanical Analysis of Polymeric Materials*, ed. T. Murayama, Elsevier, New York, 1978, pp. 40–68.
25. S. Chuayjuljit, P. Siridamrong and V. Pimpan, *J. Appl. Polym. Sci.*, 2004, **94**, 1496.
26. M. M. Jayasuriya and D. J. Hourston, *J. Appl. Polym. Sci.*, 2012, **124**, 3558.
27. A. P. Mathew, S. Packirisamy, H. J. Radusch and S. Thomas, *Eur. Polym. J.*, 2001, **37**, 1921.
28. J. C. Salamone, in *Polymeric Materials Encyclopedia*, ed. J. C. Salamone, CRC Press, Florida, 1998, pp. 8486.
29. Y. Tanaka, T. Sakaki, A. Kawasaki, M. Hayashi, E. Kanamaru and K. Shibata, *US Patent* 5,856,600, 1999.
30. V. Rao and J. Johns, *J. Appl. Polym. Sci.*, 2008, **107**, 2217.
31. C. Nakason, A. Kaesaman and K. Eardrod, *Mater. Lett.*, 2005, **59**, 4020.
32. A. I. Khalaf and E. M. Sadek, *J. Appl. Polym. Sci.*, 2012, **125**, 959.
33. J. Suksaeree, P. Boonme, W. Taweepreda, G. C. Ritthidej and W. Pichayakorn, *Chem. Eng. Res. Des.*, 2012, **90**, 906.
34. W. Pichayakorn, J. Suksaeree, P. Boonme, T. Amnuaikit, W. Taweepreda and G. C. Ritthidej, *J. Membr. Sci.*, 2012, **411–412**, 81.
35. W. Pichayakorn, J. Suksaeree, P. Boonme, T. Amnuaikit, W. Taweepreda and G. C. Ritthidej, *Ind. Eng. Chem. Res.*, 2012, **51**, 8442.
36. M. Kowalczyk and E. Piorkowska, *J. Appl. Polym. Sci.*, 2012, **124**, 4579.

CHAPTER 21

Scattering Studies on Natural Rubber Based Blends and IPNs

VALERIO CAUSIN

Dipartimento di Scienze Chimiche, Università di Padova, via Marzolo 1, 35131 Padova, Italy
Email: valerio.causin@unipd.it

21.1 Introduction

At first glance, the amorphous nature of natural rubber (NR) could seem to preclude the application of diffraction or scattering studies, which are traditionally associated to the investigation of ordered and crystalline materials. This is only partly true. First of all, rubber, when stretched, can crystallize, and this is indeed one of the reasons why NR is still undoubtedly the material of choice for very demanding, heavy-duty applications such as truck tyres. Moreover, when NR is blended with other polymers, scattering methods can probe the microscopic heterogeneity of the system, yielding information on the morphology and extent of phase separation. Scattering methods can therefore contribute to a thorough understanding of NR based blends and IPNs. This achievement is very important, because it is known that structure and morphology determine the macroscopic properties of the material, not only mechanical ones, but also functional ones such as gelation time kinetics in elastomer/thermoset blends[1] or gas permeability.[2]

Neutron scattering detects fluctuations in the density of nuclei in the sample. X-ray scattering is sensitive to inhomogeneities in electron density whereas light scattering depends on fluctuations in polarizability. Changing the characteristics of the light/particle beam, different scale lengths and morphological features can be probed, allowing us to obtain a very complete picture of the material.

A key advantage of scattering methods is that they sample the whole bulk of the material, thus giving a generalized picture of its morphology, differently from microscopy which images only small portions of the specimen, not necessarily representative of the entire sample. On the other hand, microscopy has the advantage that data are acquired in direct (real) space whereas scattering methods measure in reciprocal space, and are thus much more demanding in terms of data interpretation. When all these methods are used complementarily, fully exploiting their respective advantages, a very rich amount of information can be obtained, and a really complete picture of the several levels of morphological organization in polymeric blends and IPNs can be drawn.

21.2 Wide Angle X-Ray Diffraction

X-rays are surely the light source which found more extensive application in scattering techniques in the polymer science field.

The propagation of X-rays in space is accompanied by a periodically changing electric field. When X-rays impinge on a sample, the electrons of the atoms of the sample are excited to periodic vibrations by this changing field, and become themselves sources of spherical electromagnetic waves of the same frequency and wavelength, *i.e.* they scatter the original beam. The scattered X-ray waves from the atoms can interfere constructively or destructively along certain directions of space, provided that certain geometrical conditions are met. Diffraction is observed when the scattered waves, along a certain direction, have a difference in phase equal to an integer number of wavelengths. When this is not true, destructive interference occurs and scattered radiation is cancelled along that direction. Amorphous materials produce diffuse X-ray patterns, whereas crystalline ones give rise to sharp and neat signals due to diffraction.

A crystal is a regular arrangements of atoms, produced by the periodic repetition of a unit cell. Within the crystalline lattice, many families of planes can be identified, which are defined by three integer numbers (the Miller indices, *hkl*). The basic relationship at the basis of X-ray diffraction is the Bragg equation:

$$n\lambda = 2d_{hkl}\sin\theta \qquad (21.1)$$

The *hkl* family of planes, equally spaced by a d_{hkl} distance, impinged by an incident beam of wavelength λ, will give a diffraction signal of order n only at a precise angular value θ given by Equation (21.1). Along the other directions, destructive interference will occur and no reflections will be observed.

In wide-angle X-ray diffraction (WAXD), diffracted intensities at angles wider than about 2 °2θ are detected, allowing us to study features of characteristic size of about 30–40 Å or less. In other words, WAXD is useful to study the crystalline unit cell, the polymorphism and the size of crystallites of semi-crystalline polymers.

Polymers may be divided in amorphous and semicrystalline. Amorphous polymers, due to the nature, the shape and/or the mobility of their molecules are not able to form crystals, but they have a liquid-like disorder at the solid state. NR is an example of amorphous polymer. Semicrystallinity, on the other hand, is a characteristic feature of polymers. Due to entropic constrains, large macromolecular chains, even when they can in principle crystallize, are not able to extend their crystalline lattice to the whole sample. The structure and morphology of polymers can be described as semicrystalline, because it consists of an alternation of crystalline and amorphous domains. Both regions of the semicrystalline framework contribute to the XRD pattern of a polymeric sample: crystalline regions originate sharp reflections, whereas the amorphous zones produce a wide and diffused halo. The degree of crystallinity is the fraction of sample which is arranged in the crystalline domains, and it can be obtained by a fitting procedure of the WAXD patterns, by which the contributions of the crystalline and amorphous domains are deconvoluted.[3] Crystallinity can thus be evaluated as the ratio between the area of crystalline peaks over the total area of the diffractogram. It is worth remarking that such an assessment of the degree of crystallinity is strictly valid only in the hypothesis of a perfect two-phase system. The actual situation is far from ideal, though, because an interphase can be identified at the boundary between crystalline and amorphous domains, where a gradual transition from the high degree of order associated to crystalline domains to the disorder characteristic of the amorphous regions happens. Most of the modelling of scattering data is based on the simplifying hypothesis of ideal two-phase systems, even though correcting functions have been devised for taking transition layers into account. Figure 21.1 shows an example of the deconvolution procedure of WAXD patterns, applied to polypropylene, a typical semicrystalline polymer.

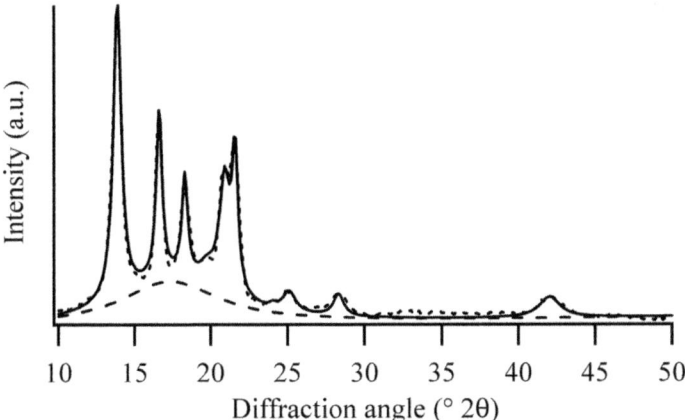

Figure 21.1 Example of the deconvolution procedure applied to calculate the degree of crystallinity from the XRD trace of polypropylene. The dotted line represents the experimental pattern, the solid line is the fitting curve, the dashed line is the amorphous halo.

The choice of the fitting functions and of the width and position of the amorphous halo introduces a certain degree of subjectivity in such data treatment, which urges to use, in the comparison of samples, always the same width and position of the amorphous halo, changing just its intensity, and the same kind of fitting functions throughout the sample series.

Polymers, depending on crystallization conditions, can order themselves according to different crystal unit cells, giving rise to the phenomenon of polymorphism. WAXD is the technique of choice for studying the polymorphism of polymers, because different crystalline phases display different diffractometric patterns. Sometimes, more than one phase coexist in the same material, and the areas of the characteristic peaks of each polymorph allow to quantify the abundance of each phase.[4] By WAXD, the size of crystallites can also be measured from the width of the peak applying the Scherrer equation,[5] and the characteristic dimensions of the crystalline cell can be assessed on the basis of the angular position of the peaks.

As said in the introduction, WAXD is usually associated to the study of crystalline materials. NR by itself is amorphous in nature, so it yields just a broad amorphous halo by WAXD. However, blending rubber to polymers which are able to crystallize may have severe consequences on the semicrystalline framework of such materials. The presence of a rubber component may favour the appearance of particular phases, as reported by Sun et al., who found that introduction of 20 vol% poly(cis-butadiene) rubber promoted the formation of β crystals in isotactic polypropylene.[6,7] Different phases have different physical and mechanical characteristics, so being able to control the polymorphism of semicrystalline polymers by a blending process is extremely promising under an applicative point of view. Blending of rubber to semicrystalline polymers is in fact usually done in a toughening perspective. If rubber is also able to modify the semicrystalline framework, a further degree of tuning of the structure and morphology of the material can be added, for an optimal design of its final properties.

The most common modification to the structure of semicrystalline polymers brought about by rubber is a hinder to crystallization, with a consequent depression of the degree of crystallinity.[8–10]

The presence of crystalline domains is not strictly necessary for acquiring information by WAXD. Halasa et al.[11] and Nigam et al.[12] extracted useful information from the amorphous halo of completely amorphous blends. They evaluated the average interchain separation in the system, $\langle R \rangle$, from the position of the maximum of the amorphous peak, and they interpreted the half-width of the WAXD amorphous halo as a qualitative expression of the distribution of $\langle R \rangle$. Although quite limited in their informative content, such data may help in assessing the morphology of completely amorphous blends with a widely available apparatus like WAXD.

Many of the properties of vulcanized NR, especially its toughness, high tensile stress and large hysteresis loss, have been ascribed to strain-induced crystallizability. The crystallization of NR has been extensively studied by many researchers[13–21] since the 1940s, using a variety of techniques.

Undeformed NR forms spherulitic crystallites below 0 °C.[17–21] The temperature/time induced crystallites generally form as folded chain lamellae, whereas the morphology of the strain/stress induced crystallites has been reported to be various: fibrils, fibrils and folded lamellae, and shish-kebabs.[14,17–22] Strain-induced crystallites and their orientation have been indicated as the key reasons for the sharp increase in modulus that accompanies strain-induced crystallization.

Most of the early experiments on these topics were done with sequential measurements in which the sample was extended to a desired strain, fixed at that strain, and subsequently removed from the dynamometer and mounted on the diffractometer to be examined. However, when the stretching was stopped, stress was relaxed and microstructures often changed, thereby significantly altering the system of interest. In order to understand what happens during stretching and retraction, it is necessary to examine the sample simultaneously with the strain. The use of synchrotron WAXD during the process of stretching has allowed us to directly follow the development of structure and stress–strain relations during deformation of rubbers in real time.[23–30] An advantage of this technique is that of enabling structural investigations on a very short time scale. For example it was possible to observe that NR responds to fast deformation by forming strain-induced crystallites more rapidly than its synthetic analogues.[24,27] This determines the superior and inimitable toughness of NR. An example of the quality of data yielded by this approach is shown in Figure 21.2.

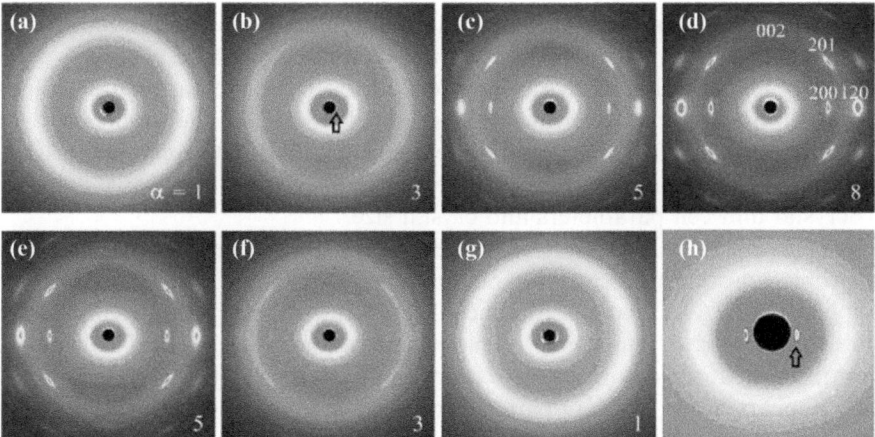

Figure 21.2 Sequential change of WAXD patterns from a NR sample. Stretching direction is vertical. Corresponding strain values are indicated at the right bottom in parts (a) to (g) of the figure. Indices of crystalline reflections of NR are indicated in part (d). Part (h) shows the enlarged image of the centre of part (b). Sharp reflections from fatty acids (indicated by arrows) corresponding to ca. 4.2 nm in real space are recognized on both sides of the beamstop.
Reprinted with permission from M. Tosaka *et al.*, *Macromolecules*, **37**, 3299–3309. Copyright (2004) American Chemical Society.

Figure 21.2 shows WAXD patterns from an NR sample during the stretching (Fig. 2a–d) and retracting (Fig. 2d–g) processes. At small strains during stretching, highly oriented reflections of fatty acids[31] were recognized on both sides of the beamstop (arrows in Fig. 2b,h). Increasing strain, NR started to crystallize (the indicization of the signals is also shown). In this figure also the orientation of nascent crystallites (Figure 21.2(c)) and the persistence of a strong isotropic amorphous halo (Figure 21.2(d)) can be observed.

This figure allows also to follow the retracting process, in which the crystalline reflections disappeared gradually, returning to the initial isotropic pattern after total relaxation of the sample. Ringlike (unoriented) crystalline reflections of NR were not observed during the cyclic deformation process, allowing to conclude that strain-induced crystallites always appear in the form of oriented crystals.[26,28]

The crystal structure of NR is orthorhombic (space group *Pbca*).[32] The (002) diffraction line, that is originated by lattice planes perpendicular to the *c* chain axis, is particularly informative. The corresponding average crystallite dimension, L_{002}, assessed by the Scherrer equation,[5] is often called in the literature the stem length and is used to describe the size of rubber crystallites.[30] The (200) and (120) reflections correspond to lattice planes parallel to the chain direction. A very widespread procedure for measuring the crystallinity or rubber is based on the Mitchell method:[30,33–36]

$$\chi = 100 \cdot \frac{I_a - I_a^*}{I_a} \tag{21.2}$$

where I_a and I_a^* are the scattered intensities of the amorphous halo in the completely amorphous and in the semicrystalline material, respectively.

A very remarkable deconvolution procedure, schematized in Figure 21.3, has been proposed,[25,26,28,37,38] by which it is possible to extract additional structural information (*i.e.*, fractions of oriented crystal, unoriented crystal, oriented amorphous, and unoriented amorphous phases) besides conventional data such as crystal unit cell parameters and crystal size.

As previously seen, the WAXD pattern of the unstretched sample exhibits an isotropic amorphous halo with no preferred orientation. As deformation is applied, the WAXD pattern becomes the superposition of a diffraction pattern due to oriented crystals and of a residual amorphous halo. The isotropic contribution at the deformed state corresponds to the original amorphous halo that appears in the WAXD pattern before stretching. The anisotropic contribution is composed of oriented crystal reflection peaks and a small amount of oriented amorphous phase.[25] After having separated the two contributions, *i.e.* isotropic and non-isotropic, ratioing of the crystal diffraction intensity over the total scattered intensity is necessary to measure the mass fraction of the strain-induced crystals. To obtain the total scattered intensity volume, a quite laborious procedure must be carried out,[25,39,40] but afterwards a normal peak fitting procedure can be applied to deconvolute the intensity profile into crystalline peaks and an amorphous halo. On the basis of the data so obtained, it is therefore possible to estimate the amounts of total isotropic and anisotropic

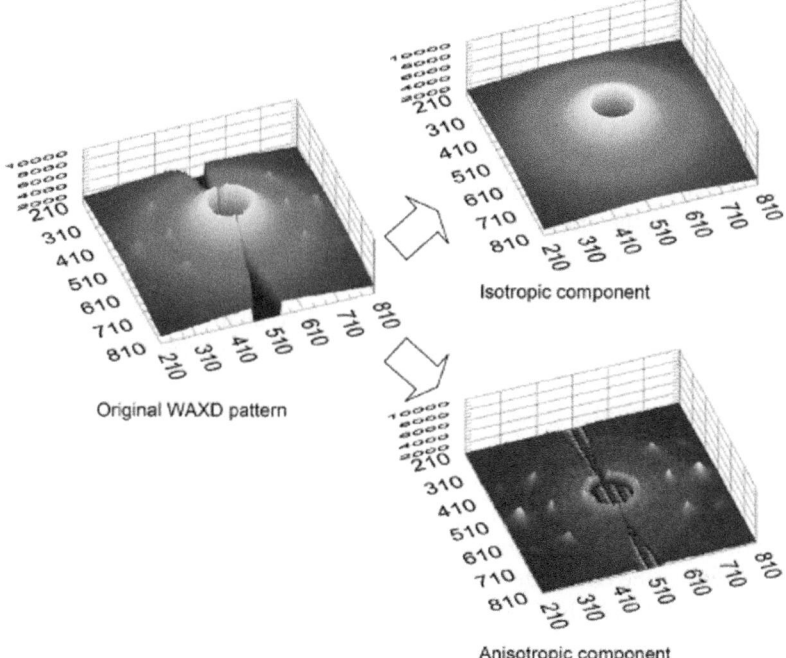

Figure 21.3 Schematic representation of the decomposition of the WAXD pattern into isotropic and anisotropic components.
Reprinted with permission from M. Tosaka *et al.*, *Journal of Applied Physics*, **101**, 084909. Copyright (2007) American Institute of Physics.

mass fractions. The isotropic fraction is composed by the unoriented amorphous fraction Φ_{UA}; the anisotropic fraction is given by the sum of oriented amorphous fraction Φ_{OA} and oriented crystalline fraction Φ_{OC}.[25] As can be observed in Figure 21.2, no unoriented crystalline phase is present during deformation experiments.

This deconvolution of the contributions due to the oriented and unoriented polymer allowed to show that, during the stretching of NR, only a small fraction of the amorphous chains become oriented and subsequently crystallize, while the majority of amorphous chains remain unstretched even at very large strains. The presence of cross-links is the main reason why chains are recalcitrant to rearranging and realigning according to the stress imposed.[26]

Probably due to the difficulty in accessing synchrotron facilities, very few such studies exist for blends containing NR,[41] although they would be extremely interesting, especially for evaluating their performance under stress.

21.3 Small-Angle X-Ray Scattering

As may be easily assessed by applying Bragg law, the analysis of aggregates of size over the tens of Å requires to limit the investigation in an angular range between 0 and 2° 2θ.

Shifting to SAXS means focusing on larger dimensional scales, neglecting the minute details of the crystallographic cell, which are the domain of WAXD, to obtain a wider view of the entire system. SAXS is observed whenever, in the material, regions of different electron density with a size between tens and thousands Å exist. Differently from WAXD, then, the ability of the polymer to crystallize is not a prerequisite to obtain a significant and informative SAXS pattern. When a component A is immersed in another component B, like in blends, the parameter that determines the scattered amplitude, and therefore the scattered intensity, is the difference in electron density between the "solute" and the "solvent": $\Delta\rho$. If the dispersed particles had the same electron density than the matrix, the X-ray beam would not distinguish between the two components and the scattered waves would be extinguished in all directions.

SAXS does not differ from WAXD from a physical point of view: when hit by X-rays, the sample's electrons become spherical emitters of coherent secondary waves, which interfere with each other.

In the case of a single particle, the scattered intensity $I(q)$ (where $q = \frac{4\pi \sin \theta}{\lambda}$, θ being half the scattering angle, according to the glancing angle used in crystallography) can be calculated taking into account the distribution function of the electrons $p(r)$, which in turn is obtained by geometrical considerations on the particle shape:[42]

$$I(q) = 4\pi \int_0^\infty p(r) \frac{\sin qr}{qr} dr \qquad (21.3)$$

$p(r)$ can be obtained by Fourier inversion of the scattering curve:[42]

$$p(r) = \frac{1}{2\pi^2} \int_0^\infty I(q) \cdot qr \sin qr \cdot dq \qquad (21.4)$$

If the dispersion of the scattering particles is homogeneous and the system is diluted (so that interference between the particles can be neglected), the total scattered intensity is given by the sum of the intensities scattered by each individual particle. When this assumption holds, then, the system may be simplified treating all the independent contributions to the diffracted intensity as coming from one single particle, representative of all the others. SAXS data analysis in such cases thus consists in developing adequate models that allow to reproduce the experimental traces, as a function of morphological parameters, such as size, shape or mass of the particles.

Theoretical $I(q)$ and $p(r)$ have been derived for a number of different geometries, so a fitting based on these models allows to obtain a good picture of the morphology of the system.[42] More complex shapes are usually approximated by a collection of primary particles, such as spheres.

If the system is not diluted, for example in cases where particles are densely packed within a matrix, interparticle interference effects need to be taken into account. Another assumption that is not always valid is that the particles are homogenous and monodisperse in size. Particle anisotropy and polydispersity are very common factors that bring about severe deviations of the system from ideality. A distribution of sizes must therefore usually be included in the models used to reproduce the experimental SAXS patterns.

As said before, many intensity functions have been calculated for a number of different shapes, e.g. spheres, ellipsoids, parallelepipedons, cylinders. A universal approximation exists for the central part of SAXS traces. Guinier[43] proposed an exponential function only dependent on the radius of gyration R:

$$I(q) = (\Delta\rho)^2 V^2 \exp\left(-\frac{q^2 R^2}{3}\right) \quad (21.5)$$

where V is the scattering volume. R is the root-mean square of the distances of all electrons from their centre of gravity and is related to the physical dimensions of the particles by geometrical relationships.

Equation (21.5) is easily linearized, allowing to obtain the radius of gyration of the particles R from the slope of a $\ln I(q)$ vs q^2 plot:

$$\ln I(q) = \ln K_0 \frac{q^2 R^2}{3} \quad (21.6)$$

where K_0 is a constant. The Guinier formula holds surprisingly well for a wide variety of cases, failing only for very anisometric particles.

V can be directly inferred from the diffraction pattern by:

$$V = \frac{2\pi^2 I(0)}{\int_0^\infty q^2 I(q) \mathrm{d}q} \quad (21.7)$$

where $I(0)$ is the scattered intensity at zero angle and the integral at the denominator is the invariant Q, which is directly related to the mean square fluctuation of electron density, irrespective of special features of the structure. In other words, deformations of the structure of the system just alter the shape of the diffraction pattern, without changing the invariant.

A further model-independent parameter can be found in the final slope of the SAXS pattern. In this region, which depends mainly on the fine structure of the particle and not on the mutual arrangement of particles, the intensity $I(q)$ can be approximated by the so-called Porod's law,[44,45] $I(q)_{q\to\infty} \frac{(\Delta\rho)^2 2\pi S}{q^4}$, where S is the surface area of the particle. This dependence on q^{-4} holds very well in most of the systems, also densely packed ones, whereas deviations from this law mostly appear in composites.[46–49] The practical application of this relationship requires the measurement of absolute intensity, but introducing the invariant this problem can be avoided and the specific interfacial area, defined as the ratio of interfacial surface area, S, to the volume, V, can be assessed from the diffraction pattern alone, without any additional data:

$$\frac{S}{V} = \frac{\pi \lim_{q \to \infty} I(q)q^4}{Q} \qquad (21.8)$$

A further approach stems from the Debye–Bueche[45,50] description of scattering from random heterogeneous media, which gives, for spherically symmetrical systems:[50]

$$I(q) = K\langle \eta^2 \rangle \int_0^\infty \gamma(r) \frac{\sin qr}{qr} r^2 dr \qquad (21.9)$$

where K is a proportionality constant and $\langle \eta^2 \rangle_{av}$ is the power fluctuation of the scattering system, which equals the difference between the electron density at a specific scattering angle and the average electron density. $\gamma(r)$ is the correlation function. For a two-phase structure with sharp interfaces where the size and shapes of the phases are random, $\gamma(r)$ may be represented by an empirical equation such as :[45]

$$\gamma(r) = \exp(-r/a_c) \qquad (21.10)$$

where the parameter a_c is the so-called correlation distance and defines the size of the heterogeneity. The correlation distance a_c is related to the particle size in dilute systems, whereas in more concentrated systems, a_c depends upon both interparticle and intraparticle distances.

Combining Equations (21.9) and (21.10) and rearranging, Equation (21.11) can be finally obtained:

$$\frac{1}{[I(q)]^{1/2}} = \frac{1}{(K'a_c^3)^{1/2}} (1 + q^2 a_c^2). \qquad (21.11)$$

where K' is the product of K and $\langle \eta^2 \rangle_{av}$.

As a consequence, a plot of $I(q)^{-1/2}$ versus q^2 should lead to a straight line having a ratio of the slope to the intercept of a_c^2.

Also by this approach, the specific interfacial area, S/V, can be obtained and it is given by:[51]

$$S/V = 4\varphi_1 \varphi_2 / a_c$$

where φ_1 and φ_2 are the volume fractions of the phases.

Three main approaches can be individuated for the treatment of SAXS data of polymer blends.

1. Qualitative inspection of the pattern;
2. Application of the Guinier, Porod or Debye-Bueche approximations for the determination of the size of dispersed particles and of the interfacial area between the phases of the blend;
3. Fitting methods based on functions, dependent on morphological features, which reproduce the experimental intensity.

A simple inspection of the SAXS patterns can by itself yield interesting preliminary information on the morphology of the samples. The appearance of a peak in the SAXS spectrum is indicative of a strong spatial correlation between the scattering particles embedded in the system, *i.e.* that the particles are arranged in an ordered array yielding a repetitive and regular pattern of electron density. If a semicrystalline polymer is present in the blend, a SAXS peak can be ascribed to the presence of lamellae, an example is reported by Li *et al.*[52] This may remind what happens in WAXD, where diffraction signals of a semicrystalline component of a blend are superposed to the amorphous halo of a rubber matrix.

In completely amorphous blends, differences in electron density that originate peaks in the SAXS pattern are caused by the particles of a polymer A dispersed within a matrix polymer B. Measuring the position of the peak and applying the Bragg relationship (Equation (21.1)), it is possible to determine the average distance between these scattering domains, as exemplified by Salgueiro *et al.*[53] The width of the peak is informative as well, since broader peaks indicate a less homogeneous and less consistent correlation length.

There are a number of works on rubber-based blends which exploited a SAXS data analysis based on approximations such as the Guinier, Porod or Debye-Bueche.[54–57] These approaches are very interesting because they offer valuable information on the size of dispersed domains within the matrix of a blend, without the need of intensive calculation and without having to develop complex theoretical models for the fitting of SAXS patterns.

The application of the Porod equation or of the Debye-Bueche approach are particularly attractive because they offer the possibility to evaluate the interfacial area between the phases of the blend, and they are probably the only way to quantify such feature in polymer blends and composites.[56–58] In fact, when the two polymers are mixed together in a blend, traditional methods based on the adsorption of small molecules, *i.e.* the BET approach, are inapplicable. Image analysis of TEM micrographs can in principle be an option, but it is extremely time consuming and it suffers from a number of limitations, such as dependence on sample preparation, on projection effects, and on image defocus. The validity of SAXS for the study of interpenetrating networks has been shown for several systems.[59–61]

A further evolution of these approaches is the Beaucage unified fit model,[62–64] which was ideated to describe scattering functions containing multiple length scales separated by power-law regimes. Scatterings from a hierarchy of structural levels are merged considering only the four parameters (G, R_i, K_p, m) required to define the single structural levels, each with its radius of gyration, R_i, and power-law exponent $-m$. The actual function used for each level is:

$$I(q) = G\exp(-q^2 R_g^2/3) + K_p \left[(\text{erf}(qR_g/6^{1/2}))^3 / q \right]^m \tag{21.12}$$

where G is the Guinier prefactor, K_p is the Porod prefactor and erf is the error function. This model is effective when it is necessary to extract the parameters associated with each hierarchical level in complex structures that yield many power-law regimes. The unified model works best with well-separated length scales, but it shows its limits when different structural regimes overlap.[65] Soares et al.[66] showed the potential of this approach in an epoxy resin blended with polybutadiene. The application of the Beaucage method allowed to describe this system as composed by primary polybutadiene particles with a 0.8 nm diameter which further aggregate in 6 nm particles, dispersed in the epoxy matrix.[66] Another example of application of the Baucage unified approach was given by Botti et al.,[67] albeit on composites and not on blends.

Deconvolution of the intensity $I(q)$ into a form factor $P(q)$ and a structure factor $S(q)$ is another possible approach to the quantitative analysis of SAXS data towards the determination of their morphology:[68–73]

$$I(q) = AP(q)S(q) \qquad (21.13)$$

where A consists of both instrument and sample dependent terms and can be treated as a scaling factor. $P(q)$ is a function that accounts for the interference effects between X-rays scattered by different parts of the same scattering body (microdomain) and is dependent upon both the size and shape of the scattering body.[74] $S(q)$ is a function that describes the interference effects between X-rays scattered by different scattering bodies in the sample and depends on their relative positions.[74]

The success of such data analysis approach is necessarily linked to the reliability of the model chosen to describe the system. This limited the use of this method of interpretation in the study of blends, in favour of more model-independent methods, like the Porod and Debye-Bueche described above. However, some examples of the use of Equation (21.13) may be found in the literature. Micellar systems of block copolymers dispersed in a polyisoprene matrix were modelled by Pavlopoulos et al.[75] with the form factor of a homogeneous sphere, multiplied by a function accounting for the polydispersity in the micelles. In this case, the structure factor was neglected, due to the extreme dilution of the system.

In analogy with what has been previously said in the context of WAXD, the use of synchrotron radiation allowed also to perform real-time SAXS measurements during tensile testing, investigating the mechanisms of deformation of the materials, and thus understanding how to improve their mechanical performance. For rubber-modified amorphous polymers, the occurrence of processes such as crazing can be identified by SAXS, as it results in typical cross-like scattering patterns consisting of two perpendicular streaks, one along the tensile direction and the other perpendicular to it.[76] The intense streak in the tensile direction is the result of scattering caused by reflection from the craze surface.[76] The less intense streak, which develops parallel to the craze plane, is attributed to the scattering of the craze fibrils and can be used to calculate the amount of crazing and the craze fibril diameter.[76] Although not focused on blends containing NR, the work of Jansen et al.[76,77] is an iconic example of how it is possible to obtain good quality data on the deformation behaviour by appropriate interpretation of SAXS data.

Figure 21.4 shows the real-time SAXS patterns acquired during tensile testing of several polymethylmethacrylate/epoxy blends. The perpendicular streaks indicating the formation of crazes can be seen in most of the data related to higher strains, on the right side of Figure 21.4.

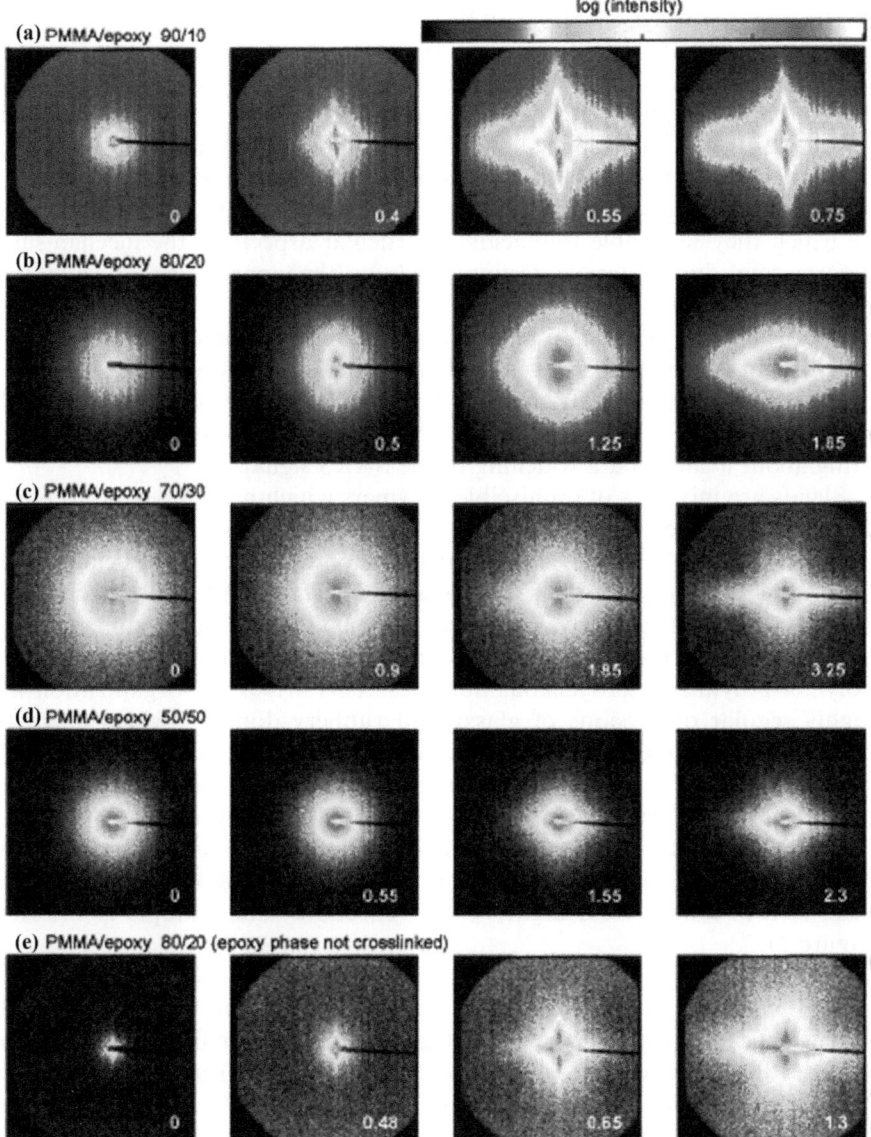

Figure 21.4 Deformation SAXS patterns of PMMA/epoxy blends measured at different clamp displacements, x [mm], indicated in the right bottom corner. Tensile direction is vertical.
Reprinted with permission from B. J. P. Jansen *et al.*, *Macromolecules*, **34**, 4007–4018. Copyright (2001) American Chemical Society.

The interpretation of the results of such experiments starts from a visual inspection of the 2-dimensional patterns. Before elongation, the patterns are isotropical rings. As a consequence of the formation of crazes which are due to the micromechanical deformation mechanism, the isotropicity of SAXS patterns is lost and intense streaks appear along certain directions.

The total scattered intensity can be obtained by integrating different regions of the 2-D pattern. Usually, focus is posed on the diameters along the equatorial and meridional directions. This allows to monitor the development of structural modifications during the deformation. Jansen et al.[76] integrated the 2-D patterns and then applied the Porod equation, quantitatively characterizing the size of the crazes which form as a consequence of tensile stress.

Several other authors reported data acquired in real-time SAXS experiments, by which they were able to elucidate particular aspects of the mechanism of deformation during tensile testing, such as the appearance of cavitation, crazing and debonding.[76-80]

SAXS can also be used to investigate the expected outcome of preparation processes of rubber-based materials. One example is the study of the evolution of the crosslink density of natural and styrene-butadiene rubbers, reported by Salgueiro et al.,[81] who observed that changes in the vulcanization conditions bring about a shift and a widening of the SAXS signal.

Blends are mixtures of immiscible polymers which naturally tend to phase-segregate; block copolymers are actually very similar systems, in which the mutually insoluble blocks are covalently bonded instead of being simply mixed together. Such systems are remarkably fit for SAXS characterization, because the presence of covalent bonds between the chemically different parts of the molecules prevents a macroscopic phase separation like that seen in polymer-polymer heterogeneous mixtures. This drives self-organization into highly regular dispersions of glassy and rubbery domains which make the performance of block copolymers similar to that of composites. A very large number of block copolymers contain polyisoprene chains. Sometimes the formation of self-organized structure is detectable also by WAXD,[82] but SAXS is surely the method of choice for such a study. Typical SAXS patterns of phase separated block copolymers consist of an intense principal peak followed by several higher order reflections. An example is shown in Figure 21.5.

The position of such higher order signals is dependent on the morphology of phase separation: in the case of a lamellar morphology, the higher order signals appear at positions that are integer multiples of the first order reflection, i.e. $1:2:3...$; the sequence of a body-centred cubic (bcc) packing of spherical microdomains is according to the ratios $1:2^{\frac{1}{2}}:3^{\frac{1}{2}}:4^{\frac{1}{2}}:6^{\frac{1}{2}}:10^{\frac{1}{2}}:12^{\frac{1}{2}}:14^{\frac{1}{2}}$ with respect to the first order peak,[83] and the secondary maxima in hexagonally packed cylinders, relative to the principal peak, follow the scaling $1, 3^{\frac{1}{2}}, 4^{\frac{1}{2}}, 7^{\frac{1}{2}}, ...$[83,84] Even though these are the most common morphologies encountered with block copolymers, many others have been reported, among which the most notable are those of the gyroid family.[85-89] Figure 21.6 illustrates some of the most common morphologies which can be found in block copolymers.

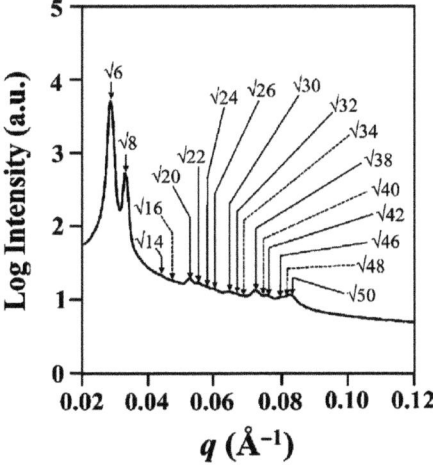

Figure 21.5 Synchrotron small-angle X-ray scattering (SAXS) data at 30 °C for a poly(isoprene-*block*-styrene) copolymer. In all SAXS profiles, peaks are indexed according to the space group characteristic of the double-gyroid morphology; solid (clearly observed) and dashed (expected) lines indicate the peak locations.
Reprinted with permission from R. Roy *et al.*, *Macromolecules*, **44**, 3910–3915. Copyright (2011) American Chemical Society.

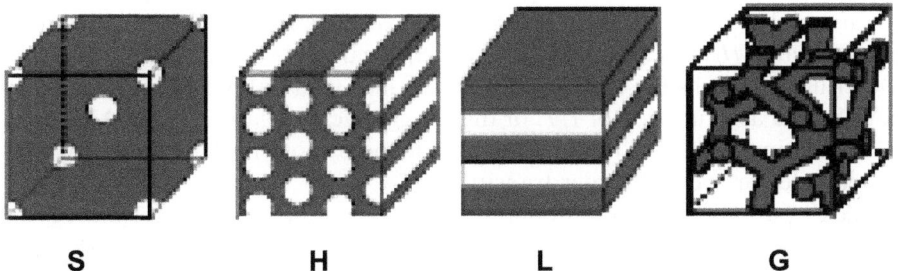

Figure 21.6 Body-centred cubic (bcc) lattice of spheres (S), hexagonally packed cylinders (H), lamellae (L), gyroid phase (G).

A large number of secondary maxima in the SAXS pattern allows to unambiguously assign the correct morphology of the block copolymer. However, this is not always the case, and the few reflections experimentally obtained may be consistent with more than one spatial group, *i.e.* to more than one arrangement of the domains in space. In such instances, for a correct phase identification it is necessary to use different and complementary characterization techniques.[88,90–92] A particularly informative example of how to use complementary techniques such as TEM to discriminate among different hypothetical space groups was offered by Avgeropoulos *et al.*[88] A further aid for a more accurate assignment of the phase-separated morphology can be attained also by inspecting the relative intensities of the peaks, as maintained by

Zha et al., who inferred, on the basis of an anomalous intensity of the primary signal with respect to the higher order ones, that a polystyrene-*block*-polyisoprene-*block*-poly(2-vinylpyridine) copolymer had a core-shell cylindrical morphology rather than a normal hexagonally packed cylindrical phase.[93]

Sometimes, for a more thorough understanding of the phase-separated structures in blends containing block copolymers, it is necessary to apply more complex approaches, which consist in developing theoretical models, depending on morphological features, which allow to reproduce a calculated SAXS pattern. A fitting of the SAXS traces is therefore possible, with the possibility of obtaining a quantitative description of the system.[91,94,95]

Systems like ternary blends of immiscible A and B homopolymers and of a macromolecular surfactant such as an AB diblock copolymer are especially well described by the Teubner-Strey equation,[96–98] which reproduces scattering experimental data by just three parameters, a_2, c_1 and c_2:

$$I(q) = \frac{\text{constant}}{a_2 + c_1 q^2 + c_2 q^4} \tag{21.14}$$

Once these parameters are obtained by the best fitting function of the experimental SAXS trace, they are used for determining the correlation length and the size of the domains.[97]

Temperature-dependent SAXS experiments allow to follow the development of the microstructure of the copolymer. The determination of the phase diagram is aided by plotting the reciprocal of the intensity of the main primary SAXS peak, $1/I_{hkl}$, as a function of the reciprocal of absolute temperature, $1/T$. Other useful graphs for this purpose are the half-width at half maximum of the SAXS peak as a function of $1/T$, or the Bragg spacing related to the SAXS peak as a function of $1/T$. Abrupt decreases in these plots reveal the onset of an order-order or order-disorder transition.[97,99–102]

SAXS patterns yield additional quantitative results on the morphology of microphase-separated block copolymers. The interdomain distance D is the domain identity period in the case of lamellae, or the nearest neighbour distance between the microdomains for the cylinders and spheres. D can be determined on the basis of the Bragg spacing of the principal peak (Equation (21.1)), according to the following equations:

$$D_{lamellae} = d_{001} \tag{21.15}$$

$$D_{cylinders} = (4/3)^{1/2} d_{100} \tag{21.16}$$

$$D_{spheres} = (3/2)^{1/2} d_{110} \tag{21.17}$$

d_{001}, d_{100}, d_{110} are the Bragg spacings of the first order peak in morphologies with a lamellar, hexagonal cylindrical and bcc spherical packing, respectively. The indicated $D_{spheres}$ is valid for spherical microdomains in a body centred cubic lattice, while in the case of a simple cubic symmetry $D = d_{100}$, where d_{100} is the Bragg spacing of the first order peak in this morphology.

The radius R of the cylinders or of the spheres can be determined on the basis of space filling considerations, based on the geometry of the microphase, *i.e.* hexagonal cylindrical or cubic, and the volume fraction of the minor phase Φ_B:

$$R_{cylinders} = [d_{100}(3^{1/2}\Phi_B/2\pi)^{1/2}] \tag{21.18}$$

$$R_{spheres} = [(d_{110}/2^{1/2})(3\Phi_B/8\pi)^{1/3}] \tag{21.19}$$

21.4 Small-Angle Neutron Scattering

Scattering by neutrons as well can be exploited for obtaining precious structural and morphological information on polymeric materials. An excellent textbook covering all the aspects of neutron scattering, from instrumentation to theory to data interpretation, has recently been written by Hammouda, and is freely available on the world wide web.[103]

Even though neutron scattering studies require expensive facilities, neutrons are very advantageous as probes for condensed matter because they are very penetrating and they allow for a non-destructive investigation of the samples. Moreover, their wavelengths are comparable to atomic sizes and inter-distance spacings. After being generated, neutrons are monochromatized and collimated. Detection of the neutrons scattered by the sample, which can be investigated in various physical forms, is usually obtained by two-dimensional area detectors.

Neutron scattering is a very flexible technique, which allows to probe structures with sizes from the near atomic to the near micrometre scale. This made SANS a method of choice when the hierarchical structure of complex materials must be elucidated, especially in conjunction with other techniques such as WAXD, SAXS or light scattering.[67,90]

Analogously to other scattering methods, SANS yields a picture of the sample in the reciprocal space. SANS data must therefore be interpreted basically with one of the following approaches, which are similar to those enumerated for SAXS:[103]

1. Qualitative inspection of the SANS pattern. Similarly to SAXS, many SANS spectra show a broad peak, whose position is related to the characteristic distance between the scattering inhomogeneities.
2. Application of the Guinier equation or of Porod-like power laws;
3. Nonlinear least-squares fitting to appropriate models. Many models have been developed, both for macromolecular scattering and for particulate scattering. Most of them are based on the formalization of appropriate structure and form factors.

Due to the high cost and poor accessibility of neutron scattering experiments, most of the researchers tend to take full advantage of the information that can be acquired by SANS, therefore the first approach, which yields at best semi-quantitative information, is often overlooked in favour of other data analysis methods which offer a richer array of details.

The application of the second approach, the one exploiting the Guinier and Porod approximations, is particularly suited for systems with a hierarchical structure, and is applied both on SAXS and on SANS data of such materials. Most of the systems of this type are elastomeric materials filled with carbon black, carbon nanotubes or silica, and are normally investigated by SAXS,[46,48,65,104–111] even though occasional studies applied also SANS.[112] If SANS is combined with ultra-small-angle neutron scattering (USANS), a very wide range of length scales can be probed, from 10 Å to microns. The scattering patterns so obtained, represented on a log-log plot, appear as an alternation of rectilinear segments connected by curved portions.[112] An example is shown in Figure 21.7.

The fitting is achieved by a combination of Guinier-type equations $\left(I = I_0 e^{-(1/3)q^2 R_g^2}\right)$, by which the radius of gyration of the scattering particles originating the signal can be obtained, and by Porod-like power laws ($I = Aq^{-\alpha}$). The exponent α of the Porod power law vary between 1, 2, 3 (for rod-shaped, lamellar or spherical objects, respectively) and 4 (objects with a smooth surface). It is informative about the dimensionality of the scattering objects and on their surface ($3 < \alpha < 4$ is due to rough interfaces, characteristic of surface fractals). The different segments of the pattern are due to particles of different size and with different shapes, so hierarchical structures composed, for example, of small particles aggregating in bigger ones, which in turn coalesce into even bigger clusters, can be effectively described by this method. Since this method is more typical of filled elastomers, and therefore of composites rather

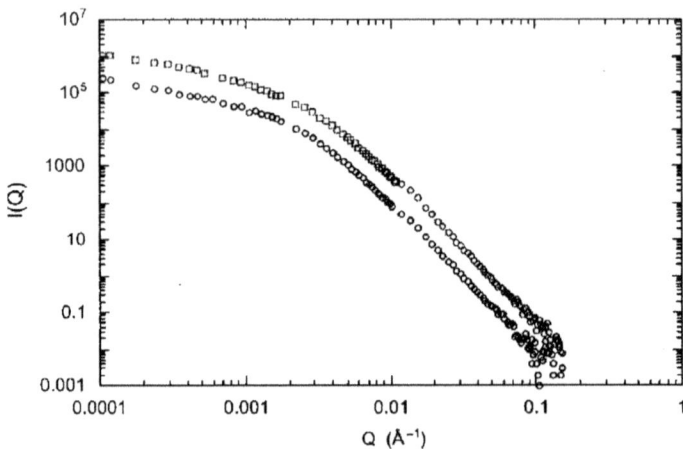

Figure 21.7 A log–log plot of scattering intensity as a function of q for a brominated poly(isobutylene-*co-p*-methylstyrene) blended with carbon N660. The filler volume fraction is 0.025 (upper curve) and 0.005 (lower curve). Reprinted with permission from Y. Zhang *et al.*, *Macromolecules*, **34**, 7056–7065. Copyright (2001) American Chemical Society.

than blends, the interested reader is directed towards the cited literature for details about this technique.

SANS has been extensively applied for elucidating the thermodynamic interactions in blends containing NR. The starting point of almost all the works appearing in the literature is the Flory-Huggins theory, which states that the Gibbs free energy (ΔG) of binary blends of monodisperse polymers is:[113–116]

$$\Delta G = k_B T \left(\frac{\varphi_1}{N_1 \left(\frac{v_1}{v_0} \right)} \ln \varphi_1 + \frac{\varphi_2}{N_2 \left(\frac{v_2}{v_0} \right)} \ln \varphi_2 + \chi \varphi_1 \varphi_2 \right) \quad (21.20)$$

where k_B is the Boltzmann constant, T is the absolute temperature, v_i and N_i are the volume per repeat unit and the number of repeat units in a chain of component i; φ_i is the volume fraction of component i; and χ is a thermodynamic interaction parameter.

The model most often used for interpreting SANS data is based on the incompressible random phase approximation (RPA), because it allows to obtain the thermodynamic interaction parameter χ. Once χ is obtained, other features of the blend can be obtained, such as for example the concentration fluctuation of the single-phase state,[117] and the thermal correlation length.[118]

In this instance, the coherent scattering intensity for a binary blend is:[119,120]

$$I(q) = \left(\frac{b_1}{v_1} - \frac{b_2}{v_2} \right)^2 S(q) \quad (21.21)$$

where b_i is the scattering length per repeat unit for species i. The structure factor, $S(q)$, is:[121]

$$\frac{1}{S(q)} = \left(\frac{1}{N_1 v_1 \varphi_1 P_1(q)} + \frac{1}{N_2 v_2 \varphi_2 P_2(q)} - 2 \frac{\chi}{v_0} \right) \quad (21.22)$$

where P_i is the weight average single-chain form factor for species i.

There is substantial agreement between the authors that this approach is the most effective for the study of issues on the miscibility and on the thermodynamic interactions in polymer blends.[117,118,122–130]

The form factor $P(q)$ can be expressed by a Debye function. One of the most used analytical expressions for $P(q)$ is:[122,123,126,127]

$$P(q) = \frac{2}{(R_{g,i}^2 q^2)^2} [\exp(-R_{g,i}^2 q^2) + R_{g,i}^2 q^2 - 1] \quad (21.23)$$

where $R_{g,i}$ is the radius of gyration of the component i.

As an alternative proposed by some authors, $P(q)$ may be formalized as:[117,125,128–130]

$$P(q) = \frac{2}{x_i^2} \left[\left(\frac{h_j}{h_j + x_i} \right)^{h_j} - 1 + x_i \right] \quad (21.24)$$

where

$$x_i = q^2 R_{g,i}^2 = q^2 N_{n,i} a_i^2 / 6 \qquad (21.25)$$

and

$$h_j = [(N_{w,i}/N_{n,i}) - 1]^{-1} \qquad (21.26)$$

$N_{w,i}$ and $N_{n,i}$ indicate the weight and number averaged degrees of polymerization for the ith component. a_i is the statistical segment length for the ith component.

When particular morphological features need to be characterized, alternative nonlinear fitting models may be used to interpret SANS data. The rationale behind such approach is, analogously to the case of the incompressible RPA described above, based on the fact that the scattering intensity can be modelled as the product of the number density of the scattering particles, of the particle volume, of the contrast factor, of the single particle form factor, $P(q)$, and of the interparticle structure factor, $S(q)$. The form and structure factors have the same meanings described in the paragraph on SAXS (Equation (21.13)), and their accurate modelling is paramount for obtaining a good fitting of the experimental scattering pattern, and a significant data interpretation. The mathematical expression is dependent on the particular system which is investigated and on the particular features that are of interest, mainly the size of the scatterers,[67,131,132] and the parameters of interaction between the components of the blend or of block copolymers.[91,131,133]

The advantage of SANS over other small-angle scattering methods (such as SAXS) is that scattering contrast depends on electron density for X-rays, and on neutron scattering length density for neutrons. This allows to use both techniques as very complementary methods to obtain a really detailed picture of complex samples. For example, Yamauchi and coworkers studied the structure of a blend of NR and high density polyethylene (HDPE).[134] The X-ray contrast factors for polyethylene/NR and amorphous/crystalline polyethylene pairs are of the same order of magnitude. Therefore, both phase-separated structures and crystalline lamellar structures contributed to the SAXS intensity.[134] On the opposite, the neutron contrast factor for amorphous/crystalline polyethylene pair is negligibly small compared with that for the polyethylene/NR pair. In other words, SANS is sensitive just to the phase-separated structure and is not influenced by the crystalline structure of HDPE.[134] The two contributors to the morphology of the sample can be thus decoupled and studied separately, obtaining a very accurate picture of the composite.

Full exploitation of this advantage can be obtained by the deuteration method. Scattering lengths for hydrogen and deuterium are widely different, so introduction of deuterium labels into the analyte will greatly increase the contrast. In order to perform the same contrast enhancement with X-rays, heavy atoms must be introduced, with the risk of unacceptable modifications of the sample morphology. Deuterium labelling is much less invasive under this point of view. SANS is capable to measure density fluctuations and composition (or concentration) fluctuations, whereas SAXS is only sensitive to density fluctuation.

The potential of SANS for probing the various morphological details of the samples and the high flux apparatus recently made available, allowed to perform several time resolved experiments, in which the structure and morphology development in response to external stimuli could be followed. An example which shows the potential of time resolved SANS study regarding blends containing NR was reported by Jinnai et al., who focused on elucidating the intermediate and late stage spinodal decomposition of perdeuterated polybutadiene/protonated polyisoprene.[118] Other applications which are very suited for time resolved SANS studies are in situ deformation studies, which are very useful to investigate the behaviour under stress and strain of materials, and therefore to understand their failure mechanisms and to devise reinforcement strategies.

When the energy and momentum of the impinging neutrons are the same as those of scattered neutrons, the observed phenomenon is elastic scattering, which is most useful when studying the structure of the samples. If the dynamics of the system is of interest, quasi-elastic or inelastic neutron scattering is exploited. Inelastic scattering is encountered whenever there is a transfer of both momentum and energy. Quasi-elastic neutron scattering (QENS) is a form of inelastic scattering where the energy transfer peak is located around $E=0$.[103] This experimental technique has been often applied to the study of polymer blends dynamics, because it is able to provide space-time information about the geometry and speed of the molecular motions.[135–138] Moreover, exploiting the very different scattering cross sections of hydrogen and deuterium, it is possible to highlight the dynamic response of one of the components in the blend by protonation, because neutron scattering intensity is dominated by the incoherent scattering of the hydrogen atoms. This allows to study the response of two components of the blend separately.[135] Broad-band dielectric spectroscopy is the natural complement of QENS studies.[135,137]

21.5 Small-Angle Light Scattering

In addition to X-rays or neutrons, also visible light can be used as the incoming radiation in scattering experiments. When a laser light impinges on a colloidal particle suspension, on a polymer blend or on an otherwise dishomogeneous sample, the corresponding oscillating electric field of the incoming electromagnetic wave deforms the electronic cloud of the atoms present in the system. As a result, the oscillating electrons, emit, *i.e.* scatter, electromagnetic radiation. The total scattered intensity at a certain angle θ with respect to the direction of the incoming laser light will obviously be the sum of the individual contributions of each small volume illuminated by the laser light in the sample. If the sample were homogeneous, no scattering would occur, because the contribution from each domain of the sample would be annihilated by the destructive interference due to another domain. However, if the sample were not homogeneous, different volumes within the sample would have different properties in terms of dielectric constant, such as their polarizability, and thus they would scatter the light with the same phase but different amplitudes, thus interference would not be completely destructive.

When the dispersed or dissolved particles are very small compared to the wavelength of the light, the intensity of the scattered light is uniform in all directions (Rayleigh scattering), whereas when the size of particles increases (above approximately 250 nm in diameter), the intensity is angle dependent (Mie scattering).

Static light scattering, in which the scattered intensity is measured as a function of the angle around the sample, is a mainstream technique for the determination of the molecular weight of polymers. Methods similar to those illustrated for X-ray or neutron scattering can moreover be applied to light scattering, such as the Guinier relationship[139] or the fitting by appropriate theoretical models developed formalizing a shape and structure factors.[140]

Dynamic light scattering focuses on the observation of time-dependent fluctuations in scattered intensity, by using coherent and monochromatic light from a laser source and an appropriate photon counting device. Dynamic light scattering (also known as quasi-elastic light scattering and photon correlation spectroscopy) is particularly suited for determining small changes in mean diameter such as those due to adsorbed layers on the particle surface or slight variations in manufacturing processes.

Light scattering allows to gather interesting information on the morphology of polymeric composite materials, and a number of examples exist in the literature on its exploitation in the investigation of polymeric blends based on NR.

Hashimoto et al.[117,141,142] used time resolved small angle light scattering to monitor in real time the phase separation of polyisoprene/polybutadiene blends. Just a visual inspection of the obtained LS curves (see for example Figure 21.8) allowed to determine that phase-separated structures grew with time. As may be seen the SALS peak is composed by several superposed signal, and their deconvolution allowed the authors of this study to reconstruct the time evolution of phase separation in the studied sample.

In addition to such qualitative interpretation of SALS data, quantitative data could be extracted, i.e. the reduced wave number and the reduced maximum intensity, which allowed to elucidate a nonlinear pathway according to which the initial structures, which had developed in a first-step process, relaxed and transformed toward an equilibrium structure after a second step.[141] A similar work was reported by Narh[143] who also worked on polybutadiene/polyisoprene blends.

Using the Debye-Bueche approach (Equation (21.11)), Sheng et al. were able to obtain from SALS data the ratio of interface surface area to the volume, S/V, along with the average chord lengths of each phase, i.e. the average length of randomly drawn vectors passing through the two phases.[144,145] The same authors were able to obtain the same amount of information, by the same data analysis approach, applied on SAXS patterns.[56,57]

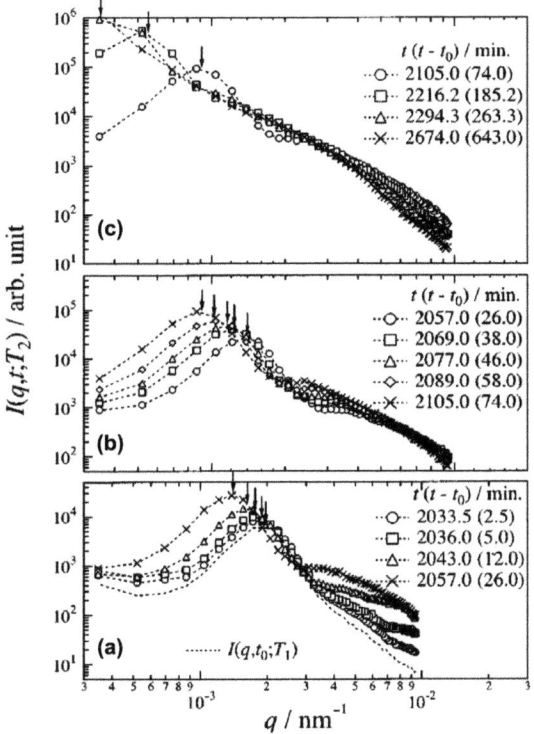

Figure 21.8 Time evolution of the LS profile of a polybutadiene/polyisoprene blend. Reprinted with permission from T. Hashimoto et al., *Journal of Chemical Physics* **112**, 6886–6896. Copyright (2000) American Institute of Physics.

References

1. D. Vlassopoulos, I. Chira, B. Loppinet and P. T. McGrail, *Rheol. Acta*, 1998, **37**, 614.
2. P. L. Drzal, A. F. Halasa and P. Kofinas, *Polymer*, 2000, **41**, 4671.
3. A. M. Hindeleh and D. J. Johnson, *J. Phys. D: Appl. Phys.*, 1971, **4**, 259.
4. A. Turner Jones, *Polymer*, 1971, **12**, 487.
5. H. P. Klug and L. E. Alexander, *X-Ray Diffraction Procedures: For Polycrystalline and Amorphous Materials*, Wiley, New York, 1974.
6. H. Sun, J. Feng, J. Wang, B. Yu and J. Sheng, *J. Macromol. Sci. Phys.*, 2012, **51**, 328.
7. H. Sun, C. Gong, X. Sun and J. Sheng, *Polym. Plast. Technol. Eng.*, 2006, **45**, 1175.
8. O. Grigoryeva, A. Fainleib, O. Starostenko, A. Tolstov and W. Brostow, *Polym. Int.*, 2004, **53**, 1693.
9. M. Baboo, M. Dixit, K. Sharma and N. S. Saxena, *Thermochim. Acta*, 2010, **502**, 47.

10. M. Baboo, M. Dixit, K. Sharma and N. S. Saxena, *Polym. Bull.*, 2011, **66**, 661.
11. A. F. Halasa, G. D. Wathen, W. L. Hsu, B. A. Matrana and J. M. Massie, *J. Appl. Polym. Sci.*, 1991, **43**, 183.
12. V. Nigam, D. K. Setua and G. N. Mathur, *J. Appl. Polym. Sci.*, 1998, **70**, 537.
13. R. Treloar, *The Physics of Rubber Elasticity*, Oxford University Press, Oxford, 1975.
14. P. Flory, *J. Chem. Phys.*, 1947, **15**, 397.
15. A. N. Gent, *Trans. Faraday Soc.*, 1954, **50**, 521.
16. K. J. J. Smith, A. Greene and A. Ciferri, *Kolloid Z. Z. Polym.*, 1963, **194**, 49.
17. E. H. Andrews, *Proc. R. Soc. London, Ser. A*, 1962, **A270**, 232.
18. E. H. Andrews, *Proc. R. Soc. London, Ser. A*, 1964, **A272**, 562.
19. E. H. Andrews, P. J. Owen and A. Singh, *Proc. R. Soc. London, Ser. A*, 1971, **A324**, 79.
20. D. Luch and G. S. Y. Yeh, *J. Appl. Phys.*, 1972, **43**, 4326.
21. D. Luch and G. S. Y. Yeh, *J. Macromol. Sci. Phys.*, 1973, **B97**, 121.
22. A. Suzuki, H. Oikawa and K. Murakami, *J. Macromol. Sci. Phys.*, 1985, **B23**, 535.
23. S. Toki, T. Fujimaki and M. Okuyama, *Polymer*, 2000, **41**, 5423.
24. S. Toki, I. Sics, B. S. Hsiao, M. Tosaka, S. Poompradub, Y. Ikeda and S. Kohjiya, *Macromolecules*, 2005, **38**, 7064.
25. S. Toki, I. Sics, S. Ran, L. Liu and B. S. Hsiao, *Polymer*, 2003, **44**, 6003.
26. S. Toki, I. Sics, S. Ran, L. Liu, B. S. Hsiao, S. Murakami, K. Senoo and S. Kohjiya, *Macromolecules*, 2002, **35**, 6578.
27. M. Tosaka, D. Kawakami, K. Senoo, S. Kohjiya, Y. Ikeda, S. Toki and B. S. Hsiao, *Macromolecules*, 2006, **39**, 5100.
28. M. Tosaka, S. Murakami, S. Poompradub, S. Kohjiya, Y. Ikeda, S. Toki, I. Sics and B. S. Hsiao, *Macromolecules*, 2004, **37**, 3299.
29. M. Tosaka, K. Senoo, S. Kohjiya and Y. Ikeda, *J. Appl. Phys.*, 2007, **101**, 084909.
30. S. Trabelsi, P.-A. Albouy and J. Rault, *Macromolecules*, 2003, **36**, 7624.
31. A. Schallamach, *Trans. Faraday Soc.*, 1942, **38**, 376.
32. E. Benedetti, P. Corradini and C. Pedone, *Eur. Polym. J.*, 1975, **11**, 585.
33. G. R. Mitchell, *Polymer*, 1984, **25**, 1562.
34. J. H. Dumbleton and B. B. Bowles, *J. Polym. Sci. A-2 Polym. Phys.*, 1966, **4**, 951.
35. D. J. Lee and J. A. Donovan, *Rubber Chem. Technol.*, 1987, **60**, 910.
36. S. Trabelsi, P. A. Albouy and J. Rault, *Macromolecules*, 2002, **35**, 10054.
37. S. Toki, I. Sics, B. S. Hsiao, S. Murakami, M. Tosaka, S. Poompradub, S. Kohjiya and Y. Ikeda, *J. Polym. Sci., Part B: Polym. Phys.*, 2004, **42**, 956.
38. S. Toki, I. Sics, S. Ran, L. Liu, B. S. Hsiao, S. Murakami, M. Tosaka, S. Kohjiya, S. Poompradub, Y. Ikeda and A. H. Tsou, *Rubber Chem. Technol.*, 2004, **77**, 317.

39. R. D. B. Fraser, T. P. Marae, A. Miller and R. J. Rowlands, *J. Appl. Crystallogr.*, 1976, **9**, 81.
40. W. Ruland, *Colloid Polym. Sci.*, 1977, **255**, 833.
41. A. Manzur and L. Rubio, *J. Macromol. Sci., Part B: Phys.*, 1997, **B36**, 103.
42. O. Glatter and O. Kratky, *Small Angle X-ray Scattering*, Academic Press, London, 1982.
43. A. Guinier, *Ann. Phys.*, 1939, **12**, 161.
44. G. Porod, *Kolloid Z. Z. Polym.*, 1951, **124**, 83.
45. P. Debye, H. R. J. Anderson and H. Brumberger, *J. Appl. Phys.*, 1957, **28**, 679.
46. T. P. Rieker, M. Hindermann-Bischoff and F. Ehrburger-Dolle, *Langmuir*, 2000, **16**, 5588.
47. P. W. Schmidt, *J. Appl. Crystallogr.*, 1991, **24**, 414.
48. T. Koga, T. Hashimoto, M. Takenaka, K. Aizawa, N. Amino, M. Nakamura, D. Yamaguchi and S. Koizumi, *Macromolecules*, 2008, **41**, 453.
49. D. W. Schaefer, T. Rieker, M. Agamalian, J. S. Lin, D. Fischer, S. Sukunaran, C. Chen, G. Beaucage, C. Herd and J. Ivie, *J. Appl. Crystallogr.*, 2000, **33**, 587.
50. P. Debye and A. M. Bueche, *J. Appl. Phys.*, 1949, **20**, 518.
51. M. Moritani, T. Inoue, M. Motegi and H. Kawai, *Macromolecules*, 1970, **3**, 433.
52. Y. J. Li, Y. Kadowaki, T. Inoue, K. Nakayama and H. Shimizu, *Macromolecules*, 2006, **39**, 4195.
53. W. Salgueiro, A. Somoza, A. J. Marzocca, I. Torriani and M. A. Mansilla, *J. Polym. Sci., Part B: Polym. Phys.*, 2009, **47**, 2320.
54. Y. Hsu and C. Liang, *J. Appl. Polym. Sci.*, 2007, **106**, 1576.
55. J. Sheng, J. Hu, X. B. Yuan, Y. P. Han, F. K. Li and D. C. Bian, *J. Appl. Polym. Sci.*, 1998, **70**, 805.
56. J. Sheng, L. Y. Qi, X. B. Yuan, N. X. Shen and D. C. Bian, *J. Appl. Polym. Sci.*, 1997, **64**, 2265.
57. G. Q. Ma, X. B. Yuan, J. Sheng and D. C. Bian, *J. Appl. Polym. Sci.*, 2002, **83**, 2088.
58. C. Marega, V. Causin, R. Saini, A. Marigo, A. P. Meera, S. Thomas and K. S. Usha Devi, *J. Phys. Chem. B*, 2012, **116**, 7596.
59. S. Tan, D. Zhang and E. Zhou, *Polym. Int.*, 1997, **42**, 90.
60. X. Yu, G. Gao, J. Wang, F. Li and X. Tang, *Polym. Int.*, 1999, **48**, 805.
61. J. H. An, A. M. Fernandez and L. H. Sperling, *Macromolecules*, 1987, **20**, 191.
62. G. Beaucage and D. W. Schaefer, *J. Non-Cryst. Solids*, 1994, **172**, 797.
63. G. Beaucage, *J. Appl. Crystallogr.*, 1995, **28**, 717.
64. G. Beaucage, *J. Appl. Crystallogr.*, 1996, **29**, 134.
65. C. N. Suryawanshi, P. Pakdel and D. W. Schaefer, *J. Appl. Crystallogr.*, 2003, **36**, 573.
66. B. G. Soares, K. Dahmouche, V. D. Lima, A. A. Silva, S. P. C. Caplan and F. L. Barcia, *J. Colloid Interface Sci.*, 2011, **358**, 338.

67. A. Botti, W. Pyckhout-Hintzen, V. Urban, J. Kohlbrecher, D. Richter and E. Straube, *Appl. Phys. A: Mater. Sci. Process.*, 2002, **74**, S513.
68. M. Hernandez, B. Sixou, J. Duchet and H. Sautereau, *Polymer*, 2007, **48**, 4075.
69. K. Varlot, E. Reynaud, G. Vigier and J. Varlet, *J. Polym. Sci., Part B: Polym. Phys.*, 2002, **40**, 272.
70. R. A. Vaia, W. Liu and H. Koerner, *J. Polym. Sci., Part B: Polym. Phys.*, 2003, **41**, 3214.
71. M. Gelfer, C. Burger, A. Fadeev, I. Sics, B. Chu, B. S. Hsiao, A. Heintz, K. Kojo, S.-L. Hsu, M. Si and M. Rafailovich, *Langmuir*, 2004, **20**, 3746.
72. W. Ruland and B. Smarsly, *J. Appl. Crystallogr.*, 2004, **37**, 575.
73. M. Y. Gelfer, C. Burger, B. Chu, B. S. Hsiao, A. D. Drozdov, M. Si, M. Rafailovich, B. B. Sauer and J. W. Gilman, *Macromolecules*, 2005, **38**, 3765.
74. S. M. King, in *Modern Techniques for Polymer Characterisation*, ed. R. A. Pethrick and J. V. Dawkins, Wiley, New York, 1999.
75. E. Pavlopoulou, S. H. Anastasiadis, H. Iatrou, M. Moshakou, N. Hadjichristidis, G. Portale and W. Bras, *Macromolecules*, 2009, **42**, 5285.
76. B. J. P. Jansen, S. Rastogi, H. E. H. Meijer and P. J. Lemstra, *Macromolecules*, 2001, **34**, 4007.
77. B. J. P. Jansen, S. Rastogi, H. E. H. Meijer and P. J. Lemstra, *Macromolecules*, 1999, **32**, 6283.
78. M. Kowalczyk and E. Piorkowska, *J. Appl. Polym. Sci.*, 2012, **124**, 4579.
79. R. A. Bubeck, D. J. Buckley, E. J. Kramer and H. R. Brown, *J. Mater. Sci.*, 1991, **26**, 6249.
80. R. M. A. l'Abee, M. van Duin, A. B. Spoelstra and J. G. P. Goossens, *Soft Matter*, 2010, **6**, 1758.
81. W. Salgueiro, A. Somozaa, A. J. Marzocca, G. Consolati and F. Quasso, *Radiat. Phys. Chem.*, 2007, **76**, 142.
82. B. D. Olsen and R. A. Segalman, *Macromolecules*, 2005, **38**, 10127.
83. C. H. Lee, H. B. Kim, S. T. Lim, H. S. Kim, Y. K. Kwon and H. J. Choi, *Macromol. Chem. Phys.*, 2006, **207**, 444.
84. I. W. Hamley, *The Physics of Block Copolymers*, Oxford University Press, Oxford, 1998.
85. R. Roy, J. K. Park, W. Young, S. E. Mastroianni, M. S. Tureau and T. H. Epps III, *Macromolecules*, 2011, **44**, 3910.
86. V. F. Scalfani and T. S. Bailey, *Chem. Mater.*, 2010, **22**, 5992.
87. J. Suzuki, M. Furuya, M. Inuma, A. Takano and Y. Matsushita, *J. Polym. Sci., Part B: Polym. Phys.*, 2002, **40**, 1135.
88. A. Avgeropoulos, B. J. Dair, N. Hadjichristidis and E. L. Thomas, *Macromolecules*, 1997, **30**, 5634.
89. J. H. Ahn and W. C. Zin, *Macromol. Res.*, 2003, **11**, 152.
90. J. Holoubek, J. Baldrian, F. Lednicky and J. Lal, *Polym. Int.*, 2009, **58**, 81.

91. S. Koizumi, H. Hasegawa and T. Hashimoto, *Macromolecules*, 1994, **27**, 7893.
92. T. P. Lodge, K. J. Hanley, B. Pudil and V. Alahapperuma, *Macromolecules*, 2003, **36**, 816.
93. W. Zha, C. D. Han, H. C. Moon, S. H. Han, D. H. Lee and J. K. Kim, *Polymer*, 2010, **51**, 936.
94. H. Takeshita, Y. Gao, Y. Takata, K. Takenaka, T. Shiomi and C. Wu, *Polymer*, 2010, **51**, 799.
95. K. C. Daoulas, D. N. Theodorou, A. Roos and C. Creton, *Macromolecules*, 2004, **37**, 5093.
96. M. Teubner and R. Strey, *J. Chem. Phys.*, 1987, **87**, 1395.
97. L. Messe, L. Corvazier and A. J. Ryan, *Polymer*, 2003, **44**, 7397.
98. C. J. Ellison, A. J. Meuler, J. Qin, C. M. Evans, L. M. Wolf and F. S. Bates, *J. Phys. Chem. B*, 2009, **113**, 3726.
99. S. Choi, K. M. Lee, C. D. Han, N. Sota and T. Hashimoto, *Macromolecules*, 2003, **36**, 793.
100. C. D. Han, N. Y. Vaidya, D. Yamaguchi and T. Hashimoto, *Polymer*, 2000, **41**, 3779.
101. S. Lee, W. Zin and J. Ahn, *J. Nanosci. Nanotechnol.*, 2009, **9**, 7499.
102. C. J. Lai, W. B. Russel and R. A. Register, *Macromolecules*, 2002, **35**, 841.
103. B. Hammouda, *Probing Nanoscale Structures – The SANS Toolbox*, available online at http://www.ncnr.nist.gov/staff/hammouda/the_sans_toolbox.pdf, 2009.
104. J. A. Chaker, K. Dahmouche, C. V. Santilli, S. H. Pulcinelli and A. Craievich, *J. Appl. Crystallogr.*, 2003, **36**, 689.
105. P. Mélé, S. Marceau, D. Brown, Y. De Puydt and N. D. Albérola, *Polymer*, 2002, **43**, 5577.
106. L. Matejka, O. Dukh and J. Kolařík, *Polymer*, 2000, **41**, 1449.
107. B. J. Bauer, E. K. Hobbie and M. L. Becker, *Macromolecules*, 2006, **39**, 2637.
108. D. W. Schaefer, J. Zhao, J. M. Brown, D. P. Anderson and D. W. Tomlin, *Chem. Phys. Lett.*, 2003, **375**, 369.
109. W. Zhou, M. F. Islam, H. Wang, D. L. Ho, A. G. Yodh, K. I. Winey and J. E. Fischer, *Chem. Phys. Lett.*, 2004, **384**, 185.
110. J. J. Hernández, M. C. García-Gutiérrez, A. Nogales, D. R. Rueda and T. A. Ezquerra, *Compos. Sci. Technol.*, 2006, **66**, 2629.
111. R. S. Justice, D. H. Wang, L. S. Tan and D. W. Schaefer, *J. Appl. Crystallogr.*, 2007, **40**, s88.
112. Y. M. Zhang, S. Ge, B. Tang, T. Koga, M. H. Rafailovich, J. C. Sokolov, D. G. Peiffer, Z. Li, A. J. Dias, K. O. McElrath, M. Y. Lin, S. K. Satija, S. G. Urquhart, H. Ade and D. Nguyen, *Macromolecules*, 2001, **34**, 7056.
113. P. J. Flory, *J. Chem. Phys.*, 1941, **9**, 660.
114. P. J. Flory, *J. Chem. Phys.*, 1942, **10**, 51.
115. M. L. Huggins, *J. Chem. Phys.*, 1941, **9**, 440.
116. M. L. Huggins, *J. Phys. Chem.*, 1942, **46**, 151.

117. H. Hasegawa, S. Sakurai, M. Takenaka, T. Hashimoto and C. C. Han, *Macromolecules*, 1991, **24**, 1813.
118. H. Jinnai, H. Hasegawa, T. Hashimoto and C. C. Han, *Macromolecules*, 1991, **24**, 282.
119. H. Benoit, J. F. Joanny, G. Hadziioannou and B. Hammouda, *Macromolecules*, 1993, **26**, 5790.
120. B. Hammouda, *Adv. Polym. Sci.*, 1993, **106**, 87.
121. M. Shibayama, H. Yang, R. Stein and C. Han, *Macromolecules*, 1990, **23**, 451.
122. M. Harada, T. Suzuki, M. Ohya, D. Kawaguchi, A. Takano and Y. Matsushita, *Macromolecules*, 2005, **38**, 1868.
123. M. Harada, M. Ohya, T. Suzuki, D. Kawaguchi, A. Takano and Y. Matsushita, *J. Polym. Sci., Part B: Polym. Phys.*, 2005, **43**, 1214.
124. G. C. Reichart, W. W. Graessley, R. A. Register, R. Krishnamoorti and D. J. Lohse, *Macromolecules*, 1997, **30**, 3363.
125. S. Sakurai, H. Jinnai, H. Hasegawa, T. Hashimoto and C. C. Han, *Macromolecules*, 1991, **24**, 4839.
126. R. N. Thudium and C. C. Han, *Macromolecules*, 1996, **29**, 2143.
127. K. Yurekli and R. Krishnamoorti, *J. Polym. Sci., Part B: Polym. Phys.*, 2004, **42**, 3204.
128. M. Takenaka, H. Takeno, T. Hashimoto and M. Nagao, *J. Chem. Phys.*, 2006, **124**, 104904.
129. M. Takenaka, H. Takeno, H. Hasegawa, S. Saito, T. Hashimoto and M. Nagao, *Phys. Rev. E*, 2002, **65**, 021806.
130. V. S. Ramachandrarao, B. D. Vogt, R. R. Gupta and J. J. Watkins, *J. Polym. Sci., Part B: Polym. Phys.*, 2003, **41**, 3114.
131. S. S. A. Rahman, D. Kawaguchi, D. Ito, A. Takano and Y. Matsushita, *J. Polym. Sci., Part B: Polym. Phys.*, 2009, **47**, 2272.
132. M. Takenaka, H. Takeno, T. Hashimoto and M. Nagao, *J. Appl. Crystallogr.*, 2003, **36**, 642.
133. B. R. Chapman, M. W. Hamersky, J. M. Milhaupt, C. Kostelecky, T. P. Lodge, E. D. von Meerwall and S. D. Smith, *Macromolecules*, 1998, **31**, 4562.
134. K. Yamauchi, S. Akasaka, H. Hasegawa, S. Koizumi, C. Deeprasertkul, P. Laokijcharoen, J. Chamchang and A. Kornduangkaeo, *Composites, Part A*, 2005, **36**, 423.
135. A. Arbe, A. Alegria, J. Colmenero, S. Hoffmann, L. Willner and D. Richter, *Macromolecules*, 1999, **32**, 7572.
136. M. Doxastakis, M. Kitsiou, G. Fytas, D. N. Theodorou, N. Hadjichristidis, G. Meier and B. Frick, *J. Chem. Phys.*, 2000, **112**, 8687.
137. S. Hoffmann, D. Richter, A. Arbe, J. Colmenero and B. Farago, *Appl. Phys. A: Mater. Sci. Process.*, 2002, **74**, S442.
138. S. Hoffmann, L. Willner, D. Richter, A. Arbe, J. Colmenero and B. Farago, *Phys. Rev. Lett.*, 2000, **85**, 772.
139. R. Prasad, R. K. Gupta, F. Cser and S. N. Bhattacharya, *Polym. Eng. Sci.*, 2009, **49**, 984.

140. B. Braun, J. R. Dorgan and J. P. Chandler, *Biomacromolecules*, 2008, **9**, 1255.
141. T. Hashimoto, M. Hayashi and H. Jinnai, *J. Chem. Phys.*, 2000, **112**, 6886.
142. T. Hashimoto, J. Kumaki and H. Kawai, *Macromolecules*, 1983, **16**, 641.
143. K. A. Narh, *Adv. Polym. Technol.*, 1996, **15**, 245.
144. G. Ma, X. Yuan, Y. Han, L. Cui and J. Sheng, *J. Vacuum Sci. Technol. B*, 2009, **27**, 1454.
145. G. Q. Ma, Y. H. Zhao, L. T. Yan, Y. Y. Li and J. Sheng, *J. Appl. Polym. Sci.*, 2006, **100**, 4900.

CHAPTER 22

Transport of Penetrant Molecules Through Natural Rubber Based Blends and IPNs

ISAAC O. IGWE

Department of Polymer and Textile Engineering, Federal University of Technology, Owerri, Imo State, Nigeria
Email: izikigwe@yahoo.com

22.1 Introduction

The study of transport of penetrant molecules (liquids, vapours and gases) through polymer materials is of interest to polymer scientists, and the transport generally involves the processes of sorption, diffusion and permeation. The interest provoked by these studies emanates mainly from the fact that the knowledge gained is utilized in the concentration of sea water and treatment of effluent water,[1] food packaging,[2] microelectronics,[3] controlled drug release[4,5] and separation science.[6,7] The understanding of the diffusion of penetrant molecules through polymer materials helps polymer scientists in the selection of polymer materials, and processing considerations to minimize sorption level in potential barrier polymers.

Diffusion is a molecular process. In this process, a concentration gradient induces a flux of penetrant molecules from regions of high concentration to regions of low concentration. The flux of penetrant is calculated as:

$$J = -M \frac{d\mu}{dz} \qquad (22.1)$$

where J is flux, M is a constant, μ is chemical potential, and z is position. The diffusion of small molecules which are smaller than the monomeric units of a given polymer are interpreted in terms of free volume concepts. Generally, polymers absorb penetrant molecules to various degrees. The absorption of a penetrant molecule (liquid or vapour or gas) causes a polymer to increase in volume, as a result, the physical properties deteriorate. The swelling of plastics is a diffusion process. Factors that affect diffusion in polymers have been identified, and they include: the degree of penetrant–polymer interactions, penetrant properties (in case of liquids or solvents) such as the solvent size and shape, hydrogen bonding, polarity, solubility parameter, *etc.*, and temperature.

Fick's first and second laws were developed to describe the diffusion process in polymers. Fickian or case I transport is obtained when the local rate of change in the concentration of a diffusing species is controlled by the rate of diffusion of the penetrant. For most purposes, diffusion in rubbery polymers typically follows Fickian law.[8] This is because these rubbery polymers adjust very rapidly to the presence of a penetrant. Polymer segments in their glassy states are relatively immobile, and do not respond rapidly to changes in their conditions. These glassy polymers often exhibit anomalous or non-Fickian transport. When the anomalies are due to an extremely slow diffusion rate as compared to the rate of polymer relaxation, the non-Fickian behaviour is called case II transport. Case II sorption is characterized by a discontinuous boundary between the outer layers of the polymer that are at sorption equilibrium with the penetrant, and the inner layers which are unrelaxed and unswollen.

The velocity of the advancing penetrant front characterizes case II transport. Here, the unswollen polymer lies ahead of the front, while the equilibrium swollen material exists in the region behind the front position.

Diffusion of liquids in glassy polymers is often anomalous. Anomalous behaviour is evidenced by significant deviations from the linear relationship between absorption and the square root of time, and can be attributed to a time dependence of polymer chain relaxation in the presence of a penetrant.

The time dependence of the weight uptake of penetrant by a polymer in a sorption experiment may exhibit a rich spectrum of behaviour. For example, the penetrant may sorb into the polymer in two stages; an initial Fickian-like stage followed by a protracted drift towards a final equilibrium uptake.[9] The penetrant weight uptake may be linear with contact time until equilibrium is reached, and the penetrant concentration profiles inside the polymer may be very sharp. This behaviour is an example of case II sorption to distinguish it from Fickian, or case I sorption.[10,11] The initial rate of penetrant uptake may also lie between the limits of Fickian, and case II diffusion.[12] Additionally, oscillations in the weight uptake have been observed as the weight uptake approaches its equilibrium value.[13] These examples illustrate the extent of responses which may be observed. More detailed descriptions of non-Fickian phenomena are available elsewhere.[14,15] Sorption data for much of the uptake process in Fickian systems can be analysed using the relationship:

$$\frac{Q_t}{Q_\infty} = kt^n \qquad (22.2)$$

where Q_t is the mole per cent of penetrant sorbed at time t, Q_∞ is the mole per cent of penetrant sorbed at equilibrium, k and n are system parameters. The value of n indicates the nature of the transport mechanism while k depends upon the structural characteristics of the polymer, and the polymer-solvent interaction. When Equation (22.3) is rewritten as:

$$\log \frac{Q_t}{Q_\infty} = \log k + n \log t \tag{22.3}$$

the slope of a log–log plot is n (according to Equation (22.3)). For planar geometry, if $n = \frac{1}{2}$, the diffusion is classified as Fickian, while a value of $n = 1$ indicates case II. Values of n between $\frac{1}{2}$ and 1 suggests anomalous transport behaviour.[16] A value of $n > 1$ indicates a super case II transport.

Diffusion and solvent uptake are the limiting factors affecting polymer end-use applications. These processes might change the mechanical properties, and sometimes, cause destruction in polymer properties. An understanding of penetrant uptake in polymers requires a thorough knowledge of the following sorption parameters: the diffusion coefficient (D), solubility (S), and permeability (P) of the penetrant. These parameters can be calculated from sorption data by using the following expressions.[17]

$$D = \Pi[h\theta/4\theta_\infty]^2 \tag{22.4}$$

$$P = D \times S \tag{22.5}$$

$$S = M_\infty/M_0 \tag{22.6}$$

where θ is the slope of the plot of Q_t against $t^{\frac{1}{2}}$.

Diffusion coefficient (D) is the ability of the penetrant to move among the polymer segment. It is a kinetic parameter which depends on the polymer segment mobility. Sorption is a surface phenomenon, and it is an indication of the tendency of the penetrant to dissolve into the polymer. Permeation on the other hand, can be considered as the combined effect of sorption and diffusion process. From equation 22.5, it can be inferred that the sorption process controls the permeability.

The molecular size of a gas molecule is the major factor affecting D. the larger the size of the gas molecule, the more difficult it is for it to penetrate through the polymer. For molecules with high solubilities in the polymer, the swelling action helps its transport through the polymer membrane.

Penetrant diffusion coefficients in rubbery polymers are typically observed to decrease with increasing penetrant size, and approaches a plateau for large penetrant molecules. In natural rubber (NR), for example, this plateau region is attained by penetrants such as n-butane, benzene or n-pentane. In glassy polymers, penetrant diffusivity is lower than in rubbery materials, with the difference increasing markedly with penetrant size.[16]

Some studies have indicated that the weight update curves for polymers immersed in different solvents show a decrease after reaching a maximum.[18,19] This phenomenon is attributed to a leaching-out effect. In some polymeric systems, the presence of solvents produces an overshoot effect followed by a steady decline to an equilibrium position.[20–22] Such effects are the result of polymer rapid relaxation during solvent migration.

The sorption and diffusion of gases in polymers are greatly influenced by the extent of crystallinity in the polymer. Generally, small gas molecules diffuse through the amorphous regions of the polymer, and are excluded from the crystalline region of the material. The crystalline regions in these polymers seem to act as non-sorbing, impermeable barriers dispersed in the amorphous polymeric medium. The presence of impermeable crystallites can have two effects on the diffusion process. Firstly, the effective path length, or tortuosity (τ), will increase due to the fact that the diffusing molecules will have to bypass the crystallites. Secondly, the presence of the crystallites can have an effect on chain mobility in the amorphous phase. Thus, the polymers segmental mobility will be restricted in the amorphous phase due to the presence of the crystallites.[23] A two-phase model has been developed to describe the penetrant sorption and transport in the semi-crystalline polymer, polyethylene, and is well described by the relation:

$$C = k_d \varnothing_a P \tag{22.7}$$

where $\varnothing_\alpha$ is the volume fraction of the polymer which is amorphous, and k_d is the Henry's law constant for the purely amorphous polymer.[23] Good descriptions of the changes in the diffusion coefficients of gases in semi-crystalline polymers as a result of polymer processing operations are available elsewhere.[14,24]

The permeation of vapours in polymeric systems is emerging as a new industrial membrane technology. Thus, in vapour permeation, the transport of a condensable vapour through a dense membrane consecutive to an activity gradient takes place. This process offers the unique feature of studying the transport process of a single permeant through a dense membrane under various upstream activities.[25] The calculation of upstream solvent activity demands the use of complicated vapour liquid equilibria methods. In the case of pure solvent vapour permeation, upstream activity can be calculated easily, provided upstream pressure is precisely monitored.[26] This process is capable of offering direct practical conclusions for the understanding and rational design of volatile organic compounds (VOCs) vapour recovery from contaminated air streams.[26–28] This technique can also offer significant opportunities for energy saving and solvent release, when compared to classical VOC control processes such as oxidation or active carbon absorption. A method for simultaneous determination of diffusivity (D), permeability (P) and sorptivity (S) of organic vapour through polymers from the mass transport was developed by Salwinski et al.[29]

22.2 Natural Rubber: Properties and Applications

NR, cis-1,4-polyisoprene, is an elastomer. Commercially, NR is almost exclusively obtained from *Hevea brasiliensis*.[30] It exists in two isomeric forms, the cis-, and trans-, because of the presence of double bond in its chains. The NR obtained from the *Hevea brasiliensis* tree is of the cis-isomer. NR occurs in nature in varying molecular weights, and which depend on the plant source. The physical properties of NR depend on the rubber structure, especially its configuration. For instance, the most striking property of NR is its elasticity. The ability to exhibit this property is due to the cis-configuration, and free rotation about single bonds which allow the rubber molecules to coil up and become tangled with one another, in a more or less random manner. X-ray diffusion studies obtained for stretched rubber shows that molecular order increases under tension. When the cohesion between the rubber chains is broken by various means, NR has some characteristics of a thick oil; on the other hand, under certain conditions, especially those of intense cooling, rubber behaves in many respects like a crystalline material.

NR has low compression set, high tensile strength, resilience, abrasion, electrical resistance, tear, and wear resistant properties. It has a good frictional surface, and exhibits excellent adhesion to metals. It also exhibits good resistant to flex cracking.

NR crystallizes easily at low temperature and so, in sub-zero temperatures, it becomes necessary to thaw rubber by stoving it in a hot room at 40–50 °C before use in winter.[31]

When compared to synthetic rubbers, NR requires lower curing temperatures, and therefore, longer vulcanization time. It has also poorer abrasion resistance and ageing properties, and cracks easily in tread grooves and tyre side walls.

NR is water repellent, and resistant to alkalis and weak acids. It has moderate resistance to ozone and oxidation, and poor resistant to light. The limitations of NR include attack by petroleum oils and deterioration on exposure to sunlight and ozone, and hence, the use of specific rubbers in the seal industry.

The uses of NR are almost entirely in the tyre sector. The ratio of NR to other rubbers used is very high and stands at about 78:22.[31] The non-tyre sector makes use of NR to produce tubes, footwear, belts, hoses, *etc*. NR also finds applications in the transportation, industrial, consumer, hygiene, and medical sectors.

22.3 Natural Rubber Based Blends

In recent years, there have been increasing interest in polymer blending since it is possible to obtain desirable properties by simple blending. Polymer blends are being used extensively in numerous applications. A blend can offer a set of properties that may give it the potential of entering application areas not possible with either of the polymers comprising the blend. The desired property improvements obtainable through polymer blending include impact strength, heat distortion temperature, flame retardancy, permeability characteristics, and processability, in addition to cost reductions. Since no rubber (an elastomer) has all the characteristics required in many application areas, rubbers are commonly blended to improve their performance. The blending of NR, an elastomer, not only leads to a reduction in the cost of the compound, it also makes it easier to fabricate complex shapes during production.[31,32] Blends of NR have been reported to be compatible with desired mechanical properties.[33–37]

The blending of natural rubber with thermoplastics, and other rubbers have been reported in the literature. Thus, blends of NR with (i) ultra-low density polyethylene,[38] (ii) styrene-butadiene rubber (SBR),[39–43] (iii) epoxidized natural rubber,[44–46] (iv) acrylonitrile butadiene rubber,[47] (v) chloroprene rubber[48] and (vi) dichlorocarbene modified styrene-butadiene rubber (DCSBR)[49] have been prepared, characterized and reported in the literature.

Since the combination of NR with other polymeric materials provides the possibility of effectively producing advanced multi-component polymeric systems with new properties, more NR based blends will continue to be studied because of the unique properties these blends offer.

22.4 Natural Rubber Based Interpenetrating Polymer Networks (IPNs)

An interpenetrating polymer network (IPN) is a polymer that comprises of two or more networks which are at least partially interlaced on a polymer scale but not covalently bonded to each other.[50] It could also be any material containing two polymers, each in network form.[51–54] These networks form a special kind of nanocomposites in which segments or units of each component are dispersed among units of a second component but cannot separate because of the network structures. Like NR blends, IPNs have gained widespread acceptance in industrial applications.

NR based IPNs are relatively, a novel type of multiphase system, where compatibility and certain degree of phase mixing is induced by crosslinking and interpenetration of component polymer chains. The preparation and characterization of some NR based IPNs have been reported in the literature.

Thus, while Mathew et al.[55] studied the effects of blend ratio, crosslinking level, and initiating system on the thermal degradation of NR/polystyrene IPNs. Results from thermogravimetric analysis (TGA) showed that the IPNs were more stable than the pure components, and full IPNs had better stability than semi-IPNs. The later observation was attributed to higher entanglement density of full IPNs. Data from thermal ageing studies of the samples showed that IPNs aged for 72 h at 100 °C; an indication of enhanced mechanical strength due to crosslinking on post curing. The preparation of semi-IPN composites composed of NR and condensed tannin by means of enzyme-mimetic crosslinking of condensed tannin catalysed by haematin was reported by Kadokawa et al.[56]

The mechanical properties and failure topography of NR/polystyrene nanostructured full IPNs was investigated by Mathew et al.[57] while a series of NR/polystyrene IPN membranes were also synthesized by Mathew et al.[58]

22.5 Transport of Penetrant Molecules through Natural Rubber Based Blends and IPNs

The transport of liquids, vapours, and gases though polymer blends and IPN's is of fundamental importance to a polymer scientist. The driving force behind the transport process is the concentration difference between the two polymer phases or the chemical potential of the penetrant in the phases separated by the membrane. The transport process involves the sorption, diffusion, and permeation of the penetrant into the polymer system.

For a polymer blend, the study of sorption, diffusion, and permeation of penetrant molecules provides additional means for its characterization.[59] The diffusion and transport through polymer blends depend upon the blend composition, miscibility, and phase morphology. Most polymer blends, however, are heterogeneous, and consists of a polymeric matrix in which the second phase is embedded.[60]

Reports in the literature on the transport of penetrant molecules through NR based blends and IPN's are few when compared to the existing literature on most common/commercial polymers/blends. The sorption and diffusion of aromatic penetrants into different NR blends such as NR/BIIR, NR/CIIR, NR/neoprene, NR/EPDM, NR/polybutadiene, and NR/SBR were studied by Siddaramaiah et al.[61] The diffusion coefficient (D) of the penetrants was found to range from 6.8 to 84.3 × 10^{-8} cm^2/s at a temperature range of 25–60 °C. Results indicated that the transport data were affected by the nature of the interacting solvent molecule rather their sizes, and also by the structural variations of the elastomers blended with NR. The activation parameters for the diffusion of the penetrants ranged from 4.16 to 30.48 kJ/mol.

Mathai et al.[62] investigated the diffusion and permeation properties of substituted benzenes (benzene, toluene, and p-xylene) through a variety of nitrile rubber and NR blend membranes in the temperature range 28–70 °C. The equilibrium solvent uptake was found to decease with increase in nitrile rubber content, and this was attributed to the inherent solvent resistance of the nitrile rubber. The diffusivity values were found to decrease with increase in molecular size of the penetrant, while the transport coefficients increased with increase in temperature. The mechanism of diffusion was found to be anomalous. The sorption data were used to estimate the activation energies or the processes of diffusion and permeation, while the van't Hoff relationship was used to determine the thermodynamic parameters. The study also correlated the morphology of blends with the difference in equilibrium solvent uptake values of the various blend composition. From Figure 22.1 as presented, it was evident that at high NR content (N_{70}), NBR was dispersed in the continuous NR matrix, and at low NR content (N_{30}), NR was dispersed in the continuous NBR matrix. At low NBR content (N_{30}), the continuous NBR phase acted as a tortuous path for the diffusion process of solvent, and the uptake was less. At high NBR content (N_{70}), the reverse was obtained, and the solvent uptake for the blend was high.

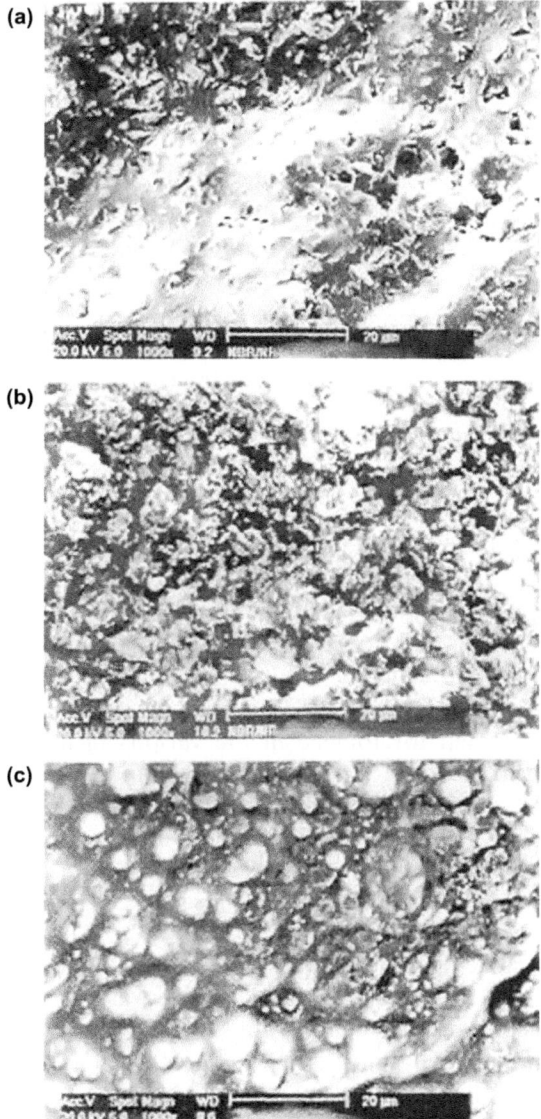

Figure 22.1 Scanning electron micrographs of (a) N_{30}, (b) N_{50} and (c) N_{70}. Reproduced with permission from *J. Membr. Sci.*, 2002, **202**, 35.

The sorption and diffusion of toluene through NR/linear low density polyethylene (LLDPE) blends at 35, 55 and 65 °C were investigated by Obasi *et al.*[63] The molar percentage uptake (Q_t) at any particular temperature plotted against the square root of time ($\sqrt{t}$) showed initial increases in the mass of toluene sorbed until the maximum absorption was reached at which time, the mass of the absorbed toluene remained constant (Figure 22.2).

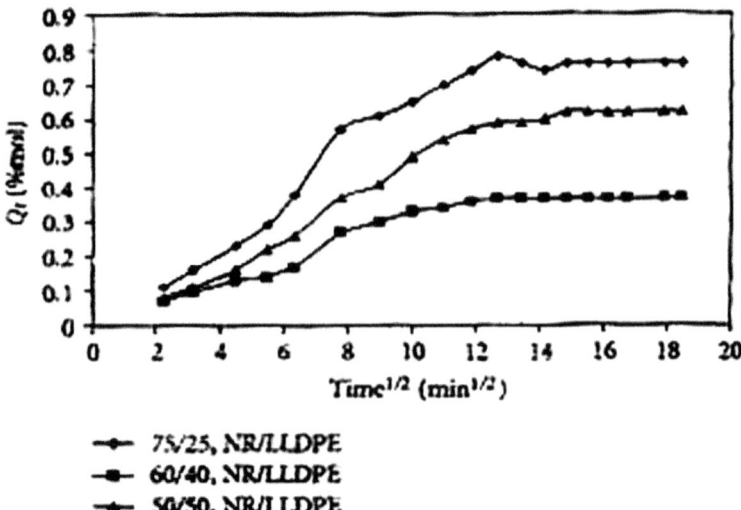

Figure 22.2 Plot of sorption data of NR/LLDRE blends at 35 °C.
Reproduced with permission from Int. J. Polym. Sci., 2009, vol 2009, Article ID, 140 682, 6 pages, doi: 10. 1155/2009/140 682.

The 75/25 NR/LLDPE blend exhibited the highest amount of Q_t of toluene at the temperatures studied. The diffusion, permeation, and sorption coefficients were obtained for toluene in the various blends. The transport of toluene through the 60/40 NR/LLDPE blend was found to be Fickian at 55 °C, pseudo-Fickian at 35 °C, while for the other blends at the temperatures studied, it was anomalous. The calculated activation energies of diffusion and permeation in the rubber blends were all positive, and were found to decrease with an increase in the amount of NR in the blends. The calculated heat of sorptions for the blends were all positive.

The effects of annealing ageing time at 70 °C on the swelling behaviour of NR/butyl rubber/SBR rubber blends in kerosene was investigated by Nasr and Gomaa.[64] The degree of swelling ($Q\%$) in kerosene was found to decrease with both physical ageing and butyl rubber content in the blend. Johnson and Thomas[65] studied the transport of pentane, hexane, heptanes and octane through NR/epoxidized natural rubber (ENR) blends at 27–60 °C. Transport parameters such as the rate constant, diffusion, sorption and permeation coefficients were calculated, and the temperature dependence of the diffusion process was used to estimate the activation parameters. Also, transport studies based on SBR/NR,[66] and NR/polystyrene[67] have been reported in the literature.

The diffusion of some aromatic solvents (benzene, toluene, and p-xylene) through microcomposites of NR/carboxylated styrene-butadiene rubber (XSBR) latex blend membranes (70/30) was investigated by Stephen et al.[68] Results indicated that the blend membrane exhibited unexpected diffusion behaviour, and this was attributed to the immiscibility of the two blend components. The blend exhibited a non-Fickian anomalous transport behaviour. The swelling behaviour of NR/reclaimed NR (STR VS60 and STR20CV) blends in toluene was investigated by Kumnuantip and Sombatsompop.[69] It was found that in all cases, the rate of toluene uptake was relatively fast in the initial stage and reached a plateau value at an equilibrium state, and that blends having greater reclaim content required less time to reach an equilibrium state (Figure 22.3).

The sample containing 80% reclaimed rubber was observed to reach the equilibrium first, and exhibited the lowest value of toluene uptake. The decrease in Q_∞ due to the reclaimed rubber was attributed to the increase in crosslink

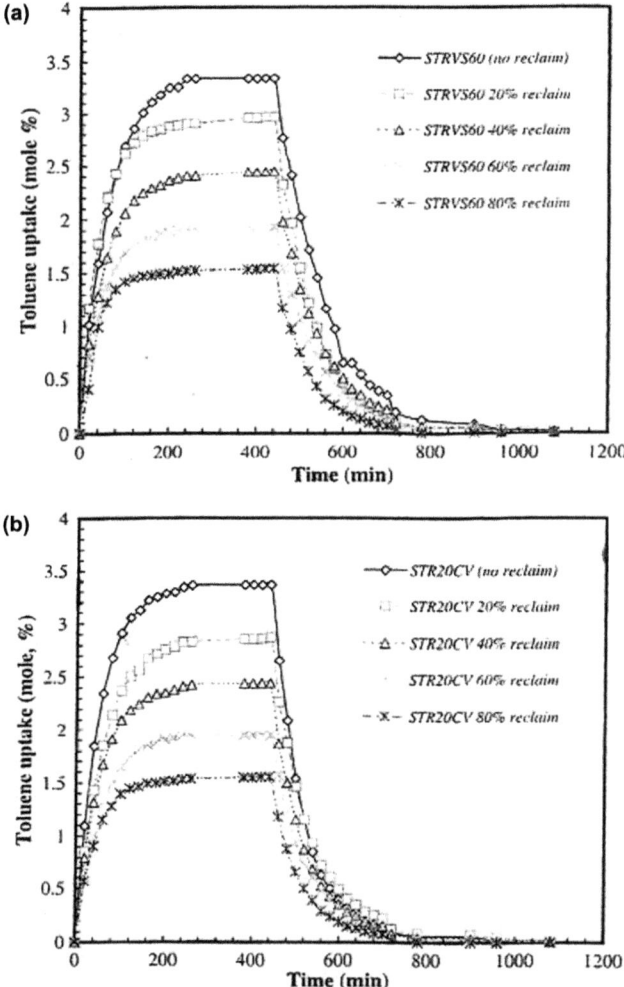

Figure 22.3 Effect of reclaimed rubber content on toluene sorption and desorption curves for (a) STRVS 60 vulcanizates and (b) STR 20CV vulcanizates. Reproduced with permission from *Mater. Lett.*, 2003, **57**, 316.

density and carbon black content in the reclaim. These restricted molecular movement in the polymer, and made it difficult for toluene to penetrate through the rubbers, and thus, decreasing the swelling. According to the authors, the decrease in Q_∞ could also indicate higher molecular interactions between carbon black in the reclaim and NR molecules. The calculated values of n were in the range of 0.27 to 0.40 for the blends, and n was observed to decrease slightly with an increase in reclaim content. The values of k were found to increase with increase in reclaimed rubber content, and this suggested a higher rubber–solvent interaction probably arising from the increased crosslink density. During desorption, it was found that the rate of loss of toluene was rapid and reached the original state (unswollen state) within 6 to 7 h. The study also found that there was no difference in the swelling behaviour of STRVS 60 and STR 20CV, and this was attributed to the similarity in the crosslink density for any given reclaim contents.

The transport of substituted benzenes through blends of thermoplastic polyurethane (TPU) and NR at 30, 50, and 70 °C was studied by Al Minnath et al.[70] Results indicated that the equilibrium solvent uptake in the blends decreased with an increase in the concentration of TPU. The mechanism of diffusion of the solvents was found to deviate from the normal Fickian trend. The sorption data obtained were used to estimate the transport coefficient, and various sorption kinetic parameters.

Few-studies on the sorption, permeation, and diffusion of vapours into NR based-blends have been reported in the literature. The transport study can provide information on the interaction of the vapours with the blends. For example, when the interaction between the vapour and each of the component of the blend is compared, information on the interaction between the blend components can be obtained.

The study of solvent vapour permeation is capable of offering direct practical conclusion for the understanding, and rationale design of VOC vapour recovery from contaminated air. Thus, Johnson and Thomas[71] investigated the sorption, diffusion, and permeation of chlorinated hydrocarbon vapours through NR/ENR blends and reported that the sorption coefficient increased with increase in epoxy content in the blend while the permeability coefficient showed a reverse trend. The permeability coefficient was minimum for the 70/30 composition, and maximum for the 50/50 composition. Similarly, the permeation of chlorinated hydrocarbon vapours through SBR/NR blends was investigated by George et al.[72] The permeation of chlorohydrocarbon vapours through the blends was compared with the permeation of oxygen and nitrogen gases through the blend membranes.

Sansone et al.[73] have calculated the permeability coefficients of benzene vapours in air from the measured values of solubility and diffusivity for NR/neoprene blend membrane.

Gas transport in polymer blends is directly affected by the morphology, and chemical uniformity of the maternal.[74,75] An extensive investigation on the permeability behaviour of gases in polymer blends is reported by Ranby et al.[76–78]

Johnson and Thomas[79] have studied the permeabilities of nitrogen, and oxygen gases in NR/ENR blends. In NR/ENR blends, the permeability of the penetrant decreased with volume fraction of ENR, whereas selectivity increased with volume fraction of ENR. Similar studies on the diffusion of oxygen and nitrogen gases through nano and microcomposites of XSBR blend membranes were investigated by Stephen et al.[80] Stephen et al.[81] also investigated the permeability of NR/XSBR latex blend membranes with special reference to the effects of blend ratio, pressure, and nature of permeants with oxygen and nitrogen gases. The experimentally determined permeability values were theoretically correlated with the Maxwell, Bruggeman, and Bottcher models. Results indicated that the oxygen-to-nitrogen selectivity decreased with an increase in XSBR content in the blend. The permeation of nitrogen and oxygen gases through SBR, NR and SBR/NR blend membranes was studied by Georgea et al.[82] The effects of blend composition and morphology, the nature and degree of crosslinking in both SBR and SBR/NR blends on gas permeation behaviour were investigated. Theoretical models proposed by Maxwell, Bruggemen and Bottcher were used to investigate the relationship between the blend morphology and permeation properties. Barrie et al.[83] studied the sorption and diffusion of propane in a series of NR-graft-polystyrene copolymers for which the polystyrene grafts were of relatively low molecular weight, and had a heterogeneity index close to unity. Results showed that the glass transition temperature and the sorptive capacity of the polystyrene domains were reduced relative to the high molecular weight polystyrene. The dependence of the permeability on copolymer composition was not well represented by expressions developed for transport in a continuum with an impermeable disperse phase. Furthermore, the results were interpreted as being consistent with a morphology which changed with copolymer composition.

The gas permeability coefficient of NR/XSBR blend (70 : 30) membranes was investigated by Stephen et al.[84] The oxygen/nitrogen selectivity of these membranes were determined. The diffusion of gas molecules through the membrane was determined by time-lag method, and diffusion selectivity of the membranes was computed.

It is important to point out that the transport of gases through polymer materials is an area of growing interest since materials with unique transport properties continue to find use in new, specialized applications ranging from extended life tennis balls to natural gas separation systems. Presently, not much research efforts have been devoted to the study of transport of gases through NR blends in spite of the potentials that these materials hold in the field of separation technology.

The study of transport phenomena into IPNs is one of the ways of characterizing them. As was pointed out earlier, IPNs have gained widespread acceptance in industrial applications. The industrial applications and properties of IPNs have been reviewed by a number of authors.[85–87]

Presently, not much report could be found in the literature on the transport of penetrants through NR based IPNs. These networks differ in their swelling properties from other crosslinked polymers since they consist of two components, one of which is rubbery, and the other, glassy. The rubbery, and glassy polymers behave differently towards penetrant molecules since there is a difference in their free volume, and molecular mobility.

For NR based IPNs, Mathew et al.[58] had investigated the sorption and diffusion of three aromatic solvents (benzene, toluene, and xylene) through NR/polystyrene semi-, and full IPNs, and reported that as the polystyrene content increased, the solvent uptake value decreased. This was attributed to the fact that the introduction of plastic phase decreased the chain flexibility of the network. The study further showed that the nature and size of the penetrant molecule affected the transport behaviour. Temperature, it was observed, affected the transport properties. The solvent uptake was found to increase with temperature up to 65 °C, and at 70 °C, a decrease in uptake was observed (Figure 22.4). The values of sorption and permeation coefficients obtained showed a direct dependency on sample characteristics, blend composition, and crosslink level. The study further showed that as the number of crosslinks increased in the blends, the resistance offered to solvent uptake increased since the solvent molecules have to overcome the dense barrier of polymer crosslinking and entanglements to diffuse into the blend.

The swelling behaviour of a membrane reflects the capacity of the membrane to specifically interact with the absorbed molecules. Thus, Amnuaypanich et al.[88]

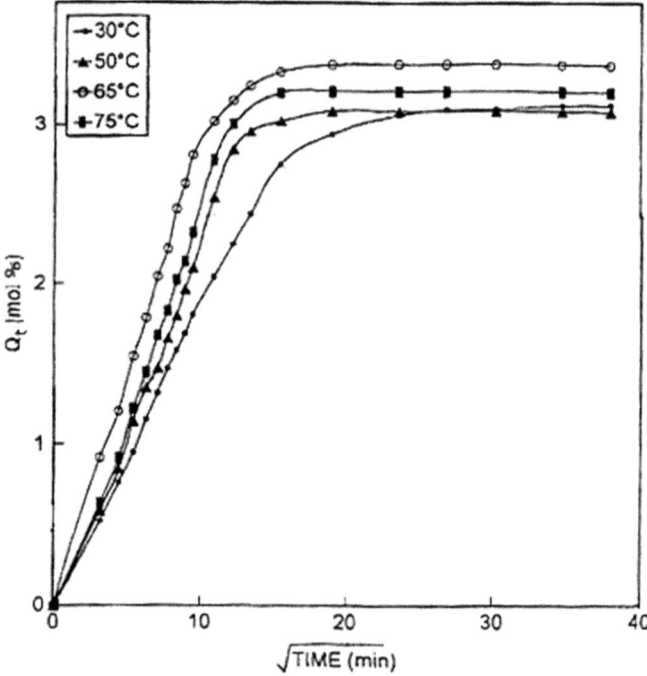

Figure 22.4 Effect of temperature on solvent uptake (Q_t) vs. $\sqrt{t}$ in toluene for D1N50 samples.
Reproduced with permission from J. Membr. Sci., 1989, **44**, 161.

investigated the swelling behaviour of NR/crosslinked poly(vinyl alcohol) (PVA) semi-IPN membrane filled with zeolite 4 A in water-ethanol mixture, and reported that the NR/PVA membrane swelled greatly in water when compared to the swelling in pure ethanol. According to the authors, this was primarily due to the specific interactions between water molecules and the –OH groups of PVA in the membranes. Consequently, the water swelling was enhanced as PVA content in the membranes was increased. When the zeolite 4 A was introduced to the NR/PVA membrane, the mixed membranes exhibited a slight decrease in the water swelling (at 10 and 20 wt% zeolite) as shown in Figure 22.5(b). The swelling of the mixed membranes was observed to fall drastically when the amount of zeolite was increased. The suppression of the membrane swelling in the presence of zeolite was attributed to the crystalline structure of zeolite which is not liable to the interacting molecules and the restricted movement of polymer chains near the zeolite particles. The temperature dependence of the partial water and ethanol fluxes was found to follow the Arrhenius relationship, and the estimated activation energies for water flux were lower than those of ethanol flux, and this, according to the authors, suggested that the developed membranes were highly water selective.

Despite the few reported research being carried out on the transport of penetrant molecules through NR based IPNs, data obtained from the transport studies can be used to advantage in determining parameters of importance as they relate to rubber based networks. The following parameters can be determined: (i) The molecular weight between crosslinks (M_C) can be calculated as:[89,90]

$$M_c = \frac{-\sigma_p V_r \varnothing^{1/3}}{\ln(1-\varnothing) + \varnothing + x\varnothing^2} \quad (22.8)$$

where σ_p is the density of polymer, V_r is the molar volume of solvent, $\varnothing$ the volume fraction of polymer, and χ, the interaction parameter, (ii) the interaction parameter (x) is given by the relation:[88]

$$x = \frac{\beta + V_r}{RT(\delta_A - \delta_B)^2} \quad (22.9)$$

where V_r is the molar volume of solvent, δ_A and δ_B are the solubility parameter of solvent and polymer respectively, R is the universal gas constant, and T is the absolute temperature. β is the lattice constant, and its value is equal to 0.34, (iii) the crosslink density (μ) is obtained using the equation:[91]

$$\mu = \frac{1}{2}M_C \quad (22.10)$$

Two models, the affine and phantom network models have been proposed to predict the structure of the IPNs from swelling studies. Details of these models can be found elsewhere.[90,92] The molecular weights between crosslinks for affine unit ($M_{c(aff)}$), and phantom unit ($M_{c(ph)}$) are then calculated using equations as given by Treloar,[93] and Mark and Erman[94] respectively.

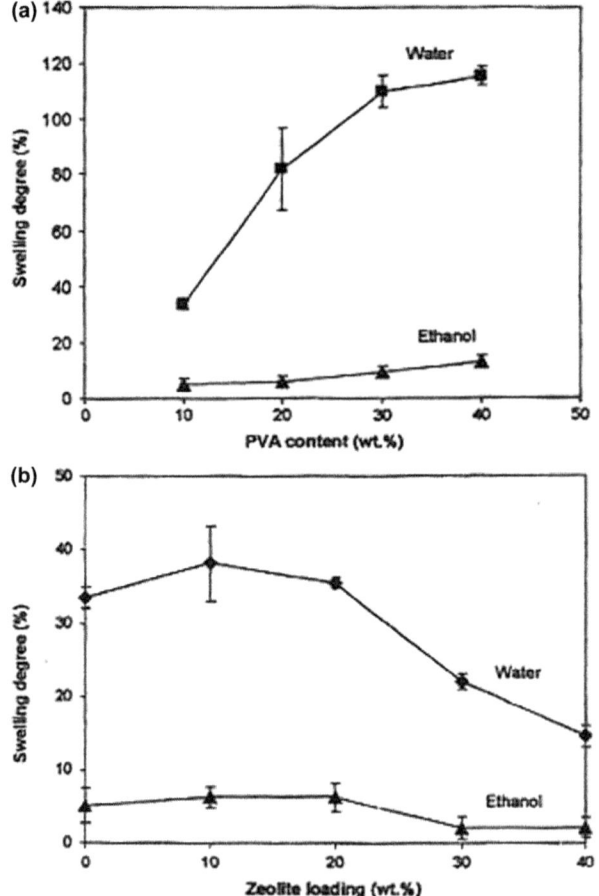

Figure 22.5 Swelling degree of NR/PVA membrane (a) and MM membranes (b) in pure water and ethanol.
Reproduced with permission from *Chem. Eng. Sci.*, 2009, **64**, 4908.

22.6 The Effects of Penetrant Absorption on the Properties of Natural Rubber Systems

Generally, the interaction of penetrant molecules with polymeric materials is a problem from both the academic and technological points of view.[95–98] For example, crosslinked polymers when brought into contact with different solvents during service applications usually exhibit swelling. The sorption and diffusion of solvents into polymer system might cause changes in the mechanical properties, and sometimes, destruction in polymer structure. For example, Johnson and Thomas[99] who investigated the effect of epoxidation on the transport of n-alkanes on NR reported that the mechanical properties of swollen NR samples were inferior to the unswollen samples. It was also reported that the crosslink density values of the swollen state were much lower

Table 22.1 Effects of swelling on tensile properties.[55]

Sample code	Unswollen		Swollen	
	Tensile strength (MPa)	Elongation at break (%)	Tensile strength (MPa)	Elongation at break (%)
D_0N_{50}	16.5	378	0.18	55
D_1N_{50}	13.0	474	0.07	35
D_2N_{50}	16.7	134	0.11	19
D_2N_{50}	12.8	290	0.06	52
D_2N_{70}	9.0	350	0.18	58
D_3N_{50}	14.0	197	0.13	19

when compared to the unswollen state. According to the authors, the plasticization of the polymer matrix by the solvent affected the physical interactions between the polymer chains. This resulted in the decreased values of crosslink density in the swollen state.

For NR systems, Ramesan[50] who investigated the oil resistance properties of NR/DCSBR blends reported that the tensile strength, modulus, tear strength, and hardness of the rubber blends after immersion in ASTM oil were progressively decreased with increase in NR content in the blends.

Also, Mathew et al.[55] who studied the transport of aromatic solvents through NR/polystyrene IPN membranes found that the tensile strength, and elongation at break of swollen samples were drastically reduced when compared to unswollen samples. The extent of drastic reductions in swollen sample properties is shown in Table 22.1.

In coding the samples, D indicates that the polymer networks were initiated by dicumyl peroxide. The blend compositions are denoted as N_{30}, B_{50} and N_{70}; where the subscripts indicate the weight per cent of NR in the network membrane.

22.7 Conclusions

The characteristic transport of penetrants across polymeric materials have been presented. The transport of penetrant molecules through polymeric systems depends on several factors among which are the nature of the polymer, penetrant properties, temperature, nature of crosslinks, and crosslink density. The transport properties of various NR based blends and IPNs studied using different penetrants have also been presented. For NR based blends, these properties are of interest for different practical applications since they are capable of giving insight into the morphology of the NR blends. The equilibrium sorption of a vapour by NR based blends is capable of not only providing information regarding the interactions between the components of the blend, but also, is utilized in practical separation of VOCs in vapour recovery from contaminated air. The transport of gases through polymer blends is nowadays assuming importance since materials having unique transport properties are used in new specialized applications. The reported transport properties of interpenetrating NR based networks have been scanty despite their wide industrial applications. These networks differ in their swelling from other crosslinked polymers. The presence of absorbed solvents in NR based blends and IPNs has been shown to adversely affect the mechanical properties of the NR systems investigated. As the penetrant (solvent) invades and swells the NR system, local stresses are built up. These stresses may be quite high, and can cause mechanical failure in the system.

References

1. N. N. Li, R. B. Long and E. J. Henley, *Ind. Eng. Chem.*, 1965, **57**, 18.
2. G. Strandburg, P. T. Delassus and B.A. Howell, in *ACS Symposium Series 423*, ed. W. J. Koros, American Chemical Society, Washington DC, 1990, p. 333.
3. H. Coll and C. G. Searls, *Polymer*, 1998, **29**, 1266.
4. P. V. Kulkarni, S. B. Rajur, P. Antich, T. M. Aminabhavi and M. I. Aralaguppi, *Rev., Macromol. Chem. Phys.*, 1990, **30**, 441.
5. N. A. Peppas, P. Bures, W. Leobandung and H. Ichikawa, *Eur. J. Pharm. Biopharm*, 2000, **50**, 27.
6. T. M. Aminabhavi, R. S. Khinnavar, S. B. Harogoppod, U. S. Aithal, Q. T. Nguyen and K. C. Hansen, *J. Macromol. Sci., Rev. Macromol. Chem. Phys.*, 1994, **34**, 139.
7. M. O. David, Q. T. Nguyen and E. J. Neel, *J. Membr. Sci.*, 1992, **73**, 129.
8. R. G. Carbonell and G. C. Sarti, *Ind. Eng. Chem. Res.*, 1990, **29**, 1194.
9. A. R. Berens and H. B. Hopfenberg, *Polymer*, 1978, **19**, 489.
10. H. B. Hopfeng, in *Membrane Science and Technology*, ed. J. E. Flinn, Plenum Press, New York, 1970, p. 16.
11. N. L. Thomas and A. H. Windle, *Polymer*, 1981, **22**, 627.
12. A. R. Berens, *Research Tech.*, 1985, **57**, Nov.
13. N. M. Franson and N. A. Peppas, *J. Appl. Polym. Sci.*, 1983, **28**, 1299.
14. W. J. Koros and M. W. Hellums, in *Encyclopedia of Polymer Science and Engineering*, ed. J. I. Kroschwitz, Wiley, New York, 1990, Supplement volume, pp. 724–802.
15. R. G. Carbonell and G. C. Sarti, *Ind. Eng. Chem. Res.*, 1990, **29**, 1194.
16. B. D. Freeman, *Macromolecules*, 1999, **32**, 375.
17. J. Crank, *The Mathematics of Diffusion*, 2nd edn, Clarendon, Oxford, 1975.
18. N. A. Peppas and K. G. Urdahl, *Polym. Bull.*, 1986, **28**, 96.
19. K. G. Urdahl and N. A. Peppas, *Polym. Eng. Sci.*, 1988, **28**, 96.
20. C. M. Walker and N. A. Peppas, *J. Appl. Polym. Sci.*, 1990, **39**, 2043.
21. D. Kim, J. M. Carothers and N. A. Peppas, *Macromolecules*, 1993, **26**, 1841.
22. M. J. Smith and N. A. Peppas, *Polymer*, 1985, **26**, 569.
23. A. S. Michaels and H. J. Bixler, *J. Polym. Sci.*, 1961, **50**, 413.
24. D. H. Weinkauf and D. R. Paul, in *Proceedings of the ACS Symposium on Barrier Polymers*, Washington, DC, 1989, p. 3.
25. C. E. Rogers, V. Stannet and M. Szwarc, *J. Polym. Sci.*, 1960, **45**, 61.

26. J. G. A. Bitter, *Transport Mechanisms in Membrane Separation Systems*, Plenum Press, New York, 1991.
27. R. W. Baker, N. Yoshioka, J. M. Mohr and A. Khan, *J. Membr. Sci.*, 1987, **31**, 259.
28. E. C. Moretti and N. Mukhopadhyay, *Chem. Eng. Prog.*, 1993, **7**, 20.
29. J. Salwinski, J. Izydorezyk and J. Podkowka, *Polymer*, 1979, **19**, 621.
30. S. F. Cheng, *Types, Composition, Properties, Storage, and Handling of Natural Rubber*, Rubber Research Institute of Malaysia, Kuala Lumpur, 1988.
31. Indian Rubber Institute, *Rubber Engineering*, Tata McGraw-Hill, New Delhi, 1998.
32. D. R. Paul (ed.), *Polymer Blends*, Academic Press, New York, 1976, vol. 11, Chapter 12.
33. Y. M. Lee, D. Bourgeois and G. Belfort, *J. Membr. Sci.*, 1989, **44**, 161.
34. A. T. Koshy, B. Kuriakose and S. Thomas, *Polym. Degrad. Stab.*, 1992, **36**, 137.
35. A. T. Koshy, B. Kuriakose, S. Thomas and S. Varghese, *Polymer*, 1993, **34**, 3428.
36. A. T. Koshy, B. Kuriakose, S. Thomas and S. Varghese, *Polym.-Plast. Technol. Eng.*, 1994, **33**, 149.
37. P. Jansen, A. S. Gomes and B. G. Soares, *J. Appl. Polym. Sci.*, 1996, **61**, 591.
38. P. Jansen and B. G. Soares, *Polym. Degrad. Stab.*, 1996, **52**, 95.
39. V. Tanrattanakul and W. Udomkichdecha, *J. Appl. Polym. Sci.*, 2001, **82**, 650.
40. K. Kim, *J. Appl. Polym. Sci.*, 1996, **61**, 431.
41. S. C. George, K. N. Ninan, G. Groenincky and S. Thomas, *J. Appl. Polym. Sci.*, 2000, **78**, 1280.
42. S. P. Basu, *J. Elastomers Plast.*, 1991, **23**, 152.
43. S. C. Goerge, J. P. Misra, V. S. Kishan Prasad and S. Thomas, *J. Appl. Polym. Sci.*, 1999, **74**, 3059.
44. S. C. Goerge, K. N. Ninan and S. Thomas, *J. Membr. Sci.*, 2000, **176**, 131.
45. B. T. Poh and G. K. Khok, *Polym.-Plast. Technol. Eng.*, 2000, **39**, 151.
46. H. Ismail and B. T. Poh, *Eur. Polym. J.*, 2000, **36**, 2403.
47. T. Johnson and S. Thomas, *J. Appl. Polym. Sci.*, 1999, **171**, 236.
48. P. Kumari, C. K. Radhakrishnan, S. Goerge and G. Unnikrishnan, *J. Polym. Res.*, 2008, **15**, 97.
49. H. Ismail and H. C. Leong, *Polym. Test.*, 2001, **20**, 509.
50. M. T. Ramesan, *React. Funct. Polym.*, 2004, **59**, 267.
51. A. P. Mathew, S. Packirisamy and S. Thomas, *J. Appl. Polym. Sci.*, 2000, **78**, 2327.
52. D. R. Paul (ed.), *Polymer Blends*, Academic Press, New York, 1976, vol. 11, Chapter 12.
53. Y. M. Lee, D. Bourgeous and G. Belfort, *J. Membr. Sci.*, 1989, **44**, 161.
54. T. M. Aminabhavi and U. S. Aithal, *J. Macromol. Sci., Rev. Macromol. Chem. Phys.*, 1991, **C31**, 117.

55. A. P. Mathew, S. Packirisamy and S. Thomas, *Polym. Degrad. Stab.*, 2001, **72**, 423.
56. J. Kadokawa, K. Kodzuru, S. Kawazoe and T. Matsuo, *J. Polym. Environ.*, 2011, **9**, 100.
57. A. P. Mathew, S. Packirisamy, H. J. Radusch and S. Thomas, *Eur. Polym. J.*, 2001, **37**, 1921.
58. A. P. Mathew, S. Packirisamy, R. Stephen and S. Thomas, *J. Membr. Sci.*, 2002, **201**, 213.
59. H. B. Hopfenberg and D. R. Paul, *Transport Phenomena in Polymer Blends*, Academic Press, New York, vol. 1, 1978.
60. L. Y. Lai, G. J. Wu and S. Shyu, *J. Appl. Polym. Sci.*, 1987, **34**, 559.
61. S. Siddaramaiah, Roopa and U. Premakumar, *Polymer*, 1998, **39**, 3925.
62. A. S. Mathai, R. P. Singh and S. Thomas, *J. Membr. Sci.*, 2002, **202**, 35.
63. H. C. Obasi, O. Ogbobe and I. O. Igwe, *Int. J. Polym. Sci.*, 2009, **2009**, Article ID 140682, doi: 10.1155/2009/140682.
64. G. M. Nasr and A. S. Gomaa, *Polym. Degrad. Stab.*, 1995, **50**, 249.
65. T. Johnson and S. Thomas, *J. Mater. Sci.*, 1999, **34**, 3221.
66. S. C. George, K. N Ninan, G. Groeninckx and S. Thomas, *J. Appl. Polym. Sci.*, 2002, **78**, 280.
67. R. Asaleta, M. G. Kumaran and S. Thomas, *Polym. Polym. Compos.*, 1998, **6**, 1.
68. R. Stephen, K. Joseph, Z. Oommen and S. Thomas, *Compos. Sci. Technol.*, 2007, **67**, 1187.
69. C. Kumnuantip and N. Sombatsompop, *Mater. Lett.*, 2003, **57**, 3167.
70. M. Al Minnath, G. Unnikrishnan and E. Purushothaman, *J. Membr. Sci.*, 2011, **379**, 361.
71. T. Johnson and S. Thomas, *Polym.-Plast. Technol. Eng.*, 2000, **39**, 363.
72. S. C. George, K. N. Ninan, G. Groeninck and S. Thomas, *J. Appl. Polym. Sci.*, 2000, **78**, 280.
73. E. B. Sansone and Y. B. Tewari, *Am. Ind. Hyg. Assoc. J.*, 1980, **41**, 170.
74. J. Sax and J. M. Ottino, *Polym. Eng. Sci.*, 1983, **23**, 165.
75. J. M. Ottino and N. Shah, *Polym. Eng. Sci.*, 1984, **24**, 153.
76. B. J. Ranby, *J. Polym. Sci., Part C: Polym. Symp.*, 1975, **51**, 89.
77. Y. J. Shur and B. Ranby, *J. Appl. Polym. Sci.*, 1976, **20**, 3121.
78. W. J. Shur, J. F. Rubek and B. Ranby, *Inter. Symp. Macromol.*, IUPAC 24th, Jerusalem, 1975, p. 18.
79. T. Johnson and S. Thomas, *Polymer*, 1999, **40**, 3223.
80. R. Stephen, C. Ranganathaiah, S. Varghese, K. Joseph and S. Abu, *Polymer J.*, 2006, **47**, 858.
81. R. Stephen, S. Thomas and K. Joseph, *J. Appl. Polym. Sci.*, 2005, **98**, 1125.
82. S. C. George, K. N. Ninan and S. Thomas, *Eur. Polym. J.*, 2001, **37**, 183.
83. J. A. Barrie, P. Sagoo and A. G. Thomas, *J. Membr. Sci.*, 1989, **43**, 229.
84. R. Stephen, C. Ranganathaiah, S. Varghese, K. Joseph and S. Thomas, *Polymer*, 2006, **47**, 858.
85. G. M. Jordhamo, J. A. Manson and L. H. Sperling, *Polym. Eng. Sci.*, 1986, **26**, 8.

86. R. Hu, V. L. Dimonie, M. S. El-Aasser, R. A. Pearson, A. Hiltner, R. A. Pearson, S. G. Mylonaki and L. H. Sperling, *J. Polym. Sci., Part A: Polym. Chem.*, 1997, **35**, 2193.
87. R. Hu, V. L. Dimonie, M. S. El-Aasser, R. A. Pearson, A. Hiltner, S. G. Mylonaki and L. H. Sperling, *J. Polym. Sci., Part B: Polym. Phys.*, 1997, **35**, 1501.
88. S. Amnuaypanich, J. Patthana and P. Phinyocheep, *Chem. Eng. Sci.*, 2009, **64**, 4908.
89. P. J. Flory, *Principles of Polymer Chemistry*, Cornell University Press, Ithaca, NY, 1953.
90. P. J. Flory and J. Rehner Jr, *J. Chem. Phys.*, 1943, **11**, 521.
91. S. H. Morrel, in *Rubber Technology and Manufacture*, ed. C. M. Blow, Butterworths, London, 1975, p. 162.
92. H. M. James and E. Guth, *J. Chem. Phys.*, 1947, **15**, 669.
93. L. R. G. Treloar, *The Physics of Rubber Elasticity*, Clarendon Press, Oxford, 1975.
94. J. E. Mark and B. Erman, *Rubber Like Elasticity, A Molecular Prime*, Wiley, New York, 1988.
95. P. G. de Gennes, *Sealing Concepts in Polymer Physics*, Cornell University Press, London, 1979.
96. A. Beamish, R. A. Goldberg and D. J. Hovrston, *Polymer*, 1977, **18**, 49.
97. K. Ito, *J. Polym. Sci., Part A: Polym. Chem.*, 1978, **16**, 497.
98. J. O. Barton, *Polymer*, 1979, **20**, 1018.
99. T. Johnson and S. Thomas, *Polymer*, 2000, **41**, 7511.

CHAPTER 23

Life Cycle Analysis, Ageing and Degradation Behaviour of Natural Rubber Based Blends and IPNs

CRISTINA RUSSI GUIMARÃES FURTADO* AND
MÁRCIA CHRISTINA AMORIM MOREIRA LEITE

Universidade do Estado do Rio de Janeiro, Instituto de Química, Rua São Francisco Xavier, 524, Pavilhão Haroldo Lisboa da Cunha, Sala 310, Rio de Janeiro, Brazil
*Email: russi@uerj.br

23.1 Introduction

The term 'sustainability' originates from silviculture and means that only wood is removed from the forest since it grows again in the long run. Cultivation and care are prerequisites of sustainable forestry. Hans Carl von Carlwitz, the founder of this practice, recognized well the economic and social implications of his idea. He therefore may also be considered as the father of sustainability in the modern sense of the word. In the global political arena, the term sustainability became renowned by its use in the report *Our Common Future* by the World Commission on Environment and Development. In this report, environmental protection is linked with global development and the responsibility that humankind has for future generations is emphasized. This can be recognized in the famous definition: "Sustainable development is development that meets the needs of present without compromising the ability of future generations to meet their own needs."[1]

At the United Nations Conference of Environment and Development (UNCED) in Rio de Janeiro, 1992, sustainable development was laid down as the most important task of the 21st century. Moreover, in Agenda 21 many political and industrial areas were analysed with regard to sustainable development and possible improvements were presented. It is now clear that three aspects, the three 'pillars' of sustainability, have to be considered and brought together: ecology (environment), economy, and social aspects.[1]

Product-related life cycle assessment with an emphasis on energy, resources and waste started around 1970. It was the time of *The Limits to Growth, a report to the club of Rome* and the first oil crisis soon afterwards showed, if not the shortage of oil, but at least the vulnerability of the global economic system. Twenty years later, life cycle assessment (LCA) was developed by the Society of Environmental Toxicology and Chemistry (SETAC) and later was standardized by the International Organization for Standardization (ISO 14040-43). LCA can be considered the first internationally standardized environmental assessment method.[1]

The polymer industry has brought great benefits to modern society, although the disadvantages appeared when polymer-based products were discarded at the end of their useful life and in particular when they appear as litter in the environment. In this way, it becomes very important to study the ageing and degradation of these polymers.

The purpose of the present chapter is to give an overview of recent works carried out in the field of LCA, ageing and degradation of natural rubber (NR) based blends and interpenetrating polymer networks (IPNs). As NR is an unsaturated polymer, NR will gradually degrade at a high temperature or when exposed to oxygen, ozone or ultraviolet light, leading to a negative effect on its special application. To overcome these limitations, polymer blends have attracted the attention of researchers due to the possibility of tailoring their properties by blending different polymers. These blends can improve the ageing properties, thermal stability, oil and solvent resistance and even produce a biodegradable material.

23.2 Life Cycle Assessment

A life-cycle assessment, LCA, also known as life-cycle analysis, ecobalance and cradle-to-grave,[2] is a process for evaluating the environmental burdens associated with a product, process or activity by identifying and quantifying the energy and materials used and wastes discharged to the environment, and assessing the impact on the environment of the energy and materials uses and waste released, to identifying ways to reduce the environmental impacts.[3] LCA, as described by SETAC,[4] addresses environmental impacts in the areas of ecological health, human health and resource depletion. It does not address other consequences of manufacturing activities, such as economic or social effects. Like all scientific models, the LCA method involves simplification of the physical system and cannot claim to provide an absolute and complete representation of every environmental interaction.[5]

The prime objective of carrying out an LCA is to provide information which shows the full effects that an activity has on the environment and thereby to identify opportunities for making changes to reduce the environmental impact of the activity.[5]

If a study is described as an LCA, it should cover the entire life cycle of the system, from extracting raw materials, through manufacturing, distribution and usage, to final disposal. The LCA method is often used for more limited studies, but this should always be made clear in reporting and applying the results of such limited applications.[5]

The method originated in the late 1960s and was developed in a number of countries, including Sweden, the UK, Switzerland and the USA. In this initial period, which lasted until the end of the 1980s, numerous studies were performed, using different methods and without a common theoretical framework. As a consequence, the results differed greatly, although the subjects of the studies were often the same, thus preventing LCA from becoming a more generally accepted and applied analytical tool.[6]

Since 1990, the European and North American organizations of Society for Environmental Toxicology and Chemistry (SETAC) have arranged a series of seminars and workshops to promote the development of methods for conducting life-cycle assessment with more scientific rigour and the proper use of the results. Widespread interest in the use of LCA for assessing the environmental impacts of products and manufacturing systems has led to the need for guidance on the most appropriate way of structuring such studies and reporting the results. In response of this need, SETAC organized a workshop of 50 experts from 13 countries who pooled their knowledge and experience over 4 days in Sesimbra, Portugal, in April 1993. The outcome of that workshop was the report *Guidelines for Life-Cycle Assessment*, a "Code of Practice" published by SETAC in September 1993.[5] The LCA methodological framework was described in detail by SETAC[4] and comprises four principal components: goal definition and scoping; inventory analysis; impact assessment; and improvement assessment.[5]

During the second half of the 1990s, the LCA started to gain a wider acceptance in many industrial sectors, particularly in processing industries such as detergent manufacturing, painting and emulsion preparation.[7] Some other segment where LCA has been used in corporate decision making include energy (including nuclear), water supply and eletronics.[7] In 1999, Adisa Azapagic[8] reviewed the state of the art of methodological development and uses of LCA. The review focused on the application of LCA in process selection, design and optimisation as a tool for identifying clean technologies. The procedures incorporated environmental criteria along with economic and technical criteria into the system optimization framework. It was shown that this approach can provide a potentially powerful decision-making tool for managers, process engineers and designers.[8] In 2001, Burgess and Brennan also published a review about the application of LCA to chemical processes.[9] In the same year, Khan *et al.* presented a brief review of the use of LCA and pollution prevention in the process industries.[10] They subsequently detailed the proposed methodology for a green and clean process design. The applicability of the method was demonstrated through a case study of the production of the monomer vinyl chloride.

Recently, some papers were published showing the need for conducting LCA of nanoproducts.[11-13] A group of Brazilian researchers studied the LCA of cellulose nanowhiskers.[14] Vegetal fibres are an important source of cellulose for the extraction of nanowhiskers, which can be used to enhance the mechanical properties of different polymers. The study contributes to the environmental performance of cellulose nanowhisker production processes in the development stage. Environmental aspects and related impacts of two cellulose nanowhiskers product systems are evaluated: nanowhiskers extracted from unripe coconut fibres (EUC system) and from white cotton fibres (EC system). The comparison between the two systems showed that nanowhiskers produced in the EC system required less energy and water, emitted fewer pollutants, and contributed less to climate change, human toxicity and eutrophication than those produced in the EUC system.

In 1996, Yoda presented a review about LCA in the polymer industry.[15] To date, however, only a few studies have been reported for polymers,[16-21] and even less for rubber.[22]

A limited number of LCA studies have been conducted on composite polymer materials, more precisely with polyester,[23] high-density polyethylene (HDPE)[24] and polypropylene (PP).[25,26] The latter two polymers were tested virgin or recycled form.

Polymeric materials have a potential to generate a significant impact on the environment throughout their life cycle, including depletion of finite, non-renewable resources and generation of solid waste. Therefore, it is essential to identify more sustainable ways of using and managing polymers. Thus, it becomes quite important to evaluate the ageing and degradation of these polymers.

In the next section we address to the ageing and degradation of natural rubber based blends and interpenetrating polymer networks (IPNs).

23.3 Ageing and Degradation of NR Based Blends and Interpenetrating Polymer Networks (IPNs)

23.3.1 Polymer Blends and IPNs

Natural rubber (NR) containing 93–95% cis-1,4-polyisoprene is an elastomeric material, produced from latex of the rubber tree. It is a renewable natural resource and has many excellent properties, such as outstanding resilience and high strength.[27] However, as an unsaturated polymer, NR will gradually degrade at a high temperature or when exposed to oxygen, ozone or ultraviolet light, leading to a negative effect on its special application. To overcome these limitations, the modification of NR is crucial. Various methods can be employed to modify the properties of NR. One way is by chemical modification, in which other groups or atoms are introduced into the NR molecular chains producing, for example, epoxidized NR, hydrogenated NR and grafted NR. Another way to enhance the NRs properties is blending it with two or more polymers.[27,28]

In recent years, polymer blends have attracted the attention of researchers due to the possibility of tailoring their properties by blending different polymers. Also, polymer blends have been extensively used in a large number of applications, because a certain polymeric blend offers properties that allow its application where the individual constituting polymers would not.[28]

IPNs are a class of blends where at least one polymer is crosslinked. When only one of the phases is crosslinked, pseudo- or semi-IPNs are formed. The properties of the IPNs are dependent on phase behaviour, blend ratio and crosslinking levels.[29] IPNs, regardless of their composition, have better thermal properties than corresponding homopolymers. This is usually attributed to the networking of the phases.

23.3.2 Ageing and Degradation

Solid polymeric materials undergo physical and chemical changes when heat is applied. This usually results in undesirable changes to the material's properties. The American Society for Testing and Materials describes thermal degradation as a process whereby the action of heat or elevated temperature on a material, product or assembly causes a loss of physical, mechanical or electrical properties and describes thermal decomposition as a process of extensive chemical species change caused by heat.[30]

As Pielichowski and Njuguna describe, the breaking of chemical bonds under the influence of heat is the result of the overcoming the bond dissociation energies. Organic polymers are highly thermally sensitive due to the limited strength of the covalent bonds that make up their structures. Scission can occur either randomly or by a chain-end process, often referred to as an unzipping reaction.[31]

Volatile products can be clipped from the end of a polymer chain during the very start of a reaction, with a distribution that is not random, or by a process of end scission or backbiting – a process of unzipping can regenerate the monomer. Thermal degradation of polymers can follow three major pathways: side-group elimination, random scission and depolymerisation that can occur randomly and/or at the end of the polymer chain.[31]

Thermooxidative degradation is characterized by random scission in the polymer backbone. The most important issues in thermooxidative degradation of polymers are where oxidation takes place, which structure fragments are most vulnerable, how they should be protected, and what the main principles of protection are.[31]

In photochemical degradation is necessary to generate an excited state, which may occur due to light incidence on the polymer. When the polymer is irradiated with light energy corresponding to the electronic transition of the chromophore existing as part of the chain or as a contaminant, of free radicals form and initiate free radical reactions.[32]

Biodegradation (*i.e.* biotic degradation) is a chemical degradation of materials induced by the action of microorganisms such as bacteria, fungi and algae. The most common definition of a biodegradable polymer is "a degradable polymer wherein the primary degradation mechanism is through the action of metabolism by microorganisms". Biodegradable materials degrade into biomass, carbon dioxide and methane.[33]

Polymers are subjected to destructive factors such as mechanical stress, the presence of different chemicals, ultraviolet light, ablation and high temperatures throughout shelf and service lives. These factors cause degradation and ultimately change the performance and lifetime of the polymers, which are sometimes stored for long periods. Therefore, it is important to know how long and under what conditions the polymers can best be stored with minimum deterioration of their properties. According to the lifetime stages of polymers, the relevant processes are classified as melt degradation, long-term heat ageing and weathering based on the mechanisms involved, *i.e.* thermomechanical, thermal, catalytic and radiation-induced oxidations and environmental biodegradation. The products are different low molecular weight (low molar mass) additives or degradation products from the additives or the polymer itself. The diffusion of these low molecular weight products changes the properties of the material and shortens the lifetime. For safety reasons, it is necessary to have a good understanding of the thermal resistance of polymeric materials and to identify precisely the products likely to be formed.[31]

The effect of heat on the performance of a polymeric system can be studied by thermal ageing. During ageing, scission of the main chain, crosslink formation or crosslink breakage can take place.[34] Sometimes, existing crosslinks may be replaced by more stable crosslinks. All these factors contribute to the properties of the aged samples.

23.3.3 NR/Thermoplastic Blends and IPNs

Mathew *et al.*[29] prepared PS/NR IPNs by the sequential technique, in which NR was first crosslinked using DCP followed by the polymerization and crosslinking of the PS phase (divinyl benzene acted as crosslinking agent). During IPN synthesis, the physical entanglements (due to networking of two phases) makes the system tough and thermally stable. The thermal stability of the IPNs was evaluated by thermogravimetric analysis and the vulcanized NR and pure uncrosslinked PS exhibited almost comparable thermal stability and both degraded around 360 °C. The NR/PS semi- or full IPN presented enhanced stability compared to the homopolymers. About 90% of the tested samples degraded at about 450 °C and as the rubber content increased, the thermal decomposition temperature decreased. The effect of heat on the performance of a polymeric system can be studied by thermal ageing. During ageing, scission of the main chain, crosslink formation or crosslink breakage can take place.[34] Sometimes existing crosslinks may be replaced by more stable crosslinks. Ageing resistance increased with increasing PS crosslinking.[35,36]

De et al. studied 30:70 and 70:30 NR/HDPE blends, which showed plastic or rubbery behaviour predominantly, depending on the blend ratio. When a very small amount of crosslinking agent was introduced during milling, it enhanced the properties of the blends, particularly of the high rubber blends. They also investigated the hot air and acid ageing behaviour of the blends and the high plastic blend showed better resistance to acid corrosion and hot air ageing than high rubber blends. Crosslinks had a favourable effect on hot air ageing resistance in the high plastic blend, but the acid resistance slightly decreased in the presence of crosslinks.[37] Thermoplastic elastomers (TPE) play an important role in the polymer industry due to their good processability and their elastomeric properties. TPEs based on NR and thermoplastics in various proportions can be called thermoplastic natural rubber blends (TPNR).

Among TPNRs, polypropylene (PP)/NR blends have been widely studied due to the high melting temperature of PP, high softening temperature (150 °C) and low T_g with dimensional stability at high temperatures and low cost due to the abundant supply of NR.[38,39] Ageing behaviour of PP/NR blends has been studied and in comparison to vulcanized rubber, TPNRs are remarkably resistant to heat ageing, even at 100 °C for 7 days.[40]

Kuriakose et al.[41] mentioned that the reduction of the molecular weight of NR by mastigation[42] improved the processability of NR/PP thermoplastic blends. It was also observed that quinoline imparts better ageing resistance than imidazole when used as antioxidants in the blend. Another study observed that the highest thermal stability is attained with the addition of ethylene-propylene diene rubber to the NR/PP blend.[43] TPEs can also be developed from recycled rubber (PP/RR) and Ismail and Suryadiasnsyah observed that its thermal stability was higher than that of PP/NR.[44,45]

The thermooxidative ageing of NR and polyethylene TPEs has been studied for various periods of ageing at temperatures up to 100 °C by Choudhury and Bhowmick.[40] In general, at a particular ageing time the properties decrease with the increase in ageing temperature. The control NR/PE system showed a marked drop in strength only after 75 °C. The changes in properties on ageing may be due to the following factors: change in morphology; degradation of rubber and crosslinking of PE and change in the level of interaction between the components at high temperature. Most of the systems showed a decrease in strength at higher temperature because of degradation of the rubber and a change in the level of interaction.[40] The photodegradation of blends of NR and polyethylene was also studied with laboratory ultraviolet exposure in the unstrained state and under tensile strain (25 and 50%). Strained exposure caused reduction of the strain to failure in subsequent tensile tests. The blends were more resistant to degradation than the NR homopolymer. The introduction of crosslinks (at a low concentration so the thermoplastic nature of the blends was retained) changed the resistance to photo-oxidation. Two different crosslinking systems were used. When dicumyl peroxide was used as the crosslinking agent, the resistance to degradation was reduced, whereas the compound containing a sulphur curing system showed improved resistance to photodegradation. Photooxidation rather than ozone degradation was found to be the major cause of breakdown, even with samples held in tension.[46]

Blends of NR and PS form a class of TPEs of great importance. NR has good elastic properties, good resilience and damping behaviour while PS has better processing properties but is very brittle. TPEs from NR and PS are expected to exhibit good processability and impact strength with good flexibility and a rubbery nature.[47] The thermal behaviour of NR/PS blends with and without compatibilizer agent (NR-g-PS) was studied by TGA and it was found that blending improves the overall thermal behaviour of the blend. Also, the addition of the compatibilizer agent improved the thermal stability of the blend.[48]

Poly(vinyl acetate) (PVAc) is a 53 polymer with excellent adhesion properties, but it is brittle and rigid. Blends of PVAc and NR were prepared by low shear blending of both lattices and were immiscible. Thermogravimetric analysis of degradation in air and nitrogen atmospheres indicated independent degradation of the parent polymers.[49]

Copolymers of ethylene-vinyl acetate (EVA) with low levels of vinyl acetate are considered thermoplastics and have a high degree of crystallinity. EVA blends with NR have been studied aiming to improve the processing characteristics and increase the material's resistance to the action of degradation agents such as ozone and γ-radiation as well as to thermal ageing. Blends of NR and EVA have been developed where EVA forms a continuous phase when its proportion is 40% or more. The resistance of the blends to thermal ageing, ozone and radiation is better for those that contain higher amounts of EVA.[50] As these properties are highly dependent on the type of curing system and also the compatibility of the blend, some studies have examined the addition of a compatibilizer agent to NR/EVA blends. Poly(ethylene-co-vinyl alcohol-co-vinyl mercaptoacetate) (EVASH) was employed as compatibilizer agent and dicumyl peroxide (DCP) as curing agent. The resistance to thermal ageing was found to be better for those blends containing EVASH/DCP, probably due to the ability of the SH groups to capture free radicals formed during the ageing process.[51] The introduction of EVASH promotes a crosslinking of NR phase, since SH groups located along the EVASH main chain are capable of reacting chemically with the elastomeric phase promoting good dispersion of components.

This crosslinking can be attributed to bonding between sulfhydryl groups along the EVASH backbone and double bonds in the rubber phase and can be responsible for the improved hardness of most of the studied blends.[52]

Polyamides are a very attractive class of engineering polymers because of their excellent tensile properties, chemical and abrasion resistance, high melting point and fatigue resistance. However, polyamides are very notch sensitive and brittle at low temperatures. Blends of different rubbers and polyamides have been studied to obtain good impact properties.[53] Some requirements are an appropriate range of rubber particle size and a uniform distribution of it. These can be achieved by controlling the level of the interfacial adhesion between the phases.

For non-functionalized rubbers, addition of a compatibilizer agent is necessary in order to improve adhesion between the rubber and polyamide phases of these immiscible blends. The addition of a reactive compatibilizer such as maleic anhydride grafted rubber resulted in improved rubber/polyamide blend properties. Grafted copolymers were formed *in situ* during processing. The resultant blend had a finer dispersion of NR in a polyamide 6 matrix.[54] The use of maleic anhydride (MA) was also studied by adding it in NR/polyamide 6. During processing, MA can react with both NR and polyamide 6, leading to graft copolymers formation.[53] Higher polyamide main chain mobility was also expected due to the presence of the rubber particles. Although higher mobility was expected to contribute to the increase of the decomposition temperature, grafting reduced the intensity of degradation due to the reduction of the chain mobility. Thermal degradation of NR (with or without MA) presented a weight loss around 400 °C, whereas polyamide showed degradation weight loss at higher temperatures of around 500 °C. Rattannape *et al.*[55] developed NR/polyamide with the induced reactivity in a single-screw extruder based on a mixture of peroxide (Perkadox 14), reactive monomer (MA) and activators (ZnO and stearic acid). The final compatible blend had very good oil resistance and can be used at a temperature of 200 °C. TPNR based on polyamide-12 blend were also studied and it was found that the simple blends with a proportion of rubber around 60 wt% exhibited a co-continuous phase structure while the dynamically cured blends showed dispersed morphology. Unmodified NR/PA12 did not show improvements in thermal resistance, which was achieved only with modified NR.[56]

23.3.4 NR/Synthetic Rubber Blends and IPNs

Natural rubber (NR) is a classic elastomer with good processability and excellent mechanical properties. It is widely used in various applications such as automobiles, gloves, tyres and seals. With the increasing demand for NR composite materials for applications at high temperatures with high performance, for example in the automobile industry, there has long been increasing interest in NR-based composites blended with other rubber systems.[57]

In 2011, Akinlabi and Okieimen[58] investigated how heat, oxygen, ozone, water and some organic solvents (toluene, carbon tetrachloride, acetone, ethanol, cyclohexane, mineral oil and brake fluid) affect the physico-mechanical properties of vulcanizates prepared from blends of NR and low molecular weight natural rubber (LMWNR). The results generally have shown that LMWNR is capable of improving the ageing properties and the solvent resistance of NR.

Rattanasom *et al.*[59] studied the morphological features and all important properties of NR and bromobutyl rubber (BIIR) blends containing hybrid fillers and varied BIIR content (from 0 to 100%). They showed a marked improvement of thermal stability of the blends with increasing BIIR loading.

A research group in Korea[57] used a new method to incorporate fluoroelastomer rubber (FR) in NR matrix. FR is the designation given to about 80% of fluoroelastomers with vinylidene fluoride (VDF), which provides extraordinary resistance to chemicals, oil, and to temperatures above 200 °C with long service life as well. The NR/FR blends including various additive fillers showed good interfacial bonding and enhanced thermal stability with substantially improved mechanical properties of strength as well as elongation at break.

NR is normally blended with ethylene-propylene-diene rubber (EPDM) to improve the ageing resistance of the former without losing its good mechanical properties.[60,61] However, due to the difference in unsaturation level between these components, a mutual incompatibility can exist, which decreases the mechanical performance. In addition to the poor interfacial adhesion caused by the thermodynamic incompatibility, these blends usually present cure rate incompatibility because of the differences between the reactivity of the elastomers with the curing agents and/or differences in solubilities of the curatives in each phase. In the case of NR/EPDM blends, the curing system can be consumed by the vulcanization of the NR phase, which is more rapidly vulcanizable because of the higher unsaturation level.[60,61]

Several strategies have been reported to improve the compatibility of NR/EPDM blends, including the addition of a third component with low molar mass[60,62] or a third polymer, such as polybutadiene (BR), styrene butadiene rubber (SBR), chlorinated rubber, chlorosulfonated PE or polyvinylchloride (PVC),[62,63] the incorporation of an accelerator moiety in the less unsaturated phase,[60] and the functionalization of EPDM with maleic anhydride.[60,63]

Natural rubber possesses good mechanical properties such as high tensile and tear strengths due to its ability to crystallize upon stretching. The elasticity and dynamic properties of NR are also excellent. However, due to the existence of numerous reactive double bonds on the molecular backbone, NR is highly susceptible to degradation by thermal ageing and ozone attack. In addition, the oil resistance of NR is relatively poor compared to some polar synthetic rubbers such as chloroprene rubber (CR) or acrylonitrile butadiene rubber (NBR). To overcome such shortcomings, NR is frequently blended with synthetic rubbers such as NBR or CR. Recently, blends of NR/CR have been extensively studied.[64] The incorporation of CR into NR helps to improve oil and thermal resistance of NR. Sae-oui *et al.* studied the silica-filled CR/NR blends and found that the mechanical properties and the resistance to degradation of the blends were mainly governed by blend morphology. They also found that good mechanical properties in association with adequately high resistance to degradation from thermal ageing and oil were obtained when CR remained the matrix in the blends. Even though the ozone cracks are found in all blends, a thorough look at the results reveals that considerable improvement in ozone resistance is achieved with increasing CR content.[64]

Freitas *et al.* investigated the effect of light and $FeCl_3 \cdot 6H_2O$ on polychloroprene/NR blends in toluene solution to discover the influence of each polymer on the degradation process. Previous investigations carried out in the author's lab showed that NR degrades in the presence of $FeCl_3 \cdot 6H_2O$, even in the absence of light. On the other hand, CR degrades only in the presence of $FeCl_3 \cdot 6H_2O$ and light. The authors said that despite the existence of many reports in the literature on the photodegradation of polymer blends, they did not find studies on the photodegradation of CR/NR blends. In their study, they suggested that the by-products formed from the degradation of CR increase the degradation rate of NR.[28]

Hydrogenated nitrile rubber (HNBR) can be blended with NR to improve its oil and ageing resistance without drastically affecting the dynamic properties. HNBR is widely used to make vibration dampers, timing belts, power transmission belts and bearings because of its ideal balance of properties like excellent heat and oil resistance coupled with good mechanical properties. It combines the oil and fuel resistance of NBR with the heat and oxidation resistance of EPDM rubber. The polarity difference between the polymers could be reduced by the incorporation of small quantities of dichlorocarbene modified NR (DCNR), which is formed during the alkaline hydrolysis of chloroform in the presence of NR. Some studies in this respect have been reported in the literature.[65–67]

The reaction of carbenes with olefins to form cyclopropyl derivatives has been used to modify elastomers. Pinazzi and Levesque[68] and Berentsvich et al.[69] found that carbene addition had a significant influence on the properties of polydienes. Thermogravimetric analysis (TGA), flammability and oil resistance in NR and dichlorocarbene modified styrene butadiene rubber (DCSBR) blends were investigated by thermogravimetric analysis as a function of different composition. The TGA plots confirmed the better thermal stability and flame resistance of DCSBR as well as its blends with NR.[70] The amount of DCSBR in the blend significantly affected the properties of blends.

The thermal stability of NR/SBR blends has been also studied. The research showed that the thermal stability of the blends chemically vulcanized increased as the SBR content in the blends increased,[71,72] and the same happened with blends vulcanized by irradiation of gamma rays with varying doses up to 250 kGy.[73]

Manshaie et al.[74] studied the physico-mechanical properties of NR/SBR blends cured by electron beam irradiation and sulphur. They showed that the irradiated blends have better mechanical properties than those cured by sulphur system. They also showed that the irradiation cured samples exhibited better heat stability than the sulphur cured samples.

The thermal stability of NR and carboxylated styrene butadiene rubber (XSBR) lattices and their blends were studied by thermogravimetric methods by Stephen et al.[75] The thermal degradation and ageing properties of these individual lattices and their blends were investigated with special reference to blend ratio and vulcanization techniques. As already described, as the XSBR content in the blends increased, their thermal stability was also found to increase. Among sulphur and radiation-vulcanized samples, radiation cured possessed higher thermal stability due to the higher thermal stability of carbon–carbon crosslinks.

Several experiments have been reported in the literature involving rubber-nanoclay composites, with the observation of improved mechanical and thermal properties.[76,77] The addition of small amounts of the nanoclay greatly improved the thermal stability and swelling behaviour, which was attributed to the good barrier properties of the dispersed and partially exfoliated organo clay particles.[77]

R. Stephen et al. also studied the effect of microfillers on the thermal stability of NR, carboxylated styrene-butadiene rubber (XSBR) lattices and their 70/30 NR/XSBR blend. Microcomposites of XSBR and their blend were found to be thermally more stable than unfilled samples.[78]

Natural rubber/cis-1,4-polybutadiene (NR/BR) blends (70/30 mass ratio) have been widely used in the tire industry. Many nanocomposites based on organo-montmorillonite (OMMT)/rubber blends have been investigated.[79] However, relatively little attention had been paid to binary rubber hybrids/ montmorillonite nanocomposites, and according to Zheng Gu et al., no studies existed dealing with OMMT/NR/BR nanocomposites. So, the authors described the preparation of OMMT/NR/BR nanocomposites by direct mechanical blending and determined the cure characteristics, static mechanical properties, dynamic mechanical properties, and thermal stability of the nanocomposites. OMMT/NR/BR nanocomposites had exactly the same onset decomposition temperature and lower thermal degradation rate as the NR/BR blends.

M. Abdollahi et al. prepared NR/BR blend/clay nanocomposites via a combined latex/melt intercalation method. The TGA results indicated an improvement in main and end decomposition by increasing the clay loading.[80]

There also are some studies about ageing and mechanical properties of NR and polybutadiene rubber blends at different blending ratios. Those studies show that both tensile stress and strain of NR/BR blends decrease after prolonged ageing.[81]

Recently, a group of Indian researchers[82] developed a new elastomer product based on NR/BR blend with reclaim rubber (RR) from ground rubber tires (GRT). The reclaiming was carried out by tetra methyl thiuram disulfide (TMTD) in the presence of spindle oil, a paraffin-based rubber process oil. Thermogravimetric analysis of RR, NR/BR and different NR/BR/RR vulcanizates was carried out in order to measure the thermal stability of the vulcanizates. Isothermal ageing test of fresh rubber RR composites showed that the ageing performance of RR containing vulcanizates are superior than that of the fresh rubber vulcanizates, which do not contain any reclaimed rubber.

Chlorinated polyethylene elastomer (CPE) has been produced by the introduction of chlorine atoms onto the polyethylene backbone in order to reduce the crystallization ability of polyethylene. In addition, the enhancement in resistance to hydrocarbon oil, heat and weathering is also achieved. To gain desired properties of the final products, CPE has been blended with many polymers, including polyvinyl chloride, styrene-acrylonitrile and polyurethane. Compared to NR, CPE is relatively expensive and therefore the blending of CPE/NR is one of methods to reduce the production cost of the final products requiring CPE properties. Some researchers from Thailand are working with this kind of blend.[83,84]

23.3.5 NR/Biopolymer Blends and IPNs

Most disposed plastic materials resist biological breakdown at disposal sites, resulting in environmental pollution. The recognition of the environmental pollution problem caused by synthetic plastics has led to the search for alternative materials – biodegradable polymers, a class of biopolymers.[85,86]

Biopolymers can be obtained from agriculture or biotechnological processes and are therefore, in principle, available from renewable sources.

Some works have reported NR as a biodegradable polymer because several species of bacteria as well as fungi are capable of degrading rubber.[87] NR (cis-polyisoprene) as it comes from the rubber tree is bioassimilated into the environment initially by peroxidation followed by biodegradation of the low molar mass oxidation products. However, NR becomes resistant to biodegradation when made into industrial products because of the presence of antioxidants adding during manufacture.[88]

Many blends of NR and biopolymers have been developed to combine different properties to open a spectrum of new biodegradable materials, which are described here.

Polyhydroxyalkanoate (PHAs) are polyesters that accumulate as spherical granules in cell cytoplasm and serve as storage compounds for energy and carbon.[85] They are synthesized by a wide range of bacteria *via* different pathways from coenzyme acid metabolism. Although PHAs are water insoluble, hydrophobic and partially crystalline polymers, they can be degraded by a large variety of microorganisms. Thus, blending PHAs with NR was studied considering three aspects: (i) chance of obtaining polymers with new properties; (ii) increasing the application range of PHA'S and (ii) increasing the degradation rate of rubber. Bhatt *et al.*[89] developed blends of PHA with NR in solution and reported that the incorporation of NR increased the thermal decomposition temperature of the blend. Besides this, they obtained a single melting temperature (around 90 °C) confirming blending at molecular level. The changes in crystallization and melting properties of PHA can be helpful in solving the problem of tacky products during processing and their adherence to equipment. Considering the biodegradation, the authors concluded that blending reduces the cost and leads to biodegradable material whose degradation can be tailored by adjusting the rubber/PHA ratio.

Starch is one of the main natural polymers studied for the production of biodegradable materials, since it is one of the major components of cereal and tubers already widely used in the food, paper and textile industries. Starch is composed by amylose and amylopectin, which are both polysaccharides, and it cannot be processed as a thermoplastic material because it decomposes before melting. When the granular structure of starch is completely disrupted by the use of plasticisers under heating, it can processed by injection moulding or extrusion. These starch compounds are commonly known as thermoplastic starches (TPS). TPS present low degradation temperatures, poor mechanical properties and high water susceptibility.[90] Carvalho *et al.*[91] prepared a blend of TPS and NR latex. The dispersion of the rubber in the thermoplastic starch matrix was homogenous thanks to the presence of the aqueous medium. Also the non rubber constituents of the latex were responsible for the latex stability. NR latex has many attractions as a suitable polymer to be blended with starch. The latex form facilitates the blending with the starch solution. It also contains natural stabilizers (proteins and lipids), is a renewable source and is biodegradable.[92] Recently, the biodegradation of a blend of starch and NR was evaluated[93] and it was concluded that *S. coelicolor* CH13 has the ability to degrade both polymers, causing the biodegradation of the blend. The authors also tested strips of commercial rubber gloves and observed that starch was a competent remover and polyisoprene chains were broken down to produce aldehyde and/or carbonyl groups. After 6 weeks, *S. coelicolor* CH13 reduced the weight of the starch/NR biopolymer by 92% and that of the rubber gloves by 14.3%. This polymer blends weight loss showed that the bacterial isolate could metabolize NR after starch disappeared. The results of molecular weight determination and FTIR confirmed the degradation of NR, indicating that *S. coelicolor* CH13 was able to cleave the carbon backbone of poly(*cis*-1,4-isoprene). Although the basic molecular mechanism of the polymer is not fully known, it is assumed that the degradation of the backbone initiated by oxidative cleavage of the double bonds in the polymer chain.[94] Roy et al.[95] reported that NR can be degraded by both bacteria and fungi and the presence of aldehyde group as proved by Schiff staining and IR spectra, indicated NR degradation, leading them to assume that the scission of the *cis*-1,4-isoprene chain by oxygen is common among all *cis*-1,4-isoprene degrading bacteria, irrespective of their colonization strategy.

Poly(lactic acid) (PLA) is a biodegradable thermoplastic polyester derived from biomass that has high brittleness and poor crystallization behaviour. The ductility of PLA has been increased by blending it with NR.[96] NR was considered a very good toughening agent increasing impact resistance in PLA/NR blends.[97,98] Impact strength and elongation at break of PLA/NR blends increased with increasing NR content up to 10% (w/w). The thermal degradation of PLA and NR showed only a single weight loss step during thermal degradation of PLA, and NR also showed only a single weight loss step. The degradation onset temperatures of PLA and NR were 290.76 °C and 305.33 °C, respectively. The final degradation temperatures of PLA and NR were found to be 409.67 °C and 490.62 °C, respectively. This indicated higher thermal stability of NR compared to that of PLA. The degradation onset temperature of PLA/NR was 294.20 °C, which is about 4 °C higher than PLA. The slight increase in thermal stability of PLA/NR was attributed to the higher thermal stability of NR. Moreover, the PLA/NR blend with NR-g-GMA showed a higher degradation onset temperatures (301.59 °C) than that of PLA/NR. The increase in thermal stability of PLA/NR when using NR-g-GMA as compatibilizer agent might be attributed to the higher interaction and better dispersion of NR in PLA matrix.[99,100]

Chitosan (CS) is a partially acetylated glucosamine obtained by deacetylation of chitin, one of the most abundant natural polymers. As a polysaccharide of natural origin, chitosan has many useful features such as nontoxicity, biocompatibility, biodegradability, good mechanical strength and antimicrobial properties. Blends of chitosan (CS) and NR obtained by solvent casting improved the thermal resistance of chitosan.[101] Johns and Rao have also developed chitosan and NR blends vulcanized with dicumyl peroxide (DCP) and as NR is thermally more stable than CS, blending improves the thermal resistance of chitosan. In dynamic vulcanization with DCP, the thermal stability of blends increases due to the formation of crosslinks in the NR phase. In the temperature range 200-270 °C, in rubber both scission and crosslinking occurs. NR undergoes thermal degradation in the range of 287–400 C to yield 39% isoprene, 13.2% dipentene and small amounts of p-menthene.[29,102]

23.4 Conclusions

Degradation of polymeric materials is an important issue from both the academic and the industrial points of view. The analysis of the natural rubber blends degradation process is very important. Thermal and thermooxidative degradation evaluation is necessary due to the use of high temperatures in engineering applications. The study of biopolymers degradation is important because biodegradation is a process that affects the product's lifetime and the number of products made of biopolymers has increased. Degradation of NR blends is therefore important to develop suitable technology for polymer processing, usability, storage and biodegradability.

Degradation is likely responsible for several damages to any rubber blend and is related to environmental impacts. Analysis of the life cycle and control of different degradation processes that blends, can be submitted to allows understanding of how many different phenomena occur. This understanding is essential to develop new blends where the properties of two or more polymers can be combined. This work covers some developments in life-cycle assessment, ageing and degradation behaviour of natural rubber-based blends and IPNs.

References

1. W. Klöpffer, *Int. J. Life Cycle. Assess.*, 2003, **B**(3), 157.
2. Life-cycle assessment [internet]. Available from: http:en.wikipedia.org/wiki/Life-cycle_assessment [accessed 14 August 2013].
3. F. J. Dennison, A. Azapagic, R. Clift and J. S. Colbourne, *Water Sci. Technol.*, 1999, **39**, 315.
4. *Guidelines for Life Cycle Assessment: A Code of Practice*, 1993, Pensacola: Society of Environmental Toxicology and Chemistry, Sesimbra, Portugal.
5. R. J. Perriman, *J. Cleaner Prod.*, 1993, **1**, 209.
6. H. A. U. de Hass, *J. Cleaner Prod.*, 1993, **1**, 131.
7. A. Azapagic and R. Clift, *Comput. Chem. Eng.*, 1999, **23**(10), 1509.
8. A. Azapagic, *Chem. Eng. J. (Amsterdam, Neth.)*, 1999, **73**, 1.
9. A. A. Burgess and D. J. Brennan, *Chem. Eng. Sci.*, 2001, **56**, 2589.
10. F. I. Khan, B. R. Natrajan and P. Revathi, *J. Loss Prev. Process Ind.*, 2001, **14**, 307.
11. T. Fleischer and A. Grunwald, *J. Cleaner Prod.*, 2008, **16**, 889.
12. C. Som, M. Berges, Q. Chaudhry, M. Dusinska, T. F. Fernandes, S. I. Olsen and B. Nowack, *Toxicology*, 2010, **269**, 160.

13. V. K. K. Upadhyayula, D. E. Meyer, M. A. Curran and M. A. Gonzalez, *J. Cleaner Prod.*, 2012, **26**, 37.
14. M. C. B. Figueirêdo, M. F. Rosa, C. M. L. Ugaya, M. S. M. Souza Filho, A. C. C. S. Braid and L. F. L. Melo, *J. Cleaner Prod.*, 2012, **35**, 130.
15. N. Yoda, *J. Macromol. Sci., Chem., Suppl.*, 1996, **A33**, 1807.
16. G. G. Smith and R. H. Barker, *Resour., Conserv. Recycl.*, 1995, **14**, 233.
17. A. Zabaniotou and E. Kassidi, *J. Cleaner Prod.*, 2003, **11**, 549.
18. E. T. H. Vink, K. R. Rábago, D. A. Glassner and P. R. Gruber, *Polym. Degrad. Stab.*, 2003, **80**, 403.
19. K. G. Harding, J. S. Dennis, H. Von Blottnitz and S. T. L. Harrison, *J. Biotechnol.*, 2007, **130**, 57.
20. S. M. Al-Salem, P. Lettieri and J. Baeyens, *Prog. Energy Combust. Sci.*, 2010, **36**, 103.
21. J. C. Arnold and S. M. Alston, *J. Environ. Manage.*, 2012, **94**, 1.
22. A. Corti and L. Lombardi, *Energy*, 2004, **29**, 2089.
23. C. L. Simões, L. M. C. Pinto and C. A. Bernardo, *Mater. Des.*, 2012, **39**, 121.
24. C. A. Bolin and S. Smith, *J. Cleaner Prod.*, 2011, **19**, 620.
25. X. Xu, K. Jayaraman, C. Morin and N. Pecqueux, *J. Mater. Process. Technol.*, 2008, **198**, 168.
26. S. Rajendran, L. Scelsi, A. Hodzic, C. Soutis and M. A. Al-Maadeed, *Resour., Conserv. Recycl.*, 2012, **60**, 131.
27. S. Riyajan, Y. Sasithornsonti and P. Phinyocheep, *Carbohydr. Polym.*, 2012, **89**, 251.
28. A. R. Freitas, A. F. Rubira and E. C. Muniz, *Polym. Degrad. Stab.*, 2008, **93**, 601.
29. A. P. Mathew, S. Packirisamy and S. Thomas, *Polym. Degrad. Stab.*, 2001, **72**, 423.
30. American Society for Testing and Materials, *Standard Terminology of Fire Standards*, ASTM E 176, ASTM International, West Conshohocken, PA, 1999.
31. K. Pielichowski and J. N. Juguma, *Thermal Degradation of Polymeric Materials*, Rapra Technology Ltd, Shawbury, 2005.
32. M.-A. De Paoli, *Degradação e Estabilização de Polímeros*, ed. J. C. Andrade, 2008, available from Chemkeys, http://www.chemkeys.com/blog/wp-content/uploads/2008/09/polimeros.pdf.
33. G. Scott, *Polym. Degrad. Stab.*, 1990, **29**, 135.
34. S. Varghese, S. Kuriakose and S. Thomas, *Polym. Degrad. Stab.*, 1994, **44**, 55.
35. C. Kim, D. Klempner, K. C. Frisch and H. C. Frisch, *J. Appl. Polym. Sci.*, 1977, **21**, 1289.
36. D. Klempner, *Angew Chem.*, 1978, **90**, 104.
37. S. K. De, S. Akhatar and P. P. De, *Mater. Chem. Phys.*, 1985, **12**, 235.
38. J. S. Oh, A. I. Isayev and M. A. Rogunova, *Polymer*, 2003, **44**, 2337.
39. C. Nakason, S. Saiwari and A. Kaesaman, *Polym. Test.*, 2006, **25**, 413.

40. N. R. Choudhury and A. K. Bhowmick, *Polym. Degrad. Stab.*, 1989, **25**, 39.
41. S. Varguese, R. Alex and B. Kuriakose, *J. Appl. Polym. Sci.*, 2004, **92**, 2063.
42. G. Scott, *Polym. Degrad. Stab.*, 1995, **48**, 315.
43. N. R. Choudhury, T. K. Chaki and A. K. Bhowmick, *Thermochim. Acta*, 1991, **176**, 149.
44. H. Ismail and Suryadiansyah, *Polym.-Plast. Technol. Eng.*, 2004, **43**, 319.
45. H. Ismail and Suryadiansyah, *Polym. Test.*, 2002, **21**, 389.
46. A. Bhowmick, J. Heslop and J. R. White, *J. Appl. Polym. Sci.*, 2002, **86**, 2393.
47. R. Asaletha, M. G. Kumaran and S. Thomas, *Eur. Polym. J.*, 1999, **35**, 253.
48. R. Asaletha, M. G. Kumaran and S. Thomas, *Polym. Degrad. Stab.*, 1998, **61**, 431.
49. S. S. Ochigbo, A. S. Luyt and W. W. Focke, *J. Mater. Sci.*, 2009, **44**, 3248.
50. A. T. Koshy, B. Kuriakose and S. Thomas, *Polym. Degrad. Stab.*, 1992, **36**, 137.
51. B. Soares, P. Jansen and A. S. Gomes, *J. Appl. Polym. Sci.*, 1996, **61**, 591.
52. B. Soares, P. Jansen, M. Amorim and A. S. Gomes, *J. Appl. Polym. Sci.*, 1995, **58**, 101.
53. E. Carone Jr, U. Kopcak, M. C. Gonçalves and S. P. Nunes, *Polymer*, 2000, **41**, 5929.
54. F. H. Axtel, P. Phinyocheep and P. Kriengchieoncharm, *J. Sci. Soc. Thailand*, 1996, **22**, 201.
55. T. Navarat, M. Seadan and S. Rattanapane, *Chiang Mai J. Sci.*, 2007, **34**, 79.
56. R. Magaraphan, C. Totanapoka and A. M. Jamieson, *Des. Monomers Polym.*, 2004, **7**, 165.
57. D. Lee, J. Hong, C. Lee, J. Oh, S. Kwak, T. Han, J. Lee, Y. Lee, B. Kim, H. R. Choi, T. Kim and J. Nam, *Macromol. Res.*, 2012, **20**, 673.
58. A. K. Akinlabi and F. E. Okieimen, *J. Appl. Polym. Sci.*, 2011, **121**, 78.
59. N. Rattanasom, S. Prasertsri and K. Suchiva, *J. Appl. Polym. Sci.*, 2009, **113**, 3985.
60. A. S. Siqueira and B. G. Soares, *Eur. Polym. J.*, 2003, **39**, 2283.
61. T. Zaharescu, V. Meltzer and R. Vîlcu, *Polym. Degrad. Stab.*, 2000, **70**, 341.
62. S. H. El-Sabbagh, *Polym. Test.*, 2003, **22**, 93.
63. H. Zhang, R. N. Datta, A. G. Talma and J. W. M. Noordermeer, *Eur. Polym. J.*, 2010, **46**, 754.
64. P. Sae-oui, C. Sirisinha and K. Hatthapanit, *eXPRESS Polym. Lett.*, 2007, **1**, 8.
65. K. I. Elizabeth, R. Alex and S. Varghese, *Plast., Rubber Compos.*, 2008, **37**, 359.
66. K. I. Elizabeth, R. Alex, B. Kuriakose, S. Varghese and N. R. Peethambaran, *J. Appl. Polym. Sci.*, 2006, **101**, 4401.

67. R. Alex, K. I. Elizabeth and B. Kuriakose, *Prog. Rubber, Plast. Recycl. Technol.*, 2006, **22.2**, 127.
68. C. Pinazzi and G. Levesque, *Compt. Rend.*, 1965, **260**, 3393.
69. E. N. Barentsvic, L. S. Breshler, E. L. Rabinerzon and A. E. Kalans, *Vysokomol Soedin Ser*, 1979, **A-21**, 1289.
70. M. T. Ramesan, *React. Funct. Polym.*, 2004, **59**, 267.
71. M. M. Hassan, A. A. El-Megeed and N. A. Maziad, *Polym. Compos.*, 2009, **30**, 743.
72. K. Pal, R. Rajasekar, T. Das, D. J. Kang, S. K. Pal, J. K. Kim and C. K. Das, *Plast., Rubber Compos.*, 2009, **38**, 302.
73. A. B. Moustafa, R. Mounir, A. A. El Miligy and M. A. Mohamed, *Arabian J. Chem.*, 2011, in press, http://dx.doi.org/10.1016/j.arabjc.2011.02.020.
74. R. Manshaie, S. N. Khorasani, S. J. Vershare and M. R. Abadchi, *Radiat. Phys. Chem.*, 2011, **80**, 100.
75. R. Stephen, S. Jose, K. Joseph, S. Thomas and Z. Oommen, *Polym. Degrad. Stab.*, 2006, **91**, 1717.
76. T. P. Mohan, J. Kuriakose and K. Kanny, *J. Ind. Eng. Chem.*, 2011, **17**, 264.
77. Z. Gu, G. Song, W. Liu, P. Li, L. Gao, H. Li and X. Hu, *Appl. Clay Sci.*, 2009, **46**, 241.
78. R. Stephen, A. M. Siddique, F. Singh, L. Kailas, S. Jose, K. Joseph and S. Thomas, *J. Appl. Polym. Sci.*, 2007, **105**, 341.
79. Z. Gu, L. Gao, G. Song, W. Liu, P. Li and C. Shan, *Appl. Clay Sci.*, 2010, **50**, 143.
80. M. Abdollahi, H. H. Khanli, J. Aalaie and M. R. Yousefi, *Polym. Sci., Ser. A*, 2011, **53**, 1175.
81. H. Chiu and P. Tsai, *J. Mater. Eng. Perform.*, 2006, **15**, 88.
82. D. De, P. K. panda, M. Roy and S. Bhunia, *Mater. Des.*, 2013, **46**, 142.
83. C. Sirisinha, P. Saeoui and J. Guaysomboon, *Polymer*, 2004, **45**, 4909.
84. P. Prasopnatra, P. Saeoui and C. Sirisinha, *J. Appl. Polym. Sci.*, 2009, **111**, 1051.
85. A. Steinbuchel, *Macromol. Biosci.*, 2001, **1**, 1.
86. S. Karlsson and A. Albertsson, *Polym. Eng. Sci.*, 1998, **38**, 1251.
87. K. Rose and A. Steinbüchel, *Appl. Environ. Microbiol.*, 2005, **71**, 2803.
88. G. Scott, *Polym. Degrad. Stab.*, 2000, **68**, 1.
89. R. Bhatt, D. Shah, K. C. Patel and U. Trivedi, *Bioresour. Technol.*, 2008, **99**, 4615.
90. H. Röper and V. Koch, *Starch/Stärke*, 1990, **42**, 123.
91. A. J. F. Carvalho, A. E. Job, N. Alves, A. A. S. Curvelo and A. Gandini, *Carbohydr. Polym.*, 2003, **53**, 95.
92. A. Rouilly, L. Rigal and R. Gilbert, *Polymer*, 2004, **45**, 7813.
93. S. Watcharakul, K. Umsakul, B. Hodgson, W. Chumeka and V. Tanrattanakul, *Electron. J. Biotechnol.*, 2012, 15.
94. R. Braaz, P. Fischer, D. Jendrossek, A. Rouilly, L. Rigal and R. Gilbert, *Appl. Environ. Microbiol.*, 2004, **70**, 7388.

95. R. V. Roy, M. Das, R. Banerjee, A. K. Bhowmick, A. Rouilly, L. Rigal and R. Gilbert, *Process Biochem.*, 2006, **41**, 181.
96. N. Bitinis, R. Verdejo, P. Cassagnau and M. A. Lopez-Manchado, *Mater. Chem. Phys.*, 2011, **192**, 823.
97. R. Jaratrotkamjorn, C. Khaokong and V. Tanrattanakul, *J. Appl. Polym. Sci.*, 2012, **124**, 5027.
98. B. Suksut and C. Deeprasertkul, *J. Polym. Environ.*, 2011, **19**, 288.
99. P. Juntuek, C. Ruksakulpiwat, P. Chumsamrong and Y. Ruksakultiwat, *J. Appl. Polym. Sci.*, 2012, **125**, 745.
100. P. Juntuek, C. Ruksakulpiwat, P. Chumsamrong and Y. Ruksakultiwat, *Adv. Mater. Res.*, 2010, **123–125**, 1167.
101. J. Johns and V. Rao, *J. Mater. Sci.*, 2009, **44**, 4087.
102. J. Johns and V. Rao, *J. Therm. Anal. Calorim.*, 2008, **92**, 801.

CHAPTER 24

Application of Natural Rubber Based Blends and IPNs in Tyre Engineering and other Fields

MIR HAMID REZA GHOREISHY*[a] AND
MOHAMMAD ALIMARDANI[b]

[a] Iran Polymer and Petrochemical Institute, PO Box 14965/115, Tehran, Iran;
[b] Department of Polymer Engineering, Tarbiat Modares University,
PO Box 14115/111, Tehran, Iran
*Email: M.H.R.Ghoreishy@ippi.ac.ir

24.1 Introduction

24.1.1 Properties of Natural Rubber

24.1.1.1 General Mechanical Properties

As described in previous chapters, natural rubber (NR) is traditionally considered as a general-purpose rubber and is used in the production of virtually all rubber products, because of its well-balanced range of physical properties. NR has a very high structural regularity, providing it with unique and valuable characteristics. In particular, strain-induced crystallization of NR can result in very high tensile strength, even in gum vulcanizates, and resistance to tearing and abrasion. On the other hand, compared to some elastomers NR has potential drawbacks, such as limited resistance to heat, oils and some chemicals, and also its susceptibility to oxidation and ozonation, mainly because of its unsaturated structure. Similarly to polychloroprene (CR), it is one of the few

Table 24.1 The basic engineering properties of NR.[4]

Characteristics of natural rubber	
Advantage	Disadvantage
Tear strength	Uniformity of quality
Wear resistance	Ageing resistance
Impact resistance	Fatigue resistance
Low heat generation	Ozone resistance

elastomers that show high strength in gum vulcanizates (cured, low hardness rubber, containing no fillers) and also good resilience, which makes the gum excellent for fine particle impact applications. NR rubber gum vulcanizates have very high elasticity and so most of the kinetic energy of an impacting particle can be converted into deformation of the vulcanizate, which is capable of releasing the energy by returning to its original undeformed state. NR also has very good low temperature resistance, down to around $-57\,^\circ\text{C}$, at which its stiffness shows a considerable increase. Thick NR products can remain in service for long periods; for example, NR bridge bearings can remain in service and function well for 100 years. NR is a frequent choice for slurry pump liners and impellers as well as for tank linings because it has very good abrasion resistance and also low relative cost. As a result of very good dynamic mechanical properties it is widely used in tyres, rubber springs and vibration mounts. It is estimated that 70% or more of world production of NR is consumed in tyres, mostly large truck tyres, off-road giant tyres and aircraft tyres. It is used mainly because of its low heat generation properties in service, together with its low rolling and high cut growth resistance. NR is used in almost every component of tyres. The only exception is the tread of modern passenger car radial tyres, which generally use a blend of butadiene rubber (BR) and styrene-butadiene rubber (SBR). The main characteristics of NR as the chief component of tyres have been recorded in Table 24.1. NR lasts a long time and tends to run cool. It is an important component in treads and any other parts of the tyre that need to flex a great deal.[1–4]

24.1.1.2 Viscoelastic Properties

Carbon black or silica filled NR generally demonstrates viscoelastic behaviour that is usually evaluated by dynamic viscoelastic measurements. One of the most widely discussed in the literature is the strain dependency of dynamic modulus known as the Payne effect. In such cases a high dynamic modulus at low strains ($< 1\%$) is measured, which decreases at higher strains ($> 10\%$), as shown in Figure 24.1. The reason for this phenomenon is the formation of a network created by filler–filler interactions. For carbon black the interactions are Van der Waals forces and for silica, much stronger hydrogen

Application of Natural Rubber Based Blends and IPNs 571

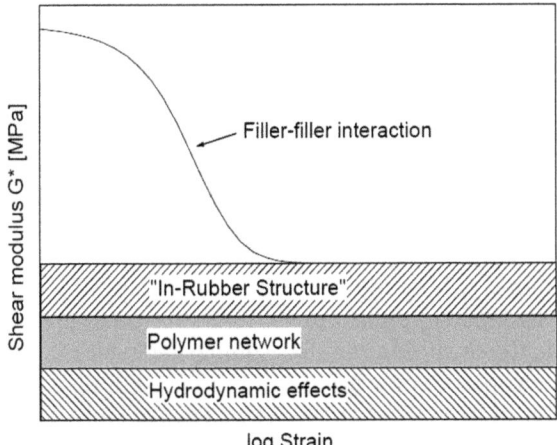

Figure 24.1 Schematic illustration of G' and G'' variations with strain.

bonding. The Payne effect can be explained by a number of mechanisms such as: adsorption–desorption of polymeric chains at the filler interface, the destruction–reformation of a filler network, disentanglement of bulk polymer from the rubber bound to the surface, and strain-softening of the glassy polymer shell surrounding the particle surfaces. In real life this effect means an energy loss and the intention is to minimize this effect in filled rubber (but with the reinforcement retained).[4]

The assessment of G' and G'' in the range of 0.1 to 0.5 strain is of great practical importance because this range corresponds to the most common solicitations of filled rubber compounds, for example in tyre tread applications. Another common viscoelastic property of rubbers is the well-known phenomena of hysteresis. In rubbers, when a load is imposed on a viscoelastic material it deforms as the stress increases. However, when the load is removed, stress and strain will not follow the initial path to return to the original position. This phenomenon happens because some of the energy used to deform the material has been dissipated as heat. In other words, hysteresis is the build-up of heat in elastomers resulting from continuous cyclic flexing. Press-on tyres and load wheels compress and deflect every revolution, which generates heat. It is also worth referring here to the Mullins effect as a phenomenon typically observed in filled polymers. It is characterized by a decrease in material stiffness during loading and is readily observed during cyclic loading as the material response along the unloading path differs noticeably from the response along the loading path. Although the details of the mechanisms responsible for the Mullins effect are not yet understood, they may be related to debonding of the polymer from the filler particles, separation of particle clusters, and rearrangement of the polymer chains and particles.

24.1.2 Elastomer Blends

Generally, elastomer blends are of technological and commercial importance because they provide an opportunity to access a combination of properties that could not be achieved by the use of individual elastomers. In 1846 Parkes patented the first polymer blend, which interestingly was a elastomer blend of shellac, NR (amorphous *cis*-1,4-polyisoprene), gutta-percha (GP; a semi-crystalline *trans*-1,4-polyisoprene) and cellulose.[5] Blending of NR with GP partially co-dissolved in CS_2 isomers resulted in partially crosslinked (co-vulcanized) materials whose rigidity was controllable by composition. The resulting blends had many applications ranging from picture frames, tableware and ear trumpets, to sheathing the first submarine cables.

The enhanced properties of a blend can be chemical, physical or mechanical, or they can offer processing benefits. In practice, all blends exhibit compositionally correlated changes in all of these properties compared to the blend components. In elastomers composed of a single monomer in a single addition mode, for example 1,2-polybutadiene, there are no other ways of changing the properties apart from blending. In the case of copolymer elastomers like SBR, changes in molecular composition, such as formation of a block polymer instead of a random copolymer, may be effective in achieving the required properties. However, intramolecular changes are usually restricted by synthesis processes. Similarly to other polymer blends, elastomers are generally immiscible. As a result of the inhomogeneous phase structure of the two component elastomers existing in immiscible blends, they show complex changes in their properties. The two separate phases would have different tendencies in the retention of fillers and plasticizers as well as vulcanization in the presence of the curative. On the other hand, incompatible polymer blends tend to improve fatigue resistance. For example, NR and polybutadiene show good resistance to fatigue, crack initiation and growth because of the formation of heterogeneous polymer phases; a crack growth in one polymer phase is arrested at the boundary with the adjacent polymer phase.[4,6]

24.1.3 General Aspects of Compounding of NR and Blends

24.1.3.1 Black/Non-Black Fillers

Fillers such as carbon black, clays and silicas are commonly added to NR formulations to meet specific property targets such as tensile strength and abrasion resistance. Carbon blacks are undoubtedly one of the most used fillers, and an extensive range is available, each of which brings a specific set of properties to a compound. For this reason, the correct choice of carbon black may be as important as the development of a formulation for a polymer system in meeting a product performance specification. Carbon blacks can be classified qualitatively by a series of properties including particle size (and surface area); particle size distribution; structure (particle aggregates); and surface activity (functional groups such as carboxyl). Increasing carbon black aggregate size or

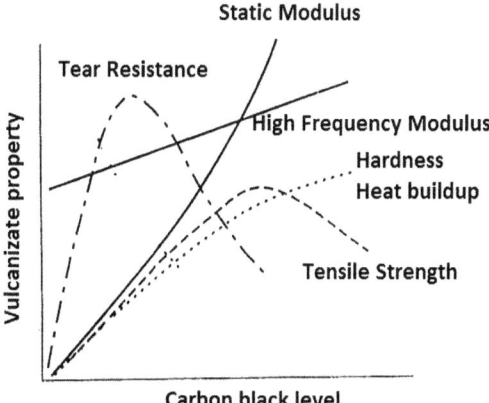

Figure 24.2 Effect of carbon black level on the properties of rubbers.

structure can result in an improvement in cut growth and fatigue resistance. A decrease in the particle size of carbon black may lead to an increase in abrasion resistance and tear strength, a drop in resilience, and an increase in hysteresis and heat build-up. The effect of carbon black loading on the mechanical and viscoelastic properties of rubbers is shown in Figure 24.2.

Due to the major impact of carbon black loading on determining the applicability of finished products, the following points are important:[4]

1. A reduction in carbon black loading leads to decreased rolling resistance of tyres. At a constant carbon black loading, an increase in oil level will give rise to rolling resistance but also improve traction (at low oil levels, an increase in oil level may decrease compound hysteresis by improving carbon black dispersion).
2. Increasing fineness of carbon black increases both traction and rolling resistance.
3. An increase in the broad aggregate size distribution decreases the tyre rolling resistance.
4. Tread-grade carbon blacks can be selected to meet defined performance parameters of rolling resistance, traction, wear, *etc.*

Addition of silica to a rubber compound offers a number of advantages, such as reduced heat build-up, improved tear strength, and increasing compound adhesion in multi-component products such as tyres. Two fundamental properties of silica and silicates influence their use in rubber compounds: ultimate particle size and the extent of hydration. Silica can also change physical properties such as pH, chemical composition and oil absorption, which are of secondary importance. Compared to carbon blacks of the same particle size, silicas do not provide the same level of reinforcement, but this deficiency can largely disappear when coupling agents or other surface modifiers are used with silica. According to Wagner, addition of silica to a tread compound leads to a loss in tread wear, even though improvements in hysteresis and tear strength are obtained.[7] The tread wear loss can be modified by the use of silane coupling agents.[8] There is also a series of additional filler systems such as kaolin clay (hydrous aluminium silicate), mica (potassium aluminium silicate), talc (magnesium silicate), limestone (calcium carbonate) and titanium dioxide which have reinforcement qualities.[4]

24.1.3.2 Plasticizers

In rubber industries the term plasticizer is used more frequently to describe the class of materials that includes esters, pine tars and low molecular weight polyethylene. Phthalates are among the most frequently used esters. Dibutyl phthalate (DBP) tends to give soft compounds with tack; dioctyl phthalate (DOP) is less volatile and tends to produce harder compounds because of its higher molecular weight. Polymeric esters such as polypropylene adipate can also be used when low volatility is needed along with good heat resistance.[4]

24.1.3.3 Antidegradants

The unique viscoelastic properties of elastomers are mainly due to their unsaturated nature. However, the presence of carbon–carbon double bonds renders elastomers susceptible to attack by oxygen and ozone, and also to thermal degradation. Oxidation of elastomers can be accelerated by various factors such as heat, heavy metal contamination, sulfur, light, moisture, swelling in oil and solvents, dynamic fatigue, oxygen and ozone. There are some variables in the compounding of elastomers such as polymer type, cure system and antidegradant system which influence the degradation resistance. Generally, there are four types of antidegradant: non-staining antioxidants, staining antioxidants, antiozonants and waxes.[4]

24.1.3.4 Vulcanizing Agents and Curing Systems

The vulcanization system constitutes the fourth component in an elastomeric formulation and functions by inserting crosslinks between adjacent polymer chains in the compound. A typical vulcanization system in a compound consists of three components: (i) activators; (ii) vulcanizing agents, typically sulfur; and (iii) accelerators.[4]

> (i) Activators. The vulcanization activator system consisting of zinc oxide and stearic acid is widely used at levels of 2.0 and 5.0 phr, respectively. Using this amount is accepted throughout the rubber industry as being adequate to achieve optimum compound physical properties when in combination with a wide range of accelerator classes and types and also accelerator-to-sulfur ratios.[4]
> (ii) Vulcanizing agents. Three vulcanizing agents find extensive use in the rubber industry: sulfur, insoluble sulfur and peroxides. Sulfur is

soluble in NR at levels up to 2.0 phr. Above this concentration, insoluble sulfur must be used to prevent migration of sulfur to the compound surface.[4]

(iii) Accelerators. Accelerators are products that increase both the rate of sulfur crosslinking in a rubber compound and crosslink density. Most accelerators fall into one of the eight groups: aldehyde amines, sulfenamides, thioureas, dithiocarbamates, guanidines, thiurams, thiazoles and xanthates.[4]

24.2 NR and its Blends for Tyre Components

NR usage in modern radial tyres has substantially increased. The amount of NR used is about 21 kg per tyre for a radial construction compared with about 9 kg in a bias truck tyre. Bernard and coworkers have compared the NR levels of heavy-duty radial truck tyres to those of the equivalent bias tyre and noted the following increase:[4]

Natural rubber (%)	Bias	Radial
Tread	47	82
Skim coat	70	100
Sidewall	43	58

The reasons for the increase have been attributed to improved green strength, increase in component-to-component adhesion, improved tear strength, lower tyre temperature generated under loaded dynamic service conditions, and lower tyre rolling resistance to improve vehicle fuel efficiency. NR compounds also tend to find use in covers of high-performance conveyor belts where a similar set of performance parameters such as those of a truck tyre tread compound are found. Low hysteretic properties, high tensile strength and good abrasion resistance are required for both products. In the following sections the use of NR blends in different parts of tyres including tread, sidewall, body plies, beads and inner tubes will be mentioned.

As mentioned earlier, resistance to environmental conditions and dynamic failure (such as fatigue) are the main drawbacks of NR in tyres. Thus, synthetic rubbers such as BR and SBR are used as the second rubber during compounding, especially in passenger tyres, to compensate for these shortcomings. The ratio of natural to synthetic rubber depends on the particular component of the tyre. Typically, blends of NR and BR are used for tread cap and base in non-passenger tyres and sidewalls. On the other hand, due to the higher stiffness and extreme resistance to tearing required for bead fillers and belt skim compounds, these components are mainly designed based on 100% of NR. In the formulation of body ply compounds and also chafers, blends of NR with SBR and BR are used.

In practice rubber blends are used in tyre rubber compositions for three reasons: reduced cost, processing improvements, and modifying final product

Table 24.2 Use of natural rubber blends in tyres.[4]

Component	Passenger tyres	Commercial vehicle tyres
Tread	SBR-BR	SBR-BR or NR-BR
Carcass	NR-SBR-BR	NR-BR
Sidewall	NR-BR or NR-SBR	NR-BR
Liner	NR-SBR-IIR	NR-IIR

performance. Table 24.2 classifies the use of blends in various sections of a tyre.[8]

Tread. Tread is the wear resistance component of the tyre and is in direct contact with the road. It must provide traction, wet skid and good cornering characteristics with minimum noise generation and also low heat build-up. Tread components can consist of blends of NR, polybutadiene (BR) and SBR, compounded with carbon black, silica, oils and vulcanizing chemicals.[4] Among recently reported formulations for tyre tread with economic and environmental merits is the work of Rattanasom, in which a blend of NR and tyre tread reclaimed rubber (RR) was prepared and mechanically characterized. Their results showed that the blends prepared with different curing systems, *i.e.* conventional vulcanization (CV) and efficient vulcanization (EV), exhibit an increase in their hardness and modulus with increasing RR content, while other mechanical properties were adversely affected.[9]

Sidewall. Tyre sidewalls that protect the casing from side scuffing, assist in tread and control vehicle–tyre ride characteristics, are usually formulated using NR, SBR and BR along with carbon black and a series of oils and organic chemicals.[4] In this application use of polyolefin elastomers such as IIR (isobutylene isoprene rubber) and EPDM (ethylene-propylene-diene rubber) as substantial components in blends of unsaturated elastomers is a rapidly developing area. Because a tyre sidewall in dynamic service conditions can easily be damaged by ozone, good ozone resistance and fatigue are required to guarantee a long service life. However, as a result of additive migration, the use of conventional antiozonants results in an unfavourable surface discoloration. Moreover, the amount of active antiozonant is reduced over time due to a reaction with ozone. The well known antiozonant 6PPD (N-(1,3-dimethylbutyl)-N'-phenyl-p-phenylenediamine) is also a toxic and environmentally unfriendly substance that the use of redoubles the difficulties in using ordinary antiozonants. The increasing demand for extending the life of tyres, and the amounts of second-hand tyres, leads to a need to improve the ozone and thermo-oxidative resistance of tyre sidewall compounds. One of the approaches recently applied to circumvent this problem is to blend NR/BR with EPDM. EPDM is a highly saturated elastomer, commonly used in applications that require good ozone resistance. Addition of EPDM into NR/BR blends improves ozone resistance, but leads to cure rate mismatch and heterogeneous filler distribution in each of the rubber phases. Consequently, many attempts have been made to overcome this problem. A comprehensive study of the preparation, cure characteristics, mechanical properties and assessment of the blend ozone and fatigue resistance

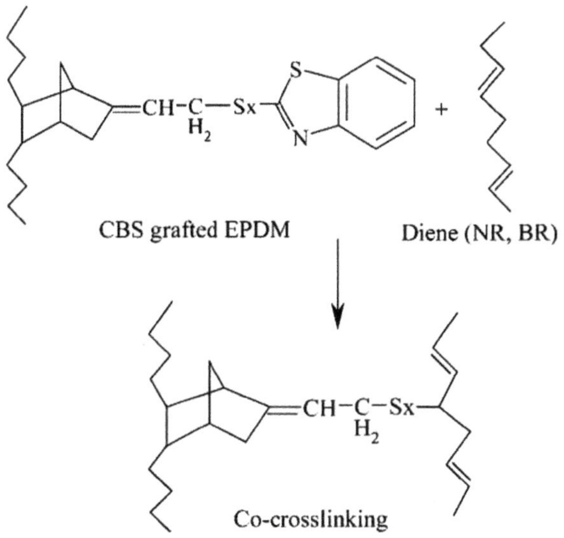

Scheme 24.1 Crosslinking between pretreated EPDM and diene rubbers (NR/BR) [10]

of EPDM/NR/BR was carried out by Sahakaro et al.[10–12] They used the common N-cyclohexyl-2-benzothiazole sulfonamide (CBS) to be grafted onto EPDM before blending with NR/BR as shown in Scheme 24.1.[10] The conclusion of the work showed that the tensile strength and elongation at break of the NR/BR/EPDM blends and the NR/BR filled blends are more or less equivalent. In addition, the physical properties of the blend under optimum conditions were also equivalent or even superior to conventional NR/BR tyre sidewall compounds, thereby indicating that the reactive procedure has led to a homogeneous blend due to either a better crosslink distribution, or more homogeneous filler distribution, or both.

24.3 NR in Seismic Isolation Bearings

Base isolation is the process of decoupling a structure from earthquake forces, which provides protection for both the structures and contents. The building or structure is decoupled from the horizontal components of the earthquake ground motion by interposing a layer with low horizontal stiffness between the structure and the foundation. This layer gives the structure a fundamental frequency that is much lower than its fixed-base frequency and also much lower than the predominant frequencies of the ground motion. Among numerous designs of base isolation systems available today, high-damping natural rubber bearings (HDNR) offer perhaps the simplest and the most cost-effective method of isolation and are relatively easy to manufacture. Rubber–steel laminated bearings for seismic isolation are designed to provide sufficient vertical stiffness to support the load of the structure and at the same time be

Figure 24.3 A rubber base isolator installed between structure and foundation.

flexible in the horizontal direction so as to accommodate the ground motion due to the earthquake. Rubber–steel laminated bearings are designed to give a mounted structure a horizontal natural frequency down to 0.4 Hz (corresponding to a period of 2.5 s). Figure 24.3 shows a giant base isolator mounted to protect the building structure from unwanted vibrations imposed by earthquake forces. In Japan in particular, the use of seismic isolation bearings has rapidly increased. These include low damping (natural) bearings, high-damping bearings and lead–rubber ones. The application of seismic bearings to nuclear facilities, however, requires more intensive research and rubber bearings with extraordinarily high reliability need to be produced.[13–16]

24.3.1 Why NR as Seismic Isolation?

Experience has shown that rubber bearings in civil engineering are more durable than ordinary constructional material such as steel or concrete. Steel plates in rubber bearings are usually embedded within the rubber to prevent deterioration of the steel. The normal cause of deterioration in rubber, such as oxidation or absorption of oil, produces only surface effects in components as large as bearings. Ozone attack, which can occur in NR subjected to tension, cannot be carried out in compression mode. Even small surface tensile strains, which can lead to unsightly superficial cracking, can be eliminated by design. A stress well in excess of 10 tons/in^2 is required to cause failure of rubber in compression. At low temperatures rubbers crystallize, resulting in progressive stiffening and loss of elasticity. For NR this does not become significant until temperatures below -20 °C are reached. However for CR the onset of this phenomenon is at nearly 0 °C. Dynamic stiffness of rubber is another important design parameter that needs to be considered in this field. Figures 24.4 and 24.5

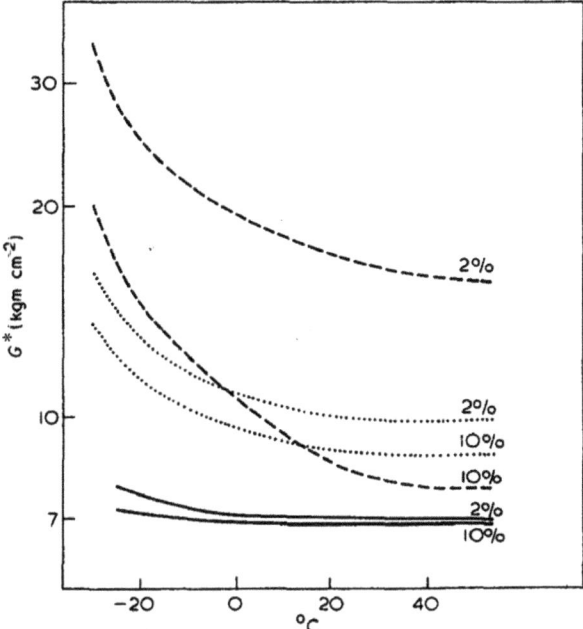

Figure 24.4 Variations of complex shear modulus with temperature and amplitude at frequency of 0.1 for three rubber samples, *i.e.* butyl (····), polychloroprene (—) and natural rubber (solid line).[21]

show the variation in complex shear modulus (G^*) with temperature and amplitude at two frequencies, 0.1 and 3 Hz, for three rubber samples, *i.e.* butyl, CR and NR. The point to note is that the dynamic stiffness does vary with temperature, amplitude and frequency and this must be allowed for in design if the variation is significant. Fortunately the dependence for NR is rather slight and if this polymer is used it will normally be possible for practical purposes to assume that its properties are constant. Moreover, strong rubbers like NR can tolerate very large deformation in shear mode without failure either in the rubber or at the rubber to metal bond that is formed chemically during the vulcanization process.[17–21]

24.4 Toughened Thermoplastics and IPNs of NR in the Automotive Industry

The automotive industry is becoming increasingly technological and is thriving, and the plastics industry has a major role to play in this revolution. The use of plastics in the automotive industry dates back to 1950 when thermoplastics began to be widely introduced in industry, starting with ABS (*acrylonitrile-butadiene-styrene*) and followed by polyamide, polyacetal and polycarbonate. This progress was timed to coincide with the advent of alloys and blends of various polymers. Promising applications for advanced and high-performance

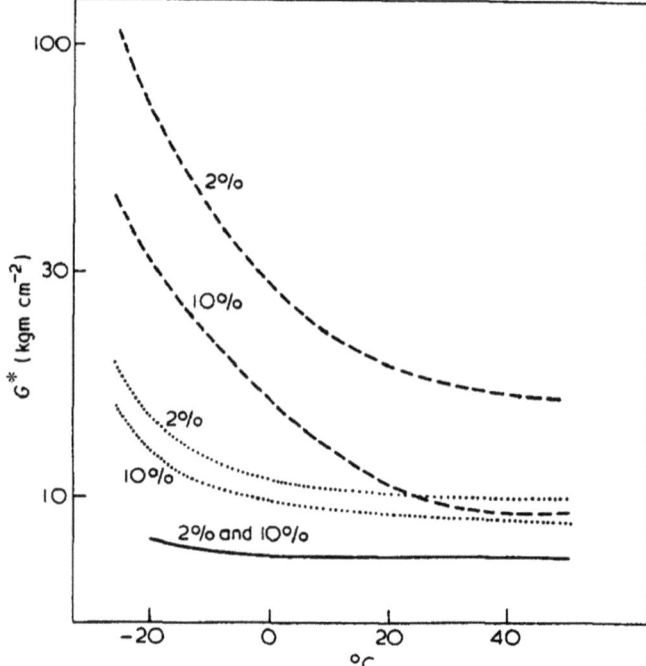

Figure 24.5 Variations of complex shear modulus with temperature and amplitude at frequency of 3 Hz for three rubber samples, *i.e.* butyl (····), polychloroprene (—) and natural rubber(solid line).[21]

polymers led to their high usage in the automotive industry. Plastics were increasingly specified because of their excellent mechanical properties in conjunction with their outstanding appearance, including the possibility of self-colouring. Recently plastics have been mainly used for the manufacture of highly energy-efficient cars, due to the fact that they can reduce weight, as well as providing durability, corrosion resistance, toughness, design flexibility, resilience and high performance at low cost. It is worth noting that the average weight of plastics and plastic composites used in vehicles is about 150 kg, much less than the total weight of steel and iron, estimated to be 1163 kg. This weight makes up around 10–15% of the total weight of a car. Table 24.3 shows the common uses for plastics in exterior and interior components such as bumpers, doors, windows, headlights and side view mirrors.[22,23] In spite of the unlimited usage of plastics in the automotive industry, they cannot be simply applied for specific applications and in some cases their properties need to be modified. For instance, the first application of plastics in the automotive industry was for bumpers; they offered an alternative to chromium-plated metal which was heavy, expensive and too stiff. The main feature required in this application is impact resistance. The plastic bumper needs to remain undamaged when subjected to an impact of 4 km/h at −25 °C (according to European regulations). High stiffness, good aesthetic appearance and environmental resistance are also

Table 24.3 Application of plastics in the automotive industry, with the weight used in an average car.[23]

Component	Main types of plastics	Weight in average car (kg)
Bumpers	PS, ABS, PC/PBT	10.0
Seating	PUR, PP, PVC, ABS, PA	13.0
Dashboard	PP, ABS, SMA, PPE, PC	7.0
Fuel systems	HDPE, POM, PA, PP, PBT	6.0
Body (incl. panels)	PP, PPE, UP	6.0
Under-bonnet components	PA, PP, PBT	9.0
Interior trim	PP, ABS, PET, POM, PVC	20.0
Electrical components	PP, PE, PBT, POM, ASA, PP	7.0
Exterior trim	PC, PBT, ABS, PMMA, UP	4.0
Lighting	PVC, PUR, PP, PE	5.0
Upholstery	PP, PE, PA	8.0
Liquid reservoirs		1.0
Total		**105.0**

required. Toughness is a measure of resistance to fracture. It is an important requirement in most load-bearing applications of materials. When a polymer with no secondary transitions is subjected to temperatures well below its glass transition temperature, deformation will be very limited before fracture occurs. Nevertheless because of the high modulus rather high tensile strengths will be recorded, in the order of 8000 lbf/in^2 (55 MPa). Therefore, the energy to break, defined as the area under the stress–strain curve, will not be very large. A common approach is to use a rigid plastic that has been toughened by using a rubber. The elastomer-modified polymer blends have several advantages. They can bring high impact strength and ductile behaviour throughout the range of the product and operating temperatures. They also have dimensional stability, and good aesthetic and mechanical properties. Generally, approaches to improving impact resistance are as follow:[24]

1. Using a semi-crystalline polymer having a T_g well below the expected service temperature.
2. Block copolymerization, so that one component of the block copolymer would have a T_g well below the expected service temperature range (*e.g.* polypropylene with small blocks of polyethylene or preferably polypropylene with small amorphous blocks of ethylene-propylene copolymer).
3. Blending of plastics with semi-compatible materials which have a T_g well below the expected service temperature range like most rubbers.
4. Plasticization. This effect also reduces the T_g of the polymer. It should, however, be noted that in the case of PVC small amounts of plasticizer actually reduce the impact strength.
5. Use of IPNs (interpenetrating polymer networks).

Due to the unique arrangement of the two phases in an IPN, these materials often exhibit good mechanical strength and toughness. There have been literature reviews on the properties of semi-IPN polymers based on a rubber mixed with different polymers such as polystyrene, polyamide and methyl methacrylate.[25] The advantages of IPNs are higher solvent resistance compared to the individual polymers and their ability to creep and flow compared to the individual polymers. Another aspect which should be considered as an advantage of toughening by IPNs is the fact that the crosslinking of the two phases can decrease the domain size of the phase. Sperling *et al.* reported that in SBR/PS IPNs, crosslinking of the phases had decreased the domain size drastically.[25] A smaller and more uniformly distributed rubber phase can effectively control the initiation and termination of crazes. In the case of IPNs both phases are intimately mixed and a high interfacial adhesion is developed, making the material impact resistant.

Toughening of plastics by using NR based IPNs can bring some special properties. The advantages of NR products are their elasticity, flexibility and resistance to splitting.[26] In NR/polystyrene IPNs, the impact measurement is of great importance in the application level as these materials can be used successfully in automobile components, footwear and various moulded articles. To prepare an impact-modified plastic by using NR based IPNs the following scheme can be taken into consideration.[25] Scheme 24.2 shows the procedure by which a semi-IPN composed of crosslinked NR and polystyrene is prepared.

As can be seen, NR should firstly be compounded and vulcanized, then subsequently the cured sheets are weighed and allowed to swell in inhibitor-free styrene monomer containing DCP (dicumyl peroxide) as initiator and divinyl benzene (DVB) if necessary. The procedure is followed by polymerization of styrene through heating in a closed mould at high temperatures for many hours. If the preparation of a semi-IPN is the aim, the polymerization of styrene is carried out in the absence of crosslinking agent (DVB). On the other hand, to prepare a full IPN, a small amount of crosslinking agent should also be added when the NR is being swollen by styrene. This is illustrated in Scheme 24.3.

The characteristic properties of an IPN depend upon many factors such as blend ratio, morphology, crosslink density and nature of the component polymers. For instance, Mathew and coworkers followed the above mentioned procedure and studied the impact behaviour of semi- and full IPNs based on NR and polystyrene (PS).[25] Their results showed that the impact properties are immensely improved by the addition of NR to pure PS and they also found that impact resistance increases with crosslinking level of an IPN. However, increase in crosslinking beyond a certain limit was found to be undesirable and

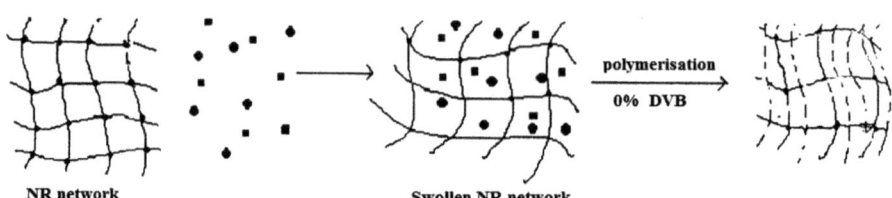

Scheme 24.2 Typical method of preparation of natural rubber based semi-IPNs.[25]

Application of Natural Rubber Based Blends and IPNs

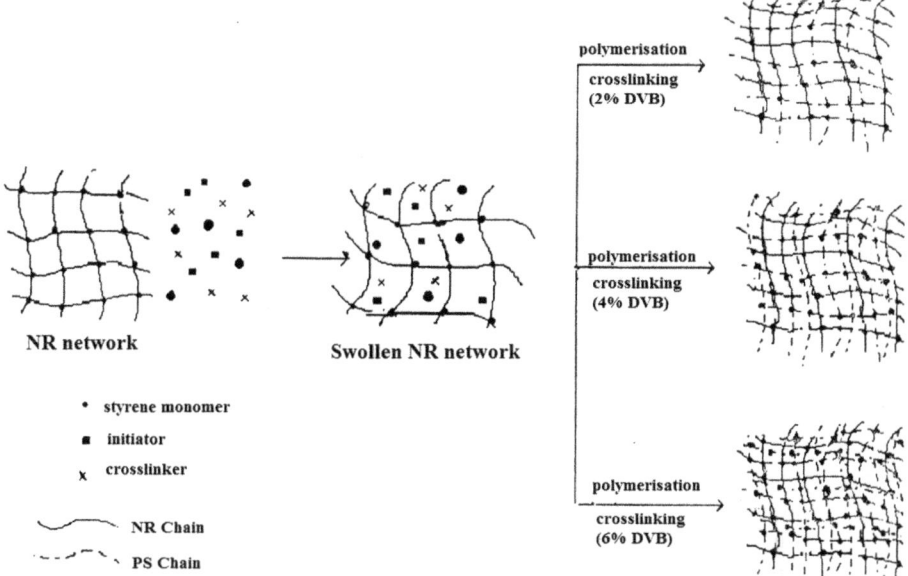

Scheme 24.3 Typical method of preparation of natural rubber based full-IPNs.[25]

Table 24.4 Effect of NR and ENR on the mechanical properties of PA6.[28]

Rubber	σ_y (MPa)	ε_y (%)	σ_b (MPa)	ε_b (%)	Impact strength (kJ/m^2)
–	81.97 ± 4.56	19 ± 1	74.37 ± 11.59	45 ± 13	6.38 ± 2.43
ENR	23.64 ± 0.82	11 ± 1	28.35 ± 4.00	60 ± 10	34.51 ± 6.97
NR	39.47 ± 2.09	12 ± 1	45.26 ± 2.43	51 ± 8	6.80 ± 1.55

therefore, high levels of PS and PS crosslinking strongly influences the impact performance. They also proposed and related toughening of semi-IPNs to a shear yielding mechanism; however, in full IPNs, the toughening mechanism was attributed to crazing followed by shear yielding.

In order to obtain desired impact properties, the polymer blends should generally be miscible, therefore phase separation in the blends should not occur.[27] Although NR could be used as a toughening agent for many thermoplastics, it is deficient in that role for polar polymers like nylons because dissimilarity in polarity would cause incompatibility. The limitation is largely overcome by using epoxidized NR (ENR), which contains epoxy groups that can be converted to hydroxyl groups. As a result it would be potentially partially miscible with polymers like nylon. Tanrattanakul et al.[28] have recently studied the toughening of nylon (PA6) with NR particles. They also evaluated the effect of compatibility of polymers on the resulting impact properties. To accomplish this, plausible reactive blending of PA6/ENR blend, PA6/NR and blends were prepared under the same method and tensile properties and also impact strength of the blends were measured as tabulated in Table 24.4.

PA6 is an engineering plastic with a wide variety of applications. However, it is well known that it is a notch sensitive plastic and is sensitive to crack propagation. It can be deduced from the table that the PA6/ENR blend has had a six-fold substantial improvement in the impact strength, while the impact improvement induced by using NR is almost negligible. According to their report, the addition of rubber also decreased yield stress (σ_y), yield strain (ε_y) and stress at break (σ_b) of PA6. Elongation at break (ε_b) of the blends was higher than that of neat resin.

It should be noted that the difficulties associated with brittleness of plastics can become more severe when they are reinforced by fillers.[29] NR is also capable of toughening the plastic composite; this was studied by Nematezadeh[29] where toughening of a PA6-montmorillonite (MMT) nanocomposite was performed by blending with ENR. As can be seen in Figures 24.6 and 24.7, the presence of ENR and MMA in PA6 was similarly examined by measuring elongation at break and by Izod impact test.

The presence of MMT decreases the impact strength of PA6, which can be improved by blending with ENR. In spite of the superior mechanical properties gained by applying these kinds of reinforcements, they can impair the ductility of polymers. One of the reasons for decreasing the ductility is the fact that a significant volume fraction of polymer which was responsible for dissipating stress through the shear yielding or crazing mechanism is now replaced by the filler, which is not capable of deforming under stress or dissipating stress.[29] Moreover, in the case of fillers that are miscible with the polymer matrix, filler may hinder the motion of the polymer molecules and therefore it sharply decreases the impact resistance of polymers. In fact, debonding of reinforcement is an important mechanism in promoting fracture toughness of polymers and in the present case ENR can act as a debonding agent by increasing plastic stretching of the matrix.

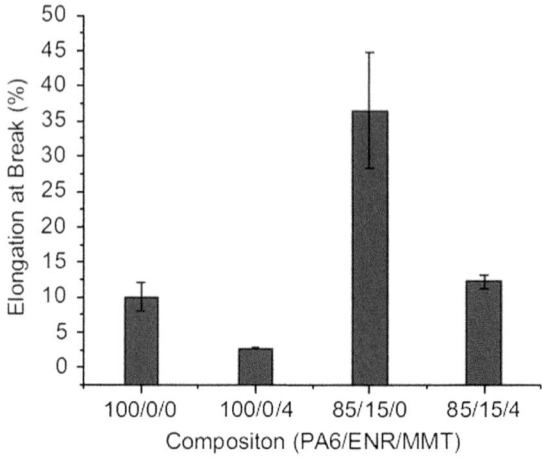

Figure 24.6 Variations of elongation at break by adding ENR and MMT into PA6.[29]

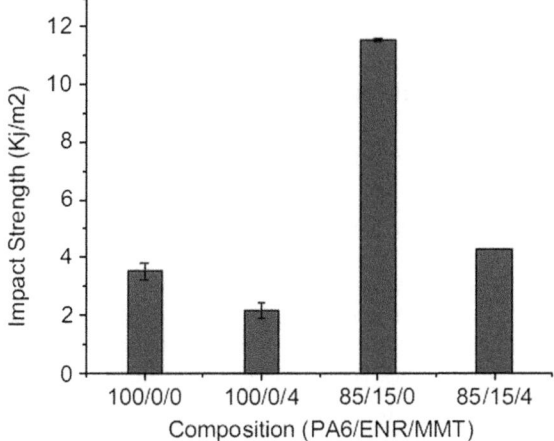

Figure 24.7 Variations of impact strength by adding ENR and MMT into PA6.[29]

Polylactic acid (PLA) is another versatile plastic that is biodegradable and therefore environmentally friendly. Compared with petroleum-based plastics, PLA has high modulus (3 GPa) and strength (50–70 MPa), but is brittle. Somdee carried out a comprehensive study on the toughening of PLA through blending with NR.[30] The results of the work showed that using 10% of NR can increase the toughness of the PLA-NR blend up to two-fold compared to pure PLA. They also investigated the effect of NR particle size on the resulting impact strength and concluded that average NR particle size of 2.30 µm can lead to the highest values for enhancement of toughness. In addition, the effects of various parameters of mixing, such as temperature and rotor speed, on the impact strength were precisely measured and optimized.[31]

24.5 Membrane Technology

24.5.1 Introduction

Membranes can be essentially defined as a barrier that separates two phases and selectively restricts transport of various chemicals.[32] Examination of the rate of diffusion of small molecules through polymeric materials is relevant for many engineering applications. Nowadays, polymer membranes are used as materials for cable coating, food packaging, electronic circuits, *etc*.[33] Therefore, the transport properties of organic solvents and gases through polymers are of great technological importance.[34] Moreover, the improved properties of polymer blends over their individual components have made them useful materials in aerospace, transportation and packaging applications, and in food industries.[35,36] Polymers in different applications are exposed to a variety of chemical environments during their lifetime. The presence of solvents in polymers or blends becomes significant because most polymers after swelling in the solvent show a reduction in properties.

24.5.2 Recent Achievements in the Field of NR Blends as Membranes

The high flexibility of the rubbery chain provides free volume 'openings' which permit gas diffusion. For instance, due to low intermolecular forces and relatively unhindered single bonds that link the silicon and oxygen backbone chain atoms together, silicone rubber has a higher than normal amount of free volume and a high degree of chain mobility, making it the most permeable elastomer (see Table 24.5).

When a rubber is exposed to a gas, solution occurs at the surface and the dissolved gas molecules diffuse into the interior. The diffusion of gas molecules into the rubber membrane is a process in which the gas molecules migrate from 'holes' (free volume) to 'holes' (free volume). The permeation of gas through a membrane involves three stages including solution on one side, diffusion through the membrane to the other side, and finally evaporation out of the membrane.[37]

Studies on the use of NR and its blends as a membrane probably date back to the work carried out by Barbier, where the permeability of mixtures of NR with other elastomers was studied; it was finally concluded that the lower permeability obtained by the addition of a synthetic elastomer does not depend solely on the permeability coefficient itself of the particular synthetic elastomer. It also depends on another factor, which, in all probability, is a function of the internal structure of the mixture of NR and synthetic elastomer.[38] Among recent studies, the barrier properties of SBR, NR and their blends to oxygen and nitrogen gases were investigated by Thomas.[39] Since the permeation properties are strongly dependent on changes in membrane structure such as crystallinity, crosslinking, additives and phase morphology, the effects of the nature and the degree of crosslinking in both SBR and SBR/NR blends on gas permeation behaviour were examined. In order to understand the impact of the curing system on the permeability properties, membranes having an equal content of SBR and NR were prepared using four different crosslinking systems including conventional (CV), efficient (EV), peroxide (DCP) and a mixture consisting of sulfur and peroxide (mixed). Using these curing systems leads to different bond lengths and various bond energies of the linkages, *i.e.* C–C, C–S and S–S bonds. The bond length is the shortest for peroxide rubber membranes and highest for CV membranes. Therefore, based on the flexibility of crosslinks it was predicted that the gas permeability would be in the order CV > mixed > EV > DCP.

Table 24.5 Oxygen permeability of rubber.[37]

Polymer	Permeability $\times 10^9$, $cm^3 \times cm/(s \times cm^2 \times cmHg)$
Dimethylsilicone rubber	60.0
Fluorosilicone	11.0
Nitrile rubber	8.5
Natural rubber	2.4
Butyl rubber	0.14

They evaluated this order and thus the crosslink densities were determined and showed that the highest crosslink density possessed by the DCP membrane results in the lowest gas permeability and the lowest crosslink density possessed by the EV membrane exhibited the highest gas permeability. In fact as the number of crosslinks per unit volume of the polymer molecules increases, it becomes very difficult for the gas molecules to pass through the tightly cross-linked system. Table 24.6 shows the permeability (P) and crosslink density (v) values for a blend of 50/50 SBR/NR.

In addition, as illustrated in Figure 24.8, dependence of the O_2/N_2 selectivity of the NR/SBR blend membrane to the type of curing system was also evaluated. It was found that the peroxide membrane exhibits a high selectivity,

Table 24.6 Permeability (P) and crosslink density (v) values for 50/50 SBR/NR blends.[39]

Systems	$P \times 10^8$ cm^3 cm/cm^2 s $cmHg$		$v \times 10^4$ (mol/ml)
	O_2	N_2	
EV	16.00	6.93	1.40
Mixed	15.43	6.39	2.08
CV	15.30	6.09	2.22
DCP	14.57	5.10	9.25

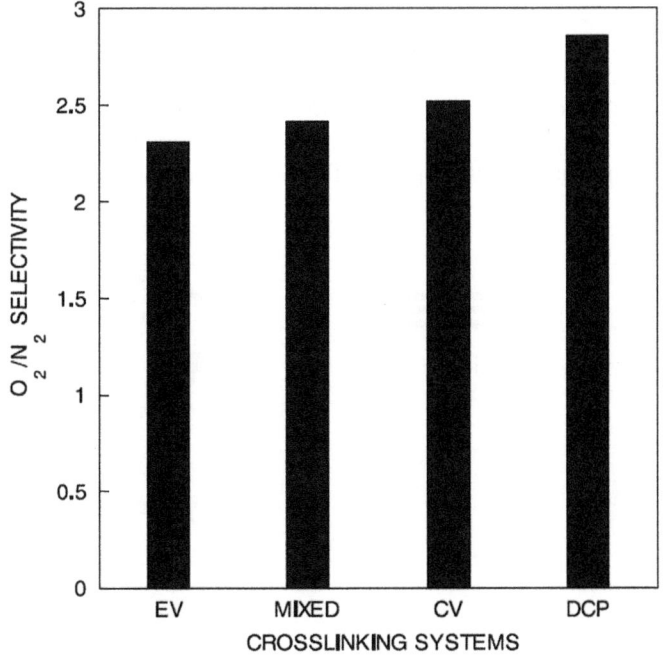

Figure 24.8 O_2/N_2 selectivity of the NR/SBR blend membrane obtained by using different crosslinking systems.[39]

and the EV vulcanized membrane exhibits a low selectivity. In other words samples that exhibit high permeability show only a low selectivity and vice versa. In principle, there is an approximately linear relation with negative slope existing between the oxygen-to-nitrogen selectivity and oxygen permeability for a series of polymers such as silicon rubber, polystyrene, poly(vinyl chloride) or poly(ethyl methacrylate).[40] A similar trend was observed for the present case.

Polymer membranes are widely used in many engineering applications and hence the study of transport of solvent molecules through these membranes is also very important. To minimize the transport of solvents through membranes, the use of polymers containing layered silicates as a barrier is well documented.[41,42] The platelet structure of layered silicates has the ability to improve the barrier properties of polymer materials according to a tortuous path model in which a small amount of platelet particles significantly reduces permeant diffusion, as shown in Figure 24.9. It has been reported that the NR reinforced with 3 phr organoclay resulted in a 50% reduction of the oxygen permeability compared to the neat matrix.[42] Solvent resistance properties of nanostructure layered silicates filled NR, carboxylated styrene butadiene rubber (XSBR) and their latex blend were examined by Stephen.[42] The effect of concentration of filler, temperature and penetrant size on the transport properties of aromatic solvents through layered silicate reinforced latex membranes were analysed. The layered nanostructure sodium bentonite and sodium fluorohectorite were used in this case and the variation in solvent uptake with increasing concentration of layered silicates was examined using benzene as the solvent. In such experiments, it is common to plot the curves as the fractional uptake of penetrant against $t^{1/2}/h$, where h is the thickness of the sample. This type of plot is often referred to as a reduced sorption curve, because the sample thickness is included in the abscissa. Figure 24.10 shows the changes in benzene uptake for the NR/XSBR blend membrane at various concentrations of silicates. It is found that the absorption of solvent is more reduced for nano-filled

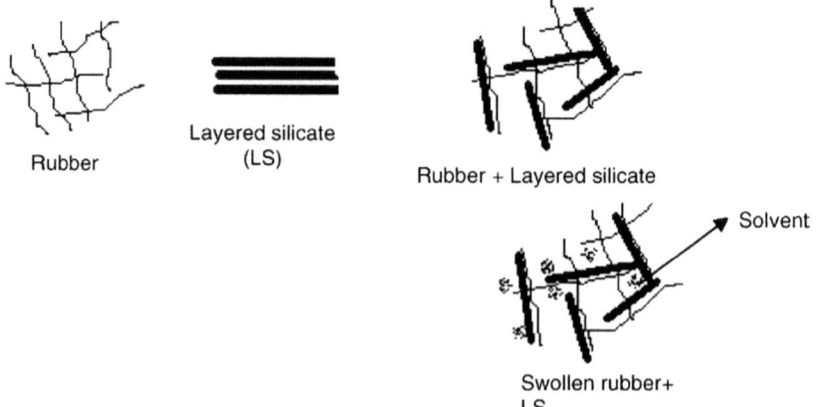

Figure 24.9 Schematic representation of swelling behaviour of rubbers filled with layered silicate.[42]

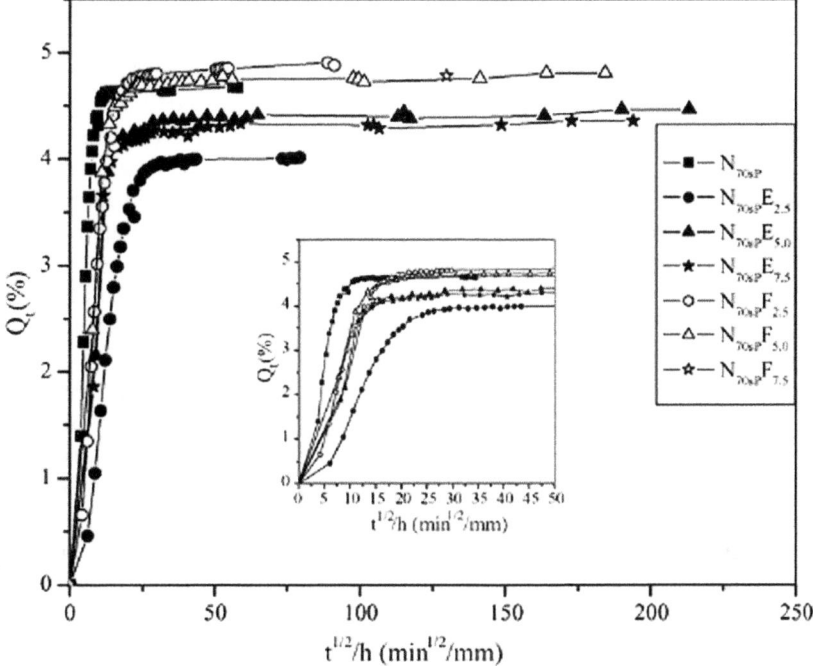

Figure 24.10 The solvent uptake (Q_t) vs. $t^{1/2}/h$ plot of nanofilled 70/30 NR/XSBR latex membranes in benzene at varying percentage of filler at 30 °C. In the sample coding, N stands for NR, the subscript 70 represent the weight percentage of NR, sP for sulphur pre-vulcanization, E and F denote sodium bentonite and sodium fluorohectorite, respectively. The subscript numbers 2.5, 5.0 and 7.5 indicate the weight percentage of fillers used.[42]

samples than pristine polymer. This is due to the platelet-like morphology of layered silicates embedded in the polymer matrix, which increases the tortuosity of the path. Moreover, two distinguishable zones can be seen in the graph. The first zone is related to the initial solvent uptake and thus the swelling rate is too high in this area owing to the large concentration gradient. The polymer sample is under severe solvent stress. The second zone indicates a reduced swelling rate due to the decrease in concentration gradient, which finally reaches at equilibrium swelling. This translates that the concentration gradient is approaching zero. The sodium bentonite filled blend displays reduced solvent uptake at lower concentration. Surprisingly, however, at higher concentrations of bentonite the solvent uptake is higher due to the aggregation of filler. As a result some portions of the silicates are not available in the rubber matrix to enhance the tortuosity of the path. The swelling rate of the fluorohectorite filled blend decreases constantly with filler loading because of the more uniform distribution of layered silicate in the matrix. Even though the more exfoliated structure can be achieved in the fluorohectorite filled blend, the solvent uptake is higher in this case than in the bentonite-reinforced system. This was attributed to the improved elasticity in this case arising from the plasticizing effect of the gallery ions.[42]

The nanocomposites of NR-XSBR were also investigated in another work and the gas transport through their membranes was assessed.[41] The oxygen and nitrogen permeability of the composites prepared using different fillers including sodium bentonite, sodium fluorohectorite, clay and silica was measured as shown in Figure 24.11 for the case of oxygen. First it should be noted that, compared to nitrogen, oxygen has more permeability than nitrogen due to the lower covalent radii of oxygen. The lowest permeability is obtained by using sodium fluorohectorite. Because the chain segments become immobilized in the presence of layered silicates, the free volume decreases and so the gas permeability coefficient is reduced.

A series of IPN membranes were synthesized using NR and polystyrene through the sequential polymerization technique and the transport of aromatic hydrocarbons through semi- and full IPN membranes was comprehensively evaluated by Mathew.[43] The sorption was carried out using a number of aromatic solvents, namely benzene, toluene and xylene. In addition to the conventional measurements, effect of temperature on swelling was studied by implementing the experiments in toluene in the temperature range of 30 to 75 °C. It was found that $Q\infty$-values increase with temperature in all cases. However, at 75 °C, a decrease in $Q\infty$-values was observed as compared to those obtained at 65 °C (Figure 24.12). This was related to the fact that at 75 °C

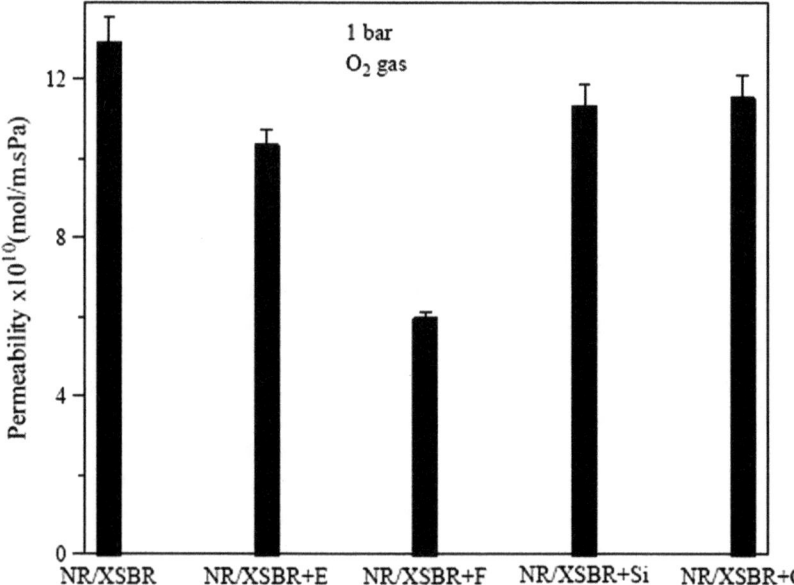

Figure 24.11 Variation in permeability of oxygen gas through 70:30 NR/XSBR latex membranes towards different fillers (2.5 phr) at 1 bar; E, F, C and Si represent sodium bentonite, sodium fluorohectorite, clay and silica, respectively.[41]

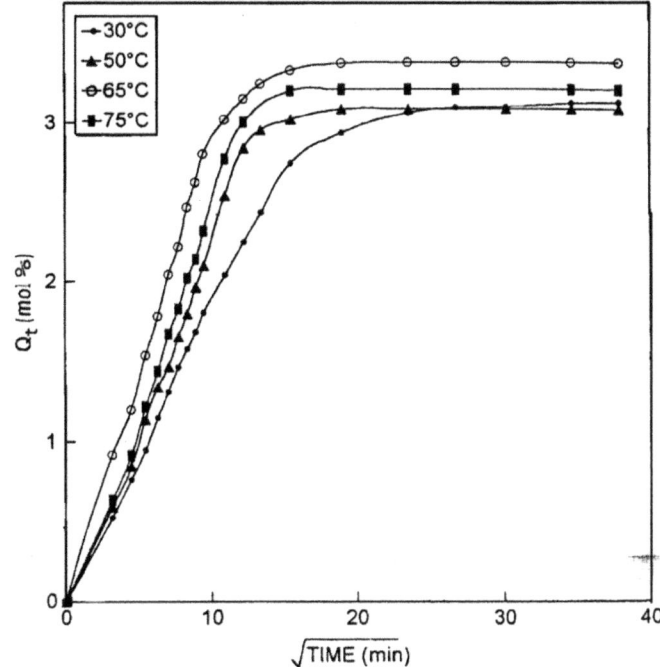

Figure 24.12 Effect of temperature on solvent uptake (Q_t vs. $\sqrt{time}$ in toluene for PS/NR IPN membrane).[43]

desorption occurs rapidly compared to sorption. The effects of blend ratio, crosslinker content and nature of initiator on the diffusion of various solvents were also analysed. It was found that in all cases, the uptake value increased by about 50% as the PS content decreased from 70% to 30%. Moreover, as the crosslink density was increased, the uptake decreased by 40%. The influence of swelling on the mechanical performance of the membranes was also investigated by conducting tensile testing of swollen specimens.[43]

Recently, preparation of a mixed matrix (MM) membrane of NR latex and crosslinked poly(vinyl alcohol) (PVA) semi-IPN filled with zeolite was studied by Amnuaypanich.[44] To do so, PVA powder was dissolved in water and the mixture was heated until a complete solution was obtained. Then, zeolite was added into the PVA solution and agitated vigorously to prepare a fine zeolite dispersion. Subsequently, NR latex was poured into the zeolite dispersion. Once completely mixed, the NR–zeolite dispersion was cast and dried in the oven where PVA was allowed to be crosslinked with sulfosuccinic acid (SSA). In principle, the swelling behaviour of a membrane can demonstrate the capability of the membrane to specifically interact with the absorbed molecules. As illustrated in Figure 24.13, the NR/PVA membranes swelled greatly in water as compared with the swelling in pure ethanol, meaning that the membranes have more affinity to water. This is due primarily to the specific interactions between water molecules and the –OH groups of PVA in the membranes. Therefore, the

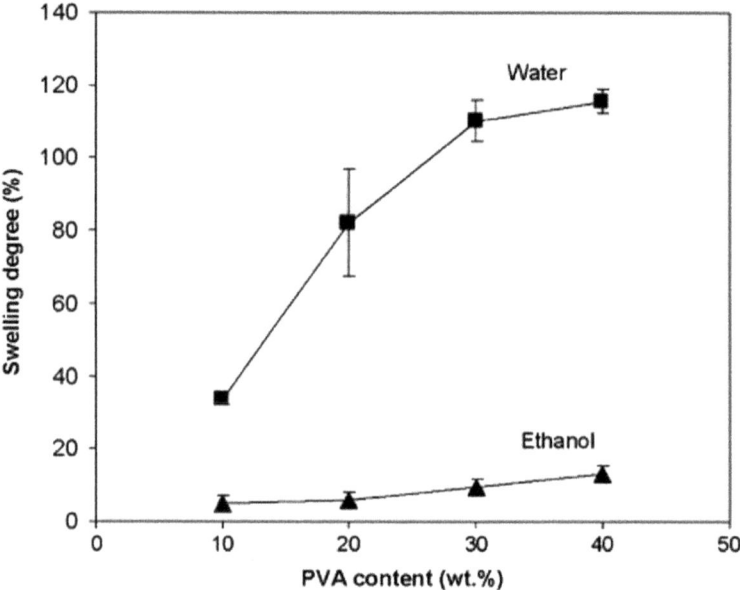

Figure 24.13 Swelling degree of NR/PVA membranes in pure water and ethanol.[44]

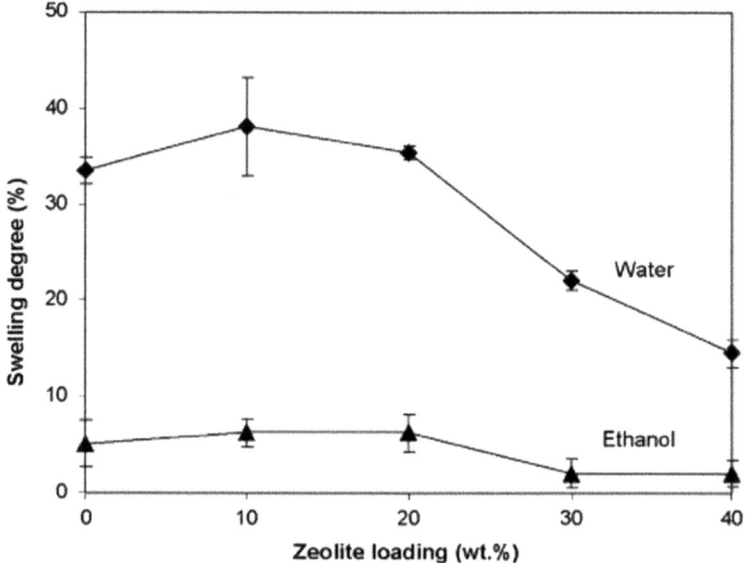

Figure 24.14 Swelling degree of mixed matrix membranes in pure water and ethanol.[44]

water swelling increased as PVA content in the membranes increased. Upon introducing zeolite to the NR/PVA membrane, the MM membranes exhibited a slight increase in water swelling, as shown in Figure 24.14; since the pore size of zeolite is about 0.4 nm, which is larger than the kinetic diameter of a water molecule (0.264 nm), it is possible that water molecules enter and are retained in the zeolite pores. This is while the swelling of MM membranes declined for higher zeolite loading. The suppression of the membrane swelling as observed by the presence of zeolite arises from the crystalline zeolite structure, which is not liable to the interacting molecules and the restricted movement of polymer chains near the zeolite particles; in other words, the zeolite acts as the physical crosslinker for the neighbouring polymer molecules. Compared to the swelling in water, the extent of swelling in ethanol of the MM membranes was less, with a slight decline at 30–40 wt% zeolite; this ensures the molecular sieving effect of zeolite which selectively excludes the ethanol molecules.[44]

Transport properties of thermoplastic polyurethane/NR (TPU/NR) blends were studied by Minnath and the diffusion of organic solvents such as benzene, toluene and p-xylene through TPU/NR blend membranes was similarly investigated.[45] The sorption behaviour of TPU/NR blend membranes in xylene indicated the equilibrium solvent uptake rises with increasing the NR content in the blends (Figure 24.15). The effect of the molecular weight of the solvents on the mol% uptake was observed as in Figure 24.16. Increasing the molecular weight of the penetrant leads to a decrease in the Q_t (moles of solvent sorbed by 0.1 kg of the sample) values in all the systems. Thus, the variation was in the order benzene > toluene > xylene. This behaviour was explained in terms of free volume theory, which states that the diffusion rate of a molecule depends primarily upon the ease with which the polymer segments exchange their positions with penetrant molecules. Moreover, the mobility of the polymer depends on the amount of free volume in the matrix. As the penetrant size increases, the exchange process becomes difficult, resulting in a decrease in sorption.

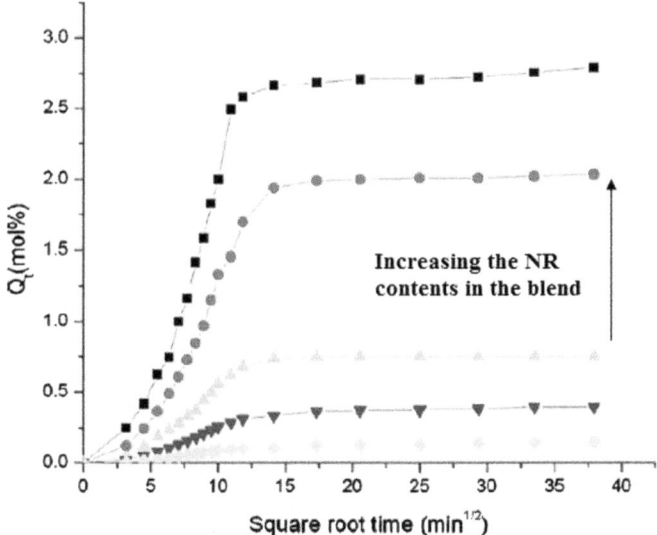

Figure 24.15 Sorption curves of the NR/PU blends in xylene at 30 °C.[45]

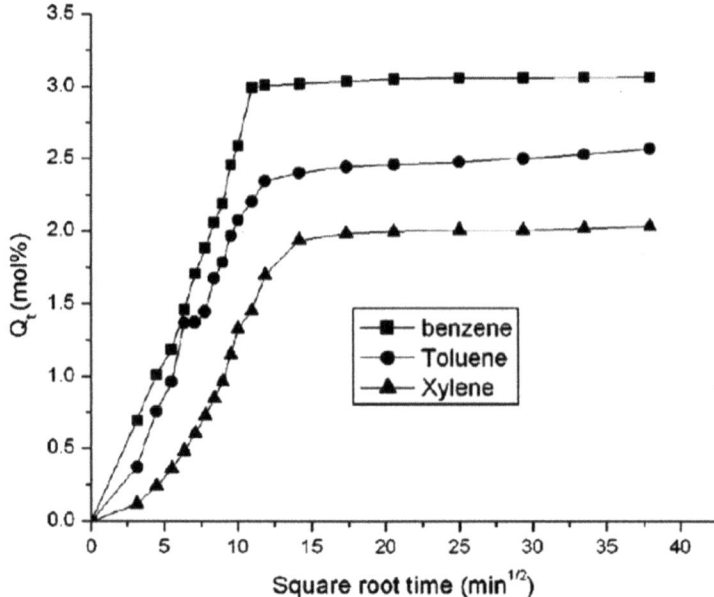

Figure 24.16 Mol% uptake of different solvents by the blend at 30 °C.[45]

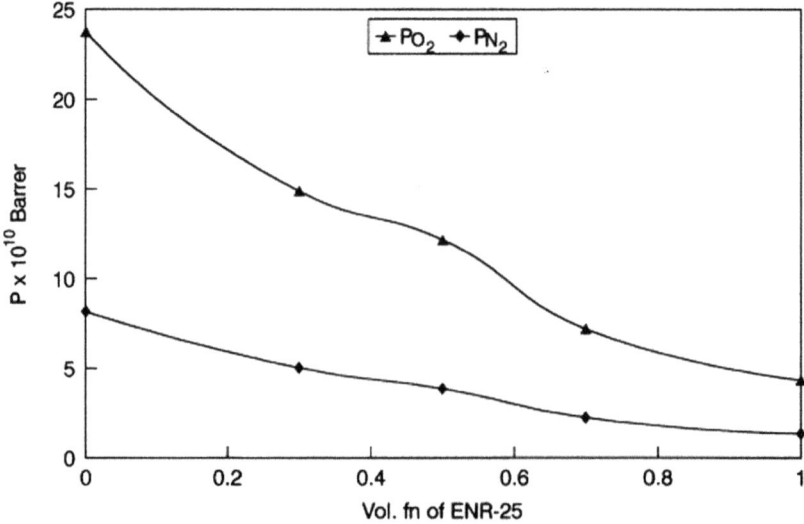

Figure 24.17 Permeation coefficients of NR/ENR blends to oxygen and nitrogen in various concentrations of ENR in the blend.[46]

Among other NR blends used as a membrane it is worth referring to the study carried out by Johnson on NR/ENR blends.[46] The permeabilities of the pure polymers and blends were determined using nitrogen and oxygen. The result of this experiment showed that gas permeability decreases as ENR concentration increases (Figure 24.17). Compared to oxygen, nitrogen exhibits lower permeability. The permeability of the polymer appears to be a very sensitive function of penetrant size. With increasing covalent radii of nitrogen, the permeability coefficient decreases. Their results also revealed that the O_2/N_2 selectivity is almost constant at low ENR fractions. With increasing concentration of ENR in the blend, however, the O_2/N_2 selectivity increases and finally at high ENR concentration, the selectivity becomes almost constant. The change in selectivity with blend composition was related to phase inversion occurring over a narrow concentration range.[46]

24.6 Miscellaneous Applications of Natural Rubber Based Blends

As mentioned previously, NR is a general-purpose elastomer and it has a wide variety of applications. For instance latex concentrates of NR are widely used in the production of gloves, condoms, balloons, catheters, baby soothers and dental bridges.[47] In addition blending can pave the way to improving the properties of this elastomer and increasing the versatility of the resulting compound. In the previous sections the most common applications were reviewed. Here a few miscellaneous applications of NR blends reported in the literature are given.

24.6.1 Retreading of Tyres

Since tyres used in mining vehicles are very expensive and need regular maintenance, it is impossible to accept its replacement expense within very short term. There are several types of damage that can occur in dump-truck tyres such as tread detachment, sidewall cuts, impact ruptures, bead damage, *etc.* Therefore, Kaushik Pal and coworkers investigated the use of carboxylated nitrile rubber and NR blends as retreading compounds for this type of tyre. The carboxyl group of XNBR can introduce new curing reactions to the rubber. The group also improves the abrasion properties of NBR significantly while retaining excellent oil and solvent resistance. Thus, XNBR provides outstanding physical properties, tensile strength and abrasion resistance. Also, XNBR generally exhibits poor hysteresis properties and reduced cold temperature flexibility, but in regard to chemical resistance XNBR is considerably superior. Blending of XNBR and NR combines all these features with the unique properties of NR. This study showed that the resulting blend has very good mechanical properties when compared with some collected rubbers from different tyre retreading industries. It was also found that the XNBR has potential as an effective modifier for NR because its incorporation into rubber enhances the plasticization of NR during mastication in the mill. Vulcanizate properties of NR such as modulus, resistance to abrasion and compression set are improved on incorporation of XNBR at a blend ratio of 80 phr XNBR with NR.[48–51]

24.6.2 NR as an Insulator

Blending of chitosan (CS) with a non-polar polymer such as NR can improve the dielectric performance of chitosan by reducing the polarity; therefore blends of NR/CS have recently been investigated. Due to excellent electrical properties such as low dielectric constant, loss factor and high volume resistivity, NR can be used as high-frequency insulators. CS has moderate insulating properties and good oil resistance. Therefore, by blending NR with CS, a material with improved electrical properties can be achieved. Johns *et al.* studied the effect of blend ratio, dynamic vulcanization with dicumyl peroxide and maleic anhydride compatibilization on dielectric constant, loss factor, volume resistivity and AC conductivity of NR/CS blends.[52–54]

24.6.3 Use of NR for Modification of Plastic Properties

Mahapram prepared a blend of LDPE (low-density polyethylene) and NR latex so as to improve the properties of heat-shrink films produced by LDPE. Plastics are widely used in packaging applications, particularly heat-shrink films, owing to their outstanding physical, mechanical and chemical properties. Shrink film is usually processed by a blown film extrusion process. The molecules of plastic processed by this method tend to become aligned in the direction of the orienting force, and thus the blown film is biaxially oriented. After that, shrink wrapping starts with loosely sealing the plastic film around the products, such as food and beverage cans, followed by passing the wrapped package through a shrink tunnel where it is exposed to heat. At present, the use of LDPE shrink films is increasing and it is predicted to continue to do so in the near future. Thus, the development of high-quality LDPE heat-shrink films is of increasing interest. Elastomers, such as epoxidized NR, carboxylated nitrile rubber and vulcanized NR, have attracted a great deal of attention in recent years from a heat shrinkability point of view. Among these elastomeric types, NR is more interesting due to its unique properties, such as elasticity, flexibility and mechanical properties, which are more suitable for use in improving the shrinkability of the blended films. Additionally, the important property required for heat-shrink film production is the crystallization of the polymer during the stretching process because this crystallization controls the shrinkability of the finished product. The outstanding properties of unvulcanized NR also include strain-induced crystallization during the stretching process.[55–66]

References

1. J. M. Vergnaud and I. D. Rosca, *Rubber Curing and Properties*, CRC Press, FL, 2004, p. 135.
2. R. C. Klingender, *Handbook of Specialty Elastomers*, CRC Press, FL, 2008, p. 493.
3. L. W. McKeen, *Fatigue and Tribological Properties of Plastics and Elastomers*, Butterworth-Heinemann, Waltham, MA, 2009, p. 215.
4. J. E. Mark and B. Erman, *Science and Technology of Rubber*, Academic Press, Salt Lake City, UT, 2005, p. 183.
5. D. R. Paul, *Polymer Blends*, Academic Press, Salt Lake City, UT, 1978, p. 5.

6. A. I. Isayev, *Encyclopedia of Polymer Blends: Volume 1: Fundamentals*, John Wiley & Sons, New York, 2010, p. 153.
7. M. Wagner, Fine-particle silicas in tire treads, carcass, and steel-belt skim, *Rubber Chem. Technol.*, 1977, **50**, 356.
8. P. T. Hao, H. Ismail, A. S. Hashim, Study of two types of styrene butadiene rubber in tire tread compounds, Polym Test, 2001, 20, 539.
9. N. Rattanasom, A. Poonsuk and T. Makmoon, *Polym. Test.*, 2005, **24**, 28.
10. K. Sahakaro, N. Naskar, R. N. Datta and J. W. M. Noordermeer, *J. Appl. Polym. Sci.*, 2007, **103**, 2538.
11. K. Sahakaro, A. G. Talma, R. N. Datta and J. W. M. Noordermeer, *J. Appl. Polym. Sci.*, 2007, **103**, 2547.
12. K. Sahakaro, R. N. Datta, J. Baaij and J. W. M. Noordermeer, *J. Appl. Polym. Sci.*, 2007, **103**, 2555.
13. H. Hamaguchi, Y. Samejima and N. Kani, A study of aging effect on rubber bearings after about twenty years in use, *11th World Conference on Seismic Isolation, Energy Dissipation and Active Vibration Control of Structures, Guangzhou, China, 17-21 November 2009*.
14. Y. Ohtori, *J. Struct. Div., Am. Soc. Civ. Eng.*, 1997, **43B**, 125.
15. T. Fujita, S. Fujita, S. Suzuki and T. Yoshizawa, *Trans. Jpn. Soc. Aeronaut. Space Sci*, 1987, **53**, 71.
16. S. Nagarajaiah and K. Ferrell, *J. Struct. Div., Am. Soc. Civ. Eng.*, 1999, **125**, 15573.
17. M. Abe, J. Yoshida and Y. Fujino, *J. Struct. Div., Am. Soc. Civ. Eng.*, 2004, **130**, 1119.
18. F. Naeim and J. M. Kelly, *Design of Seismic Isolated Structures*, John Wiley & Sons Inc., New York, 1999, p. 45.
19. A. W. Taylor, A. N. Lin and J. W. Martin, *Earthq. Spectra*, 1992, **8**, 279.
20. R. I. Skinner, W. H. Robinson and G. H. McVerry, *An Introduction to Seismic Isolation*, John Wiley & Sons Inc., New York, 1993, p. 239.
21. C. J. Derham and A. G. Thomas, *Eng. Struct.*, 1980, **2**, 171.
22. I. K. Szeteiová, *Automotive Materials Plastics in Automotive Markets Today*. A report submitted to the Institute of Production Technologies, Machine Technologies and Materials, Faculty of Material Science and Technology in Trnava, Slovak University of Technology, Bratislava, 2011, unpublished.
23. M. S. Flynn and B. C. Smith, *Automotive Plastics Chain: Some Issues and Challenges*. A report prepared for the automotive plastics recycling project, Report Number: UMTRI 93-40-6, 1993.
24. J. A. Brydson, *Plastics Materials*, Butterworth-Heinemann, Waltham, MA, 1999, p. 189.
25. A. P. Mathew and S. Thomas, *Mater. Lett.*, 2001, **50**, 154.
26. M. Schneider, T. Pith and M. Lamblas, *J. Mater. Sci.*, 1997, **32**, 6343.
27. L. A. Utracki, *Polymer Blends Handbook*, Springer, New York, 2003, p. 417.
28. V. Tanrattanakul, N. Sungthong and P. Raksa, *Polym. Test.*, 2008, **27**, 794.
29. N. Nematzadeh, *Key Eng. Mater.*, 2011, **471**, 518.

30. P. Somdee, Natural rubber toughened polylactic acid, PhD thesis, Suranaree University of Technology, 2009.
31. C. K. Riew and A. J. Kinloch, *Toughened Plastics I: Science and Engineering*, American Chemical Society, USA, 1993, vol. 23, p. 61.
32. M. T. Ravanchi, T. Kaghazchi and A. Kargari, *Desalination*, 2009, **235**, 199.
33. R. Stephen, K. Joseph, Z. Oommen and S. Thomas, *Compos. Sci. Technol.*, 2007, **67**, 1187.
34. S. C. George and S. Thomas, *Prog. Polym. Sci.*, 2001, **985**, 1017.
35. A. E. Mathai, R. P. Singh and S. Thomas, *J. Membr. Sci.*, 2002, **202**, 35.
36. A. P. Mathew, S. Packirisamy and S. Thomas, *Polym. Degrad. Stab.*, 2001, **72**, 423.
37. H. Zhang, The permeability characteristics of silicone rubber, in Proceedings of Global Advances in Materials and Process Engineering, Coatings and Sealants Section, 6–9 November 2006.
38. J. Barbier, *Rubber Chem. Technol.*, 1955, **28**, 814.
39. S. C. George, K. N. Ninan and S. Thomas, *Eur. Polym. J.*, 2001, **37**, 183.
40. M. Kajiwara, *Synthetic Polymeric Membranes*, Walter de Gruyter, Boston, MA, 1987, p. 347.
41. R. Stephen, C. Ranganathaiah, S. Varghese, K. Joseph and S. Thomas, *Polymer*, 2006, **47**, 858.
42. R. Stephen, S. Varghese, K. Joseph, Z. Oommen and S. Thomas, *J. Membr. Sci*, 2006, **282**, 162.
43. A. P. Mathew, S. Packirisamy, R. Stephen and S. Thomas, *J. Membr. Sci.*, 2002, **201**, 213.
44. S. Amnuaypanich, J. Patthana and P. Phinyocheep, *Chem. Eng. Sci.*, 2009, **64**, 4908.
45. M. A. Minnath, G. Unnikrishnan and E. Purushothaman, *J. Membr. Sci.*, 2011, **379**, 361.
46. T. Johnson and S. Thomas, *Polymer*, 1999, **40**, 3223.
47. E. Yip and P. Cacioli, *J. Allergy Clin. Immunol.*, 2002, **110**, S3.
48. K. Pal, T. Das, S. K. Pal and C. K. Das, *Polym. Eng. Sci.*, 2008, **48**, 2410.
49. K. Pal, S. K. Pal, C. K. Das and J. K. Kim, *Tribol. Int.*, 2010, **43**, 1542.
50. K. Pal, S. K. Pal, C. K. Das and J. K. Kim, *J. Appl. Polym. Sci.*, 2011, **120**, 710.
51. S. Thomas and R. Stephen, *Rubber Nanocomposites: Preparation, Properties and Applications*, John Wiley & Sons, New York, 2010, p. 233.
52. J. Johns and C. Nakason, *J. Non-Cryst. Solids*, 2011, **357**, 1816.
53. J. Johns and V. Rao, *Int. J. Polym. Anal. Character.*, 2008, **13**, 280.
54. J. Johns and V. Rao, *Fibers Polym.*, 2009, **10**, 761.
55. S. Mahaprama and S. Poompradub, *Polym. Test.*, 2011, **30**, 716.
56. J. K. Mishra, Y. W. Chang and D. K. Kim, *Mater. Lett.*, 2007, **61**, 3551.
57. S. R. Chowdhury, J. K. Mishra and C. K. Das, *Polym. Degrad. Stab.*, 2000, **70**, 199.
58. H. A. Knonakdar, J. Morshedian, M. Mehrabzadeh, U. Wagenknecht and S. H. Jafari, *Eur. Polym. J.*, 2003, **39**, 1729.

59. M. H. Senna, A. Abdel-Fattah and Y. K. Abdel-Monem, *Nucl. Instrum. Methods Phys. Res., Sect. B*, 2008, **266**, 2599.
60. A. Ibrahim and M. Dahlan, *Prog. Polym. Sci.*, 1998, **23**, 665.
61. H. M. Dahlan, M. D. Zaman and A. Ibrahim, *Radiat. Phys. Chem.*, 2002, **64**, 429.
62. C. Sirisinha, P. Saeoui and J. Guaysomboon, *Polymer*, 2004, **45**, 4909.
63. H. Azman, U. W. Mat and Y. C. Ching, *Polym. Test.*, 2003, **22**, 281.
64. H. Ismail and S. Suryadiansyah, *Polym. Test.*, 2002, **21**, 389.
65. S. H. El-Sabbagh, *Polym. Test.*, 2003, **22**, 93.
66. M. A. Adhha, *Preparation and Characterization of Natural Rubber-Polyethylene- and Natural Rubber/Polyethylene-Clay Nanocomposites*, PhD thesis, University Putra Malaysia, 2007, FS 2007 57.

Subject Index

abnormal groups
    aldehyde groups, 65–66
    bonded proteins and amino groups, 67
    epoxide groups, 64–65
    ester groups/fatty acids and phospholipids, 64
    trans-isoprene and dimethylallyl (DMA) groups, 64
accelerators, 575
acrylic plastics
    applications of, 320–321
    natural rubber, history of, 301–302
    natural rubber, preparation methods of
        acrylate blends, 302–306
        acrylate IPNs, 306–307
    properties and characterization techniques
        mechanical properties, 313–315
        morphological properties, 307–313
        rheological properties, 318–320
        thermal and thermomechanical properties, 315–318
acrylic rubber, 182
acrylonitrile butadiene rubber, 559
acrylonitrile butadiene styrene (ABS), 579
adhesives, 45–46
ageing, and degradation, 554–555
agglomerates, 375
air-dried sheet (ADS), 29, 95, 96
aldehyde groups, 65–66
alkylammonium, 391
allylic diphosphate (APP), 31
American Synthetic Rubber Research Program, 77
amino acids, **60,** 60–61
amino groups, 67
antidegradants, 44, 574
ash, 62–63
atomic force microscopy (AFM), 58, 142–143
attenuation total reflection (ATR), 234
autoclave vulcanization, 127

Banbury mixer, 120
battery-powered tapping knives, 81
Bifidobacterium bifidum, 67
biodegradable polymers, 183
biodegradable properties, 366
biodegradation, 555
biopolymers
    natural rubber/lignin blends
        blends and their applications, 351–353
        general information, 350–351
    natural rubber/polyester blends
        blends and applications, 361–366
        general information, 360–361

natural rubber/polysaccharide
    blends
        blends and applications,
            358–360
        cellulose, 358
        starch, 357–358
natural rubber/protein blends
    biosynthesis and
        biodiversity of,
            353–354
    blends and applications,
        356–357
    indigenous proteins of,
        355–356
    materials applications of,
        354–355
bis-alkylphenol disulfide (BAPD),
    219
black incorporation time (BIT), 148
black/non-black fillers, 572–574
blended matrices, 309
blended membranes, 316
blend morphology, 255–258
bonded proteins, 67
branch points, 37–38
brominated EPDM (BEPDM), 412
brominated isobutylene-co-
    p-methylstyrene (BIMS), 155
bromobutyl rubber (BIIR), 558
butadiene rubber (BR), 570

calendered products, 45
camontmorillonite, 372
carbohydrates, 61–62
carbon black (CB), 218
carbon black localization, rubber
    blends
        polarity and viscosity effect,
            157–159
        rubber blends with defined filler
            localization, using
            masterbatch technology,
            159–161
carbon black masterbatch latex, 89
carbon black-reinforced full natural
    rubber, 491

carbon nanotube (CNT) localization
    dispersion in polychloroprene
        and ionic liquid, 162–165
    SBR/NR blends, dispersion and
        localization in, 162
    selective wetting of, 165–168
carbonyl group, 328
carboxylated styrene butadiene
    rubber (XSBR), 495, 538, 560,
    588
cassava starch (ST) blend, 454
casting, 112–113
catalytic hydrogenation, 40
cationic NR latex, 89
cellulose, 358
cellulose derivatives, 498
centrifuged high ammonia, 84, **85**
chain dipping lines, **103**
chemical analysis, 477
chemically modified rubber latices
    cyclized NR latex, 92–93
    epoxidized NR latex, 91
    methyl methacrylate-grafted
        (MG) rubber latex, 91–92
    radiation-vulcanized latex
        (RVNRL), 93
chitosan, 497, 563, 595
chitosan (CST) blend, 454–456
chlorinated NR (CNR), 41, 102
chlorinated polyethylene elastomer
    (CPE), 401, 406, 561
chlorination, 41
chloroprene rubber (CR), 214, 559
chlorosulfonated polyethylene
    rubber, 216, 453, 494, 495
clay reinforcement
    crosslinking techniques
        chemical crosslinking,
            386–389
        irradiation crosslinking,
            389–391
    nanocomposites,
        characterization of
            flammability, 386
            mechanical properties,
                379–384

clay reinforcement (*continued*)
  morphology, 375–379
  thermal properties, 384–386
  preparation methods, 372–374
    ethylene vinyl acetate/natural rubber/organoclay ternary blends, 374–375
    recent developments, 371–372
colloid stability, 79
combinatorial entropy term, 6
commercial dry natural rubbers, 66
compatibility, 179–180
compatibilized blends, 2
compression moulding, 127
constant viscosity (CV), 66, 100, 101
continuous mixers, 122–126
continuous vulcanization, 127–128
conventionally mills, 117
conventional vulcanization (CV), 426, 576
covalent semi-IPNs, 301
crack energy propagation, 343
crosslinking techniques
  chemical crosslinking, 386–389
  irradiation crosslinking, 389–391
cross polarization (CP), 461
crystalline behaviour, 477
crystallization nuclei, 56
cup lump, 98
cured properties, of rubber blend, 477
curing systems, 219–220, 574–575
cyclization, 43–44
cyclized NR latex, 92–93
β-cyclodextrin (β-CD)
  graft copolymerization, 325, **326**
  inclusion complex formation of, 327–330
N-cyclohexyl-2-benzothiazyl sulfenamide (CBS), 216, 577

Debye-Bueche approach, 510, 511, 522
deconvolution, 512
degradation, 554–555
  mechanism, 477

1-deoxy-D-xylulose-5-phosphate (DXP), 31
depolymerized natural rubber (DPR), 342
deproteinized NR (DPNR), 34–36, **36**, 88, 100
dibutyl phthalate (DBP), 574
dichlorocarbene modified NR (DCNR), 233, 408
dichlorocarbene modified styrene-butadiene rubber (DCSBR), 535
dicumyl peroxide (DCP), 15, 452
dielectric properties, 230–232, 278–279
dielectric thermal analysis (DETA), 278
differential scanning calorimetry (DSC), 12, 248, 260–262, 276, 363
diffusion coefficient, 532
digalactosyl diglycerides (DGDG), 56
diglycidyl ether of bisphenol A (DGEBA), 337
dimethylallyl diphosphate (DMAPP), 31
N,N dimethyl benzyl amine, 493
dimethylolpropionic acid (DMPA), 214
2,4-dinitrophenylhydrazine (2,4-DNPH), 65
dioctyl phthalate (DOP), 574
diphenyl guanidine (DPG), 216, 220
2,2-diphenyl-1-picrylhydrazyl (DPPH), 61
dipping, 112
dispersion, 108–109
dissipation mechanisms, 343
divinyl benzene (DVB), 582
dry polyacrylate pellets, 305
dry rubber content (DRC), 29, 214
ductility, 484
dynamic light scattering, 522
dynamic mechanical analysis (DMA), 276, 382, 487
dynamic-mechanical behaviour, 225–229
dynamic mechanical properties, 488

Subject Index

dynamic mechanical thermal analysis (DMTA), 248, 487
dynamic properties, 477

earth scrap, 99
efficient sulphur vulcanization (EV), 426, 427
elasticity, 483
elastomer blends, 572
electronic impedance spectroscopy (EIS), 278
emulsion polymerization techniques, 304
energy dispersive X-ray analysis (EDX), 142
ensured epoxidized rubber (ENR), 288
EPDM *see* ethylene-propylene-diene rubber (EPDM)
epoxidation, 40–41
epoxide groups, 64–65
epoxidized natural rubber (ENR), 41, 100–101, 288–289, 342, 538
    latex, 91
epoxy-poly(2-ethyl hexyl acrylate), 13
epoxy resins, 337, 491, 493
ester groups, 64
esterified steryl glucosides (ESG), 56
ethylene glycol dimethacrylate, 307
ethylene-propylene-diene rubber (EPDM), 559, 576
ethylene thiourea (ETU), 216
ethylene vinyl acetate copolymer (EVA), 16, 396, 397
ethylene vinyl acetate (EVA), 292–294, 374, 557
ethylene vinyl acetate/natural rubber (EVA/NR), 374–375
2-ethylhexyl acrylate (EHA), 290
extraction experiment, 143
extruded products, 45

fatty acids, 64
filler migration
    equilibrium state using Z-Model, theoretical prediction of, 133–137
    equipment and experimental methods
        characterization, 142–144
        preparation of blends, 141
    experimental determination of, wetting concept for, 137–141
    results and discussion
        carbon black localization in rubber blends, 157–161
        carbon nanotube (CNT) localization, 162–168
        nanoclay transfer in rubber blends, 168–171
        silica localization in rubber blends, 144–157
fillers, 44
film form, 446
fluoroelastomer rubber (FR), 558
fluorohectorite filled blend, 589
foaming, 113
Fourier transform infrared (FTIR) spectroscopy, 141, 143, 232, 233, 247–248, 331, 362
fresh latex, composition of, **54**
Frey–Wyssling (FW) particles, 30

Gardiner mixed toluene, 214
gas form, 446
gas permeability coefficient, 541
gel, 37–38
gelatin blend, 459–460
gel content, 247, 253–254
general mechanical properties, 569–570
glass transition temperature studies, 191–192
glutamic acid, 60
glycidyl methacrylate (GMA)-grafted polymers, 362
glycolipids, 56–57
graft copolymerization, 41–42, 313, 345
grafting maleic anhydride, 305
Green Book, 29
ground rubber tires (GRT), 561

guayule, 75
Guinier formula, 509

heat-treated lignin, 352
heterogeneous polymer blends, 6
Hevea brasiliensis, 533, 534
Hevea brasiliensis latex, natural rubber (NR) from
    colloidal properties of, 78
    industrial elastomer, 75–78
    major industrial applications of, 103–104
    production methods, 94–95
        block rubbers, 98
        crepe rubbers, 97
        field coagula, rubber products from, 98–100
        sheet rubbers, 95–97
        skim rubbers, 100
        specialty rubbers and chemically modified rubbers, 100–102
    production of, 80–81
        chemically modified rubber latices, 90–93
        commercial concentration methods for, 84–87
        latex purification and concentration processes, 82–84
        latex technology, recent advances in, 93–94
        NR Latex, preservation of, 81–82
        specialty latices, 87–90
    sources, 74–75
1,6-hexaediol diacrylate (HDDA), 290
high-density polyethylene (HDPE), 397–399, 404
higher order Tyndall spectrum (HOTS), 78
hydrocarbon oils, 90
hydrogenated nitrile rubber (HNBR), 163, 168–171, 229, 560
hydrogen bonding, 477
hydrogen bromide (HBr), 64

hydrolysed proteins (HP), 355
hydroxy ethyl methacrylate, 21

immiscibility, 179
immiscible polymers, 6
incorporation, 108
infrared absorbance, 232–235
injection moulding, 127
inositols, and carbohydrates, 61–62
interfacial polarization (IP), 231
internal batch mixers, 118–120, **119**, 129
internal mixers, 120–122
International Standards Organisation (ISO), 29
International Union of Immunological Societies (IUIS), 59
interpenetrating elastomer networks (IENs), 11
interpenetrating polymer network (IPN), 535
    NR/synthetic rubber blends and, 558–561
    NR/thermoplastic blends and, 555–558
    recent trends and developments in, 14–20
    thermodynamics of, 185–186
interpenetrating polymer networks (IPN), 178–179, 553–554
    applications and the potential market for, 20–21
    environmental impact and recycling, 22–23
    introduction and history, 1–9
        development of, 9–11
        glass transition and viscoelastic behaviour, 12–13
        interpenetrating polymer networks, 9
        morphology, 13–14
        properties of, 11–12
    NR/biopolymer blends and, 561–564
ionization state, 477

IPNs *see* interpenetrating polymer network (IPN)
isobutylene isoprene rubbers (IIR), 78
isopentenyl diphosphate (IPP), 30, 31
isophorone diisocyanate (IPDI), 214

lap shear strengths (LSS), 342
latex based methods
    latex curing processes, 112–114
    latex mixing, 111–112
    maturation, 112
latex curing processes, 112–114
latex foam, 113
latex mixing, 111–112
layered silicate (LS), 225
LCA *see* life cycle assessment (LCA)
LENR *see* liquid epoxidized natural rubber (LENR)
life cycle assessment (LCA), 551–553
light scattering, 522
lignin blends
    blends and their applications, 351–353
    general information, 350–351
lignin-reinforced rubbers, 352
linear low density polyethylene (LLDPE), 436–438, 537
lipids, 54–55, **55**
    glycolipids, **55**, 56–57
    neutral lipids, **55**, 55–56
    phospholipids, **55**, 57
liquid epoxidized natural rubber (LENR), 343
liquid form, 446
liquid natural rubber (LNR), 102, 223, 397
LMWNR *see* low molecular weight natural rubber (LMWNR)
long-chain fatty acid soaps, 85
low-density polyethylene (LDPE), 596
lower critical solution temperature (LCST), 5, **196,** 199–203
low molecular weight natural rubber (LMWNR), 558
low protein natural rubber (LPNR), 395
low rubber protein latices, 88

low unsaturated ethylene propylene diene monomer rubber, 494
low viscosity (LV), 101
lysophosphatidyl choline (LPC), 57
lysophosphatidyl inositol (LPI), 57

magic angle spinning (MAS), 461
Malaysian Standard Rubber (SMR) scheme, 95
maleated depolymerized natural rubber, 493
maleated depolymerized natural rubber (MDPR), 342
maleated natural rubber, 306
maleated natural rubber (MNR), 417, 419, 422–426
maleic anhydride (MA), 190, 233, 558
manufacturing methods
    advantages and disadvantages, 128–130
    latex based methods
        latex curing processes, 112–114
        latex mixing, 111–112
        maturation, 112
    solid natural rubber based methods
        continuous mixers, 122–126
        internal batch mixers, 118–122
        solid rubber curing processes, 126–128
        two-roll mills, 115–118
    solution based methods
        solution manufacturing processes, 115
        solution mixing, 115
maturation, 112
Maxwell–Wagner–Sillars (MWS), 231
measuring online conductance, 143
mechanical evaluations
    instruments and techniques for mechanical properties, 482–485
    viscoelastic properties, 485–487

mechanical evaluations (*continued*)
  of natural rubber blends and IPNs
    natural rubber/
      biopolymers, 496–498
    natural rubber/synthetic
      rubbers, 494–496
    natural rubber/
      thermoplastics, 487–491
    natural rubber/
      thermosets, 491–493
mechanical properties, 273–276, 364–366
melt flow rate (MFR), 397, 404
membrane technology, 585
2-mercapto benzothiazole (MBT), 216, 217
metal phosphatidates (MP), 57
methacryloyl group, 325
methyl ethyl ketone (MEK), 215
methyl methacrylate-grafted (MG) rubber latex, 91–92
methyl methacrylate (MMA), 42
mevalonate (MVA) pathway, 31
micellar systems, 512
Microcyclus ulei, 74
miscibility, 179
  of NR blends and IPNs
    determination,
      identification
      parameters for, 181–182
    NR blends and IPNs, 182–184
    techniques for preparing blends, 180–181
    techniques for measuring
      glass transition
      temperature studies, 191–192
    infrared spectroscopy, 192
    morphological studies, 192
    scattering studies, 192
miscible blends, 185
mixed matrix (MM), 591, **592**, 593
mixing, 268–269

modified hydrogenated natural
  rubber/poly(methyl methacrylate-co-styrene) blended sheets, 310
modified montmorillonite (MMT), 344
molecular dynamics, 477
molecular weight distribution (MWD), 36
monodispersed poly(methyl methacrylate) latex, 302, 303
monogalactosyl diacylglycerols (MGDG), 56
monomer vinyl chloride, 553
PA6-montmorillonite (MMT), 584
Mooney viscosity, 217, 318, 319
morphological properties, 364–366
morphological studies, 271–273
morphology, 222–225
moulded products, 45
moulding, 112–113
moulding vulcanization, 127
moving die rheometer (MDR), 467
Mullins effect, 571
multifunctional vinyl monomers (MFA), 286
multivalent metal ions, 62

nanoclay greatly, 560
nanocomposites, characterization of
  flammability, 386
  mechanical properties, 379–384
  morphology, 375–379
  thermal properties, 384–386
natural rubber, 33
  applications of, 45–46
  biosynthesis of, 30–33
  branch points/gel and storage hardening, 37–38
  chemical modification of, 39–40
    chlorination, 41
    cyclization, 43–44
    epoxidation, 40–41
    grafting copolymerization, 41–42
    hydrogenation, 40
    oxidative degradation, 42–43

Subject Index

processing of, 44–45
properties of, 38–39
rubber molecule
  initiating terminal of, 33–35
  terminating end of, 35–37
natural rubber/cis-1,4-polybutadiene (NR/BR), 561
natural rubber-graft-methyl methacrylic acid, 304
natural rubber-graft-polydimethyl-(methacryloyloxymethyl), 489
neutral lipids, 55–56
nitrile butadiene rubber (NBR), 213, 275, 465
1,9-nonanediol dimethacrylate (NDMA)
  graft copolymerization, **326**
  inclusion complex formation of, 327–330
non-crystalline polymers, 205
non-intermeshing rotors, 121
non-polar lipids, 55
non-polar synthetic rubber blends
  immiscible NR blends
    background, 203–204
    NR/SBR blends, characterization of, 204–205
    tear energy, 205–210
  miscible NR blends
    background, 196–198
    LCST phase behaviour, 199–203
    NR-Sol and NR-Gel, 196–198
non-rubbers
  amino acids, **60**, 60–61
  ash, 62–63
  inositols and carbohydrates, 61–62
  lipids, 54–55
    glycolipids, 56–57
    neutral lipids, 55–56
    phospholipids, 57
  nitrogenous compounds, 61

  proteins, 57–60
  volatile matter, 63
non-vulcanized (gum) rubber, 114
NR latex based products, 46
nylon (PA6), 583

offline conductivity, 143
oil-extended NR (OENR), 90
optical microscopy, 142
organoclay ternary blends, 374–375
organomodified montmorillonite (OMMT), 220
organo-montmorillonite (OMMT), 561
OsO_4 staining, 271
oxidative degradation, 42–43

partial miscibility, 179
Payne effect, 571
penetrant molecules transport
  natural rubber based blends, 534–535
  natural rubber based blends and IPNs, 536–544
  natural rubber based interpenetrating polymer networks (IPNs), 535
  natural rubber systems properties, absorption on, 544–545
  properties and applications, 533–534
phase separation
  and compatibilization
    block/graft copolymers, 187–189
    functional/reactive polymers, 189–190
    in situ graft polymerization or reactive blending, 190–191
    thermodynamic miscibility, achievement of, 187
N-phenyl-N-isopropyl-p-phenylene diamine, 352

phosphatidic acid (PA), 57
phosphatidyl choline (PC), 57
phosphatidyl ethanolamine (PE), 57
phosphatidyl inositol (PI), 57
phospholipids, 64
photochemical degradation, 554
photoreactive nanomatrix
    inclusion complex in DPNR, graft copolymerization of, 330–334
    NDMA and β-CD, inclusion complex formation of, 327–330
plasticity retention index (PRI), 58
plasticizers, 44, 574
plastic properties, modification of, 596
polar synthetic rubber blends
    applications, 236–237
    blend characteristics
        ageing and other properties, 235–236
        curing, 219–220
        dielectric properties, 230–232
        infrared absorbance, 232–235
        mechanical and dynamic-mechanical behaviour, 225–229
        morphology, 222–225
        rheology, 217–219
        swelling and oil resistance, 220–222
        thermal properties, 229–230
    preparation methods
        latex, 214–215
        melt blending, 215–217
        solution mixing, 215
polyacrylate, 305, 306
polyaniline (PANI) blend, 445, 474–475
poly(butylene adipate-co-terephthalate) (PBAT), 354
poly(caprolactone) diol (PCL), 214
polychloroprene, 163
poly(ethylene oxide) (PEO), 356
poly(ethylene terephthalate) (PET)
    blend, 345, 461–462
polyhydroxyalkanoate (PHAs), 360–362, 562
poly(3-hydroxybutyrate-co-3-hydroxyvalerate) (PHBV), 361
poly-3-hydroxybutyrate (PHB), 360
polylactic acid (PLA), 360–366, 563, 585
polylactide, 184, 498
polymer alloy blends (PABs), 2
polymer blends, 178
    characterization techniques for, 9
    environmental impact and recycling, 22–23
    introduction and history, 1–9
        glass transition and viscoelastic behaviour, 12–13
        interpenetrating polymer networks, 9
        IPN development, 9–11
        morphology, 13–14
        properties of, 11–12
    and IPN, 553–554
    methods of, **8**
    recent trends and developments in, 14–20
    thermodynamics of, 185–186
    types of, **10**
polymeric systems, 533
polymers, radiation crosslinking of, 285–286
polymers segmental mobility, 533
poly(methyl methacrylate), 187, 490
    particles, 311
    pellets, 304, 305
polypropylene (PP), 90, 356, 416
poly(styrene-b-isoprene-b-styrene) (SIS), 272
polystyrene-nSiO$_2$, 496
polystyrene (PS), 401
polyurethane/polymethyl methacrylate (PU/PMMA), 13

polyurethane/polystyrene (PU/PS), 13
polyurethane rubber (PUR), 214
poly(vinyl acetate) (PVA), 557
poly(vinyl alcohol) (PVA), 19, 543, 591
polyvinyl chloride (PVC), 290–292, 389, 431
Porod equation, 511
potassium persulfate, 359
pressure-sensitive adhesive (PSA), 280
pre-vulcanized latices, 114
proteins, 57–60
pyrolysis of sample, 446

QENS see quasi-elastic neutron scattering (QENS)
quasi-elastic neutron scattering (QENS), 521
quebrachitol, 61

radiation crosslinked NR, 287–288
radiation effects, 285
radiation processing, vinyl plastics
    epoxidized natural rubber (ENR), radiation crosslinking of, 288–289
    natural rubber, radiation crosslinking of, 286–287
    NR based blends, radiation crosslinking of
        EVA/ENR blends, 292–294
        PVC/ENR blends, 290–292
        PVC/NR blends, 294–297
    polymers, radiation crosslinking of, 285–286
    polymers, radiation effects on, 285
    radiation crosslinked NR, properties of, 287–288
    radiation sensitizers, as crosslinking agents, 286
radiation sensitizers, as crosslinking agents, 286
radiation vulcanized natural rubber latex (RVNRL), 61, 93

random phase approximation (RPA), 519
rapid strain-induced crystallization, 204
reclaimed rubber (RR), 561, 576
residual soluble proteins, 84
rheological behaviour
    chemically modified natural rubber blends, 416–438
    natural rubber–synthetic rubber blends, 405–416
    natural rubber–thermoplastic blends, 396–405
rheological studies, 270–271
rheology, 217–219
ribbed smoked sheet (RSS), 29, 95, 96
rubber-*graft*-glycidyl methacrylate, 498
rubber-*graft*-methyl methacrylic acid, 304
rubber processing analyser (RPA), 395
rubber recovery, 22
Rubber Research Institute of Malaysia (RRIM), 101
Rubber Reserve Company, 77
rubber tapping, 81
rubber toughening, 337
RuO_4 staining, 271
RVNRL see radiation vulcanized natural rubber latex (RVNRL)

scanning electron microscopy/energy dispersive X-ray analysis (SEM/EDX), 167
scanning electron microscopy (SEM), 142, 223, 248
scanning transmission electron microscopy (STEM), 223
scattering analyses, 273
scattering studies
    small-angle light scattering, 521–523
    small-angle neutron scattering, 517–521

scattering studies (*continued*)
    small-angle X-ray diffraction, 507–517
    wide angle X-ray diffraction, 502–507
seismic isolation bearings, 577–578
semicrystallinity, 503
SETAC *see* Society for Environmental Toxicology and Chemistry (SETAC)
shear viscosity, 320
silica localization in rubber blends, 144–157
    equilibrium state, prediction of, 145–147
    filler surface tension changes, quantification of, 153–154
    kinetics, experimental determination of, 147–149
    processing and curing additives effects, 149–151
    SBR/NBR/NR and SBR/BR/NR, ternary blends of, 154–157
    silane effect, quantification of, 151–153
    silica-filled blend preparation, 144–145
single-screw extrusion, 123–125, **124**
single-walled carbon nanotubes (SWCNTs), 167, 168
small-angle light scattering, 521–523
small-angle neutron scattering (SANS), 273, 517–521
small-angle X-ray diffraction (SAXD), 507–517
small-angle X-ray scattering (SAXS), 144
smallholders lump, 99
Society for Environmental Toxicology and Chemistry (SETAC), 552
sodium bentonite filled blend, 589
solid rubber curing processes, 126–128
solution casting, 269

solution structure, 477
solvent etching, 271
solvent vapour permeation, 540
South American leaf blight (SALB), 74
spectroscopy
    applications, 477–478
    electron spin resonance (ESR) spectroscopy, 475–476
        polymer blends analysis, 476–477
        sample preparation and typical conditions for, 476
    Fourier transform infrared spectroscopy (FTIR), 445
        polymer blends analysis, 447–460
        sample preparation and typical conditions for, 446
    nuclear magnetic resonance (NMR) spectroscopy, 460
        polymer blends analysis, 461–471
        sample preparation and typical conditions for, 460–461
    Raman spectroscopy, 472–473
        polymer blends analysis, 473–475
        sample preparation and typical conditions for, 473
    UV-Vis spectroscopy, 442
        polymer blends analysis, 443–445
        sample preparation and typical conditions for, 442–443
spherulitic crystallites, 505
spraying, 113–114
spreading, 113
Standard Malaysian Rubber (SMR) scheme, 98
starch, 357–358

steryl glucosides (SG), 56
storage hardening, 37–38
storage modulus, 436
strain-induced crystallization, 207, 504
styrene butadiene rubber (SBR), 13, 154–157, 162, 449, 535
styrene-isoprene-styrene (SIS), 273
sulfosuccinic acid (SSA), 591
superior processing (SP) rubber, 101–102
surface energies, 143–144
swelling, and oil resistance, 220–222
swelling behaviour, 542
synthetic proteins, 357
synthetic rubber, 77
synthetic rubbers, 357

tactoid, 375
talalay process, 113
Taraxacum koksaghyz (TKS), 74
tear energy, 205–210
technically specified rubbers (TSR), 29, 95
technological compatibilization, 302
tensile modulus, 346
tensile properties, 246–247, 250–253
α-terminal group, 36, **37**
N-tertbutyl-2-benzothiazole sulfenamide (TBBS), 246
tetramethylthiuram disulfide (TMTD), 30, 38, 82, 561
tetramethylthiuram monosulfide (TMTM), 219
thermal analyses, 276–278
thermal properties, 229–230, 363–364
thermogravimetric analysis (TGA), 143, 248, 258–260, 316, 363, 535, 560
thermooxidative ageing, 556
thermooxidative degradation, 554
thermoplastic blends, 555–558
thermoplastic elastomers
    materials and methodology
        differential scanning calorimetry (DSC), 248
        dynamic mechanical thermal analysis (DMTA), 248
        Fourier transform infrared spectroscopy (FTIR), 247–248
        gel content, 247
        HVA-2 and sulfur curative agent, dynamic vulcanization with, 246
        scanning electron microscopy (SEM), 248
        tensile properties, 246–247
        thermogravimetric analysis (TGA), 248
        thermoplastic tapioca starch (TPS), 246
    results and discussion
        blend morphology, 255–258
        differential scanning calorimetry, 260–262
        gel content, 253–254
        processing characteristics, 248–250
        structural analysis, 254–255
        tensile properties, 250–253
        thermogravimetric analysis, 258–260
thermoplastic elastomers (TPE), 89, 90, 556
    applications of, 279–280
    characterization of
        dielectric properties, 278–279
        mechanical properties, 273–276
        morphological studies, 271–273
        rheological studies, 270–271
        scattering analyses, 273
        thermal analyses, 276–278
    preparation of
        mixing, 268–269
        solution casting, 269
    recent developments in, 267–268

thermoplastic natural rubber blends (TPNR), 89, 90, 301, 487–491, 556
thermoplastic polyurethane (TPU), 540, 593
thermoplastic starches (TPS), 562
thermoplastic tapioca starch, 246
thermosetting polymers, 491–493
 elastomer-modified epoxy resin systems, 337–344
 elastomer-modified unsaturated polyester resin systems, 344–346
thin white crepe (TWC), 97
p-toluenesulfonic acid (p-TSA), 40
total scattered intensity, 514
transfer moulding, 127
trans-isoprene and dimethylallyl (DMA) groups, 64
transmission electron microscopy (TEM), 142, 375
trans-polyoctylene rubber (TOR), 216
tree lace, 98
triallyl cyanurate (TAC), 433
triethylamine (TEA), 214
trimethylolpropanetriacrylate (TMPTA), 290, 397
twin-screw extruder, 125–126, **126**
two-roll mills, 115–118
tyre engineering fields
 automotive industry, toughened thermoplastics in, 579–585
 elastomer blends, 572
 membrane technology, 585
  recent achievements in, 586–595
 natural rubber, properties of
  general mechanical properties, 569–570
  viscoelastic properties, 570–571
 natural rubber based blends, miscellaneous applications of
  as insulator, 595–596
  plastic properties, modification of, 596
  tyres retreading, 595

NR and blends compounding, general aspects of
 antidegradants, 574
 black/non-black fillers, 572–574
 plasticizers, 574
 vulcanizing agents and curing systems, 574–575
NR and blends for, 575–577
seismic isolation bearings, 577–578
 NR as, 578–579
tyre products, 45
tyre rubber, 102
tyres retreading, 595

ultra-small-angle neutron scattering (USANS), 518
ultra-small-angle X-ray scattering (USAXS), 273
universal testing machine (UTM), 274
unsaturated polyester, 337, 344, 345
unsaturated polyester resins, 492
unsaturated polyesters/natural rubber blends, 492
unsmoked sheet (USS), 95–96
upper critical solution temperature (UCST), 5

vanadium ion, 304
vegetal fibres, 553
VFAs *see* volatile fatty acids (VFAs)
vinylidene fluoride (VDF), 558
viscoelastic behaviours, 486
viscoelastic evaluations
 instruments and techniques for
  mechanical properties, 482–485
  viscoelastic properties, 485–487
 of natural rubber blends and IPNs
  natural rubber/biopolymers, 496–498
  natural rubber/synthetic rubbers, 494–496

Subject Index 613

natural rubber/
 thermoplastics, 487–491
natural rubber/
 thermosets, 491–493
viscoelastic properties, 570–571
volatile fatty acid (VFA), 54, 82
volatile matter, 63
volatile organic compounds (VOCs), 533
vulcanization activator, 575
vulcanizing agents, 44, 574–575, 575–576

washed bottom fraction membrane (WBM), 32
water-soluble polymers (WSP), 355
water-swollen natural rubber, 318

wheat gluten, 355
wide angle X-ray diffraction (WAXD), 372, 502–507
Williams–Landel–Ferry rate–temperature equivalence, 204, 205
World Health Organization (WHO), 59

X-ray diffraction (XRD), 273, 373, 375
xylene extraction, 254

yield point, 483

zinc diethyldithiocarbamate (ZDEC), 94
zinc oxide (ZnO), 38